T0391229

Handbook of Communication for Development and Social Change

Jan Servaes
Editor

Handbook of Communication for Development and Social Change

Volume 2

With 83 Figures and 60 Tables

Editor
Jan Servaes
Department of Media and Communication
City University of Hong Kong
Hong Kong, Kowloon, Hong Kong

Katholieke Universiteit Leuven
Leuven, Belgium

ISBN 978-981-15-2013-6 ISBN 978-981-15-2014-3 (eBook)
ISBN 978-981-15-2015-0 (print and electronic bundle)
https://doi.org/10.1007/978-981-15-2014-3

This Springer imprint is published by the registered company Springer Nature Singapore Pte Ltd.
The registered company address is: 152 Beach Road, #21-01/04 Gateway East, Singapore 189721, Singapore

Introduction

This *Handbook of Communication for Development and Social Change* provides a single reference resource regarding communication for development and social change. Increasingly, one considers communication crucial to effectively tackle the major problems of today. Hence, the question being addressed in this handbook is, "Is there a right communication strategy?" Perspectives on sustainability, participation, and culture in communication have changed over time in line with the evolution of development approaches and trends, and in response to the need for effective applications of communication methods and tools to new issues and priorities.

In essence, the coherence of Communication for Development and Social Change (CDSC) is expressed in its different common underlying premises. In all its diversity, such as in theory, method, medium, and region, it is characterized by a number of underlying common values and starting points.

Therefore, the *Handbook of Communication for Development and Social Change* starts from the following premises and assumptions:

- *The use of a culturalist viewpoint*
  By means of such a viewpoint, specific attention is given to communication in social change processes. By putting culture centrally from a user's perspective, other social science disciplines can significantly contribute to the field of Communication for Development and Social Change.
- *The use of an interpretative perspective*
  Participation, dialogue, and an active vision of human beings as the interpreters of their environments are of the utmost importance. A highly considered value is the showing of respect and appreciation for the uniqueness of specific situations and identities in social change environments.
- *The preference for a transdisciplinary approach*
  While interdisciplinary collaborations create new knowledge synthesized from existing disciplines, a transdisciplinary approach relates all disciplines into a coherent whole. Transdisciplinarity combines interdisciplinarity with a participatory approach.

  The field of "sustainability" is, in essence, a transdisciplinary one.

- *The use of integrated methods and theories*
  In the field of Communication for Development and Social Change, it is considered important that the chosen methods should be connected with the used theoretical perspective. This implies that openness, diversity, and flexibility in methods and techniques are valued. In practice, it generally means triangulation and a preference for qualitative methods. This does not mean, however, that quantitative methods are excluded, and indeed an emphasis is placed on evidence-based scientific methodologies.
- *To show mutual understanding and attach importance to formal and informal intercultural teaching, training, and research*
  Tolerance, consciousness-raising, acceptance, and respect can only be arrived at when members of different cultures not only hear but also understand each other. This mutual understanding is a condition for development and social change. In order to prevent all forms of miscommunication, intercultural awareness, capacity building, and dialogue is deemed very important.

A short-hand definition of Communication for Development and Social Change (CDSC) reads as follows:

Communication for development and social change is the nurturing of knowledge aimed at creating a consensus for action that takes into account the interests, needs and capacities of all concerned. It is thus a social process, which has as its ultimate objective sustainable development/change at distinct levels of society.
Communication media and ICTs are important tools in achieving social change but their use is not an end in itself. Interpersonal communication and traditional, group, and social media must also play a fundamental role.

Divided into prominent themes comprising relevant chapters written by experts in the field and reviewed by renowned editors, the book addresses topics where communication and social change converge in both theory and praxis. Specific concerns and issues include climate change, poverty reduction, health, equity and gender, sustainable development goals (SDGs), and information and communication technologies (ICTs).

The *Handbook of Communication for Development and Social Change* integrates 11 clusters of concepts and practices. The parts are as follows:

1. An historic cluster which looks at the origins of the Communication for Development and Social Change field both from a theoretical as well as political/economic perspective
2. A *normative cluster of concepts* (e.g., on sustainability, democracy, participation, empowerment, equity, social inclusion, human rights, and accountability)
3. A *cluster of concepts that sets an important context for communication activities for development* (e.g., globalization, gender, social movements, cultural diversity, and Sustainable Development Goals (SDGs))

4. A *cluster of strategic and methodological concepts* (e.g., diffusion of innovations, social marketing, advocacy, entertainment-education, social mobilization, knowledge management)
5. A *cluster of concepts that relate to methods, techniques, and tools* (e.g., (digital) storytelling, (participatory) mapping, monitoring and evaluation, intercultural value differences, empowerment evaluation, social media, oral history, film, video, drama, and art)
6. A *cluster of cases that focus on climate change and sustainable development* (e.g., on youth participation in Indonesia, emergency communication during the 2015 Nepal earthquake, innovative education in Kenya)
7. A *cluster of cases that focus on Information and Communication Technologies (ICTs) for development* (e.g., on ICTs for learning in rural communication, strategic social media management for NGOs, online activism in Central Asia)
8. A *cluster of cases that focus on health communication* (e.g., on sustainable health, maternal health in Africa, countering sexual harassment in Bangladesh)
9. A *cluster of cases that focus on participatory communication* (e.g., on natural resource management, conflict and power, capacity building, participatory video)
10. A *set of regional cases and overviews* (e.g., on Africa, Europe, Western journalism education, ASEAN)
11. A *set of specific case studies* (e.g., on entertainment-education in radio in Africa, participatory communication in Afghanistan, Egyptian media, protest in South Africa)

This book shows how communication is essential at all levels of society. It helps readers understand the processes that underlie attitude change and decision-making and the work uses powerful models and methods to explain the processes that lead to sustainable development and social change.

This handbook is published in multiple formats: print, static electronic, and live electronic.

It is essential reading for academics and practitioners, students and policy makers alike.

World Rainforest Day
22 June 2019

Jan Servaes

Acknowledgments

The *Handbook of Communication for Development and Social Change* is the collective effort of many distinguished researchers, scholars, and professionals in the field of development communication or communication for social change.

At the editorial level, I was assisted by two excellent associate editors: Dr. Rico Lie, assistant professor at Wageningen University in The Netherlands, and Dr. Patchanee Malikhao, director of the consultancy agency Fecund Communication in Chiang Mai, Thailand.

A number of renowned experts assisted with the selection and peer review of the manuscripts. The nature of their work requires that they remain anonymous. However, I wish to single-out Professor emeritus Royal Colle from Cornell University in Ithaca NY, USA, who has kindly assisted me with the section on ICTD, Information and Communication Technologies for Development, and Dr. Jessica Noske-Turner, lecturer at Loxbourough University London, who assisted as external reviewer.

We also wish to acknowledge and thank Ms. Jayanthie Krishnan, then Publishing Editor at Springer Singapore, who initially invited me to coordinate this project. Once the project was on rails Ms. Alexandra Campbell, Editor for Social Sciences and Humanities at Springer Nature Singapore, took over. The capable Springer Reference Editorial Team was coordinated by Ms. Mokshika Gaur and her hardworking assistants Dr. Lijuan Wang and Ms. Sunaina Dadhwal.

Special thanks go to all the contributing authors, researchers, and students who have made this handbook possible.

Important Note

In all cases where gender is not implicit, we have attempted to combine feminine and masculine pronouns. However, as English is not the mother tongue of the editors and many authors in this handbook, we apologize to the reader if she/he still finds errors.

Contents

Volume 1

About the Editor

Jan Servaes (Ph.D.) was UNESCO Chair in Communication for Sustainable Social Change. He has taught International Communication and Communication for Social Change in Australia, Belgium, China, Hong Kong, The Netherlands, Thailand, and the United States, in addition to several teaching stints at about 120 universities in 55 countries.

Servaes is Editor of the Lexington Book Series Communication, Globalization and Cultural Identity (https://rowman.com/Action/SERIES/LEX/LEXCGC) and the Springer Book Series Communication, Culture and Change in Asia (http://www.springer.com/series/13565) and was Editor-in-Chief of the Elsevier journal *Telematics and Informatics: An Interdisciplinary Journal on the Social Impacts of New Technologies*" (http://www.elsevier.com/locate/tele).

Servaes has been President of the European Consortium for Communications Research (ECCR, later www.ecrea.eu) and Vice President of the International Association of Media and Communication Research (IAMCR, www.iamcr.org), in charge of Academic Publications and Research, from 2000 to 2004. He chaired the Scientific Committee for the World Congress on Communication for Development (Rome, 25–27 October 2006),organized by the World Bank, FAO, and the Communication Initiative.

Servaes has undertaken research, development, and advisory work around the world and is the author of more than 500 journal articles and 25 books/monographs, published in Chinese, Dutch, French, German, Indonesian, Portuguese, Russian, Spanish, and Thai, on topics such as as international and development communication, ICT and media policies, intercultural communication, participation and social change, and human rights and conflict management. He is known for his "multiplicity paradigm" in *Communication for Development: One World, Multiple Cultures* (1999).

Some of his recent book titles include: (2017) Servaes, Jan (ed.) *The Sustainable Development Goals in an Asian Context*, Singapore: Springer; (2016) Servaes, Jan & Oyedemi, Toks (Eds.) *The Praxis of Social Inequality in Media: A Global Perspective*, Lanham, MD: Lexington Books, Rowman and Littefield; (2016) Servaes, Jan & Oyedemi, Toks (Eds.) *Social Inequalities, Media, and Communication: Theory and Roots*, Lanham, MD: Lexington Books, Rowman and Littefield; (2014) Jan Servaes (Ed.). *Technological Determinism and Social Change*, Lanham: Lexington Books; (2014) Jan Servaes and Patchanee Malikhao. *Communication for Social Change* (in Chinese), Wuhan: Wuhan University Press; (2013) Jan Servaes (Ed.). *Sustainable Development and Green Communication: African and Asian Perspectives*, London/New York: Palgrave/MacMillan; (2013) J. Servaes (Ed.). *Sustainability, Participation and Culture in Communication. Theory and Praxis*, Bristol-Chicago: Intellect-University of Chicago Press; (2008) J. Servaes. *Communication for Development and Social Change*, Los Angeles, London, New Delhi, Singapore: Sage; (2007) J. Servaes & Liu S. (Eds.). *Moving Targets: Mapping the Paths Between Communication, Technology and Social Change in Communities*, Penang: Southbound; (2006) P. Thomas & J. Servaes (Eds.). *Intellectual Property Rights and Communications in Asia*, New Delhi: Sage; (2006) J. Servaes & N. Carpentier (Eds.). *Towards a Sustainable European Information Society*, ECCR Book Series, Bristol: Intellect; (2005) Shi -Xu, Kienpointner M. & J. Servaes (Eds.). *Read the Cultural Other: Forms of Otherness in the Discourses of Hong Kong's Decolonization*, Berlin: Mouton De Gruyter; (2003) J. Servaes (Ed.). *The European Information Society: A Reality Check*, ECCR Book Series, Bristol: Intellect; and (2003) J. Servaes (Ed.). *Approaches to Development: Studies on Communication for Development*, Paris: UNESCO Publishing House.

About the Associate Editors

Rico Lie (Ph.D. 2000, Catholic University of Brussels, Belgium) is a social anthropologist working at the research group Knowledge, Technology and Innovation, Wageningen University & Research (WUR), The Netherlands. He previously worked at the University of Brussels in Belgium and the Universities of Nijmegen and Leiden in The Netherlands. At WUR, he is an assistant professor in international communication with an interest in the areas of communication for development and intercultural learning.

Patchanee Malikhao (Ph.D. 2007, The University of Queensland, Australia) is a Sociologist with competencies in mass communication research and graphic arts research. She received her education in two prestigious universities in Thailand, Chulalongkorn and Thammasat universities, and higher education in the United States (RIT, Rochester, NY) and Australia (UQ). Dr. Malikhao has worked in Australia, the United States, Belgium, the Netherlands, and Thailand in the fields of Communication for Social Change research, Health Communication research, teaching and research in the graphic arts, journalism, and communication. She has been an academic writer since the 1980s. Dr. Malikhao has a wide range of research interests from communication for sustainable social change, over health related areas such as HIV/AIDS to mindful journalism, Thai Buddhism and social communication, globalization and Thai culture, intercultural communication, and Thai culture and communication. Currently she is the director of Fecund Communication Consultancy in Thailand.

She was the recipient of a Fulbright Scholarship, the Australian Postgraduate Award Scholarship, and a distinguished award in printing studies from the Tab Nilanidhi Foundation, Thailand.

She is the single author of the following books:

Malikhao, P. (2017). *Culture and Communication in Thailand*. Singapore: Springer. (https://www.springer.com/gp/book/9789811041235)

Malikhao, P. (2016). *Effective Health Communication for Sustainable Development*. New York: Nova Publishers, Inc. (https://www.novapublishers.com/catalog/product_info.php?products_id=58305&osCsid=98458b3794e19851cf28ac915a8c232b).

Malikhao, P. (2012). *Sex in the Village. Culture, Religion and HIV/AIDS in Thailand.* Southbound & Silkworm Publishers: Penang-Chiang Mai, 238 pp. (ISBN: 978-983-9054-55-2).

Contributors

Mario Acunzo FAO Communication for Development Team, Rome, Italy

Shahla Adnan Department of Communication and Media Studies, Faculty of Social Sciences, Fatima Jinnah Women University, Rawalpindi, Pakistan

Nova Ahmed North South University, Dhaka, Bangladesh

Rukhsana Ahmed Department of Communication, University at Albany, SUNY, Albany, NY, USA

Alfred Okoth Akwala Department of Language and Communication Studies, Faculty of Social Sciences and Technology, Technical University of Kenya, Nairobi, Kenya

Oana Brindusa Albu Department of Marketing and Management, University of Southern Denmark, Odense M, Denmark

Colin Alexander School of Arts and Humanities, Nottingham Trent University, Nottingham, UK

José Luis Aguirre Alvis Department of Social Communication, Universidad Católica Boliviana "San Pablo", La Paz, Bolivia

Antonio Arreaga Juarez & Associates, Inc., Guatemala City, Guatemala

S. Claude-Gervais Assogba Faculté d'Agronomie, Université de Parakou, Abomey-Calavi, Benin

Hina Ayaz Berlin, Germany

Silvia Balit Rome, Italy

Kriss Barker International Programs, Population Media Center, South Burlington, VT, USA

Alejandro Barranquero Universidad Carlos III de Madrid, Madrid, Spain

Hans Begeer BMC Consultancy, Brussels, Belgium

Guy Bessette Gatineau, Canada

Tanja Bosch Centre for Film and Media Studies, Cape Town, South Africa

Andrea Botero Interact Research Group, University of Oulu, Oulu, Finland

Valerie M. Campbell University of Prince Edward Island, Charlottetown, PE, Canada

Nico Carpentier Uppsala University, Uppsala, Sweden

Vrije Universiteit Brussel (VUB), Brussels, Belgium

Charles University, Prague, Czech Republic

Vidhi Chaudhri Department of Media and Communication, Erasmus University Rotterdam, Rotterdam, The Netherlands

Yang Cheng Department of Communication, North Carolina State University, Raleigh, NC, USA

Saik Yoon Chin Southbound, George Town, Penang, Malaysia

Logan Cochrane International and Global Studies, Carleton University, Ottawa, ON, Canada

Jon Corbett University of British Columbia, Vancouver, BC, Canada

Rudrajit Das UNICEF, Kathmandu, Nepal

Satarupa Dasgupta Communication Arts, School of Contemporary Arts, Ramapo College of New Jersey, Mahwah, NJ, USA

Rozane De Cock University of Leuven, Leuven, Belgium

Alvito de Souza Co-Creative Communication, Brussels, Belgium

Khyati Deesawala Department of Extension and Communication, The Maharaja Sayajirao University of Baroda, Vadodara, India

Nava Derakhshani Cape Town, South Africa

Gloria Gómez Diago Department of Communication Sciences and Sociology, Rey Juan Carlos University, Fuenlabrada, Madrid, Spain

Sandra Diehl Department of Media and Communications, Alpen-Adria-Universitaet Klagenfurt, Klagenfurt, Austria

Bruktawit Ejigu Kassa Department of Communication, University of Vienna, Vienna, Austria

Department of Journalism and Mass communication, Haramaya University, Dire Dawa, Ethiopia

P. Eric Louw School of Communication and Arts, University of Queensland, Brisbane, QLD, Australia

Michael Andreas Etter Marie Curie Research Fellow, Faculty of Management, Cass Business School, City, University of London, London, UK

Ana Fernández-Abállí Altamirano Department of Communication, Universitat Pompeu Fabra, Barcelona, Spain

Cees J. Hamelink University of Amsterdam, Amsterdam, The Netherlands

Kefa Hamidi Institute communication and media Studies, Leipzig University, Leipzig, Germany

Xiao Han Mobile Internet and Social Media Centre, Communication University of China, Beijing, China

Usha S. Harris Macquarie University, Sydney, Australia

Tom Hogan Sydney, NSW, Australia

Minna Aslama Horowitz St. John's University, New York City, NY, USA

University of Helsinki, Helsinki, Finland

Alessandro Inversini Henley Business School, University of Reading, Greenlands Henley-on-Thames, UK

Thomas Jacobson Department of Media Studies and Production, Lew Klein College of Media and Communication, Temple University, Philadelphia, PA, USA

Fatou Jah International Programs, Population Media Center, South Burlington, VT, USA

Claudia Janssen Danyi Department of Communication Studies, Eastern Illinois University, Charleston, IL, USA

Alankar Kaushik EFL University, Shillong Campus, Shillong, India

Marius N. Kêdoté Institut Régional de Santé Publique, Comlan Alfred Quenum, Université d'Abomey-Calavi, Abomey Calavi, Benin

Zafer Kıyan Department of Journalism, Ankara University, Ankara, Turkey

Lauren Kogen Department of Media Studies and Production, Temple University, Philadelphia, PA, USA

Isabell Koinig Department of Media and Communications, Alpen-Adria-Universitaet Klagenfurt, Klagenfurt, Austria

Bahtiyar Kurambayev Department of Media and Communications, College of Social Sciences, KIMEP University, Almaty, Kazakhstan

Teresa La Porte International Political Communication, University of Navarra, Pamplona, Navarra, Spain

Yuhong Li Hong Kong International New Media Group, Hong Kong, China

Rico Lie Research Group Knowledge, Technology and Innovation, Wageningen University, Wageningen, The Netherlands

Elizabeth Lubinga Department of Strategic Communication, School of Communication, University of Johannesburg, Johannesburg, South Africa

Patchanee Malikhao Fecund Communication, Chiang Mai, Thailand

Avani Maniar Department of Extension and Communication, The Maharaja Sayajirao University of Baroda, Vadodara, India

Linje Manyozo School of Media and Communication, RMIT University, Melbourne, Australia

Víctor Manuel Marí Sáez Faculty of Communication and Social Sciences, Universidad de Cádiz, Jerez de la Frontera, Spain

Paolo Mefalopulos UNICEF, Montevideo, Uruguay

Stefan Mertens University of Leuven, Leuven, Belgium

Mulatu Alemayehu Moges School of Journalism and Communication, Addis Ababa University, Addis Ababa, Ethiopia

Soledad Muñiz InsightShare, London, UK

Paul Clemens Murschetz Faculty of Digital Communication, Berlin University of Digital Sciences, Berlin, Germany

Tokunbo Ojo Department of Communication Studies, York University, Toronto, ON, Canada

Toks Dele Oyedemi Communication and Media Studies, University of Limpopo, Sovenga, South Africa

Fay Patel International Higher Education Solutions, Sydney, Australia

Tamara Plush Royal Roads University, Victoria, BC, Canada

Emily Polk Stanford University, Stanford, CA, USA

Pornpun Prajaknate Graduate School of Communication Arts and Management Innovation, National Institute of Development Administration, Bangkok, Thailand

Kiran Prasad Sri Padmavati Mahila University, Tirupati, India

Hosai Qasmi Institute of Feminism and Gender Studies, University of Ottawa, Ottawa, ON, Canada

Isabella Rega Centre of Excellence in Media Practice, Faculty of Media and Communication, Bournemouth University, Bournemouth, UK

Tony Roberts Institute of Development Studies, University of Sussex, Sussex, UK

Johanne Saint-Charles Département de communication sociale et publique, axe santé environnementale, CINBIOSE, Université du Québec à Montréal, Montréal, QC, Canada

Alamira Samah Saleh Faculty of Mass Commination, Cairo University, Giza, Egypt

Sol Sanguinetti Universidad de Lima, Lima, Peru

Programme, Environment and Technology Institute, Lima, Peru

Paula Santos Universidad Católica del Uruguay, Montevideo, Uruguay

Katharine Sarikakis Department of Communication, University of Vienna, Vienna, Austria

Jan Servaes Department of Media and Communication, City University of Hong Kong, Hong Kong, Kowloon, Hong Kong

Katholieke Universiteit Leuven, Leuven, Belgium

Hemant Shah School of Journalism and Mass Communication, University of Wisconsin-Madison, Madison, WI, USA

Thea Shahrokh Centre for Trust Peace and Social Relations at Coventry University, Coventry, UK

Song Shi Department of East Asian Studies, McGill University, Montreal, QC, Canada

Jakob Svensson School of Arts and Communication (K3), Malmö University, Malmö, Sweden

Swan Ti PannaFoto Institute, Jakarta, Indonesia

Nurcan Törenli Department of Journalism, Ankara University, Ankara, Turkey

Cleofe S. Torres College of Development Communication, University of the Philippines Los Baños, Laguna, Philippines

Micaela Trimble South American Institute for Resilience and Sustainability Studies (SARAS), Bella Vista-Maldonado, Uruguay

Ravindra Kumar Vemula Deptartment of Journalism and Mass Communication, The English and Foreign Languages University, Shillong, Meghalaya, India

Luisa Veronis Department of Geography, Environment and Geomatics, University of Ottawa, Ottawa, ON, Canada

Rahel Vetsch UNICEF, Kathmandu, Nepal

Silvio Waisbord School of Media and Public Affairs, George Washington University, Washington, DC, USA

Junis J. Warren York College, City University of New York, New York, USA

Richard Wecker UNICEF Indonesia, Jakarta, Indonesia

Franzisca Weder Department of Media and Communications, Alpen-Adria-Universitaet Klagenfurt, Klagenfurt, Austria

Joanna Wheeler University of Western Cape, Cape Town, South Africa

Loes Witteveen Research Group Communication, Participation and Social Ecological Learning, Van Hall Larenstein University of Applied Sciences, Velp, The Netherlands

Research Group Environmental Policy, Wageningen University, Wageningen, The Netherlands

Arpan Yagnik Department of Communication, Penn State, Erie, Erie, PA, USA

Stéphanie Yates Département de communication sociale et publique, Université du Québec à Montréal, Montréal, QC, Canada

Part VII

Climate Change and Sustainable Development

Communicating Climate Change: Where Did We Go Wrong, How Can We Do Better? 43

Emily Polk

Contents

E. Polk (✉)
Stanford University, Stanford, CA, USA
e-mail: empolk@stanford.edu

J. Servaes (ed.), *Handbook of Communication for Development and Social Change*,
https://doi.org/10.1007/978-981-15-2014-3_26

Abstract

This chapter traces the historical trajectory of climate change communication, theory, and practice, including the cultural and social context for attitudes and beliefs around climate change and intersections between development communication and climate change communication. It also identifies particular challenges to the field and maps out the current landscape for scholars and practitioners. The main findings support previous development communication research that suggests the distribution of information is not enough to generate sustained engagement with either climate change advocacy or individual behavior change. Climate change communicators are most effective when they focus on positive solutions and appeal to specific local communities. Scholars must do a better job sharing their work across disciplines and with practitioners in the field, and finally, all communication efforts and their evaluation processes would benefit from sustained long-term engagement.

Keywords

Climate change communication · Media framing · Advocacy and social movements · Effective evaluation

In the summer of 2017, right in the middle of the planet's hottest non-el Niño year and one of the three hottest years on record (World Meteorological Association 2017), *New York Magazine* published a controversial and widely read article "The Uninhabitable Earth" with the subhead: "Famine, economic collapse, a sun that cooks us: What climate change could wreak – sooner than you think" (Wallace-Wells 2017). The article paints a terrifying and apocalyptic picture of an inevitably doomed planet. It was controversial for many reasons, but critics and supporters primarily focused on the content and the tone. Some scientists said it was hyperbolic, referring to it as "climate disaster porn" (Cohen 2017). Still others, focusing on how different publics respond to information about climate change, noted that people respond better to hopeful messages, not fatalistic ones (O'Neill and Nicholson-Cole 2009). Acting becomes harder, they argue, when people think there is nothing they can do to change the current situation.

It is not hard to see why readers might come away from the article believing there is nothing they can do about climate change. "The Uninhabitable Earth" cites science-supported research that documents a 50-fold increase in the number of places experiencing dangerous or extreme heat over the last 40 years with bigger predicted increases to come. Droughts may cause massive food shortage and forced migration as a result of some of the world's most arable land turning to desert. According to the article:

> By 2080, without dramatic reductions in emissions, southern Europe will be in permanent extreme drought. . .The same will be true in Iraq and Syria and much of the rest of the Middle East; some of the most densely populated parts of Australia, Africa, and South America; and

> the breadbasket regions of China. None of these places, which today supply much of the world's food, will be reliable sources of any.

The apocalyptic consequences of unmitigated climate change also include increased risk of diseases that thrive in warmer climates, increasingly unbreathable air from small particles emitted from fossil-fuel burning and wildfire smoke, and more conflicts as a result of economic devastation. Let's not even get into the ruined oceans.

The article clearly wanted to communicate the urgency and gravity of climate change, but was it effective? How do we determine effectiveness? What communication processes and practices result in positive social change?

This is not a new question. Development experts, policy makers, researchers, advocates, and communication scholars have been trying to figure out how positive sustainable social change happens and the role communication can play for the better part of the last century. This chapter asks if there are any useful intersections between development communication for social change and emerging scholarship on climate change communication. What can development stakeholders learn from the people on the frontlines of climate change? And from the evolving practices and strategies of communicating it?

43.1 Current Attitudes Around Climate Change

It is important to understand the cultural and social context for attitudes and beliefs around climate change before diving into the emerging literature on climate change communication. Despite large-scale scientific consensus, and apocalyptic articles in popular media like the "Uninhabitable Earth," public attitudes and beliefs, particularly around whether or not the issue should be a priority, have fluctuated extensively. Unfortunately, the majority of current research is focused on developed countries with a notable dearth of research in developing countries (Wolf and Moser 2011; Wibeck 2013; Ming-Lee et al. 2015). Nevertheless researchers have identified compelling reasons for fluctuating attitudes despite scientific consensus. These reasons include experiences of economic insecurity, growing political polarization, media bias, personal exposure to visual images of climate change, and experience with extreme weather events, with the first two being the most significant causes of fluctuating attitudes and public concern (Brownlee et al. 2013; Weber and Stern 2011).

According to a recent Pew Research Center poll (Motel 2014), a majority of Americans believe that the earth has been getting warmer in recent decades, but rank it as a low actionable priority. It is thus important to distinguish between "awareness of" and "concern about" the issue. The poll revealed sharp partisan divides consistent with previous research with Democrats being more likely than Republicans to see it as an actionable issue. Most Americans and Congresspeople, however, still rank several other issues as greater threats than global climate change.

After evaluating and documenting significant fluctuations in over 30 years of public opinion data about global warming and the environment, Scruggs and Benegal (2012) suggest, perhaps unsurprisingly, that the decline in belief about climate change is most likely driven by economic insecurity. They argue that economic recessions lead people to demand behaviors by governments and others to increase the economy and improve the market which conflicts with their beliefs about what is needed to mitigate the negative impacts of climate change: limiting economic activity. One implication for climate change communicators is that framing to address the issue might benefit from more acutely addressing economic concerns (IE: the promise of the renewable energy industry).

It is important to note that opinion on climate change is, for some portion of the public, increasingly becoming more stable and largely fixed by ideological and social identities, identities shaped and codified by elite cues – that is, leaders of the political and cultural identity group – in which different individuals align (Brulle et al. 2012). The fluctuations in belief and growing political divides have concerned climate change communicators, researchers, and advocates who believe that there must be a bigger mobilization of engagement in order to coordinate the responses needed to create the sustained change necessary. The historical trajectory of development communication scholarship may offer some useful lessons regarding how to mobilize and evaluate sustained engagement.

43.2 Intersections Between Development Communication and Climate Change Communication

Development communication scholarship precedes the emerging literature on climate change communication by about 10 years though there are some interesting intersections, namely, around the top-down approach to communicating messages in order to inspire behavioral change. Nora C. Quebral, the first person to coin the phrase "Development Communication" in 1971, called it "the science and art to change society in a planned way." She since updated it to "the science and art of human communication linked to transform society from a state of poverty to one of socio-economic growth that makes for greater equity" (2005). Servaes (2009) defines it as a social process that involves the sharing of knowledge aimed at reaching a consensus for action that takes into account the interests, needs, and capacities of all concerned.

Lennie and Tacchi (2013) note that currently Communication for Development (C4D) "encompasses all forms and modes of communication, including community radio, community based information and communication technology (ICT) initiatives, processes such as community dialogue, participatory video, and digital storytelling activities, and the use of various combinations of new and traditional media." They highlight, however, that it is about people first and foremost.

Traditionally the communication process was seen as a one-way transmission from sender to receiver. Beginning in the 1950s, early approaches assumed a one-way progression from an agricultural to an industrial society, from the premodern to the modern, and from the nondemocratic to the democratic. Mass media was the channel through which these "modern" ideas were transmitted, and the success of their "transmission" was measured in economic terms (Schramm 1964;

Lerner 1958; Rogers 1962). Similar ideas were centered in early climate change communication: however, perhaps ironically, success in economic terms might be determined by how much people are willing to give up (relative to their consumption practices), rather than gain. Initially, in the 1980s, climate change was communicated primarily by scientists and narrowly focused on scientific findings and synthesis reports (such as the Intergovernmental Panel on Climate Change, IPCC) and sometimes by high-level conferences or policy meetings (Moser 2010).

Following on the heels of Schramm, Lerner, and Rogers, climate change communication efforts focused on the role of mass media with the expectation that an increased amount of coverage would bring an increased awareness and thus behavioral change and, perhaps more importantly, a sense of urgency with public engagement. The scientific facts were assumed to speak for themselves with their relevance and policy significance interpreted by all audiences in similar ways. However, the reality was actually much different. Different audiences interpret the news in different ways based on their own cultural, social, and political ideologies (Nisbet 2009). To complicate matters, the US news media, perhaps in an effort to garner more views or to subscribe to a "balancing" norm, initially publicized news about climate change as something that was up for debate, giving equal weight to the scientists and the "climate deniers." This gave the erroneous impression that there was limited expert agreement, contributing to a public with varying perceptions and levels of awareness that rose and fell with the attention cycles in the media (Moser 2010; Nisbet 2009).

This strategy of providing the public with information and assuming that it would lead to active and sustained social change – a strategy that has become known as the "information-deficit" model has also informed a majority of climate change communication campaigns outside of the mass media. Over the past three decades, countries, provinces, and supranational institutions have launched top-down climate change and energy-related communication campaigns pursuing a range of goals (education, awareness raising, behavior change) – with the exception of the United States, whose communication mobilizations have not been centrally coordinated by the government but by different organizations and advocacy groups (Wibeck 2013).

Alongside these information transfer campaigns, however, there has more recently been a move toward more participatory and dialogical approaches aimed at increasing public engagement. Such approaches have been the subject of communication research, as a growing number of recent studies illustrate how dialogic, deliberative processes can deepen understanding and empathy, change attitudes, and increase receptivity to policy alternatives (Moser 2016). This move parallels the historical trajectory of development communication. The multiplicity framework in development communication emerged in the 1980s in many ways as a response to the failures of the transmission approach. The framework emphasized plurality and dialogue along and in between all levels of society (Servaes 1999) while suggesting that there is no universal path or standard to development but rather that each culture and community must decide the best path based on a grass-roots, bottom-up approach: self-development of local communities.

This framework centered on participatory communication, based largely on Brazilian author, activist, and teacher Paulo Friere's work (1970) which stipulated that dialogic communication was essential for conscientization – the autonomy of each individual to realize their own self-worth. A participatory approach first recognizes that the point of departure must be the community. The viewpoints of local groups must be considered before resources for projects are allocated and distributed. Secondly, social equity and a democratic process are best fostered through a horizontal process of information exchange.

Participatory communication in theory sounds idyllic: however, the reality in practice tells a more complicated story for both the development communication field and climate change communication efforts. Waisbord (2008) notes that institutional dynamics often undercut potential contributions of participatory communication, while others argue that the language of "participation" has, in fact, been co-opted and used in the rhetoric of bureaucratic organizations that continue to operate under the top-down modernization paradigm. Another critique is that while the theory takes a cultural-specific approach, it cannot, in practice, account for the ways that different cultures structure their hierarchies of power and gender differences, since the theory advocates equality and horizontal processes of communication. Lennie and Tacchi (2013) note that a consideration of power dynamics must be prioritized in participatory efforts. "Opening up spaces for invitational participation is necessary but not enough to ensure effective participation. Supportive processes are needed because invited spaces for participation in development are often structured by those who provide them, rather than created by the people themselves. . . Invited participation in spaces created for this purpose by those in positions of power can diminish the spaces where people set their own agendas on their own terms" (p. 11).

Similar complexities with regard to participation and dialogue around climate change efforts have emerged, even as evidence suggests that participatory communication processes are more effective. For example, public participation exercises where the community is invited to deliberate on a local climate change problem or attend a workshop on global warming, while in stark contrast to top-down climate change information campaigns, have proven to generate behavioral change. "Carbon Conversations Groups" in the United Kingdom, for example, offer a supportive group experience that helps people cut their carbon footprint by connecting to group members' values, emotions, and identities. The groups are based on a psychological understanding of how people change and offer: *space* for people to explore what climate change means for them, their families, and their aspirations; *permission* to share hopes, doubts, and anxieties; *time* to work through the conflicts between intention, social pressure, and identity; *reliable, well-researched information* and practical guidance on what will make a difference; and *support* in creating a personal plan for change (Rayner and Minns 2015). The wider communication challenge is to deploy this at a scale beyond local support groups who are already "carbon capable" (Whitmarsh 2009).

Wibeck (2013) notes that while public participation in climate science and policy implies mutual learning between lay people, decision makers and scientists in deliberating on appropriate responses, in order for these participatory exercises to be successful, people need to be able to contribute their time and ideas. When public participation events can only include a limited group of participants, there is a risk that only people with the financial and social capital will attend. Vulnerable members of the population – that is, those who live in poverty, or are otherwise marginalized will not necessarily have the capacity to participate. In countries where a large proportion of the population struggles simply to survive each day, those who do end up participating are less likely to be representative of the general population and especially of the people who might benefit the most from policies that might emerge from such public deliberation (Wibeck 2013). Similar findings were echoed in Polk's long-term ethnographic study of a Transition Town in Amherst, Massachusetts, where it was found that the people who were able to participate in the initiative aimed at "transitioning" their community toward greater resiliency were the ones with the most resources – including time and money – to engage. Thus outreach to and actions involving the most vulnerable in the community were not necessarily prioritized (Polk 2015).

A lack of consideration of power not only risks the success of climate change initiatives, but it has also informed (as noted earlier) the very sites where climate change communication research has taken place (i.e., almost all in developed countries).

43.2.1 Challenges to Effective Climate Change Communication

It is important to note that climate change communication presents its own series of unique challenges and obstacles to creating sustainable social change. Indeed some of the most pressing challenges include the fact that climate change is not visible for many people, especially people in positions of privilege to set policy and mobilize sustained movement; the lack of immediacy (i.e., it is difficult to sustain mobilization around preventing the earth's temperature from heating up two degrees in the future as opposed to organizing a rally against police violence); delayed or absent gratification for taking action (this is related to the former – if one cannot see the immediate fruits of one's labor, it becomes difficult to prioritize and sustain massive action); and finally the complexity and uncertainty of a problem that affects people all over the world in different ways (Moser 2010; Marquart-Pyatt et al. 2011; Weber and Stern 2011; Whitmarsh 2011).

McAdam (2016) outlines several reasons why people have not yet mobilized around climate change. His ideas provide important considerations for climate change communicators. He suggests that climate activists have failed in the United States at creating a collective identity frame for the climate movement. Because there is no clear identity or ownership of the climate change issue, sustained collective mobilization is difficult. McAdams notes, for example, that the only segment of the American public for whom climate change is highly salient is, ironically, extreme conservatives, whose views on the issue are central to their political and ideological commitments. For the rest of the population – the majority of whom acknowledge that climate change is happening – the salience of the issue is highly variable, depending on numerous factors, as noted above. Further climate change communication research might explore the processes of other advocacy groups in creating effective identity frames as one way to inspire massive sustained mobilization.

43.2.2 Communication Challenges for Scientists

Scientists, as the gatekeepers of knowledge, will always play a crucial role in communicating climate change, and they face their own set of challenges. Lupia (2013) notes that one of the biggest challenges is that people have less capacity to pay attention to scientific information than many communicators anticipate. Secondly, even if they do pay attention, people often make different choices about whom to believe based on political ideology and cultural worldviews (Akerlof et al. 2013), and as noted earlier these choices are growing increasingly sharp among partisan lines, reinforced by a bifurcated flow of conflicting information from the media and from elites on both sides of the spectrum (McCright and Dunlap 2011). These challenges – in addition to the lack of perceived immediacy and visibility – cause policy makers and the public to be less responsive to scientific information.

Lupia (2013) suggests that scientists might be more effective if they understood two communication-related concepts: attention and source credibility. New science information is always in competition with all other phenomena to which that person can potentially pay attention. Science communicators can benefit by obtaining information about what an audience initially believes about the new information they are conveying, since people assign meaning to new information by comparing it with what they already believe. Source credibility, the extent to which an audience perceives a communicator as someone they would benefit from believing, is key for science communicators. In order to achieve source credibility, the listener must perceive the speaker to have sufficiently common interests, and the listener must perceive the speaker to have relative expertise (Lupia 2013).

43.3 Characterizing the Current Climate Change Communication Field

Despite the above challenges, the field has grown considerably for both practitioners and researchers in the last decade in ways that mirror development communication. In fact, one might note a similar trajectory – including, as noted earlier, the movement in practice and research from studying top-down transmission flows to more dialogic flows, the impact of culture and values on belief and action, as well as intersections between the climate movement and other current social movements (Caniglia et al. 2015; Luers 2013; Hadden 2014).

The year 2015 brought one of the most significant climate change communication moments, when Pope Francis released his encyclical on the human-earth relationship with its particular focus on climate change (Francis 2015). A global religious figure of his statute choosing to focus on climate change brought a heightened moral immediacy to the issue – the long-term effects remain to be documented. Such an event is situated within a larger landscape of a growing field that, much like development communication, is defined largely by its interdisciplinarity and by the depth and breadth of research and practice possibilities. Moser (2016) divides the climate change communication landscape into six categories: (1) the climate itself, which has provided ample research opportunities with increased superstorms, droughts, flooding, and acidification of the oceans, among others; (2) scientific advances, notable discoveries, and climate change assessments; (3) climate policies and actions including annual international meetings of policy-makers; (4) climate communicators and practitioners, who use the former three categories to practice their craft; (5) climate communication science as a multidisciplinary branch of academic research; and (6) important contextual and foundational factors that are relative to the site of research. This includes the social and political culture of a nation; "political destabilizations. . .heightened terrorism fears, pandemics (e.g., Ebola), or the ongoing refugee crisis in the Middle East and Europe; as well as larger economic, technological or cultural shifts and events in specific industries, nations, or regions of the world." (Moser 2016).

Within this climate change communication landscape, several central themes emerge, themes which again mirror the turns in development communication a decade ago. For example, the role of values, beliefs, worldviews, identity, and meaning-making has become one of the most central foci of climate communication researchers, which has informed significant research on framing, messaging, and language (see more below). A second theme involves the importance of storytelling and narrative formats to convey climate change. A third theme centers communication channels and forms and explores the relationship between the media (in all its forms) and the science and policy aspects of climate change. This theme is connected to a fourth theme, familiar to most development communication scholars: how best to move audiences to action. This question has informed considerable research into human motivation and how best to communicate it. Finally, there is a growing focus in practice and research on how best to communicate climate change mitigation and adaptation as well as new research on the intersectionality of the climate movement with other social movements (Moser 2016).

43.4 Moving Forward: Focus Areas for the Future of Climate Change Communication

43.4.1 Framing as a Concept, Practice, and Process

A deeply fragmented media and political landscape demands that climate change communications continue to focus on framing. Framing – as a concept and an area of research – spans several social science disciplines. Frames are interpretive storylines that communicate why an issue might be a problem, who or what might be responsible for it, and what should be done about it (Nisbet 2009).

Framing as a theory emerges from the relationship between the media landscape and relevant cultural and social forces that shape the opinions of multiple publics and these publics' wills to act. Climate change communication frames that seek to appeal to the general public have included appeals to morality and ethics, as featured in Al Gore's WE campaign (short for "We can solve it") which launched in 2008 in an attempt to unify US citizens by framing climate change as a solvable and shared moral challenge. Other effective frames include public health and economic development. Such frames have been used to both negate and suggest policy changes as well as shape public understanding of the issues.

One example is the use of the word "dangerous" in the framing of climate change. Research suggests that public support or opposition to climate policies (e.g., treaties, regulations, taxes, subsidies) is greatly influenced by public perceptions of the risks and dangers inherent in climate change (Leiserowitz 2005). However, the word "dangerous" is defined differently depending on the stakeholder. Expert definitions will be informed by scientific efforts to identify, describe, and measure thresholds in physical vulnerability to natural ecosystems or to critical components of the current climate system. Expert definitions of dangerous can also be derived from scientific efforts to define thresholds in social vulnerability to climate change and can also include efforts to identify particular ceiling levels of atmospheric greenhouse gas concentrations or global temperatures. Lay public perceptions and interpretations of what constitutes "dangerous climate change," however, are based on psychological, social, moral, institutional, and cultural processes, including trust, values, worldviews, and personal experiences. These perceptions are made all the more complex by climate change's "invisibility" for many. Leiserowitz (2005) found that while experts tend to narrowly define risks using two dimensions (e.g., probabilities and severity of consequences), the general public tends to utilize a much more multidimensional and complex set of assessments. Thus the media's role in choosing how to define and frame "danger" is directly implicated in the translation of scientific information and the way in which this information relates to the aforementioned social and psychological factors. Relatedly, it is important to note that these factors are subject to change and can be manipulated by the messenger. For example, an economic frame used by conservatives to denounce climate change mitigation policies might also be employed by environmental policy makers as a way to persuade people that the economy will be revitalized by investing in clean energy technology (Nisbet 2009).

Previous research on framing has looked at the ways in which people have responded to different terminologies to describe the changes they are seeing in the environment. The term "global warming," for example, focuses attention on temperature *increases*, which makes it a less convincing word for climate skeptics during the winter. For example, US President Donald Trump tweeted on Dec. 28, 2017:

> In the East, it could be the COLDEST New Year's Eve on record. Perhaps we could use a little bit of that good old Global Warming that our Country, but not other countries, was going to pay TRILLIONS OF DOLLARS to protect against. Bundle up!

Such idiotic comments are not new. The Drudge Report ran a headline in 2004, "Gore to warn of 'global warming' on New York City's coldest day in decades!." The implication, however, with regard to the fact that the same rhetoric is being used and distributed to the mass public suggests that climate change communication scholars and practitioners still have much work to do. Perhaps the term "global warming" will always remain too confusing for some.

Research suggests that the term "climate change," in contrast, may recruit more general associations of temperature *changes*, which can more easily accommodate cold temperatures and record snowfalls. Whitmarsh (2009) found that "global warming" evokes stronger connotations of human causation, whereas "climate change" evokes stronger connotations of natural causation. This implication is important to consider, especially in American politics, where conservatives tend to be more skeptical about the phenomenon and particularly its human origins (McCright and Dunlap 2011).

43.5 Framing Considerations for Different Stakeholders

43.5.1 Individual Attitudes, Beliefs, and Personal Experiences

A recent line of scholarship focuses on the effects of personal experience on climate change beliefs, focusing on the relationship between long- and short-term exposure to temperature anomalies and extreme weather events, such as flooding, droughts, and hurricanes (Akerlof et al. 2013; Hamilton 2011; Pidgeon 2012).

Akerlof et al. (2013) found that perceived personal experience of global warming appears to heighten people's perception of the risks, likely through some combination of direct experience, vicarious experience (e.g., news media stories), and social construction. They suggest this could be an indication that a small percentage of the general public are able to tap into aspects of direct experience that influence their local perceptions of climate change risk, apart from their political and cultural identities. These findings are particularly meaningful given the influence of increasing political polarization.

For those individuals for whom climate change remains largely invisible, communication scholars and practitioners must think of how to make the impacts resonate on a personal level. First communicators must understand that any information they are offering about an issue must compete with the immediate challenges involved with supporting one's family. Most individuals (even scientists) cannot and will never fully grasp and hold this amount of scientific complexity and uncertainty in their minds, much less be able to process it (Moser 2010). Even if it were to be accepted, mitigation and adaptation responses offer another slew of complexities to understand and engage in. Climate change communication scholars and practitioners must craft communication that takes not only the complexities into account but the lack of personal resources to accommodate them as well.

Given these complexities, Sander van der Linden (2014) suggests that successful climate change campaigns must meet three criteria: (1) interventions should design integrative communication messages that appeal to cognitive, experiential, *as well as* normative dimensions of human behavior; (2) the context and relevance of climate change needs to be made explicit; and (3) specific behaviors should be targeted, paying close attention to the psychological determinants of the behaviors that need to be changed.

43.6 The Influence of a Fragmented Media Landscape

In framing climate change, the media play many crucial roles. Brulle et al. (2012) conducted a study that included coverage of climate change in *The New York Times*, major broadcast television nightly news coverage, and weekly magazine coverage. They found perhaps not surprisingly, the greater the quantity of media coverage, the greater the level of public concern. The media reflects and shapes worldviews and values that inform public opinion and the public's understanding of options to mitigate climate change, shifts attention to specific areas the public should focus on and care about, suggests linkages between events, and proposes which actors should be seen as responsible and accountable.

The issue now is that while an increase in and fragmentation of media – in both the possibilities of consumption (via social media, mainstream media, blogs, computer games, etc.) and platform distribution (via computers, mobile app technology, printed paper, television, radio) – brings more possibilities for education and awareness, people can also choose to read information that is already aligned with and supports their political and cultural worldviews and beliefs. One implication of this fragmentation is the magnification of existing social divisions and the spread of misinformation (Moser 2010). Another potential complication is that while attention to climate change at news outlets such as *The New York Times* and *Washington Post* reaches record highs, this coverage may actually reach a proportionally smaller audience now than it would have in the past (Nisbet 2009).

Moser (2016) articulates the paradox of the current moment in the climate change communication media landscape well. "The tension between climate change as a scientifically, politically, socioeconomically and culturally complex phenomenon often requiring expert communication and interpretation on the one hand and a media landscape that is in the hands of the variably educated, motivated and ideologically leaning many on the other, however, could not be starker."

It is important to note amidst this tension that recent studies have shown that media focusing on climate change *MUST* not only reach lay audiences, but must be effective at persuading political elites. This is because, as noted earlier, science-based information is actually limited in shaping public concern, while other, more political communications appear to be more important to the public. Climate change communicators would do well to focus on getting the attention of political and social elites, not just the general public (Brulle et al. 2012).

The same might be said for engaging and sustaining participation in social movements around climate change. Some theorists and advocates argue that we must do more than mobilize massive engagement and get the attention of elites; we must also hold elites accountable for the ways in which they are culpable for the grievances caused by climate change. This next section explores how climate change communicators might be more effective at doing this.

43.7 Climate Advocacy and Social Movements Must Adhere to Frames That Hold Perpetrators Accountable, Including Those Within the Movement

The United Nations Framework Convention on Climate Change (UNFCCC), with its myriad meetings and activities from the Rio Earth Summit in 1992 to Kyoto in 1997, Copenhagen in 2010, and Paris in 2015, has inspired and helped to sustain multinational mobilization around climate change. In 2010, Brulle (2014) identified 467 unique organizations that comprised the US climate change movement alone, and the numbers are growing.

At the core of all social change processes lie cultural and political disputes to maintain or redefine a field of practice involving a number of actors, including industry organizations and their trade associations, professional bodies, government actors, social movements, and countermovements. The most basic way that social movements change the social landscape is by framing grievances in ways that resonate with members of civil society (Caniglia et al. 2015). Social movements focus members of civil society on particular dimensions of social problems of concern and provide clear definitions of those problems, along with arguments regarding who is at fault and what options exist for solving their grievances (Caniglia et al. 2015). To create and sustain a social movement on climate change, however, Brown (2016) argues that it is not enough for advocates to counter the false scientific and economic claims of climate change policy opponents; they must constantly seek to educate about the causes of the injustices that climate change is causing if they seek to build and sustain a social movement.

But accountability must extend to those within the climate and environmental movements as well and the particular framing choices they made, the consequences of which have contributed to partisan and cultural divides. Historically, large environmental organizations, which were largely funded by white members and run by men, developed fear-based public messaging around controlling consumption ("Reduce, reuse, recycle!") and "saving the planet." Framing the issue in these terms has worked to center certain cultural values, perhaps at the expense of alienating others. According to Corner et al. (2014), one unintended result is that public engagement with climate change has become polarized along values-based lines: individuals and groups that tend to strongly endorse certain values have come to view climate change as a serious problem requiring immediate action, while those who more strongly endorse different values have come to view action on climate change as an (implicit) attack on or deliberate ignoring of their own values. Recently, the environmental justice movement, which aims to center marginalized voices that have largely been on the outside of mainstream environmental movements, has made meaningful intersections with climate change movements. Such intersectionality accompanied with an awareness of and sensitivity to different cultural and social beliefs, values, and worldviews must continue if the movement is going to grow and sustain itself. Climate change communication experts can play a key role in facilitating those connections.

43.8 Measuring the Effectiveness of Climate Change Communication

Currently, there is a dearth of sustained research that has measured the effectiveness of climate change communication. This is partially due to the complexity of the word "effectiveness" and the trouble with developing generalizable frameworks to measure it – a difficulty quite familiar to the development communication field. Most concerted communication campaigns on climate change have not been guided or carefully assessed by follow-up evaluation studies to discern whether the goals set initially had been achieved and, if not, why. Historically, the success of a communication campaign has been measured by the numbers of printed pamphlets delivered, media hits, or website visits. Some follow-up studies have assessed changes in attitudes before and after specific events focused on climate change, such as viewing Al Gore's film *An Inconvenient Truth*, attending or watching the *Live Earth Concerts* in 2007, or viewing the action thriller *The Day After Tomorrow* (Moser 2010).

Researchers might fill a gap in evaluation by working with organizations to develop ways to document and evaluate their communication and engagement practices and their subsequent short- and long-term impacts. Kahan (2015) argues that much like the measurement problem of quantum physics, the measurement problem of the science of science communication involves the intrusion of multiple competing forces including education, democratic politics, and other cultural and social domains that force individuals to engage information from one of these perspectives only. Thus, the people whose orienting influence evaluators need to observe are not necessarily scientists but rather the thought leaders within specific communities. "They are the people in their everyday lives whose guiding example ordinary members of the public use to figure out what evidence of scientific belief they should credit and which they should dismiss. We can through these interactions measure *what they* [the public] *know* or *measure who they are*, but we cannot do both at once."

Kahan's insights are crucial for communication scholars seeking to evaluate the effectiveness of their campaigns. One can measure how much somebody knows about climate change in terms of the information they have, but one must also figure out a way to measure what they will *DO* with the information, and this is based on their understanding of who they are – their identity-protective selves which research suggests will be prioritized as a way of preserving their particular "group status." Greater scientific knowledge can actually increase one's capacity for and facility with explaining away the evidence relating to their groups' positions.

Climate change communication researchers interested in developing effective ways to evaluate climate campaign efforts might be interested in borrowing frameworks from development communication scholars. Recently, communication scholars have tried to address the social, cultural, institutional, and organizational complexities of measuring the success of their efforts by developing more holistic frameworks for evaluating projects – that is, shifting the focus from measurement-oriented approaches, which have historically been used as a way to be accountable to large donors, to one grounded in a systems approach and complexity theory. This approach emphasizes long-term engagement in all evaluation stages of a project. While project constraints don't always make long-term engagement possible, by grounding an assessment framework in a wider lens, it becomes possible for the assessment itself to be seen as an ongoing learning experience and less about managing a set of specific deliverables within a specific timeframe. Lennie and Tacchi (2013) argue that social change needs to be seen as dynamic, nonlinear, and unpredictable. They developed seven interrelated framework components for evaluation. They suggest the framework must be participatory (for trust and mutual learning), holistic (must take into account interrelationships and networks), complex (outcomes unknowable in advance), critical (focus on gender and other differences), realistic, learning-based, and emergent. Such an approach shifts the focus away from proving impacts to improving initiatives, using evaluation to support the development of innovation, and redirecting the focus to internal and community accountability. Climate change campaign efforts – especially those that center a dialogic approach – would be well suited to this framework.

43.9 Conclusion: Lessons Learned and Looking Ahead

This chapter has attempted to provide a brief overview of the landscape of climate change communication, including the ways in which it mirrors the historical trajectory of development communication as well as mapping out the field's own unique and particular challenges. Several important themes have emerged which may be important for both development and climate change scholars.

First unlike development communication, more climate change communication research is needed in developing countries. Although there are some studies that have been done outside of the United States and Europe, the majority continue to center the thoughts and behaviors of the Western world. Ming-Lee et al. (2015) note that academics in developed countries tend to assume widespread awareness of climate change. However, these awareness and risk perception are unevenly distributed around the world. While high levels of awareness (over 90%) were reported in developed countries, the majority of populations in developing countries from Africa to the Middle East and Asia – including more than 65% of respondents in countries such as Egypt, Bangladesh, Nigeria, and India – have never heard of climate change (Ming-Lee et al. 2015).

43.9.1 Distribution of Scientific Information Is Not Enough for Sustained Behavior Change

Research from multiple disciplines suggests that promulgation and distribution of scientific information to the public on climate change has a minimal effect. Information-based science advocacy has had only a minor effect on public concern, while political mobilization by elites and advocacy groups has been critical in influencing climate change concern (Scruggs and Benegal 2012). Thus communication scholars and practitioners *MUST* address the complexities of socially embedded uses of scientific knowledge by multiple stakeholders. The knowledge and mechanisms available (or not) to translate understanding and concern into practice must be the central focus of relevant communication efforts.

Some of these complexities are most evident in the political sphere, where climate change beliefs have become deeply partisan, particularly in the United States where recent research suggests that education and self-reported understanding of global warming have little effect on the views of climate change held by Republicans and conservatives. Thus "bombarding citizens with more information doesn't diminish polarization but instead aggravates it by amplifying the association between competing identities and competing positions on climate change." (Kahan 2015). Rather than focusing on how to get out more information, researchers would do well to focus on the circumstances that make recognizing valid scientific information hostile to the identities of certain populations.

43.9.2 Focus on Positive Solutions, Local Stories

In terms of communication that motivates and inspires action around climate change, research suggests that to enhance engagement, positive feedback on individual actions with a focus on solutions as well as locally and personally relevant framings is most effective. This includes increasing visibility of the issue that will connect locally to individuals as well as making climate-related issues more concrete as opposed to only focusing on the negative consequences (Whitmarsh 2011; Wibeck 2013). A supportive narrative might also evoke "climate protection" as a source of a socially desirable identity which might signal to a population the need for behavior and policy change (Moser 2010).

Kahan et al. (2012) note that communicators should endeavor to create a deliberative climate in which accepting the best available science does not threaten any group's values. They suggest that effective strategies can include use of culturally diverse communicators, whose affinity with different communities enhances their credibility, and information-framing techniques that invest policy solutions in ways that will resonate with diverse groups.

43.9.3 Interdisciplinary Sharing and Dialogic Communication Between All Stakeholders Must Be Central

Two-way information flows do not just mean dialogues in community; it means that opportunities must be created for all stakeholders to share information from multiple perspectives, fields, and disciplines. Information must flow from and be shared between community members, political elites, media, scientists, advocates, and researchers.

It is not yet clear at this moment just how much is shared between climate communication practitioners and climate communication scholars, but more institutionalized opportunities for sharing findings, data, insights, and experiences would benefit all stakeholders involved in communicating the issue. Moser (2016) notes that relatively few communication researchers actively interact with those who do the majority of climate communication. Academics typically are not rewarded for such outreach; it is time-consuming; researchers are not trained to do so effectively, and given the often polarized atmosphere around climate change, many shy away from it (Moser 2016). However, if researchers want climate communication to be as effective and impactful as it could be, their work must connect more effectively with those who do most of the talking – climate scientists, policy makers, advocates in all sectors of society, journalists, editors, and public intellectuals.

43.9.4 Communication Efforts and Their Evaluation Need to Be Long-Term and Sustained

The effects of communication on the general public regarding climate change are short-lived. Brulle et al. (2012) found that a high level of public concern was seen only during a period of both high levels of media coverage and active statements about the issue from political elites. It rapidly declined when these two factors declined. Thus, if public concern is to be sustained, there must be continuous and sustained public communications efforts to maintain public support for climate change action, especially in the face of opposing messaging campaigns.

Despite increased global temperatures, record-breaking storms, flooding, and droughts around the world, there are still majorities of people in many countries for whom the word "climate change" is not familiar and/or who do not have the power to play an active role in developing the adaptation and mitigation policies that will best serve them. Despite predictions from scientists regarding climate change's serious and often devastating impacts, which include massive human migrations, poisoned oceans, a range of extinctions, food shortages, and increased conflict, many people – particularly in developed countries – continue to rank climate change as a low priority. Until there is a sustained reconciliation between climate change and the willingness of stakeholders, particularly those in power, to prioritize necessary action, the need for effective climate change communication will remain urgent and critical.

References

Akerlof K, Maibach EW, Fitzgerald D, Cedeno AY, Neuman A (2013) Do people "personally experience" global warming, and if so how, and does it matter? Glob Environ Chang 23(1):81–91

Brown D (2016) What advocates of strong government action on climate change should learn from sociology. Ethics Clim. Retrieved from https://ethicsandclimate.org/category/social-movements-and-climate-change/. 18 Jan 2018

Brownlee MTJ, Powell RB, Hallo JC (2013) A review of the foundational processes that influence beliefs in climate change: opportunities for environmental education research. Environ Educ Res 19:1–20

Brulle RJ (2014) Institutionalizing delay: foundation funding and the creation of U.S. climate change countermovement organizations. Clim Chang 122:681–694

Brulle RJ, Carmichael J, Jenkins JC (2012) Shifting public opinion on climate change: an empirical assessment of factors influencing concern over climate change in the U.S., 2002–2010. Clim Chang 114(2):169–188

Caniglia BS, Brulle RJ, Szasz A (2015) Civil society, social movements, and climate change. In: Climate change and society: sociological perspectives. Oxford University Press, New York, pp 235–268

Cohen D (2017) New York Mag's climate disaster porn gets it painfully wrong. Retrieved from https://jacobinmag.com/2017/07/climate-change-new-york-magazine-response. 18 Jan 2018

Corner A, Markowitz E, Pidgeon N (2014) Public engagement with climate change: the role of human values. WIREs Clim Chang 5:411–422

Francis P (2015) Praise be to you (Laudato Si'): on care for our common home. Ignatius Press, San Francisco

Freire P (1970) Pedagogy of the oppressed. Continuum, New York

Hadden J (2014) Explaining variation in transnational climate change activism: the role of inter-movement spillover. Glob Environ Polit 14:7–25

Hamilton LC (2011) Education, politics and opinions about climate change evidence for interaction effects. Clim Chang 104(2):231–242

Kahan DM (2015) Climate-science communication and the measurement problem. Polit Psychol 36:1–43

Kahan DM, Peters E, Wittlin M, Slovic P, Ouellette LL (2012) The polarizing impact of science literacy and numeracy on perceived climate change risks. Nat Clim Chang 2(10):732–735

Leiserowitz A (2005) American risk perceptions: is climate change dangerous? Risk Anal 25(6):1433–1422

Lennie J, Tacchi J (2013) Evaluating communication for development: a framework for social change. Routledge, New York

Lerner D (1958) The passing of traditional society: modernizing the Middle East. Free Press, New York

Luers A (2013) Rethinking US climate advocacy. Clim Chang 120:13–19

Lupia A (2013) Communicating science in politicized environments. Proc Natl Acad Sci 110(Suppl 3): 14048–14054

Marquart-Pyatt S, Shwom R, Dietz T, Dunlap R, Kaplowitz S, McCright A, Zahran S (2011) Understanding public opinion on climate change: a call for research. Environment 53(4):38–42

McAdam D (2016) Social movement theory and the prospects for climate change activists in the United States. Ann Rev Polit Sci 20:189–208

McCright AM, Dunlap RE (2011) The politicization of climate change and polarization in the American public's views of global warming, 2001–2010. Sociol Q 52(2):155–194

Ming-Lee T, Markowitz EM, Howe PM, Ko CY, Leiserowitz AA (2015) Predictors of public climate change awareness and risk perception around the world. Nat Clim Chang 5:1014–1020

Moser SC (2010) Communicating climate change: history, challenges, process and future directions. WIREs Clim Chang 1:31–53

Moser SC (2016) What more is there to say? Reflections on climate change communication research and practice in the second decade of the 21st century. WIREs – Clim Chang 7:345–369. https://doi.org/10.1002/wcc.403

Motel S (2014) Polls show most Americans believe in climate change, but give it low priority. Pew Research Center. Retrieved from http://www.pewresearch.org/fact-tank/2014/09/23/most-americans-believe-in-climate-change-but-give-it-low-priority/

Nisbet MC (2009) Communicating climate change: why frames matter for public engagement. Environment 51(2):12–23

O'Neill S, Nicholson-Cole S (2009) "Fear won't do it": promoting positive engagement with climate change through visual and iconic representations. Sci Commun 30(3):355–379
Pidgeon N (2012) Public understanding of, and attitudes to, climate change: UK and international perspectives and policy. Clim Pol 12:85–106
Polk (2015) Communicating global to local resiliency: a case study of the transition network. Lexington Books, Lanham
Quebral N (1971) Development Communication in the Agricultural Context. In: Gumucio-Dagron-A, Tufte T (eds) Communication for Social Change Anthology: Historical and Contemporary Readings. Communication for Social Change Consortium, South Orange, 2006
Rayner T, Minns A (2015) The challenge of communicating unwelcome climate messages. Tyndall working paper no.162, University of East Anglia
Rogers E (1962) Diffusion of innovations. The Free Press, New York
Schramm W (1964) Mass media and national development: the role of information in the developing nations. UNESCO/Stanford University Press, Paris
Scruggs L, Benegal S (2012) Declining public concern about climate change: can we blame the great recession? Glob Environ Chang 22(2):505–515
Servaes J (1999) Communication for development: one world, multiple cultures. Hampton, Cresskill
Servaes J (2009) Communication policies, good governance and development journalism. Commun: South Afr J Commun Theory Res 35(1):50
van der Linden SL, Leiserowitz AA, Feinberg GD, et al. (2014) Climatic Change. 126:255. https://doi.org/10.1007/s10584-014-1190-4
Waisbord S (2008) The institutional challenges of participatory communication in international aid. Soc Identities 14(4):505–522
Wallace Wells D (2017) The uninhabitable earth: famine, economic collapse, a sun that cooks us: what climate change could wreak – sooner than you think. New York Magazine. Retrieved from http://nymag.com/daily/intelligencer/2017/07/climate-change-earth-too-hot-for-humans.html
Weber EU, Stern PC (2011) Public understanding of climate change in the United States. Am Psychol 66(4):315–328
Whitmarsh L (2009) What's in a name? Commonalities and differences in public understanding of "climate change" and "global warming." Public Underst Sci 18(4):401–420
Whitmarsh L (2011) Skepticism and uncertainty about climate change: dimensions, determinants and change over time. Glob Environ Chang 21:690–700
Wibeck V (2013) Enhancing learning, communication and public engagement about climate change – some lessons from recent literature. Environ Educ Res 20:387
Wolf J, Moser SC (2011) Individual understandings, perceptions, and engagement with climate change: insights from in-depth studies across the world. WIREs Clim Change 2:547–569
World Meteorological Organization (2017) Retrieved from https://public.wmo.int/en/media/news/2017-remains-track-be-among-3-hottest-years-record. 18 Jan 2018

Bottom-Up Networks in Pacific Island Countries: An Emerging Model for Participatory Environmental Communication

44

Usha S. Harris

Contents

Abstract

As interrelated human actions converge to produce devastating results on both human and natural environment, new frameworks are needed to enable a diversity of voices in bottom-up environmental communication. Communication is necessary in engaging people and entities in the process of identifying the problems and collectively finding solutions to a broad range of environmental and social change issues. In Pacific Island countries, where people are already facing existential threat from the many impacts of climate change, participatory media is increasingly being employed to share stories and influence power brokers. Based on the

U. S. Harris (✉)
Macquarie University, Sydney, Australia
e-mail: usha.harris@gmail.com

J. Servaes (ed.), *Handbook of Communication for Development and Social Change*,
https://doi.org/10.1007/978-981-15-2014-3_138

author's research in the Pacific, this chapter offers a model for participatory environmental communication (PEC) that brings together diverse human (and nonhuman) networks in a process of dialogue and collaboration to find solutions to the many environmental challenges.

Keywords

Pacific · Oceania · Participatory environmental communication · Climate change · Bottom-up · Networks · Diversity · Agency · Dialogue · Collaboration

44.1 Introduction

The village elders sit on the rough sandy ground near the deconsecrated church in the abandoned village of Tebunginako on Abaiang atoll in Kiribati. Children play on the soggy ground where the seawater has formed streams on what once was the village green and farmland. Salt water inundation has turned this once productive soil into barren land. People mill around the once busy Maneaba (community hall), keen to share their stories with the visitors. One of the village elders tells us why his people had to leave their village. As the seawater rose inundating the ground, they lost their taro pit, then the banana trees, and then the road access to the village. Slowly people started to find alternative sites for their home and taro pit. He is not sure why the water is rising and destroying their village. (This story was received by the author when she visited Kiribati in 2017 as part of an Australian delegation with Pacific Calling Partnership.)

Human-induced climate change and environmental degradation impact the lives and livelihoods of people in many parts of the world. Climate change is a crosscutting issue that makes its presence felt at the intersections of culture, politics, economy, and ecology. With natural disasters increasing in regularity and intensity, a range of information and communication tools is necessary as a lifeline in emergencies and long-term sustainability of vulnerable communities. Access to media and communication is essential to people's ability to tell their stories in an effort to bring about change, as many years of research in communication for development (C4D) tells us (Wilkins et al. 2014; Servaes 2008). Yet, the most affected communities are also those that struggle to have their voices heard.

Environmental communication research and analysis have centered on the top-down dissemination model, which uses mainstream media or strategic communication to disseminate messages from policymakers, scientists, and other interest groups to the wider population. Receiving less attention in this subfield of communication are bottom-up initiatives, which are critical in fostering whole of community participation. In this regard, environmental communication scholars and practitioners can benefit from experiences in the communication for development and social change fields, which have more than three decades of literature on alternative communication paradigms that prioritize bottom-up channels.

This chapter firstly introduces a conceptual framework for participatory environmental communication (PEC), a model that encourages connections among diverse stakeholders at local, national, and international levels. The second part of the chapter presents an overview of the Pacific region and discusses C4D approaches that communities in Pacific Island countries (PICs) employ to share their knowledge and experiences of climate change.

44.2 Framework

44.2.1 Participatory Environmental Communication (PEC)

International agencies, local NGOs, scholars, and practitioners all recognize the need for concepts and processes that strengthen both human and nonhuman interconnections. It is increasingly clear now that diverse forms of knowledge – inclusive of scientific, experiential, and traditional – are essential in order to understand vulnerabilities and resilience of local communities in the face of environmental changes and to identify effective adaptation methods on the ground. A participatory model of communication has an important role in disrupting centers of power in the face of draconian laws which inhibit citizen engagement in environmental protection. With a need to find new ways of thinking about and practicing environmental communication, this chapter offers a conceptual framework for participatory environmental communication (PEC). In introducing the term participatory environmental communication, the author brings together epistemic knowledge and practice-based insights from the two areas: participatory communication and environmental communication.

Participatory environmental communication is a process-oriented approach, which engages ordinary people in dialogue about environmental concerns so that they are able to identify the problems and are collectively empowered to make decisions to improve their situation. PEC incorporates three interrelated strands: *diversity, network, and agency.* I refer to these as the *DNA* in participatory environmental communication because each of these strands is important to the survival of both natural and social systems (Harris 2019). They are the essential building blocks of a resilient society. Embedded in these elements are the core principles of participatory communication which White and Patel so eloquently describe as "equalitarian, transactive and dialogic" in nature (1994: 363).

Through collaboration and dialogue, the PEC approach encourages cross-fertilization, forges new relational possibilities, and inspires problem-solving. In participatory environmental communication, the voice of affected communities is integral to finding solutions. PEC seeks to bring in traditional methods of adaptation alongside scientific understanding of climate change. The three elements – *diversity, network, and agency* – are interconnected and integral to participatory environmental communication processes that create the condition for greater sustainability (Fig. 1).

Diversity enables innovative and transformative thinking. The term here means both difference and inclusion of a broad range of factors – different knowledge systems, sociocultural values and beliefs, abilities, talents, demographic variables (age, gender, class, ethnicity), and the nonhuman world (ecosystems, technologies, texts).

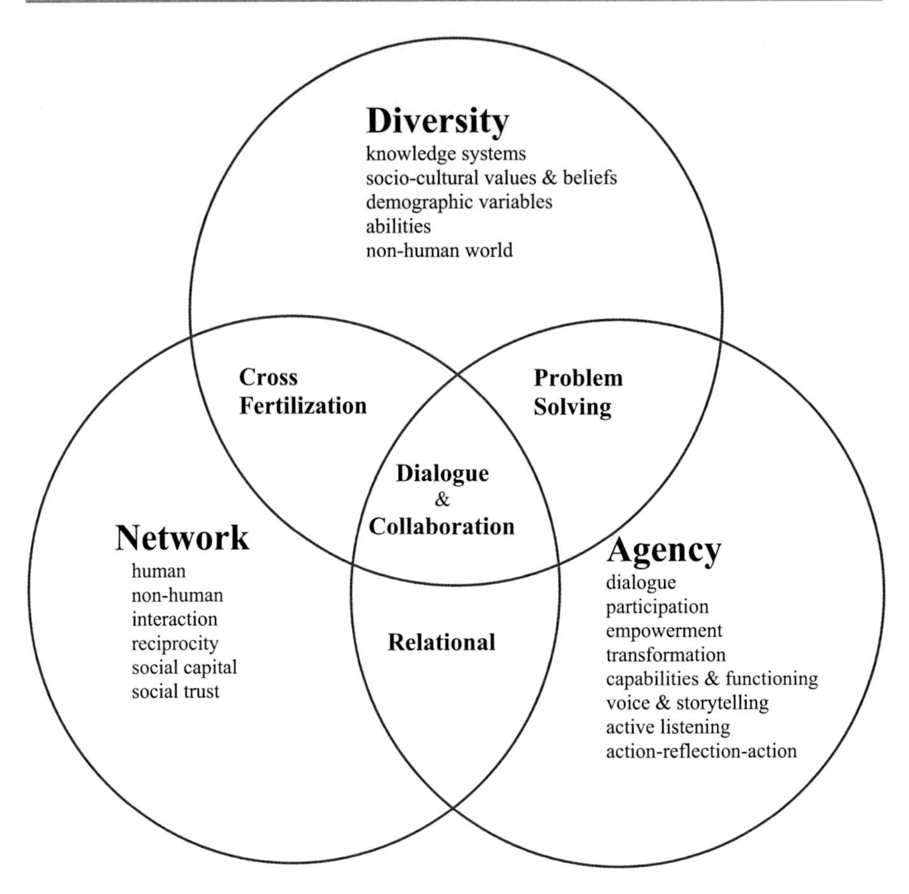

Fig. 1 The DNA framework and its attributes. (Source: Adapted from Harris 2019)

Networks are a complex system of relationships connecting both human and nonhuman worlds. They include vertical and horizontal human networks as well as nonhuman networks. The understanding that humans are but one part of this intricate web of creation, and not separate from it, is essential in the action-reflection-action cycle of the PEC process.

Agency is an action or a doing of human and more-than-human entities, which leads to an effect or outcome. Agency in the natural world contributes to the efficient working of an ecosystem. Human agency results from a realization of our own power and potential through dialogic encounters that act as the catalyst for change.

At its core, participatory environmental communication framework is designed to advance collaboration and dialogue and empower people to become engaged environmental citizens. A detailed discussion of the framework and concepts can be found in this author's book titled *Participatory Media in Environmental Communication* (Harris 2019).

Table 1 *DNA* essentials

Elements	*Diversity*	*Network*	*Agency*
Definition	*Diversity* describes both difference and inclusion of a range of entities, abilities, and ways of knowing	*Networks* are complex systems of relationships and interactions (links) between human and nonhuman worlds (nodes) vital for exchange of information and resources	*Agency* is the capacity of an actor to have the power to act in a given context
Attributes	Knowledge systems Sociocultural values and beliefs Demographic variables (age, gender, race, identities, education, place disabilities, etc.) Interface Nonhuman world Biosphere Holistic	Human Nonhuman Relational (form and content) Reciprocity Social capital Social trust Social cohesion	Dialogue Participation Empowerment Transformation Capabilities and functionings Voice and storytelling Active listening Agents of change Participatory action research
Manifestations	Communication Cross-fertilization New connection Innovation Intergenerational Intercultural Interdisciplinary Interfaith Interagency Cross-sector	Communication Local or endogenous network Heterogeneous network Trusted network Top down Horizontal Bridging and bonding Collaboration Cooperation	Communication Catalyst for change Mutual learning Problem-solving Gather information, analyze, design, knowledge sharing Transformative thinking Awareness raising Active participation Marginalized voices Action-reflection-action cycle

Source: Harris (2019), p. 37

Manifestation of communication produces a range of responses as illustrated in Table 1. Diversity in communicative practices leads to new connections and innovation and encourages intergenerational/intercultural/interdisciplinary/interagency dialogue; communication forges a variety of new networks or strengthens existing networks of collaboration; agency of individuals in communication processes becomes a catalyst for change, mutual learning, problem-solving, and transformative thinking.

PEC projects are designed as bottom-up initiatives which enable integration of local narratives and cultural practices. Through dialogue with other community members, discussions with environmental experts and by researching traditional adaptation methods, vulnerable communities learn how to plan for the future.

Participatory projects may include anything from folk theatre production to everyday media tools such as smartphones, digital cameras, and audio recording devices. Using these tools communities collectively identify problems and gather and analyze information to find viable solutions through the process of content creation. This process of production enables communities to make that critical link between (say) the impacts of climate change and why it is necessary to adapt and change, be it in relation to the types of crops they plant, fishing methods, or coastal management. In the next section, examples of participatory communication approaches used by NGOs and communities in the Pacific region are discussed.

44.3 The Pacific Region

44.3.1 Background

The Pacific Ocean spans 30% of the planet's surface and is home to 2.3 million people. Pacific Islanders inhabit more than 2,000 islands, which range from coral atolls to mountainous volcanic formations, extending from New Zealand in the south to Hawaii in the north. The three geo-cultural subregions are Melanesia, Micronesia, and Polynesia with each group having its own diversity in cultural and social practices (Smith n.d.).

More than 50% of the region's population lives within 1.5 km of the shore, and many of these countries are barely a few meters above sea level, putting them at extreme risk from sea level rise. Despite the fact that Pacific Island countries (PICs) emit less than 0.01% of total global greenhouse gas responsible for climate change, the region is among the most vulnerable in the world to its negative impacts (Oxfam 2015; WWF Pacific 2015). Changes are already being recorded in the region's climate and ocean. Parts of the Western Pacific have experienced the fastest rate of sea level rise in the world at 10 mm per year (SPREP 2013). This is already affecting communities living along the coasts as well as the coastal ecosystems and natural resources that people depend on for their survival. In Fiji several coastal communities are being relocated to higher ground with 40 other villages planning for relocation in the near future as a result of flooding from seawater inundation and storm surges. Permanent cross-border resettlement is one of the projected impacts of climate change for some small island states, among them are Tuvalu, Kiribati, and the Marshall Islands, where their populations face the prospect of becoming climate refugees (McNamara and Gibson 2009; Yamamoto and Esteban 2014). Yet government officials and communities in these countries consider that an option of last resort because displacement not only means loss of ancestral homes but also loss of their cultural identity, sovereignty, and connection to land and sea (Richards and Bradshaw 2017).

Rising sea levels are just one of the many effects of climate change in the Pacific Islands. Associated impacts of climate change, both currently experienced and projected, threaten communities' livelihoods and long-term viability. These include an increase in the frequency and intensity of cyclones, coral bleaching, coastal erosion, changing patterns of pests and diseases, saltwater intrusion, storm surges and flooding, increased temperatures (affecting taro production on lowlands, increasing risk of fire), and drought. Irregular rainfalls resulting in floods and droughts will have a devastating impact on agriculture production, forcing reliance on imported foods and threatening the livelihood of local farmers (SPREP 2013).

Collectively the Pacific Islands are the most vulnerable in the world. They suffer floods and droughts controlled by the El Niño-Southern Oscillation (ENSO) phenomenon. Drought and flood risks change during La Niña (normal cool years) and El Niño (warmer years) affecting different areas of the Pacific in different ways. For example, La Niña brings flood risk in Western Pacific and drought in Central Pacific, and during El Niño it is reversed with flood risk in Central Pacific and drought in Western Pacific. Climate change affects the frequency and intensity of natural hazards, consequently impacting on people's ability to respond effectively. Hydrometeorological hazards most common in the Pacific are ocean acidification and temperature increase, which means more warm days and warm nights and fewer cold days and cold nights. Two recent super cyclones devastated homes, infrastructure, and the economies of Fiji and Vanuatu, affecting the livelihood of almost half of their populations. The region faces other natural hazards not linked to climate change – volcanoes, tsunamis, and earthquakes – due to its location in the Pacific Ring of Fire.

Pacific Islanders face numerous homegrown ecological challenges. These include poor urban planning, waste disposal, energy inefficiency, and overuse of natural resources. Social and economic factors facing urban populations in Pacific Island countries include urban drift leading to overpopulation in cities and towns, insecure housing, and high cost of living leading to poor nutrition and health, which add to the burden of recovery from hazards. The Secretariat of the Pacific Regional Environment Programme (SPREP) has identified four main regional priorities in its 10-year strategic plan (SPREP 2017: 10): climate change resilience; ecosystem and biodiversity protection; waste management and pollution control; and environmental governance.

44.3.2 Nuclear and Mining Catastrophes

Between 1946 and 1996, Oceania was the site for nuclear testing by the United States, Britain, and France. A total of 315 atmospheric and underground tests were carried out in the colonial territories of the Marshall Islands, Gilbert and Ellis Islands (now Kiribati and Tuvalu), and the Mururoa Atoll. The colonial powers had little regard for the well-being of Pacific Island inhabitants who were under their jurisdiction. Instead the Oceanic region, with 96% sea surface and thousands of uninhabited islands, was seen as a "safe zone" because of its distance from the metropolitan centers of the world. Smith (1997: 1) explains that during the postwar period, the microstates of the Pacific were generally dismissed as "political backwaters"... and Pacific Islanders portrayed as passive actors – as opposed to recognizing their abilities to influence events.

The Pacific region has experienced other conflicts between the local population and foreign interests arising from environmental mismanagement of land and water in the extraction of mineral resources. Papua New Guinea endured ecological destruction on a huge scale as a direct result of bad mining practices in the Bougainville and Ok Tedi mines (Kirsch 2007, 2014). One of the worst human-caused ecological disasters occurred at the Ok Tedi copper-gold mine in the western province of Papua New Guinea operated by the Australian mining giant then called Broken Hill Proprietary (BHP, now BHP Billiton). An average of 50 million tonnes of tailings and mine waste per year were dumped in the Ok Tedi and Fly Rivers, causing large-scale damage to the surrounding forests and wildlife, poisoning the river system and village gardens, and leading to the displacement of villagers from their traditional lands. A class action by Papua New Guinean landowners forced BHP to commission a report into its operations at Ok Tedi mines. In 1996 the company reached an out-of-court settlement with the landowners for an estimated US $500 million. With BHP's withdrawal from the mine, its commitment to clean up the polluted areas remains to be met (Troubled Waters 2012).

In Bougainville, conflict ensued between landowners and mine operators as a result of pollution from mine tailings to the surrounding environment and its impact on the islanders' traditional way of life. Panguna mine, one of the world's biggest open-cut copper mines, was owned and operated by a subsidiary of the mining giant Rio Tinto. While the mine generated billions of dollars of revenue for the PNG government and huge profits for the mine operators, Bougainville landowners received little compensation for their land and rivers, which were being poisoned as reported in the *Sydney Morning Herald*: "...millions of tonnes of acid-laced mine tailings killed the Jaba and Kawerong rivers. The rivers had been a source of water and food for thousands, but large sections now resemble a moonscape, forcing people to leave their homes" (Flitton 2016). With their demands unmet, a rebel group made up of young Bougainville landowners formed the Bougainville Revolutionary Army (BRA). They fought a bloody civil war which drove out both the PNG government forces and Rio Tinto Zinc (RTZ) leading to the closure of the mine (see Robie 2014; Phillips 2015). The decade-long conflict, which began in 1988 and cost 20,000 lives, became known as the first successful eco-revolution.

In West Papua, similar concerns are recorded in the operation of the Freeport copper mine, where 200,000 tonnes of tailings per day (over 80 million tonnes per year) are dumped in the river systems, causing floods, deforestation, and grave impacts on community health and livelihoods (Troubled Waters 2012). In the case of West Papua, which is under Indonesian rule, the mining operations not only imperil the environment but also the indigenous population who are considered to be at risk of becoming "an anthropological museum exhibit of a bygone culture" (Schulman 2016). The West Papua resistance to Indonesian rule has claimed thousands of lives; however the international communities including Australia and the United States continue to recognize Indonesian sovereignty in the region. Many small island nations rely exclusively on their river systems for fresh water. Protection of rivers and streams from long-term pollution is more critical than ever before, given the increased presence and severity of droughts due to climate change.

Nauru and Banaba Island in Kiribati are other examples of Pacific nations that have borne the consequences of decisions made by colonial powers with vested commercial interest, which disregard the rights of indigenous people under their protection. Both these countries have suffered irreversible environmental damage from intensive phosphate mining with the loss of 80% of island soil from strip mining. Much of the mined phosphate has become part of Australian farmland as agricultural fertilizer. After having experienced the ups and downs of its mining fortunes, Nauru now earns its income by hosting the much-maligned offshore refugee detention center for the Australian government.

44.4 Networks

44.4.1 Pacific Regionalism

Pacific scholars have searched for their unique Pacific identity borne out of Pacific indigenous histories in order to counteract the neocolonial imagination. The imperialists envisioned the Pacific Island countries as small, remote, disconnected, and dependent. These nation-states were in need of salvation by their economically powerful neighbors, especially "Australia which saw itself as the natural leader of the post-colonial South Pacific" (Fry 1996: 2).

Pacific scholars emphasize the interconnectedness of islands as an important framework of analysis in Pacific Island studies. In 1994 eminent Pacific scholar Epeli Hau'ofa argued that the Oceanic region should be imagined as "Our sea of islands" with the Pacific Ocean as their common home (Hau'ofa 1994). In reimagining the region, Hau'ofa recalled the pre-colonial links of Pacific cultures, their mastery of the ocean, and the myths and legends that guided them. As he observed:

> But if we look at the myths, legends, and oral traditions, and cosmologies of the peoples of Oceania, it becomes evident that they did not conceive of their world in such microscopic proportions. Their universe comprised not only land surfaces, but the surrounding ocean as far as they could traverse and exploit it, the underworld with its fire-controlling and earth-shaking denizens, and the heavens above with their hierarchies of powerful god and named stars and constellations that people could count on to guide the ways across the seas. (1994: 154)

Hau'ofa's call to reframe the Pacific included not only dry land surfaces but the oceans that surrounded them because "There is a world of difference between viewing the Pacific as 'islands in a far sea' and as 'a sea of islands'" (Hau'ofa 1994). A new generation of Pacific leaders asserts this revised regional identity of an "Oceanic continent" echoing Hau'ofa's imaginary identity, which favors the interconnections of the Pacific Island states.

New Pacific regionalism has emerged despite pressure from the metropolitan neighbors. Maureen Penjueli, coordinator of the Pacific Network on Globalization, observes:

> Continued attempts by the Forum Secretariat to frame the region consistently as small, vulnerable, fragile, aid-dependent is no longer a working narrative for our people, let alone our countries. The wind of self-determination has set our countries on a journey to reduce our dependency on foreign aid, to be more self-sufficient, to negotiate and demand a fair share of the wealth of our resources, to redefine security and find newer partners (rightly or wrongly). (Penjueli 2017)

Instead of isolation, the surrounding sea presents new economic opportunities for the ocean states, especially the benefits generated under the exclusive economic zones (EEZ). Some of the smallest islands today command some of the largest ocean resources under the terms prescribed by the United Nations Convention on the Law of the Sea. This includes the 370 km boundary over which a state has special rights regarding the exploration and use of marine resources, including energy production from water and wind (Rona 2003). Kiribati today controls one of the largest EEZ in the world, embracing fisheries stock and future prospects of mining the ocean floor. Ironically this brings with it new threats of domination and exploitation by more powerful states which are already maneuvering for favorable positions through political and soft power diplomacy. It also raises the ugly specter of a new frontier for environmental disasters with growing interest in seabed mining and exploration.

In the twenty-first century, Pacific Island nations are strengthening their networks across a wide spectrum of sociocultural, economic, political, and scientific endeavors (Harris 2014). This solidarity has been most pronounced in the area of climate change negotiations. In 2015 Leaders of Pacific Islands Development Forum signed the Suva Declaration on Climate Change which called for an agreement "to stabilise global average temperature increase to well below 1.5°C above pre-industrial levels" (Pacific Islands Development Forum 2015: 1). The Declaration reflects the concerns of all Pacific Islands nations by noting that "climate change poses irreversible loss and damage to our people, societies, livelihoods, and natural environments; creating existential threats to our very survival and other violations of human rights to entire Pacific Small Island Developing States" (Pacific Islands Development Forum 2015: 1). The main positions put forward in the Suva Declaration were reflected in the final Paris Climate Change Agreement of the 21st session of the Conference of Parties (COP21) in Paris in 2015. The Suva Declaration for the first time enabled Pacific Island nations to put a cohesive Pacific voice in climate change negotiations. The strong-arm tactics of its more powerful neighbor, Australia, which had tried to water down the Pacific leaders' statement on climate change, failed to work. Fiji's presidency at COP23 in Bonn, Germany, in 2017 is further proof of a strengthening Pacific diplomacy on the world stage.

44.4.2 Bottom-Up Networks

At another level, regional civil society organizations (CSO) are proactive in environmental advocacy. The Pacific Island Climate Action Network (PICAN) advocates for climate justice at regional and international forums. As part of a three-tiered network, PICAN is made up of smaller country-based networks – e.g., VCAN in Vanuatu, KiriCAN in Kiribati, and TCAN in Tuvalu – and is linked to the worldwide Climate Action Network (CAN) which has over 900 nongovernment organizations "working to promote government and individual action to limit human-induced climate change to ecologically sustainable levels" (Climate Action Network 2017; Pacific Island Climate Action Network 2017). Pacific Climate Warriors from 350.org Pacific enlist youth to take global action on climate change. In 2014, Pacific Climate Warriors paddled out in traditional canoes to blockade the world's largest coal port in Newcastle, Australia. Through their banner "we are not drowning, we are fighting," Pacific Climate Warriors reframe the narrative to show that Pacific Islanders are not passive victims but proactive civil society actors demanding climate justice.

The "bottom-up regionalism" which once forged the nuclear-free movement is now seeing a resurgence around climate change advocacy led by civil society actors (Titifanue et al. 2017: 142). CSOs organized protests at the annual climate change discussions during the COP23 in Bonn and the crucial Paris Agreement at COP21. Social media platforms allow groups to "share information, communicate, and organize" using multimedia content, event invitations to protest sites, and use of hashtags using local and global networks. Titifanue et al. observe that after the antinuclear movements of the 1980s:

> Regional solidarity seemed to take a hiatus at the grassroots level and became simply something that decision makers and academics concerned themselves with. In recent times however, through digital interconnectivity, there has been resurgence in grassroots interest on key issues that have captivated the public imagination. (2017: 142)

Like the nuclear testing of 50 years ago, Pacific peoples are once again angry at the injustice of being at the receiving end of environmental impacts which are not of their making. A Pacific regionalism at both top and bottom levels of society has emerged to network, organize, and protest against the excesses of the developed world.

44.4.3 Faith-Based Networks

People in the Pacific are predominantly Christian. This places a responsibility on churches to communicate the growing threat of climate change to their constituencies. Churches are an integral part of the support networks as villagers are uprooted from their coastal homes and moved to higher ground. Member churches are trained to offer psychological trauma healing to communities who have been displaced from their ancestral homes. This includes workshops which explain the science of climate change to help people have a better grasp of the changes in their natural environment and the new challenges to social cohesion as communities move to locations which are closer to transport and mobile networks.

Belief in Christian narratives, such as God's promise to Noah to never flood the earth, often complicates public awareness. Some people demonstrate blind faith in terms of divine protection from natural disasters. This creates barriers to communication of knowledge of emergency preparedness, due to conflict between disaster information and religious beliefs (Tacchi et al. 2013). Religious leadership is important in breaking this cycle by reeducating both the clergy and the congregation as Rev. Seforosa Carroll, manager of Church Partnerships Pacific at UnitingWorld, explains:

> How do we as a community and people care for the earth the way we are called to? What we hope to do is to help people to engage and to see it as part of their spirituality. You've got to start from the grassroots; you've got to start with things as simple as Sunday school. Also, what is said from the pulpit; that is a retraining and a reforming of ministers. (Carroll 2017)

In Australia, Pacific Calling Partnership (PCP), an initiative of the Edmund Rice Centre, a Catholic NGO, creates spaces for I-Kiribati and Tuvaluan voices in international forums. PCP prepares the next generation of Pacific leaders by training them in advocacy and community education. The aim is to raise awareness of the situation in these countries by facilitating people-to-people exchange with politicians, policymakers, private industry, and community groups.

44.5 ICT and C4D Initiatives

44.5.1 Improving Connectivity

Pacific Island countries have experienced uneven access to broadband and mobile connectivity, with some countries like Fiji achieving high-speed Internet, while countries such as Kiribati experience slow and expensive connection. Connectivity is delivered through mobile and fixed networks with connections to satellites and undersea cables. While more Pacific nations are investing in undersea cable connections, mobile networks are favored in driving connectivity and Internet access, due to the dispersed nature of PICs, many of which are archipelagos comprised of a large number of islands. This is consistent with global trends in mobile broadband usage, which has outpaced fixed broadband subscriptions, with prices dropping substantially in least developed countries where 35% of individuals using the Internet are young people aged 15–24 (ITU 2017). The World Bank (2017) projects that in the next 25 years, most people in PICs will have access to faster and cheaper Internet services with increasing potential for information and communication technology (ICT) in environmental management, including monitoring, grievance redress, and citizenship engagement processes.

The Pacific Region Infrastructure Facility (PRIF) highlights the following changes in the Pacific (PRIF 2015):

- Mobile coverage across Fiji, Samoa, Solomon Islands, Tonga, and Vanuatu has jumped from less than half of the population in 2005 to 93% of the population in 2014.
- The cost of mobile calls fell by one third between 2005 and 2014.
- International Internet bandwidth jumped over 1500% between 2007 and 2014, rising from less than 100 megabits per second to over 1 gigabits per second (excluding Fiji which enjoyed high bandwidth from 2000 when it was connected to a submarine cable).

The Southern Cross Cable Network links the West Coast of the United States, Hawaii, Fiji, Australia, and New Zealand using high-capacity fiber-optic cable, with new connections to Tonga, Vanuatu, and Samoa. This has resulted in a phenomenal rise in international Internet bandwidth. Most island nations are connected by a single submarine cable, which makes them vulnerable in the case of cable damage. Some mobile network operators encourage the adoption of hybrid solutions, including both satellite and fiber-optic cable in the Pacific, thus reducing the cost of connection to many sparsely populated islands (UN-OHRLLS 2017).

44.5.2 Cultural Concerns

How does the new technology fit into the broad information ecosystem that already exists within communities? Access to improved Internet and social media brings both positive and negative impacts. The number of Facebook users in the region grew over 250% between 2011 and 2014, with 80% accessing the social media site via mobile phones (Minges and Stork 2015). In 2015 the Government of Nauru banned access to Facebook, blaming the social media site for creating instability in the nation, citing cyberbullying and offensive posts. The president of Nauru, Baron Waqa, said the action was taken to protect the small, tightly knit community of 10,000 people from unregulated use of the social networking site, which had created tensions among friends and families (ABC News 2015).

ICT use in the Pacific Islands must be seen in terms of challenges faced and within the wider communication context, including the culture and belief systems of Pacific Islanders. Effective use of ICT should not only aid information dissemination but also invite local participation in communicative processes. It can facilitate the preservation and sharing of indigenous knowledge, provide an essential network for information and communication exchange among the community sector and government agencies, and enable learning and skill building for the growing youth population. Digital narratives are a powerful way for people to capture the everyday life and concerns of a community. For example, by using participatory media tools, village elders can share the important history and heritage of their village with the younger generation and the wider Pacific community.

Digital technologies encourage greater participation in political dialogue. People in both urban and rural communities are now able to participate in debates from which they previously felt excluded. Facebook discussion groups, such as Yumi TokTok Stret in Vanuatu, and blogs provide a forum for exchange of information and opinion where all users can participate. There is a greater need for understanding how to integrate ICTs like mobile phones into media and communication plans for disaster-response technologies such as broadcast radio (Tacchi et al. 2013). Radio remains a key media platform for communication across vast distances and audiences. More research is needed in the Pacific on how radio can be successfully integrated as a communication strategy in climate change education and community-based adaptation.

The communication challenges presented above point to the potential that ICTs offer in addressing issues of local participation and content that are inclusive of gender, youth, and marginalized communities. As female-headed households are most affected by climate change impacts, Ospina and Heeks (2010) suggest that ICTs could play a key role by providing relevant information, capacity building, and empowerment to strengthen women's adaptive capacities.

In adaptation processes, ICTs are most effective in networking and knowledge exchange of good practices, inclusion of vulnerable communities in decision-making for best policy outcomes, and integration of information and knowledge that have local relevance in reducing risk and vulnerability (Ospina and Heeks 2010; Heeks and Ospina 2018). Message creation and dissemination must be strategically integrated with stakeholder inclusion using conventional media forms (participatory video and community radio) and new media (mobile and online platforms). As tools for empowerment, ICTs provide opportunities for local communities to raise awareness of their situation, demand action and accountability from their leaders, build coalitions, and share knowledge using peer-to-peer networks (Kalas and Finlay 2009).

44.5.3 Participatory Initiatives

Given the popularity of social media in the region, community-led initiatives are more likely to make it easier for communities to actively shape and take part in their own awareness process, introduce their own solutions to climate change mitigation, and share these solutions with other Pacific communities. Project Survival Pacific, Fiji's youth climate change movement, is an example of youth-based social media, which mainly shares stories through its Facebook page to reach out to young people. It encourages youth engagement in climate change communication and provides users with opportunities to share and co-create. This offers a good example of ICT use for interactive social dialogue, youth engagement, and participatory communication.

Participatory media training projects have been implemented regionally as part of the PACMAS strategic activity on climate change awareness. They have adopted innovative participatory approaches that highlight local issues through community involvement. Notable projects include mentoring of high school students across the Pacific in the production of media content to raise climate change awareness and capacity-building activities which provide training and mentoring of journalists in order to produce local content and improve coverage of regional and international events about climate change (PACMAS n.d.). Contents also focus on gendered aspects of climate change impact and adaptation. A series of participatory videos made with women living in a coastal village in Fiji shows how climate change disrupts women's lives on a daily basis and provides ways to find sustainable livelihoods (see Harris 2014).

The NGO 350.org, which coordinates global climate advocacy, offers its Pacific teams the opportunity to share their own perspectives on climate-related issues. Participatory media has become an important element of its work as it ensures stories about climate impacts coming out of the islands are shaped and told by Pacific Islanders living with the impacts of climate change and not by someone else. This included their use of Twitter during tropical cyclones Pam in Vanuatu and Winston in Fiji.

Two creative NGOs in Vanuatu have been using arts and theatre to educate communities. Wan Smolbag Theatre works at community level to raise awareness about a range of social and environmental issues. Through their travelling theatre groups and film productions, Wan Smolbag teaches about waste management, climate change, and the ecological importance of turtles and reefs in Vanuatu and other Pacific Islands. Further Arts, a community arts organization, runs a range of arts, media, and cultural projects to develop long-term social and commercial enterprises that are culturally, socially, environmentally, and financially sustainable. It uses participatory methodology in a peer-to-peer knowledge exchange through its media hub in Port Vila, Vanuatu.

Community radio is an important forum that invites participation by and gives voice to women and youth. femLINKpacific, a women-led community radio station in Fiji, has documented women's experiences with floods and other natural disasters in their communities since 2004. It uses rural networks such as women's clubs with 165 community-based women-led groups through their community media network. The Women's Weather Watch program has a network of women weather reporters on the ground in different towns and villages who update listeners about the local condition and imminent threat to community from a rain event or other weather-related disasters. femLINKpacific campaigns for gender-based disaster risk management-response strategies "because women are often left out of the formal discussions and decision-making process in any disaster although they are the ones that are responsible at household level" (Tacchi et al. 2013). Its online presence includes Facebook, Twitter, SoundCloud, and a website. While ICTs offer possibilities for a younger generation, issues of illiteracy and capacity often remain overlooked. For these reasons more face-to-face and radio initiatives are favored because they involve oral cultures and invite participation by those who cannot read or write.

The short life span of participatory media activities means that the effectiveness of these projects is not properly evaluated. Instead of single one-off projects that are sporadically funded, a more networked approach to participatory projects would encourage greater citizen engagement and improve community understanding of the issues. However it must be noted that short-lived participatory initiatives do not necessarily mean they have been ineffective. On the contrary, many participatory projects end because they have been successful in achieving their initial objectives, and the community feels no further need for them to be continued. With assistance from governments and involvement of community-based environmental groups, a focused campaign that strategically incorporates participatory activities would incorporate local perspectives, cultural knowledge, and concerns which are specific to the area.

44.6 Conclusion

As interrelated human actions converge to produce devastating results on the environment, communication tools that encourage greater collaboration and dialogue among diverse networks are essential. Pacific Island communities interact with various forms of ICT at the local level, using participatory approaches to develop content which is closer to their own realities and which reflects local cultures, values, and individual aspirations. It is important to gain greater understanding of the mediated implications of ICTs in climate change communication and disaster risk reduction at the community level. An evaluation tool, such as the DNA framework, helps to build a pool of knowledge from diverse stakeholders. The framework provides opportunities for new alliances that engender cross-fertilization of ideas and new ways of problem-solving. Participatory environmental communication initiatives go beyond raising awareness. The value of interpersonal dialogue, the importance of networks in social cohesion, and the agency of communities to bring about change are inextricably linked in resilience building.

References

ABC (2015) Nauru's president defends Facebook ban, says social media has 'power to create instability'. ABC News. Retrieved from http://www.abc.net.au/news/2015-05-29/nauru-president-baron-waqa-defends-facebook-ban/6507240

Carroll S (2017) Pacific stories – courageously encountering God in climate change. Pacific Calling Partnership, Edmund Rice Centre, Homebush West, NSW

Climate Action Network (2017) About CAN. Retrieved from http://www.climatenetwork.org/about/about-can

Flitton D (2016, 21 August) Rio Tinto's billion-dollar mess: 'unprincipled, shameful and evil'. Sydney Morning Herald. Retrieved from http://www.smh.com.au/world/billiondollar-mess-a-major-disaster-the-people-do-not-deserve-to-have-20160817-gquzli.html

Fry G (1996) Framing the islands: knowledge and power in changing Australian images of 'the South Pacific'. National Library of Australia, Canberra

Harris US (2014) Communicating climate change in the Pacific using a bottom-up approach. Pac Journal Rev 20(2):77–96. https://doi.org/10.24135/pjr.v20i2.167

Harris US (2019) Participatory Media in Environmental Communication: Engaging Communities in the Periphery. Routledge, London
Hau'ofa E (1994) Our Sea of Islands. Contemp Pac 6(1):148–161
Heeks R, Ospina AV (2018) Conceptualising the link between information systems and resilience: A developing country field study. Inf Syst J 1:1–27. https://doi.org/10.1111/isj.12177
ITU (2017) ICT facts and figures 2017. Retrieved from Geneva, Switzerland: http://www.itu.int/en/ITU-D/Statistics/Pages/stat/default.aspx
Kalas PP, Finlay A (2009) Planting the knowledge seed – adapting to climate change using ICTs: concepts, current knowledge and innovative examples. Retrieved from Melville, South Africa: https://www.apc.org/sites/default/files/BCO_ClimateChange.pdf
Kirsch S (2007) Indigenous movements and the risks of counterglobalization: tracking the campaign against Papua New Guinea's Ok Tedi mine. Am Ethnol 34(2):303–321. https://doi.org/10.1525/ae.2007.34.2.303
Kirsch S (2014) Mining capitalism: the relationship between corporations and their critics. University of California Press, Berkeley
McNamara KE, Gibson C (2009) 'We do not want to leave our land': Pacific ambassadors at the United Nations resist the category of 'climate refugees'. Geoforum 40(3):475–483. https://doi.org/10.1016/j.geoforum.2009.03.006
Minges M, Stork C (2015) Economic and social impact of ICTs in the Pacific [Online]. Sydney, Australia: The Pacific Region Infrastructure Facility. Retrieved from http://www.theprif.org/index.php/resources/document-library/121-prif-ict-study-report-2015
Ospina AV, Heeks R (2010) Unveiling the links between ICTs and climate change in developing countries: a scoping study. Retrieved from Manchester, UK: http://www.fao.org/fileadmin/user_upload/rome2007/docs/ICT%20and%20CC%20ScopingStudy.pdf
Oxfam (2015) A question of survival: Why Australia and New Zealand must heed the Pacific's calls for stronger action on climate change. Retrieved from Port Moresby: https://www.oxfam.org.au/wp-content/uploads/
Pacific Island Climate Action Network (2017) Vision. Retrieved from https://pacificclimateactionnetwork.wordpress.com/about-us-2/vission/
Pacific Islands Development Forum (2015) Suva Declaration on Climate Change. Retrieved from Suva, Fiji: http://pacificidf.org/wp-content/uploads/2013/06/PACIFIC-ISLAND-DEVELOPMENT-FORUM-SUVA-DECLARATION-ON-CLIMATE-CHANGE.v2.pdf
Pacific Media Assistance Scheme (PACMAS n.d.) Climate change communication impact briefing. Retrieved from https://www.abc.net.au/cm/lb/9580202/data/pacmas-climate-change-communication-impact-briefing-data.pdf
Penjueli M (2017) State of Pacific Regionalism: is the current Regional Architecture fit for purpose for the complexities of the 21st century? Retrieved from http://www.pina.com.fj/index.php?p=pacnews&m=read&o=139927374559af6542482380179b6d
Phillips K (2015) Bougainville at a crossroads: independence and the mine. Retrieved from http://www.abc.net.au/radionational/programs/rearvision/bougainville-at-a-crossroads/6514544
PRIF (2015) Economic and Social Impact of ICT in the Pacific. Retrieved from https://theprif.org/index.php/news/53-media-releases/169-prif-ict-study
Richards J-A, Bradshaw S (2017) Uprooted by climate change: responding to the growing risk of displacement. Oxfam Briefing Paper. Oxfam GB for Oxfam International, Cowley, Oxford
Robie D (2014) Don't Spoil my Beautiful Face: Media, Mayhem & Human Rights in the Pacific. Little Island Press, Auckland
Rona PA (2003) Resources of the sea floor. Science 299(5607):673–674
Schulman S (2016) The $100bn gold mine and the West Papuans who say they are counting the cost. The Guardian. Retrieved from https://www.theguardian.com/global-development/2016/nov/02/100-bn-dollar-gold-mine-west-papuans-say-they-are-counting-the-cost-indonesia
Servaes J (2008) Communication for Development and Social Change. Sage, Thousand Oaks
Smith RH (1997) The nuclear free and independent Pacific movement after Muroroa. Tauris Academic Studies, London
Smith A (n.d.) Thematic essay: the cultural landscapes of the Pacific Islands. Retrieved from https://whc.unesco.org/document/10062

SPREP (2013) Climate change. Retrieved from Apia, Samoa: http://www.forumsec.org/resources/uploads/attachments/documents/3-%20Climate%20Change%20SDWG%20Brief%20(8Mar13).docx

SPREP (2017) SPREP Strategic Plan 2017–2026. Retrieved from Apia, Samoa: http://www.sprep.org/attachments/Publications/Corporate_Documents/strategic-plan-2017-2026.pdf

Tacchi J, Horst H, Papoutsaki E, Thomas V, Eggins E (2013) PACMAS State of media and communication regional report 2013. Retrieved from Melbourne, Australia: http://www.pacmas.org/profile/pacmas-state-of-media-andcommunication-report-2013/

The World Bank (2017) Long-term economic opportunities and challenges for Pacific Island countries. Retrieved from Washington, DC. http://documents.worldbank.org/curated/en/168951503668157320/pdf/ACS22308-PUBLIC-P154324-ADD-SERIES-PPFullReportFINALscreen.pdf

Titifanue J, Kant R, Finau G, & Tarai J (2017) Climate change advocacy in the Pacific: the role of information and communication technologies. Pac Journal Rev, 23(1), 133–149

Troubled Waters (2012) Troubled waters: how mine waste dumping is poisoning our oceans, rivers, and lake. Retrieved from https://www.earthworksaction.org/files/publications/Troubled-Waters_FINAL.pdf

UN-OHRLLS (2017) Meeting report: regional meeting for the Asia-Pacific least developed countries. Retrieved from Port Vila, Vanuatu: http://unohrlls.org/custom-content/uploads/2017/11/Regional-Meeting-for-Asia-Pacific-LDCs-and-Pacific-SIDS-Broadband-Connectivity_LowRes1.pdf

White SA, Patel PK (1994) Participatory message making with video: Revelations from studies in India and the USA. In: White SA, Nair SK, Ascroft J (eds.) Participatory Communication: Working for Change and Development. New Delhi: Sage Publications. New Delhi: Sage Publications

Wilkins KG, Tufte T, Obregon R. (2014) Handbook of development communication and social change. Retrieved from http://mqu.eblib.com.au/patron/FullRecord.aspx?p=1602764

WWF Pacific (2015) Time is running out for low-lying islands in the South Pacific. Retrieved from http://www.wwfpacific.org/what_we_do/climatechange/

Yamamoto L, Esteban M (2014) Atoll Island States and international law: climate change displacement and sovereignty. Springer, Heidelberg

Youth Voices from the Frontlines: Facilitating Meaningful Youth Voice Participation on Climate, Disasters, and Environment in Indonesia

45

Tamara Plush, Richard Wecker, and Swan Ti

Contents

Abstract

In Indonesia children and youth are often impacted the most by disaster, environmental degradation, and a changing climate in ways unique to them. Yet in discussions to address specific impacts, they are rarely part of formal decision-making even when their participation is promoted as a foundational expression of child rights. The reason for such absence is rarely that young people have little to contribute as youth are often passionate advocates for change interested to identify, explain, and act on issues important to them. What is more likely to be lacking are opportunities to learn youth-friendly techniques and tools for exploring and sharing their concerns, as well as access to decision-makers open to hearing and

T. Plush (✉)
Royal Roads University, Victoria, BC, Canada
e-mail: Tamaraplush@gmail.com

R. Wecker
UNICEF Indonesia, Jakarta, Indonesia
e-mail: rwecker@unicef.org

S. Ti
PannaFoto Institute, Jakarta, Indonesia
e-mail: swanti@pannafoto.org

J. Servaes (ed.), *Handbook of Communication for Development and Social Change*,
https://doi.org/10.1007/978-981-15-2014-3_134

responding to what they have to say. In support, there is a growing interest by development organizations in using participatory media as a collaborative, exploratory storytelling process that can amplify youth voices in policy spaces as a way to rebalance decision-making power. Such interest was at the core of a 2016–2017 initiative in Indonesia entitled *Youth Voices from the Frontlines.* This paper offers insight from the initiative for promoting meaningful youth participation in governance such as ensuring adult participatory media facilitators have sufficient training, mentorship, and organizational support and that social mobilization projects are adequately resourced for meaningful dialogue and listening interactions critical for building influential youth voice. Creating such environment can support youth in tackling the issues they decide are most pressing to them – in this case, flooding, polluted rivers, fire, pollution haze from peat fires, and more.

Keywords

Disasters · Climate change · Youth engagement · Capacity building · Community digital storytelling · Child rights · Advocacy · Voice · Listening

45.1 Facilitating Meaningful Youth Participation

In Indonesia children and youth are impacted by disaster, environmental degradation, and a changing climate in ways contextually relevant to their subject positions and surroundings (UNICEF 2015). As one example, children breathe at twice the rate of adults, so they face a disproportionate health risks from the impacts of the air pollution from the peatland fires that occur in Kalimantan, Indonesia (UNICEF 2016a). Similarly, due to their age and development, they are more vulnerable to hunger and malnourishment during floods and droughts (ibid). In response, multiple development strategies are required not only to secure child rights to safe environments but to promote their meaningful participation in decisions that directly affect their lives (Ruiz-Casares et al. 2017; UNICEF 2017). Participation is used here in relation to Article 12 of the United Nations Convention on the Rights of the Child (UNCRC), which highlights children's rights for their views to be seriously considered in governance decision-making (United Nations General Assembly 1989). The focus on meaningful child participation as a human right means that youth-engaged initiatives go beyond merely consulting youth and amplifying their concerns. Youth must also be adequately supported with the appropriate tools, knowledge, and resources to actively engage with issues, offer their own solutions, and be sufficiently heard, valued, and responded to (Lundy 2007; Plush 2015; UNICEF 2017). Youth, in this case, refers to young people aged 15–24 as per the United Nations universal definition (United Nations 2018a).

Globally there is a growing interest in and use of participatory media to engage youth and connect their voice to policymakers, especially as a strategy for disaster risk reduction (DRR) and reducing climate change impacts (Fletcher et al. 2016; Haynes and Tanner 2015; Plush 2009). As an example, UNICEF's Disaster Risk

Fig. 1 Flood in a participating youth's Badung neighborhood. (Photo by Nurul)

Reduction (DRR) and Climate Change Unit developed the 2016–2017 *Youth Voices from the Frontlines: Community Digital Storytelling (CDST) for Social Change* initiative in Indonesia to enhance dialogue and response between young people, communities, and decision-makers on topics of climate, disasters, environment, child rights, health, and well-being (UNICEF 2016b) (Fig. 1).

This article explores the *Youth Voices* initiative to offer insights into using participatory media as a strategy to increase youth participation in decision-making. The initiative was implemented as part of the Children in a Changing Climate (CCC) coalition in Indonesia (Globally CCC is a partnership of five child-centered development and humanitarian organizations: ChildFund Alliance, Plan International, Save the Children, UNICEF, and World Vision International). The workshop focused on strengthening youth voice at scale across the region by training 11 CCC network members and 14 young people from 10 organizations to apply the CDST methodology (including ChildFund; Plan International Indonesia; Yayasan Sayangi Tunas Anak, Save the Children; Climate Warriors; Pramuka [Scouts]; Wahana Visi Indonesia [WVI, World Vision]; National Youth Forum [FAN]; Youth for Climate Change Indonesia [YFCC]; Youth Network on Violence Against Children [YNVAC]; and Sinergi Muda). The initiative was co-designed and facilitated by an international consultant and PannaFoto Institute staff in Jakarta, delivered in Bahasa, and supported through PannaFoto staff mentoring and a CDST training manual in English and Bahasa developed specifically for the initiative participants.

45.2 Using Story for Social Change

The *Youth Voices* network initiative used Community Digital Storytelling (CARE 2015b) as a process to engage and mobilize youth affected by the impacts of disaster, environmental degradation, and a changing climate. This included directly working with more than 300 youth across Indonesia in rural areas affected by air pollution caused by land and forest fires and carbon emissions, as well as urban areas affected by flood, drought, and river pollution (Fig. 2).

CDST is a creative participatory media process – directed by young people in this case – aimed at catalyzing action within a community. Photo-based videos are produced through collaborative processes in which facilitators and youth participants generate stories together using different creative forms of communication and expression such as photography, narration, and music. As a core value of the CDST process, storytellers own the stories they produce and have the ability to make informed choices about the content, production, and use of their work. In the Youth Voices project, ownership processes for youth participants were guided and agreed upon through action plans developed and reviewed as part of ongoing mentorship support to facilitators.

The intention of CDST is that story development and sharing processes of research, reflection, dialogue, engagement, and mobilization strengthens the knowledge and self-confidence of participants on a pathway for more engaged citizenship (CARE 2015a). The process is built on core values that align with participatory development and communication for social change literature, including:

Fig. 2 A West Kalimantan youth photographs her environment as part of her CDST photo-video. (Photo by Ng Swan Ti)

- It is *community-driven* where youth are equal production partners in storytelling (Milne et al. 2012; Servaes 2013; Shaw 2015).
- It is *flexible and embedded* within organizational structures and networks to enhance ongoing empowering initiatives (Gaventa and Barrett 2010; Waisbord 2008).
- It is *aware and respectful* of cultural, social, and political contexts to support complex processes of social change (Cornwall et al. 2011; Walsh 2016).
- It supports *dialogue and listening* as key components for meaningful youth participation (Dutta 2014; Dobson 2014; Waller et al. 2015).
- It is mindful of making *appropriate technology choices* in the type of media to use to enhance local capacity and minimize power imbalances so the least heard can participate (Askanius 2014).
- It is vigilant about ensuring *informed consent* as well as an environment free of harm for people sharing their stories (Cornwall and Fujita 2012; Wheeler 2011).

Participatory media initiatives like CDST hold promise for meaningful youth participation in decision-making, which can be especially empowering for young people living in poverty or vulnerable situations who may struggle to have their voices heard due to multiple factors (Gidley 2007; UNICEF 2017). For instance, they may not speak the official language being used in decision-making forums (CARE 2015a). They may have restricted access to local, regional, or international policymaking spaces or lack the confidence to raise their voice with more influential people in society (Couldry 2007; Sparks 2007). They may be restricted by gender, social, or cultural norms or affected by stigma (Cornwall and Rivas 2015; Dasgupta and Beard 2007). Here, participatory media offers potential to build youth awareness, increase self-confidence, and mobilize young people when positioned alongside strategic, comprehensive, and supported activities for youth engagement (Milne et al. 2012). Such actions are globally understood as imperative for addressing disasters and climate change, as illustrated in the child-centered adaptation approaches promoted by the United Nations Joint Framework Initiative on Children, Youth and Climate Change as part of the *United Nations Framework Convention on Climate Change* (United Nations 2018b; article 6; United Nations General Assembly 2016).

45.3 Developing CDST as an Adaptable Approach

Supporting CDST facilitators to implement participatory media processes both in the long-term and in their own contexts was deemed critical in the design of the *Youth Voices* initiative – especially in training multiple organizations focused on youth, climate change, and DRR. Thus, in addition to providing ongoing mentorship after the initial training workshop, the workshop trainers developed a CDST guide in English and Bahasa that responded to how participants planned to use the methodology. It included steps so participants could adapt the process to the various contexts in which they work (see Table 1).

Table 1 Youth voices from the frontline: steps in the guide for using community digital storytelling for social change (UNICEF 2016b)

Prepare for CDST	Facilitators analyze issues of risk for youth to determine if CDST is appropriate Facilitators ensure they will have sufficient time, resources, and management support for a CDST initiative Facilitators and youth participants determine how to monitor and evaluate the initiative
Use photography to identify key themes	Facilitators build youth photography and media ethics skills Youth take photos on the topic they are exploring Facilitators print the photos for story development (printing helps to equalize power in the group) Youth sort the photos together to develop key themes for further exploration for the final story
Develop the story	Youth determine who should hear their story, including what they want viewers to see, think, feel, discuss, and do? Youth take more photos to expand their story; ensuring the photos adhere to ethics and have proper consent Youth prioritize the top photos Facilitators print the photos for collaborative storyboarding Youth develop a script to match their storyboard Youth attain photo and script approvals from key stakeholders
Produce photo-videos	Based on feedback, youth finalize the script and record the narration Youth select emotive copyright-free music, or record music Facilitators and/or the youth technically produce the photo-videos Youth review and approve the final photo-videos, and seek final approval from stakeholders
Strategically use photo-videos to engage youth and decision-makers	Facilitators support youth in connecting them to decision-makers Youth develop questions to accompany the photo-videos Youth use the photo-videos to spark dialogue on the topics with people they want to influence Youth use social media to engage audiences as appropriate Facilitators and the youth use the photo-videos for advocacy as per a youth-approved strategy for dissemination Facilitators and the youth review the photo-video use against monitoring and evaluation objectives to assess impact, making changes as needed

The CDST steps are designed to explore and address youth concerns through young people telling their own stories in their own voice and language (Fig. 3). As a driver of social change, they incorporate techniques to negotiate power within the group that might marginalize some youth. For example, the CDST process encourages facilitators to print the photos youth have taken to create the story structure rather than using digital photos on a computer for story development. This is strategic to ensure that less technically adept youth are able to fully engage during story development as the group sorts, prioritizes, and arranges the printed photos into a storyboard – which may not occur if the photo selection is done on a computer (Fig. 4).

Fig. 3 Young people narrate their story as part of a Pramuka initiative in Bogor. (Photo by Edy Purnomo)

Fig. 4 Youth learn the CDST storyboard process with photos at the Jakarta workshop. (Photo by Ng Swan Ti)

As another example, CDST prioritizes strategically working with young people to reach different audiences according to whom they want to influence. The focus on generating meaningful response is critical, as merely holding an event to show the youth photo-videos cannot guarantee sufficient response to youth concerns and could even be harmful if decision-makers fail to respectfully and responsively listen (Kindon et al. 2012; Ruiz-Casares et al. 2017). Considered effort by supportive adults is often required for youth to enter the complex arenas of local governance and be heard in ways that can lead to meaningful social change (UNICEF 2017).

45.4 Championing Youth and Their Stories

Through the Youth Voices CDST initiative, youth across Indonesia created 16 independent stories on the themes of climate change, disaster risk, and issues of child rights, health, and well-being. The photo-videos focused on flood prevention and warning systems, cleaning up polluted rivers, fire detection, and responding to air pollution from slash-and-burn practices. The young people also shared their concerns and ideas with their peers and decision-makers locally and nationally at various events. A few examples include a youth storytelling peer learning event in Jakarta (Plan International Indonesia); a screening with health, education, and child protection government representatives (Wahana Visi Indonesia); a photo-video showing with health, agriculture, education, and social affairs government representatives (ChildFund); a school discussion event (Climate Warriors) (Fig. 5); and story sharing at the national Pramuka (Scouts) Jamboree.

Fig. 5 Youth attend a Climate Warriors photo-video screening in Bandung. (Photo by Deden Iman)

Within participatory media initiatives, skilled facilitation, and a contextually informed vision are necessary to ensure meaningful engagement and influential voice in policymaking (Plush 2016; UNICEF 2017). Accordingly, the *Youth Voices* initiative took steps early to ensure organizational support for the initiative and the trained facilitators. For instance, the training included developing action plans with both the facilitators being trained and their managers. Even so, only six of the nine implementing organizations who participated in the workshop were able to fully complete all steps within the CDST process. Time availability, sufficient resources, and ongoing organizational prioritization for CDST were identified as the key challenges for CDST facilitators. In one case, for example, the trained facilitators were unable to complete the CDST activity and handed it over to other staff members. The untrained facilitators struggled to understand and fully implement the storytelling process according to the CDST values, specific steps, or in adherence to media ethics and copyright law taught in the workshop.

Other organizations struggled with the initiative's flexible timeframe that allowed organizations to fit CDST into ongoing program activities as appropriate. Here, the non-binding commitment appeared to work best within established, child-focused, nongovernment organizations (NGOs) with a wide base from which to allocate resources as needed. In addition, internationally supported NGOs were often well-equipped to connect the youth participants to local, provincial, and national decision-makers in Indonesia with potential to respond due to internal communication and advocacy efforts. The implication in the *Youth Voices* initiative was that CDST proved the most sustainable when embedded into ongoing youth programs with dedicated facilitators equipped to support the entire storytelling process from creation to dialogue and response. That said, complex participatory media processes that aim to politically engage youth (such as CDST) are not *inherently* better suited for NGOs. For even in the initiative, one of the youth trained in the workshop from a youth-centered organization – with limited means and supervision – was able to create, share, and advocate for a concerning issue through the CDST photo-video story she developed.

Rather, the learning highlights that, from the start, initiatives must readily acknowledge that sufficiently engaging youth in participatory decision-making may require additional capacity-building investment and initiative funding when working with less-resourced and/or connected organizations. This finding in particular was highlighted by one organization that integrated use of additional tools into a broader *Adolescent Kit for Expression and Innovation* (UNICEF 2016c). The kit was designed to build capacity of adolescent groups to be better prepared before, during, and after an emergency and to find solutions to problems faced in their communities.

45.5 Focusing on Responsive Listening

As the initiative showed and scholars argue, public screenings themselves require considered attention by supportive adults (Kindon et al. 2012). Ensuring trained facilitators have the capacity to work with youth from a story's telling to realizing its influence is critical as young people negotiate complex processes of both speaking and listening. For instance, sharing their stories in public can be an affirming act of representation for aspiring youth citizens (Milne et al. 2012). However, public screenings can also bring harm if they lead to embarrassing, intimidating, or even emotionally scarring situations that might occur if those listening are not in a position to respect youth views and provide meaningful response (Wheeler 2011). What this means is that extra efforts and resources are required to ensure young people stories can be told without harm and that they are shown in places of influence where decision-makers are not only open to hearing their ideas but able to provide direct response to the issues raised (UNICEF 2017) (Fig. 6).

To foster greater listening and response, the *Youth Voices* initiative emphasized that the act of storytelling is only one part of social and political change. Attention is especially required for understanding the politics of listening that might keep youth voices from being heard within the differing contexts in which they live (Dreher 2009). Issues of concern to youth can be complex or politically sensitive and can put youth at risk in telling their stories. For instance, a group of young people in the *Youth Voices* initiative investigated and shared stories about the negative impacts of air pollution from fires intentionally started to clear land for palm plantation planting – activities which are often illegal (Varkkey 2012). They did so through an organization that knew the issue well to ensure their safety in telling their story. This illustrates the importance of participatory media facilitators understanding issues of local context, media ethics, and child well-being. For on the one hand, raising issues that directly affect them through storytelling supports young people in understanding

Fig. 6 A CDST storytelling event by youth and Plan International champions youth voice in Jakarta. (Photo by Edy Purnomo)

their world. Additionally, hearing directly from affected youth holds potential to influence policymakers when they connect lived experiences to their decisions. This can help to promote constructive and changing perceptions of adults toward young people (UNICEF 2017). On the other hand, storytelling activities can increase a young person's marginalization or vulnerability if institutions fail to recognize and respond accordingly to the risky and sometimes delicate nature of the process (Plush 2016; Ruiz-Casares et al. 2017).

Adult facilitators need to be aware that telling certain stories might threaten a young person's rights to safety and security as outlined by the UNCRC (United Nations General Assembly 1989). It is thus valuable to incorporate child protection specialists into the storytelling process so they can flag concerns early in the story production, as well as during development and public sharing. Such attention on child protection was incorporated in the facilitation training workshop but was identified as a gap within individual CDST story development during individual initiatives. Involving communication and advocacy teams early in the initiative can also support youth in connecting them to decision-makers at local and national levels. In the *Youth Voices* initiative, as mentioned, NGOs proved the most successful at linking youth to audiences with the power to meaningfully respond to their issues based on their existing programs. Here their strength lie in the potential to incorporate CDST stories into larger youth participation and mobilization efforts. This supports critical arguments that influencing policy debates is a longer and more involved endeavor than celebratory, public story-sharing events may imply (Kindon et al. 2012; Wheeler 2011). Rather, strategic efforts are required to shift marginalizing powers and processes that might silence youth concerns so their voice can be sufficiently considered, valued, and incorporated into decisions that affect their lives (Cornwall and Fujita 2012; UNICEF 2017).

45.6 Strengthening Youth Voice and Influence

The Indonesia *Youth Voices from the Frontlines* case study highlights that participatory media initiatives can benefit from a strategic design that prioritizes sufficient facilitator capacity and an attention on listening for youth participation in and influence on decisions that affect their lives. Such action requires strategic visions and plans for ensuring youth are not only provided opportunities to voice opinions. They must also be supported as active participants in researching and acting on issues that concern them as they acquire the knowledge and tools to safely share their concerns and ideas. Doing so starts with strengthening individual and organizational capacities in sustainable ways, as well as positioning youth engagement and mobilization activities in large organizations and national coalitions for greater policy influence. It requires participatory media facilitators with sufficient and appropriate tools, resources, capacity, and organizational support to adeptly use participatory media to navigate and address the contextual constraints and possibilities for youth voice and influence. In the case of the *Youth Voices from the Frontlines* initiative in Indonesia, such support offered youth opportunities to identify and champion solutions to overcome the negative impacts of disaster, environmental degradation, and a changing climate. The stories not only raise critical concerns for Indonesian youth but position young people as key actors able to understand, address, and solve the country's greatest challenges today and in the future.

Disclaimer The opinions expressed in this chapter are the authors' and do not reflect the view of UNICEF.

References

Askanius T (2014) Video for change. In: Wilkins KG, Tufte T, Obregon R (eds) The handbook of development communication and social change. Wiley-Blackwell, Chichester, pp 453–470

CARE (2015a) Our valuable voices: community digital storytelling for good programming and policy engagement. CARE International in Vietnam, Hanoi, pp 1–24

CARE (2015b) Guidelines for producing community digital storytelling (CDST) videos. CARE International, Copenhagen, pp 1–32

Cornwall A, Fujita M (2012) Ventriloquising "the poor"? Of voices, choices and the politics of "participatory" knowledge production. Third World Q 33(9):1751–1765

Cornwall A, Rivas A-M (2015) From "gender equality" and "women's empowerment" to global justice: reclaiming a transformative agenda for gender and development. Third World Q 36(2):396–415

Cornwall A, Robins S, Von Lieres B (2011) States of citizenship: contexts and cultures of public engagement and citizen action. IDS Work Pap 2011(363):1–32

Couldry N (2007) Media and democracy: some missing links. In: Dowmunt T, Dunford M, van Hemert N, Fountain A (eds) Inclusion through media. Goldsmiths, University of London, London, pp 254–262

Dasgupta A, Beard VA (2007) Community-driven development: collective action and elite capture in Indonesia. Dev Change 38(2):229–249

Dobson A (2014) Listening for democracy: recognition, representation, reconciliation. Oxford University Press, Oxford

Dreher T (2009) Listening across difference: media and multiculturalism beyond the politics of voice. Continuum 23(4):445–458

Dutta MJ (2014) A culture-centered approach to listening: voices of social change. Int J Listen 28(2):67–81

Fletcher S, Cox RS, Scannell L, Heykoop C, Tobin-Gurley J, Peek L (2016) Youth creating disaster recovery and resilience: a multi-site arts-based youth engagement research project. Child Youth Environ 26(1):148–163

Gaventa J, Barrett G (2010) So what difference does it make?: mapping the outcomes of citizen engagement. Institute of Development Studies, Brighton

Gidley B (2007) Beyond the numbers game: understanding the value of participatory media. In: Dowmunt T, Dunford M, van Hemert N, Fountain A (eds) Inclusion through media. Goldsmiths, University of London, London, pp 39–61

Haynes K, Tanner TM (2015) Empowering young people and strengthening resilience: youth-centred participatory video as a tool for climate change adaptation and disaster risk reduction. Child Geogr 13(3):357–371

Kindon S, Hume-Cook G, Woods K (2012) Troubling the politics of reception in participatory video discourse. In: Milne EJ, Mitchell C, de Lange N (eds) Handbook of participatory video. AltaMira Press, Lanham, pp 349–364

Lundy L (2007) 'Voice' is not enough: conceptualising Article 12 of the United Nations Convention on the Rights of the Child. Br Educ Res J 33(6):927–942

Milne EJ, Mitchell C, de Lange N (2012) Introduction. In: Milne EJ, Mitchell C, de Lange N (eds) Handbook of participatory video. AltaMira Press, Lanham, pp 1–18
Plush T (2009) Amplifying children's voices on climate change: The role of participatory video. Participatory Learning and Action 60:119–128
Plush T (2015) Participatory video and citizen voice – we've raised their voices: is anyone listening? Glocal Times. No 22/23, pp 1–16
Plush T (2016) Participatory video practitioners and valued citizen voice in international development contexts. PhD thesis, University of Queensland, Centre for Communication and Social Change, pp 1–258. https://espace.library.uq.edu.au/view/UQ:398615
Ruiz-Casares M, Collins T, Tisdall EK, Grover S (2017) Children's rights to participation and protection in international development and humanitarian interventions: nurturing a dialogue. Int J Hum Rights 21(1):1–13
Servaes J (2013) Sustainability, participation and culture in communication: theory and praxis. Intellect, Bristol
Shaw J (2015) Re-grounding participatory video within community emergence towards social accountability. Community Dev J 50(4):624–643
Sparks C (2007) Globalisation, development and the mass media. SAGE, London
UNICEF (2015) Unless we act now: the impact of climate change on children. Division of Data, Research and Policy, United Nations Children's Fund (UNICEF), New York, pp 1–81
UNICEF (2016a) Clear the air for children: the impact of air pollution on children. Division of Data, Research and Policy, United Nations Children's Fund (UNICEF), New York, pp 1–100
UNICEF (2016b) Youth voices from the frontline: steps in the guide for using community digital storytelling for social change. United Nations Children's Fund (UNICEF) Indonesia, Jakarta, pp 1–43
UNICEF (2016c) The adolescent kit: for expression and innovation. United Nations Children's Fund (UNICEF). http://adolescentkit.org. Accessed Jan 2018
UNICEF (2017) Child participation in local governance: a UNICEF guidance note. Public Finance and Local Governance Unit of the Social Inclusion and Policy Section, Programme Division, United Nations Children's Fund (UNICEF), New York, pp 1–39
United Nations (2018a) FAQ: what does the UN mean by youth? Division for Social Policy and Development: Youth. Retrieved from www.un.org/development/desa/youth/what-we-do/faq.html
United Nations (2018b) United Nations Joint Framework Initiative on Children, Youth and Climate Change. Retrieved from https://unfccc.int/cc_inet/cc_inet/youth_portal/items/6519.php
United Nations General Assembly (1989) Convention on the Rights of the Child, 20 November 1989, United Nations, Treaty Series, vol 1577
United Nations General Assembly (2016) United Nations Framework Convention on Climate Change: the Paris Agreement, 04 November 2016, United Nations
Varkkey H (2012) Patronage politics as a driver of economic regionalisation: the Indonesian oil palm sector and transboundary haze. Asia Pac Viewp 53(3):314–329
Waisbord S (2008) The institutional challenges of participatory communication in international aid. Soc Identities 14(4):505–522
Waller L, Dreher T, McCallum K (2015) The listening key: unlocking the democratic potential of indigenous participatory media. Media Int Aust Inc Cult Policy 154:57–66
Walsh S (2016) Critiquing the politics of participatory video and the dangerous romance of liberalism. Area 48(4):405–411
Wheeler J (2011) Seeing like a Citizen: participatory video and action research for citizen action. In: Shah N, Jansen F (eds) Digital AlterNatives with a Cause? Book 2 – To think. Bangalore: Centre for Internet and Society/Hivos, 47–60

Key SBCC Actions in a Rapid-Onset Emergency: Case Study From the 2015 Nepal Earthquakes

46

Rudrajit Das and Rahel Vetsch

Contents

Abstract

In case of an emergency, various social and behavior change communication (SBCC) instruments can play an important role not only in providing immediate access to lifesaving information to affected populations but also in resilience building and strengthening accountability of government and international and national civil society organizations.

This case study provides the readers with an overview of key challenges, needs, and SBCC strategies in case of a large-scale emergency that affects beneficiaries, service providers, and humanitarian actors at the same time. It suggests a step-by-step approach, starting with functional media channels for immediate dissemination of lifesaving messages and collection of feedback on the needs and concerns of affected population, and then slowly moving on to directly reaching out to communities through a variety of SBCC strategies including mobile 'edutainment'

R. Das · R. Vetsch (✉)
UNICEF, Kathmandu, Nepal
e-mail: rdas@unicef.org; rvetsch@unicef.org

J. Servaes (ed.), *Handbook of Communication for Development and Social Change*,
https://doi.org/10.1007/978-981-15-2014-3_129

(Entertainment with the purpose of providing educational information to the target audience) shows and face-to-face community mobilization.

The chapter focuses on key elements of the SBCC strategy pursued by UNICEF: a strong coordination mechanism between the government and national and local partners both regarding communication efforts and accountability mechanisms, a rapid assessment and rehabilitation of communication channels and mechanisms to ensure meaningful participation, and opportunities for community members to provide feedback and receive mass-scale counselling.

Keywords

Association of Community Radio Broadcasters (ACORAB) · Bhandai Sundai' radio program · Communicating with Affected Communities (CWC) · Communication for Development (C4D) program · Nepal community radio stations · Social and behavior change communication (SBCC)

46.1 2015 Earthquakes in Nepal: Impact and Needs

Due to Nepal's complex geological formation, its position in an active seismic zone and heavy annual rainfall, the South Asian country is highly prone to rapid- as well as slow-onset natural disasters such as earthquakes, landslides, floods, or droughts. In the past, Nepal experienced major earthquakes every few generations (Government of Nepal, National Planning Commission (2015)) providing a historical pattern for an eventual earthquake return period of 40 till 80 years. Given such a context, two elements are key: emergency preparedness and a rapid response mechanism once an emergency hits.

The two devastating earthquakes that occurred in Nepal on the 25th of April and the 12th of May 2015 severely affected many regions within the country. Around 8,959 people lost their lives, and 22,302 people were injured, 2,661 of whom were children. In the worst affected areas, entire settlements were flattened or swept away by landslides. (Government of Nepal, National Reconstruction Authority (2016). According to the survey conducted by the Ministry of Home Affairs (MoHA), 605,254 houses were fully damaged, and 288,255 houses were partially damaged.) With a majority of houses being damaged or fully destroyed, people were forced to leave their houses and move to safe spaces or live in makeshift shelters. Many health centers and much of the community infrastructure, including water systems and latrines, were damaged or destroyed.

The ensuing landslides blocked major roads and highways, and transportation was severely curtailed for authorities and relief agencies to reach affected districts and populations. All forms of communication and transportation were severely affected, and telecommunication networks were disrupted in all the affected districts. Public service television and radio broadcasts went off air, some of them for several days. A lot of equipment and buildings belonging to community radio stations were either damaged or destroyed. Government authorities, development agencies, and aid workers in Kathmandu and at the district headquarters faced difficulties in communicating with affected communities.

Providing critical response and early recovery is a core task of the first weeks after an emergency hits. In addition to the immediate survival and protection needs, people needed critical, lifesaving information and the means to communicate with their family members and the authorities (For a detailed timeline for an effective response and a toolkit on behavior change communication in emergencies, please see UNICEF (2017b), pp. 32–35 and UNICEF (2006)). Lack of electricity, mobile and telephone connectivity, and damaged physical infrastructure made it very difficult for communities and relief responders to communicate with each other. In addition, communication service providers such as design agencies, media-buying agencies, and printers had all been affected and were working at sub-optimal capacity.

There was an imminent need for communication channels to be re-established in order to communicate with the affected communities for rescue, aid, ensuring safety from constant aftershocks, prevention of disease outbreaks, providing key lifesaving messages and psychosocial support, and for providing affected populations platforms to voice their concerns and provide feedback on the response to duty bearers.

46.2 Social and Behavior Change Communication Response

In view of the above, social and behavior change communication (SBCC) was an intrinsic part of the UNICEF response to Nepal earthquake. (DARA (2016), p. 8) It was guided by the Core Commitments for Children (CCC) in Humanitarian Actions. (UNICEF (2010). The CCC is UNICEF's central policy to uphold the rights of children affected by humanitarian crisis and are based on global standards and norms for humanitarian action.) The founding elements of the response included:

- Forging alliances with multiple stakeholders and strengthening their capacities to effectively communicate with affected populations. In order to better coordinate, plan, manage, and monitor communication initiatives and as such avoid duplication, misunderstanding, rumors, and misinformation, an inter-agency coordination group was established.
- Using various communication channels to promote dialogue with affected populations around key lifesaving messages and critical information in the areas of health, nutrition, water and sanitation, education, child protection, and relief and rehabilitation.
- To ensure participation of and accountability to affected populations, providing them with platforms and spaces to obtain relevant information, voice their concerns, provide feedback on the response to duty bearers, and receive psychosocial counselling to help deal with their situation, with a focus on women, children, and the most marginalized.

UNICEF worked with the Government of Nepal and other development partners to develop a comprehensive SBCC strategy for responding to the situation. Based on a rapid assessment of sectoral and crosscutting communication needs, key messages and content for dialogue were developed and channels identified to disseminate these in the most affected districts. Care was taken to ensure that it was not just a one-way dissemination of messages but that there were appropriate mechanisms to ensure that community feedback and voices were heard in order to make the communication and response efforts need-based and also ensure accountability to affected populations.

46.2.1 Coordination Mechanism

To ensure a convergent and coherent SBCC response to the earthquake, UNICEF established a '*Communicating with Affected Communities*' (CWC) working group. The group brought together several development partners including UN agencies, international and national nongovernmental organizations (NGOs), civil society organizations (CSOs), media organizations, and community radio operators. It ensured that partners were informed about each other's efforts, had a common vision and approach, and that there was no duplication of efforts. In collaboration with members of this group and the National Health Education Information and Communication Centre (NHEICC), which operated under the Ministry of Health and Population, a national communication response plan was developed. This was implemented through various partners.

After the first few weeks of the earthquake, the CWC was further subdivided into four subgroups in order to coordinate the work better. These included:

Subgroup	Responsibility
Messages and materials	Through this subgroup a common set of messages were developed so that all partners communicated the same messages to communities and that there was no confusion among communities over these messages. The group would also periodically update and refresh messages based on evolving needs. Action-oriented communication materials that were developed by partners were uploaded onto a drop box site, which was accessible to all members of the CWC. Members could download, view, and print these materials as per their needs, helping in avoiding duplication in terms of different partners developing similar types of materials for the same audience
Community mobilization	The subgroup on community mobilization was responsible for coordinating direct, community-based outreach work. This helped in ensuring that partners could spread themselves and cover areas in an organized manner and avoid duplication
Radio	This group comprised of representatives of community radio operators and media-based organizations. The group managed work related to assessing the status of damaged community radio stations and preparing a roadmap for their rehabilitation. They also worked with various community radio stations to ensure uniform and correct transmission of key messages
Monitoring and evaluation	Understanding the criticality of evidence-based and data-driven work, the subgroup was responsible for carrying out communication assessments, monitoring, and evaluation

46.2.2 Rapid Appraisal of Communication Channels and Resources

If an emergency strikes, it is crucial to immediately disseminate key messages to affected families and communities through a variety of communication channels. In the aftermath of the earthquakes, however, most telecommunication networks were badly damaged, and services were disrupted. National and local television and radio stations went off air because of structural damage to their buildings, damage to equipment, and the absence of human resources to run news programs.

In Nepal community radio stations (270 radios were registered as community radios in 2015.) serve as one of the most important sources of information for communities, especially in remote areas, and was the preferred channel to receive information during the aftermath of the earthquakes. (Inter-Agency Common Feedback Project (2016), p. 12. 55% of the 2100 respondents indicated to have used radio and 31% television immediately following the earthquake.) Many of these radio stations were partially or fully damaged, and many were not in a position to communicate with communities in the geographies that they serve due to breakdown in the supply of electricity.

Through the radio subgroup of the CWC, discussions were held to identify effective ways of resuming community radio services in the districts that had suffered most. In collaboration with the Asia Pacific regional office of AMARC and the '*Association of Community Radio Broadcasters*' (ACORAB), an assessment of the extent of damage suffered by community radio stations was conducted. (The Humanitarian Data Exchange (online platform), https://data.humdata.org/dataset/radio-stations-in-earthquake-affected-areas. Accessed June 2018) Based on the assessment findings, a short-term and a long-term rehabilitation plan was developed and implemented.

The short-term plan comprised of providing rehabilitation supplies to the damaged stations such as tents, zoom recorders, power generators, telephone hybrids, power backups, batteries, stabilizers, laptops, and radio receivers for distribution in communities and also providing training to local technicians on repairing damaged radio sets. As soon as these community radio stations were equipped with the minimum requirements for airing, they started to intensively broadcast critical information related to staying safe, relief efforts, and messages around health, nutrition, water and sanitation, education, and child protection. Over 100,000 min of messages were broadcasted through 191 community radio stations. Rapid assessment findings collected in July and August 2015 indicated that around 87% people could recollect key messages that were aired through radio. (Inter-Agency Common Feedback Project (2016), p. 40. The rapid assessment was done based on 222 key informant surveys across 10 earthquake-affected districts.)

The long-term plan comprised of providing training to community radio stations on strengthening disaster risk reduction and increasing emergency preparedness and developing programs to help communities better prepare and respond to future emergencies.

46.3 Community Consultation and Participation

Immediately after the earthquake, most of the relief and response activities focused on rescue, treatment of injuries, and provision of essential relief supplies to affected communities. While these were extremely important, there was also a need to provide opportunities for affected populations to voice their needs and concerns and to provide psychosocial counselling to people to get over the emotional trauma that the earthquake had caused. (Plan International, Save the Children, UNICEF, World Vision in collaboration with the Ministry of Federal Affairs and Local Development and the Central Child Welfare Board (2016), p. 5) People including children were demonstrating common signs and symptoms of mental stress such as palpitation, sleeplessness, headache, dizziness, anxiety, fear, and inability to focus on day-to-day activities. A consultation among 1,838 girls and boys done in May and June 2015 by development agencies, in coordination with the Government of Nepal, revealed that major concerns regarding the "well-being were grief and sadness at deaths of family members, friends, and acquaintances and a strong feeling of loss, fear, and other psychosocial impacts of the damage and destruction." (Idem, p. 22)

In order to reach out to communities at scale with psychosocial counselling support and entertainment-education, the radio program '*Bhandai Sundai*' (Talking-Listening) was initiated. While many of the community FM stations, which are an important source of information for communities in rural Nepal, were unable to broadcast because they had been damaged by the earthquake, the national broadcaster '*Radio Nepal*' was still broadcasting round-the-clock. It was the only means of information for people in remote areas of the country. People were constantly tuned into Radio Nepal to get more information on the situation as well as relief and response efforts. (Inter-Agency Common Feedback Project (2015), p. 18. 94% out of the 222 respondents of the rapid assessment done in July and August 2015 indicated to listen to radio.) As such, the station provided a unique opportunity to reach out to people at scale, within a week of the earthquake.

The programs quickly gained great popularity, and many calls could not be taken due to time constraints. Listeners across the earthquake-affected districts and beyond greatly benefitted from listening to the advice given by counsellors on the show as most of them were facing similar issues (Fig. 1). Thus, gradually the program turned into psychosocial counselling on a mass scale. Considering the inherent private and personal nature of counselling, the identities of all callers were kept confidential unless they wanted to identify themselves on-air. By targeting different audiences (Table 1), the program was successfully able to address the psychosocial concerns of children, women, and families who were otherwise outside the immediate reach of direct counselling services due to the devastating impact the earthquake had had on the physical infrastructure of the country. A survey conducted 1 year after the earthquake highlighted that the psychosocial benefit was ranking just behind information to relief assistance and knowledge around earthquakes, resulting in an overall recommendation to strengthen the role of communicating with communities in addressing trauma and psychosocial needs. (CDAC Network (2016), pp. 8 and 10).

Table 1 Format and audience of *Bhandai Sundai* program

Time	Duration	Audience	Content
Morning	30 min	All audiences	Situation updates and information on relief and response efforts of the government and development partners. An opportunity for people to call in and share their needs and concerns with concerned government authorities who would periodically participate in the show was provided
Afternoon	55 min	Women	Psychosocial support: Calls on the show were answered by trained counsellors who would provide counselling to callers to help them deal with their problems and trauma
Early evening	20 min	Children and adolescents	Psychosocial support and entertainment: During the show, children were encouraged to call and share their feelings with a trained adult, who would give them practical tips on how to deal with the situation. They were invited to sing songs, recite poems, share jokes, or simply talk to help them get a few lighter moments and get over the trauma. Occasionally popular celebrities and comedians would also be invited on the show to increase the entertainment quotient of the show
Evening	45 min	All audiences	Psychosocial support to everybody who wanted to talk to a trained counsellor to discuss the emotional issues that they were going through and find options to deal with their condition

Source: Author

Radio Nepal, being a credible government body, also helped ensure accountability to affected populations by directly calling concerned government authorities on getting grievances from affected communities in order to address them at the local level. The program also helped disseminate information and critical lifesaving messages and content on health, nutrition, water and sanitation, education, and child protection. Around 13,300 min of on-air psychosocial counselling, key lifesaving messages and information were provided to listeners through the '*Bhandai Sundai*' radio program.

The program steadily turned into a very strong platform to promote initiatives such as the *Back to School Campaign*, *Nutrition Week*, and the *Cash transfer scheme* (For more information, please see: https://www.unicef.org/media/media_82139.html, https://www.thelancet.com/journals/langlo/article/PIIS2214-109X(15)00184-9/fulltext, https://www.unicef.org/evaldatabase/index_100594.html.) for earthquake-affected families belonging to certain disadvantaged groups. For instance, as part of the Back to School Campaign, 'mock classes' were organized on the show to help teachers and administrators understand the facilities and services that needed to be provided in schools and how they should carry out classes in the first few days after reopening of the school so that children could overcome their fears and slowly settle down in a child-friendly environment. Several telephone conversations and interviews were also carried out with district

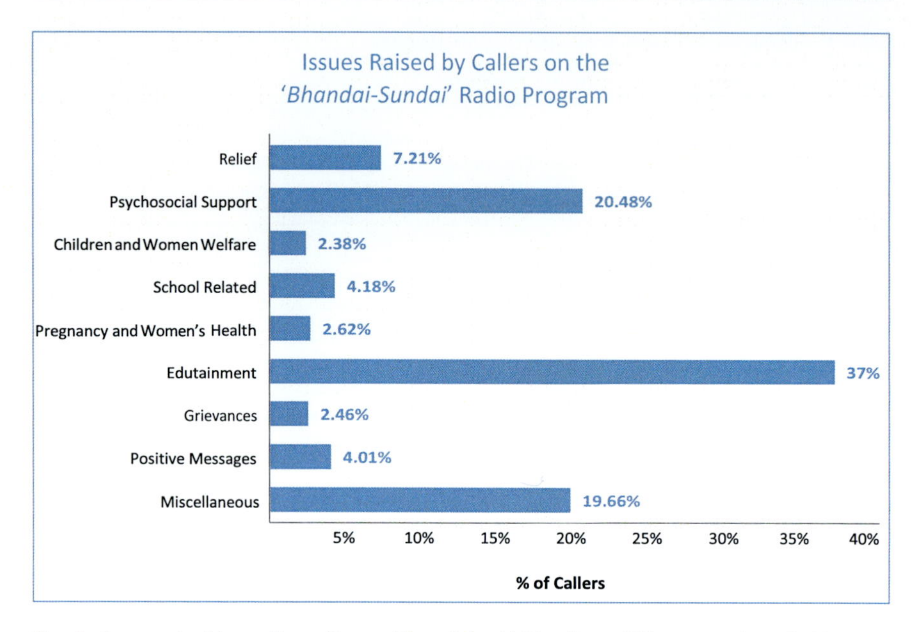

Fig. 1 Issues raised by radio audience. (Out of the 1200 callers, 68% were male and 32% were female. Although the number of female callers was lower than the male callers, an analysis of the calls revealed that many men were calling to discuss problems or issues related to their children or female members of their family. 61% of the callers were adults, whereas 39% were children. Miscellaneous included, among others, questions on rumors, other earthquakes or aftershocks, related to day-to-day life or support from government. Source: Author)

education officers, principals, teachers, parents, and students to discuss issues around reopening of schools including psychosocial and safety concerns.

Starting at a time when people were desperate for information and psychosocial support, the program was able to successfully reach out to people to satisfy both their information needs and requirements for emotional support. Further, it also created a conduit for community needs to reach duty bearers for taking necessary actions.

46.4 Mobile Edutainment Shows with Celebrities

Based on the positive response to the radio program and building on the popular brand name, '*Bhandai Sundai,*' a travelling 'edutainment' show titled '*Bhandai Sundai Gaon Gaon Ma*' '*Talking – Listening in villages*' was created.

Popular Nepali celebrities – comedians, magicians, singers and other performing artists (These included UNICEF goodwill ambassador Ani Choying Drolma, and other celebrities such as Jeetu Nepal, Kaliprasad Baskota, Komal Oli, Deepak Raj Giri, Deepashree Niraula and others.) – reached out to communities in remote areas with entertainment coupled with messaging on critical issues. The celebrities along with the messages that they carried were warmly received by people in much need of entertainment to help them forget the traumatic experiences that they had been through. The celebrities reached out to the communities and camps in severely affected districts (Rasuwa, Nuwakot, Dhading, Gorkha, Ramechhap, Dolakha, Sindhupalchowk, Kavre, Sindhuli, Kathmandu, Lalitpur and Bhaktapur) and provided them entertainment along with lifesaving messages on health, nutrition, sanitation, hygiene, and child protection – through music, comedy shows, and other entertainment-based programs. The program received tremendous response from the people, drawing in huge crowds wherever it was organized.

46.5 Youth Engagement

To reach out directly to communities in most affected and media-dark areas with critical, lifesaving messages, UNICEF partnered with a youth organization – '*Yuwalaya*' – with strong district-based networks. Hundreds of youth volunteers went door to door, community to community, and also to camp sites to talk to people and provide information to keep them safe from disease outbreaks and other effects of the earthquake such as trafficking of women and children which dramatically increased after the earthquake. These face-to-face activities were valued especially for information exchanges and discussion. Further, young volunteers also demonstrated the use of essential rehabilitation supplies and distributed communication materials.

Involving young people from local communities in the outreach activities not only ensures appropriateness and acceptance of messages but also increases identification and creates a sense of ownership. Furthermore, engaging children and youth in the design and implementation of disaster risk reduction and preparedness plans will provide them with lifesaving messages before a disaster strikes and will support them in protecting themselves and their families. (Plan International, Save the Children, UNICEF, World Vision (2016), pp. 5 and 42ff.)

46.6 Recovery and Preparedness

From a long-term perspective, the SBCC goals in the recovery phase were to help families and communities better prepare for and respond to natural disasters with a focus on women's and children's issues and to positively impact knowledge, attitudes and practices.

Given the destruction of livelihoods as well as reduction in the protection and security provided by the family and community, children and young people were at an increased risk of sexual violence, gender-based abuses, human trafficking and unsafe migration. To empower young people and their communities to make informed choices that enable them to become more resilient during and after natural disasters, a SBCC program for '*Promoting Recovery and Resilience among Earthquake-affected Communities*' was designed based on the Socio-Ecological Model and implemented in select earthquake-affected districts. The capacities of civil society organizations, community-based groups and networks and young people were built to disseminate critical information; track community perceptions and needs; develop community actions plans to address unsafe migration and human trafficking; create mechanisms for feedback generation and action by duty bearers and prepare communities for future disasters, reaching over 57,000 people.

With the aim to increase family and community preparedness for natural hazards, UNICEF designed and implemented an *edutainment* radio drama series '*Milan Chowk*' covering child survival and well-being and including episodes on health, nutrition, sanitation, education and protection. These were complemented with disaster risk reduction (DRR) messages around recurring types of natural disasters in the country. The drama, located in the imaginary village '*Milan Chowk*', was developed and broadcasted in Nepali and four widely spoken languages. It was supplemented by local content in local languages produced by 16 community radio stations from priority districts who benefited from mentoring and training. The local program capsules created community participation and ownership of the drama series and increased the effectiveness of the content. Audience feedback was generated through interviews with people from the community, so-called voxpops, group discussions held during the recap episodes, and focus group discussions (FGD) and key informant interviews (KII) held during the field and mentoring visits and fed back into the radio drama series.

46.7 Monitoring and Evaluation

The Common Feedback Project (CFP), (For further information, please see http://cfp.org.np/.) which was a member of the CWC and housed under the UN Resident Coordinator's office carried out communication assessments in collaboration with the CWC member organizations. These were carried out to understand communication needs that communities had, as well as preferences, barriers, and challenges that they were facing in accessing information. These assessments provided valuable information to UNICEF and partners to fine-tune the communication response as well as provided insights into the effectiveness of ongoing communication interventions.

The CFP also carried out community feedback and perception surveys as well as weekly rumor tracking surveys (In the initial phase, a lot of rumors where spread, including misinformation on other earthquakes, and related to government response and relief material. Reports are available on http://citizenhelpdesk.org/nepal-archive/ (previously: http://quakehelpdesk.org/).) in collaboration with Internews, Local Interventions Group, and Accountability Lab. Data from these feedback mechanisms helped inform communication efforts.

A third-party end-user monitoring system to monitor the effectiveness of the humanitarian response was established, which provided periodic reports on the performance of UNICEF emergency programs, including the SBCC initiatives. The feedback from the monitoring system helped understand program performance as well as implementation bottlenecks and course correct as required.

An independent evaluation of the UNICEF response which highlighted that the response succeeded in achieving the Humanitarian Performance Monitoring (HPM) target of reaching 1 million affected people. "The C4D (Communication for Development / SBCC) section was able as well through the CWC working group to harmonize and coordinate the messaging of different partners and different sectors. This resulted in raising the profile of communication with communities to a specific area of activity, with its own funding and objectives. Effectiveness and coverage have been outstanding."

46.8 Conclusion and Recommendations

The mega-earthquakes that struck Nepal in April and May 2015 were the biggest disasters to hit the country in a very long time. With the aim of better coordinating communication efforts among relief providers, UN agencies, international and national nongovernmental organizations (NGOs), civil society organizations (CSOs), media organizations, and community radio operators collaborated under a newly established '*Communication with affected Communities Working Group*' (CWC) in the immediate aftermath of the earthquake. The latter in collaboration with the government took a key role in providing situation updates and disseminating critical and timely lifesaving messages right from the onset of the emergency and help communities stay safe from aftershocks, clarify rumors, and recover from the effects of the earthquake.

A partnership with the main national broadcaster helped ensure accountability to affected populations by providing communities with a channel to give feedback to humanitarian responders on their concerns and needs. Intensive messaging through Nepal's widespread network of community radio stations coupled with community mobilization through youth volunteers and edutainment activities – especially in media-dark areas – ensured that affected populations had access to critical and lifesaving information. As highlighted by an external evaluation (DARA, 2016), this approach achieved significant results especially taking into account the inaccessibility of most areas. Given the increasingly high rate of mobile ownership and usage, sharing and collecting real-time information via an SMS-based technology is currently been looked into (Inter-Agency Common Feedback Project (2016), p. 12. 70% of the 2100 respondents indicated to have used the mobile phone immediately following the earthquakes. Sharecast (2017), p. 19. 67% of the respondents were listening to the radio on their mobile phones. However, special attention needs to be paid to reach women as across generations, significantly less women own phones than their male peers (22% difference for the 15–19-year-olds and 15% for the 20–24 year-olds), Government of Nepal, Ministry of Health (2017) pp. 322 and 323.).

Building capacities of community radio stations and CSOs on disaster preparedness and response can be a valuable investment so that they can prepare communities as well as immediately start local programs after a disaster. In the case of Nepal, two pilot projects were launched shortly after the earthquake, strengthening the capacity of selected community radios and CSOs in hazard-prone areas in producing programs and conducting social mobilization activities on disaster risk reduction.

Much required, on-air psychosocial counselling helped people cope with their trauma. As consequences for children and women in emergencies are likely to be more severe, special programs were created addressing their needs and concerns. A strong focus should also be put on actively involving children living with disabilities in such programs. (UNICEF 2017a)

This interactive radio program as well as community perception surveys that were carried out through the CWC provided valuable feedback on the information and rehabilitation needs of communities. However, in the absence of a SBCC cluster, this feedback could not often reach clusters in a systematic manner. Further, information on the timeframe of response efforts such as supplies, equipment, or shelter could not be fed back to communities to the extent desired as information on these were not systematically available to the CWC partners. The CWC was activated after the emergency. Given the importance of having such a group as a preparedness measure and hence to push the preparedness and prepositioning agenda, the CWC was maintained. (Renamed into the '*Community Engagement Working Group*' (CEWG), the group got reactivated during the 2017 floods in the Terai region and acted as an important coordination mechanism for all humanitarian actors in Nepal regarding community engagement and SBCC.). As an external evaluation (DARA, 2016) summarizes the SBCC response, the "creation of the CWC working group, set up of feedback mechanisms, design of communication activities, harmonization of materials, and use of radio programmes for mass psychosocial counselling were all appropriate initiatives, some of them with an innovative approach that should be highlighted as elements of good practice."

The CCC underscores the critical role of preparedness for a rapid and efficient emergency response. (UNICEF (2010), p. 5) Having a contingency plan including specific agreements with partners for community mobilization and long-term agreements for needs assessments, monitoring and evaluation, material development, and media buying greatly helps in expediting the response during a disaster.

Concurrent monitoring, evaluation, and documentation help course correct as well as establish the added value of SBCC interventions for ensuring greater investments in SBCC human resources and program budgets.

In disaster-prone countries like Nepal, it is crucial that government and developing and implementing a comprehensive SBCC strategy to reach people with critical information for disaster risk reduction and emergency preparedness, at the same time as ensure accountability to affected populations by collecting feedback and inputs from at risk and affected people, providing platforms through which children, women, youth, and populations as a whole can communicate with duty bearers on their needs and concerns. Building resilience and capacities of communities by actively involving children, adolescents, and youth will help the society as general to help them better prepare and respond to future emergencies and protect the most vulnerable.

Disclaimer The information of this document expresses the personal views and opinions of the authors and does not necessarily represent UNICEF's position.

References

CDAC Network (2016) Are you listening now? http://www.cdacnetwork.org/contentAsset/raw-data/84553f31-da55-4ce4-81a6-8c9ca61194bd/attachedFile. Accessed June 2018
DARA (2016) Evaluation of UNICEF's response and recovery efforts to the Gorkha Earthquake in Nepal. https://daraint.org/dara_evaluations/evaluation-unicefs-response-recovery-efforts-gorkha-earthquake-nepal/. Accessed May 2018
Government of Nepal, Ministry of Health (2017) Nepal Demographic and Health Survey 2016, https://www.dhsprogram.com/pubs/pdf/fr336/fr336.pdf. Accessed May 2018
Government of Nepal, National Planning Commission (2015) Nepal earthquake 2015 post disaster needs assessment, vol A: key findings. https://www.npc.gov.np/images/category/PDNA_Volume_A.pdf. Accessed May 2018
Government of Nepal, National Reconstruction Authority (2016) Post-disaster recovery framework (PDRF). National Reconstruction Authority, Nepal
Inter-Agency Common Feedback Project (2015) Information and Communications Needs Assessment. September 2015. https://bit.ly/2IHkGaz. Accessed May 2018
Inter-Agency Common Feedback Project (2016) Information and Communications Needs Assessment. March 2016. https://bit.ly/2NdDYHM. Accessed May 2018
Plan International, Save the Children, UNICEF, World Vision (2016) After the earthquake: Nepal's children speak out. https://nepal.savethechildren.net/sites/nepal.savethechildren.net/files/library/After_the_Earthquake_Nepal's_Children_Speak_Out_English.pdf. Accessed May 2018
Sharecast (2017) Nepal Media Landscape Survey 2017. https://www.slideshare.net/madhu272/national-media-landscape-nepal-2017 Accessed June 2018
UNICEF (2006) Behaviour change communication in emergencies: a toolkit. UNICEF ROSA. https://www.unicef.org/cbsc/files/BCC_Emergencies_full.pdf. Accessed May 2018
UNICEF (2017a) Guidance: including children with disabilities in humanitarian action. http://training.unicef.org/disability/emergencies/general-guidance.html. Accessed June 2018
UNICEF (2017b) Reference document for emergency preparedness and response. http://www.unicefinemergencies.com/downloads/eresource/docs/humanitarian%20learning%20resource/Reference%20Document%20-%20Full.pdf. Accessed June 2018
UNICEF Office of Emergency Programmes (2010) Core commitments for children in emergencies: framework. UNICEF. https://www.unicef.org/publications/files/CCC_042010.pdf. Accessed May 2018

Importing Innovation? Culture and Politics of Education in Creative Industries, Case Kenya

47

Minna Aslama Horowitz and Andrea Botero

Contents

Abstract

This chapter discusses the need to critically reframe the concept of innovation, especially regarding North-South development cooperation. The emerging discourse on the "Silicon Savannah" illustrates the situation: Eastern Africa is becoming more and more interesting to the Global North because of local technological innovations and emerging examples of locally inspired content that has begun to reach both local, regional, and global customers. This has evoked discussion on how ICTs, innovation, and entrepreneurship will form the core of bottom-up solutions for development. At the same time, training and mentoring programs tend to be designed to match the ideals of Western start-up

M. A. Horowitz (✉)
St. John's University, New York City, NY, USA

University of Helsinki, Helsinki, Finland
e-mail: minna@minnahorowitz.net; minskiaslama@gmail.com

A. Botero (✉)
Interact Research Group, University of Oulu, Oulu, Finland
e-mail: andrea.botero@iki.fi

J. Servaes (ed.), *Handbook of Communication for Development and Social Change*,
https://doi.org/10.1007/978-981-15-2014-3_94

culture, without broader regard to local practices. Given some of these conditions, it is fair to ask, "whose innovation?" This case study seeks to answer the question based on a case study on a creative industries' start-up incubation project in Kenya. The multi-method Living Lab study within the project (2014–2017) points to how different actors of innovation ecosystems can add value to the ecosystem and offer also value to participants through managing relationships with peers, clients, funders, and partners and thus creating new learning and development possibilities. At the same time, they tend to import Western start-up discursive culture and politics of innovation. This may be in contrast to other creative opportunities that are embedded in country- and region-specific cultural content and practices that might be overlooked. Using the case study as an example, the chapter highlights key tensions of definitions, policies, and practices of innovation and development.

Keywords

Innovation · Development policy · Digital creative industries · Discourses · Kenya · Living Labs

47.1 Introduction: Development, Creative Digital Industries, and Innovation Education

Communication for development has always had different dimensions, functions, and impact. They range from the perspectives of social relationships, of access to information and competencies, and of citizen participation to those of structures and institutions paramount for democracy and of international cooperation (e.g., Servaes 2008; Scott 2014).

In the second decade of the millennium, yet another approach seems to have entered policy-making and public discourses: that of communication as entrepreneurship. Innovation and creativity have become broadly used terms in many national development strategies. The creative industries concept in policy documents suggests added value, exports, and new jobs, implying a foundation of competitiveness. Innovation and creativity terms are used in development strategies worldwide and are included in policy documents of the UNDP, OECD, WTO, World Bank, and other large international organizations (Moore 2014).

In addition, ICT for Development (ICT4D) has evolved to broader discourses about social entrepreneurship, or the so-called Fourth Sector (e.g., Sabeti 2017), broadly referring to a variety of ventures that in some manner do good socially but also make money to sustain their operations. In tandem, due to growing digitalization, creative industries and related communication products and services have been hailed as some of the new areas that support sustainable development in the Global South (e.g., UNESCO 2012). Numerous actors involved in development, from UN programs to governments to non-governmental organizations (NGOs) and corporate donors, have embraced ICTs and digital creative industries, as significant contributors to social and economic development. A part of that strategy has been the rise of educational and entrepreneurial projects that are designed to enhance capacity for innovative products and businesses in the field.

Yet, the trend of innovation education as a part of development has at the same time evoked debates about how ICTs, innovation, and entrepreneurship can form the core of bottom-up solutions for development. Creating knowledge and skills in the Global South for participation in the global marketplace of digital creative industries seems to meet the needs of the ever-globalizing sphere of communication. At the same time, communication, training, and mentoring programs tend to be designed to match the ideals of Western start-up culture, without broader regard and engagement with local practices.

Based on a case study on a creative industries' start-up incubation project in Kenya, this chapter illustrates why asking "whose innovation?" should be key strategy when addressing communication for development projects. Using the case study as an example, the chapter highlights some of the key tensions of definitions, policies, and practices of innovation and development that these types of projects confront.

47.2 Illustration: Case Kenya

Eastern Africa is becoming more and more interesting to the Global North because of local technological innovations and emerging examples of locally inspired content that has begun to reach both local, regional, and global customers. This is also the case in Kenya. Several international rankings place the country among the ones on the rise (e.g., Digital Index 2015). Many technological innovations, such as the mobile banking system M-Pesa, and social innovations such as the crowdmapping system Ushahidi have been discussed and praised in the global media and, in many cases, have been exported to other regions. Innovators from the Global North, as well as numerous Western governments, have established and supported a variety of creative and entrepreneurial hubs and makerspaces (De Beer et al. 2017). These developments have given rise to the moniker of "Silicon Savannah" (e.g., Bright 2015) that refers to Eastern and Southern Africa.

With income levels rising, Eastern African countries and Kenya in particular are said to be on a path to become a middle-income country over the next 20 years or so. In the 2010s, Kenya's mobile phone penetration is already providing an infrastructural base for many new ventures. It is no wonder that foreign direct investment is increasing and with it innovation ecosystems that cater for local, regional, and even global markets. All these trends have thought to pave the way for growing more technologically oriented innovations as well as related cultural products and services. The pride in, and demand for, local cultural products in the region is in the rise, and this means new opportunities for digital creative industries in Kenya and elsewhere (Fleming 2015; Van der Pol 2014).

At the same time, research indicates that institutional support for cultural production and digital innovations has not developed with the same speed as the developments in the country require. A major challenge for digital creative industries is that the bulk of the applicants for work are those without qualifications or experience: Many individuals learn the skills needed on their own, because of their passion for their field (ACRI 2012). There is also a need to increase the participation of women in these sectors to fully utilize the country's human resources (UNCTAD 2012). Finally, acclaimed innovations do not automatically translate into sustainable businesses: The skills of, and educational models for, entrepreneurship in cultural production need to be further developed (Okolloh 2012).

The case in point, the GESCI-African Knowledge Exchange (GESCI-AKE) program, aims to educate young people at the beginning of their creative industries' careers with new skills in digital animation, sound production, as well as game design. The program was initiated by the Global e-Schools and Communities Initiative, aka GESCI, an international NGO in Nairobi, Kenya. As the case is with many of the Kenyan and Eastern African innovation education projects and incubator hubs (De Beer et al. 2017; Marchant 2015), the work was funded partly by foreign donors, in this case the Foreign Ministries of Finland and Ireland.

The program followed a Living Lab participatory action approach that documents the processes from curriculum development to innovation team-building and showcasing. The Living Lab research was to inspire participatory practices that would feed into the concrete teaching modules of media skills, as well as incubate ideas for future start-up businesses for the participants and reflect on the results and future of the program.

GESCI's initial focus was in ICT4D. It was founded on the recommendation of the United Nations Task Force on Information Communication Technology, at the first World Summit on the Information Society (2003), with a mandate to assist governments in the socioeconomic development of their countries through the integration of technology for inclusive and sustainable knowledge society development. Today, GESCI supports ICT and education policy-making, as well as training of teachers, but also creates educational innovation programs for the disenfranchised youth. GESCI's African Knowledge Exchange (AKE) training program belongs to that sector of GESCI's work, with the purpose of combining culture and digital media technology. The program addresses the changing employment environment driven by new technologies in the context of growing youth unemployment.

The first GESCI-AKE projects, The Sound of the City (2014–2015) and GESCI-AKE Creative Media Venture (2016–2017), were pilots to develop this approach to development, digital creative industries, and youth. They offered training for Nairobi-based young creatives, by combining digital and creative skills, and project work, with start-up business model building and learning entrepreneurial tools. GESCI-AKE's training and enterprise model is in line with the trend that believes in local entrepreneurship as one of the most empowering, and cost-effective, solutions to local systemic problems of youth employment (Schoof 2006).

In practice, these projects trained entry-level digital media creators in three different fields: animation, game and app design, and sound design. Training was done in three phases: First came technology skills training, then apprenticeship, and lastly the so-called start-up incubation. These phases included several benchmark projects: showcases, community events, roundtables, feedback and pitching sessions, as well as policy forums. The events were designed to discuss the results, progress, and insights from the projects made by participants with different stakeholders, the most important being policy-makers, educators, and industry representatives.

For most of its duration, the GESCI-AKE program employed the so-called Living Lab method, a participatory action research model in which innovation happens hand in hand with research (Feurstein et al. 2008). Living Labs have become a key tool in creative industries as well as in policy-making because they have radically mixed the roles of those who innovate. They involve user or client communities, not only as observed subjects but as cocreators (Higgins and Klein 2011). The goal was to examine how to facilitate engagement between mentors, students, digital creative industries specialists, researchers, and entrepreneurs. The secondary aim was to create new knowledge that can also be translated into policy recommendations, networking and partnership opportunities for support, and sustainability plans of the training and trainees beyond the project parameters. The Living Lab process included several methods and tools, ranging from background market and policy analyses to questionnaires, interviews, collaborative blogging, visual diaries, and participatory observation in different events. Other participants included the young creatives, their tutors and mentors, the GESCI management, as well as industry contacts and policy-makers from Kenya and the region.

47.3 Three Dimensions of Innovation in Education

Different iterations of innovation and development create challenges for digital creative industries and related education. These iterations of innovation and development can be seen in three different levels: the macro-level policy context, the meso (discursive) framing of the particular educational initiative, and the microlevel, practical reiterations of teaching innovation and pedagogy.

47.3.1 *MACRO*: Development and Innovation in the Policy Context

Organizations like GESCI operate not only in youth education in creative media but also in training, interacting, and advising teachers and policy-makers. In this case GESCI is an organization well informed by the macro-level policy goals, and it constantly interacts with Kenyan, and other African, innovation, ICT, and education policy-making as well as the related industries. Therefore, understanding the tensions requires a review of the Kenyan policy context for innovation and development.

Kenya has been hailed as one of the next hubs of technology and creative industry innovation. Specifically, cultural distinctiveness, very strong traditions, and flair across creative sectors including music, crafts, fashion, visual arts, and film define the centrality for creative industries for the economy in the region. At the same time, there are significant challenges, including aversion to risk-taking and general entrepreneurialism. Policy-making around copyright is inconsistent, as is policy support in the field of education for creative digital industries. Digital literacy still needs to be supported, as must creative and entrepreneurial educational opportunities that are few and far between (Fleming 2015).

Kenya is addressing these challenges in several fronts. Its first innovation policy, the so-called Vision 2030 initiative (2008), is the national long-term development policy that aims to transform Kenya into a newly industrializing, middle-income country by 2030. The Vision comprises three key pillars, economic, social, and political. Science, technology, and innovation (STI) are thought essential to the vision (Ndemo 2015). This Vision 2030 has been practically parallel to Kenya's new cultural policy (Kenya 2009). In general terms, Kenyan policy-making has been geared toward protecting cultural products (e.g., copyrights). The challenges lie in the livelihood of cultural creators. For instance, the draft policy recommendations for the music industry illustrates numerous mechanisms, ranging from education to incentives and funding, that can be used to not only support but actively encourage creative industries (Kenyan Ministry of Sports, Arts and Culture 2015).

These developments follow recognizable patterns identified for developing innovation ecosystems in global economies. However, Kenya's wider socioeconomic context does not match the innovation model of the Global North or even those of the Asian digital economies. In Kenya, as in many other African contexts, it is international funders and other partners who are playing bigger roles in incubating innovation, much more than venture capital does. Multinational corporations have set up research and innovation labs in Kenya to expand their own reach while getting closer to the source of unique problems and markets (Marchant 2015; Ndemo 2015). The innovation ecosystem is thus created in collaboration with for-profit and not-for-profit local and international partners.

The policy context recognizes the gaps in meeting global standards in creative media education in terms of skills, as well as in terms of entrepreneurial training. At the same time, much of this education is currently tied to international actors: foreign universities, companies, and donors, with specific agendas. This is well exemplified in the case of the emergence of innovation hubs and training labs for digital skills. Already in 2015, the country hosted 23 different ICT hubs based on innovation and learning models well established in the Global North (Fleming 2015). The emergence of these hubs is the result of research efforts in Kenyan universities, yet many have developed co-funding and other sustainability mechanisms. In those cases, international funders, either countries or businesses, are key to the sustainability of these actors (Marchant 2015).

47.3.2 *MESO*: Discursive Framing of Innovation

The ambiguity in the macro-level between local needs, national policies, and global drivers of innovation is replicated also within the meso-level in GESCI-AKE. This does not, by any means, indicate that the programs would not serve an important purpose and offer both skills training and innovation opportunities. Their experiences are illustrative of a shift that reflects the broader transformation in terms of development and communication. While GESCI started as an ICT4D-focused organization, GESCI-AKE works toward individual livelihoods and independent small business ventures.

This shift may come from the above-described global development and innovation discourses, as well as practices established in the Kenyan context at large. But GESCI-AKE's own experiences may also play a role. Indeed, the outcomes produced by participants of the 2016–2017 training are small ventures combining art and jewelry design or collaborating as freelancers in music production. The program even notes that the employment rate after the training is 100% and that some of that is self-employment. Hence, the core concept here is not local cultural products, but rather youth entrepreneurship.

It seems that the GESCI-AKE program itself has become the innovation product, rather than the framework in which its trainees learn skills and produce digital media innovations. It has transformed from organizing ICT and creative training for local youth to another innovation hub, a makerspace that facilitates meetings with investors and policy-makers. Sustainability of the existence of the programs (a key challenge for many innovation hubs in the region, CSBKE 2014) is incubated by their Living Lab approach and GESCI-AKE's multi-stakeholder policy forums.

47.3.3 *MICRO*: Practices of Innovation Education

The third tension arises at the microlevel of practices. In the GESCI-AKE, it involves between the above-described educational innovation and its concrete training curriculum. While it was clear that a significant amount of time was skill development and concrete team-building, the curriculum also entailed numerous segments that draw on typical product innovation and Western start-up practices.

GESCI-AKE has experimented with several strategies to turn the hub into an accelerator of promising start-ups. The curriculum framed entrepreneurial competences to mark a holistic training process, ranging from "opportunity-seeking behavior" to "taking calculated risks" and "self-confidence." Pitching sessions were prepared geared to help the teams gather investments. Some support was organized to help them think about business models. Discussions and training for mapping needs regarding protecting intellectual property rights were organized. Showcases of product prototypes and innovations were done in fancy hotels. The incorporation of some of these practices implied that the results of teamwork and their expected goals shifted. From the early idea of GESCI-AKE as a supportive environment for the emergence of small-sized creative ventures, a new purpose took shape, conceptualizing GESCI-AKE as an incubator or start-up accelerator. This manifested itself both in the abovementioned inserts into the training program and in the discourses about the program, including mentions of GESCI-AKE as a lab or a hub.

The initial purpose of the GESCI-AKE program placed emphasis on the underlying cultural ambition and on ventures as more open-ended aim for the entrepreneurial activities that could emerge from the work of different teams. The first pilot did not include an “incubation” phase but rather a collaborative project phase. When the approach shifted to one, in which start-ups should be formed, should begin to make money, and should grow – and therefore would initially need funding and investment – some discrepancies emerged. The ideas and prototypes were not Silicon Valley-type technological innovations, but small-scale cultural products, ranging from modest synopsis of games to singer-songwriter endeavors and jewelry. The continued reinforcement of start-up discourse ended up generating conflicting expectations for participants and for some members of the collaboration network of the program.

In the course of its transformation from an ICT training program to a hub, GESCI-AKE transformed its young creatives to innovators – but with what impact? While it is too soon to tell how the start-ups formed within GESCI-AKE are making their mark in the Kenyan digital creative industries, GESCI’s partners in industry, government, and civil society have been adamant in stressing the need for local content and opportunities for both for- and not-for-profit content creation. Recent research seems to back up the potential and urgent need as well (Bekenova 2016). Did the imported innovation discourse take over the local needs here as well?

47.4 Redefining Innovation

Does communication as entrepreneurship belong to the realm of communication for development and social change? Whose innovation is at stake here? Whose development is fostered via digital creative industries and their cultural products, as described in this case? The tension between global innovation discourse and local cultural appreciation is, in fact, expressed as an opportunity by many, ranging from market reports to policy decisions. On one hand, sub-Saharan Africa at large is seen as a vast market, and “Silicon Savannahs” are emerging, clearly indicating potential for international players. On the other hand, the reports on creative industries celebrate “Africa Rising” in terms of rekindled interest and modernization of local cultures, in ways that attract regional and international markets of creative industries. The role of culture has also been stressed globally as one of the key drivers of development, as expressed, for instance, by the United Nations Development Group: Cultural productions should “make an important contribution to poverty reduction, as a resilient economic sector that provides livelihood opportunities.” In addition, “education strategies should aim to develop cultural literacy and equip young people with the skills to live in a multicultural and diverse society, in both economic and social terms” (UNDG 2014).

Yet, absent from this discussion is the discourse of innovation as a local, communal, grassroots phenomenon. Not everyone will build a tech empire or become a global music producer. There are many skills courses and foreign-sponsored innovation hubs. But how to educate and inspire creative media producers who can service local communities and make the difference that the UN is calling for: life skills, cultural appropriation, tolerance, and diversity?

The rise of social entrepreneurship as a business model is at least a discourse, if not yet universal model, that tries to address development, social change, and livelihood by small, local, concrete steps. There is clearly a political trend to favor and support activities, big or small, that position themselves as innovators of social value. This does require specific support system, most importantly including ongoing support networks and leadership training for young social entrepreneurs (see, e.g., Brixiová et al. 2014; Ndemo and Aiko 2016). Local and international investors are ever more keen on supporting local social innovation (UNIDO 2017). But even then, who sets the parameters and frames for what innovation needs to be?

The GESCI-AKE experience highlights ways in which some new actors in innovation education ecosystems can add value by providing young entrepreneurs new brokering relationships with peers, clients, funders, and partners, creating new learning and development possibilities. At the same time, it also exemplifies how it is that it is very easy to import Western start-up discursive culture as the global standard and, with it, market-driven politics of innovation – and similar policies of development. There is a need to continue questioning and redefining innovation and related innovation education and policies. Broadening the idea of innovation, its pedagogy, and its meaning for development and social justice would not only benefit Kenya but the global North and South alike.

47.5 Cross-References

- ▶ A Threefold Approach for Enabling Social Change: Communication as Context for Interaction, Uneven Development, and Recognition
- ▶ Communication for Development and Social Change Through Creativity
- ▶ Empowerment as Development: An Outline of an Analytical Concept for the Study of ICTs in the Global South
- ▶ ICTs and Modernization in China
- ▶ Millennium Development Goals (MDGs) and Maternal Health in Africa
- ▶ New Media: The Changing Dynamics in Mobile Phone Application in Accelerating Health Care Among the Rural Populations in Kenya
- ▶ Political Economy of ICT4D and Africa

References

ACRI (2012) Unearthing the gems of culture: mapping exercise of Kenya's creative cultural industries. African Cultural Regeneration Institution – ACRI

Bekenova K (2016) Cultural and creative industries in Africa. Afr Polit Policy 2:1–12

Bright J (2015) The rise of Silicon Savannah and Africa's tech movement. In: TechCrunch. https://techcrunch.com/2015/07/23/the-rise-of-silicon-savannah-and-africas-tech-movement/. Accessed 1 Nov 2016

Brixiová Z, Mthuli N, Bicaba Z (2014) Skills and youth entrepreneurship in Africa: analysis with evidence from Swaziland. African Development Bank, Tunisia

De Beer J, Millar P, Mwangi J, Nzomo V, Rutenberg I (2017) A framework for assessing technology hubs in Africa, working paper 2, Open Air African Innovation Research
Feurstein K, Hesmer A, Hribernik K, Thoben K-D, Schumacher J (2008) Living labs – a new development strategy. In: European living labs: a new approach for human centric regional innovation. Wissenschaftlicher Verlag, Berlin
Fleming T (2015) Scoping the creative economy in East Africa. The British Council, London
Higgins A, Klein S (2011) Introduction to the living lab approach. In: Tan Y-H, Björn-Andersen N, Klein S, Rukanova B (eds) Accelerating global supply chains with IT-innovation. Springer, Berlin/Heidelberg, pp 31–36
Kenya (2009) National Policy on Culture and Heritage. The Republic of Kenya, Office of the Vice-President, Ministry of State for National Heritage and Culture, https://en.unesco.org/creativity/sites/creativity/files/activities/conv2005_eu_docs_kenya_policy.pdf
Kenya Vision 2030 (2008) Kenya Vision 2030, Vision 2030 Delivery Secretariat, http://vision2030.go.ke/
Kenyan Ministry of Sports, Arts and Culture (2015) Kenyan National Music Policy
Marchant E (2015) Who's ICT innovation for? Challenges to existing theories of innovation, a Kenyan case study. Center for Global Communication Studies, Philadelphia
Moore I (2014) Cultural and Creative Industries concept – a historical perspective. Procedia Soc Behav Sci 110:738–746
Ndemo B (2015) Chapter 9: Effective innovation policies for development: the case of Kenya. In: Dutta S, Lanvin B, Wunsch-Vincent S (eds) The global innovation index 2015: effective innovation policies for development. WIPO, Geneva
Ndemo B, Aiko D (2016) Nurturing creativity and innovation in African enterprises: a case study on Kenya. https://doi.org/10.5772/65454
Okolloh O (2012) Frustrated innovation. MIT Technology Review
Sabeti (2017) The fourth sector is a chance to build a new economic model for the benefit of all. In: World Economic Forum, https://www.weforum.org/agenda/2017/09/fourth-sector-chance-to-build-new-economic-model/. Accessed 29 Nov 2017
Schoof U (2006) Stimulating youth entrepreneurship: barriers and incentives to enterprise start-ups by young people. International Labour Organization, Geneva
Scott M (2014) Media and development. Zen Books, New York
Servaes J (ed) (2008) Communication for development and social change. Sage, Los Angeles/London/New Delhi/Singapore
UNCTAD (2012) Measuring ICT and gender: an assessment. United Nations Publications, Geneva
UNDG (2014) Delivering the Post-2015 development agenda. United Nations Development Group (UNDG)
UNESCO (2012) Culture: a driver and an enabler of sustainable development: thematic think piece. UN system Task Team on the Post 2015 Development UN Agenda and UNESCO
UNIDO (2017) Workshop on social innovation and digital currencies, Vienna (Unpublished)
Van der Pol H (2014) Key role of cultural and creative industries in the economy. UNESCO Institute for Statistics OECD, Canada

Part VIII

ICTs for Development

ICTs for Learning in the Field of Rural Communication

48

Rico Lie and Loes Witteveen

Contents

Abstract

This contribution surveys learning approaches in the field of agricultural extension, agricultural advisory services, and rural communication and explores their relationships with Information and Communication Technologies (ICTs). It makes a distinction between theory-based approaches to learning and design-based approaches to learning. The reviewed theory-based approaches are *social learning*, *experiential learning*, *collaborative learning,* and *transformative learning* and the design-based approaches are *visual learning*, *intercultural learning,* and *distance learning*. The choice for surveying these specific approaches is based on the relevance that these approaches have for the field of agricultural

R. Lie (✉)
Research Group Knowledge, Technology and Innovation, Wageningen University, Wageningen, The Netherlands
e-mail: rico.lie@wur.nl

L. Witteveen
Research Group Communication, Participation and Social Ecological Learning, Van Hall Larenstein University of Applied Sciences, Velp, The Netherlands

Research Group Environmental Policy, Wageningen University, Wageningen, The Netherlands
e-mail: loes.witteveen@hvhl.nl; loes.witteveen@wur.nl

J. Servaes (ed.), *Handbook of Communication for Development and Social Change*,
https://doi.org/10.1007/978-981-15-2014-3_89

extension, agricultural advisory services, and rural communication. It is concluded that learning itself is to be seen as social and behavioral change and that the group is much valued in existing learning processes. Furthermore, experiences and reflections are central elements in all reviewed learning processes, and the visual and the cultural play crucial roles.

Keywords

Learning · ICTs · agricultural extension · agricultural advisory services · rural communication

48.1 Introduction

Information and Communication Technologies (ICTs) cover a broad range from information leaflets and magazines to mobile communication devices with an outstanding role for radio, which has been and still is an important ICT in the agricultural sector in the so-called Global South. Agricultural advisory programs on television address large audiences in specific parts of the world and mobile phone applications have gained importance as a means of communication and information exchange in many parts of the world. ICTs have always been part of communication and information exchange in the field of agricultural extension, agricultural advisory services, and rural communication and is as such not an innovation by itself. What is new, in the past two decades or so, is that much attention has gone to so-called new ICTs as if the "T" in the abbreviation did not exist before the digital revolution. It seems limitedly recognized that e-agriculture, e-business, and e-health are becoming common terms to refer to electronic adaptations of existing concepts of knowledge exchange and learning. New ICTs claim to have the potential of easily spanning space and crossing time and reaching objectives never imagined before. Mobility and 24/7 access have created opportunities for innovative and impactful forms of communications and improved feedback qualities of communication channels.

These new features of ICTs have attracted and been given much attention. Their potential seems endless and unproblematic although currently more critical views are emerging especially in relation to the public sphere with recent Facebook scandals such as the role of Cambridge Analytica before the Brexit referendum and the Trolls used in the US electoral campaign of 2016. However, an important question that seems to be a bit under addressed is what the promising features of digital or new ICTs might mean for the quality of learning. Learning has been identified in Lie and Servaes (2015) as a central theme in communication for development and social change. Therefore, this contribution addresses the role ICTs (old and new) can play in learning in the field of agricultural extension, agricultural advisory services, and rural communication. It surveys different approaches to learning and reviews their relationship with ICTs. First, we will briefly address the historical development of the field of agricultural extension, agricultural advisory services, and rural communication and its relationship with media, ICTs, and new ICTs in learning processes. Then the contribution continues with distinguishing different approaches to learning. The theory-based approaches to learning, which are discussed first, are the approaches that are most often referred to in the context of agricultural extension, agricultural advisory services, rural communication, and the role of ICTs. These are social learning, experiential learning, collaborative learning, and transformative learning. These forms of learning have often been put on stage when dealing with complex problems in the agricultural and rural sector. Second, the design-based approaches to learning are discussed. These are learning approaches that emphasize specific aspects of an operational form and content. The aspects, which are selected for review in this contribution, are the use of visuals in learning processes (visual learning), dealing with intercultural differences (intercultural learning) and crossing space and time through distance learning and e-learning. The design-related aspects have potential in contributing to improving the quality of learning in the field of agricultural extension, agricultural advisory services, and rural communication.

Learning in this contribution is fundamentally defined as change in cognition and/or behavior and concerns adult learning in formal and non-formal contexts. This change can be individual or collective as it, for instance, relates to groups, organizations and cultures. Social change, as well as behavioral change, is thus seen as forms of learning. Learning is positioned as ranging from establishing educational and capacity building environments, "training of trainers" (ToT) activities, vocational education and training (VET) to social and behavioral transitions (see also Lie and Servaes 2015).

48.2 Historical Development of the Field

The field of agricultural extension, advisory services, and rural communication has always been about communication and learning, but it has only been in recent decades that learning has been given explicit and increased attention. "How do people learn?" and "How can we best cater to the different learning styles of people?" became new questions pushing the field forward. Röling (1988, 1989), Chambers (1993), and others have been giving quite some attention to this central concern for learning in the field that we are discussing here. Pretty and Chambers (1993, p. 182 a.f.) even talked about a new "learning paradigm."

In the 1960s, the Training & Visit (T&V) system was the dominant approach within agricultural advisory services, but as this approach depended on regular output from the research system resulting in blueprint extension messages, as it was strongly top-down and required a costly system whereby extension agents were in touch with contact farmers only, the ambitions for "a green revolution" did not materialize in many types of contexts (Moris 1991). Contrary to T&V, the emerging Farmer Field Schools (FFSs) and other participatory advisory approaches built the analytical capacity of the farmers to encourage experiments to find more locally appropriate solutions. With the general known change in paradigmatic thinking about communication, the view on learning also changed. Where previously the

T&V system emphasized the transfer of knowledge, the FFSs emphasized the interaction and co-creation of knowledge. Learning came to be a social context related process of change. Against this background, attention rose for the importance of learning. For instance, in the 1990s, Röling and Wagemakers (1998) focused explicitly on learning processes as they relate to sustainable agricultural practices. They emphasized the importance of facilitating learning through participatory approaches and appropriate institutional support and policy structures. Leeuwis and Pyburn (2002) and Wals (2007) gathered many contributions in their edited volumes, which addressed various aspects of learning as related to rural resource management and sustainability. It is also in this context that especially "social learning" became the dominant approach to learning in the field of agricultural extension, agricultural advisory services, and rural communication.

As the field shifted its focus toward the study of agricultural knowledge and information systems (AKIS) (FAO & World Bank 2000; Röling 1989; Reijntjes et al. 1992) and agricultural innovation systems (Klerkx et al. 2010) (see also Lie and Servaes 2015), the increased attention for learning continued to grow (e.g., Beers et al. 2016; Van Mierlo et al. 2010). The publication by Leeuwis and Aarts (2011) is seen as a key publication in this process as it calls for a focus on networks, power, and social learning (see also Lie and Servaes 2015). Networks and learning relate and cannot be disconnected in our interpretation. It is seen as important that people connect and that in this connection shared learning ambitions arise. In collective action lies additional value for improving the quality of learning. Another trend is the renewed focus on ICTs. As the general ICT and mobile phone possibilities expanded exponentially, all kinds of ICTs also became part and parcel of the pallet of agricultural advisory services. Some ICT-based approaches created new ways of communication, but several others just adapt ICTs to modernized traditional communication styles to transfer knowledge. Mobile applications for all kinds of data collection and social media are now considered as new communication forms and are currently being explored and are becoming established in the field. Underlying the use of these applications are quests for understanding vital roles and possibilities of ICTs in learning processes for communication and information exchange in the field of agricultural extension, agricultural advisory services, and rural communication.

48.3 Learning

48.3.1 Theory-Based Approaches to Learning

This section surveys four approaches to learning: social learning, experiential learning, collaborative learning, and transformative learning. These approaches have been termed theory-based as they originate from the minds of theoretical thinkers in an academic context. They have not originated and developed specifically in an agricultural/rural setting but have been used widely in this field. This is the reason why these approaches are discussed below. Some potentials of ICTs will briefly be addressed as to how they relate to particular learning approaches.

48.3.1.1 Social Learning

Social learning has been, and still is, one of the most influential theoretical learning approaches in the field of agricultural extension, agricultural advisory services, and rural communication. The essence of social learning theory lies in the importance of the social, collective context for learning. Learning is taking place under the influence of the environment in which the learning is taking place. Through directly observing the actions of others, through listening to stories of others, or, for instance, through role-plays, a model for aspiring behavior is constructed, which influences one's own cognition and behavior. On the one hand behavior is thus imitated; but on the other hand, learning is also based on constructing expectations based on earlier (social) experiences (cognitive learning). Based on understandings of these previous experiences combined with behavioral learning, people decide to act in a certain way in a certain situation and herewith are involved in a process of change.

Social learning theory originated from Bandura's work in the 1960 and 1970s (Bandura 1963, 1977) and has been picked up by many different scholars and practitioners since. A very useful addition to better recognize how people learn is made by Argyris and Schön (1978) and concerns the distinction between single, double, and triple loop learning. Single loop learning addresses the "what" and refers to learning outcomes such as recalling knowledge, copying behavior and applying rules. Double loop learning addresses the "how" by focusing on learning about the causes of problems; the reasons behind the way things are done as they are done. Double loop learning addresses the unknown and discovers and creates new knowledge and behavior, whereas single loop learning is about applying the known to solve problems or perfecting the way things are already done. Feedback and reflexivity play crucial roles in double loop learning. It is through these mechanisms that a deeper level of learning is reached and the "how" becomes known. Triple loop learning then refers to knowing the "why." If you know the "why" you can teach others and address the norms and values underlying second loop learning. "How do we think, behave, and learn?" "Why are we thinking in the way that we are thinking?" Seen in this way, triple loop learning is about culture. Culture underlies the assumptions and patterns of actions. Triple loop learning underlies transformational/transformative learning, which will be discussed later.

Another distinction often made is the difference between conceptual learning and relational learning (Pahl-Wostl 2006; see also Beers et al. 2016). Conceptual learning relates to knowledge and refers to changes in cognition, conceptual understandings, and ideas. Relational learning can be understood as networked learning (see also Kelly et al. 2017) through which change occurs in one's understanding about his or her own position in a network as well as of the functioning of the network as a whole.

An important distinction to better articulate that learning itself is social and/or behavioral change is the one made by Beers et al. (2016) between learning outcomes and learning impacts. Their point of departure is that learning takes place through

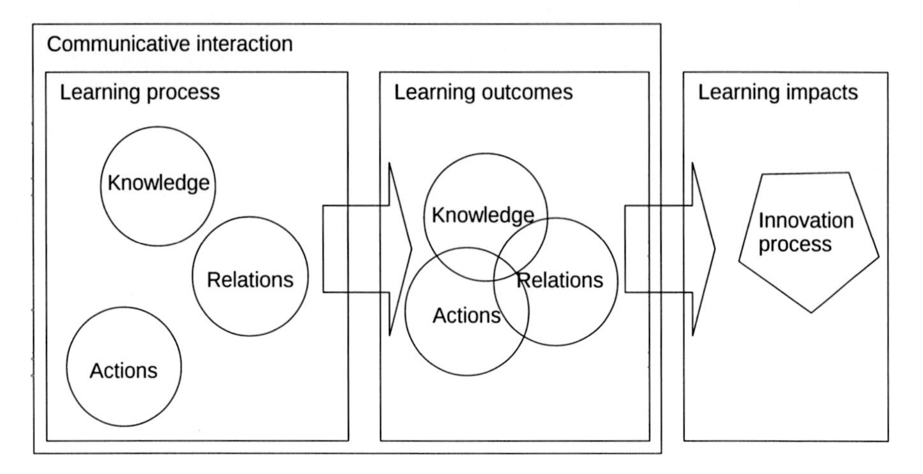

Fig. 1 Learning as a discursive process with interwoven knowledge, relations, and actions as align left outcomes

communicative interactions. They see learning as a discursive process and distinguish between knowledge, actions, and relations (Fig. 1). They emphasize that "during social learning, three dimensions of learning may become aligned: (1) new or changed knowledge (the what), (2) new or changed actions (the how), and (3) new or changed relations (the who)" (Beers et al. 2016). It is these three dimensions that become central in their view on learning. A learning outcome is then seen as a process of interweaving these three dimensions and an impact is the change achieved outside the discursive field of communicative interactions (see Fig. 1).

When it comes to ICTs, social learning theory has more than once been brought into relation with media and especially film (video) and television. The dominant thought here is that moving images can portray the behavior of socially desired models, for instance through soap operas. Besides this role of "model function" that television and film can play, ICTs also have the potential to bring people together and this is what is necessary for social learning. In these gatherings, feedback and reflection are key concepts and visually based ICTs can be strategically used to feed into the discursive process as an additional voice or portrayal of knowledge, actions, or relations.

48.3.1.2 Experiential Learning

The field of agricultural extension, advisory services, and rural communication deals with adult learning, individually or in groups, and learning is often not classroom based. Kolb's model (1984) is widely used within the field and describes an archetypal way in which people learn. Experience is the key word here. Learning is seen as a continuous interaction between observation and interpretation. Kolb distinguishes four different stages in a cycle of learning: a concrete experience, reflective observation, abstract conceptualization, and active experimentation (see Fig. 2).

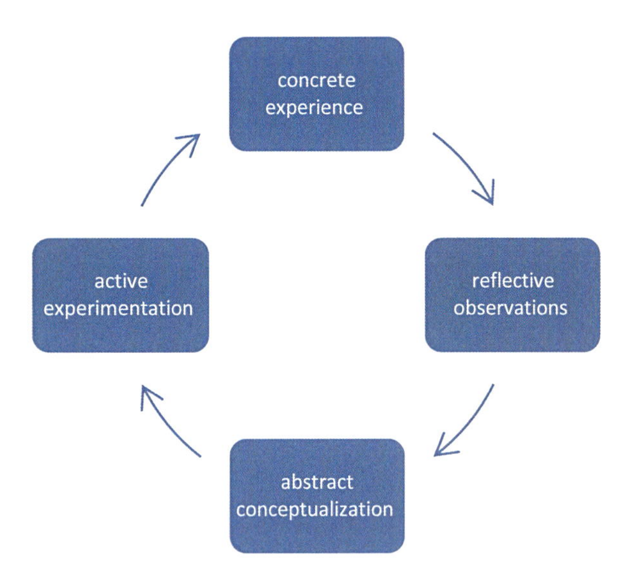

Fig. 2 Kolb's learning cycle

Experiential learning builds heavily on social learning. It is often also addressed under the terms "learning by doing" and "discovery learning." The main characteristic of experiential learning is that concrete new experiences (obtained by doing and discovery) merge with previous experiences, and such an articulated process of merging leads to change. FFSs emerged from this basic idea about learning (Anandajayasekeram et al. 2007; Braun and Duveskog 2008; Van de Fliert 1993). FFSs create space for social interactions and observations of contemporary events in the field and thus create an appropriate environment for observing others (behavioral learning) and sharing experiences and sharing and co-creating knowledge (cognitive learning). The Digital Farmer Field School (DFFS) (Witteveen et al. 2016, 2017) is an endeavor to translate some of the principles on which the FFSs are based to an ICT environment. The DFFS offers a tablet-based digital learning environment for farmers and related service providers such as extension agents. The tablet comes with tailored (instructional) films (in a social learning mode) and other educational materials and learning modules, which are then linked to features like the camera function for documenting and sharing the field experiences such as the observation and documentation of pests and diseases. It also features a telephone function to interact with an extension back-office. Besides being based on the FFS learning principles, the digital interface was also designed following the principles of responsible innovation (Stilgoe et al. 2013).

48.3.1.3 Collaborative Learning

An important aspect that is emphasized throughout the field, and which to a certain extent also underlies social learning, is the importance of the group in learning processes. A group can accomplish things beyond the capabilities of an individual, and collaborative learning, cooperative learning, team learning, multi-actor learning, networked learning, and group learning all recognize this by emphasizing collective characteristics of learning. They all share the basic idea that the old paradigm of "teacher/expert vs. student/layman" needed revision. Group interactions within Communities of Practices (CoPs), stakeholder platforms, teams, and all kinds of configurations of workable groups of people, with eventually different roles but somehow similar status, are now seen as being valuable in themselves and as having a positive influence on the process and impact of learning. Another characteristic that they all share is the view that meaning is constructed in social interaction. It is also generally accepted that this social construction has greater learning potential within heterogeneous groups than in homogenous groups. Collaborative learning is defined by Laal and Laal (2011) as "an educational approach to teaching and learning that involves groups of learners working together to solve a problem, complete a task, or create a product." Collaborative learning seems to act as an umbrella term for other group-based learning strategies. Cooperative learning can then, for instance, be seen as a specific kind of collaborative learning and is defined as "the instructional use of small groups to promote students working together to maximize their own and each other's learning" (Johnson et al. 2008).

Two concepts that stand out when discussing the importance of the group in processes of learning are "system interactivity" and "continuous reflexivity." System interactions need to be understood from a system's perspective on learning in which the value of interactions is recognized as having an important impact on learning. Within a system, learners are responsible for each other's learning as well as for their own learning. They are accountable and have responsibility towards themselves, towards each other, but also towards the whole, the functioning of the group itself. Interactions thus come with interdependencies. The basic principle here is that people depend on each other, also for their learning. How learners act with each other requires facilitation and the facilitator plays an important role in the functioning of the learning system as a whole. Learners actively participate in the group dynamics and construct the group process.

Reflexivity has already been briefly touched upon when discussing social learning and experiential learning, and it could as well be discussed under the next heading of transformative learning. Reflexivity has a central position and plays a crucial role in all forms of learning being applied in the field of agricultural extension, agricultural advisory services, and rural communication as discussed in this contribution. Although Kolb's interactional view on observations and interpretations implicitly incorporates reflection, it has been one of the critiques on Kolb's learning cycle that he did not give enough attention to reflection in his model (Boud et al. 1985; Smith 2010). Reflexivity relates to rethinking, taking a step back, overlooking the situation, and deciding about how to continue. Awareness is a fundamental condition for reflexivity. One cannot reflect without being aware and this going back and forth between observations and interpretations through reflections is the core of reflexivity. Elzen et al. (2017) following Beck, Giddens. and Lash (1994) make a distinction between reflexivity and reflection and see reflexivity as a social condition, whereas reflection refers to "reflecting at the deeper level of issues, the underlying assumptions and values" (Elzen et al. 2017, p. 245). Reflexivity is thus seen as a condition for learning (b.t.w., Elzen et al. (2017) also conceptualize it as an outcome of learning) and reflection is seen as an activity in a learning process.

48.3.1.4 Transformative Learning

Learning is bringing about change and transformative learning addresses a particular kind of change; a transformational change. Transformational change is based on a holistic perspective on change and incorporates a strategic choice for a systems change and thereby links to triple loop learning. It concerns a change in fundamental principles, underlying norms and values, and deeply rooted cultural perspectives and beliefs, focusing on questioning, imagining, and achieving radically changing routines or establishing something completely new or revolutionary. The difference between transformational change and transitional change is that transitional change replaces a practice or a condition that already exists and aims to improve the existing situation by adjusting it. Improvement and adaptation does not always require a fundamental change in norms, values, and mind-sets and when such change is not the case, it is common to talk about transitional change. Transformational change, on the other hand, does require a fundamental change, which, at the start of the transformational learning process, is open and unknown. Such learning processes align when dealing with complex problems, articulating the low predictability of future scenarios and the expected frictions or conflicts in the diverse knowledges participating in the process. One doesn't know the future outcome and needs to be flexible and adaptive during the process to be able to deal with the uncertainties, which are encapsulated in the learning process.

Mezirow is widely acknowledged as the founding father of transformative learning (1978, 1991, and 2009). To be able to change people's worldviews, Mezirow argued that people needed to be confronted with "disorienting dilemmas." Disorienting dilemmas do not fit people's existing mind-set and worldview and therefore need to be changed to be able to deal with them (see also Howie and Bagnall 2013). Other authors refer to this as "learning on the edge" or "positive dissonance." "Critical reflection" (Mezirow 1990) in a collaborative setting plays an important role in dealing with these dilemmas (see also Smith 2017). It is through confrontation – between people and between worldviews – that people learn. Transformative learning is in fact a form of experiential learning and depends on collaborative learning (see for instance Percy 2005). Collaborative learning processes in its turn is necessary to bring about transformational change. To be able to change people's frames of reference, a social learning approach, in which the social context is emphasized, is helpful to make the transformational change happen.

It is again the group that plays an important role and it is also for that reason that transformative learning formed part of the theoretical lens in the study by Taylor, Duveskog, and Friis-Hansen (2012) that explored the practice of FFSs in East Africa. But, ICTs also have the potential of bringing people together without being physically in the same room and the previously discussed DFFS is an example of that. ICTs also have the potential to offer disorienting dilemmas and offer a platform in which reflection can be facilitated.

48.3.2 Design-Based Approaches to Learning

This section surveys three design-based approaches to learning: visual learning, intercultural learning, and distance learning. In contrast to the discussed approaches above, these approaches are design-based as they do not fundamentally discuss the theoretical groundings of learning, but instead theorize design aspects of learning processes in the field of agricultural extension, agricultural advisory services, and rural communication. These design aspects relate to crossing boundaries. Visual learning crosses boundaries between people with different learning styles and searches to overcome and challenge boundaries between literacy and illiteracy. Intercultural learning crosses cultural boundaries and distance learning crosses boundaries of time and space.

48.3.2.1 Visual Learning

Visual Learning is one of the learning styles that Fleming distinguished in 1987 in his VARK model of learning. VARK stands for Visual, Aural, Read/write, and Kinesthetic and makes a distinction between the associated modes of learning or learning styles. Visual learning can be considered as demonstrating a particular modal preference for learning (Fleming and Baume 2006). Other authors focus on visual learners preferring learning with visual media such as colors, spatial organizations, (mind) maps, diagrams, drawings, photos, films, and so forth. Visual learning in such views leads to visual literacy as a skill and a reinforced preference for certain learning styles but also refers here to learning with visual media.

In this context, as far as visual literacy is concerned, Pratish (2006, p. 13) states that visual literacy refers to the competency to:

- Understand the subject matter of images.
- Analyze and interpret images to gain meaning within the cultural context the image was created and exists.
- Analyze the syntax of images including style and composition.
- Analyze the techniques used to produce the image.
- Evaluate the aesthetic merit of the work.
- Evaluate the merit of the work in terms of purpose and audience.
- Grasp the synergy, interaction, innovation, affective impact, and/or "feel" of an image.

Pratish further claims that visual literacy "involves problem solving and critical thinking and these can be applied to all areas of learning" (2006, p. 15) (see also Witteveen 2009, p. 4 a.f.). In the current digital age, visual literacy is a competence that is of crucial importance to read digital images and to be able to navigate through digital visual landscapes.

The other area of visual learning is using visual methods in learning processes and a variety of visual media (which contents of audio-visual ICTs in fact are). Visual methods are widely used in agricultural extension and advisory services. Video mediated learning is one of the forms of visual learning in which visuals are used for learning. Video and film have, in the past two decades, become often used in the field of agricultural extension, agricultural advisory services, and rural communication. In 2009, Lie and Mandler made an inventory of how film and video were used in rural development and one of the areas that the book refers to is the use of instructional videos (e.g., by organizations such as Digital Green, Video Volunteers, and Access Agriculture). In the past ten years, many studies have been conducted to get more grip on how video can be used (e.g., Bentley et al. 2015; Chowdhury et al. 2011) and how it connects to social learning (e.g., Karubanga et al. 2017). "The power of using film lies in its appropriate character and its multi-modal form of communication, and can be effective, especially in illiterate and low educated environments (e.g., Bentley et al. 2016)" (Wyckhuys et al. 2017).

Another field that is addressed in Lie and Mandler (2009) is learning about complex problems exemplified with the use of Visual Problem Appraisal (VPA). VPA is a film-based learning strategy that uses filmed narratives in collaborative learning about complex problems such as HIV/AIDS, coastal zone management, and rural livelihoods (Witteveen and Lie 2012, 2018). The essence here is that learning about a complex problem takes place by listening to, and relating to each other, the stories of various stakeholders who have a stake in a complex problem. The "system interactivity" and "continuous reflexivity," as discussed under collaborative learning, play a central role in the VPA learning strategy. Participants in the VPA workshops work in small groups of about four people and form a systemic learning unit through continuous interactivity and reflexivity. In addition, the system approach to the complex problem is being emphasized by the totality of the narratives of the various stakeholders. Complex problems can only be managed in an adaptive way as the different perspectives, which together are part of a system, are given a voice and are being heard. VPA gives a voice to these various perspectives and enhances learning in a collaborative way.

48.3.2.2 Intercultural Learning

One of the critiques on Kolb's model is that it takes very little account of different cultural experiences/conditions (Anderson 1988; Smith 2010). "As Anderson (1988, cited in Tennant 1996) highlights, there is a need to take account of differences in cognitive and communication styles that are culturally-based" (Smith 2010). Culture has indeed hardly been touched upon in the field of agricultural extension and rural communication. Even though understanding the role of culture is fundamental in implementing effective extension and communication, culture has not been given the attention that it should have been given in extension and education.

Intercultural learning, as a concept, is difficult to grasp and is broadly defined here as "a process of learning in which the role of culture is emphasized" (Lie and Witteveen 2013, p. 22). Culture always plays a role in learning. Culture is a collective phenomenon, so especially in collaborative forms of learning, which recognize culture as a determining factor, it plays an important role. In Lie and Witteveen (2013) we distinguished two different modes of intercultural learning: a sociopsychological mode of intercultural learning and an ethnographic mode of intercultural learning. The sociopsychological mode takes place in an etic way, reduces complex situations to known and identifiable variables, centralizes the individual in his or her learning about the other, emphasizes first loop learning, is short-term, and treats culture to be a context of human behavior. The ethnographic mode, on the other hand, takes place in an emic way, treats complexity as being a complex whole, values collectivity, values learning in a participatory way with the other, emphasizes second and third loop learning, is long-term, and treats culture as being text instead of context. Understanding the differences between the two modes makes us better understand how people learn in an intercultural setting. The two modes are complementary but develop differently over time. People often start learning in the first mode and over time the second ethnographic mode gains position.

Intercultural learning can be an essential part of transformative learning and touches upon aspects of triple loop learning. Intercultural learning is concerned with underlying norms and values. Transformative learning focuses on changing norms and values, but intercultural learning can also concern learning about all kinds of topics from a specific cultural perspective (conscious or unconscious). In group learning processes different cultural perspectives come together. In the field of agricultural extension and advisory services, people have to cross disciplinary cultural boundaries too. Farmers have a different culture as compared to policy makers, and extension officers have a different culture as compared to academic researchers. Recognizing and valuing these cultural differences has a positive influence on learning. Intercultural competence, which includes cultural sensitivity, is thus an important competence to acquire. Besides differences in disciplinary cultures, people can also have different culturally-related learning styles. These styles can, for instance, differ in relation to views on hierarchy, gender roles, recall capabilities, etc. Intercultural learning deals in one way or the other with diversity (e.g., UNESCO 2009).

48.3.2.3 Distance Learning

Distance learning can broadly be defined as learning away from a physically located (often formal) educational setting, like a university, a school building, or any other educational institution. Because of this spatial disconnection, distance learning is by definition connected through ICTs. Internet, e-learning, online education, MOOCs, and "synchronous learning and asynchronous learning" are all characterized by the use of ICTs of any kind. All the above terms might not all be falling strictly under the umbrella of distance learning (especially not in the traditional meaning of so-called correspondence courses), but the term distance learning nicely captures the ability of ICTs to cross distances of time, space, costs, and cultures. Distance learning thereby offers a playground to experiment and study a deliberate strategic use of ICTs, which can be overlooked or not recognized as such in place-based education.

ICTs in learning offer advantages that connect to crossing spatial and other boundaries. These include, among others, the following:

- ICTs can facilitate access to education (overcome space, time and finances).
- ICTs can create virtual learning environments.
- ICTs can facilitate individual learning as well as collaborative learning.
- ICTs can facilitate that the pace and form of learning can be defined by learners.

- ICTs can include varying combinations of content and cater for different learning styles.
- ICTs can be designed with benefits for low literate learners.
- ICTs offer wider options for using visual and film-based learning strategies.
- ICTs can cross cultural borders and facilitate intercultural learning.
- ICTs can create functional distance towards to subject of learning (e.g., film can show sensitive topics by letting people tell their story in a mediated encounter).

The above mentioned advantages can be put in a different perspective upon reviewing the so-called bold experiment in American Samoa (Schramm et al. 1981) where teachers were replaced by television sets. It was assumed that presenting the same content in the same way to all learners could enhance and level the learning experience. Although replacing teachers in the classroom with television sets is not distance learning, under the given definition above, it shows a valuable lesson for distance learning—namely, that a facilitator in person is essential for maintaining a high quality of learning. Even for relatively simple online discussion lists, moderators are important to observe and safeguard the appropriateness and the general working of the discussion list. In the beginning of distance learning there was a central focus on and a high belief in communication technology and the assumed ability for complete stand-alone learning configurations. In more recent years, also because of the success of The Open University, there is far less focus on technology and more focus on ways and qualities of learning.

Hybrid forms of learning have been introduced in the past two or three decades. Blended learning and flipped learning ("flipping the classroom") combine online and offline formats of learning, which somehow combine "traditional" classroom teaching approaches with out-of-the-class formats. Flipping the classroom is built on ideas of flipping or reviewing taxonomies of learning (e.g., Anderson et al. 2001) ensuring that more complex (cognitive) learning activities take place in the classroom and less complex tasks at home. Flipped classrooms are very specific in this by asking learners to watch an e-learning video (an instructional clip) or complete an online exercise before coming to class. This somewhat trendy interpretation of flipping the classroom meaning that "homework" tasks are presented as video recordings or other digital formats can again represent an example of framing ICTs as "new," whereas the underlying learning principles are overshadowed by new technological developments; in the case of flipping the classroom, it is built on an already existing idea of expanding the classroom to home.

The major point to make here is again, a lack of focus on the potential contribution of ICTs in a setting of combined lecturer-guided classroom session and individual homework activities. To begrime the scenario further, it deserves attention to question the feasibility of a lecturer eventually carrying a responsibility to produce a series of videos as preparations for classroom activities. If it is not about producing the videos, then the selection of visual material from a wealth of options might be time consuming. In the worst scenario, an overloaded lecturer with no qualifications, expertise, or interest has to produce videos, which leave him / her without resources for further teaching preparations. Reviewing flipped classroom evaluation creates an impression that the term attracts and motivates certain groups of teachers to rethink and innovate their teaching strategy, combining classroom activities and homework in a more coherent learning process.

Flipping the classroom is very much related to formal education, but in the field of agricultural extension and advisory services, forms of blended or hybrid learning can also be found. For instance, distance learning in the field of development in the Global South still involves radio as a main means of communication, but through a blended learning strategy radio was combined with FFSs to attend to skills and knowledge about cocoa farming with farmers in Sierra Leone (BBC Media Action 2005). Another example is provided by researchers from Johns Hopkins University who conducted research into the potential of blended learning models to reinforce learning and application of knowledge for health professionals. The researchers concluded that blended learning is an important strategy for reaching health professionals in lower- and middle-income countries (Ahmed et al. 2017).

48.4 Conclusion

A first conclusion that can be drawn from the above presented survey of approaches to learning is the basic understanding that *learning itself is social or behavioral change*. Learning not only leads to social and behavioral change but is the change process itself. Learning and change are integrated. This is an important perspective to incorporate in designing rural learning interventions, and Beers et al. (2016) made it clear by making a distinction between learning outcomes and learning impacts. The learning process and the learning outcomes together are to be seen as communicative interaction (see Fig. 1). The process and outcomes of learning are thus to be seen as social and/or behavioral change and this change leads to impact.

What further stands out from the review is *the value of the group* in adult learning processes. The group, but also the social environment and the network, contribute in a positive way to the quality of learning. Collaborative learning is indeed a much-valued approach to learning in the field of agricultural extension, agricultural advisory services, and rural communication. Relational learning is then about learning about one's own position in a social functional whole. It is also here that the system approach to learning comes into play. System approach means that the whole is considered and that interdependencies in learning processes are not only taken into account but are also seen as adding value to the learning process itself. The emphasis is on social construction.

A third point that stands out is *the centrality of experiences and reflexivity*. Kolb's experiential learning and Mezirow's transformative learning have both emphasized the importance of experiences, and reflections and reflexivity is widely seen as an important condition for collaborative learning. Having many experiences is a strong asset in adult learning processes. Experience is the key word in experiential learning, and old and new experiences lead – by the act of reflecting – to learning. Learning can therefore be seen as an interactive play between experiences and reflections and between observations and interpretations. Reflexivity is to be seen as a condition for learning.

A fourth point is that *the visual and the cultural play crucial roles* and that there is a need for addressing these aspects explicitly. In this contribution, the visual and the cultural aspects have been operationalized as design aspects of learning. Calling these aspects "design" feels a bit arbitrary as they could as well be theoretically explored. However, fact is that these two aspects have been undervalued in designing learning interventions in the field of agricultural extension, agricultural advisory services, and rural communication, whereas they do play a decisive role. Visual learning has its merits as visual literacy is becoming of increased importance in the current digital age. In addition, it can be concluded that visual ICTs – especially film and video – have a huge potential in learning processes. The case of the flipped classroom described in the section on distance education can be interpreted as a recent example of continuous fashionable uses of new technologies whereby underlying learning theories are easily deleted from educational designs.

How then are these four concluding points on learning and change related to the use of ICTs? The challenge is to create innovative learning strategies and learning environments, which are based on theories of learning and make use of the potential of new and old ICTs. In this regard, it is important to have a basic theoretical understanding about how people learn before engaging in designing, developing, and applying ICTs. The challenge is indeed not to center the technology. Simply wrapping existing strategies in innovative new technological designs is likely to underestimate the full potential of possibilities for learners, learning facilitators, and involved institutional arrangements. The focus should first and foremost be on articulating the guiding principles of the learning and change process and every professional and academic should have a vision on learning before taking ICT-based actions. Considering agricultural and rural learning needs in a contemporary context of climate change adaptation, increasing conflicts over resources, and issues of rural-urban and international migration as characterized by a sense of urgency, demands full attention on the smart use of available learning resources. Conceptualizing motivation, time, and other human capitals for learning as an ever-scarce resource instigates the notion that learning processes have to constitute a meaningful praxis of learning and learning facilitation. A further recognition of ICTs integrating a concise learning theory, creative design, and appropriate development carries the ambition to create a wider access and enhanced participation of rural learners in learning processes, which links to their rural realities and may influence their futures.

References

Ahmed N, Ballard A, Ohkubo S, Limaye R (2017) Global Health eLearning: examining the effects of blended learning models on knowledge application and retention. Johns Hopkins University, Baltimore

Anandajayasekeram P, Davis KE, Workneh S (2007) Farmer field schools: an alternative to existing extension systems? Experience from eastern and southern Africa. Journal of International Agricultural and Extension Education 14:81–93

Anderson JA (1988) Cognitive styles and multicultural populations. J Teach Educ 39(1):2–9
Anderson LW, Krathwohl DR, Airasian PW, Cruikshank KA, Mayer RE, Pintrich PR, Wittrock MC (2001) A taxonomy for learning, teaching, and assessing: a revision of Bloom's taxonomy of educational objectives, abridged edition. Longman, White Plains
Argyris C, Schön DA (1978) Organizational learning: a theory of action perspective. Addison-Wesley, Reading
Bandura A (1963) Social learning and personality development. Holt, Rinehart, and Winston, New York
Bandura A (1977) Social learning theory. Prentice-Hall, Oxford
BBC Media Action (2005) How radio and distance learning built skills and knowledge for cocoa farmers. BBC Media Action, London
Beck U, Giddens A, Lash S (1994) Reflexive modernization: politics, tradition and aesthetics in the modern social order. Stanford University Press, Stanford
Beers PJ, Van Mierlo B, Hoes A-C (2016) Toward an integrative perspective on social learning in system innovation initiatives. Ecol Soc 21(1):33. https://doi.org/10.5751/ES-08148-210133
Bentley JW, Boa E, Salm M (2016) A passion for video: 25 stories about making, translating, sharing and using videos on farmer innovation. Access Agriculture and CTA, Nairobi
Bentley JW, Van Mele P, Harun-Ar-Rashid M, Krupnik TJ (2015) Distributing and showing farmer learning videos in Bangladesh. Journal of Agricultural Education and Extension 22:179–197. https://doi.org/10.1080/1389224X.2015.1026365
Boud D, Keogh R, Walker D (eds) (1985) Reflection. Turning experience into learning. Routledge, London
Braun A, Duveskog D (2008) The Farmer Field School approach – History, global assessment and success stories. Background paper for the IFAD Rural Poverty Report 2011
Chambers R (1993) Challenging the professions: frontiers for rural development. Intermediate Technology Publications, London
Chowdhury AH, Van Mele P, Hauser M (2011) Contribution of farmer-to-farmer video to capital assets building: evidence from Bangladesh. J Sustain Agric 35:408–435. https://doi.org/10.1080/10440046.2011.562059
Elzen B, Augustyn A, Barbier M, Van Mierlo B (2017) Agroecological transitions: changes and breakthroughs in the making. https://doi.org/10.18174/407609
FAO and World Bank (2000) Agricultural knowledge and information systems for rural development (AKIS/RD). Strategic vision and guiding principles. FAO, Rome
Fleming N, Baume D (2006) Learning styles again: VARKing up the right tree! Educational Developments 7(4):4–7
Howie P, Bagnall R (2013) A beautiful metaphor: transformative learning theory. Int J Lifelong Educ 32(6):816–836
Johnson DW, Johnson RT, Holubec EJ (2008) Cooperation in the classroom, 8th edn. Interaction, Edina
Karubanga G, Kibwika P, Okry F, Sseguya H (2017) How farmer videos trigger social learning to enhance innovation among smallholder rice farmers in Uganda. Cogent Food & Agriculture. https://doi.org/10.1080/23311932.2017.1368105
Kelly N, McLean Bennett J, Starasts A (2017) Networked learning for agricultural extension: a framework for analysis and two cases. J Agric Educ Ext 23(5):399–414. https://doi.org/10.1080/1389224X.2017.1331173
Klerkx L, Aarts N, Leeuwis C (2010) Adaptive management in agricultural innovation systems: the interactions between innovation networks and their environment. Agric Syst 103(6):390–400. https://doi.org/10.1016/j.agsy.2010.03.012
Kolb D (1984) Experiential learning: experience as the source of learning and development. Prentice-Hall International, Hemel Hempstead
Laal M, Laal M (2011) Collaborative learning: what is it? Procedia Social and Behavioral Sciences 31:491–495
Leeuwis C, Aarts N (2011) Rethinking communication in innovation processes: creating space for change in complex systems. J Agric Educ Ext 17(1):21–36. https://doi.org/10.1080/1389224X.2011.536344

Leeuwis C, Pyburn R (eds) (2002) Wheelbarrows full of frogs: social learning in rural resource management. Koninklijke Van Gorcum, Assen
Lie, R., & Mandler, A. (2009). *Video in development. Filming for rural change*, Wageningen/Rome: CTA/FAO
Lie R, Servaes J (2015) Disciplines in the field of communication for development and social change. Commun Theory. https://doi.org/10.1111/comt.12065
Lie R, Witteveen L (2013) Spaces of intercultural learning. In: Mertens S (ed) International perspectives on journalism (internationale perspectieven op journalistiek). Academia Press, Gent, pp. 19–34
Mezirow J (1978) Perspective transformation. Adult Educ Q 28:100–110. https://doi.org/10.1177/074171367802800202
Mezirow J (1990) How critical reflection triggers transformative learning. Fostering Critical Reflection in Adulthood 1:20
Mezirow J (1991) Transformative dimensions of adult learning. Jossey-Bass, San Francisco
Mezirow J (2009) Transformative learning theory. In: Taylor EW, Mezirow J (eds) Transformative learning in practice: insights from community, workplace, and higher education. Jossey-Bass, San Francisco
Moris J (1991) Extension alternatives in tropical Africa. Overseas Development Institute, London
Pahl-Wostl C (2006) The importance of social learning in restoring the multifunctionality of rivers and floodplains. Ecol Soc 11(1):10
Percy R (2005) The contribution of transformative learning theory to the practice of participatory research and extension: theoretical reflections. Agric Hum Values 22(2):127–136. https://doi.org/10.1007/s10460-004-8273-1
Pratish KM (2006) Visual communication beyond word. GNOSIS, Delhi
Pretty JN, Chambers R (1993) Towards a learning paradigm: new professionalism and institutions for agriculture. Institute of Development Studies, Brighton
Reijntjes C, Haverkort B, Waters-Bayer A (1992) Farming for the future. An introduction to low-external-input and sustainable agriculture. Macmillan, London
Röling NG (1988) Extension science: information systems in agricultural development. Cambridge University Press, Cambridge
Röling NG (1989) The agricultural research-technology transfer interface: a knowledge system perspective. International Service for National Agricultural Research, The Hague
Röling NG, Wagemakers MAE (1998) Facilitating sustainable agriculture: participatory learning and adaptive management in times of environmental uncertainty. Cambridge University Press, Cambridge
Schramm W, Nelson LM, Betham MT (1981) Bold experiment: the story of educational television in American Samoa. Stanford University Press, Stanford
Smith E (2017) Transformative learning theory (Mezirow), in Learning Theories, September 30, 2017. https://www.learning-theories.com/transformative-learning-theory-mezirow.html
Smith MK (2010) 'David A. Kolb on experiential learning', the encyclopedia of informal education. http://infed.org/mobi/david-a-kolb-on-experiential-learning/. Retrieved 13 July 2018
Stilgoe J, Owen R, Macnaghten P (2013) Developing a framework for responsible innovation. Res Policy 42(9):1568–1580
Taylor EW, Duveskog D, Friis-Hansen E (2012) Fostering transformative learning in non-formal settings: farmer-field schools in East Africa. Int J Lifelong Educ 31(6):725–742
UNESCO (2009) UNESCO world report. Investing in cultural diversity and intercultural dialogue. UNESCO, Paris
Van de Fliert E (1993) Integrated pest management: Farmer field schools generate sustainable practices. A case study in Central Java evaluating IPM training. Published doctoral thesis, Wageningen University, Wageningen
Van Mierlo B, Leeuwis C, Smits R, Woolthuis RK (2010) Learning towards system innovation: evaluating a systemic instrument. Technol Forecast Soc Chang 77(2):318–334. https://doi.org/10.1016/j.techfore.2009.08.004
Wals AEJ (ed) (2007) Social learning towards a sustainable world. Wageningen Academic Publishers, Wageningen

Witteveen LM (2009) The voice of the visual: visual learning strategies for problem analysis, social dialogue and mediated participation. Eburon Uitgeverij BV

Witteveen LM, Goris M, Lie R, Ingram VJ (2016) Kusheh na minem Fatu, en mi na koko farmer, Hello, I am Fatu and I am a cocoa farmer; A Digital Farmer Field School for training in cocoa production and certification in Sierra Leone, Science Shop report 330. Wageningen UR, Wageningen

Witteveen L, Lie R (2012) Learning about "wicked" problems in the global south. Creating a film-based learning environment with "visual problem appraisal". J Media Comm Res 52:81–99

Witteveen L, Lie R (2018) Visual Problem Appraisal. An educational package, which uses filmed narratives. In: Griffith S, Bliemel M. Carruthers K (eds) Visual tools for developing student capacity for cross-disciplinary collaboration, innovation and entrepreneurship. A. Rourke and V. Rees (Series Curators), Transformative Pedagogies in the Visual Domain: Book No. 6. Common Ground Research Networks, Champaign

Witteveen L, Lie R, Goris M, Ingram V (2017) Design and development of a digital farmer field school. Experiences with a digital learning environment for cocoa production and certification in Sierra Leone. Telematics Inform. https://doi.org/10.1016/j.tele.2017.07.013

Wyckhuys KAG, Bentley JW, Lie R, Nghiem LTP, Fredrix M (2017) Maximizing farm-level uptake and diffusion of biological control innovations in today's digital era. BioControl 1–16. https://doi.org/10.1007/s10526-017-9820-1

How Social Media Mashups Enable and Constrain Online Activism of Civil Society Organizations

49

Oana Brindusa Albu and Michael Andreas Etter

Contents

Abstract

Activists of civil society organizations often use social media (SM) to organize and achieve social change by sharing content across different SM technologies. These technologies themselves can be understood as non-human actors that crucially influence how activists share content and organize. This chapter focuses on how the sharing of content, which is shaped by the interplay between human

O. B. Albu (✉)
Department of Marketing and Management, University of Southern Denmark, Odense M, Denmark
e-mail: oabri@sam.sdu.dk

M. A. Etter
Marie Curie Research Fellow, Faculty of Management, Cass Business School, City, University of London, London, UK
e-mail: michael.etter@city.ac.uk

J. Servaes (ed.), *Handbook of Communication for Development and Social Change*,
https://doi.org/10.1007/978-981-15-2014-3_80

and non-human actors, results in mashups, i.e., mutable interactions that emerge from disparate locales. Based on affordances theory and an ethnographic study, this chapter investigates how these mashups influence activist organizing of two civil society organizations. The study shows how the human-technology interplay that rests on the feature of "exporting" and "importing" content across SM connects various actors and interactions. The study furthermore shows the role and agency of non-human actors (algorithm-driven hashtags) in creating mashups and shows how these mashups can develop ordering and disordering effects.

Keywords

Social media · Digital activism · Mashups · Organizational theory · Connective affordances

49.1 Introduction

Social media (SM) are computer-based tools (such as websites and apps) that people use to create and share content with other people and organize collectively (McKenna et al. 2017). SM are important for connective action, i.e., a new form of collective engagement, whereby multiple actors come together spontaneously and informally, even if they do not all equally identify with a common cause, and engage in co-participation and co-production of content (Bennett and Segerberg 2012; Vaast et al. 2017). Civil society organizations (CSOs) are increasingly using these technologies in their efforts to promote social change by mobilizing and engaging large and loosely connected crowds (Lee and Chan 2016). Research has looked at the effects technologies have on the way people organize (for a review see Leonardi and Barley 2008; Leonardi et al. 2012). Much of the work looking at SM affordances (i.e., the relations linking the capabilities afforded by technology interactions to the actors' purposes) has extensively examined the ways in which SM affect technology-organization relationships (Treem and Leonardi 2012) and the interdependent roles of different actors that these technologies create (Vaast et al. 2017). However, the emphasis is typically placed on how actors use *one SM technology at a time* (Leonardi et al. 2013, emphasis added). As a result, there is only little known about what happens when multiple SM are used at the same time.

Furthermore, current studies usually assume that the actors in SM networks are humans. This chapter argues that more attention needs to be paid to how non-human elements, such as algorithms, play an increasingly important role in information diffusion and organizing processes, i.e., selecting what information is considered relevant to individuals for coordinating organizational tasks (Scott and Orlikowski 2014; Gillespie et al. 2014). Particularly, when SM technologies are used at the same time, these algorithms affect how different technologies interact with each other and consequently how human actors organize. Recent studies show, indeed, that algorithms change the way humans interact and organize when they are using hashtags (i.e., linguistic markers preceded by the dash sign #). One such example is the hashtag #LasVagasShooting that was used to share news and information about a mass shooting that took place in Las Vegas (Abbruzzese 2017). While it may seem like an innocent typo that can be discarded (LasVagas instead of LasVegas), it represents a larger problem. Such mistakes can lead to big misinformation problems when algorithms are involved and even cause the system to help spread easily disproven information. The algorithm takes into account, on the one hand, the number of tweets that are related to the trends when ranking and determining trends. On the other hand, the algorithm groups together trends and hashtags if they are related to the same topic. For instance, #MondayMotivation and #MotivationMonday may both be represented by #MondayMotivation (Twitter 2017). While Twitter does have some human oversight of its trends, both employees and users have little power to make changes – even correcting a misspelling since it is the algorithm that decides which hashtags will be trending (Abbruzzese 2017).

In addition, the same hashtags are often used on different SM platforms, thereby connecting interactions across platforms, which results into mashups (i.e., freely mixed and combined content, Yoo 2012). While mashups can be central to organizing by connecting platforms, actors, interactions, and content, there is still little knowledge of the subsequent implications for connective action (Bennett and Segerberg 2012). For a richer understanding, this study will explore how both human and non-human elements create mashups through the sharing of content (i.e., clustering mutable interactions that connect with each other across large communities, Asur and Huberman 2010).

To shed light on these processes, this chapter builds on complementary approaches from affordances literature and organizational communication studies (Leonardi et al. 2013). These approaches, which are well-established in the social sciences research tradition, were adopted because they work specifically within a relational and performative ontology and help us examine how human and non-human elements form a social and material nexus that is constitutive of organizing. Accordingly, in this study SM interactions are conceptualized to be created by both humans (SM users) and non-humans (i.e., algorithm-based hashtags). The analytical focus is on material agency, which allows to show that action and agency are not human beings' privileges but that humans are acted upon as much as they act. Such an approach would neither overplay technology's effects (seeing it as primarily deterministic) nor underplay its pliability (seeing it as primarily subject to human interpretations and intentions), and it would neither black-box the dynamics and entailments of technology nor diminish its workings and effects in the world (Scott and Orlikowski 2014). With this theoretical lens, the chapter identifies connective affordances that appear when SM are used in the same time and the implications of these affordances for the everyday organizing of fluid collectives.

In short, this chapter proposes that due to distributed agency, SM afford content sharing in uncontrollable manners, and because they do, SM introduce new forms of collective engagement (Bennett and Segerberg 2012). As it is discussed next, content sharing enables and constrains organizing, by introducing both order and disorder for the organizing processes. Disorder occurs when digital interactions exceed authors' full control as they leave the initial context of their creation and have negative effects for collective organizing. This happens, because digital interactions have the capacity to produce unprompted actions and collective coordination across large, varied, and uncoordinated audiences (Yoo 2012).

In sum, this chapter provides insight into the ordering and disordering effects of SM use and shows how both human and material agency shape connective action. It thereby sheds light on the hybrid use of SM, whose interrelations, implementations, and outcomes are not yet fully understood (Fulk and Yuan 2013). The chapter proceeds as follows: first it is discussed that, despite emerging studies on non-human agency in SM, a tendency to focus only on humans prevails in existing research. Then the connective affordances of SM are introduced, and it is shown how these enable forms of content sharing with both ordering and disordering effects. After describing the ethnographic methods and analytical steps employed, the chapter presents a case study that shows the implications of multiple SM use in two CSOs. Lastly, the chapter concludes by discussing the limitations of the study, which acts as a springboard for a brief future research agenda.

49.2 Social Media and Activism

49.2.1 Beyond a Human-Centered View of SM Use

There are many actors on SM that interact in far-reaching social networks, such as Twitter and Facebook. In social network studies (Wasserman and Faust 1994), the actors of networks ("nodes") are thought to be human beings that are connected through ties, such as friendships. For instance, studies on the impact of social influence on product adoption and content generation on SM platforms emphasize the human aspect of actors by conceptualizing them as individuals/people (e.g., Aral and Walker 2014). Also, in organization and communication studies, SM are typically considered to be tools used by humans for organizing processes, such as recruitment, mobilization of resources, and more generally the promotion of social change (Dobusch and Schoeneborn 2015). SM are then often understood as instruments that individuals use to disseminate information and create community engagement. In social movement studies, for example, actors are considered to be users that protest authoritarian regimes or use SM instruments to execute communication strategies (Lovejoy and Saxton 2012).

The focus on *human agency* offers of course valuable insights concerning how SM facilitate organizational processes such as information diffusion (Meraz and Papacharissi 2013), campaigning, mobilization, and engagement (McPherson 2015). However, studies about the role of non-human elements, which are increasingly present on SM, for connective action are scarce. For instance, algorithmic syntaxes have been shown to have influence over changing peoples' behavior through creating awareness and social learning. Indeed, studies indicate that collaborating for knowledge, exerting or receiving influence, purchasing or reviewing products, or diffusing information in SM do not need to be driven by people; such processes can also be guided by algorithms which modify how SM work and hold so far unexplored implications for organizational theory and practice (Salge and Karahanna 2016). In fact, SM can be seen to remove human agency from the center stage (Kallinikos et al. 2012), and it is only through these technologies that human and non-human actors are able to act in certain ways and therefore make human agency possible in the first place (Kallinikos et al. 2012).

Besides a tendency to focus strongly on human agency in SM research, an enthusiastic tone tends to pervade current studies about SM use for social change. Indeed, studies understand SM as technologies that facilitate and create dialogue, increase the access to information, and help organizing processes (see Lovejoy and Saxton 2012). While such research about the human use of SM is valuable, studies need to investigate also the increased complexities introduced by algorithms and how these shape data flows (Flyverbom 2016) and connective action. Furthermore, more knowledge is needed about the ways in which non-human elements act on SM and their ability of manipulating data sharing practices, because they may generate both order and disorder. Specifically, important issues surrounding non-human actors' agency are related to digital activism. The ways in which algorithm-based hashtags or the multiple use of SM can be used by CSOs to raise awareness for a social cause in a way that also mobilizes people to participate in offline connective action, and the constraints that arise when doing so, are still undertheorized.

For understanding such matters, it is useful to explore how the uses across, and communication through, different SM can enable content sharing in which multiple actors are tied together. To this end, it is next discussed how connective affordances facilitate different types of content sharing. Since the chapter adopts an inductive grounded theory approach (Charmaz 2006), the study iteratively compared existing data with emerging data and refined the theoretical framework accordingly. Therefore, the connective affordances of three SM (Facebook, Twitter, and Meerkat) used in the case study are next discussed.

49.2.2 Connective Affordances of SM

An established stream of literature has examined SM affordances, such as visibility, persistence, editability, and association, which influence organizing processes and their outcomes (Treem and Leonardi 2012; Leonardi et al. 2013). For example, the visibility of communications, which were once invisible, facilitates new practices of knowledge sharing and may influence actors' decision to participate in organizing processes (e.g., Yardi et al. 2009). Similarly, the association of actors with multimedia content (e.g., "tagging" users and creating hashtags on Facebook and Twitter) allows actors to enrich their social connections (e.g., Thom-Santelli et al. 2008). Recent studies have shown how SM create connective affordances, which means that users are mutually dependent upon each other's particular SM use for coordinating connective action. For example, Vaast et al. (2017) show how interrelated use of Twitter creates interdependencies and new emerging roles of different actors that are fluid and that depend on socio-material relationships and the ways in which material features are employed by multiple individuals. For these affordances to play out, individuals depend on each other's SM use through similar patterns.

To this extent, SM affordances that facilitate data sharing across spaces, times, and technologies afford connective action, while depending on multiple actors that use SM in similar ways. For example, in order to mobilize action through a hashtag that connects activists from multiple sites, actors need to use SM in similar ways, thereby creating mashups, which in turn will influence the processes of organizing, for example, during instant protests (see Rane and Salem 2012). Algorithms play a crucial role for the creation and impact of mashups, thereby facilitating new socio-material relationships that influence how individuals connect and organize. To understand how this relationship between human and non-human actors unfolds, this chapter shifts the focus away from emerging and interdependent roles of connected individuals (e.g., Vaast et al. 2017), toward the emergence of mashups that depend on socio-material practices of both humans and non-humans.

Hashtags as non-human actors develop agency because they are driven by an algorithm, which embodies a command structure enabling them to act (Goffey 2008). This is not to say that human agency is passed over in the process of using SM. Instead, the key phenomenon is hybrid association: the entanglement between humans and technologies that affords connective action in manners that could not be done otherwise because both actors exchange properties with each other leading to unprecedented fluid and messy collectives (Deuze 2012). This relationship helps one understand the extent to which combined interactions substantiate both: *order* through collectively specifying goals and targets, setting and sharing deadlines and responsibilities, and mobilizing actors in relation to sudden environmental changes and *disorder*, as the interactions misguide individuals about meeting locations; decontextualize, omit, or promote fake information on reports; and disassociate actors from texts. Both ordering and disordering are specific to each instance of organizing since digital interactions can take a life on their own, as they are co-produced by various actors in different times and spaces. For a more detailed understanding of the implications of SM use, the chapter turns the attention to trace how such interactions propagate and connect with each other across outspread communities of activists. This is important, because digital interactions are based on the transformation and accumulation of communication acts and interaction (Jackson 2007). This approach allows the examination of how such interactions "travel" crisscrossing multiple sites with underexplored ordering and disordering implications.

Specifically, when actors engage in practices that share content across different SM, content becomes combined and can produce unprompted change driven by large, varied, and uncoordinated audiences (Yoo 2012). In the case study of this chapter, content sharing is facilitated in two manners, on the one hand, by the ability to "export" interactions from one SM to another (Twitter interactions can be exported to Facebook). This allows actors to link different SM platforms and organize and act at a distance across different SM platforms. On the other hand, as the case study will show, content sharing is based on the integration of interactions from one SM to another (e.g., Facebook allows integrating live videos from Meerkat). To this extent, both human and non-human actors "import" interactions from distinct sites to coordinate with large often anonymous audiences. In this context, actors become temporarily connected with each other enabling connective action (see Dobusch and Schoeneborn 2015). However, these types of data sharing done by human and non-human actors introduce also disorder in organizing processes because they create mutable, often uncontrollable, and unpredictable mashups that constrain organizing processes (Wilkie et al. 2015).

49.2.3 Instantaneous and Integrative Content Sharing as Means for Connective Action

When interactions are combined across SM, mashups are established with visible and persistent links ("tags") between texts, photos, or videos with other users or with specific spatiotemporal locations ("check-in") that users can edit and modify at any time. For instance, in the case of the "We are all Khaled Said" Facebook page, "liking," "tagging," "hashtagging," and "reposting" or "retweeting" a Facebook page across other SM created mashups with visible links which, among other factors, enabled over 390,000 members of civil society to put into motion the Egyptian revolt of 2011 (Herrera 2014). Thus, mashups have become central elements in organizational processes, particularly in an environment of turmoil, fast changing conditions, and high uncertainty. Studies have shown that a major part of connective action that occurs in a publicly open manner is often based on visible hyperlinks (Segerberg and Bennett 2011). Mashups may therefore increase the visibility of the dynamics of connective action and mass mobilization (Milan 2015). Based on an inductive analysis, this chapter identified that mashups result from two interrelated types of content sharing (e.g., integrative and instantaneous) made possible by connective affordances of SM, which are presented next.

Integrative content sharing happens as interactions that are publicly exchanged between Twitter, Facebook, and Meerkat locales prompt actors to act as a collective based on emerging incidents. Here the dynamic integration of various interactions produces order in organizational activities since this integration can modify the direction and content of organizational conversations, circulate reports, set up meeting coordinates, and share strategic data (Segerberg and Bennett 2011). Multiple actors and their conversations are connected by being re-embedded in the situated space-time of another actor, if that actor in turn retweets or reposts (i.e., forward the tweet to the bot's or individuals' "followers" or "friends," see Murthy 2011). Disorder may happen due to different reasons, such as the limited visibility of SM interactions (Murthy 2009) or confusion caused by the intended or unintended integration of deceptive information by intervening actors which alter and edit the interactions (Salge and Karahanna 2016). For instance, in the case of the conflict in Eastern Ukraine, SM became a propaganda outlet whereby users integrated and broadcasted opposing views and misinformation about the ongoing conflict in Facebook and Twitter mashups (Makhortykh and Lyebyedyev 2015).

Instantaneous content sharing takes place when Facebook, Twitter, and Meerkat interactions permit actors to coordinate connective action at high speed. The intertwined sharing of Twitter interactions with Facebook and Meerkat ones generates mashups that can lead to order by allowing fast response to sudden changes in the environment (Agrawal et al. 2014). This happens, as combined interactions form the basis for ad hoc coordination, for example, for the coordination of meetings after significant turning of events (Aouragh and Alexander 2011). In doing so, interactions track events in real time and spontaneously mass mobilize members (Eltantawy and Wiest 2011). One example is the hashtag #SidiBouzid which was used on both Facebook and Twitter to share content (e.g., messages, photos, videos, and locations), which connected over 200,000 Tunisian members of civil society to coordinate concurrent protests in different locations that subsequently contributed to starting the Jasmine Revolution of 2011 (Rane and Salem 2012). Disorder may also be introduced because of the speed with which combined interactions move across SM. Confusion can arise because these interactions transport information faster than organizational members' abilities to keep track of the immediate developments across SM, even though they might be connected to each other.

In short, integrative and instantaneous content sharing, which connects human and non-humans actors, can have both ordering and disordering implications for organizational processes (Campbell 2005). It therefore becomes relevant to identify the distributed forms of agency present on SM and their (dis)ordering potential, which are revealed primarily in the interaction with its users (Deuze 2012). The case study presented next thus examines the ways in which data sharing through connective affordances leads to order and disorder in digital activism.

49.3 Digital Activism in Tunisia

Tunisia was chosen as a site of investigation for digital activism, because it is a representative of the MENA region where social media has a high penetration rate for civil society organizations (Rane and Salem 2012). For instance, on the list of the top 20 nations worldwide in terms of new Facebook users in 2010, half are from the MENA region, and during 2011 the number of new MENA users has increased by 78% (Rane and Salem 2012). Two CSOs were studied because these types of organizations use SM as central resources in organizing against human rights abuses and coordinating with international actors to lend their support for protests and uprisings (Lee and Chan 2016). A CSO with a longer experience in using SM for collective action (Kappa, 3 years) and a CSO who was in the beginning of using such technologies (Omega, 9 months) were selected in order to obtain a diversity of parameters for the purpose of comparison (Tracy 2013). Omega is a CSO that works to promote open government and democracy in Tunisian institutions

(Omega Annual Report 2015). Omega members use SM for coordinating their daily work (create advocacy strategies, attract supporters for anti-corruption campaigns, recruiting of volunteers, etc.). Kappa is an advocacy CSO that works to defend the freedom of access to information fundamental right by offering citizens the means to stay updated with their elected representatives and thus reposition citizens at the core of political action (Kappa Annual Report 2015). In doing so, Kappa members rely on SM for coordinating organizational tasks (e.g., disclosing the activities of Tunisia's National Constituent Assembly to the public by broadcasting, etc.).

Digital activism was analyzed based on multi-sited fieldwork (participant observation, qualitative interviews, and sourcing the SM interactions of Kappa and Omega members). The analysis focused on critical events where interactions had an impact on the ability of members to organize and collectively coordinate specific activities (i.e., campaigns, protests, conferences). The rationale of selecting critical SM events builds upon an established tradition in research that investigates connective action (Segerberg and Bennett 2011). Critical events were identified by following four criteria: (1) they involved multiple actors interacting through multiple SM use; (2) they were characterized by an identifiable, but not necessarily very precise, common cause or theme; (3) they unfolded over time; and (4) they involved specific and intended organizing tasks, taking place in physical or virtual locations (e.g., press release or virtual vote count). The case study shows what actors (i.e., users, hashtags) did (share data) in specific critical events, what this meant for collective action (order and disorder), and how this related to connective affordances (see Table 1). These results are presented in the following section on the basis of two themes.

Table 1 Connective Action and Content Sharing

Data analysis			
Content sharing	Order	Disorder	Actors and examples
Integrative			
Is facilitated through connective affordances, where human and non-human actors integrate content (i.e., reports, videos, pictures, interactions, etc.) across spaces, times, technologies	Integrative content sharing and the resulting mashups enable connective action by mobilizing and engaging large and loosely connected crowds by giving mutual access to information and interaction, thereby connecting actors	Integrative content sharing and the resulting mashups disrupt connective action by misguiding actors about meeting locations, and omitting or promoting fake information on reports, and disassociating authors from texts	Once **#Tunisia** and **#TnXYZ** started ***trending*** that day on Twitter and Facebook, we had to use them everywhere although they weren't our hashtags (Omega manager, interview) [H]ashtags ***stand*** *for something else* and *work* ***differently*** across platforms than what you'd expect (Kappa manager, interview)

(continued)

Table 1 (continued)

Data analysis			
Content sharing	Order	Disorder	Actors and examples
Instantaneous			
Is facilitated through connective affordances that rest on the socio-material feature of importing and exporting content across technologies, which makes sharing to happen at high speed and instantly	Instantaneous content sharing and the resulting mashups enable connective action by producing unprompted actions and driving collective coordination across large, varied, and uncoordinated audiences at high speed	Instantaneous content sharing and the resulting mashups disrupt connective action by confusing actors because interactions transport information faster than one's abilities to keep track of the immediate developments across SM, even though actors might be connected to each other	We have a team [. . .] ***posting*** on our social media all the sessions planned, all the votes. Like this we give those interested in the accountability of elected representatives the chance to be in ***real-time control*** of their votes in the assembly (Kappa manager, interview) @Kappa where can we find information about the plenary session of today? I follow #FL2016 but the ***info*** went ***too quickly*** among all sorts of stuff. Which is the room number? (Tweet, Kappa manager)

49.4 Hashtags and Mashups: The Ordering and Disordering Role of Non-human Actors for CSO Activism

Connective action by Omega and Kappa, as it is illustrated next, depends crucially on connective affordances of SM, which facilitate content sharing and connect human and non-human actors through dynamic socio-material relationships. In doing so, SM allowed both CSOs to generate and co-produce activist online tactics (i.e., raising digital petitioning signatures, broadcasting and live streaming protests, increasing awareness for advocacy campaigns, etc.) on platforms, such as Meerkat, Twitter, and Facebook, where visible hyperlinks connect actors and aggregate content from these different platforms. In the case of Omega, an instance indicative of the ability of SM to create connective action was the live broadcasting of a press conference in order to mobilize actors across different sites: "@Omega |LIVE NOW| Press Conference Now: Our symposium of civil society #meerkat Watch Omega |LIVE NOW| from Tunis mrk.tv" (Omega Facebook/Meerkat mashup). Meerkat interactions allowed Omega members to collaboratively broadcast through the "cameo" feature where the video stream was "taken over" for 60 seconds by another actor following the event, who could integrate his or her own content into the broadcast. Such affordance allowed a shift from simply broadcasting to a certain audience to broadcast together with other connected actors that track and contribute to the respective mashup. In such instances, mashups create order as they allow activists to engage in connective action and attract supporters for their activities, as the manager indicates:

> We managed to gather more than 20 national and more than 5 international organisations for our symposium where we gave a preliminary report of the problems encountered in our fight for human rights vis-à-vis the authorities' fight against terrorism. In order *to get more activists to show up* write a reply to our *tweets where we tag their user names like cc @user and a hashtag*. This goes across all our [social] *media* accounts as we've *put all together*. We broadcast live streams on Meerkat and use Facebook at the same time, as in Tunisia most of people that live outside of the big cities rely on Facebook. So now when we share something it *goes everywhere* and this *gets bloggers on their feet*. (Omega manager, interview, emphasis added)

However, the resulting mashups introduce also disorder to connective action. In the case of Kappa, a mashup of a Facebook event and Meerkat broadcast was mobilized for coordinating and broadcasting a press conference: "@Kappa |LIVE NOW| #meerkat #Décentralisation The questions/remarks will follow at the end of our press conf – live: https://mrk.tv/" (Kappa, Twitter/Meerkat mashup message). The mashup acted as an "action alert" (Obar et al. 2012) and mobilized actors to participate in the event. At the same time, disorder occurred because of intervening actors, whose SM interactions are at best obscure and often nonsensical (Wilkie et al. 2015). This happens because these intervening actors were able to aggregate content in routinely odd and contrary ways to the visible flow of exchanges and therewith alter the meanings of the mashups. Hence, mashups had disordering effects by taking conversations "off track," as the manager indicates:

> We get disturbing comments all the time both on [our interactions from] Facebook and Twitter that *take us a bit off track* with claims such as "you are funded by the Mossad." And they are *all the time visible, I mean everyone can see* the comments. And it's getting so absurd that no matter what I answer they say the same, so I should reply "yes the Mossad funds me" [laughs]. (Kappa manager interview)

Similarly, in the case of Omega, activists' efforts of generating connective action were disrupted because nonsensical interactions benefited of the same visibility affordances provided by SM, as an Omega member suggests:

> They [nonsensical mashups] *are* always on topics where the political debates are highly polarized such as Libya where you currently have two governments claiming legitimacy. I think Twitter as a platform makes visible all sorts of conversations and trolls are part of what it means to have open conversations. But yes, such visibility works both ways. (Omega manager interview)

Furthermore, disorder also occurred due to the fleeting visibility of the interactions: when Omega stopped streaming, the video message disappeared from the access list, leaving only a "stream over" message behind. As a result, actors could not participate in the respective activity and connect with other actors, therefore not receiving details about meeting points and other necessary information. Similarly, the importance of connective affordances that base on the feature to import and export conversations across technologies was confirmed, when looking at Meerkat and Facebook, which do not allow such combined use. Here, actors were unable to connect and coordinate with each other, when relying on these specific SM. More in detail, the visible digital interactions that exhibited ordering and disordering properties in both cases facilitated two interrelated types of data sharing that are presented next.

49.4.1 Integrative Content Sharing

The integration of various interactions between SM produced order by combining scattered content from multiple sites that facilitated connections between various actors. In one instance indicative of many similar situations in the case of Kappa, mashups of Facebook and Twitter interactions facilitate Kappa members' efforts of connective action by providing information about simultaneously occurring tasks despite their different geographical and temporal locations (e.g., following up fiscal legislation making proposals, signposting MPs' interventions). Integrative data sharing took place as interactions were retweeted and re-embedded in the organizational conversations of 30 actors simultaneously. In doing so, integrative content sharing ensured offline mobilization of key constituents (i.e., volunteers) at ad hoc events by connecting actors and indicating time and location coordinates (Segerberg and Bennett 2011). Yet, the interactions exhibited also disordering properties, when they allowed the integration of unwanted or deceptive content by unknown audiences. As the manager indicates next, mashups and hashtags disrupt the presumed one-to-many communication properties and the intended positive effects on connective action that SM are considered to have:

> We use #FL2016 #GenLeg and #TnXYZ on both Facebook and Twitter to *"signpost"* on what project *we are working now*, so #FL2016 and #GenLeg shows that we are working on the Fiscal Law 2016 that is part of the general legislative package. We also use these hashtags because citizens use them to *search, find* information and *be part* in the democratic process with their elected representatives. But we can't really predict how hashtags *work*... sometimes they are of big help and sometimes not and everyone gets confused. Now and then, the hashtags *stand for something else* and *work differently* across platforms then what you'd expect. But you can't buy or control a hashtag because social media is not a media. *It's a world and it makes you as much as you make it.* And since you can't escape it you'd better be ready to deal with it. (Interview, Kappa manager, emphasis added)

Furthermore, connective affordances create order and enable SM to act as rallying and mobilizing mechanisms (Segerberg and Bennett 2011) by connecting actors and content across technologies. But equally, such affordances create disorder given that mashups can also "stand for something else," "work differently" (manager Kappa), and enable open authorship across multiple sites. On the one hand, the #TnXYZ hashtag integrates Facebook and Twitter content (the hashtag was initially developed by Kappa managers to geo-locate their work in the house of representatives and signpost specific organizational activities in order to mobilize more support for their actions). On the other hand, the hashtag disrupts integrative sharing by permitting other actors to alter and include it in their interactions that coordinated unrelated events. Similarly, Omega members were able to use the same #TnXYZ hashtag to share their daily news coverage activities. By posting and tweeting the hashtag, Omega members together with other actors "bec[a]me part of the conversation," as it happened in the case of a terrorist attack at the Bardo National Museum in Tunisia:

> Once #Tunisia and #TnXYZ started *trending* that day on Twitter and Facebook, we had to use them everywhere although they weren't our hashtags. So you use what hashtags are used in that day. *This is what you have to do if you want to be part of the conversation* along with international channels such as Al Jazeera. When people were looking for the Bardo attack on the tourists the only thing they would find is #TnXYZ and #JesuisTunisien, which was started by Tunisian activists in our network, so I use those. But everyone else does it too, which makes things a bit chaotic sometimes. (Interview, Omega manager)

The SM used in both Omega and Kappa are not simple means or products of interactions and discourses of human collectives (Fayard and Weeks 2007). Instead, the agency of the interactions is dependent on both their material features and situated use. It is therefore the distributed use that enables connective action in specific situations, such as the algorithmic operation of "trending." Such mashups gain traction based on the non-human agency of an algorithm that promotes in some arbitrary manner hashtags that suddenly dramatically increased in volume (Wilkie et al. 2015). The agency of the interactions resides thus in their interconnected use by both human and non-human actors (hashtags, algorithms, etc.). In doing so, integrative content sharing creates order for Omega, Kappa, or new actors, such as the community of followers contributing to #TnXYZ, for that matter. However, additional to the disordering effects due to information overload and misappropriation of the combined interactions described above, Twitter and Facebook interactions cannot be exported on the Meerkat platform. Thus mashups are impeding both Omega and Kappa members from relying on combined interactions to coordinate for activism.

49.4.2 Instantaneous Content Sharing

Sharing content at high speed is a daily practice in Omega and Kappa, which aim at real-time coordination. Omega managers engage with an average of ten times a day "tweeting," "posting," "streaming videos," and "hashtagging," across multiple SM with several hundreds of followers. Mashups generate order, when they enable actors to manage ad hoc activities, share relevant data, and connect to key audiences instantly:

> We use the Facebook messenger for practical things, [be]cause it's fast and convenient. But we mix up social media for our work. People *follow us* on Twitter and like our Facebook page because they are interested in what we have to say…uhm…so *we tweet* and *at the same time post* on Facebook, *do hashtags* and *share videos* a lot across platforms…[uhm] about things people care about. This way *we show* what is going on. (Omega manager, interview, emphasis added)

Similarly, in the case of Kappa, Facebook and Twitter interactions ("sharing," "tweeting," and "posting") create mashups that facilitate instantaneous content sharing. On the one hand, mashups allow Kappa's members the ability to connect with unknown audiences "on the spot" and share information about activities such as making recommendations on bills in the same time as these are promulgated by the parliament. On the other hand, the mashups give the ability to be in "real-time" control of constituents, as the manager indicates:

> The [Tunisian] parliament only audio-records the meetings but the records are not immediately available to the citizens. So, we have a team in the parliament measuring attendance of deputies, *posting* on our social media all the sessions planned, all the votes. Like this we give those interested in the accountability of elected representatives the chance to *be in real time control* of their votes in the assembly. We basically *tweet* and *post everything* the MPs say, pictures, videos, you name it, everything from the moment the parliament session is opening in the morning until the evening when the doors close…uhm…like showing misbehavior or so on. By *doing this*…uhm… *every day*…uhm…we are *able to make recommendations on the spot* for the assembly that is preparing the bill that day. (Interview Kappa manager, emphasis added)

The capacity of quickly transporting interactions across different SM and thereby creating mashups contributed to the immediate synchronization of actions among connected activists, since it provided information about "where and when are we meeting that day" (interview manager Kappa). For instance, in the case of Kappa, a Facebook interaction mashed up with a link to the Twitter account, and #TnXYZ permits Kappa members to signal the time and location of their work in the house of representatives. In addition to the instantaneous coordination of work tasks, the mashup also upholds the mission of the organization across different physical and virtual locations: "@Kappa we are back for the plenary this evening. Our mission, we are leading to the end #TnXYZ." In doing so, the mashup communicates and generates awareness about "who we are and what we do" (interview manager Kappa).

Nevertheless, instantaneously combined interactions create also disorder for connective action. For instance, instantaneous data sharing at high speed led to data overload and confusion, which stopped individuals – even though connected – from obtaining relevant information or participating in certain events, as the manager indicates: "[w]e work hard to get the message out the right way but we often get feedback from our interns and volunteers such as 'these hashtags are flooding our timelines'" (manager Omega interview). In the case of Kappa managers, mashups exhibited disordering properties because of the overwhelming speed with which these work: "@Kappa where can we find information about the plenary session of today? I follow #FL2016 but the info went too quickly among all sorts of stuff. Which is the room number?" (tweet Kappa manager). The importance of connective affordances that enable integration of content across technologies for connective action was confirmed by the fact that Facebook interactions do not aggregate with Meerkat ones, which restricted the emergence of mashups and disrupted the ability of both Kappa and Omega to engage in connective actions when they relied only on these technologies for coordination purposes.

In sum, in both Kappa's and Omega's case, SM use is indicative of the distributed agency specific to connective action. When it comes to SM, not only what humans do with SM is important, but also the actions of non-human actors such as hashtags are highly relevant. The chapter has shown how instantaneous and integrative content sharing specific to SM combined use has ordering and disordering effects for connective action.

49.5 Future Directions

Literature that looks at the role of SM for connective action typically studies how individuals use SM as tools for disseminating information and creating new modes of participation and organizing (Lee and Chan 2016; Segerberg and Bennett 2011). By adopting a socio-material and affordances lens on connective action (Vaast et al. 2017), this chapter has shown how these connective affordances, by enabling data mashups, facilitate connections between these actors and analyzed the outcomes of data sharing practices that are shaped by both human and material agency.

Accordingly, the presented study enriches the understanding of data sharing practices and digital activism, in which human and non-human actors are tied together (Leonardi 2017). The chapter shows how connective action depends as much on similar patterns of SM use by human actors, as much as it does on material agency of non-human actors. Overall, the chapter indicates that two kinds of effects are introduced through instantaneous and integrative content sharing practices, specifically: (a) ordering effects that stem from distributed material and human agency that makes it possible for communication and organization to exist in a virtual time and space without disrupting the routines of the connected actors and (b) disordering effects which are equally generated by human and non-human agencies that omit and disassociate actors from interactions, thus disturbing connective action.

The introduced theoretical framework therewith accounts for the multiple use of intertwined technologies, which has so far only gained little attention in current research (Leonardi 2017). Indeed, research typically looks at the use of one SM technology at the time (for an overview see Treem and Leonardi 2012). The complex combination of multiple SM use, instead, requires an understanding of the relationship between these technologies and the various actors, which apply them. The proposed theoretical framework made it possible to understand these relationships and conceptualize the connective affordances that account for the interplay between the practices that humans apply on multiple technologies and the material features that these technologies offer. In doing so, the framework enabled an analysis of affordances on collective level, of how multiple SM use creates connections between different actors, and of how the resulting mashups affect connective action.

In contrast to many studies that are pervaded by a mostly positive or even enthusiastic tone about the effects of connective actions through SM use (e.g., Segerberg and Bennett 2011), the chapter has, next to ordering effects, detected the disordering effects of multiple SM in digital activism. Indeed, combined interactions across technologies help but also often restrain actors to organize across physical and virtual settings. When actors modify interactions, mashups act back simultaneously in often uncontrollable and unintended ways while being appropriated by various actors. In fact, connective affordances allow appropriation by actors that may intentionally disturb organizing processes. Furthermore, disorder may be introduced without bad intentions, when instantaneous content sharing leads to information overflow or when integrative content sharing makes a certain hashtag that takes a different direction than intended. Hence, by acknowledging agency of non-human actors, the chapter shows how disorder occurs because combined interactions produce unprompted change driven by fluid collectives (Yoo 2012).

In sum, this study contributes to existing research in two ways: on the one hand, it reveals that SM interactions are entities irreducible to humans or human communication and have both ordering and disordering effects. On the other hand, it indicates that combined interactions introduce connective action across multiple spatiotemporal localities and SM platforms. While the case study examined three of the most popular SM technologies in a specific country context, these practices are feasible between other SM. As the study shows, SM are not stable but vacillate depending on the situated interactions between both human and non-human actors. Thus, the material and organizational world created by these technologies does not exist separate from the people communicating with them. Instead, organizing happens through heterogeneous and conflicted socio-material practices mediated by both human and non-human actors.

The chapter offers a particular view of the combined use of multiple SM in a particular context (Tunisian CSOs) and for a distinct period of time. With the limitation of interconnected Facebook, Twitter, and Meerkat uses, the study covers three of the most popular SM currently in use. Future technological developments of these and other tools will possibly result in other forms of interaction. For example, Facebook recently introduced the live stream function that opens up new possibilities and at the same time renders hybrid use with Meerkat less important. Nevertheless, this chapter provides relevant and generalized knowledge concerning the role of non-human actors, such as algorithm-based hashtags, and the crucial role of connective affordances that connect actors, interactions, and content across SM. The results that emerge from the chapter can be transferred if there are common characteristics between the cases in focus. Furthermore, the presented analysis of SM uses in a politically unstable environment such as Tunisia can act as a springboard for future research to validate or challenge the results of this case through a longer examination of the struggles and dynamics introduced by SM in organizations that rely on such technologies for accomplishing everyday work, such as medium enterprises and start-ups. This is particularly important as such tensions tend to be rendered invisible by the search for coherent and unitary answers.

49.6 Conclusion

This chapter has investigated the use of intertwined SM technologies by drawing on affordances theory and an ethnographic study. The chapter advances research by providing insights concerning how the combined SM use generates connective action through mashups, in which multiple actors are tied together. Firstly, the chapter identifies connective affordances that are particular to the interplay between multiple SM technologies and human actors, and allow integrative and instantaneous content sharing, which rests on the socio-material feature of "exporting" and "importing" content across platforms. Such connective affordances allow the creation of mashups, i.e., mutable interactions, which emerge from multiple and disparate locales. Secondly, the analysis has shown how mashups develop not only ordering but also disordering effects, whereby the socio-material relationships between various human and non-human actors and their distributed agency have a crucial role for how these effects play out. The developed framework may support future research that aims at investigating organizing practices through SM, which takes human and non-human actors equally seriously.

49.7 Cross-References

▶ Communication for Development and Social Change: Three Development Paradigms, Two Communication Models, and Many Applications and Approaches
▶ Communication for Development and Social Change Through Creativity
▶ Political Engagement of Individuals in the Digital Age
▶ Protest as Communication for Development and Social Change in South Africa
▶ Strategic Social Media Management for NGOs
▶ Women's Empowerment in Digital Media: A Communication Paradigm

References

Abbruzzese J (2017) Twitter's 'LasVagas' hashtag fail shows the worst part of algorithms. http://mashable.com/2017/10/03/twitter-algorithm-fail-las-vegas/#nvynexnERPqr. Accessed 11 Oct 2017

Agrawal A, Catalini C, Goldfarb A (2014) Some simple economics of crowdfunding. Innov Policy Econ 14(1):63–97

Aouragh M, Alexander A (2011) The Arab spring| the Egyptian experience: sense and nonsense of the internet revolution. Int J Commun 5(1):1–15

Aral S, Walker D (2014) Tie strength, embeddedness, and social influence: a large-scale networked experiment. Manag Sci 60(6):1352–1370

Asur S, Huberman BA (2010) Predicting the future with social media. In: Web Intelligence and Intelligent Agent Technology (WI-IAT), 2010 IEEE/WIC/ACM international conference, vol 1. IEEE, Washington DC, pp 492–499

Bennett WL, Segerberg A (2012) The logic of connective action: digital media and the personalization of contentious politics. Inf Commun Soc 15(5):739–768

Campbell KK (2005) Agency: promiscuous and protean. Commun Crit/Cult Stud 2(1):1–19

Charmaz K (2006) The power of names. J Contemp Ethnogr 35(4):396–399
Deuze M (2012) People and media are messy. New Media Soc 14(4):717–720
Dobusch L, Schoeneborn D (2015) Fluidity, identity, and organizationality: the communicative constitution of Anonymous. J Manag Stud 52(8):1005–1035
Eltantawy N, Wiest JB (2011) The Arab spring| Social media in the Egyptian revolution: reconsidering resource mobilization theory. Int J Commun 5(1):18–40
Fayard AL, Weeks J (2007) Photocopiers and water-coolers: the affordances of informal interaction. Organ Stud 28(5):605–634
Flyverbom M (2016) Transparency: mediation and the management of visibilities. Int J Commun 10:110–122
Fulk J, Yuan CY (2013) Location, motivation, and social capitalization via enterprise social networking. J Comput-Mediat Commun 19(1):20–37
Gillespie T, Boczkowski PJ, Foot KA (2014) Media technologies: essays on communication, materiality, and society. MIT Press, Cambridge, MA
Goffey A (2008) Algorithm. In: Martin F (ed) Software studies: a lexicon. MIT Press, Cambridge, MA, pp 15–20
Herrera L (2014) Revolution in the age of social media: the Egyptian popular insurrection and the internet. Verso Publishing, London
Jackson MH (2007) Fluidity, promiscuity, and mash-ups: new concepts for the study of mobility and communication. Commun Monogr 74(3):408–413
Kallinikos J, Leonardi PM, Nardi BA (2012) The challenge of materiality: origins, scope, and prospects. Materiality and organizing: social interaction in a technological world. Oxford University Press, Oxford
Lee FL, Chan JM (2016) Digital media activities and mode of participation in a protest campaign: a study of the umbrella movement. Inf Commun Soc 19(1):4–22
Leonardi PM (2017) The social media revolution: sharing and learning in the age of leaky knowledge. Inf Organ 27:47–59. https://doi.org/10.1016/j.infoandorg.2017.01.004
Leonardi PM, Barley SR (2008) Materiality and change: challenges to building better theory about technology and organizing. Inf Organ 18(3):159–176
Leonardi PM, Nardi BA, Kallinikos J (2012) Materiality and organizing: social interaction in a technological world. Oxford University Press on Demand, Oxford
Leonardi PM, Huysman M, Steinfield C (2013) Enterprise social media: definition, history, and prospects for the study of social technologies in organizations. J Comput-Mediat Commun 19(1):1–19
Lovejoy K, Saxton GD (2012) Information, community, and action: how nonprofit organizations use social media. J Comput-Mediat Commun 17(3):337–353
Makhortykh M, Lyebyedyev Y (2015) SaveDonbassPeople: Twitter, propaganda, and conflict in Eastern Ukraine. Commun Rev 18(4):239–270
McKenna B, Myers MD, Newman M (2017) Social media in qualitative research: challenges and recommendations. Inf Organ 27(2):87–99
McPherson E (2015) Advocacy organizations' evaluation of social media information for CSO journalism the evidence and engagement models. Am Behav Sci 59(1):124–148
Meraz S, Papacharissi Z (2013) Networked gatekeeping and networked framing on # Egypt. Int J Press/Politics 18(2):138–166
Milan S (2015) From social movements to cloud protesting: the evolution of collective identity. Inf Commun Soc 18(8):887–900
Murthy D (2011) Twitter: microphone for the masses? Media Cult Soc 33(5):779–789
Obar JA, Zube P, Lampe C (2012) Advocacy 2.0: an analysis of how advocacy groups in the United States perceive and use social media as tools for facilitating civic engagement and collective action. J Inf Policy 2(1):1–25
Rane H, Salem S (2012) Social media, social movements and the diffusion of ideas in the Arab uprisings. J Int Commun 18(1):97–111

Salge C, Karahanna E (2016) Protesting corruption on Twitter: is it a bot or is it a person? Acad Manag Discov. https://doi.org/10.5465/amd.2015.0121
Scott SV, Orlikowski WJ (2014) Entanglements in practice: performing anonymity through social media. MIS Q 38:873–893
Segerberg A, Bennett WL (2011) Social Media and the Organization of Collective Action: using Twitter to Explore the Ecologies of Two Climate Change Protests. Commun Rev 14(3): 197–215. https://doi.org/10.1080/10714421.2011.597250
Thom-Santelli J, Muller MJ, Millen DR (2008) Social tagging roles: publishers, evangelists, leaders. In: Proceedings of the SIGCHI conference on human factors in computing systems. ACM, New York, pp 1041–1044
Tracy SJ (2013) Qualitative research methods: collecting evidence, crafting analysis, communicating impact. Wiley-Blackwell, USA
Treem JW, Leonardi PM (2012) Social media use in organizations: exploring the affordances of visibility, editability, persistence, and association. Commun Yearb 36(1):143–189
Vaast E, Safadi H, Lapointe L, Negoita B (2017) Social media affordances for connective action-an examination of microblogging use during the Gulf of Mexico oil spill. MIS Q 41(4):1179. https://doi.org/10.25300/MISQ/2017/41.4.08
Wasserman S, Faust K (1994) Social network analysis: methods and applications, vol 8. Cambridge university press, Cambridge, UK
Wilkie A, Michael M, Plummer-Fernandez M (2015) Speculative method and Twitter: bots, energy and three conceptual characters. Sociol Rev 63(1):79–101
Yardi S, Golder SA, Brzozowski MJ (2009) Blogging at work and the corporate attention economy. In: Proceedings of the SIGCHI conference on human factors in computing systems. ACM, Boston, MA, pp 2071–2080
Yoo Y (2012) Digital Materiality and the Emergence of an Evolutionary Science of the Artificial in Materiality and Organizing: Social Interaction in a Technological World. Oxford University Press, New York, pp. 134–154

Strategic Social Media Management for NGOs

50

Claudia Janssen Danyi and Vidhi Chaudhri

Contents

Abstract

Social media channels provide intriguing opportunities for NGOs to communicate with their audiences, raise awareness, and campaign for change. Maintaining strong social media communication that supports the organization's ability to achieve its goals is the mandate of social media management, the day-to-day maintenance and management of an organization's social media channels and communication. This chapter provides an overview of the potential, practice, process, methods, and tools of effective social media management, including strategic planning based on formative research, implementation, and communication, as well as evaluation.

C. Janssen Danyi (✉)
Department of Communication Studies, Eastern Illinois University, Charleston, IL, USA
e-mail: cijanssen@eiu.edu

V. Chaudhri
Department of Media and Communication, Erasmus University Rotterdam, Rotterdam, The Netherlands
e-mail: chaudhri@eshcc.eur.nl

J. Servaes (ed.), *Handbook of Communication for Development and Social Change*,
https://doi.org/10.1007/978-981-15-2014-3_72

Keywords

Digital communication · Engagement · Social media management · Social media · Strategy

50.1 Introduction

When the ALS Ice Bucket Challenge started in 2014, few would have thought that the campaign would end up raising more than $115 million (Sample and Woolf 2016) to fund research for a cure and support for those living with the disease. Those who accepted the challenge donated to the cause, poured a bucket of ice water over their heads, and shared a video of the spectacle on their social media accounts while challenging their friends to do the same. As social media users, including high-profile persons like George W. Bush, Heidi Klum, or Leonardo DiCaprio, challenged each other, they shared more than 2.2 million videos (Steel 2014). With this simple and cost-effective strategy, the campaign took advantage of the unique features of social media and traversed individual networks to raise awareness and funds.

Social media are commonly defined as "a group of internet-based applications that build on the ideological and technological foundations of Web 2.0 and that allow the creation and exchange of user generated content" (Kaplan and Haenlein 2010, p. 61). They bring intriguing opportunities for NGOs. At the onset of 2018, approximately 3 billion people used social media worldwide, an increase of 13% compared to the previous year. While the percentage of users varies widely by country and continent, social media accounts continue to increase globally (WeAreSocial 2018).

Differently from traditional channels, "social media [. . .] has made the ability to comment, respond, share, critique, change and add to information possible on a broad scale" (Pridmore et al. 2013). With large numbers of active users, social media platforms can thus enable even small organizations with little budgets to reach stakeholders across the world, raise awareness for issues, fundraise, recruit volunteers, and share information quickly, among others. What's more, social media allow for immediate two-way communication, thus providing opportunities for organizations to build and maintain strong relationships with stakeholders and seek their input (see Saffer et al. 2013; Saxton and Waters 2014).

At the same time, social media networks also pose managerial and communicative challenges. Organizations need to implement new processes and divert resources to their social media communication within digital environments where information can spread rapidly and widely. Organizations are often used to high levels of control over content shared via traditional channels like newsletter, press releases, flyers, or read-only websites. Control, however, is shared on social media platforms, which evolved around Web 2.0 principles of sharing, participating, and collaborating (see Constantinidis and Fountain 2008; Kietzmann et al. 2011; Trunfio and Lucia 2016). Thus, when even small social media blunders may impact the organization's reputation and any user can share criticism and information, social media communication requires thoughtful and professional management.

In response, social media management has emerged as a sophisticated organizational function (Torning et al. 2015) and "the collaborative process of using Web 2.0 platforms and tools to accomplish desired organizational objectives" (Montalvo 2016, p. 91). It is commonly located within an organization's integrated communication function, which may include public relations, marketing, and internal communication, among others. As a strategic practice, social media management includes planning based on research and listening, implementation and communication, and evaluation. In addition, it requires everyday monitoring and maintenance of social media communication with the organization's stakeholders (Kim 2016).

This chapter provides an introduction to the strategic social media management process with a particular focus on NGOs, which include "private, not-for-profit organizations that aim to serve particular societal interests by focusing advocacy and/or operational efforts on social, political and economic goals, including equity, health, environmental protection and human rights" (Teegen et al. 2004). We first provide an overview of NGO communication and social media. Subsequently, we lay out key steps of a social media process that may guide organizations when planning and implementing their social media communication.

50.2 Social-Mediated Communication: What's in It for NGOs?

Long before social networking sites became mainstream, scholars argued the potential of the Internet as a "potential equalizer" (Coombs 1998, p. 289), allowing activists (and advocacy groups) greater access to power resources in an effort to influence change in business practice. Limited access to power resources, i.e., "the number of followers, favorable media coverage, public support, money, and political champions," explained, in large part, why activists (and NGOs, more broadly) were considered secondary stakeholders (Coombs 1998, p. 291).

Social media have considerably altered this power relation, making it easier for NGOs to challenge and expose unethical and socially irresponsible business practices (Coombs and Holladay 2015). Oft-cited examples include the Greenpeace campaign against Shell's plans to drill in the Arctic and the now-classic Greenpeace challenge to Nestle's palm oil sourcing practices. In these and other cases, social media provide previously marginalized stakeholders such as NGOs a platform to make themselves salient and visible. By posing a threat to corporate reputation, an intangible organizational asset, social-mediated communication has potentially altered NGO-corporate communication and relationships.

Aside from digital activism, social media open up new avenues for NGOs to engage with their stakeholders and supporters. A survey of nearly 5000 global NGOs reports a growing digital presence on social networking sites such as Facebook (92%), Twitter (72%), YouTube (55%), LinkedIn (51%), and Instagram (39%) (27 Stats 2017). Such communication may take the form of raising awareness, providing information, engaging in dialogic communication, fundraising, and/or mobilizing support for specific causes. Current research finds that, thus far, most NGOs use social media in conventional ways; purely to disseminate information and as a form of one-way communication on focal issues and/or campaigns. Most studies of NGO communication on Facebook and Twitter find limited use of the dialogic and interactive capabilities that are unique affordances of social media (Waters et al. 2009; Waters and Jamal 2011). There are exceptions. For instance, a study found that the American Red Cross uses a variety of tools such as websites, blogs, Twitter, and Facebook to recruit and maintain volunteers, update the community on disaster preparedness and response, and engage the media (Briones et al. 2011). Even when there is recognition to use social media to build relationships with primary publics, efforts may be impeded by the lack of time, knowledge, or dedicated personnel (Ibid).

On the other hand, there is more optimism about the potential to mobilize, connect, and build community especially during humanitarian crises. As a case in point, crowdsourcing for disaster relief and humanitarian crises has been on the rise. Following the 2010 earthquake in Haiti, the explosion of content on social media and mobile messaging landed the Red Cross $8 million in donations via text messaging in under 48 hours (Gao et al. 2011). Social enterprises such as Ushahidi provide open-source tools, crowdmaps, and crowdsourcing platforms to reach marginalized populations and have, since its inception in 2008, reached 25 million people in crises across the globe. Real-time and far-reaching status updates, photo sharing, tweets and retweets, and mobilization of networks in times of crisis amplify recovery and aid efforts (Giridharas 2010). Social media also played a vital role in the campaign to secure the release of Chibok girls in Nigeria. Triggered by a grassroots campaign #BringBackOurGirls, the issue garnered worldwide attention of the abductions by Boko Haram and increased pressure on the Nigerian government to escalate efforts to secure the girls' release (Bring Back Our Girls' Rights 2015).

These illustrative cases do not suggest an unqualified position about some inherent power of social media; rather, they are situated in a larger debate about the potential of social media to foster digital engagement. Critics contend that online support ("Likes") for social causes are manifestations of slacktivism and/or arm-chair activism with no substantive offline impact. Proponents, on the other hand, provide evidence for online behavior being a springboard for engagement, moving individuals from "'knowing' to 'doing'" and possibly even reinforcing offline behavior (Mano 2014, p. 292). The Cone Communications Digital Activism Study (2014) highlights several positive developments. First, liking or following organizations on social media is not an end in itself but a "gateway to further participation." Second, donations are more likely to be made online as opposed to offline. Third, compelling content, urgency and relevance of an issue, and ease of participation drive engagement. Fourth, social media demands that organizations tailor their messages and networks to specific audiences. This implies that NGOs need to understand the motivations and usage habits of their constituents if they are to effectively use digital networks.

Despite the optimism, two points merit attention. Contextual differences are decisive in determining the relative power of NGOs as well as the outcomes of social-mediated communication. A recent report found that NGOs in developing nations lag behind their Western counterparts in the optimal use of the Internet and social platforms because "the economic and political factors and the quality of Internet infrastructure in each region affects how NGOs use these tools and how donors in each region respond to them" (Morrison 2016). NGOs are also constrained in their ability to effectively deploy social media for fundraising because of lack of dedicated personnel and the reliance on volunteers for day-to-day operations.

Second, it is naïve to view social media as separate from or supplanting traditional forms of NGO communication, be it email, petitions, or offline activism. Scholars have hypothesized that the impetus toward and adoption of social media in NGOs is contingent on organizational strategy (i.e., what strategic approaches does the NGO deploy in fulfilling its social mission), governance (e.g., membership structure, board size), and environmental considerations such that NGOs that depend on or face high pressure from external constituents may be more likely to adopt social media (Nah and Saxton 2012). There is, as yet, limited research to treat these variables as complete or universal; however, the argument that social media be treated as an integral component of overall organizational strategy remains valid.

Together, these tools are part of the strategic communication effort of any organization. For all organization types, whether commercial or nonprofit, a clear understanding of how social media fit with business and communication strategy is a key starting point. Clarity on goals is the precursor to tailoring the tactical choices (e.g., what networks to use, developing a content strategy, etcetera) to the appropriate target audience. Social media are an amplifier, as noted by a WWF executive, "I see our web site as our home base, the blog as our podium and Twitter, YouTube, Flickr and LinkedIn as our mega phone [...] provides us with the opportunity to mobilize thousands more people [...[" (Catone 2009), not a substitute for conventional channels.

50.3 A Social Media Management Process

50.3.1 Developing a Social Media Strategy

It is tempting for organizations to jump into social media as it only takes a few minutes to start a free page or profile on multiple platforms. Maintaining and implementing effective social media communication, however, takes time, commitment, and knowledge. Just as other areas of organizational operations, it should be informed by sound strategy.

Most basically, a social media strategy outlines how social media goals should be reached so that resources can be used effectively to help the organization fulfill its mission (see Clampitt 2018). It thus provides direction for the organization's social media management as it informs the development and implementation of specific social media tactics, including messages and content, among others (Kim 2016). Doing so, strategy also helps maintain consistency, for instance, when the team faces turnover or crises (Clampitt 2018).

Before developing a strategic social media plan, which provides a guiding framework for the organization's social media activities, the organization should gather and consider information on several dimensions (see Clampitt 2018; Kim 2016). First, social media management needs to contribute to the organization's overall mission, vision, and goals (Kim 2016). Second, it cannot stand isolated but needs to be planned as an integral part of the organization's overall strategic communication function. It thus ideally works in concert with other on- and offline communication tactics so that they can enhance each other.

Third, while the social media landscape is diverse, organizations do not have to be active on every platform. The selection should rather be based on the platforms' appropriateness (for instance, its users, reach, features, and norms) for the organization's social media plan as each platform provides unique features and opportunities to reach different types of audiences (Clampitt 2018). With over 2 billion active user accounts, Facebook, for instance, remains the most visited social network and attracts people of almost all ages who share a wide variety of content and information. In comparison, LinkedIn has only 260 million active accounts and a strong focus on business and career-related content. And while Qzone is hardly known in many parts of the world, it is one of the most popular social media platforms in China (WeareSocial 2018). Because social media are vibrant and constantly developing, the social media team should further monitor and evaluate changing features and policies, shifts in characteristics of users, as well as emerging platforms.

Fourth, social media management should be tailored toward defined audiences and organizational stakeholders. It is thus important to know how they use social media, what they expect from the organization, and how they might best be reached, among others (Clampitt 2018). This also includes considering different patterns of social media use in different parts of the world. Finally, the organization needs to consider its available resources (for instance, staff, budget, skills, and technology), existing social media presence, as well as its strengths and weaknesses before developing a strategic social media plan (Kim 2016).

Based on the analyzed background information, Kim (2016) suggests that organizations proceed by developing social media goals, vision statements, and objectives (see Fig. 1). Social media goals are general statements about what the organization ultimately wants to accomplish with its social media activities. At this stage, target audiences should also be defined (Kim 2016). A nonprofit organization, for example, may want to be widely known as a trusted source of support for victims of sexual abuse among Millennials. An activist group may want to gain public support for its advocacy to change a law while maintaining strong bonds with its supporters and members. A disaster relief organization may set enhancing transparency with donors and volunteers as a social media goal.

Vision statements then require the careful selection of social media platforms as they summarize the specific function and value of each platform in the pursuit of the social media goal(s) (Kim 2016). Clampitt (2018) suggests thinking about these statements as "platform job descriptions" (p. 84), which define specific duties. While the activities on each platform should be consistent and enhance each other, they may play quite different roles and focus on different target audiences. Doing so, the unique capabilities of each platform for sharing and presenting content and engaging audiences should be considered (Clampitt 2018; Kim 2016). What's more, an

Fig. 1 Structure of a strategic social media plan. (Visualization based on Kim's (2015) model)

organization may decide to consider personal accounts of organizational members as part of its strategy and/or operate multiple accounts on one platform. Multiple accounts can be effective and necessary when the organization is large and has a variety of organizational units and functions that require their own social media representation. UNICEF, for instance, operates dozens of twitter accounts for its units in different countries as well as multiple pages and groups on Facebook. A large number of different profiles, however, also increases the resources needed, may dilute a clearly defined social media presence, and poses high demands on consistency and coordination.

Vision statements for the selected platforms (and profiles) should specify the channel's purpose for the defined audience. The disaster relief organization aiming to increase transparency may, for instance, want to use Facebook for sharing insights into the organization's governance process and facilitate discussions about current issues (for instance, by live-streaming executive board meetings or hosting Q&A sessions with leadership) with donors and volunteers. Instagram, a platform that focuses on sharing pictures, short videos, and gifs, could be used to engage audiences via photos of successful projects and behind-the-scenes impressions, while Twitter may primarily function as a news channel to share information about current developments.

Finally, the social media team can set specific objectives for each channel (Kim 2016). Objectives establish specific targets as a base for evaluation and "facilitate communication planning" (Hallahan 2015, p. 251) by defining desired lower-level outcomes and outputs that will lead to achieving higher-level goals (see Hallahan 2015; Kim 2016). While output objectives focus on the quantity and quality of the social media team's posts, content, and activities, outcome objectives set targets for actual results (Kim 2016). The latter commonly aim to affect change in behavior, which also requires change in awareness and attitudes (Hallahan 2015). An

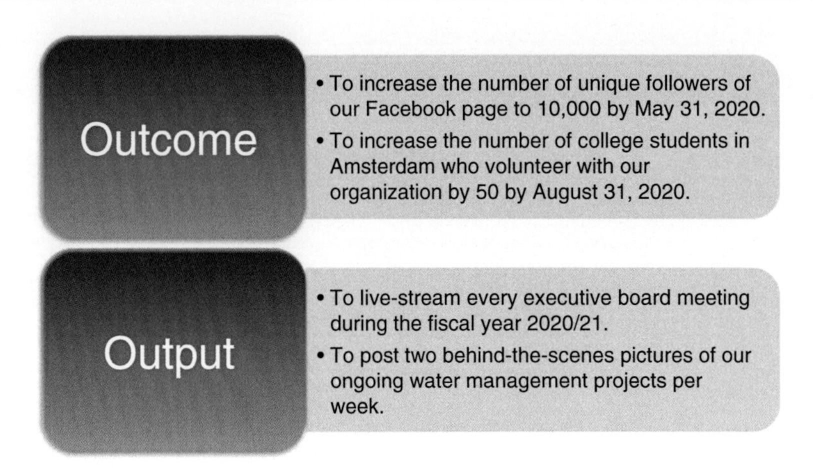

Fig. 2 Examples of SMART output and outcome objectives

organization may, for instance, want to mobilize a certain number of people to participate in a protest against hydraulic fracturing (behavior), which requires change or reinforcement of opinions (attitude) about the practice as well as knowledge about the process and environmental impacts (awareness). While the organization can control its outputs, but not its stakeholders and audiences, a strong social media strategy will balance output and outcome objectives, so that outputs are directed toward achieving outcomes (Kim 2016).

As for most strategic communication efforts, social media managers commonly define objectives following the SMART model (see Hallahan 2015; Kim 2016; Lipschultz 2017). SMART objectives are specific, measurable, attainable, relevant, and timely. Strong objectives (see Fig. 2) thus avoid any vagueness, set a measurable target to clearly establish success or failure, are realistic, relevant to the organization's mission, and include a timeframe. Together, the social media goals, vision statements, and specific objectives outline the strategic orientation that can be translated creatively into lower-level strategies (for instance, we will recruit more followers by having a contest) and tactics (the actual social media communication and activities).

50.3.2 Social Media Communication: From Strategy to Implementation

In formulating communication strategy, social media goals should be aligned with organizational objectives. Doing so offers a pathway for implementation and evaluation (more on this in the next section). Sinek (2009) contends that the most inspiring organizations and leaders "Start with Why" – the raison d'être and core purpose – that, in turn, defines the "how" (processes and tools to realize the goal) and "what" (results and outcomes of the goal) of organizational actions. By no means specific or limited to particular forms of media, Sinek's claim resonates with the idea that understanding social media should go beyond knowledge of tools and technical specifications to a broad-based appreciation of their affordances and the possibilities they create for organizational communication (Kaul and Chaudhri 2017).

Having established a clear understanding of goals – whether fundraising, engagement, growing the subscriber base, recruiting volunteers, increasing brand awareness, and the like – organizations can move from strategy to implementation. Although there is no "one size fits all," there are factors to consider in implementing social media strategy.

50.3.2.1 Listen, Scan, and Monitor

Social listening is a vital yet often ignored effort in flagging potential issues that could threaten an organization's reputation, if left unchecked. Issues management through social listening is a valuable source of information on the conversations and topics that pertain to an organization (for a difference between social listening and monitoring, see Jackson 2017). Social listening can also help categorize and prioritize organizational stakeholders relative to their importance and position on specific issues and tailor communication strategies to appropriately deal with each group, for instance, leverage supporters and make an effort to address the negative opinion of problematic stakeholders.

Scanning for posts that mention the organization further allows for identifying content the organization can engage with or needs to respond to. It can indeed be challenging to determine which mentions merit a response and which ones do not. The Chartered Institute of Public Relations (CIPR n.d.) suggests that social media teams categorize posts according to the stones, pebbles, and sand principle. Informed by the organization's goals and objectives, posts should be sorted based on whether they require immediate action (stones), can be dealt with later (pebbles), or don't require a response (sand).

Sorting through the mentions of the organization and other relevant categories, the social media team also functions as a "boundary spanner" (see Gregory 2016) who brings issues relevant to external stakeholders to the attention of leadership in the organization. In this way, monitoring and social listening also help detect emergent challenges and assist with crisis preparedness. Additionally, data and insights gained through social listening can be an invaluable source of knowledge for content creation, idea generation, and campaign focus. Hootsuite, Social Mention, Sprout Social, and Google Alerts are a few of the available tools for social listening and tracking conversations.

50.3.2.2 Select Platforms Based on Goals and Target Audience

Not all social media are created equal; instead, they offer differentiated functionalities. As an illustration, the honeycomb model of social media outlines seven constructs or building blocks (identity, presence, relationships, conversations, groups, reputations, and sharing) that, in some combination, constitute the defining features of each platform (see Kietzmann et al. 2012). Per this model, "relationships" and "sharing" are the primary building blocks of Facebook and YouTube, respectively, whereas "identity" is core to LinkedIn followed by "reputation" and "relationships." With knowledge of primary and secondary functionalities, organizations can implement informed tactics of engagement to leverage the affordances of each platform. Relatedly, all organizations need to be judicious about their digital footprint although NGOs, in particular, may have limited resources, time, and effort to cast a wide net on social media. Moreover, simply having a Facebook page or a Twitter account is not enough; these need to be regularly updated, monitored, and incorporate engaging content consistent with strategic goals and the defining features and norms on each platform.

50.3.2.3 Create Opportunities for Engagement

Contrary to the static architecture of Web 1.0, social media offer dynamic opportunities for real-time, two-way communication between an organization and its audiences. Interactive features (such as likes, shares, and comments), virtual dialogues, and visual storytelling dramatically alter relational and communicative possibilities. Realizing these benefits, however, can be an uphill task for many organizations.

For one, audiences differ widely in their motivations to engage with organizations. Whereas those that are highly aware, interested, and personally affected by an issue might be more inclined to participate in reciprocal communication, extant scholarship also notes that priorities and behaviors may change over time and latent groups may become salient (see Grunig 1997). Likewise, engagement is not a magic bullet but a process that requires incremental effort. Social engagement, then, may be conceptualized as a ladder starting with awareness and building up to interest, interaction, engagement, and, finally, evangelization. Through a series of progressive actions, this form of digital organizing cultivates "casual supporters into advocates and donors" (Schossow 2014). Barack Obama's election campaign is an illustrative case that started with a simple ask (friending or liking a Facebook page); once campaigners secured a "foot in the door," they could target followers with status updates and specific tasks (e.g., signing a birthday card). With user data and opt-ins, the campaign opened avenues for conversation and furthering the relationship (e.g., actions such as signing an issue-specific petition or making a small donation). At the very top rung, the campaign could comfortably ask supporters for bigger donations, time for voluntary work, and advocacy among friends and networks, etcetera (Ibid).

50.3.2.4 Creating Content that Connects

There is no dearth of advice on creating social media content that is compelling, unique, authentic, surprising, grabs attention, and tells a good story. Using the power of social media, The Dutch Alzheimer's Foundation launched a creative Facebook campaign to "mobilize younger people in the fight against dementia" by letting them experience what patients of the disease go through on a daily basis (The Alzheimer's Event n.d.). The idea was simple: starting with 100 Dutch influencers, people were invited to upload an image of a friend on the campaign website (http://hetalzheimerevent.nl/). The photos were digitally manipulated and, using Facebook functionalities such as photos and tagging, the altered photo was placed in a non-existing event and the participant notified. Participants could tag their friends in the photo, which appeared on the friends' timeline with a simple message that drove the point home, "Confusing right? You are now experiencing what it's like to have Alzheimer's disease." The novel approach and a simple yet powerful message went viral as more people passed on the message and posted images to get their friends to "experience 'the Alzheimer's effect'." Notably, the campaign reached 2,450,000 people only via blogs, news sites, and Twitter, and over 200,000 people experienced the Alzheimer Event directly or via a tagged friend. Indeed, there are many examples to learn from, and storytelling to forge emotional connections is perhaps an unquestioned "truth" in the digital age. As NGOs compete for attention, funds, and followers in a competitive landscape, they confront the challenge of telling their stories in a unique and distinctive manner.

Content creation, however, is more complex than simply telling a good story, as difficult as storytelling is. Empirical studies indicate that virality in social media is driven by several factors. According to Mill's (2012) SPIN framework, for instance, four key variables make virality more likely; viral content has to be likable and sharable (spreadability), easy to redistribute (propagativity), run on multiple social media platforms (integration), and "successively reinforced by virtue of sequentially releasing units of viral content" (p. 168) (nexus).

Social media managers also need to know what works from a stakeholder standpoint. A study of online reactions to NGO communication (Gregory and Richard 2014) attests to the complexity of content creation, which aims to further engagement through likes, comments, and shares. In this study, the public responded positively to all types of messages from nonprofits, including information-sharing, promotion, mobilization, and community-building. Actual engagement, however, differed significantly by message type. Community-building messages, for instance, prompted the highest number of comments, confirming a preference for dialogic forms of communication over information dissemination. Simultaneously, call-to-action messaging garnered the maximum number of likes. Ironically, however, the messages that are most likely to be shared are one-way informational updates, especially if they include photo statuses. This leads the authors to posit that "individuals will share information about nonprofits, but they are not actively encouraging their networks to participate by sharing information about fundraising, events, or calls to action" (p, 294).

In practical terms, organizations need to have realistic expectations and create content that is best suited to specific goals. By creating a social media calendar, organizations can plan and map periodic social media activity (e.g., tweets, Facebook posts, blog posts, videos, etcetera) across networks and ensure alignment with other marketing activities across the organization. Even with the planning, though, emergent issues require organizations to be adaptive and flexible, a defining paradox of the digital age.

50.3.2.5 Rules of Engagement and Social Media Governance

Irresponsible social media usage can prompt serious "social media fails." Preventing, or at least anticipating, these missteps and managing them is contextual – albeit guiding principles may come in handy. Among the first rules of engagement are maintaining a pleasant tone of voice, being responsive, and open to (especially critical) feedback. Censorship in social media can threaten its participatory and open nature and can even undermine organizational reputation.

As organizations negotiate the loss of control in the digital age, they turn to social media guidelines and policies as a mechanism for governance. Along with training programs, "formal and informal frameworks which regulate the actions of the members of an organization within the social web" (Macnamara and Zerfass 2012, p. 302) allow organizations to discipline employee use of social media in order to mitigate potential reputational, legal, and operational risks and offer a semblance of control. Most policies highlight rules of etiquette and engagement consistent with brand values, clear identification of authorship, and issues of confidentiality and privacy. They can run from one page to multiple pages of content. Most policies recognize that employees are representatives of the organization and, as brand ambassadors, they assume a powerful position as reputation agents (Rokka et al. 2014). Thus, their actions can hurt or help their employer (Opitz et al. 2018). In light of blurring boundaries, guidelines help but are not enough. Organizational success with social media is equally about enabling and empowering employees embedded in a supportive culture with latitude for mistakes.

Finally, organizations must be ethical in their use of social media and simultaneously be cognizant of the ethical challenges of irresponsible usage. Bowen (2013, p. 126) outlines several guidelines for ethical digital engagement, which include avoiding deception, verifying data and sources, and being consistent as well as transparent, among others. These guidelines also demand consideration during the formative planning stages and ongoing reflection throughout the social media management process. Social media misuse can have far-reaching effects largely because of its permanence and visibility. Organizations, NGOs included, thus need to look in the mirror regularly and assess whether all forms of strategic communication are conducted ethically and in alignment with organizational values. Once the social media plan has been implemented, social media managers can proceed to evaluate the quality and effectiveness of their efforts in order to improve and adjust their tactics.

50.3.3 Evaluation and Social Media Metrics

Social media communication should be evaluated frequently, and the social media team should assess quality and quantity at all stages of the social media process. This includes preparation and strategic design, implementation, content and audience interactions, as well as the impact of the organization's social media communication (Kim 2016). Evaluation can give confidence that strategic and tactical decisions were appropriate to meet the social media goals and objectives, and allow the social media team to document the value of its work. They can also provide opportunities for reflection and improvements as they identify where the team fell short.

Evaluating the quality of the preparation and background research, the team should consider if important information was considered, was correct, and interpreted appropriately (Kim 2016). Clampitt's (2018) general 5C assessment model suggests that evaluation focuses on five general aspects:

1. Coordinates – appropriateness of goals and how well they are met
2. Channels – social media platform selection and management
3. Content – resonance with audiences, variety, quality, and goal orientation
4. Connections – linkages to other off- and online channels (such as the organization's website, flyers, newsletters, etc.), relevant internal resources and departments, and external communities
5. Corrections – ability to change as well as respond to gaffes and mistakes based on proper evaluation protocols

More specifically, establishing the value and impact of communication is a central concern of evaluation. This question has also been a prominent challenge for strategic communication scholars and practitioners (see Macnamara 2014). Indeed, professional associations such as the Public Relations Society of America (see PRSA n.d.) or the Institute for Public Relations (see IPR n.d.) have commissioned task forces and committees with exploring frameworks and standards for measurement and evaluation. In recent years, these efforts have also focused on the unique characteristics of social media communication and metrics.

The international Association for Measurement and Evaluation of Communication (AMEC 2016) has put forth the Integrated Measurement Evaluation Framework. It suggests that organizations evaluate their communication on three dimensions:

1. *Communication activity* to be assessed by focusing on the quality and quantity of social media communication (outputs)
2. *Audience effects* to be analyzed by measuring outtakes (responses and reactions to the organization's content) and outcomes (change or reinforcement of audience awareness, attitude, or behavior)
3. *Impact* to be established by assessing how the social media communication has contributed to achieving higher-order organizational goals such as establishing a strong reputation or increasing donations

It is helpful that social media companies collect and provide an abundance of data about how followers interact with the organization's profile page and content. Most allow basic free and/or paid access to their analytics. Facebook pages or Twitter profiles, for instance, come with their own analytics feature, and Google Analytics allows for tracking how much traffic to the organization's websites was triggered from its social media platforms. In addition to tools provided by social media companies, organizations have a variety of options for free and paid third-party analytics services.

Social media data is reported based on common social media metrics. These are particularly useful to assess responses and reactions but also provide statistics about communication activity and some outcomes (see Table 1). Facebook Insights, for

Table 1 Selected social media metrics for different levels of evaluation

Level of evaluation	Metric	Measures
Communication activity	Post rate	Number of social media posts within a specific timeframe
	Response rate	Percentage of inquiries and comments the organization responded to within a specific timeframe
Outtakes	Impressions	Number of times a post could be seen by others
	Reach	Number of unique individuals a post could be seen by
	Likes/applause rate	Number of times content has been liked/average number of likes per post
	Shares/ amplification rate	Number of times content has been shared/average number of shares per post
	Comments/ conversation rate	Number of comments in response to content/average number of comments per post
	Engagements	Number of interactions with posts
	Followers/ unfollows	Number of people who follow a profile or page/number of times people unfollow a page or profile
	Mentions/share of voice	Number of times the organization has been mentioned/ percentage of mentions compared to competitors
	Click-through/ click-through rate	Number of times a shared link has been opened, click-throughs divided by impressions
Outcomes	Sentiments	Percentage of positive, neutral, and negative mentions
	Conversions	Number of times social media content interaction was converted into a desired action (subscriptions, donations, signing a petition, attending an event, etc.)

instance, tracks fluctuations in following, reach of posts by time of day, likes, reactions, and shares, among many other metrics.

While social media offer unique access to data on how audiences interact with their profiles and content, evaluation methods should also include qualitative assessments and be tailored to the unique objectives and strategies pursued. The quantity of posts and responses does not reveal much about the quality of the communication activities. To get a complete picture, aspects such as quality of writing and production (for instance, videos), readability and relevance, alignment with strategic objectives, and appropriateness of responses should be assessed as well. Outtake metrics become more meaningful when, for instance, content of comments and mentions, reasons for following/unfollowing, or whether audiences have actually understood the message are explored (see IPR 2003). What's more, social media content may lead to outtakes beyond social media, for instance, if an organization's message gets picked up by traditional news media or leads to feedback and comments by phone.

Metrics are also limited when evaluating outcomes. While conversions can provide important information for specific types of lower-level goals, such as increasing subscribers, they cannot account for other changes in awareness, attitude, and behaviors. Appropriate ways for further assessment depend on the type of change as well as on the organization's budget. If the organization, for instance, aims to increase donations or recruit volunteers, it should also collect information on what made individuals donate or sign up, in addition to tracking conversion metrics. Changes in awareness and attitude often require surveys before and after implementation to establish that a difference was made (see Macnamara 2005). This is particularly useful when organizations launch large-scale campaigns with specific timeframes (for instance, a campaign to change attitudes toward breastfeeding).

A holistic approach to evaluating multiple dimensions of social media communication quantitatively and qualitatively allows for generating important insights. This in turn helps identify where strategy and tactics need to be revisited and can be improved. Evaluation thus closes the circle of the social media process as it leads back to strategic planning and implementation.

50.4 Conclusion

In a time that some call the “era of social media” (for instance, Mutsvairo 2016), NGOs are well advised to embrace and explore the opportunities for communication that social media platforms provide. Many NGOs have shown how social media communication can significantly support efforts to increase awareness, mobilize, raise funds, build community, and engage stakeholders, among others. Navigating the social media environment so that it benefits the organization’s cause, however, is a time-consuming and challenging task. It requires resources, strategic planning, as well as expertise.

This chapter has reviewed opportunities for NGOs and outlined a strategic social media process for organizations. It includes strategic planning, implementation, and evaluation. Doing so, strategic social media management encourages organizations to implement communication creatively and oriented toward the organization’s goals and mission while embracing two-way communication, collaboration, and participation.

A strategic orientation to social media management provides focus, consistency, discipline, and orientation. While many everyday tasks of social media management may become routine, social media are vibrant, dynamic, ever-evolving, messy, collaborative, and spontaneous in nature. Besides a strategic mindset, they thus also require openness and flexibility (see Macnamara and Zerfass 2012), and one-size-fits-all recipes for communication and content will thus likely fall short. Indeed, some of the most successful social media campaigns stand out because they creatively play with the unique possibilities that social media afford and embrace user participation as well as collaboration.

50.5 Cross-References

▶ Capacity Building and People’s Participation in e-Governance: Challenges and Prospects for Digital India

▶ Communication Policy for Women's Empowerment: Media Strategies and Insights
▶ Digital Communication and Tourism for Development
▶ Empowerment as Development: An Outline of an Analytical Concept for the Study of ICTs in the Global South
▶ How Social Media Mashups Enable and Constrain Online Activism of Civil Society Organizations
▶ Online Social Media and Crisis Communication in China: A Review and Critique
▶ Strategic Social Media Management for NGOs

References

2014 Cone Communications Digital Activism Study (2014) Available via http://www.conecomm.com/research-blog/2014-cone-communications-digital-activism-study. Accessed 13 Feb 2018

27 stats about how NGOs worldwide use online technology and social (2017, February 7) Available via Nonprofit Tech for Good. http://www.nptechforgood.com/2017/02/07/27-statsabout-how-ngos-worldwide-use-online-technology-and-social-media/. Accessed 22 Feb 2018

AMEC (n.d.) Integrated evaluation framework. Retrieved from https://amecorg.com/amecframework/

Bowen SA (2013) Using classic social media cases to distill ethical guidelines for digital engagement. J Mass Media Ethics 28:119–133. https://doi.org/10.1080/08900523.2013.793523

Bring back our girls' rights in Nigeria (2015) Available via Plan International. https://plan-international.org/bring-back-our-girls-rights-nigeria#. Accessed 13 Feb 2018

Briones RL, Kuch B, Liu BF, Jin Y (2011) Keeping up with the digital age: how the American Red Cross uses social media to build relationships. Public Relat Rev 37:37–43. https://doi.org/10.1016/j.pubrev.2010.12.006

Catone J (2009, June 24) How WWF is using social media for good #FindingTheGood. Available via https://mashable.com/2009/06/24/wwf-profile/#ytwqQ28im5qr. Accessed 15 Feb 2018

CIPR (n.d.) Guide to social media monitoring. Retrieved from https://www.cipr.co.uk/content/policy/toolkits-and-best-practice-guides/social-media-monitoring

Clampitt PG (2018) Social media strategy: tools for professionals and organizations. Sage, Thousand Oaks

Constantinides E, Fountain SJ (2008) Web 2.0: conceptual foundations and marketing issues. Direct Data Digit Mark Pract 9:231–244. https://doi.org/10.1057/palgrave.dddmp.4350098

Coombs WT (1998) The internet as potential equalizer: new leverage for confronting social irresponsibility. Public Relat Rev 24:289–303

Coombs WT, Holladay SJ (2015) How activists shape CSR: insights from internet contagion and contingency theories. In: Adi A, Grigore G, Crowther D (eds) Corporate social responsibility in the digital age (developments in corporate governance and responsibility, vol 7). Emerald Group Publishing Limited, pp 85–97. https://doi.org/10.1108/S2043-052320150000007007

Gao H, Barbier G, Goolsby R (2011) Harnessing the crowdsourcing power of social media for disaster relief. IEEE Intell Syst 26:10–14 Available via https://www.computer.org/csdl/mags/ex/2011/03/mex2011030010.pdf. Accessed 15 Feb 2018

Giridharas A (2010, March 13) Africa's gift to Silicon Valley: how to track a crisis. Available via New York Times. http://www.nytimes.com/2010/03/14/weekinreview/14giridharadas.html. Accessed 15 Feb 2018

Gregory A (2016) Public relations and management. In: Theaker A (ed) Public relations handbook. Routlege, New York, pp 76–101

Gregory DS, Richard DW (2014) What do stakeholders like on Facebook? Examining public reactions to nonprofit organizations' informational, promotional, and community-building messages. J Public Relat Res 26:280–299. https://doi.org/10.1080/1062726X.2014.908721

Grunig JE (1997) A situational theory of publics: conceptual history, recent challenges and new research. Public Relat Res 3:48

Hallahan K (2015) Organizational goals and communication objectives in strategic communication. In: Holtzhausen D, Zerfass A (eds) The Routledge handbook of strategic communication. Routledge, pp 76–101

IPR (n.d.) IPR Measurement Commission. Retrieved from https://instituteforpr.org/ipr-measurement-commission/about/

Jackson D (2017, September 20) What is social listening & why is it important? Available via https://sproutsocial.com/insights/social-listening/. Accessed 15 Feb 2018

Kaplan AM, Haenlein M (2010) Users of the world, unite! The challenges and opportunities of social media. Bus Horiz 53:59–68

Kaul A, Chaudhri V (2017) Corporate communication through social media: strategies for managing reputation. SAGE, New Delhi

Kietzmann JH, Hermkens K, McCarthy IP, Silvestre BS (2011) Social media? Get serious! Understanding the functional building blocks of social media. Bus Horiz 54:241–251

Kietzmann JH, Silvestre BS, McCarthy IP, Pitt LF (2012) Unpacking the social media phenomenon: towards a research agenda. J Public Aff 12:109–119. https://doi.org/10.1002/pa.1412

Kim CM (2016) Social media campaigns: strategies for public relations and marketing. Routledge, New York

Lipschultz JH (2017) Social media communication: concepts, practices, data, law and ethics. Taylor & Francis

Macnamara J (2005) PR metrics: how to measure public relations and corporate communication. Public Relations Institute of Australia. Retrieved from http://www.pria.com.au/knowledgebank/area?command=record&id=238&cid=20

Macnamara J (2014) Emerging international standards for measurement and evaluation of public relations: a critical analysis. Public Relat Inq 3:7–28

Macnamara J, Zerfass A (2012) Social media communication in organizations: the challenges of balancing openness, strategy, and management. Int J Strateg Commun 6:287–308. https://doi.org/10.1080/1553118X.2012.711402

Mano R (2014) Social media, social causes, giving behavior and money contributions. Comput Hum Behav 31:287–293. https://doi.org/10.1016/j.chb.2013.10.044

Mills AJ (2012) Virality in social media: the SPIN framework. J Public Aff 12:162–169. https://doi.org/10.1002/pa.1418

Montalvo RE (2016) Social media management. Int J Manag Inf Syst 15:91

Morrison K (2016, April 8) How NGOs around the world use technology and social media. Available via http://www.adweek.com/digital/how-ngos-around-the-world-use-technology-and-social-media/. Accessed 13 Feb 2018

Mutsvairo B (2016) Digital activism in the social media era: critical reflections on emerging trends in sub-saharan Africa. Springer

Nah S, Saxton GD (2012) Modeling the adoption and use of social media by nonprofit organizations. New Media Soc 15:294–313. https://doi.org/10.1177/1461444812452411

Opitz M, Chaudhri V, Wang Y (2018) Employee social-mediated crisis communication as reputational asset or threat? Corporate communications. An Int J 23:66–83. https://doi.org/10.1108/CCIJ-07-2017-0069

Pridmore J, Falk A, Sprenke I (2013). New media & social media: What's the difference v2.0. Working paper on Academia.edu. [Cited with permission of first author]

PRSA (n.d.). Measurement standardization. Retrieved from http://apps.prsa.org/intelligence/BusinessCase/MeasurementStandarization/

Rokka J, Karlsson K, Tienari J (2014) Balancing acts: managing employees and reputation in social media. J Mark Manag 30:802–827. https://doi.org/10.1080/0267257X.2013.813577

Saffer AJ, Sommerfeldt EJ, Taylor M (2013) The effects of organizational twitter interactivity on organization–public relationships. Public Relat Rev 39:213–215

Sample I, Woolf N (2016). How the ice bucket challenge led to an ALS research breakthrough. In: The Guardian. Retrieved from https://www.theguardian.com/science/2016/jul/27/how-the-ice-bucket-challenge-led-to-an-als-research-breakthrough

Saxton GD, Waters RD (2014) What do stakeholders like on Facebook? Examining public reactions to nonprofit organizations' informational, promotional, and community-building messages. J Public Relat Res 26:280–299

Schossow C (2014, March 3) Using the ladder of engagement. Available via https://www.newmediacampaigns.com/blog/using-the-ladder-of-engagement. Accessed 15 Feb 2018

Sinek S (2009) How great leaders inspire action [Ted Talk]. Available via https://www.ted.com/talks/simon_sinek_how_great_leaders_inspire_action?language=en. Accessed 15 Feb 2018

Steel E (2014) Ice bucket challenge has raised millions for ALS Association. In: New York Times. Retrieved from https://www.nytimes.com/2014/08/18/business/ice-bucket-challenge-has-raised-millions-for-als-association.html? r=1

Teegen H, Doh JP, Vachini S (2004) The importance of nongovernmental organizations (NGOs) in global governance and value creation: an international business agenda. J Int Bus Stud 35:463–483

The Alzhiemer's Event (n.d.) Available via http://www.award-entry.com/thealzheimersevent/. Accessed 18 Feb 2018

Tørning K, Jaffari Z, Vatrapu R (2015) Current challenges in social media management. In: Proceedings of the 2015 international conference on social media & society. Association for computing machinery. Available via ACM Digital Library https://dl.acm.org/citation.cfm?id=2789191

Trunfio M, Della Lucia M (2016) Toward eb 5.0 in Italian regional destination marketing. Symp Emerg Issues Manag 2:60–75. https://doi.org/10.4468/2016.2.07trunfio.dellalucia

Waters RD, Jamal JY (2011) Tweet, tweet, tweet: a content analysis of nonprofit organizations' twitter updates. Public Relat Rev 37:321–324. https://doi.org/10.1016/j.pubrev.2011.03.002

Waters RD, Burnett E, Lamm A, Lucas J (2009) Engaging stakeholders through social networking: how nonprofit organizations are using Facebook. Public Relat Rev 35:102–106. https://doi.org/10.1016/j.pubrev.2009.01.006

WeAreSocial (2018) Digital in 2018. Available via https://wearesocial.com/blog/2018/01/global-digital-report-2018

ICTs and Modernization in China

51

Song Shi

Contents

Abstract

A multidisciplinary field at the intersection of communication for development and social change (CDS) studies and ICT studies, ICT for Development (ICT4D) is generally understood as the research of development and social change brought about by the application of ICTs. China, the largest developing country, with the largest population, and the biggest Internet population in the world, offered a good case for the investigation of ICT4D in developing countries. This chapter explores the ICT4D phenomena in China in relation to the modernization paradigm in Communication for Development. This chapter reveals that assumptions and theories of the modernization paradigm have significantly influenced the policies and projects on ICT4D in contemporary China. Yet, discussion on the potential of other approaches in ICT4D in China has also emerged among scholars.

Keywords

Communication for Development · ICT for development (ICT4D) · ICT studies · Modernization

S. Shi (✉)
Department of East Asian Studies, McGill University, Montreal, QC, Canada
e-mail: song.shi2@mail.mcgill.ca; shisong1973@gmail.com

J. Servaes (ed.), *Handbook of Communication for Development and Social Change*,
https://doi.org/10.1007/978-981-15-2014-3_27

51.1 Introduction

As the biggest developing country in the world, China has experienced tremendous economic development over the past three decades. It has also been widely considered as an example that how an agrarian society could be rapidly "modernized" into an industrial society. At the same time, China now has the largest Internet user base in the world, with upwards of 750,000,000 Internet users (as of June 2017), over twice the population of the United States. The use of Information and Communication Technologies (ICTs) for development and "modernization" in contemporary China has attracted worldwide interest among scholars, policy makers, and the general public. This chapter first provides a brief survey of the concept of modernization in Communication for Development and in ICT for Development. Then, it explores the phenomenon of ICT4D in relation to modernization theories in the specific context in China (PRC). The outline of the chapter is as follows: Section 51.2 will briefly summarize the concept of modernization as well as the critique of the modernization paradigm in Communication for Development. Section 51.3 will survey the concept and practices of ICTs for development. Section 51.4 will discuss the practice of ICT for development in China in relation to modernization theories and alternatives approaches to ICT for development. Section 51.5 concludes.

51.2 Communication for Development: Modernization, Dependency, and Multiplicity

Communication for Development is the study of development and social change brought about by the application of communication research, theory, and technologies (Rogers 1976; Servaes 1999, 2008). Modernization theories (e.g., Almond and Coleman 1961; Lerner 1958; Lerner and Schramm 1967; Rogers 1976; Rostow 1953) is the best established paradigm in this field. They significantly shaped the definition and understanding of the concept of "development" as well as the concept of "modern" and "modernization" among scholars, policymakers, and practitioners, as well as the practice of development from the 1950s to the 1960s (Servaes 1999, 2008; Sparks 2007). The modernization paradigm has also critically influenced the international development policies in the Western countries such as the United States. This paradigm stems from neo-liberalism, functionalism, and behaviorism traditions. It defined development and underdevelopment in terms of observable quantitative differences between rich (developed) countries and poor (developing) countries. And it used a number of quantitative economic growth indicators, such as income, per capita GDP, savings volume, and investment level to measure different stages of development. In the late 1950s and early 1960s, modernization theories incorporated non-economic indexes, such as attitudes toward change, education level, mass media adoption rate, and institutional reforms, to measure development (e.g., Almond and Coleman 1961; Lerner 1958). Modernization theorists contended that the cause of underdevelopment in developing countries and less developed communities is internal, meaning that internal factors, such as traditional social-political and cultural structures; traditional mindsets; and passive attitudes toward change in these societies and communities, lead to underdevelopment. This implies that the stimulus for change or the solution to underdevelopment in developing countries must come from the outside (Western) world (Servaes 1999). A typical example of this kind of argument is presented in Lerner's canonical book, *The Passing of Traditional Society: Modernizing the Middle East*: "From the West came the stimuli that will undermine traditional society . . . for reconstruction of a modern society . . . West is still a useful model. What the West is . . . the Middle East seeks to become" (Lerner 1958). Another characteristic of the perception of development in modernization paradigm is that it understands development as a universal linear evolutionary process. By universal, it means that the Western development model is a spontaneous, unavoidable, and irresistible process that every society must go through (Lerner 1958; Servaes 1999). On a more practical level, it believes that replicating the Western political-economic system is the only means by which developing countries can develop into modern societies. By linear evolutionary, it means that the development process is an irreversible growth process without limit.

With regard to the role and effects of communication in development projects, the modernization paradigm perpetuated the "diffusion model" (Rogers and Shoemaker 1983; Inagaki and World Bank 2007; Servaes 1999) in which communication is considered to flow one-way from the "sender" (development experts) to the "receiver" (less developed communities). In this model, communication is overwhelmingly oriented toward persuasion. Accordingly, in Communication for Development projects guided by modernization theories, less developed communities, the intended beneficiaries of these projects, are considered passive targets. And they are excluded from the policymaking and the designing of these projects as well as from developing a base of knowledge to solve underdevelopment.

On an international level, the modernization paradigm asserts that the transfer of technology, capital, values, and sociopolitical structure from developed countries (Western countries) to developing countries will solve their underdevelopment problems and "modernize" them. The modernization paradigm was criticized for being Eurocentric, overlooking local participation, focusing overwhelmingly on economic growth, and ignoring the consequences of macro and international economic and social-political factors in local development (e.g., Servaes 1999, 2008; Sparks 2007). In the 1960s, the modernization paradigm was criticized and challenged by a new paradigm in development studies-dependency paradigm emerged from Latin America (Servaes 1999). Dependency theories stem from the neo-Marxist or structuralist intellectual traditions and the Latin American researchers' study on development issues. Most dependency paradigm researchers agreed that dependency is a conditional situation in which the economies of one group of countries are conditioned by the development of others. The relationship between the two groups of countries becomes a dependency relationship when the dominant group can expand through self-impulsion while the dependent group can only expand as a reflection of the expansion of the dominant countries (Dos Santos 1970). Dependency theories argued that the dependency relation between the developing countries and the developed countries causes the underdevelopment in developing countries. Thus, the solution of underdevelopment in developing countries is removing the dependency relationship. The developing countries should disassociate themselves from the world market and opt for a self-reliant development strategy. On a technical level, dependency researchers advocated limiting the import of privately owned and Western controlled technologies, organizing import-substitution programs to design the "appropriate" indigenous technologies in developing countries, and fostering technological cooperation among all developing countries (Servaes 1999).

In the 1980s, the multiplicity paradigm (Servaes 1999) or the participatory paradigm (Inagaki and World Bank 2007) emerged as the critique of the first two paradigms from diverse origins in both Western countries and developing countries. Contrary to the modernization paradigm, it highlights that there is no universal development model and every society must pursue its own development strategy (Servaes 1999). In addition to economic development, the problem of freedom and justice also needs to be addressed. Instead of endless growth, limits of growth are considered to be inherent in the interaction between society and nature. Different from the dependency paradigm, it highlights interdependency. By interdependency, it means that all nations, in one way or the other, are dependent upon each other. Thus, development has to be studied in a global context, in which center (developed countries) and periphery (developing countries), as well as their interrelated subdivisions, have to be taken into consideration (Servaes 1999). With regard to the role of communication in development projects, the multiplicity paradigm proposed a new model – the "participatory model" (Inagaki and World Bank 2007; Ogan et al. 2009; Servaes 1999). Compared with the sender-oriented diffusion model, participatory model is more receiver-oriented. It views the ordinary people (receivers) as the controlling actors or participants in using communication for development. It believes that communication is a dialogue. This means, first, there is no sender and receiver, and all participants are equal peers; second, the meaning and the knowledge of development is not packaged knowledge sent from sender to receiver (the diffusion model); rather, it is constructed in the interactions among different participants. The advocators of participatory model in development projects stated that real participation must involve the redistribution of power, meaning that in a participatory model development project, the elites' power must be redistributed so that a community can become a full-fledged democracy (e.g., Inagaki and World Bank 2007; Melkote and Steeves 2001; Servaes 1999). Overall, in Communication for Development, although modernization theories were criticized first by dependency theories then by multiplicity paradigm, it remains a popular discourse among policymakers and a dominant approach in many development contexts, especially in projects related to ICTs (Inagaki and World Bank 2007; Ogan et al. 2009).

51.3 ICT for Development

A multidisciplinary field at the intersection of Communication for Development studies and ICT studies, ICT for Development (ICT4D) is generally understood as research that uses ICTs to bring about development and social change. The rise of ICT4D as an academic field and the large number of ICT4D projects in different countries is an indication of the strong interest among scholars and practitioners in the effects of ICTs on development and social change (e.g., Donner 2015; Heeks 2008; Hudson 2013; Kleine 2013; Mansell and Wehn 1998; Ogan et al. 2009; Torero and Von Braun 2006). Using meta-analysis and content analysis to examine in peer-reviewed journals between 1998 and 2007, Ogan et al. (2009) found that ICT4D has become the most dominant approach employed by researches concerned with the relationship between communication and development at that time, with 42.3% of the articles (85 out of 201) using ICT4D as their primary approach.

The research of Ogan et al. also reveals that a substantial number of researchers in ICT4D embraced the technological deterministic view of the modernization paradigm, believing that technology is the primary driver of development and ICTs, such as the Internet and mobile phones, are new magic solutions to development problems (2009). Yet, their research and that of others (e.g., Kleine 2013; Torero and Von Braun 2005; Servaes et al. 2006; Slater and Tacchi 2004) also indicate that an increasing amount of ICT4D scholarship considers ICTs to be tools for development and social change. For example, Torero and von Braun stated: "ICT is an opportunity for development, but not a panacea. For the potential benefits of ICT to be realized in developing countries, many prerequisites need to be put in place" (2006: 343). Servaes et al. assert that ICTs provide "a new potential for combining the information embedded in ICT systems with the creative potential and knowledge embodied in people" (2006, p. 5). The second group of scholars therefore believed that research on ICT4D should focus on how users utilize ICTs and the sociopolitical, economic, and cultural contexts of the use of ICTs, as well as the characteristics of ICTs.

51.4 ICT for Development and "Modernization" in Contemporary China

Although not among the most dominant or populous approaches in studies of China's ICTs and the Internet, ICT4D has also been employed by communication and media scholars focusing on China (e.g., Liu 2016; Shi 2013; Ting 2015; Zhao 2008a, b; Zhang and Chib 2014).

Zhao's monograph (2008b) significantly contributed to our knowledge of ICT4D in China by offering an in-depth analysis of ICT4D projects in five rural towns in China. Her research fully applied the ICT4D approach in the context of rural China. In the Communication for Development filed, Zhao's theoretical framework is more in line with the modernization paradigm especially Rogers' theory on diffusion of innovation (1983). Yet, different from most of past research on diffusion of ICTs, Zhao used a qualitative method, which offers her research a unique capacity to seek in-depth understanding of Internet diffusion in relation to political, socioeconomic, and technological contexts of rural China. Her book also explores the implications of Internet use upon various aspects of rural development and the how government, private industries, and individual users contribute to the diffusion of Internet in rural China. A significant strength of Zhao's research is that it is based on ethnographic research of five ICT4D projects in rural Chinese towns in western, central, and eastern regions in China. The rationale that Zhao selected the five projects is that they represent ICT4D projects initiated by different stakeholders in different parts of China. Two of the projects were initiated by the local government and one was initiated by local government and international development organization. The rest two projects were initiated by the private sector. Through this approach, Zhao's research explored the role of different stakeholders in promoting Internet diffusion in different parts of rural China. Zhao's findings offered a solid critique to technological determinism assumption of the modernization paradigm such as ICTs drive development and social change in less developed community. Zhao argued that the impacts of the five ICT4D projects to various development problems in rural China are very limited (2008b). Zhao's findings also show that local government, development organization, and private industry led every process of the five projects, whereas local rural communities' participation in these projects was relative passive. This indicates that although the five ICT4D projects were initiated by different stakeholders in different parts of rural China, they are all more or less in line with the modernization development theories and the diffusion model in Communication for Development. Zhao's book also shows that the effectiveness of ICT4D projects guided by the modernization theories is questionable.

Zhao's research primarily focuses on ICT4D projects in local level. Shi (2013), on the contrary, explored the state ICT4D policy in rural China. Through a textual analysis of policy documents related to the Connecting Every Village Project, a state initiative to promote telecommunication and Internet services in the nation's vast rural regions, Shi analyzes development theories and assumptions that underpin this state policy. Shi argued (2013) that the Connecting Every Village Project is significantly influenced by the modernization theories and assumptions. Shi argued that in the Connecting Every Village Project, the rural-urban dichotomy replaced the traditional-modern dichotomy of the modernization paradigm as a core concept in the policy discourse. Yet, the conceptualization of development and the solutions of underdevelopment in the project are completely congruent with the modernization paradigm. For example, the modernization paradigm assumes that development in traditional society can be stimulated by external factors–the transfer of capital and technology, expertise and technique from modern Western nations. Similarly, the Connecting Every Village Project believed that the development of the rural regions can be stimulated by external factors such as the transfer of capital and technology, expertise, and technique from more industrialized urban regions within China (2013). Shi argued that in the modernization paradigm, the "modernization" of the traditional societies in developing countries was perceived as a Westernization process, while in the the Connecting Every Village Project, the "modernization" of rural regions was perceived as an urbanization process. Moreover, like in most ICT4D projects in modernization paradigm, the Connecting Every Village Project oversimplified the ICT development in rural China to a single quantitative criterion – the percentage of villages covered by either the Internet or telecommunication networks. Shi's finding opened up the door to introduce critical research on the modernization paradigm into the study of the unbalanced development and unequal relations between rural China and urban China.

Shi's research revealed the modernization paradigm assumptions that underpin China's state ICT policy in rural China. Yet, it has not addressed the various problems and weaknesses of the modernization paradigm ICT4D initiatives led by the Chinese government. Ting and Yi (2013) offered an in-depth investigation of the weakness of state-led ICT4D projects in Guangdong Province, south China. Guangdong has the biggest GDP in China, $7,281,255 million in total as of 2015. It also has the eighth highest GDP per capita in China, reaching $ 11,143 in 2016. Yet, the economic development in Guangdong was very unbalanced. According to Ting and Yi (2013), the GDP per capita in the most developed Pearl River Delta in Guangdong has researched 67,407 Yuan whereas in the mountainous region in the province, it was just 16,726 Yuan as of 2009. The most economically developed province as well as a province with sharp rural-urban inequality, Guangdong offered a good case to examine ICT4D projects in rural China. Ting and Yi identified four major weaknesses of two provincial ICT4D projects led by the government: "(1) inefficient and wasteful spending resulting from interdepartmental rivalry (2) lack of policy continuity and institutional learning (3) lack of accountability and credible measurements (4) central planning resulting in gap between services and local needs" (2013). The first weakness of the projects is related with the top-down designing and implementation processes of Communication for Development projects that are guided by modernization theories and the diffusion model. The third weakness, lack of accountability, is related with the fundamental characteristics of the modernization theories guided projects such as overlooking local participation and the dominance of the government or elites in all process of the projects. The last weakness of the two projects revealed by Ting and Yi is actually the result of the features of the "sender" oriented diffusion model such as the intended beneficiaries of development projects are considered passive targets and are excluded from developing a base of knowledge to solve their own problems. Ting and Yi (2013) offered a concrete example of various weaknesses of ICT4D projects influenced or guided by modernization theories in rural China.

Liu's article in 2016 examined a provincial level government led ICT4D project in rural region in Sichuan province, with a focus on the sustainability risks of the project. Sichuan ranked fourth in term of population but 24th in terms of GDP per capita in China as of 2016. Its GDP per capita just reached $6022, around 54% of Guangdong in 2016. Compared with Ting and Yi (2013), Liu's research offered insights on government led ICT4D project in a relative poor region in China. Drawing upon Kumar and Best's theoretical framework (2006), Liu (2016) offered a critical analysis of Sichuan's informatization project led by the government by examining its sustainability failure from five perspectives: financial/economic sustainability failure, cultural/social sustainability failure, technological sustainability failure, political/institutional sustainability failure, and environmental sustainability failure. Liu (2016) argued that the government-led project is at risk of financial, social, and institutional sustainability failures. By institutional sustainability failure, Liu means that the lack of coordination between different government departments and the lack of policy continuity threatened the sustainability of the project in Sichuan. This is in line with the findings in Ting and Yi's research on the weakness of government-led ICT4D projects in Guangdong. In terms of social sustainability failure, Liu (2016) revealed that in the current design of the project, members of rural communities are perceived as passive receivers of what information is delivered to them. They were excluded from the discussion of what contents should be created for and delivered by the newly developed ICT systems in their own communities. Liu (2016) argued without the engagement of local representatives and local communities, the quality of the locally relevant and locally oriented information might deteriorate in the long term. According to the data in Liu article, five out of seven of the local residents interviewed in the research stated that the information from the newly developed ICT systems is not useful for them. Liu's findings reconfirmed the weakness of ICT4D projects guided by modernization theories that Ting and Yi (2013) has revealed in their research in Guangdong such as lack of local participation.

Existing research (e.g., Liu 2016; Shi 2013; Ting and Yi 2013; Zhao 2008b) indicated that modernization paradigm dominated ICT4D projects in China especially in its vast rural regions. Influenced by the increasing significance of new theories in Communication for Development, scholars investigating ICT4D in Chinese society have also begun to examine the potential to introduce new approaches in ICT4D projects and policies in China. Ting (2015), for example, examined the potential and challenges to incorporate the Capacity Approach in ICT4D in the Chinese government's rural ICT4D policy. The Capacity Approach in ICT4D is generally in line with the multiplicity paradigm or participatory paradigm in Communication for Development. The Capacity Approach stems from Indian economist and philosopher Amartya Sen's research (Sen 1993, 1999). It challenged the economic growth focused definition of development in the modernization paradigm. In this approach, development is understood as the expansion or enhancement of the capacity of individuals to live a life he or she values (Kleine 2013). Contrary to the modernization paradigm's belief that technology is the driver of development, the Capacity Approach considers technology as tools and means for individual to expand their capacities. The Capacity Approach has been effectively used to analyze ICT4D projects and policies in other countries (Kleine 2013) and has been incorporated into the Human Development Index of the United Nations Development Programme. Ting (2015) argued that the Capacity Approach has the potential to develop "a more coherent and holistic framework" for ICT4D projects in rural China. Through case studies of two ICT4D projects in rural China, Ting examined the compatibility between the Capability Approach and the conventional top-down modernization approach towards rural ICT4D projects in China and offered policy suggestions (Ting 2015).

51.5 Conclusion

This chapter explores the ICT4D phenomena in China in relation to the modernization paradigm in Communication for Development. Existing research (e.g., Liu 2016; Shi 2013; Ting and Yi 2013; Zhao 2008a, b) indicated that assumptions and theories of the modernization paradigm have significantly influenced the policies and projects on ICT4D in contemporary China. From the national level project (Shi 2013), to provincial level projects (e.g., Liu 2016; Ting and Yi 2013), to village level projects (Zhao 2008a, b), the modernization paradigm dominated in the design and implementation of ICT4D initiatives in China. Yet, discussion on the potential of other approaches has also emerged among scholars (Ting 2015). In order to bring new insights on ICT4D in China, future research could examine why modernization paradigm dominated the ICT4D field in China, the emerging practice of ICT4D that employed other approaches such as the Capacity Approach or the participatory communication model, and how to use the participatory communication model in critical analysis of ICT4D efforts led by the Chinese government or other stakeholders.

References

Almond GA, Coleman JS (1961) The politics of the developing areas: Coauthors: James S. Coleman (o. Fl.). Princeton University Press, Princeton

Donner J (2015) After access: inclusion, development, and a more mobile internet. MIT Press, Cambridge, MA

Dos Santos T (1970) The structure of dependence. The American Economic Review, 60(2). Papers and Proceedings of the Eighty-second Annual Meeting of the American Economic Association, pp 231–236

Heeks R (2008) ICT4D 2.0: the next phase of applying ICT for international development. Computer 41(6):26–33

Hudson HE (2013) From rural village to global village: telecommunications for development in the information age. Routledge, New York, USA

Inagaki N, World Bank (2007) Communicating the impact of communication for development: recent trends in empirical research. World Bank, Washington, DC

Kleine D (2013) Technologies of choice?: ICTs, development, and the capabilities approach. MIT Press, Cambridge, MA

Kumar R, Best ML (2006) Impact and sustainability of e-government services in developing countries: lessons learned from Tamil Nadu, India. Inf Soc 22(1):1–12

Lerner D 1917–1980 (1958) The passing of traditional society: Modernizing the middle east. Glencoe: Free Press

Lerner D, Schramm WL (1967) Communication and change in the developing countries. East-West Center Press, Honolulu

Liu C (2016) Sustainability of rural informatization programs in developing countries: a case study of China's Sichuan province. https://doi.org/10.1016/j.telpol.2015.08.007

Mansell R, Wehn U (eds) (1998) Knowledge societies : information technology for sustainable development. Oxford University Press, Oxford/New York

Melkote SR, Steeves HL (2001) Communication for development in the third world: theory and practice for empowerment. Sage, New Delhi

Miletzki J, Broten N (2017) Development as freedom. Macat Library

Ogan CL, Bashir M, Camaj L, Luo Y, Gaddie B, Pennington R et al (2009) Development communication the state of research in an era of ICTs and globalization. Int Commun Gaz 71 (8):655–670
Rogers EM (1976) Communication and development: critical perspectives. Sage, Beverly Hill
Rogers EM, Shoemaker F (1983) Diffusion of innovation: a cross-cultural approach, New York
Rostow WW (1953) The process of economic growth, by W.W. Rostow. Clarendon Press, Oxford
Sen A (1993) Capability and well-being73. The Quality of Life 30
Sen A (1999) Development as freedom. In: The globalization and development reader: perspectives on development and global change. Wiley, p 525
Servaes J (1999) Communication for development: one world, multiple cultures. Hampton Press, Cresskill
Servaes J (2008) Communication for development and social change. Sage, New Delhi/Thousand Oaks
Servaes, J., Carpentier, N European Communication Research and Education Association (2006). Towards a sustainable information society: deconstructing WSIS. Bristol/Portland: Intellect
Shi S (2013) Two cases and two paradigms: connecting every village project and CSO Web2.0 project. In: Servaes J (ed) Sustainability, participation and culture in communication. Intellect/ The University of Chicago Press, Bristol, pp 153–172
Slater D, Tacchi JA (2004) Research on ICT innovations for poverty reduction. UNESCO, Paris
Sparks C (2007) Globalization, development and the mass media. Sage, London
Ting C (2015) ICT4D in China and the capability approach: do they mix? Int J Info Comm Technol Human Dev 7(1):58–72
Ting C, Yi F (2013) ICT policy for the "socialist new countryside" – a case study of rural informatization in Guangdong, China. Telecommun Policy 37(8):626–638
Torero M, Von Braun J (2006) Information and communication technologies for development and poverty reduction: the potential of telecommunications. International Food Policy Research Institute, Johns Hopkins University Press, Baltimore
Zhang W, Chib A (2014) Internet studies and development discourses: the cases of China and India. Inf Technol Dev 20(4):324–338
Zhao J (2008a) ICTs for achieving millennium development goals: experiences of connecting rural China to the internet. Knowl Technol Policy 22(2):133–143. https://doi.org/10.1007/s12130-009-9071-2
Zhao J (2008b) The Internet and rural development in China: The socio-structural paradigm (Vol. 97). Peter Lang.

Online Social Media and Crisis Communication in China: A Review and Critique

52

Yang Cheng

Contents

Abstract

This chapter presents a review of the scholarship on the social media and crisis communication in China. Through a content analysis of research articles published in 11 journals listed in the Social Science Citation Index (SSCI) and in the Chinese Social Science Citation Index (CSSCI), 58 directly relevant articles are identified in the period from 2006 to 2016. The chapter examines the theoretical framework, methodological preferences, types of research, crisis communication practice, as an overview of current ongoing research trends. This research also explores how the

Y. Cheng (✉)
Department of Communication, North Carolina State University, Raleigh, NC, USA
e-mail: ycheng20@ncsu.edu

J. Servaes (ed.), *Handbook of Communication for Development and Social Change*,
https://doi.org/10.1007/978-981-15-2014-3_5

unique Chinese characteristics affect the communication practice. Finally, critiques and suggestions for future research are provided from four dimensions, which include emphasizing on the crisis phases, extending theories and models, applying a comparative logic in discussion, and unifying the research standard between articles published in the SSCI and CSSCI journals.

Keywords

Social media · new media · crisis communication · content analysis · China

52.1 Introduction

As China and India have risen into the world's most rapidly developing markets, new patterns of online communication and crisis management are emerging within the Asian area, thereby the twenty-first century has been called "the Asian Century" and attracted increasing attention from global scholars. In the previous 10 years, several major crises in China were broadcast to the world through the rapid transmission of social media such as Facebook, Twitter, Weixin, and Weibo. Among these crises (e.g., the SARS crisis in 2003, Sanlu milk contamination crisis in 2008, China's Red Cross credibility crisis in 2011, and Japan's nuclear crisis in 2013), social media as the mediator between organizations and publics have transformed the way of sending and achieving information. On the one hand, organizations can directly send information to massive subscribed stakeholders on social media. On the other hand, current online users can actively engage in crises through collectively commenting or publishing their own news stories (Bruns 2005). This social-mediated communication may assist effective crisis communication community (Howe 2008) while meantime generate the possibility for intensive crises or risks as well (Huang and Lin 2004). Hackers, viruses, and rumors on social media can easily trigger crises and lead to large amounts of tangible or intangible cost (e.g., organizational litigation cost and customer complain cost). Studies based on interviews in 16 global companies found that 36% of crises resulted from digital security failures or online negative publicities (Burson-Marsteller and PSB 2011).

This chapter proposes to explore the form and practice of online social media and crisis communication as they exist in China. Through a content analysis of 58 articles published in 11 academic journals from 2006 to 2016, the chapter presents the theoretical framework, methodological preferences, types of research, and crisis communication practice addressed in these articles.

52.2 Online Social Media and Crisis Communication

Types of social media such as Twitter, Facebook, WeChat, and Weibo are ushering in a new era of crisis communication. Previous literature has conducted numerous discussions on online social media and crisis communication, which is referred to as the "social-mediated crisis management research (SMCM)" (Cheng 2016a).

Earlier, Hearit (1999) studied the crisis management of flawed Intel Pentium chip on the Internet and found that companies were urged to use the Internet under the pressure of public movements. In turns, Taylor and Perry (2005) suggested that about half of studied 92 organizations adopted online tools in crisis management. Recently, Liu et al. (2012) constructed a blog-mediated crisis communication (BMCC) model and described how the American Red Cross uses blogs in crises. In China, Choi and Lin (2012) explored the consumer emotions posted on online bulletin boards in the crisis of Mattel product recalls. In the Chinese context, Tai and Sun (2007) found that the Internet tools particularly empower the public to challenge official claims during the SARS crisis. Wu and Yeh (2012) suggested that although the majority of Taiwan corporations rapidly respond to the crises online within the first 24 h, the crisis communication effectiveness is below the expectation. In sum, studies in the social media and crisis communication research mainly covered areas such as the use of external social media in organizational crisis communication (Huang and Lin 2004; Jin and Liu 2010; Liu et al. 2012; Taylor and Perry 2005), the management of brand image and organizational-public relationship (Moody 2011), and the crisis communication strategy and effectiveness (Utz et al. 2013; Wu and Yeh 2012).

Considering the large amount of relevant literature on crisis communication involving social media in public relations or in the field of communication, it is time that we should take a synthesized review of how global scholarship examines the realm of social media and crisis communication and provide insights for future research agendas. Research that examines the theoretical framework, methodological preferences, types of research, and public relation practice in this area is lacking (e.g., Veil et al. 2011; Ye and Ki 2012).

According to Huang et al. (2005), although crisis communication scholarship has grown rapidly over the past decades, most paradigms originated, in terms of conceptualization and operationalization, still were applied to and tested only in North America. Little research has explored the social media and crisis communication in China where cultural traits together with institutional contexts can greatly differ from those in Western countries such as the U.S.A.

Traditional Chinese society places a special emphasis on moral values such as collectivism, hierarchy, harmony, social relationships, and face-favor practice (Hwang 1987), while in Western countries, individualism, equality, freedom, and personal achievements are valued (Hofstede 2001). Contrasting with the USA, the block of international social media (e.g., Facebook, Twitter, YouTube) and the Internet censorship serving for political regimes characterize a distinctive online media landscape in China (Jiang 2013). For example, over half (61%) of 984 million Chinese Internet or mobile web users could get access to online media. The nation's top social network, QZone gained 712 million users in 2013, 84% of which participated as active users (Millward 2013). Social media in China enjoyed large numbers of users and a high level of engagement. Sixty-one percent Renren users shared personal information online, and 58% users were willing to accept friend requests from strangers, while Facebook users showed a lower degree of openness that only 36% of them were prone to disclose personal information and only 26% users would like to accept strangers (Yin and Liu 2012). Thus, how did the forms of social media and crisis communication practice present themselves in

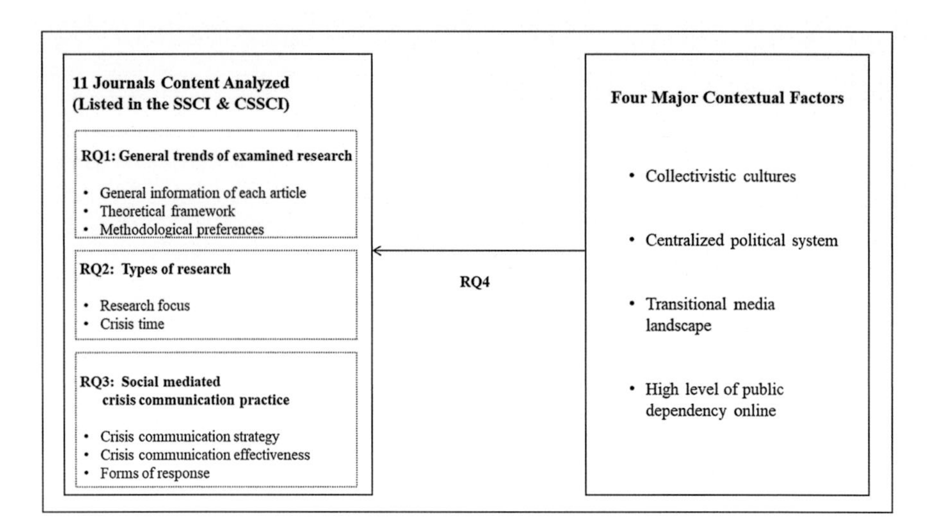

Fig. 1 The theoretical framework of the social media and crisis communication research

China? How did the contexts affect such practice? A comprehensive analysis in the Chinese communication should shed light on these questions and extend Western theories of public relations in general and the crisis communication in particular.

This chapter analyzed a total of 58 articles published by 11 academic journals from 2006 to 2016. Based upon the findings, this study presented the general trend of the social media and crisis communication research, in terms of the theoretical framework and methodological approach, what types of social media and crisis communication research have been studied and what were the forms of social media and crisis communication practice. Finally, the impact of the Chinese context on the crisis communication practice was explored. The ultimate goal is to provide conceptual clarity for future research that seeks to develop a better understanding of the social media and crisis communication research in China. Figure 1 depicts the theoretical framework for this study and suggests four research questions below:

- What is the general trend of social media and crisis communication research, in terms of the article number in each journal, the theoretical framework, and the methodological preferences in each article?
- What types of social media and crisis communication research have been studied (i.e., research focus and crisis time)?
- What are the forms of social media and crisis communication practice (i.e., crisis communication strategy, forms of response, and crisis management effectiveness) in China?
- How do cultural traits and institutional contexts influence the social media and crisis communication practice in China?

52.3 The General Trend of Social Media and Crisis Communication Research

This chapter first presented research trends of all available including general information of articles, theoretical framework, and methodological preferences. Among the total 58 articles drawn from 11 journals, only SSCI (12 articles, 21%) and CSSCI (46 articles, 79%) journals were included, with 5 articles (9%) in *Public Relations Review* [SSCI], as the leading journal in the field of public relations. Twelve articles (21%) were found in the *Journal of International Communication* [CSSCI] and 11 articles (19%) in the *Journalism and Communication Research* [CSSCI] serving as major outlets for the social media and crisis communication research in China. Findings also demonstrated an increasing attention to "social media and crisis communication" over the past 10 years, with only 9 articles (16%) published between 2006 and 2008, 17 articles (29%) published between 2009 and 2011, and 32 articles (55%) published from 2012 to 2016.

52.3.1 Theoretical Framework

Among the articles applying specific theories, the most frequently examined theory was image repair theory (38%), followed by situational crisis communication theory (16%), framing theory (11%), excellence theory (10%), media system dependency theory (8%), uses and gratifications (5%), spiral of silence (5%), diffusion of innovation (3%), contingency theory of accommodation (2%), and others (2%).

With regard to the use of research questions (RQs) or hypotheses (Hs), articles published in SSCI and CSSCI journals exhibited significant differences ($X^2 = 28.55$, $df = 3$, $p < 0.001$). For articles in the SSCI, the use of RQs or Hs increased from 2006 to 2016. As a whole, "RQ" (seven articles, 58%) was the most prevalent one, followed by "neither RQ nor H" (four articles, 34%), and "RQ and H" (one article, 8%). For those indexed in CSSCI journals, only 4 articles (9%) applied RQ; the other 42 articles (91%) applied neither RQ nor H.

52.3.2 Methodological Preferences

Data showed that articles indexed in SSCI and CSSCI journals differ significantly in research method ($X^2 = 16.60$, $df = 2$, $p < 0.001$) and data sources ($X^2 = 45.68$, $df = 14$, $p < 0.001$). Among the 12 articles in SSCI journals, quantitative research was relatively more prominent (10 articles, 83%), and quantitative content analysis (32%) was the dominant method. As for data sources, the most widely used sources were traditional media such as print newspapers (six articles, 50%). In contrast, the qualitative method (39 articles, 85%) and qualitative content analysis (37 articles, 81%) were mostly used, and archival data served as the major source (30 articles, 65%) among the 46 articles indexed in CSSCI.

52.3.3 Types of Research

Types of research discussed the research focus and crisis time. Research focus refers to the main subject of specific research. In the dataset, articles in SSCI and CSSCI journals exhibited significant difference on research focus ($X^2 = 22.22$, $df = 7$, $p < 0.01$). The 12 articles in SSCI showed great interests in the "public" and widely discussed public motivation of using online media, public emotions, engagement, identification, and generated online contents in crises (Choi and Lin 2012; Tai and Sun 2007), while the articles in CSSCI journals emphasized on the evaluation of crisis situation (43 articles, 93%), and few of them discussed public reactions (3 articles, 7%). Among the total 58 articles, the widely discussed crises included SARS (2003), the melamine-contaminated milk powder of Sanlu (2008), flu pandemic (2009), China's Red Cross credibility crisis (2011), and Japan's nuclear crisis (2013).

52.4 Forms of Crisis Communication Practice

This chapter also discussed the forms of crisis communication practice, including crisis communication strategy (CCS), crisis communication effectiveness (CCE), and forms of response.

52.4.1 Crisis Communication Strategy (CCS)

For CCS, several clear features manifested. First, although the context studied was China, the CCS spectrum proposed by Western scholars (Benoit 1997; Coombs 2007) still dominated the social media and crisis communication research. Strategies such as modification, reduction of offensiveness, correction, accommodation, and denial were frequently utilized (e.g., Cho and Cameron 2006). Second, several new patterns of crisis communication strategies have been found in the Chinese context. Different with the recommended "two-way" symmetrical relationship in Western societies (Grunig 1992), organizations in China pertained the "enclosed control model," which described a dominant and asymmetrical relationship with stakeholders (He and Chen 2010, p. 21). Strategies such as face-saving, risk communication avoidance, deception, lying, and offering briberies were utilized to cover up crises, manipulate the public, and reduce negative media exposure. For instance, Tai and Sun (2007)'s study disclosed the highly controlled information system in South China, where local officials lied and covered up the information on the SARS epidemic when this crisis erupted at the beginning. Veil and Yang (2012) found that in a corporate context, the covering up strategies were also applied: instead of admitting the quality problem of products, the company manipulated its relationship with the local government and media (i.e., China largest search engine Baidu and popular forums such as Sohu and Sina) to reduce negative publicity. Thus, the new strategies such as covering up and manipulation were forming an asymmetrical pattern of crisis communication strategies (Cheng 2016a). Hundreds of dairy farms or companies, however continually announced bankruptcy due to the crisis in the year of 2012 (Veil and Yang 2012).

Third, the social media and crisis communication research in China has realized the empowering function of social media (He and Chen 2010). A transparent and symmetrical online crisis communication strategy was suggested (Cheng, 2016a; Cheng and Cameron 2017). Scholars advised that organizations could apply multi-functions of social media to monitor stakeholder generated contents, pay attention to their desired strategies, and cultivate relationships with online opinion leaders (Cheng and Cameron 2017; Choi and Lin 2012; He and Chen 2010; Veil and Yang 2012)

52.4.2 Forms of Response

Three traditional forms of response (i.e., timely, consistent, and proactive) (Huang and Su 2009) were frequently applied in the social media and crisis communication research. Furthermore, a new type of form – interactive – was intensively discussed in 21 articles (36%). Scholars suggested that organizations should have "interactive" dialogues with key publics online to improve the effectiveness of crisis communication and utilize the social media tools for interpersonal communication and emotional support (Cheng 2016a; Cheng and Cameron 2017; Gilpin 2010; Lev-On 2011).

52.4.3 Crisis Communication Effectiveness (CCE)

It was found that media publicity was most frequently used as a measurement of CCE (50%), followed by revenue reputation (38%), reputation (31%), cost reduction (23%), and organizational-public relationship (15%).

Meanwhile, new measurement of CCE on social media emerged: on the one hand, new items such as numbers of visitors, followers, and subscribers or attributes of comments/posts were added to measure the online media publicity (e.g., Wang 2012). On the other hand, the measurement of economic value involved reduction of negative public emotion (e.g., anger, confusion, fear, and sadness), increased account acceptance, public awareness and engagement (Cheng and Cameron, 2017)

52.5 The Chinese Context and Crisis Communication Practice

This chapter explored how the unique characteristics of cultural traits and institutional systems in China affected the social media and crisis communication practice. The major Chinese contextual factors include the following four dimensions: collectivistic cultures, centralized political system, transitional media landscape, and a high level of public dependency online.

First, cross-culture scholars emphasized the differences between people who live in individualistic and collectivistic cultures (Hofstede 2001). People from individualistic countries, such as the U.S.A. tend to ask for freedom and rights, prefer dominating and integrating styles, and keep short-term relationships. In contrast, people from collectivistic countries such as China emphasize on authority and order; prefer avoiding, obliging, and compromising styles; and embrace long-term relationships (Ting-Toomey 2005). The social media and crisis communication research showed that the collectivistic culture characterized the unique Chinese crisis communication practice. One the one hand, the value of "saving face" and "the ugly things in family shall not go public" resulted in the asymmetrical CCS, leading to certain unethical and unprofessional practices such as bribing officials and manipulating news reports (e.g., Veil and Yang 2012). On the other hand, "relationships" in China were particularly important in facilitating the online crisis communication. Scholars found that if no relationships established among organizations, bloggers, and followers, no information would flow on the Chinese social media (Hu 2010).

Second, contrasting to the American political system, which promotes the democracy, human rights, and free flow of capital in the public diplomacy (Thussu 2006), current Chinese political system reflects its Soviet or Leninist origins (Oksenberg 2001). Enormous power resides in the Chinese Communist Party (CCP), whose power and influence are transmitted by local governments and reinforced through the control of education, media, and military systems. This centralized political system impedes a timely, proactive, transparent, and interactive crisis response form on social media. For instance, in the SARS crisis, local officials instead of responding the public timely concealed the real number of infectors until receiving instructions from the central government (Hong 2007). In the melamine-contaminated milk powder crisis, the first action of the accused state-owned corporation, Sanlu, was to hide the truth, rather than actively giving the public a timely and accurate response (Veil and Yang 2012).

Third, in Western societies such as the USA, media system belongs to the liberal model, which enjoys press freedom and enough autonomy, relies on marketization, and acts as a societal watchdog (Hallin and Mancini 2004). The functions of media in crises focus on the scrutiny and information providing. Previous literature discussed the way that online media served as the key information platform in natural crises (Greer and Moreland 2003; Murphree et al. 2009). However, in China, the transitional media system is facing a dilemma in taking responsibilities in crisis communication. On the one hand, the media, especially "state-owned media" under the central governmental control, have to promote and legitimate policies of the CCP. Both online and offline media can hardly escape from the censorship of the CCP. On the other hand, contemporary Chinese media system is in transition and moving toward the marketization and globalization. It is expected that the media could follow international professional standards and ethics to serve public interests in crisis (Xue and Li 2010). In the future, social media may act as a platform for rapid information dissemination and promotion of a transparent media environment in China (Tai and Sun 2007).

Last but not the least, media dependency relations vary in contexts and influence the crisis communication effectiveness (Tai and Sun 2007). In the USA, information transmitted by television messages is more trustworthy than the same message transmitted via social media, which suggest that traditional media exert a stronger influence on public communication than new media in crises (Jin and Liu 2010; Utz et al. 2013), while in China, individuals depend on the social media to a much higher extent than traditional media (Lyu 2012). People rely on social media as a depoliticized and decentralized online communication environment to seek all kinds of information, including rumors and gossips as well (Yin and Liu 2012). Under this circumstance, a rumor of shortage of dairy suppliers can easily stimulate social panic and irrational behavior of the public. For example, in the rush of salts crisis in 2013, despite the government informed that salts were useless in reducing the effects of radiation after Japan nuclear crisis, people still believed this rumor and spread the information quickly online. Crowded people rushed to the store to purchase salts, which led to an immediate out of stock of salts and an intense social panic (Yu et al. 2011). A similar situation of rush for the Radix isatidis (*Ban Langen*) also appeared in the SARS crisis in 2003.

Two major reasons below may help explain the high level of online engagement and the crises emerged on Chinese social media. First, since the traditional media is highly controlled and international social media is blocked, domestic social media (e.g., Renren and Weibo) emerged and provided active "live" reporting tools. Empowered by the high sharing and re-tweet rate with targeted dissemination via social media, large numbers of Chinese Internet users enjoyed the strong personal connection, massive discussion, and interaction in cyberspace, which created the possibilities for social movements (Tai and Sun 2007). In the post-Mao era, Chinese netizens have demonstrated their ability to accelerate the speed of information dissemination and act against the power of government authorities or corporate institutions (Wang 2008). The Wukan incident in 2012 served as a good example illustrating how Chinese villagers utilized new media tools to spread messages, solicit global supports, and act against the local government (Cheng 2016b). Second, the low level of trust toward the whole society may explain the high level of public dependency online. Although Chinese economy developed so rapidly, the level of social trust reached the lowest among recent 5 years (Wang and Yang 2012). Publics hardly trust the governments and traditional media system (Cheng, Huang and Chan 2017). Instead, they have to rely heavily on the Internet to acquire information and relieve stressed emotions and opinions.

52.6 Future Directions

In summary, both the SCCI and CSSCI journal articles showed an increasing attention to the social media and crisis communication research over the past 10 years. The literature in China has realized the empowering function of social media, and current research focus began to shift from the "organization" to the "public." Topics such as public motivation for online media uses, public emotions, or engagement were frequently discussed (Cho and Cameron 2006; Tai and Sun 2007). Scholars also identified the unique unsymmetrical crisis communication practice in China, and an accurate, transparent, and interactive response form was suggested to accommodate the strong online public agenda (He and Chen 2010).

Based on the review of relevant articles, this chapter also found distinct weaknesses in current social media and crisis communication research in China, including a lack of emphasizing on the crisis phases, narrow application of theories and models, the neglect of a comparative logic in discussion, and the different research standards between journal articles listed in SSCI and CSSCI. The following section offers critiques and suggestions for how scholars should move forward in future social media and crisis communication research.

52.6.1 To Emphasize the Crisis Phases

Previous research showed within a certain "time," organizations' monitoring and application of strategies could effectively prevent a look-like crisis (Taylor and Kent 2007). In the social media and crisis communication research, scholars found that the length of time within which organizations were expected to react has been shortened due to the rapid timeframe of social media (Cheng, 2016a; Gilpin, 2010). Given this, scholars recommended an immediate response way (Muralidharan et al. 2011; Veil et al. 2011). Specifically, an exact response time was proposed. Wang et al. (2012) suggested the response of organizations in China within 48 h was too late for effective crisis communication. Based on the analysis of two rumor storms in China, Sun (2012) found storms usually budded within 48 h and went into the climax from the third to the fifth day. Thus, immediate response within 48 h could effectively terminate rumors and prevent crises.

Depending on the inconsistence of the abovementioned findings and only 9 out of 58 articles (20%) which discussed "time" or stages in crises, it is strongly suggested that future studies could emphasize the variable of "time" in the conceptualization of research design. A longitudinal study could be used to track changes from all involved parties over time and help to identify the critical response time in each stage of crises (Cheng 2016a; Cheng and Cameron 2017).

52.6.2 To Extend Theories and Models

In the social media and crisis communication research, interesting phenomena on new media effects were observed. He and Chen (2010) suggested a reversed agenda setting in online media crises of China, during which the media agenda may lose the power of setting the public agenda on the Internet. Tian (2011) proposed a changing "spiral of silence" in the generation of rumors. However, most studies still adopted public relation theories such as the image repair theory and excellence theory. Little empirical research could challenge traditional media theories such as agenda-setting/building theories and establish new models to extend the area of crisis communication. As the new technology was changing the way of communication dramatically, the creation of theories or models became necessary in future research. Below several research questions were listed for the next waves of study to rethink traditional media effects: How does interpersonal communication such as the word-of-mouth communication affect public responses in crises? How does the interactivity of social media mediate the relationship between crisis communication strategies and outcomes? How do the online media and traditional media contents intercorrelate in crises? In what conditions, the public agenda may lead the media or policy agenda in crises?

52.6.3 To Consider a Comparative Logic

As any crisis with a certain type, stage or issue must happen within context(s), where a specific legal, media, political, or cultural system exists. Scholars in current social media and crisis communication research already realized the importance of context (s) and focused their studies on China areas (e.g., Veil and Yang 2012; Wang 2012). For example, findings showed that the unique cultural and institutional characteristics of China affect the social media and crisis communication practice and formed an asymmetrical and dominated crisis communication strategy (He and Chen 2010; Veil and Yang 2012). However, few of them applied a comparative logic in discussing multiple contexts within one study, and a cross-contextual empirical study remained lacking. Future research could consider a comparative logic and test the posited hypothesis: in the process of social-mediated crisis communication, the more likely the context has individualistic cultures and democratic political, legal, and media systems, the more frequent the symmetrical strategy and proactive, consistent, and timely response forms could be adopted by organizations.

52.6.4 To Adopt a Uniformed Research Standard

By comparing articles published in the SSCI and CSSCI journals, it was not difficult to find that articles (21%) published in the SSCI journals usually followed the same research standard by reviewing relevant literature, generating research questions or hypotheses, applying the sampling method, and conducting data analysis to make the conclusion, while articles (79%) in the CSSCI journals did not reach the same level of research standard as those in the SSCI journals. These articles written in Chinese presented interesting discussions about the updated social events such as the China's Red Cross credibility crisis but seldom specified the research question or hypothesis, and most of them drew the conclusions based on the subjective judgment of archival data or literature. For example, although He and Chen (2010) proposed a reversed agenda-setting model in the crisis of China, this failed to adopt systematic quantitative methods applied in the 1968 Chapel Hill study to test the relationship between the media and public agenda. Thus, without following the same research standard as other published agenda-setting articles in SSCI journals, it is difficult to draw the conclusion that the reversed agenda-setting effects significantly exist in the Chinese crisis context. As social media is becoming an emerging research area in the field of crisis communication, future scholars may adopt a uniformed research standard and produce more innovative empirical studies on the social media and crisis communication in China.

References

Benoit W (1997) Image repair discourse and crisis communication. Public Relat Rev 23(2): 177–186

Bruns A (2005) Gatewatching: collaborative online news production. Peter Lane, New York

Burson-Marsteller, PSB (Penn Schoen Berland) (2011) Reputation in the cloud era digital crisis communications study. Retrieved from http://www.slideshare.net/bmChina/bursonmarsteller-digital-crisis-communications-study

Cheng Y (2016a) How social media is changing crisis communication strategies: evidence from the updated literature. J Conting Crisis Manag. https://doi.org/10.1111/1468-5973.12130

Cheng Y (2016b). Activism in Wukan Incident: Power and confrontation strategies at a Chinese village. China Media Research 12(2):90–104

Cheng Y, Cameron G (2017). The status of social mediated crisis communication (SMCC) research: An analysis of published articles in 2002–2014. In L. Austin & Y. Jin (eds.), Social Media and Crisis Communication (pp. 9–20). New York, NY: Routledge

Cheng Y, Huang YH, Chan, CM (2017). Public relations, media coverage, and public opinion in contemporary China: Testing agenda building theory in a social mediated crisis, Telematics and Informatics 34(3):765–773. https://doi.org/10.1016/j.tele.2016.05.012

Cho S, Cameron GT (2006) Public nudity on cell phones: managing conflict in crisis situations. Public Relat Rev 32(2):199–201

Choi Y, Lin Y (2012) Consumer responses to Mattel product recalls posted on online bulletin boards: exploring two types of emotion. J Public Relat Res 21(2):198–207

Coombs WT (2007) Protecting organization reputations during a crisis: the development and application of situational crisis communication theory. Corp Reput Rev 10(3):163–176

Gilpin D (2010) Organizational image construction in a fragmented online media environment. J Public Relat Res 22(3):265–287

Greer CF, Moreland KD (2003) United Airlines' and American Airlines' online crisis communication following the September 11 terrorist attacks. Public Relat Rev 29(4):427–441

Grunig JE (1992) Excellence in public relations and communication management. Lawrence Erlbaum Associates, Inc, Hillsdale

Hallin D, Mancini P (2004) Comparing media systems: three models of media and politics. Cambridge University Press, Cambridge

He Z, Chen XH (2010) Dual discourse context: the interaction model of China official disclosure and unofficial disclosure within crisis communication. J Int Commun 2010(8):21–27 (In Chinese).

Hearit KM (1999) Newsgroups, activist publics, and corporate apologia: the case of Intel and its Pentium chip. Public Relat Rev 25:291–308

Hofstede G (2001) Culture's consequences: comparing values, behaviors, institutions, and organizations across nations, 2nd edn. Sage, Thousand Oaks

Hong T (2007) Information control in time of crisis: the framing of SARS in China-based newspapers and internet sources. Cyberpsychol Behav 10(5):696–699

Howe J (2008) Crowdsourcing: why the power of the crowd is driving the future of business. Crown Business, New York

Hu BJ (2010) New media, meta-discourse of public relations and moral heritage. J Int Commun 2010(8):15–20 (In Chinese).

Huang YH, Lin YX (2004) Construction of Internet crisis management model: analysis of relations between situation, Internet communications, and responses from organizations. Paper presented in the Chinese communication society annual conferences, Macao (In Chinese)
Huang Y-H, Su S-H (2009) Determinants of consistent, timely, and active responses in corporate crises. Public Relat Rev 35:7–17
Huang YH, Lin YH, Su SH (2005) Crisis communicative strategies: category, continuum, and cultural implication in Taiwan. Public Relat Rev 31(2):229–238
Hwang K (1987) Face and favor. The Chinese power game. Am J Sociol 92(4):944–974
Jiang M (2013) The business and politics of search engines: a comparative study of Baidu and Google's search results of internet events in China. New Media Soc 0(0):1–22
Jin Y, Liu BF (2010) The blog-mediated crisis communication model: recommendations for responding to influential external blogs. J Public Relat Res 22(4):429–455
Lev-On A (2011) Communication, community, crisis: mapping uses and gratifications in the contemporary media environment. New Media Soc 14(1):98–116
Liu BF, Jin Y, Briones RL, Kuch B (2012) Managing turbulence online: evaluating the blog-mediated crisis communication model with the American red Cross. J Public Relat Res 24:353–370
Lyu JC (2012) How young Chinese depend on the media during public health crises? A comparative perspective. Public Relat Rev 38(5):799–806
Millward S (2013) With 600 million social media users, this is China's Web in 2013. Retrieved from http://www.techinChina.com/social-media-and-social-marketing-china-stats-2013/
Moody M (2011) Jon and Kate plus 8: a case study of social media and image repair tactics. Public Relat Rev 37(4):405–414
Muralidharan S, Dillistone K, Shin JH (2011) The gulf coast oil spill: extending the theory of image restoration discourse to the realm of social media and beyond petroleum. Public Relat Rev 37(3): 226–232
Murphree V, Reber BH, Blevens F (2009) Superhero, instructor, optimist: FEMA and the frames of disaster in hurricanes Katrina and Rita. J Public Relat Res 21(3):273–294
Oksenberg M (2001) China's political system: challenges of the twenty-first century. China J 45:21–35
Tai Z, Sun T (2007) Media dependencies in a changing media environment: the case of the 2003 SARS epidemic in China. New Media Soc 9(6):987–1009
Taylor M, Kent ML (2007) Taxonomy of mediated crisis responses. Public Relat Rev 33(2): 140–146
Taylor M, Perry DC (2005) Diffusion of traditional and new media tactics in crisis communication. Public Relat Rev 31:209–217
Thussu D (2006) International communication: continuity and change, 2nd edn. Hodder, London
Tian XL (2011) The dissemination and digestion of "noise" in emergent public affairs: a case study of the crisis of nuclear leakage triggered by violent earthquake in Japan. Contemp Comm 2011 (3):47–49 (In Chinese).
Ting-Toomey S (2005) The matrix of face: an updated face-negotiation theory. In: Gudykunst W (ed) Theorizing about intercultural communication. Sage, Thousand Oaks, pp 71–92
Utz S, Schultz F, Glocka S (2013) Crisis communication online: how medium, crisis type and emotions affected public reactions in the Fukushima Daiichi nuclear disaster. Public Relat Rev 39(1):40–46
Veil SR, Yang A (2012) Media manipulation in the Sanlu milk contamination crisis. Public Relat Rev 38(5):935–937
Veil SR, Sellnow TL, Petrun EL (2011) Hoaxes and the paradoxical challenges of restoring legitimacy: dominos' response to its YouTube crisis. Manag Commun Q 26(2):322–345
Wang SG (2008) Changing models of China's policy agenda setting. Mod China 34:56–87
Wang X (2012) Combating negative blog posts and a negative incident: a case study of the "Mayday" incident between Juneyao airlines and Qatar airways. Public Relat Rev 38(5): 792–795
Wang JX, Yang YY (2012) Bluebook of social mentality. Social Sciences Academic Press, Beijing

Wang ZY, Chen X, Wang YR (2012) Role-play of online media in image crisis cycle. Contem Commun 2012(1):77–79 (In Chinese).
Wu YC, Yeh MH (2012) Dialogue theory of public relations and internet crisis communication-an exploratory study. Commun Soc 22:99–134
Xue G, Li Z (2010) Keeping balanced between the country and the society: the identity crisis of Chinese media. Mod Commun 9:11–15 (In Chinese)
Ye L, Ki E-J (2012) The status of online public relations research: an analysis of published articles in 1992–2009. J Public Relat Res 24(5):409–434
Yin YG, Liu RS (2012) Annual report on development of new media in China. Social Sciences Academic Press, Beijing (In Chinese)
Yu GM, Li B, Wu WX, Song MJ, Liu JY (2011). The trends of public opinion and patterns of communication in the rush of salts: Based on the intelligent analysis of network text, Journal of International Communication, 2011(7):28–39 (In Chinese)

Diffusion and Adoption of an E-Society: The Myths and Politics of ICT for the Poor in India 53

Ravindra Kumar Vemula

Contents

Abstract

The label "ICT for the poor" has been widely used in India whenever ICT moves out of urban settings. There have been many debates on the "digital divide" and reluctance on the part of the masses to adopt the "remarkable changes" being brought out by the ICT boom. In the last 15 years, many initiatives have been undertaken in the form of "information kiosks" to diffuse ICT on the pretext of making the poor e-literate and build an e-society ultimately. Most of these projects have been funded by international multilateral and bilateral organizations and have also got awarded by various reputed international and national bodies for "reaching out" to the masses. It has been observed that most of these projects start with lots of fanfare promising a "leap frogging" to an information society,

Hamelink's (1997) definition of ICTs: "Information and Communication Technologies (ICTs) encompass all those technologies that enable the handling of information and facilitate different forms of communication among human actors, between human beings and electronic systems, and among electronic systems" (Hamelink 1997: 3). This functional definition of ICTs includes both the new (i.e., Internet, e-mail) and traditional (i.e., community radio, TV) forms of ICT into its definition.

R. K. Vemula (✉)
Deptartment of Journalism and Mass Communication, The English and Foreign Languages University, Shillong, Meghalaya, India
e-mail: ravi@efluniversity.ac.in

J. Servaes (ed.), *Handbook of Communication for Development and Social Change*,
https://doi.org/10.1007/978-981-15-2014-3_53

but they falter somewhere down the line for various reasons. Though, masses approach the kiosks for land records, birth or death certificate, or any document that needs to be obtained from the Government. As a result of which, the information kiosks, after a while, are no more used by the masses, because their "temporary need" for an e-service has been fulfilled. Ultimately, all these information kiosks which have been put up by the Government, nongovernment organizations, and other philanthropic bodies end up as training centers on software/hardware for the local village youth at a price. Later on, the kiosk sustainability is solely based on the revenue that is generated by the e-courses that it offers and on the other allied services like printouts, serving as a public phone booth or may be as a cool drink center. This paper attempts to understand whether ICT is a boon or bane for India It also tries to understand how the poor is defined by the information kiosks, or are they being only catered to a particular class of people who have proximity to the kiosk operator and how comfortable they are in utilizing the services of the kiosk. It also argues that in the pretext of creating an "information society," the state has somewhere missed out on other needs of the poor, which are required for a holistic development of the society and an e-society ultimately. This article aims to understand if the information and communications technologies (ICTs) really empower poor communities and does it actually bring in the change that is anticipated of it. This article investigates this question, focusing on the role of information and communications technologies in diffusion and adoption of e-society in India through the ICT for the poor programs in the last few decades. The framework attempts to contrast with the global discourse around the "digital divide" and the holistic human development of the poor brought in by the ICT revolution.

Keywords

ICT for the poor · Diffusion · Kiosks · Village information center · Digital divide

53.1 Introduction

The advocates of ICTs (World Bank 2000; UNDP 2001; Pohjola 2002; Braga 1998) take an optimistic view and highlight the positive effects ICTs to create new economic, social, and political order for developing countries and especially for the poor, whereas the critics take a pessimistic view and claim that ICTs due to existing socioeconomic inequalities and structural factors will only favor the privileged segments within the society and not reach the economically and socially disadvantaged, thus, leading to a further widening of the gap. The impact of ICTs on poor communities cannot be understood without first understanding the role information and knowledge play for development. In the ICT *for* development debate, the emphasis is placed on providing access to ICTs to the poor before analyzing the value information and knowledge exchanges play for development at the local level (Black 1999; Mansell 1999; Norris 2001). Digital divide is actually has more to do than the adoption of technology. It is more of a social divide, an economic divide, a cultural divide, a political divide, and of course a "technological divide." Technological gap can always be bridged easily because it is market driven. The technological companies will jump at any given opportunity to intervene in the ICT market in the developing countries. However, hypothesizing on the fact that access to ICTs, computers, and Internet will lead to sustainable social development is far from being the answer. Gumucio Dagron (2003) critically approaches the development paradigms which developing countries have gone through around the world. He clearly states that technology alone is not sufficient, but states that information and knowledge are also important to help the rural population to improve their living conditions. A number of development organizations around the world have begun to understand that information and communication are not the same thing. Information alone does not generate changes, whereas communication – which implies participation, sharing of knowledge in a horizontal way, and respect for diversity and culture – is key to social change. Unfortunately, too many development programs today are still basing their approach on the diffusion of innovations theories of the 1960s, often mocking participatory approaches, but seldom really involving communities in the decision-making process, because it clashes with institutional agendas and the "annual report" syndrome.

The very first broad category of development's primary objective is to better the lives of the poor. The Millennium Development Goals (United Nations 2009) strongly focuses on targets to end poverty and hunger and to improve health and education for the poor. ICT is seen as a key prerequisite to improved healthcare delivery and the better assessment of development programs (Braa et al. 2004). Telecenters have been widely applied with the aim of bridging the digital divide for the poor (Reilly and Gómez 2001), providing them with access to information and better freedom of choice. Digital divide is not solely a matter of technology but also of the social, political, institutional and cultural contexts that shape people's access and use of ICTs (Warschauer 2003).

53.2 ICT for the Poor Initiatives in India and Learnings: Understanding the Myths and Politics

India from the past few decades has become a test-bed for innovations in information and communications technology (ICT), serving the rural people through village information center (VICs) or kiosks. Various reasons explain the emergence of VIC or kiosk. The most obvious reason may be rural area has remained undeveloped in terms of basic infrastructure and basic facility. Another reason is providing a bunch of services through a kiosk is seen as a very easy process. Rural India is the largest underserved area for banking, microfinance, e-commerce, insurance, medical services, and e-governance service. Therefore, a kiosk can work as a bridge between the outer world and a village. Rural kiosks are one manifestation of various attempts to apply information and communications technology for socio-economic development. In many cases, a kiosk can be thought of as Internet cafes for rural villages, with one or more connected PCs available for shared use by village residents. They differ greatly from urban Internet cafes, however, in that the operational challenges and user needs of remote rural villages are appreciably different from that of cities. Rural kiosks typically offer a broad range of services and applications specialized for rural areas. Kiosks are initiated by various kiosk project agencies, which identify one or more people per rural village to act as a kiosk operator. In many cases, the kiosk operator is also the kiosk owner, in which case the agency takes on a franchise model, with operators as franchisees or rural entrepreneurs. By some estimates, there are more than 150 rural kiosk projects across India, some of which already have, or are planning, thousands of installments (According to I4D Magazine, 2005). Most of these projects were started within the last 5 years. Reflecting the nation's diversity, these initiatives differ in goals, models, operating paradigms, and geographic distribution. Every sector is involved – large enterprise, entrepreneurs, government, and NGOs – with motives ranging from turning a commercial profit to driving socioeconomic growth to streamlining government bureaucracy. Early evidence indicates that rural kiosks can help villagers improve their economic standard of living by expanding livelihood options and empowering them with information, tools, goods, and services (such as education and healthcare). The true challenge is in finding ways to deliver this benefit broadly and consistently while making kiosk projects economically sustainable in the long term. The experiences of these noncommercial projects suggest that the information needs of a community should be thoroughly assessed before the launch of a project. It suggests finding out motivated and skilled grassroots intermediaries as a necessary condition for any project to succeed in bringing e-government to rural communities. Lack of awareness among the community is a major learning from the ICT for the poor initiatives in India. Either the people are completely unaware of the kiosk in their village, or they would not know about the services it offers. Many of the kiosks were put up in the homes of the operators, thus, making it difficult for the people to access it. The information center operator needs to understand the basic technology, such as how to navigate the Internet or to maximize the use of the available tools, and marketing. In most of the kiosks, the kiosk operator does have the details of the services offered. And they are lacking in basic operation of the kiosk. Information and communications technology (ICT) can reduce poverty by improving poor people's access to education, health, government, and financial services. ICT can also help small farmers and artisans by connecting them to markets. It is clear that in the rural area, realization of this potential is not guaranteed. Low-cost access to information infrastructure is a necessary prerequisite for the successful use of ICT by the poor, but it is not sufficient. The implementation of ICT projects needs to be performed by organizations and individuals who have the appropriate incentives to work with marginalized groups. These intermediaries are best suited to promote local ownership and poor people's participation. People also continuously face the languages barriers while using the Internet as it is predominantly in English. Though, off late other language interface is developing but it's far from achieving the desired impact because they lack the local language products. It was also observed that many people also lacked the motivation to use information over the Internet.

Information communication technology is to bridge the digital divide (i.e., disparity between digital have and have-nots according to their geographical location or demographical groups) and aid economic development by ensuring equitable access to up-to-date communication technologies. Information and communications technologies played a significant role in promoting social and economic development that includes improvement of individual livelihoods, community prosperity, and the achievement of national development goals related to the UN Millennium Development Goals (David et al. 2005). National ICT strategies and the programs of international donors are incorporating ICT components on this basis, with specific objectives in reaching poor rural and semi-urban as well as urban communities. The earliest of the focus by India was on the use of IT (not ICT) in the mid-1950s to late 1990s. It was predominantly used in the government and private sector organization. The Millennium Development Goals coupled with the rapid rise and spread of Internet in the late 1990s to late 2000s led to a rapid increase in investments in ICT infrastructure and projects in developing countries. The telecenter was used to bring information on development issues such as health, education, and agricultural extension into poor communities. Later, telecenters were also used to deliver government services. In the early 2000, the shift toward mobile phones and SMS came into vogue where there was less concern with e-readiness and more interest in the impact of ICTs on development. Additionally, the focus on the poor as producers and innovators with ICTs (as opposed to being consumers of ICT-based information) was more prevalent. The earlier approaches on ICT marginalized them, allowing a supply-driven focus, and ICT brought them into the core and operationalized by creating a demand-driven focus. The previous initiatives on ICT viewed the poor at the "bottom of the pyramid" concept and characterized them largely as passive consumers, whereas ICT4D viewed the poor as active producers and active innovators. Access to ICTs in the developing world has been framed through the concepts of digital divide and use/non-use. Market liberalization and competition as well as various regulatory and technical solutions are believed to be useful in closing the digital divide and ensuring the universal access to ICTs. The general perception is that people who have access to ICT will benefit from it and those who don't would not. Benefits include boundless information sharing, connectivity, and participation in the global economy. The use of mobile phones shows some positive effects in improving access to information and services. The ITC e-Choupal was an important initiative that aimed to provide computers and Internet access in rural areas across several agricultural regions of the country, where the farmers can directly negotiate the sale of their produce with ITC Limited. Online access enabled farmers to obtain information on mandi (market) prices, and good farming practices, and to place orders for agricultural inputs like seeds and fertilizers. This helped farmers improve the quality of their products and helps in obtaining a better price. ITC kiosk with Internet access run by a sanchalak – a trained farmer. The computer is housed in the sanchalak's house and is linked to the Internet via phone lines or by a VSAT connection. Each installation serves an average of 600 farmers in the surrounding ten villages within about a 5 km radius. The sanchalak bears some operating cost but in return earns a service fee for the e-transactions done through his e-Choupal. The warehouse hub is managed by the same traditional middlemen, now called samyojaks, but with no exploitative power due to the reorganization. These middlemen make up for the lack of infrastructure and fulfill critical jobs like cash disbursement, quantity aggregation, and transportation. Since the introduction of e-Choupal services, farmers have seen a rise in their income levels because of a rise in yields, improvement in quality of output, and a fall in transaction costs. Even small farmers have gained from the initiative. Farmers get real-time information despite their physical distance from the mandis. The system saves procurement costs for ITC. The farmers do not pay for the information and knowledge they get from e-Choupals; the principle is to inform, empower, and compete. market place for spot transactions and support services to futures exchange. There are 6100 e-Choupals in operation in 40,000 villages in 10 states, affecting around 4 million farmers. Gyandoot which means "purveyor of knowledge" in Hindi was a government-to-citizen, intranet-based service portal, implemented in the Dhar district of the state of Madhya Pradesh, India, in January 2000. The project was designed to extend the benefits of information technology to people in rural areas by directly linking the government and villagers through information kiosks. The kiosks provided access to a variety of government services, such as registration of complaints and submission of applications for the issuance of certificates and loans. Data on prices of agricultural crops in different markets are also available. Gyandoot pioneered the idea of rural telecenters in India. The project concept has been replicated by other information and communication technologies (ICT) development initiatives in India. Gyandoot was considered to be very successful in the early years of its implementation, and the project was awarded the Stockholm challenge information technology (IT) award in 2000 for public service and democracy. However, subsequent evaluations have reported diminishing levels of activity, placing in question on the long-term viability of the project. ICT for development has attempted many initiatives in different parts of India. This was an infiltration with or without the knowledge and consent of the people related. While some proved to be a success, others did not. This gap is yet to be studied in detail in the field of this venture.

53.3 Conclusion

ICTs under certain conditions can significantly enhance the human and social capabilities of the poor, thus empowering them at the individual and collective level. At the core of this empowerment process stands the notion that ICTs can enhance peoples' control over their own lives. Similarly, to literacy, newly acquired "informational capabilities" can act as an agent for change for individuals and communities enhancing their abilities to engage with the formal institutions in the economic, political, social, and cultural spheres of their life. In this context, the issue of whether ICTs are channeling resources away from the real priorities and needs of poor communities seems to be misguided. Instead this question should be rephrased and address the issue of how ICTs could be used to meet the "basic needs" of the poor. This however will require a shift in focus of ICT interventions to address such challenges as the fight against HIV and AIDS, helping to avoid famines and their support in the mediation of conflicts. At the same time, the case studies have demonstrated that due to the existing "hype" around the potential benefits of ICTs, the high expectations of poor communities cannot be met. The experience shows that ICTs are only able to address certain aspects of the development challenges facing poor people and that in fact they are not able to change the existing structural, social, political, and economic inequality. For instance, while ICTs can act as an effective tool in improving the access of small-scale farmers to market price information, they are not able to address the underlying structural market inequalities between small-scale farmers and agrobusinesses. Furthermore, the paper has illustrated that there is not a direct and causal relationship between ICT and poverty reduction. This relationship is much more complex and indirect in nature, whereby the issue of its impact on the livelihoods of the poor depends to a large extent on the dynamic and iterative process between people and technology within a specific local, cultural, and sociopolitical context. Frequently, the most immediate and direct effect of ICT programs seems to be the psychological empowerment of poor people, whereby newly acquired ICT skills provide poor people with a sense of achievement and pride, thus strengthening their self-esteem. A key recommendation of the paper is that the human development of people, rather than technology itself, should be the center of the design and evaluation of ICT programs. As has been shown, the important advantage of using the "capability approach" as the basis for the evaluation of ICT programs is its emphasis on the ability of ICTs to improve the daily livelihoods of poor communities, in contrast to more conventional approaches which overemphasize the significance of technology itself for social change. Furthermore, evaluations of the impact of ICT programs should focus on an analysis from the vantage point of the poor, rather than from the perspective of outside donors. In addition, the analysis provides the following concrete recommendations on the manner in which ICTs programs should be designed in order to be most effective on facilitating the empowerment of marginalized groups.

References

Black J (1999) Information rich- information poor, bridging the digital divide. International Institute for Communication and Development. Available at: www.iicd.org. Accessed on 25 Apr 2018

Braa J, Monteiro E, Sahay S (2004) Networks of action: sustainable health information systems across developing countries. MIS Q 28(3):337–362

Braga CP (1998) Inclusion or exclusion, Information for Development (InfoDev). The World Bank. http://www.unesco.org/courier/1998_12/uk/dossier/txt21.htm. Accessed on 25 Apr 2018

Dagron AG, Bleck C, Gumucio Dagron A, Dagron AG (2001). Making waves. Rockefeller Foundation, New York.

David S, Nigel S, Christopher G, Jain R, Ophelia M (2005) The economic impact of telecommunications on rural livelihoods and poverty reduction: a study of rural communities in India (Gujarat), Mozambique and Tanzania (No. WP2005-11-04). Indian Institute of Management Ahmedabad, Research and Publication Department

Ellis F (2000) Rural Livelihoods and diversity in developing countries. Oxford University Press, Oxford

Ernberg J (1998) Universal access for rural development: from action to strategies. In: First International Conference on Rural Telecommunications, Washington, DC, 30 November–2 December

Escobar A (1995) Encountering development: the making and unmaking of the Third World. Princeton University Press, Princeton

Freire P (1972) Pedagogy of the Oppressed. Penguin Books, London

Gumucio Dagron A (2003) What can ICTs do for the rural poor. World Summit for the Information Society, Geneva

Hamelink CJ (1994) Trends in World Communication, on disempowerment and self-empowerment. Southbound, Third World Network, Penang Malaysia

Hamelink CJ (2002) Social development, information and knowledge: whatever happened to communication? Development 45(4):5–9

Ippmedia (2018) https://www.ippmedia.com/en/features/how-icts-are-accelerating-sdgs. Accessed on 26 Nov 2018

Mansell R (1999) Information and communication technologies for development: assessing the potential and the risks. Telecommun Policy 23:35–50

Norris P (2001) Digital divide: civic engagement, information poverty and the Internet worldwide. John F. Kennedy School of Government (KSG), Harvard University

Panos (1998) The Internet and poverty. Panos media briefing, no. 28. Panos Institute, London

Pigato M (2001) Information and communication technology, poverty, and development in sub-Saharan Africa and South Asia. World Bank, Washington, DC

Pohjola M (2002) The new economy: facts, impacts and policies. Inf Econ Policy 14:133

Reilly K, Gómez R (2001) Comparing approaches: telecentre evaluation experiences in Asia and Latin America. Electron J Inf Sys Dev Ctries 4(3):1–17

UNDP (2001) Human Development Report 2001, “Making new technologies work for human development”. UNDP, Oxford

Vijaybaskar M, Gayathri V (2003) ICT and Indian development: processes, prognoses, policies. Econ Polit Wkly 38:2360–2364

Wade R (2002) Bridging the digital divide: new route to development or new form of dependency? Glob Gov 8(4):443

Warschauer M (2003) Dissecting the ‘Digital Divide’: a case study of Egypt. Inf Soc 19(4):297–304

World Bank (2000) Global Information and Communication Technologies Department, The Networking Revolution, Opportunities and Challenges for Developing Countries

Online Activism in Politically Restricted Central Asia: A Comparative Review of Kazakhstan, Kyrgyzstan, and Tajikistan

54

Bahtiyar Kurambayev

Contents

Abstract

This comparative study examined the Internet's role and wider social media in Kazakhstan, Kyrgyzstan, and Tajikistan, three Central Asian countries in the democratization process. Specifically, this work aims to discuss how the Internet and social media are allowing Internet users wider opportunities to access and share information in a media-restricted region as well as collectively speak up in a restricted region of Central Asia. In general, Internet penetration is relatively low compared to other parts of the world. Still, the Internet has demonstrated its power in the region when presidents of Kyrgyzstan in 2005 and in 2010 were ousted. Both times, the Internet played the key role in facilitating such drastic change. While it is true that Central Asian countries have differently related policies and practices, varying from some freedom in Kyrgyzstan and total state control in Uzbekistan, it is also true that the region is experiencing an unprecedented boom in mobile phones, which brings the Internet to citizens.

B. Kurambayev (✉)
Department of Media and Communications, College of Social Sciences, KIMEP University, Almaty, Kazakhstan
e-mail: b.kurambayev@kimep.kz

J. Servaes (ed.), *Handbook of Communication for Development and Social Change*,
https://doi.org/10.1007/978-981-15-2014-3_99

Keywords

Online activism · Censorship · Surveillance · Central Asia · Democracy · Authoritarian government · Social media · Internet · Former Soviet Union

54.1 Introduction

A great deal of research has been conducted around the world about the role of the Internet in nondemocratic contexts. Some scholars now subscribe to the belief that the Internet has a positive relationship with democratization processes in authoritarian contexts and elsewhere. They argue that the Internet is associated with regime changes in places like Egypt, Tunisia, Jordan, Algeria, Ukraine, and elsewhere, where rulers of several decades were forced from power. In other places, new communication technology has been used extensively to achieve smaller-scale positive results, such as protest mobilization and collectively speaking up against the government in Tunisia (Breuer et al. 2014), awareness campaigns of electoral fraud in parliamentary elections in Russia (Reuter and Szakonyi 2015), and coordinating and organizing uprisings after 2009 elections in Iran (Wojcieszak and Smith 2014). Supporters of positive influence of the Internet argue that it would not be possible to accomplish such successful and fundamental changes in abovementioned restrictive and authoritarian locations without the Internet. What is striking about these waves of uprisings and mass protests is that the Internet, including YouTube, Twitter, Facebook, and bloggers, played an important role in communicating, coordinating, and channeling this rising tide of opposition and variously managed to bypass state-controlled national media as they propelled images and ideas of resistance and mass defiance (Cottle 2011) (Table 1).

Others have argued that the Internet has a limited influence and even negative relationship with democratization process. For example, Vanderhill (2015) argues that much of the existing literature that is critical of the democratization process relies on how authoritarian governments can control and manipulate the Internet to prevent challenges and maintain their rule (p. 32). Vanderhill (2015) offers three arguments why the Internet has limited influence on democratization in nondemocratic and authoritarian countries. First, information communication technology does not always succeed in organizing activities and protests against the regime because of the strength of the regime. Second, some authoritarians have coercive

Table 1 Internet Penetration Rate in Central Asian Countries

Country	Internet penetration rate (percentage of population)	Facebook users (June 30, 2017)
Kazakhstan	76.8%	1,500,000
Kyrgyzstan	34.5%	360,000
Tajikistan	20.5	84,000

Note: Internet penetration rate and number of Facebook users as of June 30, 2017. Source: Internet World Stats (2017)

capacity to resist democratization demands. Third, even if information and communication technology (ICT) helps facilitate mass mobilization, such recruitment is insufficient for full democratization in the long-term changes. Furthermore, there is a growing literature to suggest that governments may use social media for carefully coordinated messages, including for the spread of misinformation (Howard and Bradshaw 2017; Marwick and Lewis 2017).

Despite the plethora of studies around the world about the role of the Internet in democratization process, almost no research has been conducted in Central Asia. This is a region, according to Freedman and Shafer (2012), that scholars have barely touched the surface from an academic perspective: "There is no shortage of potential research topics pertaining to the press, journalism and mass communication in the region, given globalization and the rapid changes in communications technologies, such efforts would be particularly timely" (p. 124). So, this chapter reviews the extent to which the Internet has the potential to play a role in facilitating democratization process in former Soviet Union Central Asia, specifically in Kazakhstan, Kyrgyzstan, and Tajikistan. This chapter also reviews recent events and developments related to the Internet and social media in these abovementioned countries.

54.2 Context of Central Asia

To understand the relevance of this chapter, it is important to understand wider socioeconomic and political context of Central Asia. Kazakhstan, Kyrgyzstan, and Tajikistan are former Soviet Union countries. All three gained their independence in 1991 when the Soviet Union formally collapsed. Since independence, these countries have seen few changes in leadership. For example, Kazakhstan President Nursultan Nazarbayev has been in power since 1991. Tajikistan President Emomali Rahmon has been president of the country since 1992. The only exception in the region is Kyrgyzstan. The new constitution adopted in 2010 in this country bars the same individual from serving as president more than one term, with a term lasting 6 years. However, the first two presidents of the country, Askar Akayev (in 2005) and Kurmanbek Bakiyev (in 2010), both were ousted from office by people's revolutions. Overall, since independence, these countries have achieved various levels of political and economic successes.

Mass media outlets in all of these three countries are under government control. "News website blocking and Russian-inspired laws to curtail free expression are spreading, while journalists are often subject to torture and unlawful imprisonment" (Reporters Without Borders 2017) in reference to all Central Asian countries. Reporters Without Borders said authoritarian tendencies in Central Asia are on the rise because there is a growing fearful environment for independent journalists to work in the region. Bowe et al. (2012) argue that Central Asian republics remain bastions of official and extralegal censorship, self-censorship, constraints on journalists and news organizations, and insufficient financial resources to support independent, and sustainable, market-based press systems (p. 145). Respect to human rights in the region remains disregarded. Human Rights Watch documents various incidences of violations of basic human rights in the region. The organization's latest report from 2017 outlines some of them. For example, Kazakhstan restricts peace protests and jail those who dare to do so. Tajikistan has imprisoned more than 150 activists on politically motivated charges since the middle of 2015. "Under the pretext of protecting national security, Tajikistan's state telecommunications agency regularly blocks websites that carry information potentially critical of the government, including Facebook, Gmail, Radio Ozodi, the website of Radio Free Europe's Tajik service, news and opposition websites" (Human Rights Watch 2017). The Kyrgyz Republic is one of Central Asia's poorest countries. According to the Heritage Foundation's 2017 report of Index of Economic Freedom, corruption is pervasive in the country, and this is why judges are reported to pay bribes to attain their positions. Despite some anticorruption efforts, the country is trapped in a cycle in which predatory political elites use government resources to reward clients, including organized crime figures, and punish opponents (Freedom House 2017).

In this restrictive region of Central Asia, Kulikova and Perlmutter (2007) argue that the Internet seems to be the only way to be read and heard in Kyrgyzstan, but their argument can be applied to wider Central Asia where free expression is curtailed and all mass media outlets are under government control. Given this context, understanding how the Internet and wider social media may have played a role in the recent changes of the region takes on great urgency.

54.3 More Government Control over the Internet and Social Media

Kazakhstan is the 9th largest country by land mass in the world and the largest one in Central Asia. The country has a population of 18 million and is the largest economy in the region. Nursultan Nazarbayev has been the leader of the county since its independence in 1991 from the Soviet Union, and since then he has been elected multiple times. However, the Organization for Security and Co-operation in Europe (OSCE) reports that none of the presidential elections held in the country were noted fair and free in compliance with democratic standards. Human Rights Watch reports Kazakh authorities tolerate little public criticism of the government or its record. While the government of Kazakhstan has long restricted the right to freedom of association and the right to carry out peaceful dissent, in recent years the government has further tightened controls over trade unions and civil society groups (Human Rights Watch 2017). According to Reporters Without Borders, Kazakhstan ranks 157th out of 180 countries in the world in terms of press freedom. In Freedom House ranking, Kazakhstan is ranked among "not free" countries in *Freedom in the World 2017*, among "not free" countries in *Freedom of the Press 2016*. The same can be said for Kazakhstan's position in *Freedom on the Net 2016*. Kazakhstan receives a democracy score of 6.61, on a scale of 1 to 7, with 7 as the worst possible score, in *Nations in Transit 2016*. In other international organizations' evaluations, Kazakhstan is ranked among the countries of authoritarian regime (Economist's Democracy

Index 2017), earning 3.06 out of 10. The higher the score a country earns, the more democratic it is. In terms of government accountability, this Central Asian nation is ranked 134th among 176 countries in Transparency International Index in its latest report.

Kazakhstan does has some features to appear that it is moving toward democratization, such as multiparty parliament and the fact that almost 80 percent of all media outlets in the country are private (IREX 2017), but in reality most of these media outlets are affiliated with the government or officials. Kazakhstan has in recent years moved further in the direction of restricting press freedom by repressive methods of limiting access to information in the country, and such methods are continuing (IREX 2017; Emrich et al. 2013). The IREX report describes some limitations for journalists, including significantly delayed responses, irrelevant answers, and statements that a question is within the competence of the responder. Some independent news outlets continue operating in the country, but they routinely face political pressure and state interference (Emrich et al. 2013). Emrich and his colleagues note that the country is characterized by lack of pluralism and prevalence of pro-government outlets, especially among the broadcast media, which often are either directly owned by the state or by highly loyal government officials or businesses affiliated with them. In this context, journalists and editors self-censor themselves, fearing administrative, civil, and criminal prosecutions, and IREX (2017) reports that media outlets have blacklists of persons unsuitable for airing and a list of forbidden subjects and names. Those who try to do independent journalism face many troubles. For example, Kazakh journalism and press freedom defender Ramazan Yesergepov was stabbed in May 2017 (Committee to Protect Journalists 2017a). Yesergepov had scarcely begun the journey of roughly 1200 kilometers (745 miles) north from Almaty to Astana to discuss threats to media freedom with foreign diplomats when he was stabbed. Zhanbolat Mamay, editor of the independent newspaper *Sayasi kalam/Tribuna*, was arrested in 2017 after allegedly receiving illegal funds. Overall, Freedom House noted that authorities in Kazakhstan continued to arrest and prosecute journalists and social media users on a range of criminal charges in 2017 restrict freedom of speech freedom of speech.

Niyazbekov (2017) notes that Arab Spring protests in 2011–2012 and similar developments elsewhere signaled to post-communist dictators to treat social media networks with caution. He cites the case of Uzbekistan when former President Islam Karimov banned social media in early 2010. He also noted that Kazakhstan did not pay much attention to controlling social media, assuming developments like the Arab Spring would never reach Kazakhstan's border. Niyazbekov also notes that the Kazakh government was wrong. An example he and many other scholars discussed is the mass protests in the city of Zhanaozen. This is reported to be the worst civic conflict in the post-Soviet history in Kazakhstan and social media played a role in it. In December 2011 during the celebration of the country's 20th anniversary of Kazakhstan's independence, oil workers clashed with the state police. These oil workers were on strike since May 2011 following disputes over pay and working conditions (Beisembayeva et al. 2013; Achilov 2016). These oil workers demanded higher salaries and better working conditions. Beishembayeva and her colleagues wrote that 16 people were killed and 100 were injured during the uprising. These authors note that YouTube videos demonstrated that police fired directly into the large crowd, prompting President Nursultan Nazarbayev to impose a state of emergency in Zhanaozen. International outlets placed casualties at 73 (Niyazbekov 2017), and all forms of communication with the rest of the country were cut off, including mobile phone and Internet services. The government cut all forms of communication with the outside with the intention of preventing the uprising from spreading elsewhere in the country. While authorities claimed that police were there to defend themselves, eyewitnesses reportedly observed the security forces opening fire indiscriminately at unarmed protestors (BBC, December 16 2011). The significance of this case is that protesters used social media (Facebook, Twitter, etc.) to mobilize resources, attract popular support, and appeal to foreign governments and international organizations (Niyazbekov 2017). Niyazbekov also notes that the Zhanaozen massacre resulted in the government curbing press freedom and heightening control over the public. In September 2011, Kazakhstan's general prosecutor was quoted as saying "the question of control over social networks, over the internet, is a question of time. . .countries must join efforts to counter this evil" (Freedom on the Net 2012). The Zhanaozen crisis marks the major challenge to the government of Kazakhstan with the use of social media.

It is important to understand the context of Kazakhstan's wider Internet usage. Specifically, Kazakhstan's government made the development of digital information technologies a national priority (Emrich et al. 2013) and aims to give 100 percent of households the opportunity to access to information communication technologies by 2020. Kazakhstan is still pushing hard to digitize the nation but only for economic growth. Most people access the Internet from their mobile devices and at home in Kazakhstan. Free access is available in many public areas. According to Internet World Stats website as of June 30, 2017, almost 77% of Kazakhstan's population has access to the Internet. Mostly, Internet connection fees range from as little as KZT 3830 ($12) to the fastest available at KZT 19,900 (roughly $57). Overall, Kazakhstan is listed among Internet "not free" countries, according to the latest 2017 Freedom House report on Internet freedom around the world. The report highlights some of the developments in the country. It said authorities imprisoned activists attempting to organize protests using social media. It also said that overall environment remains oppressive to ICT users, with continued online censorship and arrests of social media users.

"Imposing systematic censorship under the rubric of national security allows the regime to limit even further the freedom of Kazakhstan's new media: in a move that is reminiscent of China's large-scale blocking strategy" (Anceschi 2015, p. 294). Kazakhstan authorities used criminal charges against social media users in an effort to silence dissent and punish online mobilization, issuing prison sentences of up to 5 years. Opposition activists and dissidents were targeted with malware attacks likely originated from the government (Freedom House 2017). At the time of writing this chapter, Kazakhstan's government has reportedly accepted new amendments to existing law that Kazakh Internet users will no longer be able to post comments online anonymously (Zakon.kg 2017). Details about this new law are yet to be revealed. But social media users are already in trouble for their "behavior" online. Facebook user Alexander Babanov faced a libelous lawsuit for a social media post about accidents that allegedly happened in a local plant in the city of Pavlodar. The former head of the local plant filed a libel lawsuit against Babanov seeking 10 million Kazakh tenge ($30,000). Another example is when a woman was financially fined for using her Facebook to allege that criminal activity was taking place in neighboring apartments (Toleukhanova 2016). Criminal charges were brought to some social media users in the country for allegedly inciting religious and ethnic discord (IREX 2017).

Turning to the situation in Kyrgyzstan, the country has suffered from political instability and widespread corruption. Two revolutions occurred since 1991 in the country, with the deadliest ethnic clashes between Kyrgyz and minority Uzbeks occurring in 2010 when almost a half of million people were displaced, with thousands of fleeing to neighboring Uzbekistan (BBC 2010). Overall, the country is the most dynamic in terms of the frequency of collective protests among the countries in Central Asia (Achilov 2016) because the country's politics are largely shaped by clan-based loyalties, weak institutions, fragmented independent elites, and regional differences with the country. This can explain why Kyrgyzstan is one of the poorest countries among the former Soviet Union countries. Almost 1 million out of the 6 million population are abroad in search of jobs, primarily in Russia, Turkey, Kazakhstan, and elsewhere. This can be further noted by the fact that a Kyrgyz lawmaker initiated a draft law in 2016 to seek humanitarian aid.

But the country's post-independence history is frequently cited among academics who support the view that the Internet facilitates the democratization process. This is because the Internet and wider social media sites played an instrumental role in the people's first revolution in 2005 that ousted President Askar Akayev from power (McGlinchey and Johnson 2007; Kulikova and Perlmutter 2007). Specifically, on March 24, 2005, about 20,000 people demanded President Akayev's resignation (Hiro 2009; Kulikova and Perlmutter 2007) because of the manipulation of the parliamentary election held in February 2005 and because of growing public discontent toward rampant corruption, widespread poverty in the country, and failed economic policies (Achilov 2016). Specifically, Radnitz (2012) noted that election manipulation included paying cash to buy votes. This is why thousands of people showed up in the capital city of Bishkek to demand invalidation of the parliamentary election held in the previous month, February 2005. In response to public protest, the Kyrgyz government blocked all traditional media outlets but the Internet. Access to the Internet was relatively low level, and it is possible that the Kyrgyz government did not consider the Internet to be a serious threat. "Most of the domestic media were in a difficult position when reporting about the events, as no one knew whether the president was still in the country" (Kulikova and Perlmutter 2007, p. 30). In such context, domestic media outlets were providing unreliable accounts, leading people to search for alternative sources of information on the Internet (Kulikova and Perlmutter 2007). In their work, Kulikova and Perlmutter (2007) extensively discussed how the website Akaevu.net played an important role to keep people informed about what was happening in the country. Kulikova and Perlmutter argued that the Internet is the only venue for people to express themselves because mass media outlets in the country are under government control. Even though Kyrgyzstan's media freedom ranking is higher than other countries in the region, journalists are threatened, harassed, violently attacked, prosecuted, and imprisoned (Freedman and Shafer 2012; Pitts 2011; Kurambayev 2016). Journalists in Kyrgyzstan work in a fearful environment (Toralieva 2014) not only for physical threats but also large financial lawsuits. Also, journalists earn so little that they commonly sell the "news" (Mould and Schuster 1999), and some journalists have "extortion" practices while government agencies and even international organizations operating in the country bribe journalists (Kurambayev 2017). At the time of writing this chapter, a British journalist working for the French news agency AFP was denied entry to Kyrgyzstan upon arrival in December 2017 and told he was banned from the ex-Soviet republic in Central Asia (Radio Free Europe/Radio Liberty, December 9, 2017). Also, local journalist Elnura Alkanova was facing a legal threat for her independent investigative work. Specifically, a local bank urged local authorities to open a criminal case for her investigative report allegedly sharing classified information about the bank's clients. In such context and specifically during the 2005 revolution, Kulikova and Perlmutter (2007) argued the Internet encouraged people to share information, post comments, ask questions, and mobilize like-minded groups of people to group together for collective actions against the government via Akaevu.net website and some other blogging platforms. This Akaevu.net website was used to share information about planned gatherings for people to join forces to stand up against the government's election manipulation. Eventually, then-President Askar Akayev fled the country and sought asylum elsewhere. This marked the first time in the region a state leader was forced to step down and seek asylum in a foreign country.

Melvin and Umaraliev (2011) argued that this 2005 revolution contributed to the development of new media in the country. This is because international organizations felt that the Internet could facilitate positive change in the country, and therefore international organizations supplied the country with the software and extensive computer and network equipment in return for cooperation from the Kyrgyz government to keep the Internet deregulated (McGlinchey and Johnson 2007). For example, Transitions Online along with Open Society Institute (previously known as Soros Foundation) funded projects developing citizen journalism by offering free of charge training on how to use computers and the Internet to anyone who wanted to learn. Five years later, people extensively used social media sites for exchange of information during people's second revolution in Kyrgyzstan in 2010 (Melvin and Umaraliev 2011). This marked the second people's popular revolution in the country since its independence in 1991. The reasons for the second people's revolution are similar to the previous revolution, including wide-scale corruption, growing living expenses, and a deteriorating economy. On April 6, 2010, a group of people protested against the government's decision to increase tariffs significantly. Specifically, the Kyrgyz government had increased heating costs up to 400% and electricity up to 170% in February of 2010. So, on April 6, 2010, the protest turned violent in a clash with state police. Such protests spread to other cities, gradually moving more toward violent demonstrations, including taking hostages of certain high-ranking government officials. President Bakiyev declared a state of emergency and took total control of media outlets, but social media users were tweeting about the protests using the hashtag #freekg to disseminate information (Melvin and Umaraliev 2011). Social media platforms were key places to obtain and share information. As a result of this revolution, another president, Kurmanbek Bakiyev, had to flee to a foreign country to seek asylum.

Overall, almost 35% of Kyrgyzstan's population has access to the Internet (International Telecommunications Union 2016; Freedom House 2017). But the government of Kyrgyzstan is also attempting to digitize the nation like neighboring Kazakhstan by bringing Internet connections to all educational institutions, including secondary schools, with the help of international organizations and foreign governments. As of December 5, 2017, 55% of all secondary schools have been connected to the Internet (Bengard 2017, 24.kg website), and more than 4000 computers were purchased to distribute among schools. The average cost of an Internet connection can be around $15 per month in major cities of the country, while rural areas might be slightly higher. More and more Wi-Fi public locations are appearing in major cities. Private and state transportation companies have recently begun offering free Wi-Fi on public transportation.

There are some other examples where the power of the Internet and wider social media played a positive role in the context of Kyrgyzstan at a small scale where people collectively chose to speak up against Kyrgyz government's wrongdoings via the Internet, leading authorities to respond and/or take actions. For example, locally well-known singer Dilshat Kangeldieva shared her frustration about the authority's reported refusal to accept a report about stolen items from her car on December 17, 2017. She wanted the police officers to investigate the theft. The Kyrgyz singer said in her post that Kyrgyz officers on duty suggested her "forgive those thieves and move on" and that those officers on duty even accused her that because of her, they [officers] could not have a dinner and that they were hungry. As a result of social media users' collective frustration about police handling this case, an investigation was launched. This is one of thousands of cases where people are now using social media to express their frustration about government incompetence. Another recent case involved a schoolteacher beating students in class while students secretly recorded the teachers. Once shared via YouTube, the Ministry of Education fired the teacher in question and issued a statement to all teachers in the country to honor the professional code of conduct. A very notable example included when social media users in the country collectively spoke up about government's wrongdoing that led Kyrgyz authorities to focus more about controlling social media users so social media users do not challenge Kyrgyz authorities in any format. The incident began in 2015 when lawmakers announced that they wanted to buy new furniture, such as chairs, to replace old ones. Parliament announced that that it had ordered 120 new chairs at a cost of 2.6 million soms, or over US $38,000 (BBC 2015). The BBC reported that each planned chair would cost $310 apiece. Social media users inundated social media platforms with their frustration about the purchases coming during a time when government officials regularly spoke about the need to cut government spending. As a result of massive frustration, parliament dropped the plan to purchase new furniture.

This may have led the Kyrgyz government to realize that social media users in the country are having tremendous influence on government decision-making. In this context, the Kyrgyz government has been strengthening its control over the Internet and social media platforms and suppressing online criticism of the Kyrgyz government and government officials. For example, Kyrgyz lawmaker Irina Karamushkina said that Internet users often criticize the president and officials and urged authorities to pursue online critics. The deputy chief of National Security suggested in one parliamentary meeting that Internet users critical of president and officials are being identified (Mamytova 2016). Also, Privacy International reported that the Kyrgyz government purchased sophisticated Western technology to spy on the entire country's Internet communication (Radio Free Europe/Radio Liberty 2014). Since then, a foreign citizen was deported from the country for his social media post about Kyrgyzstan, and numerous locals were detained for social media behavior such as liking or sharing certain content, etc. Some Facebook users were taken to court for damaging comments about officials, while some lost their jobs as a result of their social media posts. While the Kyrgyz government may justify such control with the justification of fighting terrorism and online radicalization, they may have a chilling effect on freedom of speech and expression online. Kyrgyz lawmakers have recently proposed a law that would demand that social media users use their real name, make their contact information visible to all, and verify any information before posting online.

While Tajikistan formally boasts many features of democracy, such as multiple political parties, elections, and laws that protect civil liberties, the country has struggled with poverty and instability since its independence. Tajikistan, with a population of about 8 million people, is the smallest country in the region and one of the poorest countries of Eurasia. Upon its independence from the Soviet Union in 1991, the country faced civil war during the 1992–1997 years, disrupting economic production facilities. Since then, the country seems to have achieved some political stability. However, currently, 33% of the economically active population has left the country (like citizens of neighboring Kyrgyzstan) in search of employment opportunities abroad, primarily in Russia, Turkey, Kazakhstan, and elsewhere. The remittance from these labor migrants consists of 50% of Tajikistan's GDP (UNDP 2017). The current rate of unemployment is at 33%, according to the World Bank.

Tajikistan is an authoritarian state dominated politically by President Emomali Rahmon and his supporters, says the 2017 report of US Department of State. The report also noted that the most significant human rights problems included citizens' inability to change their government through free and fair elections, repression, increased harassment of civil society and political activists, and restrictions on freedoms of expression, media, and the free flow of information, including through the repeated blockage of several independent news and social networking websites. President Rahmon's family members hold many prominent government positions. For example, his daughter Ozoda Rahmonva served as a deputy foreign minister. She was later appointed as the president's chief of staff. Another daughter is the head of Foreign Ministry's international relations department (Putz 2017). His son is the mayor of the capital city of Dushanbe. Tajikistan's human rights record continues to deteriorate amid an ongoing crackdown on freedom of expression and the political opposition, as well as the targeting of independent lawyers, journalists, and even the family members of opposition activists abroad (Human Rights Watch 2017). The authorities persecute and arrest journalists and political opposition leaders and curtail the free flow of information (Shafiev and Miles 2015, p. 301).

Like other countries in the region, corruption in Tajikistan is widespread and at all levels of society. Rule of law is weak and most institutions lack transparency and integrity structures (Transparency International 2017). Criminal groups have considerable influence on judicial functions, while bribery of judges, who are poorly paid and poorly trained, is commonplace (Library of Congress). The 2017 Media Sustainability Index (MSI) report for Tajikistan says that more than 40% of the population lives in poverty. Freedom House reported that recent constitutional changes in 2016 formally removed presidential term limits and lowered the minimum age for presidential candidates from 35 to 30. The changes effectively allow Rahmon to rule indefinitely and also render his 29-year-old son eligible for candidacy in the 2020 presidential election.

In terms of freedom of speech and expression and the media, Tajikistan has an atmosphere of intimidation for journalists (IREX 2017). The IREX report notes that many sources in the country do not talk to journalists partly because of fear of the authorities. This is why most news outlets focus more on social issues and culture to avoid harassment, intimidation, firing, or worse (Gross and Kenny 2011). Gross and Kenny argue that censorship and self-censorship are among the paralyzing problems faced by the Tajik media and society. The media avoid covering the president's family and private life, corruption in the top ranks of government, and activities of the special services, tax services, and their business partners. The report also noted that hacking of the email and social accounts of journalists and civil society activists, phone tapping, and others forms of cyber-crime are becoming routine in the country. Citing national security in fighting terrorism, Tajikistan government requires all subscriber identity module (SIM) cards reregister in the country. Since 2012, Tajikistan has begun to repress its opposition and others through pro-active online measures (Shafiev and Miles 2015). A group of volunteers monitored websites and social media, informing relevant state authorities when national interests and the state image were in danger or when the country was being humiliated (Shafiev and Miles 2015).

Approximately 20 percent of the population has regular access to the Internet, and monthly Internet connection fees seem to range from $50 to $200, which is out of reach for many people. Daily wintertime power outages can be another obstacle for public use of the Internet. Overall, Tajikistan regularly blocks access to social media sites including YouTube, Facebook, Google Services, etc. Once, Tajikistan State Communications Service Mr. Bek Zuhurov said that hundreds of citizens requested he shut down social media access, including Facebook (Shafiev and Miles 2015), and the government's blocking of access to social media is to follow the people's request. Because of the irregularity of Internet access and the government's steps to block some of the popular social media and news websites, journalists say these sites can be unreliable in reaching out to people. There are certain moments when Tajikistan's government seems to restrict flow of information. First, blocking systematically follows an event, such as military action, which the government did not want publicized, or when articles, audio, or video leaks or discussions that made the government uncomfortable appear online (Shafiev and Miles 2015, p. 307). Second, it happens during times when groups (including political groups) are created on Facebook, the social media site Odnoklassniki, or any other social media sites where opposition political leaders use platforms to criticize the Tajik government. However, every time, the state denies that any state orders for censorship exist. For example, Shafiev and Miles discussed one case that occurred in 2013 involving Gruppa 24, a political opposition group founded in Moscow by Tajik government critic Umarali Quvvatov. He propagated on the social media site Odnoklassniki about planned a protest in Dushanbe, the capital city of Tajikistan. While online, the protest attracted the attention of hundreds of supporters. News website EurasiaNet.org reported that Quvvatov was shot and killed in 2015 in Istanbul. Shafiev and Miles also discuss a similar case when a bid in social media about gathering a protest at the High Court of Tajikistan. The purpose of this protest was to protest against the detention of new political party Zaid Saidov. However, Tajik police became aware of this planned protest ahead of the actual protest date and were well prepared for it. But the major wave of calls for protest generated via social media against President Emomali Rahmon's government occurred in 2014. Tajik diaspora from Russian cities, including Moscow, Ekaterinburg, and other cities, called for massive protests in Tajikistan. It is important to note that almost 90% of all Tajik migrants who left the country in search of employment abroad are in Russia. Such social media calls led Tajikistan government to block dozens of websites countrywide and led to a complete Internet shutdown (Shafiev and Miles 2015), all while sending armed vehicles to Dushanbe. At the time of this writing, Tajikistan's Taxation Department argued that social media platforms should be blocked because of a declining state budget and that Tajik people are using other social media (Skype, WeChat, IMO, Viber) more to communicate with each other rather than making a traditional phone call (Asia Plus, December 21 2017). Yet, government officials frequently appeal to the collective memory and fears of civil war, warning that views on the Internet could cause a resurgence in violence (Shafiev and Miles 2015, p. 308). "Authorities began to ride the Internet wave; beyond volunteers reporting slanders, a large number of users with fake names and pictures now defend state positions" (Shafiev and Miles 2015, p. 315). Tajikistan, along with other countries in Central Asia, reportedly purchased a sophisticated Western technology to spy on its citizens inside and abroad. The Privacy International report noted that:

> some [Central Asian] countries are equipped with sophisticated surveillance capabilities that allow the monitoring of communications on a mass scale. . . . These surveillance capabilities are centralized and accessed by security agencies in monitoring centers, located across the region, allowing agents to intercept, decode and analyze the private communications of thousands of people simultaneously.

In such context, social media users often self-censored their views while posting on the Internet.

54.4 Conclusions

Although this work may seem of concern to only a small group of readers of Central Asia, it should in fact concern anyone who cares about the Internet and its potential power to democratize authoritarian countries and how governments can use repression technology to undermine freedom of information online (Tucker et al. 2017). While it is true that existing literature offers mixed conclusions about whether the Internet opens previously closed societies – for example, "much of the present excitement about the internet, particularly the high hopes that are pinned on it in terms of opening up closed societies, stems from selective and, at times, incorrect readings of history" (Morozov 2012, p. xi) – others argue that the Internet facilitates democratization process by giving access to information even among those who previously paid little attention (Gil de Zúñiga et al. 2012; Hamilton and Tolbert 2012). Some other scholars regularly cite the "Arab Spring" as an example of their view that the Internet can promote democracy regardless of contexts. This chapter demonstrates from recent examples of Kazakhstan, Kyrgyzstan, and Tajikistan that social media democratizes access to information, promotes and coordinates political planning, and pursues collective activism and a group of like-minded people together, while social media can be used simultaneously to censor and manipulate information to try to silence others' voices, including through hindering access to information or threatening or employing trolls to change the conversation online (Tucker et al. 2017).

54.5 Cross-References

▶ Asian Contributions to Communication for Development and Social Change
▶ Communicative Analysis of a Failed Coup Attempt in Turkey
▶ How Social Media Mashups Enable and Constrain Online Activism of Civil Society Organizations
▶ Media and Participation
▶ Multidimensional Model for Change: Understanding Multiple Realities to Plan and Promote Social and Behavior Change
▶ Online Social Media and Crisis Communication in China: A Review and Critique

References

Achilov D (2016) When actions speak louder than words: examining collective political protests in Central Asia. Democratization 23(4):699–722

Anceschi L (2015) The persistence of media control under consolidated authoritarianism: containing Kazakhstan's digital media. Demokratizatsiya 23(3):277–295

Asia Plus (2017, December 21) Официально. Служба связи отозвала лицензии на NGN у всех таджикских операторов. Retrieved from https://news.tj/ru/news/tajikistan/society/20171221/sluzhba-svyazi-otozvala-litsenzii-dlya-realizatsii-ngn-u-vseh-tadzhikskih-operatorov

BBC (2010) BBC Q&A: Kyrgyzstan's ethnic violence. Retrieved on July 27, 2018, from https://www.bbc.co.uk/news/10313948

BBC (2011, December 16). Kazakh oil strike: 10 dead in Zhanaozen clashes. Retrieved on 22 Dec 2017, from http://www.bbc.com/news/world-asia-16221566

BBC (2015, October 14) Kyrgyzstan: online protests over new chairs for MPs. Retrieved from http://www.bbc.com/news/blogs-news-from-elsewhere-34527694

Beisembayeva D, Papoutsaki E, Kolesova E (2013) Social media and online activism in Kazakhstan: A new challenge for authoritarianism? Paper presented at the Asian conference on media and mass communications 2013

Bengard A (2017, December 5) В Кыргызстане только половина школ подключена к Интернету. Retrieved from https://24.kg/obschestvo/70149_vkyirgyizstane_tolko_polovina_shkol_podklyuchena_kinternetu_/

Bowe B, Freedman E, Blom R (2012) Social media, cyber-dissent, and constraints on online political communication in central Asia. Central Asia Caucasus 13(1):144–152

Breuer A, Landman T, Farguhar D (2014) Social media and protest mobilization: evidence from the Tunisian revolution. Democratization 22(4):764–792. https://doi.org/10.1080/13510347.2014.885505

Committee to Protect Journalists (2017a, May 15) Journalist and press freedom defender stabbed in Kazakhstan. Retrieved on 23 Nov 2017 from https://cpj.org/2017/05/journalist-and-press-freedom-defender-stabbed-in-k.php

Cottle S (2011) Media and the Arab uprisings of 2011. Journalism 12(5):647–659

Economist's Democracy Index (2017). Democracy Index 2017. Retrieved on July 27, 2018, from https://www.eiu.com/topic/democracy-index

Emrich F, Plakhina Y, Tsyrenzhapova D (2013) Mapping digital media: Kazakhstan. Retrieved from https://www.opensocietyfoundations.org/sites/default/files/mapping-digital-media-kazakhstan-eng-20131024.pdf

Freedman E, Shafer R (2012) Advancing a comprehensive research agenda for central Asian mass media. Media Asia 39(3):119–126

Freedom House (2012) Freedom on the net 2012: Kazakhstan. Retrieved on 22 Dec 2017 from https://freedomhouse.org/report/freedom-net/freedom-net-2012

Freedom House (2017) Kazakhstan: freedom on the Net 2017. Retrieved on 22 Nov 2017 from https://freedomhouse.org/report/freedom-net/freedom-net-2017

Gil De Zúñiga H, Jung N, Valenzuela S (2012) Social media use for news and individuals' social capital, civic engagement and political participation. J Comput-Mediat Commun 17:319–336

Gross P, Kenny T (2011) Journalistic self-censorship and the Tajik Press in the context of Central Asia. In: Freedman E, Shafer R (eds) After the czars and commissars: Journalism in authoritarian post-Soviet Central Asia. Michigan State University Press, East Lansing, pp 123–139

Hamilton A, Tolbert C (2012) Political engagement and the Internet in the 2008 U.S. presidential election: a panel survey. In: Anduiza E, Jensen M, Jorba L (eds) Digital media and political engagement worldwide: a comparative study. Cambridge University Press, Cambridge, UK, pp 56–79

Heritage Foundation (2017) 2017 index of economic freedom: promoting economic opportunity and prosperity. Retrieved 26 Nov 2017 from http://www.heritage.org/index/country/kyrgyzrepublic

Hiro D (2009) Inside Central Asia: a political and cultural history of Uzbekistan, Turkmenistan, Kazakhstan, Kyrgyzstan, Tajikistan, Turkey and Iran. Overlook, London

Howard P, Bradshaw P (2017) Troops, trolls and troublemakers: a global inventory of organized social media manipulation. Computational Propaganda Project, Oxford http://comprop.oii.ox.ac.uk/wp-content/uploads/sites/89/2017/07/Troops-Trolls-and-Troublemakers.pdf

Human Rights Watch (2017) World report 2017: Kazakhstan. Retrieved from https://www.hrw.org/world-report/2017/country-chapters/kazakhstan

International Research and Exchanges Board (2017) Media Sustainability Index: Development of sustainable independent media in Europe and Eurasia. IREX, Washington, DC

Internet World Stats (2017) Internet usage in Asia: Kazakhstan. Retrieved on 23 Nov 2017 from http://www.internetworldstats.com/stats3.htm

Kulikova S, Perlmutter D (2007) Blogging down the dictator: the Kyrgyz revolution and samizdat websites. Int Commun Gaz (69):29–50

Kurambayev B (2016) Journalism and democracy in Kyrgyzstan: the impact of victimizations of the media practitioners. Media Asia 43(2):102–111

Kurambayev B (2017) Bribery and extortion in Kyrgyz journalism or simply profitable profession? Asia Pacific Media Educ 27(1):170–185

Mamytova A (2016) Ирина Карамшукина: Оскорбляя президента, оскорбляете государство. Retrieved from http://24.kg/vlast/38455_irina_karamshukina_oskorblyaya_prezidenta_oskorblyaete_gosudarstvo/

Marwick A, Lewis R (2017) Media Manipulation and Disinformation Online. Retrieved on July 27, 2018, from https://datasociety.net/pubs/oh/DataAndSociety_MediaManipulationAndDisinformationOnline.pdf

McGlinchey E, Johnson E (2007) Aiding the internet in Central Asia. Democratization 14 (2):273–288

Melvin N, Umaraliev T (2011) New social media and conflict in Kyrgyzstan. SIRPI Insights on peace and security

Morozov E (2012) The Net Delusion: The Dark Side of Internet Freedom. New York: Public Affairs Publishing

Mould D, Schuster E (1999) Central Asia: Ethics–a Western luxury. In M. Kunczik (Ed.), Ethics in journalism: A reader on their perception in the Third World. Bonn: Fridrich-Ebert-Stiftung

Niyazbekov N (2017) Kazakhstan's government is using social media to tame rebellion. Retrieved from https://theconversation.com/kazakhstans-government-is-using-social-media-to-tame-rebellion-73542

Pitts G (2011) Professionalism among journalists in Kyrgyzstan. In: Freedman E, Shafer R (eds) After the czars and commissars: Journalism in authoritarian post-Soviet Central Asia. Michigan State University Press, East Lansing, pp 233–243

Putz, C. (2017). Thanks Dad! Tajik President's son gets a new job. Retrieved from https://thediplomat.com/2017/01/thanks-dad-tajik-presidents-son-gets-a-new-job/

Radio Free Europe/Radio Liberty (2014). Report: Western firms help central Asian states spy on citizens. Retrieved from http://www.rferl.org/a/26701293.html

Radio Free Europe/Radio Liberty (2017) British AFP Journalist Says Banned From Kyrgyzstan. Retrieved on July 27, 2018, from https://www.rferl.org/a/british-afp-journalist-banned-from-kyrgyzstan/28907023.html

Radnitz S (2012) Weapons of the wealthy: predatory regimes and elite-led protests in Central Asia. Cornell University Press, Ithaca

Reporters Without Borders (2017) 2017 World press freedom index. Retrieved from https://rsf.org/en/ranking

Reuter O, Szakonyi D (2015) Online social media and political awareness in authoritarian regimes. Br J Polit Sci 45(1):29–51. https://doi.org/10.1017/S0007123413000203

Shafiev A, Miles M (2015) Friends, foes, and Facebook: blocking the internet in Tajikistan. Demokratizatsiya 23(3):297–319

Toleukhanova A (2016, April 1). Kazakhstan: Facebook lands more people in trouble. Retrieved on 21 Nov 2017 from http://www.eurasianet.org/node/78051

Toralieva G (2014) Kyrgyzstan-challenges for environmental journalism. In: Kalyango Y, Mould D (eds) Global journalism practice and new media performance. Palgrave Macmillan, New York, pp 214–226

Transparency International (2017) Corruption Perceptions Index: Tajikistan. Retrieved on July 27, 2018, from https://www.transparency.org/country/TJK

Tucker J, Theocharis Y, Roberts M, Barbera P (2017) From liberation to turmoil: social media and democracy. J Democr 28(4):46–59

United Nations Development Program (2017) Tajikistan. Retrieved from http://www.tj.undp.org/content/tajikistan/en/home/countryinfo.html

Vanderhill R (2015) Limits on the democratizing Influence of the Internet: lessons from Post-Soviet States. Demokratizatsiya 23(1):31–56

Wojcieszak M, Smith B (2014) Will politics be tweeted? new media use by Iranian youth in 2011. New Media Soc 16(1):91–109. https://doi.org/10.1177/1461444813479594

Zakon.kg (2017, November 23). Оставлять комментарии на интернет-ресурсах анонимно казахстанцы теперь не смогут. Retrieved on 23 Nov 2017 from https://www.zakon.kz/4890305-ostavlyat-kommentarii-na-internet.html?utm_source=web&utm_medium=chrome&utm_campaign=notification

New Media: The Changing Dynamics in Mobile Phone Application in Accelerating Health Care Among the Rural Populations in Kenya

55

Alfred Okoth Akwala

Contents

Abstract

Millennium Development Goals (MDGs) had health-care improvement as one of the pillars; however, a majority of the African countries were not able to achieve these goals and have now embarked on Sustainable Development Goals. The third objective of SDG aims to ensuring healthy lives and promotion of the well-being of all ages.

A. O. Akwala (✉)
Department of Language and Communication Studies, Faculty of Social Sciences and Technology, Technical University of Kenya, Nairobi, Kenya
e-mail: akwala08@yahoo.com

J. Servaes (ed.), *Handbook of Communication for Development and Social Change*,
https://doi.org/10.1007/978-981-15-2014-3_84

Kenya has implemented two strategies aimed at influencing skilled facility care utilization: the provision of free maternity health services and the Beyond Zero Campaign. However, these strategies may not have achieved much due to lack of adequate and relevant information for the populations. Therefore, utilization of new media in disseminating health knowledge to populations may be an alternative to the traditional media and other channels of health knowledge dissemination which may not be in tandem with the ever-changing socioeconomic realities in the rural communities. To this end, this chapter seeks to assess the appropriation of mobile phone technology in accelerating the access to skilled facility health care, investigate the different types of health knowledge that can be accessed through the mobile phone, and assess the efficiency of the mobile phone in disseminating maternal health knowledge. The study will review available literature using a case study of Kenya.

Keywords

Maternal health · Knowledge · Communication · Skilled facility care · Mobile technology · Dissemination

55.1 Introduction

Mobile phone interventions have been found to improve utilization of antenatal and postnatal care services, which are essential for maternal and newborn health.

World Health Organization (WHO) report (2011) states that the unprecedented spread of mobile technologies as well as advancements in their innovative application to address health priorities has evolved into a new field of eHealth, known as mHealth. The report further states that the use of mobile devices to send appointment reminders is becoming more common in many countries.

Advances in technology can enable patients to receive a diagnosis and treatment plans without leaving their homes. Application of mobile phones in dissemination of maternal health knowledge can strengthen health-care systems in developing countries.

The challenges of access and utilization create disparities in the availability of health facilities and skilled maternal health professionals in the rural areas. Therefore mobile technologies can be used to deliver health behavior interventions. A report by WHO (2011) indicates that if use of mobile technology in health messages provision is implemented strategically and systematically, mHealth can revolutionize health outcomes, providing virtually anyone with a mobile phone with medical expertise and knowledge in real time.

mHealth applications can assist in achieving the SDG especially goal three which targets to ensure healthy lives and promote well-being for all at all ages. Mobile phone text message reminders for appointment attendance and further high-quality research are required to draw more robust interventions in facilitation of accessing skilled health-care uptake.

55.2 Securing Lives Through Mobile Phones

Technology has enhanced the way maternal health care is being conducted in sub-Saharan Africa. Zurovac et al. (2012) maintain that poor health indicators, restricted resources, poor infrastructure, and weak health systems in many African countries have overcome communication problems with the widespread use of mobile phone technology.

Geographically isolated communities can now have access to vital information on health in the comfort of their sitting rooms. Facility health providers can track the progress of their patients from their offices. As such, mobile phone applications have generated hope for the maternal health patients and the service providers.

UN et al. (2011) document that hundreds of thousands of women are still dying due to complications of pregnancy and/or childbirth each year. Many of these deaths occur due to haemorrhage, infection, and obstructed labor during child birth. Health providers can make an important contribution to improving maternal and newborn health of communities especially if they can have a means to disseminate health knowledge to the patients. Mobile phones have become more accessible than ever before in today's contemporary society. They are no longer tools of calling, texting, and sending or receiving money, but they are tools for access of critical information. One of the critical information is behavior change information. mHealth is proving to be part of the solution to the maternal health information, from text messages educating mothers on what to expect when they're expecting to information that assist medical professionals to provide comprehensive care. Zurovac et al. (2012) say the application of mobile technologies to health behavior interventions is an exciting and rapidly growing field as it has the ability to provide frequent and time-intensive interventions to maternal health patients. They maintain that mobile phone technology used in Kenya showed that text message communication between health workers and patients substantially improved patients' adherence to antiretroviral treatment and HIV treatment outcomes.

In Kenya, there are various mobile phone applications available to both the antenatal and postnatal mothers. The following are some of those applications:

55.2.1 Totohealth

Totohealth is a mobile technology platform in Kenya used to detect a mother's pregnancy stages and child development after birth. The aim of this mobile application is to improve the maternal and child health status of marginalized communities. Patients subscribe to the platform, and then text messages about schedules, reminders for clinic visits, and vaccinations are sent to them periodically. The application also detects development abnormalities of a child and pregnancy, maps out mothers due to deliver and avails resources to ensure safe deliveries, and analyzes data collected from patients to enhance decision and policymaking. This application has motivated both the antenatal and the postnatal patients to get relevant maternal health information without wasting time and spending much money on transport costs.

Irena et al. (2016) document that health-care provision has reduced costs by offering an alternative to costly travel to clinics or hospitals. Instead of face-to-face appointments, mHealth ventures offer patient care through doctor hotlines.

55.2.2 mHMtaani

mHMtaani mobile application is a community-based platform that is aimed to promote healthy living among the antenatal and postnatal patients in the coastal communities of Kenya. The app monitors and tracks the health of pregnant women. The rural women are assisted in knowing the pregnancy danger signs, delivery date, use of family planning, breastfeeding, and the general well-being of pregnant mothers without necessarily visiting the facility/doctor. This service has improved the quality of care for pregnant women and increased the number of hospital deliveries. Irena et al. (2016) maintain that mHealth projects educate individuals on various aspects of health-care benefits to increase the appropriate utilization of resources. The communication methods used include SMS messages, smartphone apps, website videos, Facebook pages, hotlines, online messaging, and online learning modules. The services offered include information on affordable drugs, childcare, nutrition, diseases, sexual and reproductive health, and referrals to local care centers.

Fotso et al. (2008) observe that the mere availability of health services to a population may not translate into increased utilization. This is due to many other socioeconomic factors that affect the utilization of health services. Therefore, there is a need to set up and strengthen maternal-child health promotion campaigns geared toward the marginalized population groups (ibid.).

Ransom and Nancy (2002, pp. 19–29), in discussing maternal-child health utilization, identified four obstacles that limit women from accessing and utilizing maternal-child health services, thereby leading to maternal-child deaths: delay in recognizing danger signs, deciding to seek care, reaching the appropriate health facility, and delay in receiving health care at a health facility. mHMtaani platform has created a conducive atmosphere for mothers to be able to undergo their deliveries in a facility under the supervision of a facility health staff as they able to be promptly updated on their pregnancy status via mobile phone messages. Linda (2006) documents that maternal-child health patients should be given adequate information in regard to maternal health issues. Effort should be made to ensure that all women have access to a skilled birth attendant (physician, midwife, or nurse professionally prepared for the provision of prenatal care) in order to decrease the high number of preventable deaths.

55.2.3 Baby Monitor

Baby Monitor is another mobile phone application in Kenya used for linking pregnant women with health clinics through interactive voice response technology. WHO (2011) report indicates that in Bangladesh, awareness of health campaigns is created by broadcasting SMS text messages to all mobile telephone numbers in the country, irrespective of their service providers to mobilize citizens to bring their children to get vaccinated. In Kenya, patients use Baby Monitor mobile application to receive information through SMS and voice messages once they call the hotline. The messages are disseminated in a language the patient can understand.

55.2.4 ChildCount+

ChildCount+ is yet another mHealth platform which relies on SMS data entry to create a centralized database whose aim is to register every pregnant woman and every child under the age of five in rural areas. Subramaniam (2011) documents that in Tanzania a mother who has TB and HIV and seeks antenatal care will have three identification numbers, her information residing in three different registers. This makes patient tracking and follow-up burdensome. It also makes compiling monthly reports tedious. However health-care providers can load all cluster patient data into the ChildCount+ database which can be used to track patients' progress for all the three ailments.

Fotso et al. (2008) observe that women should be encouraged to attend antenatal care where they can be given advice on delivery care and other pregnancy-related issues targeting the poorest, less educated, and high-parity rural women groups. These innovative technologies can provide better information to mothers to ensure that they have a safe delivery.

55.3 Setting of the Study

This study was conducted in Western Province, Busia County, a rural setting in Western Kenya and an area confronted with a myriad of health-related problems. The study was conducted between April and August 2012. Thirty antenatal health patients and 28 postnatal health patients were drawn from the region to participate in the study.

55.4 Methods

The study adopted the mixed method approach. The use of both qualitative and quantitative approaches enhanced the credibility of the research findings as the findings were triangulated to elicit superior evidence. The two data strands were generated concurrently but analyzed differently through thematic analysis and descriptive and inferential statistics analysis. The interpretations are mixed to establish the relationship between the variables.

In addition the use of both quantitative and qualitative approaches in combination provides a better understanding of research problems as qualitative data which is mostly a narration backed up with statistical scores from quantitative data, hence assisting the researcher to compare and contrast both qualitative and quantitative data.

55.5 Results

This study sought to identify communication strategies used in promoting skilled maternal health-care utilization in rural Busia County among antenatal and postnatal health patients. Therefore, a question on the channels of accessing maternal health knowledge was asked. Similarly, a question on the frequency of mobile phone use in accessing maternal health knowledge was also asked.

This question was intended to find out the average number of people who use mobile phones to access maternal health information and the kinds of knowledge they are able to access.

55.6 Dissemination of Maternal-Child Health Knowledge to Patients (Antenatal Patients) in Busia County

The dissemination of information on maternal-child health is a key factor to the utilization of maternal-child health services. A total of 30 antenatal patients filled the questionnaires and 8 of them were interviewed. A majority of the respondents acknowledged that chief's *barazas* (43.3%) could be utilized as key avenues for dissemination of information, especially in rural areas. One of the respondents said that:

> messages from the chief's *Baraza* or a 'community worker' can be clarified and substantiated as opposed to those we 'hear' from radios.

According to her, community outreach programs offer an opportunity for patients to ask for clarification and questions on issues that are not clear. She further said that chief's *barazas* are communal media since they act as a forum where one meets masses of people and disseminates information to them. The mobile phone was also singled out as one avenue to access maternal-child health information. Mobile phone was approved by 16.7% as an appropriate source of maternal health information.

Tables 1 and 2 show the breakdown of the methods used for information access, the number of respondents, and the percentages for each method for both antenatal and postnatal patients.

Table 1 Channels used to disseminate maternal-child health information (antenatal patients)

Method	Frequency	Percent
Mobile phone	5	16.7
Relative/friend	2	6.7
Chief's barazas	13	43.3
CHWs	6	20.0
Information at facility/interactive posters	1	3.3
Mass media	3	10.0
Total	**30**	**100.0**

Source: Research data (2018)

Table 2 Methods of disseminating maternal-child knowledge (postnatal patients)

Means	Frequency	Percent
Mobile phone	6	21.4
Mass media	7	25.1
Door to door (CHWs)	9	32.1
Chief's *barazas*	3	10.7
Churches	2	7.1
FHS	1	3.6
Total	**28**	**100.0**

Source: Research data (2018)

55.7 Methods of Disseminating Maternal-Child Knowledge (Antenatal and Postnatal Patients)

For both antenatal and postnatal patients, mass media (a channel known to inform, educate, entertain, persuade, socialize, and market health products) did not score highly. For antenatal patients, only 10% preferred the use of mass media, while among the postnatal patients, only 25.1% preferred it to any other as a channel most suitable to disseminate maternal-child health knowledge. CHWs, chief's *barazas*, relatives, and friends are seen as a means that can achieve better results in maternal-child health campaigns, especially when it relates to immunization than any other means.

One of the respondents said that the health facilities are too far away and there are a lot of tests before one is enrolled into the program; "we" would be better if all maternal health information was easily available.

This shows that women would prefer to get this information on a mobile platform so that they only visit facilities for critical issues.

The methods of knowledge dissemination that were recommended as the most preferred by the respondents included mobile phone, mass media, door-to-door campaigns, community health workers' campaigns, chief's *barazas*, churches, and facility health staffs.

For postnatal patients, door-to-door campaigns, which preferred by 32.1% of the sample population, are to be steered by community health workers who mobilize maternal health patients by informing them on the need to visit maternal health clinics. Though most patients claimed that they only used their mobile phone to call, text, and receive or send money, 21.4% of the postnatal patients acknowledged using their mobile phones to access maternal health information.

55.8 Discussions

Most maternal health patients both in the pre- and postnatal stages claimed that they got information about maternal health-care services from chief's barazas and community health workers. However, patients said that mobile phone applications could be of assistance as other patients turned to family, friends, co-workers, women who had been pregnant before, and older women in the community.

Communication about the benefits of utilizing skilled maternal-child health services is necessary. The findings corresponded with Ivanov and Flynn (1999) observation that point out that entry or early access into the health-care system is an indicator of health-care utilization (p. 374). They further say that negative experiences with health providers decrease women's satisfaction with prenatal care services (ibid., p. 383).

55.9 Strategic Communication

The source of information determines the level of trust placed on the message. Patients may trust the message from the mobile phone as opposed to one from a community volunteer worker or one from a TBA as opposed to a message from a community health volunteer/worker. Accordingly, communities need to be educated on the need to access and utilize skilled maternal-child health care, and this can only be possible if appropriate strategies of disseminating this knowledge are used.

It also emerged from the study that mass media is not the most appropriate means to disseminate maternal-child health knowledge as it is not interactive and participatory. Most mothers can rely on mobile phone text messages to access meaningful pre- and postnatal care information in SMS format that can be easily consumed according to the findings. Therefore, mass media is more effective when backed by complementary activities on the ground such as door-to-door campaigns. Oronje et al. (2009) note that the mass media has the ability to disseminate information in a broad, timely, and accessible manner; it constitutes an important source of information for the general public and policymakers. They further say that as information providers, the mass media informs, educates, entertains, persuades, socializes, and markets commercial products, among other roles (p. 2). However, most of the maternal health information disseminated through radio has been termed as a nonparticipatory. Participatory communication is usually relevant in achieving rural development according to the findings.

In view of the findings, communication strategies that embrace new technology and are illustrative and participatory, such as the use of mobile phones, should be used. Therefore, health providers can use technology to enhance maternal health knowledge dissemination. Oluwaseun et al. (2015) say that e-health interventions for MCH can improve access to health information, influence pregnant women to adopt safe MCH practices, encourage prompt patient diagnosis, and increase the utilization of health facilities.

However, Fotso et al. (2008) acknowledge that to ensure wider reach, health education programs should be channelled through a mix of avenues including the mass media, organizations working in the communities such as chiefs, community outreach activities, posters, and leaflets. In addition, Winskell and Enger (2005, p. 412) document that it is generally agreed that health communication through the mass media has great impact when it is reinforced through interpersonal channels.

These channels include discussions, educational sessions, local theaters and songs, and debates to generate emotional engagement, thereby influencing behavior change. The mass media messages cannot singly influence behavior change as Adam (2005, p. 363) documents that Bandura proposes that "social persuasion" by itself will not succeed in bringing about adoptive behavior and that the mass media is more effective when backed by complementary activities on the ground.

Infrastructural support on maternal health is necessary as some of the cases relate to access to maternal-child health facilities. In line with the findings, Pade-Khene et al. (2010), quoting McNamara (2003), point out that health infrastructure hampers the ability of rural communities to preserve good health and treat illnesses. They further note that facility health workers have to have appropriate communication channels to receive and disseminate current and new health or medical information (ibid.). Health infrastructure in terms of accessibility to the health facilities in the area under study is an issue that needs to be addressed for the realization of appropriate maternal-child health service utilization. Fotso et al. (2009) findings support the view that in developing countries patients face significant obstacles in accessing health care. Therefore, those who live nearby hospitals are most likely to access health services than those who live far away. There are those patients who prior to giving birth always felt sick and hence had to visit the clinic regularly for checkups and, even the birth of their child, had to end up in a health facility. Magadi (2000, p. 188) also observes the same that the issue of physical access to health services is a critical problem in many rural parts of Kenya, where long distance to hospitals and poor road conditions are actual obstacles to reaching health facilities and often a disincentive to even trying to seek care. However, Magadi et al (2000) aver that even though more than half of the births take place in a health facility, of the women who manage to get to the health facilities, their chances of dying from maternal-related causes are still high. According to their household survey, 46% of maternal deaths were observed to occur in a health facility (p. 6). Generally, communication strategies in rural areas have to be participatory and illustrative and have to make use of new technology and the opinion leaders. Pade-Khene et al. (2010), quoting McNamara (2003), point out that health infrastructure hampers the ability of rural communities to preserve good health and treat illness. They further observe that facility health workers have to have appropriate communication channels to receive and disseminate current and new health or medical information (ibid.).

55.10 Client-Service Provider Communication

Client-service provider communication enhances dissemination of maternal-child health knowledge and facilitates appropriate decision-making so that through participatory communication, patients can share and exchange knowledge. Once done, appropriate rating of the facility health staff could be given. Thomas (2006, p. 53) says that health-care audiences are hampered by lack of knowledge on the cost of care and other issues as well. Mobile phone health messages are likely to encourage utilization of skilled facility services.

Consumers must make judgments based on the provider's reputation or on superficial factors such as the appearance of the facilities, the available amenities, or the tastiness of the hospital's food. Consequently, education could play a major role in influencing the behavior of the maternal-child health patients. TBAs are assumed to be more hospitable and loving than the FHSs who have been branded by some patients as cruel and arrogant; as such, patients avoid utilizing skilled facility health care and resorting to using the TBAs. Oronje (2009) documents that about 10% of births in the slums are handled by traditional birth attendants (TBAs). These attendants, however, lack skills to handle delivery. Women prefer TBAs to nurses in public health facilities because the nurses are more often reported to have a bad attitude and are abusive toward them. The more negative experiences women have with health care providers, the less satisfied they become with the doctor's midwife's behavior at the health facility. Through strategic communication, this mind-set and attitude can be eradicated. Mobile phone technology could be of importance in dealing with such situations. Pade-Khene et al. (2010) observe that facility health workers have to have appropriate communication channels to receive and disseminate current and new health or medical information, whereas Fotso et al. (2008) observe that women should be encouraged to attend antenatal care where they can be given advice on delivery care and other pregnancy-related issues targeting the poor, less educated, and high-parity rural women groups.

55.11 Conclusion

Communication is very essential in health as health-related information disseminated as it enables the patients to understand the importance of accessing and utilizing skilled facility maternal-child health services. Information dissemination and behavior change strategies should be embraced in promoting the utilization of skilled maternal-child health services in rural areas. Servaes (2007) says communication should influence attitudes, disseminate knowledge, and bring about a desired and voluntary behavior change. This clearly indicates that the information reaching maternal-child health patients should be informative.

Information dissemination of maternal-child health knowledge is a key factor in the utilization of maternal-child health services. The interview with maternal-child patients revealed that the empathy in communication of the caregiver at the health facility was an indicator for facility health-care utilization by the patient. Thomas (2006) reports that communication experts indicate that "effective communications must be meaningful, relevant, understandable to the receiver, relevant to the lives of 'real' people, and stimulate the receiver emotionally" (p. 99).

Service provider-client communication is very essential in the realization of positive maternal-child health. Fotso et al. (2008) document that lack of or poor transportation facilities and poverty are also key contributing factors to non-institutional deliveries, especially those deliveries that take place en route to health facilities. Lack of equipment and personnel in health facilities hinders also some patients to access and utilize skilled maternal-child health services. Participatory communication is very vital here, and these messages could be passed through mobile phone technology.

Unfriendly health facility staff and long procedures at the facility centers that force patients to make very long queues make the patients not to utilize skilled maternal-child health services and hence opt for other services which are instant and unfortunately may not be skilled. Negligence by maternal-child patients and lack of information are also other challenges that lead to non-utilization of skilled maternal-child health services. Mobile phone applications may be a solution to all these.

Education on maternal-child health issues should be geared toward behavior change among the maternal-child health patients. Health communication is concerned with a shift from previous visions of health care and prevention which is largely dominated by high technology and hospital-based concepts of health care to searching for innovative and flexible approaches that pay greater attention to knowledge dissemination already possessed by local people.

The use of mobile technology in solving health-related matters does not imply that health-care professionals have to ignore their respective role. Rather, the goal of technology is to uncover alternative solutions that will better service for the pre- and postnatal women. It should also foster local microenterprise, creating upgraded platforms and new functionalities that will generate ongoing economic opportunities for communities. He further says that the conveyance of information can increase or decrease people's anxiety, depending on their information preferences and the amount and kind of information they are given (ibid.). Therefore, maternal health messages must be developed with knowledge of the cultural characteristics of the audience. Culture has been known to be a major determinant of maternal-child health services in many African communities. Magadi (2000) notes that the poor health outcomes (e.g., high infant and child mortality) consistently observed in Nyanza Province, for instance, may be more of a factor of cultural practices, rather than availability and accessibility of health services.

References

Adam G (2005) Radio in Afghanistan: socially useful communication in wartime. In: Oscar H, Tufte T (eds) Media and glocal change: rethinking communication for development. Nordicom, Göteborg, pp 349–365

Fotso JC, Ezeh A, Oronje R (2008) Provision and use of maternal health services among urban poor in Kenya: what do we know and what can we do? J Urban Health 85(3):428–442

Fotso JC, Ezeh AC, Essendi H (2009) Maternal health in resource-poor urban settings: how does women's autonomy influence the utilization of obstetric care services? Reprod Health 6:9. African Population and Health Research Center (APHRC), Nairobi

Irena GJ, Bram T, Sutermaster S, Eckman M, Mehta K (2016) Value propositions of mHealth projects. J Med Eng Technol 40(7–8):400–421. https://doi.org/10.1080/03091902.2016.1213907

Ivanov LL, Flynn BC (1999) Utilization and satisfaction with prenatal care services. West J Nurs Res 21(3):372–386. Sage

Linda VW (2006) Beliefs and rituals in traditional birth attendant practice in Guatemala. J Transcult Nurs 17(2):148–154. Sage
Magadi MA (2000) Maternal and child health among the urban poor in Nairobi, Kenya. Afr Popul Stud 19:179–198. Centre for Research in Social Policy (CRSP), Loughborough University
Magadi M, Diamond I, Madise N (2000) Patient socio-demographic characteristics and hospital factors associated with maternal-mortality in Kenyan hospitals (working papers). Population Council, Nairobi
Oluwaseun OI, Mabawonku I, Lagunju I (2015) A review of e-health interventions for maternal and child health in sub-Sahara Africa. Matern Child Health J. https://doi.org/10.1007/s10995-015-1695-0. Springer, New York
Oronje NR (2009) *The Maternal Health Challenge in Poor Urban Communities in Kenya: A policy brief 12*. African Population and Health Research Centre, Nairobi
Oronje RN, Undie C, Zulu EM, Crichton J (2009) Engaging media in communicating research on sexual and reproductive health and rights in sub-Saharan Africa: experiences and lessons learned. Health Res Policy Syst 9(1):S7
Pade-Khene C, Palmer R, Kavhai M (2010) A baseline study of Dwesa rural community for the Siyakhula information and ground, communication technology for development project: understanding the reality. J Inf Dev 26(4):265–288. Sage
Ransom IE, Nancy VY (2002) Making motherhood safer: overcoming obstacles on the pathway to care. Population Reference Bureau, Washington, DC
Servaes J (2007) Harnessing the UN system into common approach communication for development. Int Commun Gaz J 69(6):483–507. Sage
Subramaniam H (2011) Knowledge is power: ChildCount+, mHealth in Tanzania. Columbia University, Columbia
Thomas KR (2006) Health communication. Springer, New York
UN, UNICEF, UNFPA, WB (2011) Trends in maternal mortality: 1990 to 2010. WHO, Geneva
WHO (2011) mHealth: new horizons for health through mobile technologies: second global survey on eHealth. WHO, Geneva
Winskell K, Enger D (2005) Young voices travel far: a case study of scenarios from Africa. In: Oscar H, Tufte T (eds) Media and global [sic] rethinking communication for development. Nordicom, Göteborg, pp 403–416
Zurovac D, Talisuna AO, Snow RW (2012) Mobile phone text messaging: tool for malaria control in Africa, vol 9, no 2. Malaria Public Health and Epidemiology Group, Kenya Medical Research Institute-Wellcome Trust Research Program, Nairobi, p e1001176

ICTs for Development: Building the Information Society by Understanding the Consumer Market

56

Shahla Adnan

Contents

S. Adnan (✉)
Department of Communication and Media Studies, Faculty of Social Sciences, Fatima Jinnah Women University, Rawalpindi, Pakistan
e-mail: shahla.adnanz@gmail.com

J. Servaes (ed.), *Handbook of Communication for Development and Social Change*,
https://doi.org/10.1007/978-981-15-2014-3_93

Abstract

Ensuring that consumers have a right to choose from the innovative technological facilities, at the most cost-effective affordable price, is the main objective for building the information society. Mechanisms involving comprehensive strategies, concerned with the buildup of information societies, have been addressed in this chapter. The particular focus is on the digitally divided underserved areas where consumer market needs have to be identified in their peculiar diverse culture, technology consumption ability, as well as practically applicable mechanisms for increasing the teledensity. Technological landscapes appear on the brink of radical transformations. Consumer patterns and technological advancements are going through a consolidation and adaptation phase for devising new structures. The ICT convergent era is offering a vibrant discourse to the dynamically changing consumer needs in their native and global scenarios. In addition, the potential of the alternative avenues for addressing the consumer market needs have also been analyzed.

Keywords

Information Society · Understanding consumer rights · Use of ICT · Networked media · Bridging ICT challenges and solutions · ICT solutions · Developing countries · WLL · Increasing teledensity

56.1 Introduction

The encouragement for the use of Information Communication Technologies, or commonly referred to as ICTs, for developing an information society has grown on a global scale due to an increase in its demand. ICTs have long been in use for multi-dimensional purposes ranging from a personal palm-size gadget to extremely sophisticated equipment in space. The usage of ICT is directly linked with the consumers for whom it is created. The consumers of ICT reside in an information society that is characterized by its properties to manufacture, distribute, and even manipulate the information. Technology-based needs are first introduced in the societies. If they get diffused and accepted in the society, its influences can be seen on the lives of the people that build up an information society ultimately. Now, if this ICT is to be used for the development of the underserved areas of the world, the consumer rights need to be understood and served in their true sense. Section 56.1 of this chapter identifies the characteristics of the information society where ICT can flourish for bringing development. Certain identified analytical yardsticks have been discussed in the information society along with a focus on the understanding of the consumer rights in the information society. Section 56.2 illustrates the concept of ICT penetration in the underserved areas where role of the ICT industry has been also discussed for its fast penetration. Section 56.3 contains the strategies required to address the consumer rights in the underserved areas, in which further expression has been given from multiple dimensions to support consumer's development. Section 56.4 exemplifies the ICT-based services by mentioning the applied benefits of ICTs particularly in developing areas scenarios. Section 56.5 puts up a detailed case study for the usage of wireless local loop ICT (WLL) in Pakistan as a developing country where certain ICT alternative avenues have also been represented. Section 56.6 contains the conclusion along with recommendations to increase the opportunities for usage of ICT that can raise the teledensity of the underserved areas along with certain fruitful results for development of these areas. It also recommends on widening the freedom of the choices for the ICT consumers of the underserved areas that can enable them to benefit from the subsided rates and enhanced connectivity among the competitors in government and private companies. This can ultimately satisfy the true needs of these areas where ICTs get an actual requirement-based spread-out, while addressing the consumers' rights.

56.1.1 Information Society

Information Society is a society where the manufacturing, dissemination, and manipulation of information exist as the most significant activities. Ultimate purpose of this information is consumption by the society. Such activities are linked with the economic paradigm, access usage paradigm, and cultural paradigms. In the converging yet networked societies, the use of information and communication technologies has emerged out as a note-worthy tool. The usage of Information and Communication Technology (ICT)-based platforms and techniques has invaded the face-to-face interactions of the existing consumer societies.

In case of economic paradigm, the manufacturing of information involves the making of new information by considering human minds as apparatuses (Herman and Chomsky 2006). Consumers of social media are exposed to the social and political news content through status updates and shared links on their online social connections (Mitchell 2015; Mitchell et al. 2013). Consumers exhibit their behaviors while performing certain online or other technology-based activities. The ICT techniques help for the content curation by monitoring behavior patterns of the consumers exhibited in their expressions while using respective platforms. For example, the patterns of liking in the online space are curated by different advertising companies for future information manufacture by suggesting information for the same consumer. This manufactured information is then disseminated via various platforms of ICT such as mobile phones and online platforms. According to the attentive response patterns of the consumers during their online activities, the format of manufactured information is packaged for dissemination based on consumer choice. This is the case when the consumer has opted for Really Simple Syndication (RSS) feeds on mobile platforms Facebook or over a twitter account, etc. The manipulation of the manufactured information occurs by behavior targeting: via analyzing the browsing history of the consumers or the semantics of the web pages, as well as by keyword stuffing for the search engines. Thus, these three activities of manufacturing, dissemination, and manipulation of information play a vital role in establishing an Information Society.

In case of access and usage paradigm related with the ICT, the basic units of society are those individuals who are consuming units of ICTs. In case of underdeveloped areas, the consumers are provided access to the ICTs yet they are unable to utilize it for their personal empowerment (Alderete 2017). Consumers of developing areas do not have the appropriate skill set for the usage of the ICTs with them. Hence, as Morawczynski and Ngwenyama (2007) have suggested that corresponding investment in ICTs, healthcare and education could have major influences on the human development, in underserved areas the access to ICTs can merely have a rise in their teledensities of the population yet actual benefits of the ICTs remain missing because of the lacking skills for ICTs usage. Alderete (2017) has endorsed as confirmed by Brynjolfsson and McAfee (2015), Hargittai (2008), Spiezia (2010), and Sunkel andTrucco (2011) that ICT access has a positive and significant effect on the ICT use and skills. A suitable ICT access is a prerequisite for the beneficial usage of the ICT in developing areas. If ICT access conditions are poorer like unavailability of connections, lack of devices at convenient places, or outdated devices, it steps back the usage of ICTs at the very initial step. The situation gets declined when adequate skills to use ICTs are not with the consumers.

The third cultural paradigm related with ICT is linked with the fear of technology found in the consumers of the underdeveloped areas. The consumers are not only underserved in terms of basic infrastructure, health facilities but also are unprivileged in terms of their behavior devolvement and attitude towards ICT. The use of latest tech-based gadgets even if adopted by the youngsters is discouraged by the elders as a curse that spoils their important time. Elders of underdeveloped societies, e.g., Pakistan, tend to exhibit reluctance towards ICT as they do have a fear of ICT usage that can only detach their descendants from them by destroying their social sittings on the daily basis with their families. This psychological barrier ultimately becomes a major road blocker in the usage of the ICT especially by the seniors. A digital divide exists even within the same society in the form of rural and urban segmentation.

If issues of ICT access, usage skills and fear to use ICT are addressed properly the consumers will be enabled to generate huge amounts of information. Such consumption can not only empower ICT consumers rather generate and disseminate their content on a massive scale.

56.1.2 Understanding Consumer Rights in the Information Society

Ensuring consumer's rights in scenario of underserved areas of world is still a challenge as consumers in this scenario are served with the ultimate-manipulated information. They are forced to use the provided lenses already chosen by the manufacturer of the information. Here the agenda setting theory of communication works into play where it directs the minds of the consumers not only what to think about but also how to think about it by influencing their opinions and attitudes. The salience of the consumer pertaining issues is raised with a pre-designed frame and primed out in the minds of the consumers (McCombs and Guo 2014). This process of manufacturing, dissemination, and manipulation of information is underpinned with an economic activity where information is the prime commodity. The consumers of ICT ultimately make financial transactions that generate profits. With the flourishing use of information in societies, the need for understanding the rights of the information consumers has also emerged. Some significant rights of the consumers in the information society include the *Right to safety* where consumers of underserved areas should be protected from potential physical and mental health hazards. Moreover, the *right to be informed* applied on consumers of ICTs can be ensured by the availability of alternative weighing information which can be implemented by safeguarding protection from false and misleading advertised claims and labeled practices. In direct relation, the strategy for consumer's *right to choose* should focus on the balanced accessibility of the competing ICTs and their services along with their alternatives in terms of price, quality, and service. It further follows with the strategies for the *consumers right to be heard* where need of mechanisms, such as the forces of demand and supply, is a real challenge to administrate for consumers in underserved areas. By doing as such, a deeper analysis regarding the demands and concerns of consumers of ICTs can be carried out and obtained by the government as well as involved structures. Rights of underserved consumers to safety, to information, to exercise choices, and ultimately to be heard need a collaborative mechanism for ICTs penetration.

56.2 Contextual Factors: ICT Penetration in Underserved Areas

A precise need-based road-map is required for ICT penetration. In rural areas, understanding the consumers for conveying information to them by the means of modern technology is of great importance. Certain contextual factors need to be addressed by adopting a strong consumer gratification mechanism for imparting the technology. The term "*Technology just for the sake of technology*" never works and cannot sustain in any domain, until and unless an objective is focused. Until the usage of technology is aimed towards the actual problems of the target consumers, the real benefits of consumers at certain affordable prices cannot be achieved. Some of the impacting *contextual factors* for creating effective collaborations in this domain are as follows:

Connectedness comes first and encircles the domain of how native people are connected to each other in their social circles. Potential consumers of the specific area may face several possible limitations. Connectivity of the consumers with the ICTs with respect to their category and niche requirements needs to be established. For example, farmers need agriculture-based guidelines and solutions for their crops, women with limited logistic needs easy access for doing daily grocery or clothes shopping within their proximity. Certain telecenters can be established at the local community health centers where women usually visit or e-commerce facility can be introduced at local market shops. Recently, in 2018 in Karachi, population wise largest city in Pakistan, an online as well as mobile app facility for grocery shopping is available through a private company. Moreover, in past such successful implementation of ICTs at underserved areas in form of "E-Choupal" project in India, via establishment of specific telecenters in the rural areas, addressed farmer needs. Furthermore, such e-commerce facilities can be provided at railway stations, bus stations, schools, etc. Initial revenues may be very less in this scenario or even to nil in certain cases but can be enhanced with proper usage along with the passage of time.

The second contextual factor focuses on the relationship of the ICT industry with its potential consumers. ICT has an important role in the customer-organization relationship, where customer satisfaction, commitment and loyalty are the influencing variables (Bauer et al. 2002). In the technology-deprived areas, in initial adaptation of customers to certain technology, customer satisfaction is closely dependent on the eradication of their fears regarding usage of ICT. Meanwhile, the respective industry can be provided with certain help from Universal Service Fund (USF) for its financial soundness. When the consumers get used to the provided technology and start getting certain benefits from it, they would be happy to pay for certain advanced services and devices for their benefits in life routines.

Third contextual factor is the community, i.e., whether the consumers work cooperatively or competitively. It also influences the collaborations among the consumers of the ICT particularly in those areas where urban and rural segmentation exists. Communities with collectivist culture, where cohesion among consumers exists, have a good collaboration among its consumers. Zegers (2017), while mentioning the influencing factors for possibilities and limits of urban and rural collaboration, considers ahead mentioned factors like mutual respect, understanding, and Trust; appropriate cross-section of members for learning and complementing each other through knowledge sharing, experiences, and ideas; ability to compromise; shared vision; and open and frequent communication at the base of collaboration. Although Zegers (2017) has mainly focused on collaborations in tourism and leisure, yet they appropriately focus on the factors that influence community collaborations in urban and rural consumer segments.

The diversity existing in the community members can act as a welcoming resource for the penetration of the ICT that can enhance creative solutions for the end consumers. The consumers as well as service providers exercise the choice of the services. This choice mechanism hinges upon income status of the consumers versus their needs and cost of service by the service providers. In rural areas of the developing countries, the agricultural economy exists and is quite a cumbersome job to induce technology-based needs as preferred requirements in the people's mind where customer's decision is vital. Servon (2002) mentions factors of cost, infrastructure, discrimination, policy, and culture as interacting ones to keep certain groups from being able to participate fully in the information society. The power structures existing in consumer's communities were one land lord owns the complete resources of land and man power. The benefits of the rest of the community may get compromised. Therefore consumer's needs, income status, and behaviors directly affect the penetration of the newly disseminated technology in such power central communities. Lovink (2016) sets out a vision of an alternative pathway using "collective awareness platforms" to resist unequal power structures. While implementing ICT-based solutions, the principal interest should not be attainment of market dominance. Fostering alternatives to the commercial market such as collaborative sharing in ICT-based relationships should be encouraged.

There is a close relationship between social change and ICT that it has both positive and negative effects. ICT deployment in developing consumer markets follow a sequence where ICT advances create a new opportunity to achieve some desired goals. It is followed by an increase in demand for alterations in the existing social structure of the organizations for availing the new ICT opportunities. The real issue faced by the ICT penetration is the gradual decline of the older or existing social structures which have helped the society to feel connected on a more personal basis. However, the preceding statement doesn't intend to deny the use of ICT for building an information society, rather lays emphasis towards the emerging need for inter-disciplinary researches specially between the ICT-based subjects, economics, communications, as well as digital humanities.

56.2.1 Role of ICT Industry in Faster Penetration

Industry acts as the center of gravity for ICT penetration. Service implementation and its timely delivery that is compatible to the needs of the consumer market is the main concern particularly in rural areas. In order to win certain laurels in the underserved areas of the world, ***technology infrastructure*** needs to be cost effective for deployment as well as in terms of maintenance. For example, where the issue to provide connectivity to a larger area arises, broadband mobile connectivity such as cellular and Wireless Local Loop (WLL) services can provide the unparalleled large coverage in the wireless realm. The limitless mobility as well as the restricted mobility options provided by wireless service provider allows networks to be built with far fewer optimized cell sites. Fewer cell sites covering a larger diameter of an area can ensure reduced operating expenses, which results in savings to both operators and consumers and hence allow faster penetration. Technical support industries can also play a crucial role to produce, distribute, and maintain PCs, software, manage E-financing facilities, and administrate networking. Certain solution-oriented industries that aptly address the consumer needs in order to survive well with easiness in their lives are essential for the digital revolution that can gradually architect the information society. The mobile industry is one of those industries that are fully committed to achievement of sustainable development goals, as a universal plan has been unanimously adopted by 193 countries under United Nations (GSM 2017). Better networks of the mobile operators support disaster communication and broadcast services. This makes them to provide accurate and timely information to the affected communities. Rolling out in remote areas by making mobile services more affordable to the poorest individuals along with efforts to accelerate the digital inclusion for the women are some of the steps in progress by mobile operators. Hence, the ICT industries require appropriate human and technological resources that help in the rapid development of ICT in the society.

56.3 Strategies to Address ICTs Consumers' Right in Underserved Areas

The empathetic need for the consumer-based solutions through simple and prompt mechanisms is required in underserved areas. In the backdrop of the above parameters, certain strategies to address the ICTs consumers' rights particularly in the underserved areas have been suggested as follows.

56.3.1 Open Regulatory Framework

It can catalyze the digital connectivity by allowing innovative communication technology to be used for providing connectivity between communities particularly in rural areas. Communication infrastructure, including telecommunications, internet connections, and broadcasting should reach not only urban high-income class but also connect semi-urban as well as rural areas. Servon (2008) states unequal investment in infrastructure between poor urban areas and rural regions as a source of market failure; where private companies will invest more likely in those areas where they can yield more return on investment. Such situations warrant government intervention. It is further mentioned that "same places that are characterized by economic poverty also tend to suffer from information poverty." In developing countries, the main infrastructures are usually owned by their respective Governments. This government ownership is of an authoritative style where provision of technology-based solutions provided by private ownership is monopolized by Govt.-owned stakeholders of the ICT domain. Open regulatory frameworks can ease the invasion of the private sector for provision of services along with providing healthy competition for the govt. sector. The technological interconnects between the govt. and private stakeholders of the ICT can simplify the access of the technology-based services to the society. For example, in case of Pakistan, the biggest terrestrial network for the media broadcast, i.e., Pakistan Television, is owned by Govt. Private Operators have to go through a rigorous procedure despite de-regularization of media. At times, certain exclusive categories like govt. exclusive areas and content that may vary from the live coverage of a parliament session to the cricket match remain exclusively of govt.-owned operators. The interconnection of government and private sector producers can establish giant networks of ICT and ensure maximum utilization. Hence, the operating costs for the stakeholders, as well as the access costs for the end consumers can be lessened to a most affordable cost that maintains a considerable amount of profit.

56.3.2 Use of Universal Service (US) and Universal Access (UA)

This is a generous mechanism for addressing the consumer rights. GSM (2012) explains the difference between, though closely related yet very different terms of, Universal Service (US) and Universal Access (UA) as: "US refers to the provision of telecom's services to all households within a country. UA refers to the provision of services on a shared basis. UA programs typically promote the installation of public payphones or public access businesses in rural villages or low-income urban areas with the aim of providing basic telecoms services." The provision of US and UA ensures accessibility and usability of the ICT services that leads towards acceptability of Information Society Technologies by anyone, anywhere, at any time, and through any media and device, ensuring an ease of access. Mobile operators are providing UA at an unimaginable pace in the majority of developing markets. The ICT penetration rates of US have increased due to lower costs and high speed of ICT services deployment, thereby meeting demands for basic voice services in a rapid and flexible way. It has eliminated barriers for low-income communities to obtain subscription and usage of communication services. GSM (2012) further elaborates that the mobile operators have overcome the barriers in their services for the developing marketers like removal of virtually all bureaucratic formalities, non-monetary entry barriers to service access via simple pre pay model, low-cost initial access, and budget control concerns of low-income people. According to GSM (2018), "To support customer migration and further drive consumer engagement in the digital era, mobile operators will invest $0.5 trillion in mobile capex worldwide between 2018 and 2020." In this upcoming scenario, underserved communities have the potential to become a skilled consumer market, as their consumer interests vary largely and vividly from the high income communities. UA safeguards that all people have practical means of access to a publicly available telephone and emergency services in their communities. For US and UA, the mobile network operators enjoy dominated de facto status in ICT services because of their economical reach, as well as flexible mobility and affordable mobile handsets with an affordable tariff package for niche communities. The rapid transition of GSM has enabled technology to reach an appealing domain with very less barriers for example transition of ICT service networks in developing countries to contemporary technologies like GPRS, EDGE, 3G, and HSPA. In this way, mobile operators can also offer enhanced data, facsimile, Video Conferences, Internet, and various advance ICT services.

56.3.3 Emphasis on Content Industries

Emphasis on the content industries is the most powerful mechanism that should be flourished. Its implementation can provide relevant information that addresses the local people's need. Developing countries do access foreign content via cost-effective methods (e.g., in certain developing countries like South Asian countries cable TV operators access foreign transmission via dish head ends and then use cables for distribution to the local community at their affordable prices). All of the content produced from foreign countries is in foreign languages such as English and is further dubbed to Hindi/Urdu language, especially in case of cartoons and drama genre of the media content. Though access of the international market products to local markets can be provided by the innovative usage of technology, availability of good and culturally relevant content, a prerequisite for the development of the Information Society, is missed out. As the local people cannot develop any relevance and actual understanding of the disseminated information; hence they remain deprived from the main essence of information that can help them in developing a true information society. This ultimately creates a gap between the society and the information being provided to them. For instance, if in an area clay pottery is done, in other local area stone carving and stone decorative items are made, and in another small community the local habitants have the skill of local food-making. In all such diverse scenarios, the different usage of technology in the form of addressing the pure local needs in their native languages with a customizable, user-friendly technology is required. Once this issue of putting up technology in their native language is addressed, the technology can provide enhancement of such under developed areas. In this way, the consumers can start owning the ICT at an affordable and suitably customized way according to their specific requirements. This can ultimately bring a level of most efficient and effective way for usage of ICTs.

The following exemplary cases from Pakistan depict the worth of manufacturing and thus disseminating the information in local language indigenous content via ICT in underserved areas.

The options for local language content in the native understandable context are missed out especially in the media. This creates a hollow zone for the underserved developing areas which are already creeping with the deprivation of adequate quality of education and awareness. The usage of the local language content in the dissemination of the provided ICT-based solutions can ease their proper implementation and usage by the local community. For example, in Karachi, a city with the largest population of Pakistan, existing slums at least know about their consumer rights due to the efforts of NGOs. They have started to use basic technology gadgets like cell phones and now use them for conveying their complaints to the govt. officials for the access of clean water and basic sanitation rights in their area; they have obtained some success to information by using ICTs in this regard. In KPK province of Pakistan, the provincial govt. has initiated a smart phone-based service to promote the citizen participatory journalism, where residents of the society can put up their problems (e.g., area cleanliness, traffic issues, and health issues, etc.). This can be done via smart phone technology by capturing videos, pictures, or even texting to bring the issues in government's notice for providing solutions. This strategy can be more productive if is deployed using native language of the consumers.

The Pplice department in Punjab (the most populated province of Pakistan) has launched a traffic regulation smart technology-based system for the masses to assist them. Recently, the Govt. of Pakistan has launched a smart phone app named, "Roshan Pakistan" for creating awareness to the masses of the society about the dissemination of information relating daily utilities, e.g., load shedding schedule, billing info, net metering, and bill calculation, etc. In 2018, under active citizenship program by British Council in Pakistan, kitchen gardening mobile apps have been created for giving an awareness and training to grow and consume organic foods. Such creation of the local content in native language and delivery via ICT-based devices and services enjoys a welcoming response from the consumers. This ultimately leads towards the strong buildup of an information society.

56.3.4 Interconnects Between Government and Private Organizations

Largely self-sustainable business model for the backward rural areas is crucially required. Because by having an increase in the rural teledensity, there would be an increase in the GDP per capita. This may result in the form of the index of prosperity in these developing countries, where about 70% of the population resides in rural areas.

More project initiatives with the collaboration of private organizations and governments need to be taken that eventually result in the access of modern digital technology-based facilities with lower cost of production for producers, such as by subsidizing projects, which would resultantly be more effective and viable for the local habitants. Kshetri (2014) points towards a crucial aspect about collaboration and cooperation among ICT stakeholders for fostering a data ecosystem required for development. He further says that "creation of appropriate databases may stand out as particularly appealing and promising for some entrepreneurial firms. Governments, businesses, and individuals are willing to pay for data when they perceive the value of such data in terms of helping them make better decisions. In the meantime, policymakers, academics, and other stakeholders should make the most of what is available."

Rather than constructing new infrastructure to facilitate for such services, already established infrastructure can be leased in terms of lower inter connect costs. Several countries in Europe are using infrastructure sharing model to reduce the cost of offering telecom services. Some of the examples are white-zone concept in France and sharing of infrastructure between two mobile operators in Australia to offer 3G services. Such scenarios need to be addressed with easy terms and conditions between the Significant Market Players (SMPs) and the newly emerging service providers, particularly in the developing countries.

56.3.5 Role of Social Change Agents

To build an Information society and to utilize ICT for socially beneficial purposes, social change agents' roles are essential. In order to fuse the two ends of the rope, where technology-based solutions reside at one end and consumers of the underserved areas on the other, consumers need to go through a systematic process of changed yet beneficial and productive life styles. Humans intrinsically fear change, so social change agents can act as a suitable aid in this regard. Certain technology-based U-turns can be induced according to the needs of the local community that could lead to enhance connectivity with the rest of world at massive scales. For that sake, change agents may even be chosen from the local community. They should be trained well with ability to use and implement innovative models to bring about a change in the social process. Education, health, environment, governance, and community social problems are the typical areas where ICT will help create real change. In rural and underdeveloped areas, first of all the living customs need to be explored and analyzed to determine the ground reality. For instance, a wide gap in women empowerment can be filled using ICT. In these underserved areas of the world, uneducated housewives/ladies do embroidery over dresses, caps, bedsheets, etc. as a primary source of income or in some instances as a hobby. Moreover, some women have skills of intertwining wooden beds, chairs, making baskets, and various decorative materials. Women do make hand-woven clothes, shawls, carpets, and rugs with spun into yarn on a spinning wheel called charkha and hand-woven khadi. Such women can be provided with digital assistance such as an online educational course in textile and trained accordingly to develop a knowledge-base in modern embroidery and cloth-making leading to an increased productivity. A social change agent provided by the supplying ICT agency can be initially trained to run a digital telecenter for such women. Such a working scenario can be enhanced by gradually networking via a Wireless Local Loop (WLL) service in accordance with other areas, resulting in a financially beneficial connectivity through various ICT platforms on a national or even a global scale.

56.3.6 Use of Community Media in Connection with ICTs

In certain cases, the needs of a local community may conflict with that of its subcommunities. Such conflicts may lead towards ICT producers for diversifying their product and service range to meet consumer-specific demands. Hence local community radio stations (that played an influential role in both world wars), TV stations, and smart phone apps can be precisely programmed to provide and address the intrinsic local demand of consumers in communities of the developing world. INFOdev (2003) explains the technical challenges faced by online community radio stations as; "implementing Streaming online community radio sessions for addressing the community issues need adequate bandwidth that is required to obtain a good quality stream, streaming compatibility with media players running on all operating platforms and obtaining open source workstations (Linux, etc.)." However, in underserved areas, the community radio has to pave through an uncertain trajectory due to a lack of existing resources that can sustain an always-on internet connection with good bandwidths. Mobile service providers can link this gap by providing certain value-added services for community radio service, as mobile economy is the most committed one with the sustainable development goals.

Certain other obstacles need yet to be resolved for the optimum implementation of the ICT solutions in the neglected areas of the world. The upcoming chapter further discusses the solutions for bridging challenges faced for initiating ICT solutions in developing countries.

56.3.7 Gauging Consumer Community in the Developing Countries

Developing countries experience a lack of connectivity via broadband connections that require digital and expensive satellite equipment. To begin with, developing countries are lagging behind where technology-based industries either do not exist or if found, they function on a small scale. For example, the absence of an industry's satellite systems used for overlying the information broadcasts within the country or for the foreign exchange of information. In either scenario, real-time broadcast or buffered data in form of recorded broadcast is required. It is expensive hence much beyond viable for the masses of developing communities to opt for such options without prior knowledge.

Enabling the common people, especially the illiterate women, which are in sheer need of advocacy, may prosper by using ICTs to gauge the intensity of their problems. In this backdrop, particular strategies for influencing disseminated information and communication policies need to be worked out from a gender perspective for creating gender sensitive ICT guides, in order to prevent possible chances of discrimination. This could integrate ICT into their poverty-related issues, common conflict avenues and local governance issues. From various case studies round the globe, it became evident that an individual does not have to be fully literate to utilize ICT. This discovery has increased the impact of ICT within low-income societies, where education is not considered a necessity for existence.

Funding for ICT-based projects is another challenge, to be proactively managed in a timely and relevant way, which would cover education and technology costs. Personal/convenient and synchronous/asynchronous communication can be enabled by mobile phones. However, Chipchase (2008) has identified problems of limited formal education and lower levels of literacy in emerging markets. He further considers, "literacy is often a result of lack of opportunity rather than ability" and has thus coined the term "Textual literacy." Addressing needs of those who can afford but are textually illiterate in developing countries should also be the target of ICT-based solutions.

56.3.8 Ease of ICT Platforms Usage

ICT-based solutions are effective only if the consumers can access them with ease. Wiendenbeck (1999) explains that the perceptions of ease of use were consistently better for the icon-labeled interface. Whereas by conducting field researches with relevant participants from common masses, Chipchase (2008) describes an icon-based mobile interface is more effective for developing communities that are textually illiterate like cooks, cleaners, fuel station attendants, sweepers, uneducated yet skilled women, etc. Illiterate user's willingness to explore features on the mobile phone demands more proactive and solution-based research. Cardiac (2013) while addressing the area of accessible and assistive ICT mentions "Icons used for ICT interfaces must be easy to understand." The market for illiterate users is vast with potentially growing margins. Therefore, a minimal feature mobile set is the most adequate for such illiterate developing communities where audio feedbacks, spoken menus in local languages with native dialect support can bring very personal, easy-to-use, and meaningful ICT solutions. Effective serving for this potential market can not only improve the lives of illiterates massively but also act as a springboard for enhancing literacy. Maximizing the extent of the ICT-based solutions is a demanding task where societies dwell on the basis of the networked and converged mechanisms.

As national ICT strategy of Egypt focuses on four main goals as "Supporting democratic transformations, fostering digital citizenship, supporting sustainable social developments and fostering knowledge based national economy" (Egypt MOCIT 2012–2017). After meeting these goals in an underserved area it can lead towards the making of a smart city as defined by Kayat and Fashal (2019) as "that is able to efficiently mobilize technological innovations so as to anticipate, understand, openly discuss, act and serve many actors with a wide range of profiles."

ICT enables connectivity within and among cities (Kayat and Fashal 2019). A standalone computer can offer only minimal utility. Therefore, networking is essential to maximize gains. As interactions go digital, these can be coordinated over greater distances, creating new communities of interest and new challenges for governance. While dealing with networking, speed and security would be the prime concerns. In addition to these challenges, a skilled human resource is required as Phani (2006) concludes that "Technology works only if people have the right training to execute it effectively."

56.3.9 Presence and Solutions of the Digital Divide

Worldwide digital divide is a big challenge for developing information societies. Krizanovic Cik et al. (2017) explains about the improvement of universal access to digital information sources. The inequalities related to the levels of deployment and adoption of broadband solutions between rural and urban areas lead towards the digital divide. For lessening the digital divide, it requires an emphasized focus towards the needs and abilities required for improving the digital skills. Moreover, access to knowledge base and awareness about the underserved population also matters. Using techno-economic methodology, this digital divide can be covered. It involves finding the business strategies of ICT operators along with the relevant case studies that may also greatly help in reducing the digital divide.

Additionally, for some developing and even developed communities, all this change in the name of technological innovation pathway is singular and it is often depicted as being inevitable even if "there will be heartbreak, conflict, and confusion in addition to incredible benefits" (Kelly 2016, p. 267). However, Mansell (2017) debates that "these outcomes depend on multiple technical and nontechnical factors, and it does not follow that these developments will improve the overall quality of life or contribute to reduced social and economic inequality."

While (Park 2017) has proposed for lessening the digital divide existing between rural and urban Australia that for implementation of digital inclusion strategies, "both supply (infrastructure) and demand (education levels, industry sector, employment opportunities, socio-demographics) factors must be considered." Within the same society there exists segmentation of "haves and have nots" that ultimately widens the technology gaps. Although, such inadequacy creates, at times, certain difficult situations for both the "haves and have nots," as both have to go through the process of change, but lack alternative solutions that aptly address the differing needs.

56.3.10 Consumer Involvement in Policy and Decision Making

The non-involvement of consumers in the process of policy-making and decisionmaking in developing countries is another challenge in this domain. Siddiqui (2005), while discussing the reasons of under development, contemplates that a natural repercussion to planning and a development paradigm is that consumer needs are distantly identified and implemented by the administrative departments. She further considers Pakistan's planning and developing process as "flawed" because it excludes people from the decision-making process where the whole system is quantity oriented where targets and numbers are more important than quality or satisfaction of the beneficiaries. Involvement of the communities in the whole process of planning and development should start from identifying the needs and locations, to designing projects, and their implementation and maintenance.

56.4 ICT-Based Services

ICT can be beneficial in providing facilities that can enhance the human living experiences via usage of corresponding devices and services. Some of them are mentioned as below.

56.4.1 mHealth Services

Around the globe, the use of wireless and mobile communication technologies has a unique potential to transform the delivery of mobile health services (mHealth). There are now over five billion wireless subscribers and over 70% of them reside in low- and middle-income countries. The GSM Association reports that commercial wireless signals cover over 85% of the world's population, extending far beyond the reach of the electrical grid. For determining the status of mHealth in 114 member states, 14 categories of mHealth services were surveyed. The 14 surveyed categories of mHealth services were: health call centers, emergency toll-free telephone services, managing emergencies and disasters, mobile telemedicine, appointment reminders, community mobilization and health promotion, treatment compliance, mobile patient records, information access, patient monitoring, health surveys and data collection, surveillance, health awareness raising, and decision support systems. Successful implementation of ICTs in these above-mentioned categories particularly in case of emergencies and mobile telemedicine can be substantially beneficial for the underserved areas. Regarding barriers to implementation of mHealth services, the conflicting priorities resides at top with 52% whereas the lack of knowledge concerning the possible applications of mHealth and public health outcomes comes at second barrier with 47%.

Mechael (2008) considers usage of mobile phones within general population and among health professionals as creating new paradigms. Access to emergency and general health services can be improved by the coordination and collaboration amongst users. The increased efficiency and effectiveness of health service delivery by using ICT-trained health workers from the developing community can bring prospective results. Yet certain areas for remote patient monitoring, consultation, and infectious disease prevention with control and telemedicine are a target that can be achieved by using the ICT trained local health workers from both gents and ladies, along with participatory connections with the health professionals for consultations. This needs first to bridge up the gap between the conflicting priorities regarding mHealth as well as the knowledge required for using mHealth services.

56.4.2 E-Rate Program

The use of an E-Rate universal funding mechanism as employed in developed countries can address the technology gap by defraying the connection costs at public educational institutions. This E-Rate program can provide 20–90% discounts on the telecom services, internet net access, and internal wireless connections for schools and libraries that serve poor and rural communities.

Servon (2004) considers not only the innovative work under way by the local ICT sector but also the active engagement of the public sector for bridging the technology gap. The policy makers in favor of E-Rate program also believe in the creation of a skilled workforce that can improve educational opportunities by its usage. This can vitally bridge up the existing digital divide.

56.4.3 IoT Using ICT Devices

ICT-based devices, services, and apps in Internet of Things (IoT) do vary over a wide range as ITU (2017b) report mentions: “objects in connected homes (e.g., appliances such as refrigerators, washers, dryers; kitchen and cooking tools; and applications that manage home security) to devices supporting better energy management, heating and cooling, health monitoring devices and lifestyle devices.” Networking of these ICT-based devices for wider data collections can be ultimately mined. It can create enhanced value services as mentioned above. This ultimately leads towards the better decision-making by consumers who can potentially gain access to more accurate and detailed information. ITU (2017b) report also explains the usage of ICT-based apps for smart homes, in health domain, energy management, precision agriculture, environment, transportation, and security and emergency avenues. It further describes that "Internet of Things (IoT)," cloud computing, big data analytics, and artificial intelligence all have useful applications on a standalone basis. Hence, potentially much higher benefits can be realized if they are used jointly. Lastly, ITU (2017b) mentions that previously mentioned four technologies form a highly complementary Innovation system if IoT is combined with data analytical capability.

56.5 WLL Usage in Pakistan: A Case Study

A case study where Wireless local loop (WLL), a specific genre of ICT, has been used in Pakistan with a particular focus on increasing the teledensity in the converging era of mobile ICTs, specially. The writer, however, tends to emphasize the usage of the same WLL ICT in the digitally divided underserved areas where consumer market needs to be identified in their peculiar diverse cultures.

The consumer's ability to consume technology-based solutions for their needs can increase the teledensity in an area. WLL technology was launched in Pakistan with the vision to increase teledensity with the availability of the telecom facilities at affordable prices to the larger population stream in the country, living in rural areas. As teledensity is referred as to the ratio of telephone connections to the population. WLL is mostly making use of CDMA technology that consistently provides better capacity for voice and data communications than other technologies. Being the common platform for 3G technologies, it has outstanding voice and call quality as it filters out background noise and cross-talk. CDMA's spread spectrum signal provides an unparalleled large coverage in the wireless domain of ICT. It allows networks to be built with far fewer cell sites than is possible with other wireless technologies. Fewer cell sites covering larger areas ensure reduced operating expenses, which results in savings to both operators and consumers. These are the factors which allow fast penetration levels along with expeditious network deployments as intrinsic advantages of the WLL.

Regarding the cellular mobile industry, Pakistan has performed excellently as the most progressive economy in establishing a thriving mobile communications sector in the past years and has been recognized by the achievement of the GSM award. To win similar laurels in the field of WLL industry, certain policy matters are required to be addressed and recognized globally. There is a need of reviving regulations and licensing practices by waiving off the vague restrictions in the WLL industry that may pose to the industry groom. For example, the spectrum offered to the candidates must be cleared from the current occupants before the auction takes place. Hence, the same spectrum should not be allocated to more than one operator in the parameter regions to avoid service degradation, such as low voice quality for subscribers.

Currently, PTA (2017) conveys that there are a total of 14 WLL operators in the whole of Pakistan, operational in all 14 telecom regions or in certain selected licensed areas. Bhatti (2007) mentions the greatest share is of the state owned company PTCLV with 59% share and the rest main market share is held by the three private operators. Gradually a total of six WLL operators terminated their operation as inferred from the PTA report of WLL status of operators in Pakistan. If we have close quarters look on the WLL industry, it is under a threefold pressure. These pressures include:

(a) ***Regulatory institutions*** *which* implement policies legislated by the government, monitor performance for ensuring standards and quality of services, and enforce compliance of the industry with the licensing conditions.
(b) The second corner is ***Government*** which issues directions on the matters of general policy and procedures by imposing license conditions as well as conferring regulatory rights to the industry, for self-regulations.
(c) The third corner is of ***Consumers*** who exercise the choice of the services as well as service providers. Their choices definitely confine them with respect to their living area being either urban or rural, hinging upon then their income status as well as their need for a particular service. This service may particularly be the invasion of technologies arrayed against one another through competing services in rural areas of developing countries where agriculture holds the economy. Thus, it becomes quite a cumbersome job to induce technology-based needs as preferred requirements in the peoples' mind. Here customer's decision is vital. Therefore consumer's needs and behaviors directly impact the ICT industry.

Huysman and Wulf (2004) considers ICT nurtures knowledge sharing by those who are mainly eager to share. ICT enhances the *efficiency of the consumers* with which they can communicate their knowledge to the other group members, regardless of time and geographic location. ICT's efficiency route gives consumers the greater enhanced efficiency for communicating and sharing their knowledge with others. They further elaborate that ICT facilitates a communication style that fits the consumers' attitude towards knowledge sharing, primarily oriented towards just sending rather than aiming to create a common view. Thus the contribution of ICT towards a more collectivist climate is much less clear, as it is expected to be dependent on the degree to which the other richer means of communication are used. If no common bounds are developed, no group feel emerges that leads towards a mismatch between communication needs and the medium characteristics. But those consumers for whom efficiency to communicate matters less rather than creating a mutual understanding and common view, ICT offers the collectivist route. For its usage, the group consumers of ICT should have a common social identity or collectivist norm that is based on a common identity. PTA (2017) mentions the state of the fastest growing teledensity in Pakistan has increased from 13.7% to 71% in last decade. Whereas the WLL-based teledensity has declined visibly. This is because if the inappropriate treatment to the WLL technology, though it has the best inexpensive potential to serve the consumers of the underserved areas.

In this backdrop, by developing an understanding about the undeserved consumer market can guide for creating alternative avenues for ICT usage. This can help in ensuring the consumer rights in the underserved areas in current contemporary convergent era of ICTs.

56.5.1 ICT Alternative Avenues in Underserved Areas of Pakistan

The current ICT era is offering a vibrant discourse to the dynamically changing consumer needs in their native and global scenarios. The potential of the alternative avenues for addressing the consumer market needs is analyzed here in scenario of Pakistan's underdeveloped areas. Pakistan is very fortunate as it is currently experiencing an increase in telecom growth at a time when various technological developments are taking place. This development is set in a direction that encourages low-cost telecom technologies in access devices, access networks, switching networks, and transmission technologies. The challenge today is that the regulatory system should be such that the growth and deployment of all these technologies should receive a stimulus. According to Pakistan's Telecom Regulatory body (PTA), its annual report of 2017 with major populations living in rural areas have only a very small percentage of WLL teledensity. Therefore, there is a big market for the introduction of latest technologies in these rural areas. For ensuring the increase in the teledensity of the country, bold and prudent steps are required to be taken both generally by the ICT industry and particularly by the WLL Industry along with ICT regulators.

These steps are enumerated as follows:

- The aforementioned empirical data from PTA (2017) closely reveals the fact that the WLL industry in Pakistan (although being the most efficient, and of high quality as well as cost effective) is still deprived of the environment required for its flourishing phases. The regulatory body has limited WLL's mobility to one BTS cell site, i.e., about 60 km of its service area. Yet industry operators might intend to expand their services to establish more access areas for increasing their return on investment (ROI). Mobile sets with limited mobility options can best serve the needs of consumers in rural areas where initially, info services can be disseminated to the local community such as; educational audios, health training audio manuals, awareness about citizens' basic rights at very affordable prices. These can be categorized in the form of value-added services, besides basic audio telephony service to build small-scale info communities depending on the niche areas. To conclude, industries in this neglected region can never groom up until and unless their outputs are made easily available to the common man at affordable costs. Hence govt. and private partnership mergers to initiate and launch such ICT-based projects can be proactively used in this scenario.
- In Pakistan, teledensity ratios are low, especially amongst the provinces. These provinces may possess a lesser market potential due to low literacy rates and a lack of government interest required for its development. Moreover, regions in the vicinity of rural areas have an emerging potential for increasing teledensity. In recent events, Catco International has undergone an agreement with Telecard to start a project in order to enhance the efficiency and performance level of Govt. schools in Sindh, the second largest province of Pakistan. With the collaboration of Sindh educational ministry, a wireless network will be established using WLL connectivity. In this scenario, more project initiatives need to be taken which

would ultimately result in the access of WLL facilities at cost-effective rates to producers that are also affordable to the local habitants.

- In the scenario of women empowerment, a thorough, suitable use of ICT in developing rural areas of Pakistan, a case study of District “Umerkout” was carried out. Here the local women of “kunri” village have the potential for making “Rali,” a specific style of dress-making for women using applique work. But the local shopkeepers pay them a lesser amount for their hard work. Using ICT, an E-business initiative can provide a brand name to the local women that can bring adequate wages to match their hard work. This innovation needs to be diffused into their immediate community. Women can be empowered either by training them for certain applications over user friendly gadgets in their respective native languages; i.e., applications specifically designed for their needs or even establishing certain telecenters with a few properly trained native females. This would also lead to the empowerment of the rest of the networked women in that area.

56.6 Conclusions and Recommendations

Area of ICT is very dynamic where connectivity leaps on progressive pace. The emerging trends in ICT are interrelated, leading towards a networked convergent avenue. For complete coupling of the economic and social benefits of these ICT developments, efficient and affordable physical infrastructures, services, and more advanced (yet consumer needs’ specific skills along with public supportive) policies are required in developing areas. The presence of digital divides and inequalities affects the extent to which information societies can contribute towards the economic and social development of different regions, countries, communities, households, and individuals. Recent widespread and dominant growth in broadband networks and services, particularly mobile broadband, has started to intensify the economic support development as well as individual empowerment. Yet digital gender gaps as well as digital age gaps have emerged as new forms of digital divides in the developing societies where ICT development Index (IDI) needs monitoring in order to bring improvement.

Interventions in developing societies are at times crucial so that scrutiny by policy makers, for enabling of the commercial solutions to be groomed as well to avoid time and resource wastage. In certain areas, bottom to up approach (Consumers to ICT) is required whereas in other areas top to bottom approach (ICT to consumers) for invasion of ICT solutions is effective. It may involve the technology-first or the consumer-first approaches or vice versa.

Prices of ICT-based services as well as communication hardware that are the actual needs of the consumers need to be more affordable, especially in the digitally divided communities. Focused mediations on such areas definitely need assistance using appropriate predictive programs. Best practice, time-bound evaluation and adjustment USFs may be best executed during a limited period in a country’s development sector. Afterwards levies should be phased out as they will no longer serve the purpose. If addressed appropriately using the strategies discussed in this chapter, the developing, low-income communities will be able to create a multilayered information society with multiple audiences of the ICT services who consume as per their niche interests. Consultation between the industry and government along with involvement of local consumers who have to actually use the ICT-based solutions should be a routine process of consultation in public policy-making. For strengthening the functioning of the consumer market working collaborations between the government and industry can lessen the access gaps especially in the geographically largest and least developed regions.

ICT developments can lead towards a new ICT-based ecosystem for the proliferation of ICT by improvement in their performance, decreased costs along with networked connectivity for commercial user generated contents. The roll out and continuous addressing of the consumer needs while ensuring their rights along with upgradation of the infrastructures and joint interconnect ventures between giant and small entrepreneurship can allow co-investments with sufficient licensed and unlicensed spectrums. This allows for new smart cities that can safe guard the information provision to consumers with innovative services and solutions. Technology just for the sake of technology never works and cannot sustain in any domain, until and unless its objective is focused towards benefits and ensures rights of consumers at certain affordable prices.

The manufacturing of consumer-specific information derived from their relevant cultural origins can empathize the consumer's actual needs. Foreign impositions of ICTs can inculcate only the superficial usage of the gadgets but what and how to go with these ICT gadgets can be only obtained with the detailed analysis and understanding of the indigenous individual constituting the society. Ultimately such scenarios can build up the trust of the consumers and can prepare individuals who make better use of the opportunities of a digital future, where human mind frame of feelings do exist as well.

Building information societies demand equitable access and active participation of potentially all the citizens in the society. This requires theoretical, methodological, as well as empirical research from both technological as well as nontechnological perspectives where empirical results, reviews, case studies, and best result-oriented practices can be fruitful.

References

Alderete MV (2017a) Mobile broadband: a key enabling technology for entrepreneurship? Technol Innov Small Bus 55(02):254–269

Alderete MV (2017b) Examining the ICT access effect on socioeconomic development: the moderating role of ICT use and skills. Inf Technol Dev 23(1):42–58. https://doi.org/10.1080/02681102.2016.1238807

Bauer HH, Grether M, Leach M (2002) Building customer relations over the internet. Ind Mark Manag 31:155–163

Bhatti B (2007) WLL market overview and trends in Pakistan. http://telecompk.net/2007/08/15/wll-market-overview-and-trends-in-pakistan/. Accessed Oct 2017

Brynjolfsson E, McAfee A (2015) The second machine age. W.W. Norton, New York
Cardiac-EU.Org (2013) Pictograms icons and symbols. http://www.cardiac-eu.org/guidelines/pictograms.htm. Accessed 15 Oct 2017
Chipchase J (2008) Reducing illiteracy as a barrier to mobile communication. In: Katz JE (ed) Handbook of mobile communication studies. Massachusetts Institute of Technology, Cambridge, MA, pp 79–89
Egypt Ministry of Communication and IT (2012) National ICT strategy 2012–2017. http://www.mcit.gov.eg/Upcont/Documents/ICT%20Strategy%202012-2017.pdf. Accessed 21 Jan 2019
GSM Association (2012) How mobile can bring communication to all. GSM Association Universal Access Report. https://www.gsma.com/publicpolicy/wpcontent/uploads/2012/03/universal accessfullreport.pdf. Accessed Oct 2017
GSM (2017) Mobile industry impact report: sustainable development goals, available on https://www.gsmaintelligence.com/research/?file=622ab899f558a6ab3b7f14881f0f031e&download
GSM Association (2018) The Mobile Economy. https://www.gsma.com/mobileeconomy/wp-content/uploads/2018/02/The-Mobile-Economy-Global-2018.pdf. Accessed 14 Dec 2018
Hargittai E (2008) The digital reproduction on in-equality. In: Grusky D (ed) Social stratification: class, race, and gender in sociological perspective. Westview Press, Boulder, pp 936–944
Herman E, Chomsky N (2006) Manufacturing consent: the political economy of the mass media. Vintage, New York
Huysman M, Wulf V (2004) Social capital and information technology. In: Hooff B, Ridder J, Aukema E (eds) Exploring the eagerness to share knowledge: the role of social capital and ICT in knowledge sharing. MIT, Cambridge, MA, pp 163–183
Infodev (2003) ICT for development contributing to the millennium development goals, lessons learned from seventeen infoDev Projects. The World Bank. https://www.infodev.org/infodevfiles/resource/InfodevDocuments_19.pdf. Accessed Oct 2017
ITU (2017a) ICT facts and figures 2017. https://www.itu.int/en/ITU-D/Statistics/Documents/facts/ICTFactsFigures2017.pdf. Accessed Aug 2017
ITU (2017b) Measuring the Information Society Report 2017. https://www.itu.int/en/ITUD/Statistics/Documents/publications/misr2017/MISR2017_Volume1.pdf. Accessed Nov 2017
Jaju S (2006) Delivery of civic services online: the Saukaryam way. In: The state, IT, and development. Sage, New Delhi, pp 214–227
Kayat GA, Fashal NA (2019) Inter and intra cities smartness: a survey on location problems and GIS tools. In: Smart cities and smart spaces: concepts, methodologies and applications. IGI Global, Hershey, p 32
Kelly K (2016) The inevitable: understanding the 12 technological forces that will shape our future. Viking Press, New York
Knieps G (2016) Internet of Things (IoT), Future Networks (FN) and the Economics of Virtual Networks. Paper presented at TPRC 44: the 44th research conference on communication, information and internet policy, Arlington, 30 Sept–1 Oct 2016
Krizanovic Cik V, Zagar D, Grgic K (2017) Univ Access Inf Soc. Springer Berlin Heidelberg. https://doi.org/10.1007/s10209-017-0560-x. Accessed Nov 2017
Kshetri N (2014) The emerging role of Big Data in key development issues: opportunities, challenges, and concerns. Big Data Soc 1:1–20. https://doi.org/10.1177/2053951714564227
Lemstra W, Melody WH (eds) (2015) The dynamics of broadband markets in Europe: realizing the 2020 digital agenda. Cambridge University Press, Cambridge
Lovink G (2016) Social media abyss: critical internet cultures and the force of negation. Polity Press, Cambridge
Mansell R (2017) The mediation of hope: communication technologies and inequality in perspective. Int J Commun 11:4285–4304
Margaret R (2017) Information-Society. http://whatis.techtarget.com/definition/Information-Society. Accessed 4 Oct 2017
María Verónica Alderete (2017) Examining the ICT access effect on socioeconomic development: the moderating role of ICT use and skills

McCombs ME, Guo L (2014) Agenda-setting influence of the media in the public sphere. In: Fortner RS, Fackler PM (eds) The handbook of media and mass communication theory. Wiley, West Sussex, pp 249–268

Mechael P (2008) Health services and mobiles: a case from Egypt. In: Handbook of mobile communication studies. Massachusetts Institute of Technology, Cambridge, MA, pp 91–101

Mitchell A (2015) State of the news 2015. Pew Research Center. Retrieved from http://www.journalism.org/2015/04/29/state-of-the-news-media-2015/Google Scholar

Mitchell A, Kiley J, Gottfried J, Guskin E (2013) The role of news on Facebook: common yet incidental. Pew Research Center Report. http://www.journalism.org/2013/10/24/the-role-of-news-on-facebook/Google Scholar. Accessed 15 Dec 2018

Morawczynski O, Ngwenyama O (2007) Unraveling the impact of investments in ICT, education and health on development: an analysis of archival data of five West African countries using regression splines. Electron J Inf Syst Dev Ctries. https://doi.org/10.1002/j.1681-4835.2007.tb00199.x

Park S (2017) Digital inequalities in rural Australia: a double jeopardy of remoteness and social exclusion. J Rural Stud 54:399–407. https://doi.org/10.1016/j.jrurstud.2015.12.018

Phani KM (2006) ESeva: transforming service delivery to citizens in Andhra Pradesh. In: The state, IT, and development. Sage, New Delhi, pp 207–213

Pick JG, Azari R (2008) Global digital divide: influence of socioeconomic, governmental, and accessibility factors on information technology. Inf Technol Dev 14(2):91–115

Poushter J (2016) Smartphone ownership and internet usage continues to climb in emerging economies. But advanced economies still have higher rates of technology use. Pew Research Center, Washington, DC

PTA (2017) Pakistan Telecom Regulatory Authority. Telecom Indicators. http://www.pta.gov.pk/en/telecom-indicators, http://www.pta.gov.pk/en/telecom-indicators/2. Accessed 15 Dec 2017

Servon LJ (2002) Bridging the digital divide, technology, community and public policy, the information age series. Blackwell, UK, pp 77–106

Servon LJ (2004) Community technology centers: training disadvantaged workers for information technology jobs. https://doi.org/10.17848/9781417596317.Ch7

Servon LJ (2008) Bridging the digital divide, technology, community and public policy, the information age series. Blackwell, Boston

Siddiqui AT (2005) Dynamics of social change. SAMA Editorial and Publishing Services, Karachi, pp 18–23

Smith J, Brown B (eds) (2001) The demise of modern genomics. Blackwell, London

Spiezia V (2010) Does computer use increase educational achievements? Student-level evidence from PISA. J Econ Stud 7(1):1–22

Srinivasa S (2006) The power law of information: life in a connected world. Sage: Response Books, New Delhi

Stojmenovic I, Wen S (2014) The Fog computing paradigm: scenarios and security issues. Paper presented at the 2014 Federated Conference on Computer Science and Information Systems (FedCSIS)

Sunkel G, Trucco D (2011) New information and communications technologies for education in Latin America. Risk and opportunities. Santiago: N.U. Cepal. Google Scholar

The International Telecommunication Union (ITU) (2015) The 13th World Telecommunication/ICT indicators Symposium. Conclusions and recommendations. http://www.itu.int/en/ITU-D/Statistics. Accessed Dec 2017

Toyama K (2015) Geek Heresy: rescuing social change from the Cult of Technology. Public Affairs, New York

UNDP (2016) Human Development Report 2016: human development for everyone. United Nations Development Programme, New York

United Nations (2015) Transforming our world: the 2030 Agenda for Sustainable Development. Resolution adopted by the General Assembly on 25 September 2015. A/RES/70/1. http://www.un.org/ga/search/view_doc.asp?symbol=A/RES/70/1&Lang=E

United Nations Development Programme (UNDP) (2015) Human Development Report 2015: work for human development. United Nations Development Programme, New York

Unwin T (2017) Reclaiming information and communication technologies for development. Oxford University Press, Oxford

Wiendenbeck S (1999) The use of icons and labels in an end user application program: an empirical study of learning and retention. Behav Inform Technol 18(2):68–82. https://doi.org/10.1080/014492999119129

World Health Organization (2016) Global diffusion of eHealth: making universal health coverage achievable. Report of the third global survey on eHealth, Geneva. http://apps.who.int/iris/bitstream/10665/252529/1/9789241511780-eng.pdf?ua=1. Accessed 5 Aug 2017

Zegers M (2017) The possibilities and limits of urban–rural collaboration in tourism and leisure. Wageningen University, Wageningen. http://edepot.wur.nl/412461. Accessed 21 Jan 2019

Part IX

Health Communication

Health Communication: Approaches, Strategies, and Ways to Sustainability on Health or Health for All

57

Patchanee Malikhao

Contents

P. Malikhao (✉)
Fecund Communication, Chiang Mai, Thailand
e-mail: pmalikhao@gmail.com

J. Servaes (ed.), *Handbook of Communication for Development and Social Change*,
https://doi.org/10.1007/978-981-15-2014-3_137

Abstract

This chapter starts from a holistic perspective on health in the society. It makes a statement that, in order to become effective and sustainable, health communication needs to be studied, assessed, and practiced from a rights- or social justice-based position. Such an approach implies the use and integration of multidisciplinary perspectives that try to grasp the complexity of health issues from both global and local, individual, interpersonal, group, and community levels.

Keywords

Health for all · Wellness · Well-being · Health behavior · Communication strategies · Health literacy · Intercultural communication · Media literacy · Participatory media · Community media · New media

> Health is a core element in people's well-being and happiness. Health is an important enabler and a prerequisite for a person's ability to reach his/her goals and aspirations, and for society to reach many of the societal goals (Minister of Social Affairs and Health, Finland, 2013: 3).

Health communication has been a part of development communication or communication for development for the past five decades. Royal Colle (2003: 44–51) explains that health communication has been one of the threads of development communication together with population information, education, and communication (IEC) since 1969. Then, it was concerned with population and family planning programs, with an emphasis on reproductive health that includes family planning, maternal and infant death and disability prevention, sexually transmitted diseases (STD) and HIV/AIDS prevention, harmful cultural practices such as female genital mutilation (FGM), violence against females, human trafficking, and female health (Colle 2003: 46).In her book, *Effective Health Communication for Sustainable Development*, Patchanee Malikhao (2016: 6–7) explains how the history of health communication, as a separate field of study, has emerged from being only a part of health education and training in medical and public health to the integration of health-related aspects of individuals, communities, and organizations or their environment, with appropriate communication and mass communication theories. These communication theories borrow models and frameworks from (1) social science fields such as psychology, social psychology, anthropology, and sociology; (2) humanities subjects such as culture, linguistics, and languages; (3) ecological and environmental science; and (4) medical science fields.

Today health communication has expanded its scope from biomedical interventions at a personal level to more context-based communication about health, which includes the socials and the environment that have impacts on an individual's health. Robert Rattle (2010: 130–141) affirms that these are the social determinants of health, apart from the physical determinants, and above all the health policies that impact health behaviors.

57.1 What Constitutes and Has Impact on Health

Let's try to understand what constitutes health before we discuss what determines good health. Some scholars think that health means only the absence of disease or infirmity. That is not enough, because, during our life course, we all experience discomfort, disabilities, and pains along the way, such as child teething, babies being unable to act as adults do, women having menstruation pain or having labor pain, and certain conditions such as pregnancy (Janzen 2002: 69). These kinds of pain do not indicate that we are not healthy. Janzen (2002: 69) also includes aging, fatigue, birth impairs, and growth disorders as diseases. Our individual health is indeed hardwired with intrinsic diseases, pains, and discomforts. It is a fact of life! Each individual is subject to different healthy degrees, depending upon the physical determinants of health – the physical environment, biology and genetic endowment, and medical service (Rattle 2013: 181). A twin, who lives in a good and clean environment where medical service is well-organized, would be healthier than the other twin who lives in a polluted environment with poor medical service. New medicines and vaccines could save lives, but some people can get access to these innovations easier than others.

Some people are born with birth defects and that makes them less advantageous than others. Some people inherited genes from their ancestors which makes them prone to high blood pressure, diabetes, anemia, etc.

How we think, feel, cope with issues, and manage our life should not be neglected. That is how we differentiate the terms "well-being" from "wellness." Wellness is more related to the physical abilities of a person to perform tasks up to the full potential (Dunn 1977: 9–16), but well-being indicates that there are intangible elements that we cannot miss out. We therefore need to look at other dimensions of health apart from focusing on only our physical health: fitness, agility, body mass index, etc. That's why Hunter, Marshall, Corcoran, and Leeder (2013) categorize four aspects of health: (a) psychological/emotional, which includes positive attitudes, awareness, resilience, etc.; (b) intellectual/cognitive, which constitutes the ability to learn and to be creative and critical; (c) spiritual, which includes one's values and beliefs, a sense of meaning and purpose in life, inner peace, and an ability to transcend one's own self, and occupational, which means the satisfaction from working unpaid or paid and recreational activities; and (d) environmental.

Rattle (2010: 190) explains that there are eight factors in the socials that have impacts on our health. They are relative income and socioeconomic status, education, employment and working conditions, social support networks, health practices and coping skills, healthy child development, culture, and gender. All of these are socially constructed, which means that it is shaped by our interactions with others within the context we live.

People with higher incomes seem to have a higher education, and those with higher socioeconomic status can afford private health insurance and have regular health checkups. People with a higher income tend to live in a safer neighborhood and do not face depression that is often the result of living in unsafe public spaces. Those with better incomes can enjoy better outdoor life with access to more sports facilities and recreation. Better income guarantees hygienic sanitation and healthier living conditions that could prevent respiratory diseases, health hazards, and epidemics.

Better education would result in better personal hygiene and better access to knowledge on diseases and their prevention. Better education is closely related to meaningful job opportunities which in turn brings in satisfaction/self-contentment, inner peace, self-esteem, and self-control/self-power/self-autonomy among other life qualities (Mirowsky et al. 2000: 50–56). Moreover, good working conditions will prevent employees from hazards and bad physical conditions caused by a nonsafe environment.

People with good social networks can get more access to self-help medical knowledge, health practices, and have a better basic knowledge of hygiene and sanitation, first aid, vaccinations, etc. With social supports, an individual benefits from inner peace and is better able to cope with traumatic experiences and loss.

Healthy childhood development depends upon the socioeconomic status of the parents and the health-care services within the community. One can notice a higher mortality at birth, poorer overall health, and the development of chronic illnesses in adulthood among individuals from a lower socioeconomic status (Sarafino 2006: 158).

Culture is socially constructed and dynamic. It is what we pass on from generation to generation. It influences the way we interact and how we experience the world, including suffering, pain and its articulation, healing, mental health, and self-help. According to Brown (1995: 8–9), the tangible aspects of culture can be seen as artifacts, norms and behaviors, heroes and symbols, the media, languages and expressions, stories, myths, jokes, rites and rituals, ceremonies, and celebrations. Its intangible aspects are beliefs, values, attitudes, and the worldviews of people. The worldviews of people in different cultures are influenced mostly by their religion and beliefs (Malikhao 2016: 78). In *Caring for Patients from Different Cultures*, Galanti (2008) explains at length how different cultures, influenced by different belief systems and religions, affect the way people express pain and suffering, perceive and discuss mental health, perform healing practices, develop culturally bound syndromes, and organize support networks.

Now it comes to the last social determinant: gender. Gender is socially constructed as well. Males seem to have more social burdens than females, as more males are still expected to be the breadwinners. They tend to be more stressed than females and use more drugs, tobacco, and alcohol to cope with stress (Malikhao 2016: 72).

It is worth noting that the physical and social determinants can be intertwined, and induce cross-link effects, and that can be messed up with biological determinants such as age, sex, and more importantly the genes or our DNAs and the lifestyle of the people in this postmodern era. All of that makes an individual's health different from one another. Pollard (2008) explains in her book, *Western Diseases: An Evolutionary Perspective*, how our genes adjusted themselves when people changed their lifestyle and dietary pattern from the East to the West, and that caused the so-called Western diseases or noncommunicable diseases such as type 2 diabetes and colon and breast cancer among other diseases mostly found as the cause of death among people in the Western world – North America, Western Europe, Australia, and New Zealand. Pollard states clearly in her book that the Western diseases are also found among the rich populations in the Third World as a result of imitating the lifestyle of those living in the West. This can be explained with the term "aberrant epigenetics." It means that a change in the environment and/or change of dietary patterns in an early stage of life could have an impact on the DNA of our cells, not as much to alter the genetic makeup of our body but enough to cause certain diseases such as Alzheimer, schizophrenia, asthma, and autism (Marchlewicz et al. 2015:4). Western diseases are one of many indicators of how globalization affects the health of the world population. Postcolonial or contemporary globalization has been speeded up in both degree and kind by advances of information and communication technologies (ICT). The moving of labor, goods, production sources, technology, etc., with driving forces such as the marketing and advertising industry across the globe, results in unequal affluence and instigates new patterns of consumption and lifestyle and new diseases and the comeback of almost eradicated diseases (Lee 2005: 14; The College of Physicians of Philadelphia 2018; CDC 2010, 2014; Emedicinehealth.com 2018). To elaborate, some parts of the world or sectors within the same country benefit from globalization more than others, which causes unequal opportunities in accessing basic needs for city dwellers: clean water supply, proper sanitation, waste disposal, proper housing, clean air, public playgrounds, good schools, good working conditions, adequate primary health-care services, or a good marketplace that provides fresh, clean, and nutritious foods. Respiratory diseases; dysentery diseases; tuberculosis; plague; depression; mental health and health issues due to tobacco, drugs, and alcohol use; and other epidemics happen more in less developed, mainly crowded areas. As more and more people migrate in pursuit of better opportunities, communicable diseases could spread far and wide. Patchanee Malikhao explains at length in her 2012 book, *Sex in the Village: Culture, Religion and HIV/AIDS in Thailand*, how the pandemic HIV/AIDS occurred due to a complex factors including migration of labor, unsafe sex practices, the change of sexual norms due to globalization, discourses on safe sex education, HIV/AIDS prevention public health policies, the localization of the global HIV/AIDS prevention, discourses on the religious HIV/AIDS interventions, gender inequality, and poverty to name a few (Malikhao 2012).

As people have more sedentary lifestyles, triggered by using automations and performing white-collar work, together with eating fast food and processed food that contains too much sugar, salt, and fat, they are prone to develop non-communicable diseases (obesity, cardiovascular diseases, diabetes, among others) (Lee 2005: 14).

Globalization brings in advantages and disadvantages in disease prevention and eradication. Malikhao (2016: 51–52) reports the resurgence of diseases that broke out because either the bacteria have become antibiotic-resistant (such as avian influenza, cholera, influenza, and chikungunya) or there have been cross-links between the virus that used to cause diseases in animals in previous days with that used to cause diseases in humans resulting in new hybrid diseases such as mad cow disease, bird flu, or swine flu. At the same time, thanks to vaccinations, certain diseases such as small pox and rinderpest have been wiped out, and some diseases such as polio, measles, and mumps are on the way to die out.

Malikhao (2016: 51, 52, 55, 112) reports that, in a developed world, the haves are not much healthier than the have-nots; some people face health problems such as stress, anxiety, depression, burn out, sleep deprivation, obesity, or risk-taking behaviors such as smoking, alcohol abuse, and drug addictions due to unemployment and underemployment. The developed world can impose more health threats to the less developed world by exporting, in the name of advertising and marketing, unhealthy products such as tobacco, alcohol, fast food, chemical waste, etc. While, in the developing world, some people face starvation due to poverty, some in the developed world develop eating disorders such as anorexia nervosa and bulimia as a result of the change in perception of one's self-image, partly due to the complexity of mediatization (change of social interactions modulated by the media – Hjarvard 2013: 17), mediation, and new forms of individualization with symptoms of depression, anguish, apprehension, and anxiety found among the haves (affluenza) in the contemporary globalization period (Lemert and Elliot 2006; Twenge and Campbell 2010; James 2007).

Above all health policies of each state or country do influence the health behaviors of a population in major ways (Rattle 2010: 141). That includes the food-producing and food-marketing industry which has strong relations with the politico-economic and legal systems and the public health policy of a country. Pesticides, insecticides, hormones, antibiotics, and controls in the agriculture and food industries would benefit the health of the consumers a great deal. Moreover, the systems of health-care services are also determined by the public health policy of a country. Malikhao (2016: 80–81) uncovered four basic health-care systems in the world, as presented on the PBS Frontline documentary in 2015:

1. The Beveridge model. It is a government model that supports the health system totally with the tax payers' money. It is used in Great Britain, Hong Kong, Spain, most of Scandinavia, and New Zealand.
2. The Bismarck model. It is a nonprofit model in which health care is privately funded by payroll deduction of employees and funded by the employers. It is used in Germany, Japan, Belgium, the Netherlands, France, Switzerland, and somewhat in Latin America.
3. The National Health Insurance model. It is a model funded by a government-administered insurance program through private providers. It is used in Taiwan and South Korea.
4. The out-of-pocket model. This is used in the USA. It is the nonmedical care for the have-nots and medical care for the haves.

Having said all of that about health, one could not agree more with the definition of health principles in the constitution of the World Health Organization (WHO 2018a).

57.1.1 Constitution of the World Health Organization: Principles

Health is a state of complete physical, mental, and social well-being and not merely the absence of disease or infirmity.

The enjoyment of the highest attainable standard of health is one of the fundamental rights of every human being without distinction of race, religion, political belief, and economic or social condition.

The health of all people is fundamental to the attainment of peace and security and is dependent on the fullest cooperation of individuals and states.

The achievement of any state in the promotion and protection of health is of value to all.

Unequal development in different countries in the promotion of health and control of diseases, especially communicable disease, is a common danger.

Healthy development of the child is of basic importance; the ability to live harmoniously in a changing total environment is essential to such development.

The extension to all people of the benefits of medical, psychological, and related knowledge is essential to the fullest attainment of health.

Informed opinion and active cooperation on the part of the public are of the utmost importance in the improvement of the health of the people.

Governments have a responsibility for the health of their people which can be fulfilled only by the provision of adequate health and social measures.

The Constitution was adopted by the International Health Conference held in New York from June 19 to July 22, 1946, signed on July 22, 1946, by the representatives of 61 states and entered into force on April 07, 1948. Later amendments are incorporated into this text."

The WHO acknowledges the multidimensional nature of health, that health involves complete physical, mental, and social well-being and not just the absence of disease. Moreover, everyone has the right to maintain and enjoy the benefits of being healthy regardless of their socioeconomic-politico and religious status. That means everyone has the right to medical, psychological, and related knowledge necessary to attain health to the fullest capacity. The health inequality caused by unequal development should be attended to by each government. Each government should be responsible for the health of their population by providing appropriate health and social measures.

We can draw three important notions from the constitution of the WHO: that "rights to health care," "health inequality reduction," and "health for all" are essential to devise good communication strategies to achieve the health goal. This leads to the next topic.

57.2 Health Communication Perspectives

In the 1980s and early 1990s, health communication was known as a form of health education and health promotion and preventive medicine and focused on communication at many levels (interpersonal and organizational communication) in health-care settings. Mainly in the USA, but also in the UK, the focus was even more down to communication among the patients and the health-care provider (Lupton 1994: 56; Irwin 1989: 32, 40).

Health communication theories that originated in the USA have focused on mainstream quantitative research perspectives for about four decades. Malikhao (2016: 19–26) reports that these theories are borrowed from the fields of psychology and social psychology, which emphasize cognitive and behavioral changes at the intrapersonal, interpersonal, and group/organizational level. Popular models for intrapersonal communication are, for instance, the health belief model, theory of reasoned action (TRA), theory of planned behavior (TPB), the integrated behavioral model (IBM), the transtheoretical model and stages of change (TTM), and the precaution adoption process model (PAPM). Models used for interpersonal communication are social cognitive theory (SCT) or social learning theory (SLT). Models used for organizational communication are stage theory of organizational change, diffusion of innovation, and social marketing and edutainment.

Malikhao (2016: 16–17) reports that the signature of these kinds of models is that they rely heavily on positivism which is based on natural science models of cause and effect. They flatten the well-rounded facts of life into a linear line for prediction with inferential statistics. The models used for intrapersonal communication seem to assume direct relationships between knowledge, attitude, and behavior regardless of the context within which people live. The models used for interpersonal communication pay attention to the simplicity of the stimulus-response formula and the modifications of it, such as Laswell (1948) formula, "Who? Says what? Through which channel? To whom? With what effect?" and Katz and Lazarfeld (1955) two-step flow of communication, which relays on the spreading of messages from opinion leaders received via the mass communication to other people. Moreover, at the societal level, these models focus on the ability of humans to act in stages from being laggards to people who adopt innovation completely. However, they pay less attention to the socio-politico-cultural context that impinges on the ability to change those individuals. The organizational models assume the one-fit-all models and technology transfer from a more developed country to a less developed country.

These American-based models are built under the modernization paradigm which assumes that the Western way of living is a desirable goal for development everywhere. It is a positivist perspective that emphasizes empirical observations and statistics. Health communication within the modernization paradigm involves a high-tech, top-down, and unilinear approach from health professionals either directly or through the mass media to the receivers aiming to educate, upgrade, or train them to be informed in public health, have good attitude toward biomedical interventions or health-related advice/information, and have self-efficacy to change health-risk behaviors to a healthy lifestyle and health behaviors.

The modernization paradigm has been challenged since the 1990s in the multiplicity paradigm proposed by Servaes (1999). This paradigm is more than a many-roads-lead-to-Rome approach as one can go by foot, by plane, or by boat and one can mobilize others to join them to Rome without having to listen to the commands from Rome. Health communication within this paradigm emphasizes human rights: freedom from exploitation, the right to access adequate health care and health insurance, equity, community efficacy to come up with one's own solutions to manage resources and health4 issues, participatory democracy, and sustainability in health or health for all in a given socioeconomic and cultural system at all geopolitical levels. With the help of the new media (which is going to be elaborated in the next section), the dream of managing community health and disease prevention by the people and for the people has come true.

Malikhao (2016: 31–40) researches communication perspectives within the framework of the multiplicity paradigm at the individual level, interpersonal level, and group or community level:

At the individual level, the Self Determination Theory of SDT by Ryan and Deci (2002:5) is preferred. This theory focuses on the context where an individual lives or the extrinsic factors that motivate that person to motivate oneself to engage in behavioral change. The internalization process of a person comes from having opportunities to make a choice that is meaningful to oneself in the socialization process and receive a positive feedback to encourage the change of behavior. That means one cannot change his/her behavior by just receiving messages, one has to have an enabling environment to foster desirable health behaviors.

At the interpersonal level, life skill training and education are essential to build up intrinsic factors of an individual to prevent health risk behaviors and health hazards. Intercultural competency training is important for health care professionals to empathize patients and people in the community.

At a group/community level, some social capitals such as support groups and peers together with positive rewards can help reinforce the change of habits into desirable ones. Theories of social network and social support, community organization and community building theories as well as the PRECEDE/PROCEED Model and the Ecological Models of Health Behavior are having common characteristics of enabling environment, including advocacy communication, participatory communication, communication for structural and sustainable social change. Advocacy communication for health is about using the mass media to empower the voiceless to be heard regarding health hazards, issues on environment, and health-related issues. Participatory communication for health enables the locals, regardless of sociocultural and politico-economic status, to act and have dialogues in a democratic way to discuss and prevent diseases, hazards, and pollutions, and promote healthy life style, safety and clean environment. Participatory-based advocacy diversifies advocacy communication by adding the idea that the locals could get together and manage the content of the mass media used to advocate solutions on health issues, pollution, hazards, and clean environment. Health communication for structural and sustainable social change uses mix and match approaches, to advocate change and participation according to the situation, felt-needs of the locals, the budget, and available resources. A health communicator in the Multiplicity Paradigm can be called a social mobilizer who cultivates his/her attitude to empathize with others to achieve capacity building and empowerment and be able to mobilize the community to research, plan, and execute projects that are useful for the sustainability on health of the community (Malikhao 2016: 99).

All of these models in both paradigms have been used by many scholars for research. According to Babrow and Mattson (2003: 47–53), based on Craig (1999), historic traditions of research on health communication can be summarized under seven categories:

- The rhetoric tradition focuses on how the health communicators persuade the policymakers to act upon health projects.
- The semiotic tradition pays attention on how health communicators are mediated in intersubjectivity fashions by signs and sign systems.

- The phenomenological tradition studies an individual's and others' health experiences in the process of the patient and health-care provider communication.
- The cybernetic tradition looks at the process of health communication in linearity, which implies the encoding of a message, the transmission of the data, decoding of the data, feedback, and the impact on the environment.
- The sociopsychological tradition is about what affects the cognition, emotion, and behaviors of an individual when she/he interacts with the socials.
- The sociocultural tradition studies how health communication produces new sociocultural patterns which have been shared.
- The critical tradition relies on the critical theory used to study discursive reflection of health and illness.

The interdisciplinary research on health communication encourages to gain rich, insightful, and meaningful data to accompany the empirical observations.

57.3 Sustainability in Health and Health Communication

Sustainability is what many people talk about in this postmodern era. It seems as if one realizes that, within the limited resources on Mother Earth, we should manage, conserve, and nurture our own habitat in line with an ecological balance, so that new generations to come can enjoy it as well. That is only one part to reach sustainability. Apart from having a healthy environment, we should consider a healthy social (social justice) and healthy economy as sustainability as a whole (Cox and Pezzullo 2016: 264). Sustainable development implies five different areas that we need to pay attention to: water and sanitation, energy, human health, agricultural productivity, and biodiversity and ecosystem management (Wallington 2014: 170).

Sustainability on health is to uplift equality of the quality of life for everyone. Poverty, discrimination, less opportunities to get education and employment, living in substandard housing and environment, and less opportunity to access primary health care are the underlying assumptions of inequality in health. According to Malikhao (2016: 99), sustainability on health is a process of social mobilization empowered by both stakeholders, some of whom can be health communicators, and health communicators from outside who have empathy toward the stakeholders, to achieve two goals: first, to engage the people in the community in upgrading the health and media literacy status so that they can make an informed choice on their body and health and health care and, second, to build up community capacity and networking with other communities so that the people can solve problems related to community health, achieve social justice in health, prevent diseases, maintain well-being, and cultivate health knowledge, good attitude, ethical values, cosmopolitan worldview, and health behaviors, including advocating for structural change for a local healthy lifestyle and accommodating environment. In addition, we should call for a rights-based health communication which means everyone should have the equal right to access primary health care regardless of his/her socioeconomic status. We should cultivate intrinsic values, such as self-contentment to instill inner peace that is opposite to short-term pleasure from mindless consumption triggered by the advertisement and marketing industry. Moreover, we should economize on natural resources and energy consumption, including the reduction of carbon footprints to prevent the acceleration of global warming. Extrinsic factors are those that enable people to think, rethink, assess own and community values, empower themselves to improve one's own health, and participate to improve community health. They are families; schools; communities; the governmental sectors related to public health, energy, and the environment; the NGOs related to health and environment; the mass media and new media; man-made environment; and natural habitats.

Health communication advocating for sustainability ought to be operating under the framework of the multiplicity paradigm mentioned above. To achieve the status of "health for all," "rights to health care," and "health inequality reduction," the integration of five different approaches is needed: behavioral change communication, mass communication, advocacy communication, participatory communication, and communication for structural and communication change. The recipes for each case vary according to each setting, but we can summarize what Patchanee Malikhao suggests in the section of Communicating for/about Health for Sustainable Development in her book, *Effective Health Communication for Sustainable Development* (Malikhao 2016):

Behavioral Change Communication for Health for All is about interpersonal communication on health literacy, health control and management, disease prevention, food and nutrition, wellness and well-being, etc. The main aim is to engage the stakeholder to become a health agency who has an autonomy of own health and can make informed decision. The media used can be both the mass media of the new media that encourage social supports.

Mass Communication for Health for All is about using the mass media convergence to empower the stakeholders and ease the participation process in addressing a health problem in a community. The mass media convergence is the ability of the modern mass media, thanks to the new media, to have more platforms that can interact among one another. For instance, a newspaper can have a digital TV channel, a Facebook page, an Instagram page, a Twitter channel, and a website which allows the audience to follow life video/audio clips or stock audio files for podcast or stock video clips to be watch on a laptop, smartphones, or digital TV screen. Media literacy should be as important as health literacy for health communicators.

Participatory-Based Advocacy Communication for Health for All means empowering the grassroots to interact with key decision-makers on health issues aiming at influencing them to support policy changes at all levels (and also international) and to sustain accountability and commitment from governmental and international actors (WHO 2018b; Servaes and Malikhao 2010: 43).

Participatory Communication for Health for All employs both interpersonal communication and multi-community media as well as social media to cultivate community interests and participation, by taking diversity and pluralism into account, in health-related areas. Participatory social marketing on health is the added values on the traditional social marketing to communicate about/on health. That can be both online and offline campaigns using the social interconnectedness online and face-to-face dialogues to buzz news, information, events, and other social media entries. Self-management and production of the media and access of both media producers and stakeholders are emphasized (Berrigan 1979: 8).

Communication for Structural and Sustainable Social Change for Health for All is a combination of all of the above approaches to empower the stakeholders to upgrade and advocate for their own and community health on correct information on health, health and media literacy, disease prevention, environmental health, health behavior, and access to affordable and quality health-care system.

In order to achieve sustainability, health communicators should possess essential knowledge and skills of intercultural communication. Malikhao (2016: 102-105), adapting from the framework of Martin and Nakayama (2010: 50–52), elaborates on intercultural competency in health in five aspects: First, understanding personal and contextual way of communication. Health communicators should be able to discern personal health behavior (such as brushing teeth two times daily) from the behaviors that are results of social construction (such as sharing syringes among prisoners).

Second, understanding the differences and similarities between cultures. People from different cultures may have different ways in verbal and nonverbal communication. Pain expression, treatment option, or healing rituals can be different from culture to culture. The similarities are obvious that everyone needs respect, kindness, and a nonjudgmental attitude.

Third, understand the local cultural context in which we are operating. If the health communicators can speak the local language, it would be a plus to create rapport between them and the stakeholders. Understanding one's own culture and biases in one's own culture, but at the same time having sensitivity to other cultures and appreciation for the differences, should be the characteristics of health communicators (Galanti 2008: 2).

Fourth, understanding the privilege and disadvantage in the socials. Health communicators have advantages in the sense that they may have good connections with the community leaders and policymakers and they may already know the strengths and weaknesses of the community. But they may face resistance from the stakeholders. They may need to step back and assess the situation.

Fifth, history and past understanding. Knowing the history of the health situation of the unit of analysis is important to assess the present situation and plan ahead. The unit of analysis varies from small to large: individual, family, ethnic group, group, community, institution, or a country.

Next, we need to understand the media, especially the new media, to assist health communicators to do the right job.

57.4 Media Literacy and the Media to Assist Health Communicators for Sustainable Development

As summarized from Malikhao (2016: 109-123), it should be clarified here that the media is not the message, but the type, format, and limitation of each medium shape the content of the messages. In our mediated world, fueled by digitization, no matter whether we like it or not, we are in the process of mass self-communication, a term coined by Castells (2013: 55), as we are parts of a digital network. We are using the new media every day. According to Fuchs and Sandoval (2015: 165), new media are the media that are based on the World Wide Web, which includes social networking sites such as Line, WhatsApp,

Facebook, etc., or video-sharing sites such as YouTube, Vimeo, etc. They can be wikis or a website on which users are allowed to modify the text from the web browser in collaboration with others such as Wikipedia. They can be blogs such as Blogspot, WordPress, etc. or microblogs such as Twitter and Sina Weibo. They can be online pinboards such as Tumblr and Pinterest. They can be a photo-sharing site such as Instagram. While sending texts that we create and sharing images and motion images we create ourselves or repost from someone else, with friends and our circles or networks of contacts, we are acting as both an interpersonal and a mass communicator.

Hence, the term mass self-communication is quite handy to describe our way of communication. Mediation was the old term, that we are mediated by mostly traditional media (newspaper, radio, television, or the classic media) as well as the new media or alternative media such as Indy media (or independent media – which are the media that do not aim for commercial profits and tend to stand up for the underdogs). However, there is another term coined by Hjarvard (2013: 17): mediatization, that is, the long-term influence of mediation creates changes in the culture and institution. Mediatization makes contacts across the globe a reality, makes people think about their identity, and creates digital divides, meaning the gap between those who can access the new media and those who cannot is wider (Martin and Nakayama 2004: 6).

Let's discuss the concept of "media literacy," which any health communicator should understand apart from being a health literate. As summarized from Turow (2014: 20-21), a health communicator should know that, first of all, media are not the reflection of a reality. Media are constructed, constrained, and influenced by those who produce them and the media business owners. Second, media are related to power and profits. The larger the company, the more power it has to media convergence, more power to control the entire process of media making, and more chances to dominate other cultures with the media products as cultural products: that is called the "soft power." Third, the media contents are shaped by a political ideology, worldview, and values of the creators. Fourth, the format of the media shapes the characteristics of the media. Fifth, different audiences interpret the same media message in different ways due to their own worldview, background, and education. Sixth, each medium is unique aesthetically. Seventh, the media are the mirror of the visions in society.

Media literacy can be categorized further into four different subgroups, explains Share (2015: 192-197): visual, aural, multimedia, and alternative literacy. Do keep in mind that different lighting, lenses used, types of cameras, camera angles, distances between objects and cameras, and so forth create a different visual image. Aural literacy makes us aware that the same sentence that was spoken and heard in a different context will convey different messages. Multimedia literacy calls for the understanding and capability in the process of making multimedia from the beginning to the end. Alternative literacy calls for the ability to analyze the mainstream media how they create hegemony, how the main stream media represent the power from the dominant ideology, and how to represent underserved perspectives.

The alternative to the mainstream media is community media (Carpentier et al. 2012). They can assist health communicators to become health mobilizers. They can help raise awareness, amplify, and engage the stakeholders and solicit solutions or immediate response from the community. These media can be community radio, community television, community video, community-based telematics systems, or just a community folk media, or the same old print media that can be displayed at the places where people gather together temporarily, such as a fair, or permanently, such as in a railway station.

Community new media offer interactive responses. They can be various media, such as telemedicine technology, e-health, Internet radio, Internet television, geographic information system (GIS), and Webs 2.0, 3.0, and 4.0, which allow a community website or community social media platform where everyone can exchange information and voice his/her concerns on health-related areas, environment, and well-being of the community. This author would like to call them participatory new media.

57.5 Health Communication Strategies for Sustainability

Stakeholder analysis, recognizing and defining the public health problem, setting goals and objectives, identifying resources, and maintaining control of the problem are steps useful for participatory action research on health-related issues (Malikhao 2016: 127).

According to the WHO Europe (2013), to achieve sustainability in health, these strategies can be categorized fourfold:

1. To improve health through a life course of empowering people.
2. To tackle local major health challenges of noncommunicable diseases, injuries and violence, and communicable diseases.
3. To strengthen people-centered health systems, public health capacity, and emergency preparedness, surveillance, and response.
4. Communication strategies to create resilient communities and supportive environments, including a healthy physical environment.

Malikhao (2016: 143–164) Modifies the abovementioned strategies by introducing the integration of five different approaches: Behavioral change communication, mass communication, advocacy communication, participatory communication, and communication for structural and communication change health communication but with a stress on the last three approaches. The integration of traditional media, new media, and/or alternative media is possible depending on cases. In order to come up with tangible strategies, we should keep in mind that women development must come first in order that they can bring up their children (WHO Europe 2013: 73–74). Then, we can focus on children and adolescent, adult, and elderly health.

57.5.1 Communication Strategies to Improve Health Through a Life Course of Empowering People

In all of these age groups there are similar strategies aiming at an individual level, a group level, and structural level. The following communication strategies are summarized from Malikhao (2016: 144–153), aiming to improve on health quality at an individual level:

- Inviting stakeholders to a training program on health and media literacy via a community website or letter, emails, tweeting, phone calls, leaflet distribution, posters, and messages on social media applications and on cell phones.
- Promoting health education on sex education, communicable and non-communicable diseases, and rights and justice, for the female group and each age group in public institutions, and by using folk media and community media.
- Promoting information on female health and each age group's health on social media, community media, and the mass media. The community website can be created with built-in podcasts, vodcasts, links to YouTube videos, and links to urls that give information on gender equality, reproductive health, healthy lifestyle, parental skills, preventive information on diseases, clean environment and sanitation, dementia and mental health, etc.
- Raising awareness for routine immunization program for each age group by using emails, letters, posters, blogs, websites, twitter, and community folk media.
- Promoting the inclusion of females, people from each age group, and people from other ethnic groups in any community event.
- Promoting values for healthy lifestyles and advocating for an alcohol-, smoke-, and drug-free life on new community media such as blogs, websites, podcast, digital storytelling, and community media such as theater plays, posters, banners, walls, community radio and TV, social media, and the mass media.
- Promoting nonviolent and nondiscriminating ways to solve problems by using digital storytelling, theater plays, podcast, community radio, and TV.
- Using personal contacts can help unlearn undesirable health behavior like eating fast food and consuming sugary drinks and replace old habits with new habits of eating fresh vegetables, whole grains, and fruit, exercising, and diet control. A workshop and a follow-up program can be designed to accommodate offline discussions.
- Promoting benefits of living in good and clean housing, settings, and environment by using video clips shared in the social media as well as mobile exhibitions with videos and digital storytelling.

The following are communication strategies that aim at improving health quality at a group level:

- Advocating for a school-based curriculum on media and health literacy and courses on media and health literacy in college.
- Advocating for a school-based sex education which includes respect for female rights, gender equality, sexual debut, contraception, and STD and HIV/AIDS prevention.
- Training health-care and medical personnel and interested stakeholders media literacy, communication, and digital media skills. Promoting the training courses can be done by using traditional media as well as the new media.
- Using participatory community media to air information on many topics such as maternal health, teens and adolescent forums, family planning, parenting skills, etc.
- Facilitating collaborating among families, age groups, and members in a community by producing participatory video clips or vodcasts on healthy lifestyle for teens, tweens, adults, the elderly, pregnant women, etc.

- Forming a peer-to-peer online and offline group to exchange views and information on maternal health, mental health, adolescent health, elderly health, etc. brainstorming sessions are recommended to come up with cue, routine, and reward for building up healthy lifestyle such as eating fresh vegetable and fruit, exercising, etc. filming the sessions and upload video clips on the community website or air them on the community media.
- Organizing an online or offline group on preventing alcohol and substance abuse among women and other age groups.
- Participating in the evaluation of all programs and coming up with recommendations and plans for improvement.

The following are communication strategies that aim at improving health quality at a structural level:

- Advocating for rights of the women, children, and the elderly. Video clips made by the collaboration of community members are encouraged to tackle on a particular issue. Sending those clips to the public health authorities and writing articles about these rights on websites, social networking pages, and blogs are recommended.
- Advocating for improving of women and men education. This includes anti-domestic violence and rapes. Digital storytelling or interview clips of women in the community can be displayed on diverse social media platforms.
- Advocating for the equality of men and women in employment and organizing media campaigns to raise awareness to policymakers and business owners.
- Advocating for the inclusion of the marginalized groups. Promoting video clips participatory made by the members of the marginalized groups on websites, social media platforms, and community media.
- Advocating for intercultural communication sessions on community media programs.
- Advocating for a clean and safe environment to work and to live by producing video clips or writing articles in both traditional and new media.
- Advocating for recreation spaces in workplaces and community. Video clips can be shown and uploaded on websites.

57.5.2 Communication Strategies to Tackle Local Major Health Challenges of Noncommunicable Diseases, Injuries, and Violence

Communication strategies aiming at an individual level are the following:

- Promoting eating right and exercising right by using video clips, messages, interpersonal contacts, and social networking sites.

- Raising the awareness of monitoring weight, blood pressure, blood sugar, and blood cholesterol by a concert of campaigns and social marketing by using the new media.
- Promoting nonviolent solutions and gender respect with interpersonal communication and new media communication.
- Promoting the knowledge of noncommunicable diseases such as type 2 diabetes mellitus, heart infarct, coronary diseases, etc. by using print media, electronic media, new media, and interpersonal media.
- Engaging the stakeholders in a community to obey the traffic rules by encouraging them to report the violators in a social media site of the community.

Communication strategies aiming at a group level are as follows:

- Empowering the stakeholders to form peer-to-peer and support groups on any noncommunicable disease triggers such as smoking, lacking exercise, etc.
- Advocating for primary health-care units to monitor noncommunicable diseases with the collaboration of public health officials, the private sectors, and the civil society.
- Empowering for first aid courses at workplace, schools, and community.
- Advocating for a school-based training/education on noncommunicable diseases.

Communication strategies aiming at a structural level are the following:

- Advocating for poverty reduction to reduce mental health problems among multiple sectors. Personal media together with video clips or digital storytelling are to solicit support from policymakers.
- Advocating for the reduction of salt, sugar, trans fat, and saturated fat in processed food. This can be done with MPs, authorities, civil groups, and consumer groups.
- Advocating for monitoring, control, and penalties of pesticides and insecticides in fresh produce; and hormones, antibiotics, heavy metals, and other chemicals in meat, poultry, fresh water fish, and seafood.
- Advocating for stronger penalties for gender-based violence, including sexual harassment, and traffic rule violators.

57.5.3 To Tackle Vaccine-Preventable Communicable Diseases

In order to tackle vaccine-preventable communication diseases at an individual level, mass communication and various forms of community media can be employed to raise awareness. URL links to online information can be provided to share information on new vaccines and vaccines that bridge communicable and noncommunicable diseases such as Hepatitis B and human papillomavirus.

At a group level, advocating on any collaboration among risk groups and authorities by using participatory videos. Develop a surveillance and early-warning system in a community and monitor the outbreaks of diseases and keep the stakeholders informed by new media, traditional media, and interpersonal media.

At a structural level, advocating contacts with international organizations on sharing vaccine-preventable epidemiological data, laboratory data, evidence-based and cost-effective interventions, etc. Mobilizing for social support across the board. This includes lobbying for law enforcement on every childhood vaccination, subsidies for vaccination, food and water safety, etc.

57.5.4 Communication Strategies to Tackle Non-vaccine-Preventable or Not-Yet-Vaccine-Available Communicable Diseases Such as Malaria, Dengue Fever, Tuberculosis (TB), and HIV/AIDS and Respiratory Diseases Such as SARS, MERS, H1N1, and H5N1

Communication strategies aiming at an individual level are as follows:

- Advocating via the social and community media to emphasize the cause of the disease, the transmission, symptoms, and prevention of the disease.
- Empowering the stakeholders via personal media and community media campaigns, complemented by training courses of how important the sanitary and hygienic environment is in preventing some vector-borne diseases such as malaria and dengue fever and airborne diseases such as SARS.
- Participatory videos, digital storytelling, podcasts, and vodcasts can be uploaded to promote better education. Better education implies better occupation and thus better HIV prevention as HIV prevalence is partly related to poverty and low socioeconomic status.
- Using participatory community media to promote HIV counseling and testing, especially to mothers to be.

Communication strategies aiming at a group level are as follows:

- Advocating for an early warning and surveillance system for vector- and airborne diseases via the community media, new media, and traditional media.
- Use participatory short film of video to advocate the government for multi-sectoral collaboration, and raise funds to provide budget for bed nets in case of vector-borne diseases and budgets for helping dwellers in crowded and substandard housing to prevent airborne diseases.
- Empowering the stakeholders to support outreach and peer education activities by community and personal media.
- Social marketing by using the mass media, social media, and community media on anti-discrimination against people living with HIV/AIDS. Use personal media to include the people living with HIV/AIDS in community forums and activities.
- Advocating school-based interventions to teach students about hygiene, respiratory diseases, HIV/AIDS, TB, malaria, and dengue fever.

- Empowering the stakeholders to take the mosquito-free environment and clean environment in their own hands.

Communication strategies aiming at a structural level are as follows:

- Using the mass media to gain social support on HIV/AIDS prevention in sex education curricula. Gender education should be done in a way of role plays and brainstorming to emphasize gender respect and learn that violence toward females can lead to HIV transmission.
- Mobilize social supports for alcohol control, improving housing standard, etc. by using participatory films, clips, photos, or audio files.
- Using participatory new media to promote a political environment that supports access to antiretroviral therapy for people living with HIV and employment for them and promote a well-informed civil society and nongovernmental bodies.
- Using integrated media to promote the linkage and integration of national programs on HIV/AIDS prevention and malaria, TB, dengue fever, and airborne epidemic prevention with broader health and development agendas.

57.5.5 Communication Strategies to Strengthen People-Centered Health Systems, Public Health Capacity, and Emergency Preparedness, Surveillance, and Response

Communication strategies aiming at an individual level are as follows:

- Empowering the health-care professionals to produce participatory video or audio files to advocate better working conditions, social supports, and recognition.
- Engaging health-care professionals in dialogues and skill training to increase competencies in intercultural communication, fiscal and budget, decision-making, etc.
- Empowering the stakeholders with participatory media to let them take part in designing, implementing, and evaluating health policies and services.
- Integrated media can be used to empower patients to make informed decisions and/or shared decisions between health-care providers and the patients.
- Empowering the disables, the elderly, and people with chronic diseases with integrated media to advocate access and services for them.

Communication strategies aiming at a group level are as follows:

- Using community media and new media to advocate for quality in primary health care.
- Using integrated media to call for good governance and collaborating with other health-care services. The media can be used to engage the stakeholders in helping forming ideas for a people-centered health care.

- Contacting all actors and stakeholders to discuss indicators for success of such a health care service.

Communication strategies aiming at a structural level are as follows:

- Using integrating participatory new media and mass media to advocate national strategies for developing public health services; assessing present public health laws; revising the laws, if needed; and evaluating partnerships for their effectiveness.
- Using participatory new media to mobilize public health workers, health-care providers, patients, and the public at large to solicit more funding to pay off the operating costs of the health-care services.
- Using participatory media to promote social support for capacity building and assess good governance and efficiency in sharing data and networking with accredited public health organizations abroad.

57.5.6 Communication Strategies to Create Resilient Communities and Supportive Environments, Including a Healthy Physical Environment

Communication strategies aiming at an individual level are as follows:

- Setting up forums to discuss online on participatory media or offline to engage stakeholders in the environmental management of their community.
- Empowering each person in the community to act as a citizen journalist on outbreaks of diseases and natural disasters by engaging them in the activities of sharing video clips and audio files discussions online and offline.
- Empowering individuals by integrated media to monitor climate change and conserving the natural habitats, including reporting on pollutions.
- Engaging the entire community by integrated mass media, community media, and new media to promote garbage management and recycles within the community.
- Empowering individuals to use bicycles instead of motor vehicles by using traditional social marketing and new media social marketing.

Communication strategies aiming at a group level are as follows:

- Offering new media platforms in a community operated by peers to discuss ways to help improve the environment.
- Advocating for safe water, standard sanitation, and clean energy by integrating the mass media, community media, and new media.
- Using new media social marketing to promote the surveillance and preparedness systems for extreme weather events and disease outbreaks.
- Using traditional and new community media to promote events and engage the community dwellers to see the importance of recycling and garbage separation.

Communication strategies aiming at a structural level are as follows:

- Gaining social support on measures, policies, and strategies to monitor and mitigate climate change by using participatory media.
- Mobilizing social support on educational and awareness programs on biodiversity, ecology, and climate change by using participatory media.
- Mobilizing support for research and development on conservation of natural habitats by using integrating mass communication, community media, and new media.

All of these tangible strategies are just examples of how to reach sustainability in health. People can be empowered to take health- and environmental-related issues in their own hands to ensure capacity building and networking.

57.6 Conclusion and Recommendations

By now the reader should be familiar with health communication for sustainability. Thanks to the advances in information and communication technologies or ICT, we can engage the community by media convergence and multimodal digital communication. As the media are not the messages, health communicators should research the strengths and weaknesses in communities regarding health and environment and get to know the stakeholders in order to devise the right communication strategies to suit the needs of the community.

The World Health Organization (WHO) acknowledges the multidimensional nature of health, that health involves complete physical, mental, and social well-being and not just the absence of disease. Moreover, everyone has the right to maintain and enjoy the benefits of being healthy regardless of their socioeconomic-politico and religious status. That means everyone has the right to medical, psychological, and related knowledge necessary to attain health to the fullest capacity. The health inequality caused by unequal development should be attended to by each government. Each government should be responsible for the health of their population by providing appropriate health and social measures.

Three important notions can be drawn from the constitution of the WHO: that "rights to health care," "health inequality reduction," and "health for all" are essential to devise good communication strategies to achieve the health goal.

Health communication is no longer a top-down fashion commanding the public to listen and act. We need to consider the enabling environment that would influence the decision-making process, attitude, and health behavior of an individual. Empowerment and advocating are the two terms health communicators as social mobilizers should be acquainted to. In order to do that, a health communicator should possess empathy and intercultural communication skills, apart from being media literate and health literate. This requires training and work experience. Apart from knowing public health, journalism, and communication, a health communicator should also have good knowledge of sociology, anthropology, environmental science, and management. Thus, interdisciplinary training would be a plus for a health communicator.

References

Babrow AS, Mattson M (2003) Theorizing about health communication. In: Thompson TL, Dorsey A, Miller KI, Parrott R (eds) Handbook of health communication. Lawrence Erlbaum Associates, New York/London, pp 263–284

Berrigan FJ (1979) Community communications. The role of community media in development. UNESCO, Paris

Carpentier N, Lie R, Servaes J (2012) Multitheoretical approaches to community media: capturing specificity and diversity. In: Fuller L (ed) The power of global community media. Palgrave Macmillan, New York (paperback), pp 219–236

Castells M (2013) Communication power. Oxford University Press, Oxford

Colle R (2003) Threads of development communication. In: Servaes J (ed) Approaches to development: studies on communication for development. UNESCO, Paris, pp 22–72

Cox R, Pezzullo P (2016) Environmental communication and the public sphere, 4th edn. Sage, Los Angelis

Craig RT (1999) Communication theory as a field. Commun Theory 9(2):119–161

Dunn HL (1977) What high level wellness means. Health Values 1(1):9–16

Fuchs C, Sandoval M (2015) The political economy of capitalist and alternative social media. In: Atton C (ed) The Routledge companion to alternative and community media. Routledge, London/New York

Galanti G (2008) Caring for patients from different cultures, 4th edn. University of Pennsylvania Press, Philadelphia

Hjarvard S (2013) The mediatization of culture and society. Routledge, London/New York

Irwin H (1989) Health communication: the research agenda. Media Inf Aust 54:32–40

James O (2007) Affluenza. How to be successful and stay sane. Vermilion, London

Janzen JM (2002) The social fabric of health: an introduction to medical anthropology. McGrawHill, Boston

Katz E, Lazarfeld PF (1955 – latest ed. 2017) Personal influence. Routledge, London

Lasswell HD (1948) The structure and function of communication in society. Harper & Bros, New York

Lee K (2005) Global social change and health. In: Lee K, Collin J (eds) Global change and health. Open University Press/McGraw-Hill Education, New York, pp 13–27

Lemert C, Elliot A (2006) The new individualism. Routledge, London

Lupton D (1994) Toward the development of critical health communication praxis. Health Commun 6(1):55–67

Malikhao P (2012) Sex in the village. Culture, religion and HIV/AIDS in Thailand. Southbound & Silkworm Publishers, Penang-Chiangmai

Malikhao P (2016) Effective health communication for sustainable development. NOVA Publishers, New York

Marchlewicz EH, Anderson OS, Dolinoy DC (2015) Early-life exposures and the epigenome. Interactions between nutrients and the environment. In: Ho E, Domann F (eds) Nutrition and epigenetics. CRC Press, London/New York, pp 3–52

Martin JN, Nakayama TK (2004) Intercultural communication in contexts. McGraw-Hill, Boston

Martin JN, Nakayama TK (2010) Intercultural communication in contexts (5th ed.). McGraw-Hill, New York

Ministry of Social Affairs and Health, Finland (2013) Health in all policies. In: Leppo K, Ollila E, Pena S, Wismar M, Cook S (eds) Seizing opportunities, implementing policies. http://www.euro.who.int/__data/assets/pdf_file/0007/188809/Health-in-All-Policies-final.pdf. Accessed on December 1, 2018

Mirowsky J, Ross C, Reynolds J (2000) Link between social status and health status. In: Bird CE, Conrad P, Fremont AM (eds) Handbook of Medical Sociology. Prentice-Hall Inc, Upper Saddle River, pp 68–78

Pollard TM (2008) Western diseases. An evolutionary perspective. Cambridge University Press, Cambridge
Rattle R (2010) Computing our way to paradise? The role of internet and communication Technologies in Sustainable Consumption and Globalization. Rowman & Littlefield Publishers, Inc, New York
Ryan RM, Deci EL (2002) Overview of self-determination theory: an organismic dialectical perspective. In: Deci EL, Ryan RM (eds) Handbook of self-determination research. The University of Rochester Press, Rochester, pp 3–33
Sarafino EP (2006) Health psychology. Biopsychosocial interactions, 5th edn. John Wiley & Sons, Inc, New York
Servaes J (1999) Communication for development: one world, multiple cultures, 1st edn. Hampton Press, Inc, Cresskill
Servaes J, Malikhao P (2010) Advocacy strategies for health communication. Public Relat Rev:42–49. https://doi.org/10.1016/j.pubrev.2009.08.017
Share J (2015) Media literacy is elementary. Teaching youth to critically read and create media, 2nd edn. Peter Lang, New York
Turow JW (2014) Media today, 5th edn. Routledge, New York/London
Twenge JM, Campbell WK (2010) The narcissism epidemic: living in the age of entitlement. ATRIA, New York
Wallington SF (2014) Health disparities research and practice. The role of language and health communication. In: Hamilton HE, Chou WS (eds) The Routledge handbook of language and health communication. Routledge, London-New York, pp 168–183
World Health Organization Regional Office for Europe (2013) A European policy framework and strategy for the 21st century. World Health Organization (WHO) Regional Office for Europe, Copenhagen

Websites

Center for Disease Control and Prevention (CDC) (2010) http://www.cdc.gov/h1n1flu/cdcresponse.htm. Accessed 18 Aug 2018
Center for Disease Control and Prevention (CDC) (2014) http://www.cdc.gov/flu/avianflu/h7n9-virus.htm. Accessed 18 Aug 2018
Emedicinehealth.com (2018) https://www.emedicinehealth.com/mad_cow_disease_and_variant_creutzfeldt-jakob/article_em.htm. Accessed 18 Aug 2018
The College of Physicians of Philadelphia (2018) https://www.historyofvaccines.org/content/articles/disease-eradication. Accessed 18 Aug 2018
World Health Organization (2018a) http://www.who.int/about/mission/en/. Accessed 21 Aug 2018
World Health Organization (2018b) http://www.who.int/tb/people_and_communities/advocacy_communication/en/. Accessed 21 Aug 2018

Health Communication: A Discussion of North American and European Views on Sustainable Health in the Digital Age

58

Isabell Koinig, Sandra Diehl, and Franzisca Weder

Contents

Abstract

Following the United Nations' Sustainable Development Goals, the UN's third goal is meant to "ensure health lives and promote well-being for all at all ages" (UN 2017). Thus health is closely linked to sustainability. While progress has been made over the past decades, which have seen an increase in life expectancy and a success in combatting several diseases (e.g., children's diseases such as measles or adult diseases such as HIV and malaria), new health issues have emerged and need

I. Koinig (✉) · S. Diehl · F. Weder
Department of Media and Communications, Alpen-Adria-Universitaet Klagenfurt, Klagenfurt, Austria
e-mail: Isabelle.Koinig@aau.at; sandra.diehl@aau.at; Franzisca.Weder@aau.at

J. Servaes (ed.), *Handbook of Communication for Development and Social Change*,
https://doi.org/10.1007/978-981-15-2014-3_81

to be addressed. In this context, communication is of uttermost relevance. Broadly speaking, health communication refers to "any type of human communication whose content is concerned with health" (Rogers, J Health Commun 1:15–23, 1996) and can be directed at both individuals and organizations with the goal of preventing illness and fostering health (Thompson et al., The Routledge handbook of health communication, 2nd edn. Routledge, New York, 2011).

As a multifaceted and multidisciplinary approach, health communication draws from and combines influences from different theoretical backgrounds and disciplines, such as education, sociology, (mass) communication, anthropology, psychology, and social sciences (WHO, Health and sustainable development. Key health trends. Available via WHO. http://www.who.int/mediacentre/events/HSD_Plaq_02.2_Gb_def1.pdf. Accessed 20 Dec 2017, 2003; Institute of Medicine, Health literacy: a prescription to end confusion. Available via The National Academies of Sciences Engineering Medicine. http://www.nap.edu/openbook.php?record_id=10883. Accessed 11 Apr 2016, 2003; Bernhardt, Am J Public Health 94:2051–2053, 2004). Health communication – regardless of the form it takes (e.g., policies, patient-provider interactions, community projects, public service announcements, or advertising) – is concerned with "influencing, engaging and supporting individuals, communities, health professionals, special groups, policy makers and the public to champion, introduce, adopt, or sustain a behavior, practice or policy that will ultimately improve health outcomes" (Schiavo, Health communication: from theory to practice. Wiley, San Francisco, 2007). As such, it needs to be perceived as "a part of everyday life" (du Pré, Communicating about health: current issues and perspectives. Mayfield Publishing Company, Mountain View, 2000).

Since health communication occurs in the health communication environment (Schiavo, Health communication: from theory to practice, 2nd edn. Jossey-Bass, San Francisco, 2014), which is composed of four main domains, namely: (1) health audience; (2) recommended health behavior, service, or product; (3) social environment; and (4) political environment, it takes place on various levels (societal, institutional, and individual) which need to be studied in order to provide a comprehensible and complete picture of the subject area. The present contribution seeks to highlight the contribution of the different disciplines to effective health communication, outline changes in the health communication environment, as well as carve out future challenges that are brought about by changes in demographics, disease treatment, and communication patterns. A special focus will be put on gender-specific and digital health communication. In conclusion, limitations and directions for future research are addressed.

Keywords

Digital health technologies · eHealth · *Every Woman Every Child* project · Health communication · Health information technology · Men's health · mHealth · Pharmaceutical advertising · Pragmatism · Universal Health Coverage (UHC) · Women's health

58.1 Introduction

As part of its Sustainable Development Goals, the United Nations strive to transform the world as it is known. Its third goal emphasizes the need to improve the overall health and well-being of global citizens regardless of age (UN 2017a). Recent developments within the health environment, which took place over the last years, respectively, decades, have resulted in a broader conceptualization of health (Schiavo 2014). In consequence, the topic of health has conquered a variety of related academic disciplines, becoming multifaceted in its approach (Earle 2007a).

The present chapter will start out by defining the core terms before discussing the particularities of the health communication environment. Afterward, current trends will be examined, such as digital health offerings, gender-specific health issues, and empowerment, among others. The paper will also address ways of achieving sustainable health and will be concluded by limitations and directions for future research.

58.2 Core Terminology

To guarantee that the terms used throughout the chapter are understood as intended by the authors, they will be briefly defined beforehand.

58.2.1 Health

Listed as a fundamental human right by the World Health Organization (2013), health has become a matter of global relevance. Yet, in terms of definitions, consensus is hardly achieved on grounds of health abstract character (Earle 2007). A rather utopian and idealistic view is put forward by the WHO (2006), which conceptualizes health as "a state of complete physical, mental and social well-being" and thus the absence of disease or infirmity (Balog 1978; Boruchovitch and Mednick 2002; WHO 2006). Parsons (1951) regards health as the "state of optimum capacity of an individual for the effective performance of the roles and tasks for which s/he has been socialized." Similarly, Dubos (1972) posits that health is a "physical and mental state relatively free of discomfort and suffering that allows the individual to function as long as possible in the environment where chance or the choice have placed him or her." The latter notions are expressive of health being embedded in everyday life, rendering it a "functional capacity" (Blaxter and Paterson 1982), also taking environmental and cultural parameters into account (Ewles and Simnett 2003); it also corresponds more with the United Nations' take on health. The influence of lifestyle factors and social aspects is emphasized by a variety of other authors as well (Berry 2007; Schiavo 2007; Tones and Tilford 2001), who see it as an important determinant of individual success: "Health is the essential foundation that supports and nurtures growth, learning, personal well being, social fulfillment, enrichment of others, economic production, and constructive citizenship" (Jenkins 2003). According to a contemporary view, health – if maintained properly – makes for a good quality of life (Rod and Saunders 2004; Buchanan 2000) and is claimed to consist of three essential qualities: (1) wholeness, as it is linked to a person's immediate environment and subjective experiences, and (2) pragmatism, which pays tribute to health's relative nature in terms of experience and is closely linked to the third quality, namely, (3) individualism, according to which health is highly personal (Svalastog et al. 2017), especially in the digital age, which challenges academia to take a more individualized look at health.

58.2.2 Health Communication

Communication is a viable asset to the health domain. If conceptualized very broadly, health communication is concerned with the practice of communicating and disseminating information on health-related topics to a dispersed mass audience (US Department of Health and Human Services 2014) as well as communication encounters in the health-care setting (Thompson 2000; Dutta 2008). In practitioners' eyes, health communication is understood as "the study and use of communication strategies to inform and influence individual and community decisions that enhance health" (Center for Disease Control and Prevention; CDC 2001; US Department of Health and Human Services 2005). Health messages can take numerous forms, such as public health campaigns, health education materials, as well as doctor-patient interactions (Schiavo 2007), and are concerned with "informing, influencing, and motivating individual, institutional, and public audiences about important health issues" (US Department of Health and Human Services 2000). Thus, it also includes public service announcements (PSAs) or advertising for both prescription/direct-to-consumer (DTC) and non-prescription/over-the-counter (OTC) drugs (for more details, see Koinig et al. 2017). As such, health communication's goal is to achieve "disease prevention through behavior modification" (Freimuth et al. 2000), encouraging them to unlearn health-compromising behaviors and rather adapt and maintain health-enhancing behaviors (Bernhardt 2004; US Department of Health and Human Services 2005; Schiavo 2007).

58.2.3 Health Promotion

Health promotion alludes to "any combination of health education and related organizational, economic, and environmental supports for behavior of individuals, groups, or communities conducive to health" (Green and Kreuter 1991). Ideally, health promotion will convince individuals to learn how to activate their personal skills in order to improve their health, respectively, well-being, triggering them to seek advice from medical experts and their peers after being exposed to useful and enabling health information (McLaurin 1995). In consulting and utilizing individual, social, and structural resources, the determinants of health are changed through communication (Nutbeam 2000), e.g., invoking recipients to take up recommended actions to improve their current health situations. Thus, health promotion "recognizes the fundamental importance of environmental influences on health and the complex interplay between these factors and health-related behavior" (Green and Tones 2010) while empowering individuals and communities to take more control over their health and well-being (WHO 1996). Therefore, health promotion efforts "are essential and ubiquitous parts of the delivery of health care and the promotion of public health" (Kreps 2014).

58.2.4 Sustainable Development

"Sustainable development is development that meets the needs of the present without compromising the ability of future generations to meet their own needs" (IISD 2016). In its 2030 Agenda for Sustainable Development as well as the accompanying 17 Sustainable Development Goals, the United Nations emphasizes the need to "promote well-being for all ages," which it perceives as a precondition for a "prosperous society" (UN 2017b). In spite of the rather general formulation, the UN puts a focus on women and children, who are especially affected by health issues.

"Human beings are at the centre of concerns for sustainable development. They are entitled to a healthy and productive life in harmony with nature" (WHO 2003). Hence, the WHO's priority is to grant all individuals access to health. Proclaimed to be a "human right" (WHO 2013), health made it to the Sustainable Development Agenda to ensure that all populations, regardless of age, ethnicity, income, and geographic are enabled to live a life free from disease. (UN 2017a, b). While progress has been made in numerous areas, including maternal health, child health, HIV/AIDS, and malaria, other aspects still remain to be addressed. These include universal health coverage, vaccine and medicine access, affordable health care, and child morbidity (UN 2017). Special emphasis is put on universal health coverage (UHC) and the global action plans for prevention and control of noncommunicable diseases (NCDs; Takian and Akbari-Sari 2016).

Several initiatives have already been launched, among them the *Every Woman Every Child* project – "an unprecedented global movement that mobilizes and intensifies international and national action by governments, multilaterals, the private sector and civil society to address the major health challenges facing women and children around the world" (Sustainable Development 2015). Significant progress has already been made, as maternal deaths have dropped by 47% between 1990 and 2012 (SDSN 2014). Still, challenges remain to be tackled, among them national and socioeconomic inequalities, putting especially vulnerable groups at disadvantage (Buse and Hawkes 2015). For instance, The Elders (2016) suggest to improve health-care systems altogether instead of focusing on individuals or particular diseases. Likewise, pharmaceutical companies (e.g., Novartis) have also started to take this issue at heart to "create long-lasting solutions for global health" by combining "philanthropy, zero profit, and Social Ventures" (Novartis 2013).

58.3 The Health Communication Environment

While Dahlgreen and Whitehead (1991) base their notion of the health environment on a social ecological theory and center on the individual and its health-enhancing behavior, this theory is rather limited in its view. Hence, an alternative health communication model will be consulted. In her revised edition of *Health Communication*, Schiavo (2014) outlines the four environmental factors that constitute the so-called Health Communication Environment (see Fig. 1). Drawing from socio-ecological, behavioral, or marketing influences (Morris 1975), the author perceives health communication to be nourished by political, social, community, and service, respectively, product influences. With regard to the *political environment*, policies and laws as well as a country's or an organization's political agenda need to be borne in mind. In terms of *community*, individual health variables, such as health beliefs, attitudes, literacy rates, or behaviors together with gender and lifestyle attributes and demographic factors, play a crucial role in how people manage and communicate about health. On the *product or service level*, health products are under close scrutiny in terms of benefits and risks as well as availability and access. Finally, *social environment* aspects concern social norms, social support systems, existing projects, plus established practices (Schiavo 2014).

Fig. 1 The Health Communication Environment (Schiavo 2014: 22)

58.4 Current Trends and Developments

58.4.1 Multidisciplinary Approach

In their 2015 article, Hannawa and colleagues attempted to map out the health communication field, focusing on the different forms of communication utilized within the health field. A more discipline-focused approach was taken by Bernhardt (2004) and the WHO (2003), who regard health as multi- or transdisciplinary in its very nature, drawing from and influencing multiple (academic) disciplines. As such, it pays tribute to the diverse and complex parameters involved in shaping, respectively, attaining individual health (Schiavo 2014). In the area of *behavioral and social science theories*, health communication seeks to explain how and why individuals do (not) engage in or take up health-enhancing measures, as in the case of the Health Belief Model (Becker et al. 1977) and the social cognitive theory (Bandura 1977). *Mass media and new media theories* are not only concerned with how health messages are communicated but also intend to unravel both how media is used by and influences individuals with regard to their health. A mass media concept commonly employed in this context is McQuail's (1994) media effectiveness. *(Social) marketing theories* are applied to both commercial and nonprofit healthcare settings and scrutinize how marketing is used to make consumers use a certain kind of (health) product (Schiavo 2014). Increasingly, these theories are also linked to social change due to a growing people-centeredness (Lefebvre 2013; see also *A threefold approach for social change* by Diago, *Community Media for Social Change in South India* by Pavarala, *Films for Social Change in Hong Kong and China* by Li, *Role of Participatory Communication in Strengthening Solidarity and Social Cohesion in Afghanistan* by Ahmed, and *Institutionalization and implosion of Communication for Development and Social Change in Spain* by Saez in this volume). Medical models are, first and foremost, used for explaining patient-provider interactions (Schiavo 2014). The biomedical model regards poor health as a physical phenomenon, and health can only be restored through physical means (du Pré 2000). While it blends out selected components (e.g., individual beliefs, norms, and attitudes), this shortcoming is explicitly addressed by the biopsychosocial model of health, which integrates "people's feelings, their ideas about health, and the events of their lives" (du Pré 2000: 9). Finally, s*ociological* aspects of health are discussed as part of health policy, population health, and health services (Bourgeault et al. 2013), while, by contrast, *anthropological* notions of health look at health in the context of a particular culture, respectively, social group (Langdon and Wiik 2010) (Fig. 2).

58.4.2 Digital Health and Health Monitoring Systems

Due to the convergence of technologies, digital and genomic content is increasingly digitized, leading to the development of digital health, which is meant to not only enhance the accessibility of health information but also make health delivery more

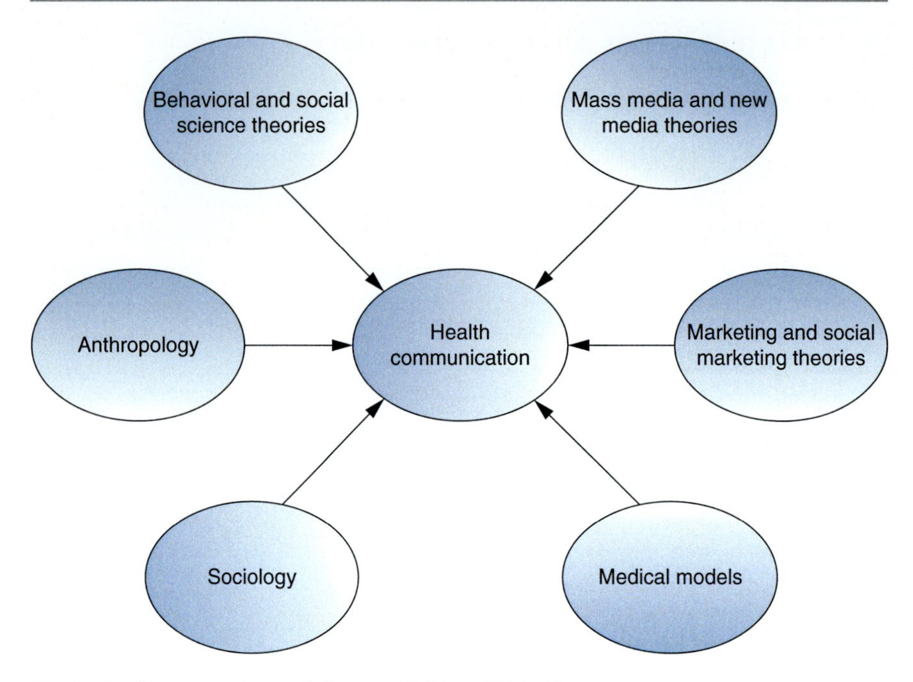

Fig. 2 Health communication influences (Schiavo 2014: 22)

efficient and precise (Bhavnani et al. 2016). Digital health technologies are driving change toward a more individually centered, self-responsible, and empowered health care (Banos et al. 2016). In addition, the last two decades have witnessed a dramatic increase in health monitoring systems (Baig and Gholamhosseini 2013). Thereby, services range from software to hardware solutions, encompassing email, text messages, and apps just to name the most prominent ones (Widmer et al. 2015). Even though digital health has not explicitly been integrated into the UN's Sustainable Development Goals, this area warrants further discussion, since it holds enormous potential to revolutionize health care. In the following, eHealth, mHealth, as well as digital health information and communication will be discussed in more detail.

eHealth: eHealth is used to refer to the integration and use of new information technologies in health-related matters (WHO 2017). "eHealth refers to the use of modern information and communication technologies to meet the needs of citizens, patients, health care professionals, health care providers as well as policy makers" (EU Ministerial Declaration of eHealth 2003; also see Eysenbach 2001). As such, eHealth promotes the use of interactive communication tools in the health context while also holding the potential to change health beliefs, policies, and outcomes lastingly (Schiavo 2014). In the advent of the Internet revolution and new technologies, eHealth is available on-demand, on-the-go, and is no longer a push, but a pull, service, also broadening its community boundaries (Abroms et al. 2011). At the same time, it allows for a new, advanced form of health literacy – defined as "the degree to which individuals have the capacity to obtain, process, and understand basic health information and services needed to make appropriate health decisions" (Selden et al. 2000). Consequently, eHealth is concerned with promoting health globally and in disadvantaged communities (Schiavo 2014), taking changed health behaviors and patterns into account (Fischer et al. 2016).

eHealth services among US physicians are constantly gaining popularity and have risen from 68% in 2012 to 84% in 2015 (Statista 2017h). Thereby, practitioners have started to rely on these electronic devices for specific/clinical content and pharmaceutical information to the largest extent (61% and 53%, respectively; Statista 2017i). Nonetheless, Americans seem to be only somewhat familiar with eHealth, as more than half of them claim not to have heard of the concept in 2017 (Statista 2017j). Among those US consumers who use eHealth, services – more precisely, medical video visits – were most frequently used to get prescription refills (78%) or for chronic care management (60%; Statista 2017k). Besides video chats, emails are the preferred form of communication with physicians (66%) and are followed by online web portals (36%), text messages (34%), and mobile apps (31%; Statista 2017l) (see also mHealth). The willingness to pay for these services is, nonetheless, only moderately pronounced (Statista 2017m). However, the potential these offerings hold is huge, as eHealth services have also become popular in outpatient care and are, for instance, employed for early-stage breast cancer patients (Ventura 2016).

mHealth: mHealth can be considered as a part of eHealth. Albeit it is often conceptualized as a part of telemedicine – a term used to describe the potential of new information technologies to bridge geographical distances and barriers to deliver and grant individual access to health-related services (WHO 2010) – mobile health (mHealth) has become used in its own right. Following the WHO, mHealth refers to "the use of mobile and wireless technologies to support the achievement of health objectives" (WHO 2011) and was already employed by more than one third of Americans in 2014 (39%; Statista 2017a). Yet, the tremendous growth of mobile devices does not only affect health in the developed world but also offers hope to vulnerable populations in resource-poor environments (United Nations Foundation 2011).

With a mHealth market share of only 6.7 billion in 2012, the value is expected to almost reach 60 billion by 2020 (58.8%; Statista 2017b), presenting a compound annual growth rate of more than 40% (Statista 2017c). Drivers of mHealth popularity are multiple and concern its potential for cost reductions (Statista 2017d) and its vast array of applicability, ranging from remote monitoring, diagnosing symptoms, consulting doctors, and complying with treatment or dietary regimes (Statista 2017e). mHealth, however, does not only benefit individuals; there are also benefits for the health-care system, as mHealth holds the potential to reduce doctor consultations, examinations, as well as the length of hospital stays (Statista 2017f). The area of application is expected to be broadened in the near future, with 71% of US Americans expressing a willingness to use mHealth services for health emergencies (Statista 2017g).

The applicability of mHealth services is constantly growing. Academia has repeatedly pointed out the applicability of social networking sites to grant individuals (online) social support, offering both informational and emotional support to both patients, survivors, and family members (Hether et al. 2016; Murthy and Eldredge

2016). In addition, health apps and mobile tracking devices are also very helpful in and often used for monitoring, detecting, and managing medical conditions, such as pulse, heart rate, or breast cancer risk assessment (Goth 2015, see also a more detailed description below). On the other hand health symptoms brought about by (excessive) (mobile) social media use might also be worth investigating (Nasirudeen et al. 2017).

Digital/Online Health Information: Health information technology (HIT, Buntin et al. 2010) has led digital health information to be available in abundance. The variety of online and mobile services, such as wireless and mobile services, online doctor review portals, self-diagnosing websites, or rating sites, among others, are not only of advantage to practitioners but can also involve individuals more strongly in their health care (Martin 2012). Thus, digital health information is also increasingly requested by consumers and was already of interest in the early 2000s (Rice and Katz 2001). For example, in Great Britain, the number of individuals consulting the Internet for health information has increased from 18% in 2007 to 43% in 2013 (Statista 2017n).

Surveys from all over the world report that consumers' education levels influence their search habits for health information; for instance, Finnish consumers with secondary and tertiary education go more frequently online to research health, nutrition, or diseases (62% and 72%, respectively) than those with primary education (43%; Statista 2017o). US college graduates were also almost twice as often looking for information on health than high school graduates (81% vs. 45%; Statista 2017p). Similar results are reported for US cell phone users (38% vs. 17%; Statista 2017q). Findings from France support the influence of gender, with women looking for health information more frequently than men (45% vs. 37%; Statista 2017r). This was also found by Osma et al. (2016), who discovered that women do not only turn to the Internet more often than men for health information but also use their smartphones more frequently to do so.

Digitized health information has empowered consumers, allowing them to get more involved in their health care. In 2015, US consumers claimed to be very interested in online medical records (29%), online cost estimators (27%), online appointment setting (24%), email access to their doctors (23%), as well as the ability to pay their medical bills online (21%; Statista 2017s). The same year, UK residents turned to the Internet for self-diagnosing purposes (73%) and health-condition management (63%) or when in need for health-enhancing information, respectively, treatment options (both 39%), while 38% researched the risks associated with a specific medical condition (Statista 2017t).

Wearables: Mobile technology has also become attached to bodies, where devices provide personalized health information and are increasingly used to identify and discuss health symptoms (Piwek et al. 2016). As such, these gadgets have become part of the "quantified self" movement, in the course of which the reporting and tracking of data can be used to improve individual health (Patel et al. 2015). Further advantages are connected information, the creation of health-care communities, and participation due to gamification (Huffington Post 2017).

While social networking sites (33%) and mobile apps (30%) have been commonly used by US consumers in 2017, wearables already come in 3rd place (28%; Statista 2017u). In an age of automated physical activity monitoring technology, which is often linked to social networking sites like Facebook, etc., some recipients meet these devices and services with reservation. Numbers are a bit inconclusive, ranging from claiming only 1% of US consumers to use wearables (Wang 2014) to even every 6th US consumer relying on wearables (Juniper Research 2013). Based on the apps' monthly users, Fitbit is the leading health and fitness app in the United States with 23.6 million users (Statista 2017v). Given the vast array of applicability by users themselves, but also by health-care organizations, employers, insurance companies, and practitioners (Patel et al. 2015), sales are predicted to increase significantly in the years to come (Wang 2014). Rising user numbers also correspond with increasing shipments, which were expected to reach 5.3 million units by 2017 (Statista 2017w); smart wristbands alone are predicted to reach sales of 63.86 million units in 2021 (Statista 2017x), while connected wearable devices are expected to achieve sales of 325 million in 2021 (Statista 2017y).

Increasingly, patients demand doctors to base health recommendations on the results produced by wearables (Cello Health Insight 2014), and they are called upon the leverage the power of wireless devices to their advantage, which include tracking, analyzing, and optimizing individual health (Waracle 2016). At the same time, these devices are expected to not only reduce health-care costs but also increase efficiency, accessibility, and quality (Waracle 2016). Moreover, they are claimed to foster health empowerment and improve individual quality of life (Waracle 2016). Nonetheless, these devices predominantly appeal to a younger generation but fail to engage people above 40 (Statista 2017z). The major reasons as to why these devices are used concern tracking purposes (30%), awareness raising (28%), and individual motivation (27%; Statista 2017aa).

Currently, the biggest challenge lies in providing smart health-care solutions and devices that do not invade individuals' health, trying to overcome established conceptualizations of u-health (ubiquitous health) and p-health (pervasive health; Custodio et al. 2012). In addition, reservations need to be tackled, which are most prominently held by women (Arigo et al. 2015).

58.4.3 Gender-Specific Health Issues

When it comes to health, men and women differ significantly in their attitudes. Men define health in rather functionalist terms of "efficiency" and "absence of disease"; they prefer offers that help them to overcome specific health problems (Tempel et al. 2013). Women, on the other hand, are assumed to proactively and publicly address their health issues (Broom and Tovey 2009). Thus, gender-specific health issues and their impact on health communication present a challenge that needs to be addressed.

Women's Health: Already in 1992, research recognized that women's health is "a patchwork quilt with gaps" (Clansy 1992). As part of the United Nations' Sustainable Development Goals, the necessity to address women's health issues is prioritized. Overall, women's health is defined as all issues that are particulate to women. According to the WHO (2017), women's health deserves specific attention because of the following reasons: an unequal distribution of power, social norms limiting women's educational and professional choices, and their reproductive roles together with an increased likelihood of women to fall victim to violence of any kind (WHO 2017).

However, in the academic context, women's health has received significant attention. Women are generally more interested in health issues than men (75% vs. 63%; Bonfadelli 2002) and are more engaged in their health, whereby more than 80% of women visit a doctor at least once a year, while only slightly more than 70% of men do (The Austrian Health Survey 2007). Women also participate more often in prevention programs. In addition, they visit their primary care provider to a significantly greater extent than men who are still underrepresented in primary care (Thompson et al. 2016). Recent research by Rowley et al. (2015) also confirmed that women are more proactive in health information seeking and check a wider range of health information sources. In addition, women are more likely to go to the Internet for health information (Tennant et al. 2015) and are more likely to consult health professionals and family and friends than men. Moreover, women are also the primary recipients of health messages: "Health promotion messages often target women in their assigned role as caregivers in the family" (Östlin et al. 2006: 27).

Despite progress has been made since then (Charney 2000), there are also numerous diseases that are unique to women and warrant further research. Present-day research has already taken up some issues, dedicating more research toward breast cancer risk assessment (Goth 2015), perinatal depression (Osma et al. 2016), preconception care, and prenatal health issues (Nwolise et al. 2017).

Men's Health: One area of research that received very limited academic attention is men's health (Weder 2014). As experts worldwide express their concern with this issue – even talking about a men's health crisis – this issue is slowly gaining recognition. In general, men's health is something that can and needs to be differentiated from both general health and women's health, since it is concerned with "the prevention and diagnosis of illnesses that afflict the male population. Men's health should also include the personal well-being and quality of life of men" (Hoon 2005: 172). From a clinical point of view, common male-specific diseases are prostate cancer, erectile dysfunction, and testicular cancer (MacDonald 2016), while mental health issues like depression receive limited attention (Branney and White 2008). Yet, there are some obstacles to overcome first: while women address their health issues openly and proactively, men feel rather indifferent toward their health and are thus reluctant to become engaged (Broom and Tovey 2009). Due to men's rather functionalist perspective on health, they prefer offers that help them to overcome specific health problems (Tempel et al. 2013). Since they often tend to underestimate physical and psychological symptoms, they wait longer to seek help from a doctor; at times, it even takes their wives to urge them to go and make an appointment or seek treatment (Raml et al. 2011). This concerning trend can be explained by the fact that masculine traits – strength and independence – "are seen as injurious to [men's] health" (MacDonald 2016: 285).

And even though there are studies in the area of strategic communication and health promotion focusing on gender-specific issues (Galdas et al. 2005), there is a clear scarcity of research on specifically male health issues like prostate cancer, baldness, cancer, diabetes, cardiovascular diseases, etc. (Hoebel et al. 2013). As such, men's health is conceptualized "as the prevention and diagnoses of illnesses that afflict the male population. Men's health should also include the personal well-being and quality of life of men" (Hoon 2005: 172).

Men's health issues and concerns have also increasingly moved online and concern safer sex, respectively, condom use (Bailey et al. 2017), as well as prostate cancer interventions.

58.4.4 Pharmaceutical Advertising as Health Communication

Also, commercial messages qualify as health communication, the most prominent one being pharmaceutical advertising. It is "defined as [paid] messages created by marketers of pharma products that attempt to inform, persuade and even entertain the target audience with the goal of influencing recipients' attitudes—and ultimately behavior—in a favorable manner" (Diehl et al. 2008: 100). Pharmaceutical advertisements are utilized to promote both prescription and non-prescription drugs. The former comprise drugs used to treat more serious diseases and are subject to a doctor's prescription since their use is associated with significant risks and adverse effects; consequently, they are only allowed to be promoted to physicians (with the exception of the United States and New Zealand, where these preparations can also be advertised to the general public; Diehl et al. 2008). The latter, on the other hand, are commonly labeled self-medication preparations, home remedies, or over-the-counter (OTC) medications, which can be promoted to the general public in almost every country throughout the world, since they hold comparatively minor risks and are only intended to treat minor medical symptoms.

Koinig's (2016) field study, which was conducted on three continents, highlighted that among four different ad appeal types (informative, emotional, mixed, and CSR), the mixed pharmaceutical promotional message was most effective with regard to ad evaluation and empowerment. Another study by Koinig et al. (2017) showed that pharmaceutical advertising can indeed empower consumers on three distinct levels, namely, message empowerment, self-medication empowerment, and health empowerment. In terms of ad appeal, findings suggest that the mixed appeal ad led to the highest degree of empowerment in all of the three empowerment categories and was followed by the informative ad. The emotional and the CSR appeal ads failed to appeal to consumers and empowered them to a lesser degree. Results thus suggest that mixed appeals were not only consumers' favorites in terms of ad liking but were also the most suitable way to aid consumers in making qualified and reasonable decisions, educating and "empowering" them by strengthening their beliefs in the product, their own self-medication capabilities, and their health.

58.4.5 Health Empowerment

The major goals associated with health promotion are equity and empowerment (Green and Tones 2010), the latter having also been taken up by the Ottawa Charter (WHO 1996). Following Tones and Tilford (2001), empowerment could "be even seen as a worthwhile health goal in its own right" which is brought about by significant changes that do not leave the health sector unaffected: "Empowered by better access to higher education, information sources like the Internet, and greater personal wealth, consumers expect to have a much greater say in their own medical treatment" (David 2001: 1). As such, empowerment is concerned with including lay perspectives – perspectives by non-expert or ordinary people – in health-related decisions (Earle 2007). Lately, the concept of empowerment has also been linked to advertising messages (Koinig 2016; Koinig et al. 2017).

Through mobile and interactive health technologies, consumers are given the opportunity to become more involved in and empowered regarding their own health care (see also ▶ Chap. 55, "New Media: The Changing Dynamics in Mobile Phone Application in Accelerating Health Care Among the Rural Populations in Kenya" by Okoth in this volume); increasingly, they also agree to disclose their online health records to receive more tailored treatments (Makovsky 2015). Furthermore, the availability of online services does not only allow for increased consumer participation but also for a faster delivery of care (see also ▶ Chap.64, Participatory Development Communication and Natural Resources Management by Bessette and ▶ 9, "Media and Participation" by Carpentier in this volume). Even against the background of already heightened health-care costs, Americans would nevertheless leverage technology, expressing a high willingness to pay for innovative services (Makovsky 2015) and the clear benefits they bring: they "have the potential to improve communication with clinicians, access to personal health information, and health education with the goal of preparing patients to take a more active role in their care" (Franklin et al. 2009: 169f.).

58.5 Working Toward Sustainable Health and Sustainable Health Communication

Sustainable health comprises initiatives taken on the individual, community, public/national, and international level (UN 2017; Buse and Hawkes 2015).

On the individual level, preventive care is the number one priority and concerns vaccinating children, as well as engaging in safe sex only – the latter being a movement that is also increasingly featured on social media (Bailey et al. 2017). Also, the necessity of involving partners in their spouse's fitness activities has been found to benefit women, specifically in the digital context (Arigo et al. 2015). Weder (2014) discovered this necessity also for men and discussed this against the background of "crabwise campaigns" – promotional campaigns that addressed women to encourage their men to get a preventive examination.

On a community level, inequalities between urban and rural or wealthier and poorer groups need to be balanced, guaranteeing that all groups are educated with regard to nutrition, health affecting risky behavior, etc. (Maeda et al. 2014; Schmidt et al. 2015). Practicing and promoting health lifestyles in groups is recommended as well, whereby schools are listed as prime examples. Moreover, the issue of tailoring information to communities of different academic and ethnic backgrounds is emphasized (Noar et al. 2007) even though the success of these tailored approaches is ambiguous (Saywell et al. 2004). Yet, tailored information holds one advantage: the topic's personal relevance is stressed, and centralized information processing is encouraged (Hawkins et al. 2008). Moreover, tailored content is also of interest to the wearables' industry (Kim et al. 2017).

Nationally, ensuring high-quality workforce training, especially in low-income or middle-income countries, has to be made a core focus (Maeda et al. 2014). Special focus should also be put on health professionals' communication competencies (Ruben 2016), following the example of the Health Professions Core Communication Curriculum (HPCCC; Bachmann et al. 2016). Where shortfalls exist, apps and eHealth services might be used, as it is the case in outpatient care (Ventura 2016). These gadgets are specifically useful in emphasizing, respectively, highlighting, patient needs and preferences (tailoring, customization; Nwolise et al. 2017): "Person-centered e-supportive systems may bridge the communication gap between the hospital setting and patients' homes by fostering a reciprocal partnership in care that acknowledges and reinforces patients' expertise and agency" (Ventura 2016).

Some countries even publish a health report to make their efforts more binding and achieve a clearly set goal, as it is the case with the Healthy People 2020 (Hesse et al. 2014), or even implement specifically designed laws, such as the Health Information Technology for Economic and Clinical Health Act of 2009, which incorporates health technology into health care in the United States (Gold et al. 2012). Ojo (in this volume) sheds some light on how ICTs are used in Africa for development purposes.

The most prominent goal on the international level is the call for universal coverage, which means that "all people can use the promotive, preventive, curative, rehabilitative and palliative health services they need" (WHO 2017). It is considered to be "the single most powerful concept that public health has to offer. It is our ticket to greater efficiency and better quality. It is our savior from the crushing weight of chronic noncommunicable diseases that now engulf the globe" (Chan 2012). If accompanied by a sustainable financing plan, this move is expected to reduce health-care costs (Schmidt et al. 2015).

Another international goal aims at fostering health education and health literacy (Cappelletti et al. 2015), also through interactive offers with appealing designs (see also ▶ Chap.29, Visual Communication and Social Change by Loes in this volume). In the area of digital health, calls to improve the knowledge about usability, design, and acceptance of mHealth and eHealth services as well as wearables should be taken seriously (Hether et al. 2016; Nwolise et al. 2017), which cannot only increase individual involvement but also foster empowerment in the long run.

58.6 Limitations and Directions for Future Research

The present paper sets out to outline some topics that are of high importance to the health-care sector and build upon the United Nations' Sustainable Development Goals. Yet, due to the limited scope of the paper, only a few, selected aspects could be discussed. We pointed out the potentials of digital and tailored health communication – based on their growing acceptability. Here, empirical research is still quite scarce, so it would be interesting to delve deeper in this topic, on a global, national, as well as glocal level (see also ▶ Chap.26, Glocal Development for Sustainable Social Change by Patel in this volume). The gender debate should be intensified as well, taking into account gender-specific health topics and usage patterns. Moreover, the term of empowerment should be examined in the context of communities, wearables, and (advertising) message reception (see also ▶ Chap. 10, "Empowerment as Development: An Outline of an Analytical Concept for the Study of ICTs in the Global South" by Svensson in this volume). While the previously discussed theories on models of health are hardly cross- or transdisciplinary in nature, more recent approaches set out to overcome these segregated notions (Schiavo 2014) and might require further elaboration. Thus, while the present research looked at the microlevel of health communication, further research should not neglect the broader, macro perspective. Especially the role of media advocacy and public health advocacy warrants future considerations, as the role of both the media and health practitioners in shaping health policy development must not be underestimated (Ingram et al. 2015).

The digitization of health information – starting with electronic health records (EHRs) – is expected to continue in the future, with countries (e.g., Austria) considering the introduction of electronic prescriptions, electronic pregnancy passports, and electronic immunization cards. It is worthwhile to investigate not only how these new digital offerings are perceived and used, but also inasmuch they are able to improve patient-provider interactions and reduce disparities in access to health care.

References

Abroms LC, Padmanabhan N, Philips T, Thaweethai L (2011) iPhone apps for smoking cessation: a content analysis. Am J Prev Med 40:279–285

Arigo D, Schumacher LM, Pinkasavage E, Butryn ML (2015) Addressing barriers to physical activity among women: a feasibility study using social networking-enabled technology. Digit Health 1:1–12

Bachmann C, Kiessling C, Härtl A, Haak R (2016) Communication in health professions: a European consensus on inter- and multi-professional learning objectives in German. GMS J Med Educ 33:1–13

Baig MM, Gholamhosseini H (2013) Smart health monitoring systems: an overview of design and modeling. J Med Syst 37:9898

Bailey J, Tomlinson N, Hobbs L, Webster R (2017) Challenges and opportunities in evaluating a digital sexual health intervention in a clinic setting: staff and patient views. Digit Health 3:1–20

Balog JE (1978) The concept of health and the role of health education. J Sch Health 9:462–464

Bandura A (1977) Self-efficacy: toward a unifying theory of behavioral change. Psychol Rev 84:191–215

Banos O, Amin MB, Khan WA, Afzal M, Hussain M, Kang BH, Lee S (2016) The Mining Minds digital health and wellness framework. Biomed Eng Online 15:76

Becker MH, Maiman LA, Kirscht JP, Haefner DP, Drachman RH (1977) The health belief model and prediction of dietary compliance: a field experiment. J Health Soc Behav 18:348–366

Bernhardt JM (2004) Communication at the core of effective public health. Am J Public Health 94:2051–2053

Berry D (2007) Health communication. Theory and practice. Open University Press, New York

Bhavnani SP, Narula J, Sengupta PP (2016) Mobile technology and the digitization of healthcare. Eur Heart J 37(18):1428–1438

Blaxter M, Paterson S (1982) Mothers and daughters: a three-generation study of health, attitudes and behavior. Heinemann Educational, London

Bonfadelli H (2002) The Internet and knowledge gaps: a theoretical and empirical investigation. Eur J Commun 17(1):65–84

Boruchovitch E, Mednick B (2002) The meaning of health and illness: some considerations for health psychology. Psico-USF 7:175–183

Bourgeault IL, Benoit C, Bouchard L (2013) Towards a sociology of health and healthcare. Health Policy 9:10–11

Branney P, White A (2008) Big boys don't cry: depression and men. Adv Psychiatr Treat 14(4):256–262

Broom A, Tovey P (eds) (2009) Men's health: body, identity and social context. Wiley-Blackwell, Chichester

Buchanan DR (2000) An ethic for health promotion: rethinking the sources of human well-being. Oxford University Press, New York

Buntin MB, Jain SH, Blumenthal D (2010) Health information technology: laying the infrastructure for national health reform. Health Aff 29:1214–1219

Buse K, Hawkes S (2015) Health in the Sustainable Development Goals: ready for a paradigm shift? Glob Health 11:1–8

Cappelletti ER, Kreuter MW, Boyum S, Thompson T (2015) Basic needs, stress and the effects of tailored health communication in vulnerable populations. Health Educ Res 30:591–598

CDC (2001) What is health communications? http://www.cdc.gov/healthcommunication/healthbasics/whatishc.html. Accessed 21 June 2014

Cello Health Insight (2014) Digital health debate 2014. Available at Cello Health Insight. http://cellohealthinsight.com/digital-health-debate/. Accessed 20 Dec 2017

Chan M (2012) Universal coverage is the ultimate expression of fairness. Acceptance speech at the 65th World Health Assembly, Geneva. 23 May 2012. Available via WHO. http://www.who.int/dg/speeches/2012/wha_20120523/en/index.html. Accessed 29 May 2015

Charney P (2000) Women's health. An evolving mosaic. J Gen Intern Med 15:600–602

Clansy SM (1992) American Women's Health Care: a patchwork quilt with gaps. JAMA 268:1919–1920

Custodio V, Herrera FJ, López G, Moreno JI (2012) A review on architectures and communications technologies for wearable health-monitoring systems. Sensors 12:13907–13946

Dahlgren G, Whitehead M (1991) Policies and strategies to promote social equity in health. Institute for Futures Studies, Stockholm

David C (2001) Marketing to the consumer: perspectives from the pharmaceutical industry. Mark Health Serv 21(1):5–11

Diehl S, Mueller B, Terlutter R (2008) Consumer responses towards non-prescription and prescription drug advertising in the US and Germany: They don't really like it, but they do believe it. Int J Advert 27(1):99–131

Du Pré A (2000) Communicating about health: current issues and perspectives. Mayfield Publishing Company, Mountain View

Dubos R (1972) L'homme ininterrompu? Denoel, Paris

Dutta MJ (2008) Communicating health: a culture-centered approach. Polity, Malden

Earle S (2007a) Exploring health. In: Earle S, Lloyd CE, Sidell M, Spurr S (eds) Theory and research in promoting public health. Sage, London, pp 37–66

Earle S (2007b) Promoting public health: exploring the issues. In: Earle S, Lloyd CE, Sidell M, Spurr S (eds) Theory and research in promoting public health. Sage, London, pp 1–36

EU Ministerial Declaration of eHealth (2003) Retrieved from http://bme2.aut.ac.ir/~towhidkhah/MI/seminar83/Eslami/HIS%20%20%20%20EHR%2D%2D-%20Documents%20of%20Classmates/F.R.Eslami/HIS%2D%2DIn%20%20Uropean%20Countries%20%2D%2D%20E-References%2D%2D%20%20Eslami/min_dec_22_may_03.pdf

Ewles L, Simnett I (2003) Promoting health: a practical guide, 5th edn. Bailliere Tindall, Edinburgh

Eysenbach G (2001) What is eHealth? J Med Internet Res 3(2)

Fischer HH, Fischer IP, Pereira RI, Furniss AL, Rozwadowski JM, Moore SL, Durfee MJ, Raghunath SG, Tsai AG et al (2016) Text message support for weight loss in patients with prediabetes: a randomized clinical trial. Diabetes Care. https://doi.org/10.2337/dc15-2137

Franklin PD, Farzanfar R, Thompson D (2009) E-health strategies to support adherence. In: Shumaker SA, Ockene JK, Riekert KA (eds) The handbook of health behavior change. Springer, New York, pp 169–190

Freimuth V, Linnan HW, Potter P (2000) Communicating the threat of emerging infections to the public. Emerg Infect Dis 6:337–374

Galdas PM, Cheater F, Marshall P (2005) Men and health help-seeking behaviour: literature review. J Adv Nurs 49(6):616–623

Gold MR, McLaughlin CG, Devers KJ, Berenson RA, Bovbjerg RR (2012) Obtaining providers' 'buy-in' and establishing effective means of information exchange will be critical to HITECH's success. Health Aff (Millwood) 31:514–526

Goth G (2015) Older Low-Income Women Liked mHealth Breast Care App. Healthdatamanagement.com:4–4

Green L, Kreuter M (1991) Health promotion planning: an educational and environmental approach. Mayfield, Mountain View

Green J, Tones K (2010) Health promotion: planning and strategies, 2nd edn. Sage, Thousand Oaks

Hannawa AF, García-Jiménez L, Candrian C, Schulz PJ (2015) Identifying the field of health communication. J Health Commun 20:521–530

Hawkins RP, Kreuter MW, Resnicow K, Fishbein M, Dijkstra A (2008) Understanding tailoring in communicating about health. Health Educ Res 23:454–466

Hesse BW, Gaysynsky A, Ottenbacher A, Moser RP, Blake KD, Chou W-YS, Vieux S, Beckjord E (2014) Meeting the healthy people 2020 goals: using the Health Information National Trends Survey to monitor progress on health communication objectives. J Health Commun 19:1497–1509

Hether HJ, Murphy ST, Valente TW (2016) A social network analysis of supportive interactions on prenatal sites. Digit Health 0:2–12

Hoebel J, Richter M, Lampert T (2013) Social status and participation in health checks in men and women in Germany – results from the German Health Update (GEDA), 2009 and 2010. Dtsch Arztebl Int 110:679–685

Hoon W (2005) The coverage of prostate cancer and impotence in four popular men's magazines (1991–2000). Int J Mens Health 4(2):171–185

Huffington Post (2017) Wearable technology: the coming revolution in healthcare. Available via Huffington Post. https://www.huffingtonpost.com/vala-afshar/wearable-technology-the-c_b_5263547.html. Accessed 20 Dec 2017

IISD (2016) Sustainable Development Goals. IISD perspectives on the 2030 agenda for sustainable development. In: Sustainable development. International Institute for Sustainable Development. Available via IISD. http://www.iisd.org/sites/default/files/publications/sustainable-development-goals-iisd-prespectives.pdf. Accessed 20 Dec 2017

Ingram M, Sabo SJ, Gomez S, Piper R, de Zapien JG, Reinschmidt KM, Schachter KA, Carvaja SC (2015) Taking a community-based participatory research approach in the development of methods to measure a community health worker community advocacy intervention. Prog Community Health Partnersh 9(1):49–56

Jenkins DC (2003) Building better health: a handbook for behavioral change. PAHO, Washington, DC

Juniper Research (2013) Smart wearable devices. Fitness, healthcare, entertainment & enterprise. Available via Juniper Research. http://www.juniperresearch.com/reports/Smart_Wearable_Devices. Accessed 20 Dec 2017

Kim KJ, Shin D-H, Yoon H (2017) Information tailoring and framing in wearable health communication. Inf Process Manag 53:351–358

Koinig I (2016) Pharmaceutical advertising as a source of consumer self-empowerment: evidence from four countries. Springer, Wiesbaden

Koinig I, Diehl S, Mueller B (2017) Are pharmaceutical ads affording consumers a greater say in their health care? The evaluation and self-empowerment effects of different ad appeals in Brazil. Int J Advert 36(6):945–974. https://doi.org/10.1080/02650487.2017.1367353
Kreps GL (2014) Evaluating health communication programs to enhance health care and health promotion. J Health Commun 19:1449–1459
Langdon EJ, Wiik FB (2010) Anthropology, health and illness: an introduction to the concept of culture applied to the health sciences. Rev Lat Am Enfermagem 18:459–466
Lefebre RC (2013) Social marketing and social change: strategies and tools for improving health, well-being, and the environment. Jossey-Bass, San Francisco
MacDonald J (2016) A different framework for looking at men's health. Int J Mens Health 15(3):283–295
Maeda A, Araujo E, Cashin C, Harris J, Ikegami N, Reich MR (2014) Universal health coverage for inclusive and sustainable development: a synthesis of 11 country case studies. World Bank, Washington, DC
Makovsky (2015) Fifth annual "Pulse of Online Health" survey finds 66% of Americans eager to leverage digital tools to manage personal health. Retrieved from http://www.makovsky.com/insights/articles/733
Martin T (2012) Assessing mHealth: opportunities and barriers to patient engagement. J Health Care Poor Underserved 23:935–941
McLaurin P (1995) An examination of the effect of culture on pro-social messages directed at African-American at-risk youth. Commun Monogr 62:301–326
McQuail D (1994) Mass communication theory: an introduction. Sage, London
Morris JN (1975) Towards prevention and health. J Intern Med 197(January/December):13–17
Murthy D, Eldredge M (2016) Who tweets about cancer? An analysis of cancer-related tweets in the USA. Digit Health 2:1–16
Nasirudeen AMA, Lau Lee CA, Wat NJ, Lay S, Wenjie L (2017) Impact of social media usage on daytime sleepiness: a study in a sample of tertiary students in Singapore. Digit Health 3:1–9
Noar SM, Benac CN, Harris MS (2007) Does tailoring matter? Meta-analytic review of tailored print health behavior change interventions. Psychol Bull 133:673–693
Novartis (2013) Making healthcare sustainable in the developing world. Available via Novartis. https://www.novartis.com/stories/access-healthcare/making-healthcare-sustainable-developing-world. Accessed 20 Dec 2017
Nutbeam D (2000) Health literacy as a public health goal: a challenge for contemporary health education and communication strategies into the 21st century. Health Promot Int 15:259–267
Nwolise CH, Carey N, Shawe J (2017) Exploring the acceptability and feasibility of a preconception and diabetes information app for women with pregestational diabetes: a mixed-methods study protocol. Digit Health 3:1–11
Osma J, Barrera AZ, Ramphos E (2016) Are pregnant and postpartum women interested in health-related apps? Implications for the prevention of perinatal depression. Cyberpsychol Behav Soc Netw 19:412–415
Östlin P, Eckermann E, Mishra US, Nkowane M, Wallstam E (2006) Gender and health promotion: a multisectoral policy approach. Health Promot Int 21(Suppl_1):25–35. https://doi.org/10.1093/heapro/dal048
Parsons T (1951) The social system. Free Press, New York
Patel MS, Asch DA, Volpp KG (2015) Wearable devices as facilitators, not drivers, of health behavior change. JAMA E1–E2 313(5):459–60
Piwek L, Ellis DA, Andrews S, Joinson A (2016) The rise of consumer health wearables: promises and barriers. PLoS Med 13:e1001953
Raml R, Dawid E, Feistritzer G (2011) 2. Österreichischer Männerbericht (unter Mitarbeit von Radojicic, N und Seyyed-Hashemi S) (im Auftrag des Bundesministeriums für Arbeit, Soziales und Konsumentenschutz – BMASK). Vienna
Rice RE, Katz JE (2001) The Internet and health communication: experiences and expectations. Sage, Thousand Oaks

Rod M, Saunders S (2004) The informative and persuasive components of pharmaceutical promotion: an argument for why the two can coexist. Int J Advert 28(2):313–349
Rowley J, Johnson F, Sbaffi L (2015) Gender as an influencer of online health information-seeking and evaluation behavior. J Assoc Inf Sci Technol. Advance online publication. https://doi.org/10.1002/asi.23597
Ruben BD (2016) Communication theory and health communication practice: the more things change, the more they stay the same. Health Commun 31:1–11
Saywell RM Jr, Champion VL, Skinner CS, Menon U, Daggy J (2004) A cost-effectiveness comparison of three tailored interventions to increase mammography screening. J Womens Health 13:909–918
Schiavo R (2007) Health communication: from theory to practice. Wiley, San Francisco
Schiavo R (2014) Health communication: from theory to practice, 2nd edn. Jossey-Bass, San Francisco
Schmidt H, Gostin LO, Emanuel EJ (2015) Public health, universal health coverage, and Sustainable Development Goals: can they coexist? Lancet 386:928–930
SDSN (2014) Health in the framework of sustainable development. Technical report for the post-2015 development agenda. Available via UN. http://unsdsn.org/wp-content/uploads/2014/02/Health-For-All-Report.pdf. Accessed 20 Dec 2017
Selden CR, Zorn M, Ratzan SC, Parker RM (2000) Health literacy. Retrieved from http://www.nih.gov/clearcommunication/healthliteracy.htm. Accessed 20 Mar 2014
Statista (2017a) Availability of selected telemedicine practices to U.S. patients as of 2014. Available via Statista. https://www.statista.com/statistics/419533/availability-of-selected-telemedicine-practices-to-us-patients/. Accessed 20 Dec 2017
Statista (2017aa) Primary reasons for U.S. Internet users to access mobile health and fitness apps as of March 2014. Available via Statista. https://www.statista.com/statistics/298033/us-health-and-fitness-app-usage-reasons/. Accessed 20 Dec
Statista (2017b) mHealth (mobile health) industry market size projection from 2012 to 2020 (in billion U.S. dollars)*. Available via Statista. https://www.statista.com/statistics/295771/mhealth-global-market-size/. Accessed 20 Dec 2017
Statista (2017c) Projected CAGR for the global digital health market in the period 2015–2020, by major segment. Available via Statista. https://www.statista.com/statistics/387875/forecast-cagr-of-worldwide-digital-health-market-by-segment/. Accessed 20 Dec 2017
Statista (2017d) Cost drivers where mobile health will have the highest positive impact worldwide in the next five years, as of 2016*. Available via Statista. https://www.statista.com/statistics/625219/mobile-health-global-healthcare-cost-reductions/. Accessed 20 Dec 2017
Statista (2017e) Mobile health app categories that will offer the highest global market potential in the next five years, as of 2016*. Available via Statista. https://www.statista.com/statistics/625181/mobile-health-app-category-market-potential-worldwide/. Accessed 20 Dec 2017
Statista (2017f) Potential cost savings in health care by mHealth in 2014. Available via Statista. https://www.statista.com/statistics/449430/potential-mhealth-cost-savings-in-health-care/. Accessed 20 Dec 2017
Statista (2017g) Percentage of U.S. adults who would be willing to use an app for health emergencies as of 2017. Available via Statista. https://www.statista.com/statistics/698331/us-adults-that-would-use-an-app-for-health-emergencies/. Accessed 20 Dec 2017
Statista (2017h) Physicians' usage of smartphones for professional purposes in the U.S. from 2012 to 2015. Available via Statista. https://www.statista.com/statistics/416951/smartphone-use-for-professional-purposes-among-us-physicians/. Accessed 20 Dec 2017
Statista (2017i) Most common types of health-related content used on mobile devices among U.S. physicians in 2015. Available via Statista. https://www.statista.com/statistics/416957/health-content-viewed-on-mobile-devices-among-us-physicians-by-type/. Accessed 20 Dec 2017
Statista (2017j) Percentage of U.S. adults who have ever heard about e-health as of 2017. Available via Statista. https://www.statista.com/statistics/697317/us-adults-who-ever-heard-about-e-health/. Accessed 20 Dec 2017

Statista (2017k) Major purposes of having medical video visits among U.S. consumers as of 2016. Available via Statista. https://www.statista.com/statistics/667623/purpose-of-video-visits-in-us-consumers/. Accessed 20 Dec 2017
Statista (2017l) Willingness to use selected technologies to communicate with healthcare providers among U.S. adults as of 2017. Available via Statista. https://www.statista.com/statistics/297827/healthcare-providers-and-consumers-use-of-communication-technolog/. Accessed 20 Dec 2017
Statista (2017m) E-health services Americans would be willing to pay for as of 2017. Available via Statista. https://www.statista.com/statistics/242961/mhealth-doctors-and-payers-changes-2012/. Accessed 20 Dec 2017
Statista (2017n) Share of individuals seeking health related information online in Great Britain from 2007 to 2013*. Available via Statista. https://www.statista.com/statistics/286275/medical-web site-use-in-great-britain/. Accessed 20 Dec 2017
Statista (2017o) Internet usage for searching for information on health, diseases or nutrition in Finland in Q1 2016, by level of education. Available via Statista. https://www.statista.com/statistics/552022/internet-usage-for-searching-for-information-on-health-by-education-level/. Accessed 20 Dec 2017
Statista (2017p) Percentage of U.S. internet users looking online for health information in 2010, by education. Available via Statista. https://www.statista.com/statistics/194584/us-internet-users-looking-online-for-health-information-by-education/. Accessed 20 Dec 2017
Statista (2017q) Percentage of cell phone owners who looked up health or medical information via mobile phone in 2010 and 2012, by education. Available via Statista. https://www.statista.com/statistics/247511/use-of-cell-phones-to-access-health-info-by-income/. Accessed 20 Dec 2017
Statista (2017r) Proportion of people searching for health information on the Internet in the last 12 months in France in 2007, 2011 and 2015, by sex. Available via Statista. https://www.statista.com/statistics/768794/research-health-sure-internet-la-france-by-sex/. Accessed 20 Dec 2017
Statista (2017s) New technologies that are important to U.S. health care consumers as of 2015. Available via Statista. https://www.statista.com/statistics/504001/important-new-technologies-for-health-care-consumers-in-us/. Accessed 20 Dec 2017
Statista (2017t) Share of individuals who have used the internet to search for health care information in the United Kingdom (UK) in 2015. Available via Statista. https://www.statista.com/statistics/505053/individual-use-internet-for-health-information-search-united-kingdom-uk/. Accessed 20 Dec 2017
Statista (2017u) Percentage of consumers who used select sources for health and wellness in the United States from 2014 to 2017. Available via Statista. https://www.statista.com/statistics/727647/sources-used-for-health-and-wellness-in-us/. Accessed 20 Dec 2017
Statista (2017v) Most popular health and fitness apps in the United States as of July 2017, by monthly active users (in millions). Available via Statista. https://www.statista.com/statistics/650748/health-fitness-app-usage-usa/. Accessed 20 Dec 2017
Statista (2017w) Forecast unit shipments of smart garments worldwide from 2015 to 2017 (in million units). Available via Statista. https://www.statista.com/statistics/385757/smart-gar ments-worldwide-shipments/. Accessed 20 Dec 2017
Statista (2017x) Forecast unit shipments of smart wristbands worldwide from 2016 to 2018 and in 2021 (in million units). Available via Statista. https://www.statista.com/statistics/385749/smart-wristbands-worldwide-shipments/. Accessed 20 Dec 2017
Statista (2017y) Number of connected wearable devices worldwide from 2016 to 2021 (in millions). Available via Statista. https://www.statista.com/statistics/487291/global-connected-wearable-devices/. Accessed 20 Dec 2017
Statista (2017z) Share of respondents monitoring their health or fitness via applications, fitness band, clip or smartwatch in the United Kingdom (UK) in 2016, by age group. Available via Statista. https://www.statista.com/statistics/684876/share-of-health-or-fitness-monitoring-devic e-users-in-the-uk-by-age-group/. Accessed 20 Dec
Sustainable Development (2015) Health and population. Available via UN. https://sustainablede velopment.un.org/topics/healthandpopulation. Accessed 20 Dec 2017

Svalastog AL, Donev D, Jahren Kristoffersen N, Gajović S (2017) Concepts and definitions of health and health-related values in the knowledge landscapes of the digital society. Croat Med J 58(6):431–435

Takian A, Akbari-Sari A (2016) Sustainable health development becoming Agenda for Public Health Academia. Iran J Public Health 45:1502–1506

Tennant B, Stellefson M, Dodd V, Chaney B, Chaney D, Paige S, Alber J (2015) eHealth literacy and Web 2.0 health information seeking behaviors among baby boomers and older adults. J Med Internet Res 17(3):e70. https://doi.org/10.2196/jmir.3992

The Austrian Health Survey (2007) Austrian health survey 2006/07 – results on health in Vienna. Available at Stadt Wien. https://www.wien.gv.at/gesundheit/einrichtungen/planung/pdf/gesundheitsbefragung-2006-2007.pdf. Accessed 21 Dec 2017

The Elders (2016) UHC explained: universal health coverage and the Sustainable Development Goals. Available via The Elders. https://theelders.org/article/faqs-uhc-and-sustainable-development-goals. Accessed 20 Dec 2017

Thompson TL (2000) The nature and language of illness explanations. In: Whaley BB (ed) Explaining illness: research, theory and strategies. Erlbaum, Mahwah, pp 3–40

Thompson AE, Anisimowicz Y, Miedema B, Hogg W, Wodchis WP, Aubrey-Bassler K (2016) The influence of gender and other patient characteristics on health care-seeking behaviour: a QUALICOPC study. BMC Fam Pract 17:38. https://doi.org/10.1186/s12875-016-0440-0

Tones K, Tilford S (2001) Health promotion: effectiveness, efficiency and equity, 3rd edn. Nelson Thornes, Cheltenham

U.S. Department of Health and Human Services (2000) Healthy People 2010: understanding and improving health and objectives for improving health. US Department of Health and Human Services, Washington, DC. Available via CDC. http://www.cdc.gov/nchs/data/hpdata2010/hp2010_final_review.pdf. Accessed 10 Feb 2016

U.S. Department of Health and Human Services (2005) Making health communication programs work. Available via U.S. Department of Health and Human Services. http://www.cancer.gov/publications/health-communication/pink-book.pdf. Accessed 10 Feb 2016

U.S. Department of Health and Human Services (2014) What is health literacy. Available via U.S. Department of Health and Human Services. http://www.health.gov/communication/literacy/. Accessed 10 Feb 2016

UN (2017a) Goal 3: ensure healthy lives and promote well-being for all at all ages. Available via UN. http://www.un.org/sustainabledevelopment/health/. Accessed 20 Dec 2017

UN (2017b) Good health and well-being: why it matters. In: Goal 3: ensure healthy lives and promote well-being for all at all ages. Available via UN. http://www.un.org/sustainabledevelopment/wp-content/uploads/2017/03/ENGLISH_Why_it_Matters_Goal_3_Health.pdf. Accessed 20 Dec 2017

United Nations Foundation (2011) mHealth for development. The opportunity of mobile technology for healthcare in the developing world. Available via United Nations Foundation. https://web.archive.org/web/20121203014521/http://vitalwaveconsulting.com/pdf/2011/mHealth.pdf. Accessed 20 Dec 2017

Ventura F (2016) Person-centred e-support. Foundations for the development of nursing interventions in outpatient cancer care. Available via University of Gothenburg. http://hdl.handle.net/2077/39524. Accessed 20 Dec 2017

Wang Y, Wang L, Li X, Zang X, Zhu M, Wang K, Wu D, Zhu H (2014) Wearable and highly sensitive graphene strain sensors for human motion monitoring. Adv Funct Mater 24(29):4666–4670

Waracle (2016) How wearable tech is transforming digital health. https://waracle.com/wearable-transforming-digital-health/. Accessed 20 Dec 2017

Weder F (2014) Frauen als Zielgruppe von Männerkampagnen: 'Crabwise campaigns' als Antwort auf gender-spezifische Herausforderungen im Bereich der Gesundheitskommunikation. Medienjournal 45:103–159

WHO (1996) Ottawa Charter for Health Promotion. First international conference on health promotion. Available via WHO. http://www.who.int/healthpromotion/conferences/previous/ottawa/en/. Accessed 15 Feb 2016

WHO (2003) Health and sustainable development. Key health trends. Available via WHO. http://www.who.int/mediacentre/events/HSD_Plaq_02.2_Gb_defl.pdf. Accessed 20 Dec 2017

WHO (2006) Constitution of the World Health Organization. Basic documents, 45th edn, Geneva. Available via WHO. http://www.who.int/governance/eb/who_constitution_en.pdf. Accessed 11 June 2016

WHO (2010) Telemedicine. Opportunities and developments in Member States. Report on the second global survey on eHealth. Global Observatory for eHealth series, vol 2. Available via WHO. http://www.who.int/goe/publications/goe_telemedicine_2010.pdf. Accessed 20 Dec 2017

WHO (2011) mHealth. New horizons for health through mobile technologies. Based on the findings of the second global survey on eHealth. Global Observatory for eHealth series, vol 3. Available via WHO. http://www.who.int/goe/publications/goe_mhealth_web.pdf. Accessed 20 Dec 2017

WHO (2013) The right to health. Available via WHO. http://www.who.int/mediacentre/factsheets/fs323/en/. Accessed 1 Mar 2016

WHO (2017) eHealth at WHO. Available via WHO. http://www.who.int/ehealth/about/en/. Accessed 20 Dec 2017

Widmer RJ, Collins NM, Collins CS, West CP, Lerman LO, Lerman A (2015) Digital health interventions for the prevention of cardiovascular disease: a systematic review and meta-analysis. Mayo Clin Proc 90(4):469–480

Millennium Development Goals (MDGs) and Maternal Health in Africa

59

Alfred Okoth Akwala

Contents

Abstract

Millennium Development Goals (MDGs) were adopted at the 2000 Millennium Summit. Improving maternal health was one of the eight goals; Millennium Development Goals four (MDG 4) and five (MDG 5) are related to maternal and child health. The fourth goal (MDG 4) aimed at reducing child mortality by two thirds between 1990 and 2015, while the fifth goal (MDG 5) aimed to improve maternal health by reducing the maternal mortality ratio (MMR) by three quarters between the same years (WHO, Trends in maternal mortality

A. O. Akwala (✉)
Department of Language and Communication Studies, Faculty of Social Sciences and Technology, Technical University of Kenya, Nairobi, Kenya
e-mail: akwala08@yahoo.com

J. Servaes (ed.), *Handbook of Communication for Development and Social Change*,
https://doi.org/10.1007/978-981-15-2014-3_136

(1990–2008). WHO, UNICEF, UNFPA, and World Bank, Geneva, 2010, p. 3). However, they did not meet these promises because the essential role of information and communication was given an afterthought. The United Nations has now embarked on Sustainable Development Goals in which goal 3 aims to ensure healthy lives and promote well-being for all at all ages, while goal 5 aims to achieve gender equality and empower all women and girls. To this end, the objectives of the chapter were investigation of the maternal-child health patients' knowledge, influence of provider-patient communication, and maternal health communication strategies. The study used the setting of Busia County in Kenya and a review of available literature. Data was analyzed thematically. Findings indicate that lack of maternal-child health knowledge leads to non-utilization of skilled facility health. From the study, we conclude that appropriate communication strategies if mainstreamed in maternal health campaigns can enhance utilization of skilled facility health care, thus accelerating the achievement of MDGs 4 and 5. The study recommends that communication strategies in Africa should foster participation and strengthening of community structures to enhance attaining MDGs 4 and 5. The study was grounded in the social cognitive theory postulated by Albert Bandura. This theory posits that although environmental stimuli influence behavior, individual personal factors such as beliefs and expectations also influence behavior change.

Keywords

Millennium Development Goals · Sustainable Development Goals · Communication strategies · Maternal health campaigns

59.1 Introduction

It is estimated that each year, 529,000 maternal deaths occur globally and that the burden of maternal mortality is greatest in the sub-Saharan Africa and South Asia (WHO 2000). WHO et al. (2004) reports that the country with the highest estimated number of maternal deaths is India (136,000), followed by Nigeria (37,000), Pakistan (26,000), Democratic Republic of Congo and Ethiopia (24,000 each), the United Republic of Tanzania (21,000), Afghanistan (20,000), Bangladesh (16,000), Angola, China, Kenya (11,000 each), Indonesia and Uganda (10,000 each). These 13 countries account for 67 per cent of all maternal deaths. Improving maternal health was one of the eight Millennium Development Goals (MDGs) adopted at the 2000 Millennium Summit. Millennium Development Goals four (MDG 4) and five (MDG 5) are related to maternal and child health. The fourth goal (MDG 4) aimed at reducing child mortality by two thirds between 1990 and 2015, while the fifth goal (MDG 5) aimed in improving maternal health by reducing the maternal mortality ratio (MMR) by three quarters between the same years (WHO 2010, p. 3). These goals were not met by most developing countries. The World Health Organization (WHO) has now embarked on achieving the 11 Sustainable

Development Goals (SDGs), in which goal 3 aims at ensuring healthy lives and promoting well-being for all at all ages. A report by the UN et al. (2012) gives estimates of maternal mortality across the world indicating countries that have made progress, those with insufficient progress and those with no progress in meeting MDG 4, to reduce the under-5 mortality by two-thirds between 1990 and 2015.

Maternal-child health is one of the biggest problems as lack of access to health facilities especially in the sub-Saharan Africa contributes to as high as 1000 deaths in 100,000 live births in some countries. Therefore information of different kinds needs to be used to raise health awareness which in turn would lead to skilled health service utilization and thus realization of MDGs 4 and 5. The information should entail health education to convey messages about preserving health care. A majority of developing countries have poor health outcomes as a result of long distances to facilities, the cost, and lack of knowledge. A report by WHO (2010) indicates that more than 800 women globally die each day from preventable causes related to childbirth and pregnancy. Of these 99% live in developing countries.

59.2 Maternal-Child Health Care

According to WHO et al. (2005), maternal death is the death of a woman while pregnant or within 42 days of termination of pregnancy, from any cause related to or aggravated by the pregnancy or its management but not from accidental or incidental causes. A significant number of women face life-threatening complications during pregnancy, childbirth, and in the period after birth, and, therefore, it is highly important that they seek skilled maternal-child health care during these periods. This is only possible if essential elements of communication for social change are identified. These may range from the use of interpersonal communication to the use of community forums such as churches. Pandey et al. (2011), quoting the United Nations (2000, 2010), report that in 1987, international organizations sponsored a global conference and adopted the Safe Motherhood Initiative to reduce the high rate of women dying during pregnancy and childbirth. The initiative recommended that all countries provide three types of maternity care services for all pregnant women: prenatal care, delivery care, and postnatal care (p. 556). Effective participatory and advocacy communication if used can enhance the utilization of skilled maternal-child health services and thus reduce the maternal and infant deaths.

Complications at pregnancy and childbirth are among the leading causes of morbidity and mortality. Medical experts point out that the direct causes of maternal death include hemorrhage, infection, and obstructed labor and unsafe abortion, while child mortality is caused by infections, baby asphyxia, birth injuries, complications of prematurity, low birth weight, and birth defects. These could be controlled if detected early enough as some of the infections are immunizable. Figure 1 shows the distribution of the causes of child mortality across the sub-Saharan Africa (see file on Figures and Tables).

The Democratic Republic of the Congo (DRC), Ethiopia, and Nigeria account for more than 43% of the total under-five deaths in all of Africa, and of the 46 countries

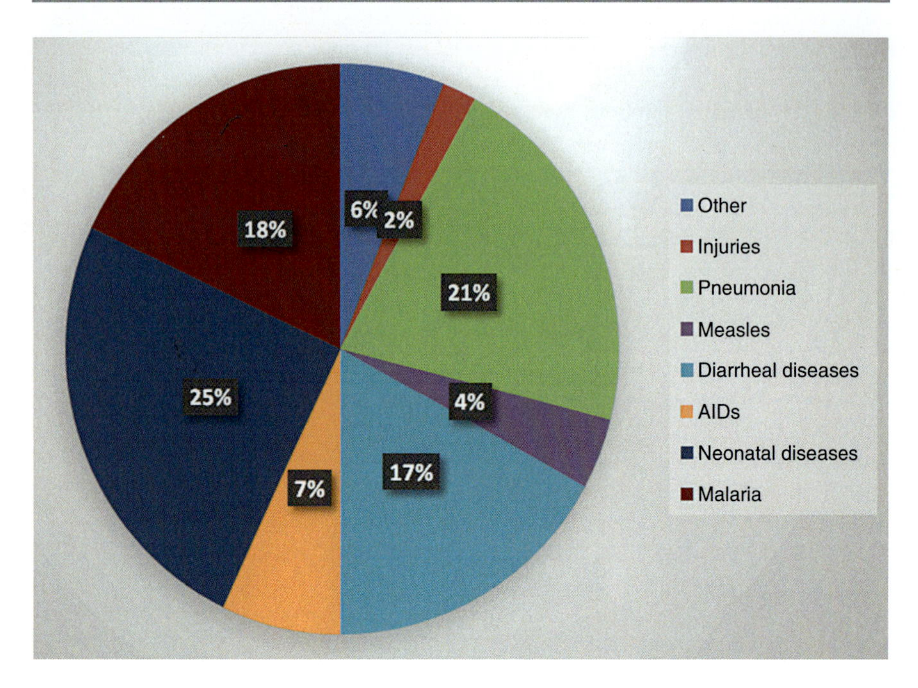

Fig. 1 Major causes of deaths among children under 5 years in the sub-Saharan Africa as rated by WHO. (Source: UNICEF 2008)

in the sub-Saharan Africa, only Cape Verde, Eritrea, Mauritius, and the Seychelles are on track to attain MDG 4 (UNICEF 2008, p. 7). In the ratings of countries according to those which were making progress toward improving maternal health between the years 1990 and 2008, Kenya was among the seven countries marked as having made no progress in improving maternal health, the other countries being Zimbabwe, Zambia, Swaziland, Somalia, South Africa, and Congo. According to UNICEF (ibid.), nearly 100 children die every day from preventable diseases, and at least 8 women die every day due to pregnancy-related complications in Zimbabwe; a majority of these women are residing in rural areas where health facilities are not easily accessible. The report further indicates that the issue of user fees charged by facilities giving maternal health services is one of the biggest barriers to poor women and children's accessing life-saving and critical health care in Zimbabwe.

59.2.1 Maternal Health in Africa

Africa has been among the highest worldwide, representing about 50% of all maternal deaths. For African countries to achieve the target of Millennium Development Goal (MDG) 5, concerted effort is needed in maternal and newborn health (MNH) interventions. Most African countries developed their national maternal and newborn health (MNH) road map. This included developing strategies to tackle maternal, newborn, and child health issues. However, the challenges of inadequate access to quality MNH health care; inequitable distribution of quality MNH services; inadequate financial and human resources; huge burden of HIV/AIDS, malaria, and other infections; and weak community involvement and participation are stumbling blocks to these efforts. Oronje (2009) says that Kenya needs to prioritize in national planning because failing to reach this rapidly growing sub-population with health services may result in lack of progress towards the health-related Millennium Development Goals. Africa has other determinants such as poverty, gender inequity, and poor communication systems that contribute to delays in timely accessing MNH services.

59.2.2 Maternal Health in Kenya

In Kenya, maternal health has been a critical issue with due reference to the vision 2030 goal of reducing infant mortality rate and accelerating skilled birth attendance (KPSA 2011, p. 111). Of concern is the fact that a number of expectant mothers still lose their lives due to maternity and childbirth complications. For instance, it is estimated that Kenya loses 488 mothers out of 100,000 births per year mainly because women do not give birth under the care of skilled health providers. A report by NCAPD (2010) indicates that infant mortality rate (IMR) in Kenya is high with Western Province recording comparatively higher percentages. These high rates are attributed to well-known and preventable (both direct and indirect) causes. These causes would easily be managed when a woman delivers in a health facility under the care of a trained facility health staff who ensure that mothers have access to a facility that is equipped and staffed with personnel who could assist in case of an emergency such as cesarean. A report by (UNESCO 1990) indicates that education is linked to child health through its influence on the ability of women being autonomous in decision making within their families; it determines how much power a woman wields in the marital relationship. Usually, women are the primary caregivers in households, and in most cases, the first person in the home to recognize that a child is sick is usually the mother.

Dissemination of information and utilization of health-care services in rural areas is thus very important for the preservation of good health among the rural population. Magadi (2000) points out that apart from mere attendance of antenatal care, the quality of care received (in terms of the timing and frequency of visits as well as the content of the care) plays a key role for outcomes (p. 96). Lack of maternal-child knowledge leading to failure in seeking and utilizing skilled maternal-child health services has been the major cause of infant and child mortality, especially in rural areas. There is a need to empower maternal health patients in Kenya and specifically in rural areas to gain information on how to improve their health status and meet this Sustainable Development Goal. This is only possible if appropriate communication systems are developed to transmit information among the patients. Berry (2007) confirms that good communication contributes virtually to all aspects of healthcare. Berry further argues that a substantial body of evidence shows that patients who are attended to by health providers with good communication skills have better health outcomes (p. x).

Mefalopulos (2005, p. 248) says that achieving sustainability in rural development depends largely on the way stakeholders perceive the proposed change and the way they are involved in assessing and deciding about how that change should be achieved. Implementing strategic communication approaches to the rural communities could be one of the essential ways to change the status of a rural community. There are different ways in which maternal-child health patients receive appropriate health knowledge, thereby influencing their perception and reaction to maternal-child health utilization. Acquisition of maternal-child health knowledge influences a great deal of the decision a patient makes and hence pregnancy outcome and the survival of the newborn. Bandura (1986) states that human learning can occur through observation. He says that much of the human learning takes place through other people modeling various behavior that they are able to encounter in their day to day life.

59.3 Dissemination of Maternal Health Knowledge

Information about maternal health is crucial to a patient as it can either hamper or influence the usage of skilled health services depending on how the patients interpret, discuss, and make sense of the maternal health messages received. Communication and education is very important, especially to the rural women seeking maternal health care, given that it is a population with the greatest disparities in pregnancy outcomes. Campaigns in Kenya about maternal-child health have had little effect in reducing maternal and child mortality. This may be because most of these campaigns have been through the mass media which may not be accessible to the rural population. This corroborates Kamali's documentation that for communication to be improved successfully in rural areas, participatory communication must be embraced (Kamali 2007); this involves the consumers of communication also being the designers of that communication. Williams (2006) also acknowledges that in poor communities, informal communication strategies such as street theater can serve to inform the marginalized about community health issues.

59.3.1 Maternal-Child Health and Maternal-Child Health Knowledge

Information about maternal-child health is crucial to a patient as it can either hamper or influence the usage of skilled health services depending on how they interpret, discuss, and make sense of the maternal-child health messages received. In a report by the Kenya Demographic Health Survey (KDHS) (2009), rural women are less likely than their urban counterparts to get antenatal care from a doctor, and they are more likely to get no care at all. The report further points out that the proportion of babies delivered in health facilities reduce the health risks to both the mother and the baby. This is because proper medical attention and hygienic conditions during delivery reduce the risk of complications and infection that cause morbidity and mortality either to the mother or child.

59.4 Maternal Health Communication

Maternal health communication is aimed at changing knowledge, attitudes, and behavior of maternal-child health patients, and this communication's success depends largely on the partnership between the caregiver and the client. The strategy should be able to meet the needs of the patients so that it is economically, socially, and culturally sensitive to the needs of the patients. The caregivers should involve the patients and the community in developing behavior and social change communication strategies. Rosato et al. (2009) document that Malawi has a neonatal mortality rate of 27 per 1,000 live births, an infant mortality rate of 76 per 1,000 live births, and an under-five mortality rate of 111 per 1,000 live births. They further point out that to scale up the progress toward achieving the millennium goal 4, behavior change communication should play a central position in efforts to improve health in rural areas by disseminating health education messages to the local communities. Kenya Demographic Health Survey (2009), maintains that rural women in Kenya are less likely than their urban counterparts to get antenatal care from a doctor, and they are more likely to get no care at all.

There has been a growing focus on communication as a vehicle that can ensure good health among populations as opposed to the treatment of illnesses. Thomas (2006) confirms that the process of communication is critical in the implementation of health communication since it is often through the process that the right influence can be attained to improve health behavior.

Heavy investment in health may not be a prerequisite for success in maternal health; however, appropriate communication between the service provider and the client plays an important role. Communication as a tool should be developed to bring about meaningful behavior changes through interpersonal communications between the maternal-child health patients and health providers.

The health communication strategy used should attempt to increase women's power to choose and access quality maternal-child health services. It should ensure that critical understandable health information is disseminated from the relevant sources to the audiences. The communication process involves dialogue and exchange of information so that the participants build a platform of understanding, negotiation, and decision-making. Dagron (2001) acknowledges that establishing dialogue with beneficiaries of maternal health care during a health campaign builds up a sense of "ownership" by the community, hence contributing to behavior change among the users.

An appropriate communication must therefore make sure that the focus shifts from mere information-giving approach to an information-sharing approach so that there is a continuous exchange of ideas, feelings, or information between health providers and clients.

59.4.1 Maternal Health Campaigns

Effective health promotion and communication initiatives should often adopt an audience-centered approach so that the patients' needs are taken care of. This further ensures that patients are not just passive recipients of information but that they are also active contributors to health information through dialogue. Dialogue creates an opportunity for the patients to share their ideas, perceptions, and attitudes during meetings. The traditional forms of communication have not served well in disseminating maternal-child health, and there is a need to adopt new approaches in maternal health campaigns. According to Snyder (2007), campaigns may use a variety of communication strategies to try to change the behavior of the target populations, including strategies that attempt to change the political and economic context in which people are making decisions, those aimed directly at the populations and those aimed at people who may have influence with the target population. Behavior change communication (BCC) strategies used in health campaigns endeavor to break the societal barriers to behavior change; they use communication activities to help change the community's perceptions and beliefs about childbirth.

59.5 Results

The section presents results of an investigation that we did in one of the counties in Kenya to assess level of maternal-child health knowledge among the rural population, establish the extent to which health provider-patient communication influences the outcome, and find out the different strategies that can be used to promote maternal-child health-care utilization.

59.5.1 Levels of Maternal Health Knowledge Among the Patients

Empowering women by removing barriers to accessing skilled facility services would call for improvement of maternal-child health knowledge among the patients. This knowledge can be disseminated through appropriate communication which would function as a catalyst for behavior change. Health education campaigns should be aimed at creating public awareness. The communication strategies need to stress and consider social, economic, and cultural factors that inhibit utilization of skilled facility care.

The respondents were asked to state why they felt it was important to access and utilize facility maternal-child health services. Various reasons were given in regard to this. Table 1 shows some of the reasons given by the respondents.

It is evident that community health workers (CHWs) play a vital role in terms of maternal-child health service utilization. It was established that 23.3% of the respondents visited the clinic simply because of the advice from a CHW. The findings revealed that while mass media has been one of the ways of disseminating maternal information through mobilizing pregnant women to access and utilize maternal-child health services, it is not the best method as it was more likely to be utilized by the urban population than the rural population afflicted by issues of poverty, ignorance, and lack of infrastructure.

Table 1 Reasons for attending antenatal clinic, respective frequencies, and percentages

Reason for attending clinic	Frequency	Percent
Advice from CHW	7	23.3
Advice from FHS	3	10.0
Know HIV status	2	6.7
Know status of child	9	30.0
Mandatory	1	3.3
Seeing other women going	1	3.3
Sickness feeling	7	23.3
Total	**30**	**100.0**

Source: Research data. This is the data that was generated by the researcher during the field study. The researcher is the author of the article.

59.5.2 Influence of Provider-Patient Communication on Maternal-Child Health Outcomes

Communication between the health provider and the patient enhances dissemination of important maternal-child health knowledge. This facilitates appropriate decision-making so that the health provider and the patient can share and exchange views. The health provider needs to be knowledgeable about the patient's community and understand their cultural context in order to make communication successful. Table 2 shows a summary of the methods respondents used to receive maternal-child health information.

According to the survey, information at the facility through the facility health staff and posters which are placed at the waiting bays of the various health facilities scored highly with 50% of the respondents approving them as appropriate channels to receive information.

59.5.3 Communication Strategies in Disseminating Maternal-Child Health Knowledge

According to the findings, improving maternal health in rural areas may not be achievable unless appropriate communication strategies are mainstreamed in maternal health campaigns. This could be through community-based health education, local media, chiefs'*mabaraza*, or door-to-door campaigns. The dissemination of information on maternal-child health is a key factor to the utilization of maternal-child health services. A total of 30 antenatal patients filled the questionnaires and 8 of them were interviewed. A majority of the respondents acknowledged that chiefs'*mabaraza* (50%) could be utilized as key avenues for dissemination of information, especially in rural areas. This was followed by CHWs constituting 23.3%. These two strategies were more accessible to the rural women. One of the respondents said that:

Table 2 Channels used to disseminate maternal-child health information

Method	Frequency	Percent
Relative/friend	3	10.0
Chief barazas	1	3.3
CHWs	7	23.3
Information at facility/interactive posters	15	50.0
Mass media	4	13.3
Total	**30**	**100.0**

Source: Research data. This is the data that was generated by the researcher during the field study. The researcher is the author of the article.

Table 3 Channels used to disseminate maternal-child health information

Method	Frequency	Percent
Relative/friend	3	10.0
Chief barazas	15	50.0
CHWs	7	23.3
Information at facility/interactive posters	1	3.3
Mass media	4	13.3
Total	**30**	**100.0**

Source: Research data. This is the data that was generated by the researcher during the field study. The researcher is the author of the article.

> Messages from the chief baraza or a "community worker" can be clarified and substantiated as opposed to those we "hear" from radios.

According to her, community outreach programs offer an opportunity for patients to ask for clarification and questions on issues that are not clear. She further said that chiefs'*mabaraza* are communal media since they act as a forum where one meets masses of people and disseminates information to them.

The church was also singled out as one avenue to disseminate maternal-child health messages. In churches, clinics, schools and at funerals or traditional ceremonies, people often get information on the current affairs in the community, providing them with a platform to communicate and discuss their views, hence shaping the sociocultural and economic status of the community.

Table 3 shows a summary of the methods respondents preferred for disseminating maternal-child health information.

59.6 Conclusion

The quantity and quality of information that a maternal-child patient can access greatly influence the decision taken in terms of accessing and utilizing maternal-child health services. It is evident from the data analysis in this chapter that various communication strategies influence the utilization of skilled facility health care. It is only upon embracing appropriate communication strategies that both maternal and child mortality in Africa can be reduced. These strategies have to be interactive and participatory. Communication between the health provider and the patient enhances dissemination of important maternal-child health knowledge. Patients, if given appropriate maternal-child health service information, tend to access and utilize the health facilities. Education on maternal-child health issues should be geared toward behavior change among the maternal-child health patients. The social cognitive theory is instrumental here as it dictates the behavior change strategy. Communication messages should therefore be generated from the maternal-child health patients themselves or those they can identify with. Messages from Community Health Volunteers previously referred to as community health workers are likely to be believed than the ones from the media. Mainstreaming strategic communication could ensure dissemination of significant maternal-child health knowledge which will in turn influence behavior change. This hence means that communication interventions are responsible for individual and communal behavior change as they help people acquire the knowledge and skills they need to improve their condition and that of society and to improve the effectiveness of institutions.

Lack of information, ignorance, poor attitudes toward utilization of skilled maternal-child health services, and poor infrastructure contribute to non-utilization of skilled maternal-child health services including defaulting on immunization. These impact on preservation of health and thus failure to achieve MDG 4.

References

Bandura A (1986) Social foundations of thought and action: a social cognitive theory. Prentice-Hall, Englewood Cliffs

Berry D (2007) Health communication: theory and practice. McGraw Hill Education, New York

Dagron GA (2001) Making waves; stories of participatory communication for social change. Rockefeller Foundation, New York

Kamali B (2007) Critical reflections on participatory action research for rural development in Iran. Action Res J 5:103. Sage

Kenya Demographic and Health Survey (KDHS) (2009) Kenya demographic and health survey 2008–2009. ICF Macro, Calverton

Kenya Service Provision Assessment Survey 2004–2005 (2011) Maternal and child health, family planning and STIs. NCAPD, MOH, CBS and ORC Macro, Nairobi

Magadi MA (2000) Maternal and child health among the urban poor in Nairobi, Kenya. Centre for Research in Social Policy (CRSP), Loughborough University, Loughborough

Mefalopulos P (2005) Communication for sustainable development: applications and challenges. In: Oscar H, Tufte T (eds) Media and glocal [sic] rethinking communication for development. Nordicom, Gothenburg, pp 247–259

National Coordinating Agency for Population and Development (NCAPD) (2010) Facts and figures on population and development 2010. National Coordinating Agency for Population and Development (NCAPD), Nairobi

Oronje NR (2009) The maternal health challenge in poor urban communities in Kenya: a policy briefs no 12, 2009. African Population and Health Research Centre, Nairobi

Pandey S, Lama G, Lee H (2011) Effects of women's empowerment on their utilization of health services: a case of Nepal. J Int Soc Work. Accessed on 12 Apr 2013 from http://isw.sagepub.com/. Sage
Rosato M, Lewycka S, Mwansambo C, Kazembe P, Costello A (2009) Women's groups' perceptions of neonatal and infanthealth problems in rural Malawi. Malawi Med J 21(4):168–173
Snyder LB (2007) Health communication campaigns and their impact on behaviour. J Nutr Educ Behav 39(2S):S32
Thomas KR (2006) Health communication. Springer, New York
UN, UNICEF, UNFPA, WB (2012) Trends in maternal mortality: 1990 to 2010. WHO, Geneva
UNESCO (1990) Literacy and civic education among rural women. UNESCO, Bangkok
UNICEF (2008) The state of Africa's children 2008: child survival. UNICEF, Nairobi
United Nations (2000) World's women: trends and statistics. United Nations, New York
United Nations (2010) The Millennium Development Goals report. United Nations, New York
WHO (2000) The world health report 2000-health systems: improving performance. WHO, Geneva
WHO (2010) Trends in maternal mortality (1990–2008). WHO, UNICEF, UNFPA and World Bank, Geneva
WHO, UNICEF, UNFPA (2004). Maternal Mortality in 2000: Estimates. WHO, Geneva.
WHO, UNICEF, UNFPA, WB (2005) Maternal mortality in 2005. WHO, Geneva
Williams JJ (2006) Community participation: lessons from post-apartheid South Africa. J Policy Stud 27:197. Routledge

Impact of the Dominant Discourses in Global Policymaking on Commercial Sex Work on HIV/STI Intervention Projects Among Commercial Sex Workers

60

Satarupa Dasgupta

Contents

Abstract

This chapter examines how dominant discourses in policymaking in the realm of commercial sex work and international aid affect health communication practices in HIV/STI intervention among commercial sex workers. The policy documents of global HIV/STI research and aid organizations often actively conflate commercial sex work and trafficking. The choice of the profession of sex work is depicted as an outcome of coercion facilitated by trafficking. Sex workers are portrayed as victims of abuse as well as sexual servitude. Volition on part of sex workers in executing their profession is not acknowledged by many US and global donor organizations. Delegitimization and eradication of sex work, and rescue and rehabilitation of the sex workers, are propositions supported by some

S. Dasgupta (✉)
Communication Arts, School of Contemporary Arts, Ramapo College of New Jersey, Mahwah, NJ, USA
e-mail: satarupadasgupta@gmail.com

J. Servaes (ed.), *Handbook of Communication for Development and Social Change*,
https://doi.org/10.1007/978-981-15-2014-3_107

of the international donor organizations. The conflation of trafficking and sex work can problematize health promotion among sex workers and jeopardize HIV/STI intervention projects. The chapter explores how the delegitimization of sex work, rescue, and rehabilitation propositions offered to sex workers and conflating trafficking with sex work affect HIV/AIDS intervention programs among commercial sex workers. The chapter also looks at a case study conducted among commercial female sex workers in a red light district in India and takes into account the voices of commercial female sex workers on the proposed equation of trafficking and sex work and how it affects HIV/STI intervention project in that population.

Keywords

Commercial sex work · Trafficking · Global policymaking · HIV/AIDS intervention programs · Coercion · Sexual servitude · Rescue and rehabilitation · Delegitimization · Criminalization

60.1 Introduction

This chapter examines how dominant discourses in policymaking in the realm of commercial sex work and international aid affect health communication practices in HIV/STI intervention among commercial sex workers. The policy documents of global HIV/STI research and aid organizations often actively conflate commercial sex work and trafficking. The choice of the profession of sex work is depicted as an outcome of coercion facilitated by trafficking. Sex workers are portrayed as victims of abuse as well as sexual servitude. Volition on part of sex workers in executing their profession is not acknowledged by many US and global donor organizations. Delegitimization and eradication of sex work, and rescue and rehabilitation of the sex workers, are propositions supported by some of the international donor organizations. The conflation of trafficking and sex work can problematize health promotion among sex workers and jeopardize HIV/STI intervention initiatives. The chapter explores how the delegitimization of sex work, rescue, and rehabilitation propositions offered to sex workers and conflating trafficking with sex work affect HIV/AIDS intervention programs among commercial sex workers. The chapter also looks at a case study conducted among commercial female sex workers in a red light district in India and takes into account the voices of commercial female sex workers on the proposed equation of trafficking and sex work and how it affects HIV/STI intervention in that population.

The conflation of trafficking and sex work has been a recurrent phenomenon observable in national and global HIV/AIDS intervention and policy formulation circuits. Whether it is the Trafficking in Persons (TIP) report (2009) brought out by the United States government or the HIV/AIDS intervention documentation published by United Nations development Fund for Women (UNIFEM), trafficking is equated with sex work (termed prostitution in the aforesaid documents) and recognized as one of the socio-behavioral factors that exacerbate health conditions by facilitating transmission of sexual diseases and HIV/AIDS. The sex workers are portrayed in the policy documents of international donor agencies as victims who are coerced into their profession by trafficking and have little agency on their own fates. The result is the proposition of eradication of sex work and rehabilitation of sex workers by most international donor agencies. The Global AIDS Program's "prostitution pledge" which specifically aims to purge trafficking and sex work imposes a restriction on organizations that strive for legalization, unionization, or organization of sex workers. Consequently, health intervention projects among commercial sex workers that also strive for the legalization of sex work are deprived from major global funding.

60.2 The Global Discourse on Sex Work

60.2.1 Impact on HIV/AIDS Policy Formulation and Legislation: The Global Conflation of Trafficking and Sex Work

The distinction between sex work and trafficking is not a well-defined one in the arena of global HIV/AIDS healthcare. "By some definitions, trafficking can be understood to involve coercion and forced labor, while prostitution infers the voluntary sale of sex. However, there is still not an agreed taxonomy of terms which renders discussion amongst different stakeholders concerned with HIV/AIDS, and the health and rights of people engaged in sex work, difficult" (UNAIDS issue paper, 2003, p. 2). Although the term trafficking in persons can refer to both genders and children, the emphasis put by policymakers on HIV/AIDS communication intervention lies on the trafficking of women and children for sexual purposes.

The United States Victims of Trafficking and Violence Protection Act. Section 103 (8), defines several forms of trafficking in persons such as "sex trafficking in which a commercial sex act is induced by force, fraud and coercion" and "the recruitment harboring, transportation, provision, or obtaining of a person for labor or services, through the use of force, fraud or coercion for the purpose or coercion for the purpose of subjection to involuntary servitude, peonage, debt bondage or slavery" (TIP report 2009, p. 1). The TIP report (2009, pp. 21–22) further notes that:

> Sex trafficking comprises a significant portion of overall human trafficking. When a person is coerced, forced, or deceived into prostitution, or maintained in prostitution through coercion, that person is a victim of trafficking. All of those involved in recruiting, transporting, harboring, receiving, or obtaining the person for that purpose have committed a trafficking crime.

The TIP report does not make a strong distinction between sex work and trafficking. The articulation of sex work as intrinsically related to be an offshoot of trafficking is a process undertaken by several global HIV/AIDS donor agencies. Sex workers are portrayed as victims of abuse who are duped or coerced into their profession by the menace of trafficking. The conflation of trafficking and sex work can problematize development and healthcare activism among sex workers and jeopardize HIV/AIDS intervention initiatives (Ditmore 2003). Such an equation of trafficking with sex work is promoted vigorously by anti-trafficking organizations active in the United States and globally.

A UNAIDS issue paper (2003) on HIV/AIDS and human rights notes that there has been an "alarming shift" in policy formulation and programmatic support for HIV/AIDS intervention geared toward sex workers (p. 1). The issue paper says that the distinction between trafficking and sex work has become increasingly blurred as a consequence of which protection of the rights and health of the sex workers are being jeopardized. Restrictive policies by governments, such as stringent immigration laws, manage to keep the sex workers underground, especially those who have been trafficked or else living as illegal migrants:

> The vulnerability of women in prostitution and sex work is heightened because they are often subjected to sexual abuse at the hands of authorities, petty political leaders, immigration and police officials, as well as local criminal gangs. Forcible detention, lack of access to redress, police corruption, and the invisibility of women engaged in sex work only compound vulnerability to HIV infection, and once infected, hinder the ability to access needed care and support. (UNAIDS issue paper 2003, p. 1)

The UNAIDS issue paper (2003) emphasizes how the repressive nature of most strategies intended to combat trafficking neglects to address the issue of vulnerability of the trafficked people, some of whom may engage in prostitution. Policy formulation and programmatic responses to HIV/AIDS among sex workers are often motivated by a moralistic standing on sex work itself and a lack of consideration of the socioeconomic compulsions of the trafficking victims (Butcher 2003; Wolffers and Beelan 2003). Sex workers are routinely pathologized as conduits of virus transmission rather than as being vulnerable to the virus.

An example is the moralistic stance adopted by the United States on HIV/AIDS policymaking. In January 2003, the then United States President George W. Bush announced the President's Emergency Plan for AIDS Relief (PEPFAR). It comprised $15 billion dollars for programs to combat the global HIV/AIDS epidemic. Within the detailed plan, Congress noted its concern about the sociocultural, economic, and behavioral causes of HIV/AIDS. Prostitution and sex trafficking were specifically named as being the behavioral forces behind the spread of the virus (Masenior and Beyrer 2007). "This legislation advanced a new policy goal for the US: the global eradication of prostitution" (Masenior and Beyrer 2007, p. 1158).

The conditions for receiving grants for countering HIV/AIDS were based on an explicit relationship between HIV/AIDS prevention and the abolition of prostitution. The requirement for receiving AIDS funds from the United States needed the intended recipient to have "a policy explicitly opposing prostitution and sex trafficking and certification of compliance with the 'Prohibition on the Promotion and Advocacy of the Legalization or Practice of Prostitution or Sex Trafficking,' which applies to all organization activities, including those with funding from private grants" (Masenior and Beyrer 2007, p. 1158). Hence one of the requirements of the Global AIDS Program – "the prostitution pledge" – rendered organizations that strive for legalization, unionization, and organization of sex workers ineligible for obtaining much needed funds from USAID.

60.2.2 A Pejorative Framing of Sex Work

The Global AIDS Program's usage of the term prostitution, as Masenior and Beyrer (2007) observe, is in itself controversial as people associated with the profession generally tend to refer to themselves as sex workers rather than prostitutes. The latter term is widely considered as stigma-inducing and derogatory. "The core debate is that for many stakeholders, the category of sex workers includes consenting adults who sell sex of their own volition, who are not trafficking victims, and who have called for recognition of their rights as worker" (Masenior and Beyrer 2007, p. 1159).

However such a volition on part of sex workers in executing their profession is not acknowledged by many US and global donor organizations. The articulation of sex workers as passive victims of trafficking, abuse, and slavery-like practices has been a mainstay of global health policymakers especially those pertaining to HIV/AIDS prevention and intervention. The TIP report (2009) which equates sex work with trafficking refers to both activities as exploitation and servitude. The TIP report (2009) notes that "there can be no exceptions and no cultural or socioeconomic rationalizations that prevent the rescue ..from sexual servitude" (p. 22). UNIFEM which explicitly portrays sex workers as victims of trafficking and coercion vocalizes the need for "control and suppression of prostitution through the legal system," "rescue and rehabilitation for women and girl victims of trafficking," and "supply reduction through the provision of alternative employment and income-earning opportunities for women and girls" (UNIFEM Factsheet 2009, p. 5).

Sex work can be exploitative and remains illegal in many countries. The population of sex workers remains a heavily deprived sector in dire need of services and support from health sectors to reduce their risk of venereal diseases and HIV/AIDS infection (Masenior and Beyrer 2007, p. 1159). But the propositions of global aid organizations like the Global AIDS Program has laid a funding freeze on those initiatives that strive for decriminalization or legalization of sex work. Thus, successful HIV/AIDS intervention projects that strive to vocalize the rights of the much maligned sex workers can lose out on funds from USAID for advocating legalization of prostitution.

60.2.3 The Discourse of "Rescue" and "Rehabilitation"

Combating sex work through "rescue" and "rehabilitation" of the women involved is a complex proposition. As Cohen (2005, p. 12) notes in a report published by Guttmacher report:

> The moral imperative to rescue women from brothels is compelling when young girls are involved or there is clear evidence of duress, but 'rescuing' adult women from brothels against their will can mean an end to their health care and economic survival. In countries and situations in which basic survival is a daily struggle, the distinction between free agency and oppression may be more a gray area than a bright line.

Cohen (2005) notes that sex workers may resist rehabilitation not because there may not be viable economic alternatives to sustain themselves and their families. Proposed rehabilitation of sex workers or coercive measures such as mandatory examination for HIV/AIDS and venereal diseases are often not feasible as shown by previous research. "Mandatory HIV testing of people who are or are assumed to be engaged in sex work, detention and specialized health and 'rehabilitation' services all may be understood to push the people engaged in this work further underground" (UNAIDS issue paper 2003, p. 2).

60.3 A Study of Sex Workers' Voices

The researcher sought the opinions of commercial sex workers themselves concerning the relationship of trafficking and sex work and volition in the trade. Interviews were conducted with 37 commercial female sex workers in a red light district called *Sonagachi* in Calcutta, India. With a population of more than 50,000 commercial sex workers, *Sonagachi* is one of the largest red light districts in South and Southeast Asia. The sex workers of *Sonagachi* are unionized and conducting a health intervention project by carrying out peer outreach-based campaigns to increase condom usage compliance and reduce rates of HIV and sexually transmitted infections (STI) among their colleagues.

The HIV/STI intervention project was originally started by the sex workers themselves to disseminate awareness information about STIs including HIV and arrest the infection incidence. A HIV infection incidence of 10% has been achieved which is significantly lower than 50% to 90% among similar red light areas in India. The usage of condoms among sex workers also improved from 3% to 90% during implementation of the project. Few of the additional outcomes of the project include the attainment of healthcare facilities for the sex workers and their children, creation of literacy programs and vocational training centers for the latter, and unionizing of the sex workers. The sex workers' union is called the *Durbar Mahila Samanwaya Committee* (DMSC), and it has offices in the heart of *Sonagachi* itself.

Structured and semi-structured interviews were conducted with the subjects at the DMSC offices and at their places of work. In *Sonagachi*, there are concentrated pockets of sex work zones amid regular neighborhoods. At the entrance of each neighborhood, the sex workers, who wear colorful garb and makeup, solicit for clients. This practice is colloquially referred to as "standing at the gates" and signifies the presence of sex work sites within the neighborhood. It needs to be noted that all of the interviewees were active members of the DMSC and were also peer outreach workers in the HIV/STI intervention project conducted by the union.

Most of the interviewees were heavily involved in the daily operations of the DMSC and various other projects. The subjects interviewed cannot be said to be representative of all the sex workers plying their trade within the borders of the red light district of Calcutta. However, their voices offer glimpses of the opinions of sex workers which traditionally go unheard and unrecognized in the public sphere.

Prior permission was obtained from the DMSC central governing committee before the commencement of any research and interviews. This permission was obtained by sending an application letter drafted in Bengali and addressed to the Central Management Committee of the DMSC headed by its president. Institutional Review Board (IRB) research approval was also received (Protocol # 13456). The narratives of the sex workers as gathered from the interviews will be discussed in the following sections.

60.3.1 Sonagachi Women: Disarticulating Trafficking and Sex Work

The interviews with the women at *Sonagachi* raise a question on the conflation of trafficking and sex work in global policy documents and the underlying assumption of coercion and sexual servitude in sex work. The USAID policy of refusing funding to organizations which try to legalize sex work is also rendered questionable by the interviewees' statements.

All of the 37 interviewees at the *Sonagachi* emphasized the distinction between trafficking and sex work. Eight of the interviewees noted the conflation between sex work and trafficking in general discourse about sex work. Three of the interviewees also pointed out how sex work is perceived as trafficking and hence considered a punishable criminal offence by the media and the intelligentsia in India. Sapna excitedly noted, "We are considered to be sex slaves and sold or trafficked into this profession by the media, by all the important people and by the entire society. Why doesn't anybody ask us?"

When questioned whether women are coerced or trafficked into this profession, 36 of the 37 interviewees noted that they had entered the profession of their own accord and have not been trafficked or sold or coerced into sex work. One of the interviewees however stated that she was introduced to sex work against her will by a close family member. She noted that after initial abuse, she managed to break ties with an exploitative brothel-owner and conducted the profession independently by renting rooms within the red light area. She continued in the profession since, according to her, it provided her a steady source of income. She added that her forced introduction into the profession occurred almost 40 years ago, and the current mechanism of self-regulatory boards, which DMSC had installed, prevented cases like hers today.

The reason for the *Sonagachi* women engaging in sex work appeared to be primarily economic. 19 of the interviewees, which is 54.2% of the sample size, mentioned financial exigencies as a reason for entering the profession. Twelve of the interviewees noted that they could afford a comfortable standard of living by sex work. Nine of the interviewees observed being subjected to violence by spouse or a family member prior to entering sex work. Sapna said, "My partner beat me up to the point of breaking my limbs. I had nobody in this world. The red light area was my refuge for it gave me the income to survive and escape my partner."

Thirteen of the interviewees reported sexual assault in prior professions which included domestic sector work, construction work, brick kiln labor, and secretarial jobs. Krishna who had previously worked a secretarial job recalled how she had been sexually harassed by her employer during her tenure. "I was compelled to offer sexual favors to my boss. Yet my salary was a pittance. I was finding it really hard to make ends meet. Well, then I decided I might as well get paid for sex."

Interviewees also noted the mechanism of sexual harassment in place in construction and menial labor industries in India. "To get a job and maintain it in bricklaying, masonry and construction, you have to sleep with the *rajmistri* (*Rajmistri* refers to head mason in Bengali), his assistant and respective subordinates. You need to keep them happy and also do back-breaking physical work to earn your meager salary," noted Purnima. Her views echoed that of Krishna: "You see after being exploited by a string of supervisors I decided if I give my sex I might as well get paid for it." Sapna recounts being raped by her employer while working as a domestic helper. "No more free and forced sex," she quipped.

All of the 37 interviewees emphasized the harmful effects of trafficking and the emergent need to curb trafficking. Seventeen of the interviewees acquiesced that women are often trafficked into sex work. The porous border between Bangladesh and India was noted to be a reason for easy trafficking of women for sex trade in the Indian side. The interviewees also noted that women from Nepal are trafficked into India to work in red light areas in and around Eastern India. Sadhana notes:

> I will not say that sex workers are not trafficked, yes sometimes they are. But many times they are not trafficked against their will. It often happens that they will pay an agent for trafficking them across the border for work. Many girls come in from across the (Indian) border to work in *Sonagachi*. Why do you think they come? It is because they have heard of this place. They know they can earn some money from sex work and feed their family. What is a girl who is extremely poor and has no education and has five or ten mouths to feed work in? She knows sex work can provide her a steady source of income. She needs sex work to survive, to provide for herself and for her family. When it comes to hunger there is no good profession or bad profession.

According to the interviewees, a mechanism for paying traffickers and sneaking across the Indian border appear to be in place in many places of Bangladesh and Nepal. But it is noted that these women aid in their own trafficking and come voluntarily – their motivation is earning money through sex work in India and sustaining their families in the neighboring countries. The Indian government's attempt to curb trafficking across borders and stem the flow of potential sex work seekers has been futile so far according to Rama:

> The government has tried to seal the borders before and it is still trying to do so now. What is the result of that? Has it succeeded in reducing trafficking? No, it has not. Thousands have crossed the border seeking to work in *Sonagachi* in spite of government efforts. Trafficking will not stop and the entry of women into this profession will not stop. Do you know why? It

> is because there is a huge demand for this profession. And then there is hunger. Who will feed them and their children? The government does not provide food, it only provides laws and bans.

The sex workers' entry into this profession is thus noted to be voluntary and sometimes a result of their own collusion with traffickers. This is in direct disagreement with current policy research documents on sex work and HIV/AIDS intervention most of which characterize sex workers as victims of trafficking, coercion, and sexual servitude.

The single interviewee who said she was forced into this profession noted that she did not leave sex work even when she got a chance as it gave her a steady and secured source of income. Seven of the interviewees stated that sex work was not the primary source of income and they had additional means of sustenance such as small businesses. Eight of the interviewees did not live within the perimeters of the red light district and were daily and weekly commuters to *Sonagachi*. The presupposition of coercion into sex work and sexual servitude in the profession as advocated by HIV/AIDS global policy documents appears to be in direct contradiction to the testimonies of the interviewed women of *Sonagachi*. The interviewed sex workers enter the profession of their own volition – this is emphasized by Bisakha:

> I am here working in this profession out of my own free will. Nobody has forced me into this profession. If I go to work as a domestic help in somebody's house I am going there voluntarily, right? Similarly when I come here to work as a sex worker I come voluntarily. I am not compelled by anyone. Why do people think that sex workers are sex slaves? I think it is because they make their own ideas, they do not care to ask the opinion of sex workers.

60.3.2 Re-articulation of Sex Work: Demanding Legitimacy

One of the important findings from the interviews was the sex workers' re-articulation of issues related to sex work – including the status of sex work itself – and demanding for legitimization. Sex work was asserted to be a valid form of employment whose legalization and decriminalization were demanded. The rights of the sex workers to demand benefits such as healthcare for themselves and educational opportunities for their children were emphasized by the interviewees.

The interviewees also laid importance on the rights for self-determination which included the right to choose the occupation of a sex worker as a livelihood without interference from legal, social, and moral authorities. As the manifesto produced by DMSC (2009) asserted, "Sex work needs to be seen as a contractual service, negotiated between consenting adults. In such a service contract there ought to be no coercion or deception. DMSC is against any force exercised against sex workers, be it by the client, labor contractors of the sex sector, room owners, pimps, local goons, the police or the traffickers" (p. 1). Legalization of the profession was noted to be a necessary step for protecting the health and securing the rights of the sex workers.

The initiatives of the *Sonagachi* women to promote sex work as a legitimate form of labor are in direct opposition to the agenda of many global HIV/AIDS policymakers and international legal resolutions that intend to curb trafficking. The latter portray sex work as a profession interrelated to and produced by trafficking itself. Hence sex work is often articulated by these agencies as a profession incompatible to human dignity and welfare and which needed to be delegitimized on an urgent basis.

The USAID grant policy explicitly stresses the need to eradicate sex work to counter the impact of HIV/AIDS and ensure global health. Some feminist scholars supported policies seeking a legal ban on sex work and endeavoring to rehabilitate the sex workers and eliminate the profession all together. An example may be cited in the words of Hughes (2000):

> Prostitution and trafficking are extreme forms of gender discrimination and exist as a result of the powerlessness of women as a class. Sexual exploitation is more than an act; it is a systematic way to abuse and control women that socializes and coerces women and girls until they comply, take ownership of their own subordinate status, and say, "I choose this." Legalization of this violence to women restricts women's freedom and citizenship rights. If women are allowed to become a legitimate commodity, they are consigned to a second-class citizenship. No state can be a true democracy, if half of its citizens can potentially be treated as commodities

The interviews obtained by the researcher make it apparent that the women of *Sonagachi* might disagree with Hughes and many international policymakers. All of the 37 interviewees noted sex work to be a valid form of labor. Two of the interviewees admitted that their profession was morally questionable according to social norms but added that sex work should be considered to be a legitimate vocation. All of the interviewees wanted their profession to be given legal status. Decriminalization of sex work was noted to be an emergent need to protect the rights of the sex workers and to establish and maintain the reduction of HIV/AIDS and STI infection incidence.

All of the 37 interviewees emphasized sex work to be like any other job. Santana noted:

> Our profession helps us to sustain our families, our children. We work hard, use our bodies to make our clients happy and earn money to survive. How are we different from workers who use manual labor to make an income? How are we different from government workers who work hard in their offices to feed and educate their children? At least we do not engage in corruption like some of the government officials do. The money we make comes in exchange of hard physical labor to make our clients happy. If that is not a valid form of labor, then tell me what labor is?

Bishakha added:

> Our job involves its own sort of physical and mental exertion. Tell me which job does not create exertion? You are an interviewer, you have come here to interview me, for that you have woken up early in the morning. Haven't you undergone some exertion for your job of interviewing? But your work serves your interest for doing your study. Similarly my work

also has its stresses but it looks after my interest—my interest of keeping myself and my family in comfort. Plus you like what you do, right? I also like what I do. So how is my work any less valid than yours or anybody else's? The people who want to ban sex work.. they are rich people. They sit high up, in air-conditioned rooms. They need to climb down a bit, to our level, to think like us.

Criminalization of sex work was questioned and strongly castigated by the interviewees. The latter noted the propensity among media and sections of society in general to equate sex work with criminal activity in an endeavor to ban it. Krishna observed:

Sometimes people tell us sex work is criminal, it is akin to stealing, robbing and murder. They say if sex work is to be legalized then one should legalize stealing, robbing and murder too. I would like to remind them that stealing, robbing and murder cause fear, anxiety and grief. But the clients of sex workers come to them to get sexual service, and pleasure. Nobody can say that the sex workers cause fear, anxiety and grief to the clients. If they had, the clients would not have been coming back again and again. Well somebody from the media once asked me that one may get happiness by taking heroin and marijuana, in that case should the seller of these drugs be given legal status? I would like to add here that heroin and marijuana ruin a person's health. It is a proven fact, is it not? But show me one study which says a client having sex with a sex worker using condoms is ruining his health.

One of the primary demands of DMSC was reiterated in the words of the interviewees – "*gautore khatiye khai, taai sromiker adhikar chaai*" which translates as "we use our bodies to work hard, so we need to get legitimate worker rights." Eight of the interviewees noted their work to be a part of service industry. "We provide service to our clients" and "we are part of the service industry" were reiterated by a number of interviewees.

Of the 37 interviewees, 15 noted sex work to be akin to entertainment work. Kalavati said, "We entertain our clients. We provide them happiness. Our job is similar to that of actors, singers, artists and other entertainment professionals." Bishakha noted, "We give pleasure and we take money for it, that is our job. Our profession is similar to that of entertainment workers."

All of the 37 interviewees emphasized the drawbacks of delegitimization of sex work. Fourteen noted how rendering sex work illegal created an unsafe environment for the sex workers and hindered the implementation of safe sex practices. Kalavati said:

See when our profession is illegal what will the women do? They will have to earn a living after all. Else who will feed the kids? So they go in hiding. They do their trade in hidden and dark lanes and allies. The chances of violence against them in such hidden locations increase. The anti-socials will target these women. Rape the women, and no money paid. And safe sex, condoms. . .there is no safe sex in rape.

Kajal adds:

Our profession is illegal. We are criminals according to law. So if we get raped what justice can we get? The police can tell us, in fact the police had told us in the past that a criminal

> cannot get raped. And sex workers who live on their sex cannot get raped. And yes, the police has raped us in the past. But now with *Durbar* we have learnt how to live with our heads high. The police is wary of us because of our union, they register our complains, treat us with respect. But what happens to the sex workers in other states of India? They are raped regularly, by clients, pimps, police. There is no justice, for where is the crime? The sex workers are criminals, their profession is illegal. Do you think rapists use condoms? I tell the government that if they want to stop HIV then make sex work legal. Give these women a solid ground to stand on, to protest and make their demands known.

From the interviews it becomes apparent that delegitimization and criminalization compel sex workers to operate in subterfuge, such as in dark and isolated geographical spots. Their attempt at concealment in order to escape from detection and prosecution by law enforcement agencies exacerbated their vulnerability to rape and assault. Such sexual violence rarely includes safe sex measures, and hence the chances of HIV/AIDS and STI infection transmission are greatly increased. Legalization and decriminalization of sex work are noted by the interviewees to be a step in the right direction for HIV/AIDS harm reduction and securing the rights and health of the average sex worker.

The interviewees stressed that unionization enabled them to fight the violence and stigmatization imposed by their illegal and criminal status. A stronger sense of self-belief and confidence in their chosen profession reinforced the demand for formal legalization. The futility of the logic that upheld banning sex work for moral and health reason was also discussed by the interviewees. Sapna said:

> We believe that banning sex work is not the answer. Banning sex work will not be a solution to any problem. Tell me why there is sex work? It is because sex work has always been in demand through the ages. It is in demand now and it will be in demand in the future. More so in the future perhaps. Whenever there is a demand there is a supply. Sex work cannot be banned and it cannot be wished away. We are here to stay. Sooner people understand this, the better.

The emphasis is thus on the welfare of the women who are "here to stay" by legalizing and decriminalizing them. As Sadhana observed, "Our clients are not criminals and neither are we. They come in pursuit of physical pleasure and happiness which we give to them. We are adults and there is no crime in that."

60.3.3 Rescue and Rehabilitation of Sex Workers

The TIP report of 2009 which equated sex work with trafficking and sexual servitude noted that there can be no exception to rescue of sex workers and rehabilitation from sex work. The USAID funding policy guided by the Global AIDS Act propounds rescue and rehabilitation of sex workers to be one of the targeted goals in HIV/AIDS intervention projects. In order to achieve this objective, the Global AIDS Act deprives organizations that do not strive for the rescue and rehabilitation of sex workers from receiving any funding.

Indian legislation on sex work criminalizes sex workers, their clients, pimps, and brothel owners subjecting them to a fine and imprisonment of 3 to 5 years on prosecution. The clause 2(f) of the Immoral Traffic (Prevention) Act (ITA) of India defines sex work as prostitution which is "sexual exploitation or abuse of persons for commercial purposes or for consideration of money or in any other kind" (Government of India publication 2009, p. 1). The clause 4(a) of the ITA also criminalizes sustenance of a sex worker's earnings which creates a precarious situation for the offspring and family members of the sex workers. The articulation of sex work as an unlawful and oppressive vocation leaves the sex workers as illegal beings themselves with little rights or opportunities.

However the interviews with the sex workers of *Sonagachi* showed that propositions of rescue and rehabilitation were not considered to be feasible options. Rather the interviewees considered the rehabilitation approach impracticable for several reasons. One of the reasons voiced against the rescue and rehabilitation option was that such a proposition violated the rights of the sex workers as a legitimate labor group. Fourteen of the interviewees subscribed to this notion that rescue and rehabilitation entailed the violation of the rights and dignity of a sex worker. The interviewees noted that rehabilitation is applicable for the poor and destitute, the homeless, and the dispossessed. The sex workers were noted to belong to none of the aforesaid categories. The interviewees also noted that individuals in dangerous or coercive situation have to be rescued. The sex workers were pursuing their vocation voluntarily and were not in distressed circumstances. Hence the proposition of rescuing the sex workers was questionable. Santana noted:

> Well, if they have to rehabilitate, why don't they rehabilitate the homeless people, the people displaced by floods, the street dwellers who are starving? We are not starving, we have a job. Why don't they rehabilitate the poor beggars? Does the government classify us as beggars. We are not beggars. We have a job. We can look after ourselves and our families. We don't need rehabilitation.

The rescue proposition was also repudiated by the interviewees. Rama remarked:

> What are they rescuing us from? From our professions? We are adult women who have engaged in this profession willingly. Nobody has sold us into this, nobody is forcibly keeping us in this. Why do we need to be rescued then? Is it because the government or the outside agencies cannot accept the fact that we follow our profession voluntarily? Does it pain them to accept that we sell sex willingly?

Most of the interviewees seemed to hold the assumption that the rescue and rehabilitation proposition for sex workers had moralistic undertones. Such schemes framed by moralistic motives were noted to be in violation of the dignity of the sex worker and her profession.

Another reason for repudiating the rescue/rehabilitation proposition was economic. It was noted by 16 interviewees that rehabilitation was not feasible for financial reasons, for the income generated by sex work was often greater than that gained from suggested alternate professions. These women noted that sex workers were mostly illiterate and lacked educational skills that would enable them to be placed in anything other than minimum wage jobs. The alternative professions suggested to the sex workers by government agencies were generally domestic help jobs and menial labor work. Alternative vocations like handicrafts and domestic labor which were considered "honorable" rehabilitation options for sex workers by aid agencies, government, and NGOs were not financially viable to be considered feasible. The interviewees were also extremely wary about the chances of sexual violation in the suggested alternative professions and hence questioned the moral ground of such rehabilitation proposition.

Another reason for dismissing the rehabilitation proposition was the prospect of sexual violation and exploitation in the process of rehabilitation itself. Eleven of the sex workers noted that such rehabilitation projects had failed previously because rehabilitated sex workers were often sexually harassed by the concerned officials engaged in the process.

No data could be obtained on sex workers who were rehabilitated by government or NGO initiatives to judge the success or the lack of it of such rehabilitation projects. It appeared from the interviewee's statements that such rescue and rehabilitation endeavors were often motivated by ulterior factors and were futile. However no relevant data on the subject could be found to verify or nullify these observations. The interviewees may be motivated by a sense of skepticism against external intervention that prompted them to view any such rescue and rehabilitation endeavor with skepticism. Logistical realities such as stigmatization, ostracization, and sexual harassment of rehabilitated sex workers might render such efforts futile. And sexual exploitation of sex workers by rehabilitation workers might be a reality in red light areas that unravel the impact of such external interventions.

60.4 Commercial Sex Work and Ground Realities

The study conducted with the sex workers in *Sonagachi* show that the interviewees' statements are in direct contradiction to some of the policies of global HIV/AIDS aid organizations. All of the interviewees noted the distinction between trafficking and sex work. Thirty-six of the thirty-seven interviewees had entered their vocation of their own accord. They have not been duped or coerced into sex work, neither were they kept in a state of servitude. The primary motivation of engaging in sex work was found to be economic. Trafficking was noted to be a present and persistent problem by the interviewees. But the latter also noted it to be a process aided and abated by desperately poor individuals striving to get into a vocation like sex work to sustain their family members.

The chapter does not purport to conclude that sex workers are not trafficked or coerced into their profession, but the study shows that not all sex workers are victims of trafficking and sexual servitude. In such a case, the application of policies and legislatures that uniformly treat them as victims of trafficking and sexual slavery might be impracticable. The interviews with the *Sonagachi* women indicate that they chose this profession under financial exigencies. Yet the interviewees asked their choice of livelihood to be respected and not subjected to moralistic evaluation. They demand legalization and decriminalization of their vocation and acquisition of labor rights guided by international labor regulations.

Previous research shows that delegitimization and criminalization of sex work can jeopardize the health and safety of sex workers and increase unsafe sex practices and consequently HIV/AIDS infection rates. Heeding the cries for legalization and being granted labor rights can therefore ensure the sustainability of an already successful health outreach initiative among the sex workers. The rescue and rehabilitation propositions for sex workers are often not feasible options as shown by previous research. The interviews with the sex workers of *Sonagachi* show that the women emphasize the impracticability of these propositions and repudiate them vigorously. The ground realities of the trade, as indicated by the interviews, ensure that such rescue and rehabilitation schemes continue to fail. The interviewees also expressed serious reservations about the moral compunctions that precipitated such rescue and rehabilitation schemes. Evidently rescue and rehabilitation propositions are not always the appropriate objectives in health intervention projects among sex workers.

It appears that the population of *Sonagachi* would not be the suitable target population for rescue and rehabilitation schemes. Similarly there might be comparable groups of sex workers in India and globally among whom the application of rescue and rehabilitation options would not succeed. In such a situation, one is led to question the policies of the Global AIDS Act which deprives sex workers groups from funding if rescue and rehabilitation propositions are not implemented. A change in the policy of the Global AIDS Act and the USAID funding policy appears to be essential for the sake of many sex workers' groups across the globe.

References

Butcher K (2003) Confusion between prostitution and sex trafficking. Lancet 361(9373):1983

Cohen S (2005) Ominous convergence: sex trafficking, prostitution and international family planning. Guttmacher Rep Public Pol 8(1):12–14

Ditmore M (2003) Morality in new policies addressing trafficking and sex work. In: Conference paper of IWPR's seventh international women's policy research conference, June 2003

Durbar Mahila Samanway Committee (2009) Durbar Bhabona. Calcutta: Durbar Prokashoni.

Government of India (2009) The Immoral Trafficking (Prevention) Act 1956. Available at https://indiacode.nic.in/bitstream/123456789/1661/1/1956104.pdf. Accessed on 1 December 2017.

Hughes D (2000) Men create the demand; women are the supply. Safety at Work, 7:10–14.

Masenior NF, Beyrer C (2007) The US anti-prostitution pledge: first amendment challenges and public health priorities. PLoS Med 4(7):1158–1161

Trafficking in Persons Report (2009) United States Government. Available at http://www.state.gov/documents/organization/123357.pdf. Accessed on 1 Dec 2017

UNAIDS Global Reference Group on HIV/AIDS and Human Rights (2003) Sex Work and HIV/AIDS. Available at http://data.unaids.org/Topics/Human-Rights/hr_refgroup2_01_en.pdf. Accessed on 1 Dec 2017

United Nations Development Fund for Women (2009). Factsheet on trafficking and commercial sexual exploitation. Available at http://www.unifem.org/attachments/products/339_Chapter_4.pdf. Accessed on 1 December 2017

Wolffers I, van Beelan N (2003) Public health and the human rights of sex workers. Lancet 6:1981

Designing and Distribution of Dementia Resource Book to Augment the Capacities of Their Caretakers

61

Avani Maniar and Khyati Deesawala

Contents

Abstract

As per the estimates of the WHO, aging population is increasing in developing countries, and dementia is going to be epidemic among elderly in the coming decades. This demands early action to prevent the disease and treatment of the affected persons, which is poorly existed in middle- and low-income countries.

The need of the hour to tackle dementia in India is to estimate the disease burden in the community, search for risk and protective factors, and undertake measures to provide social benefits to the sufferers and those who are at risk. Raising awareness among the civil society is an important task ahead. Apart from that, family members and caregivers of dementia sufferers have to be very sensitive and resourceful while

A. Maniar (✉) · K. Deesawala
Department of Extension and Communication, The Maharaja Sayajirao University of Baroda, Vadodara, India
e-mail: avanimaniar@gmail.com; khyatideesawala@yahoo.com

J. Servaes (ed.), *Handbook of Communication for Development and Social Change*,
https://doi.org/10.1007/978-981-15-2014-3_36

handling them. If they are well equipped to deal with the issue, positive results can be seen in their efforts.

Without proper information about dementia, the person and the family may respond in different ways which may aggravate the problems in dealing with dementia patients. Hence, to prepare family members and caregivers of dementia sufferers, a resource book in simple, nonmedical language with a lot of illustrations and variety of information is designed to help them and gain support for various issues while dealing with the patients.

The book chapter discusses the need for designing the resource book, the approach adopted while selecting and designing the book, and also the communication strategies used in distributing the resource book among its caretakers.

Keywords

Dementia · Resource book · Caretakers · Leaflet · Awareness · Old age · Capacity building

61.1 Introduction

The World Health Organization (1999) recognized that the developing world often defines an old age not by years but by new roles, loss of previous roles, or inability to make an active contribution to society. The age of 60 or 65, roughly equivalent to retirement ages in most developed countries, is said to be the beginning of old age. The WHO predicts that by the year 2025, about 75% of the estimated 1.2 billion people aged 60 years and older will reside in developing countries.

https://www.scribd.com/document/190077600/WHO-Definition-of-an-Older-or-Elderly-Person

The aging of the society comes along with several specific problems, i.e., physical and mental problems. The top five causes of death among older adults are heart disease, cancer, osteoporosis, cerebrovascular disease (relating to the blood vessels that supply the brain), pneumonia and flu, and chronic obstructive pulmonary disease. Hearing impairment among older adults is often moderate or mild, yet it is widespread; 48 percent of men and 37 percent of women over age 75 experience hearing difficulties. Visual changes among aging adults include problems with reading speed, seeing in dim light, reading the small print, and locating objects. By age 80, it's common to have lost as much as 2 inches (5 cm) in height. This is often related to normal changes in posture and compression of joints, spinal bones, and spinal discs. With age, the skin becomes less elastic and more lined and wrinkled (Brock 2014).

(http://hemispherespublishing.com/social-security-qualitative/)

He further mentioned that for most of the older adults, good health ensures independence, security, and productivity as they age. Yet millions struggle everyday with emotional and psychological stress closely related to loneliness, isolation, or loss of a loved one, along with emotional problems. Elderly also go through health and safety challenges such as chronic disease, dementia/Alzheimer's disease, diabetes, Parkinson's disease, and much more, and at times these dilemmas are not taken into account by the medical or healthcare system.

Das et al. (2012) while analyzing dementia scenario in India described about the number of people who may be affected by it. It is estimated that the number of people living with dementia will almost double every 20 years to 42.3 million in the year 2020 and 81.1 million in the year 2040. The rate of growth will be the highest around 336% in India, China, South Asia, and Western Pacific regions and 235–393% in Latin America and Africa and the lowest (100%) in developed regions. Based on the year 2001 global population, about 24.3 million have dementia, and 4.6 million incidents or new cases are added yearly.

As per global burden of disease study by WHO and World Bank, dementia contributes 4.1% of all disability-adjusted life years (DALYs). According to the Alzheimer's and Related Disorders Society of India (2010), in India, the number of people with the AD and other dementias is increasing every year because of the steady growth in the older population and stable increment in life expectancy. Thus, an estimated twofold increase by 2030 and threefold by 2050 can be expected. According to the Neurological Society of India, It is expected that the burden of dementia will be increasing in developing countries due to increase in longevity and increasing prevalence of risk factors such as hypertension and stroke and lifestyle changes.

http://www.neurologyindia.com/article.asp?issn=0028-3886;year=2012;volume=60;issue=6;spage=618;epage=624;aulast=Das

61.1.1 What Is Dementia?

According to ***Alzheimer's society***, the word dementia is defined as a set of symptoms that may include memory loss and difficulties with thinking, problem-solving, or language. These changes are often small to start with, but for someone with dementia, they become severe enough to affect their daily life. A person with dementia may also experience changes in their mood or behavior.

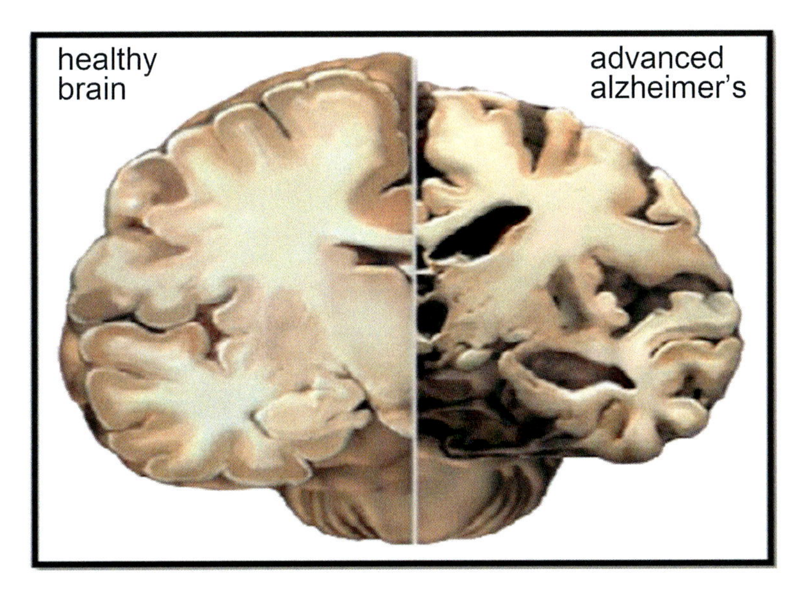

Source: https://goo.gl/329PY1

The people affected by dementia become easily confused and restless and perform repetitive actions. The patients can become irritable, agitated, and tearful. This scenario is stressing for both the patient and their family. Because of the frustration the patients feel about their condition, they can also develop depression and aggression and show improper sexual behavior and incontinence at times.

There is 1 out of 20 people over the age of 65 who can develop dementia. When the elderly are over the age of 85, there is a higher risk of developing dementia: one out of four elderly can develop the disorder. People who have been diagnosed with dementia before the age of 65 are very rare cases. This type of dementia is called presenile or early onset dementia as mentioned on a mental health information website of Hamlet Trust, UK (www.hamlet-trust.org.uk/articles/dementia.html).

61.1.1.1 What Are Signs and Symptoms of Dementia?

WHO factsheet (2017) highlighted that problems linked to dementia can be best understood in three stages:

- Early stage – in first year or 2 years of its onset
- Middle stage – during second to fourth or fifth year of its suffering
- Late stage – in its fifth year and after

Early Stage

The early stage is often overlooked. Relatives and friends (and sometimes professionals as well) see it as "old age," just a normal part of the aging process.

The onset of the disease is gradual and difficult to recognize.

- Have problems in talking properly (language problems)
- Have some memory loss, particularly of recent events
- Have difficulty in making decisions
- Become inactive and unmotivated; show mood changes, depression, or anxiety; and may react angrily or aggressively
- Show a loss of interest in hobbies and activities

Middle Stage

As the disease progresses, limitations become pronounced and more restricting.

The person with dementia has difficulty with day-to-day living.

- May become very forgetful – especially of recent events and people's names
- May become extremely dependent on their family and caretaker; unable to cook, clean, or shop; needs help with personal hygiene

(continued)

- Have wandering and other behavior problems such as repeated questioning and calling out, clinging, and disturbed sleep
- Unable to recognize familiar and unfamiliar places at home or outside
- May have hallucinations (seeing or hearing things which are not really there)

Late Stage

This stage is one of near-total dependence (confined to a wheelchair or bed). Memory disturbances are very serious with more physical complications. The person may:

- Have difficulty eating and walking and be incapable of communicating
- Not recognize relatives, friends, and familiar objects
- Have bladder and bowel incontinence, breathing difficulties, and respiratory infections.

http://www.who.int/mediacentre/factsheets/fs362/en/#Factsheets

Below snapshots (Figs. 1 and 2) are from the Reference Book depicting the signs and symptoms of dementia and also what the affected patient feels like.

61.1.1.2 What Are Causes of Dementia?

Alzheimer's Society (2007) elaborated about the causes of dementia. It is caused by damage to brain cells. When the brain cells are damaged and become nonfunctional, thinking, behavior, and feelings get affected. The human brain is divided into regions, and each region is responsible for different functions associated with the body. For example, one region controls memory; another is responsible for psychomotor controls and still another for judgement.

Different types of dementia causes are associated with particular type of brain cell damage in particular regions of the brain; some of the causes are simpler to understand in terms of how they affect the brain and lead to dementia:

A. **Alzheimer's disease** – This is the most common cause of dementia. High level of certain proteins inside and outside brain cells makes it hard for brain cells to stay healthy and to communicate with each other in the brain region called the hippocampus. This region is concerned with memory and learning; hence, the problems with day-to-day memory are often noticed first, but other symptoms may include difficulties with finding the right words, solving problems, making decisions, or perceiving things.
B. **Vascular dementia** – If the oxygen supply to the brain is reduced because of narrowing or blockage of blood vessels, some brain cells become damaged or die. This causes vascular dementia. The symptoms of vascular dementia vary and

Signs and Symptoms of Dementia

The signs of dementia are different for different people. These signs can also be the signs of other illnesses or of getting older. The most common signs of dementia are:

Forgetfulness: This is usually the earliest and most noticeable symptom. For example, trouble recalling recent events or recognizing people and places.

Finding it hard to plan or do things that you used to be able to do. Difficulties making decisions, solving problems or carrying out a sequence of tasks (eg cooking a meal)

Not feeling sure of things, even when you are somewhere you know and difficulty recalling events that happened recently. For eg: Losing track of the day or date, or becoming confused about where they are.

Being moody or depressed. For example, they may become frustrated or irritable, withdrawn, anxious, easily upset or unusually sad.

Fig. 1 Signs and symptoms of dementia

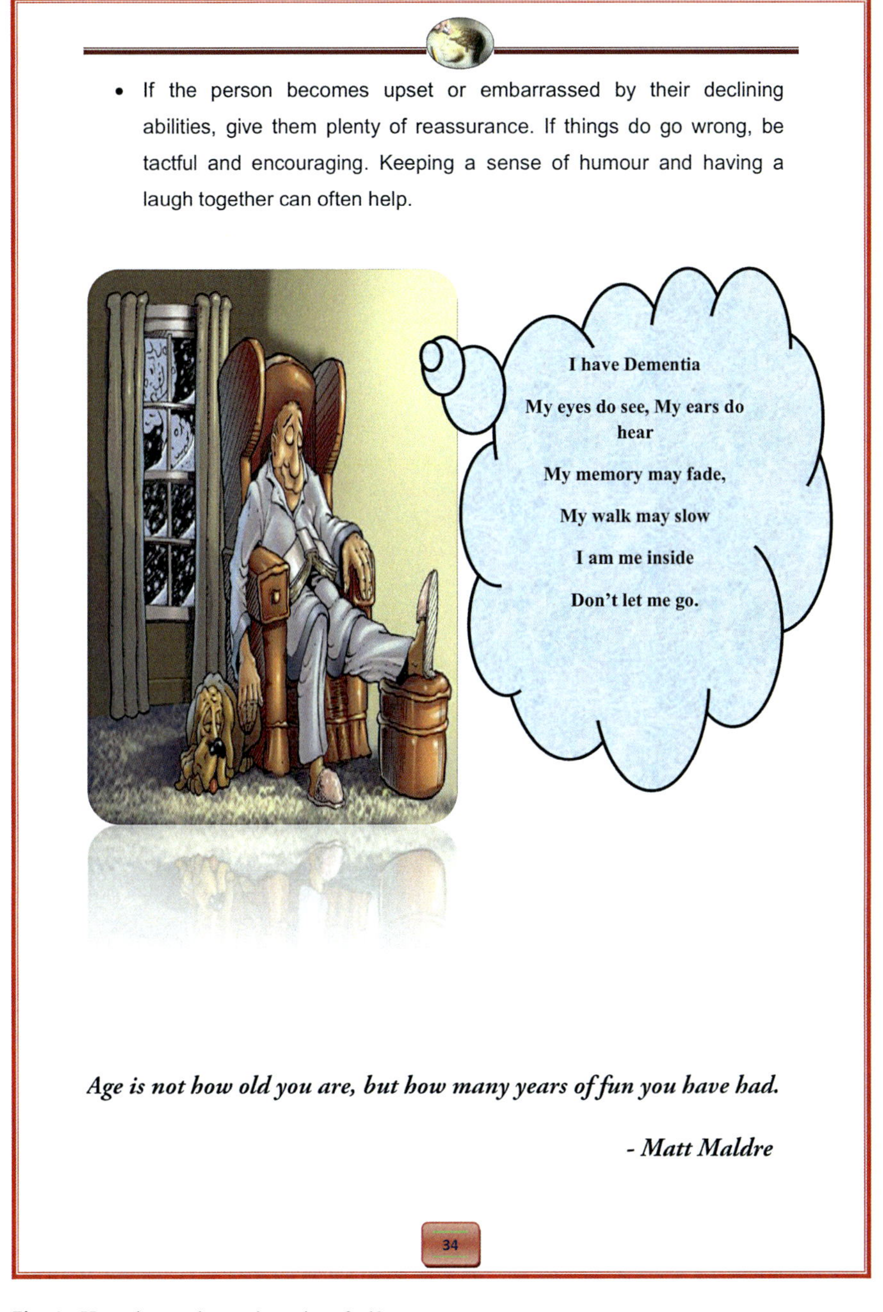

Fig. 2 How does a dementia patient feel?

may overlap with those of Alzheimer's disease. Many people have difficulties with problem-solving or planning, thinking quickly, and concentrating.

C. **Mixed dementia** – This is when someone has more than one type of dementia and a mixture of symptoms. It is common for someone to have Alzheimer's disease and vascular dementia together.
D. **Dementia with the Lewy bodies** – This type of dementia involves tiny abnormal structures (Lewy bodies) developing inside brain cells. They disrupt the brain's chemistry and lead to the death of brain cells. Early symptoms can include fluctuating alertness, difficulties with judging distances, and hallucinations.
E. **Frontotemporal dementia (including Pick's disease)** – In frontotemporal dementia, the front and side parts of the brain are damaged. Clumps of abnormal proteins form inside brain cells, causing them to die. At first, changes in personality and behavior may be the most obvious signs. Depending on which areas of the brain are damaged, the person may have difficulties with fluent speech or forget the meaning of words (Fig. 3).

https://www.alzheimers.org.uk/info/20007/types_of_dementia/1/what_is_dementia/3

61.1.1.3 How Can Dementia Be Treated?

According to Alzheimer's Society (2007), the vast majority of causes of dementia cannot be cured, although research is continuing into developing drugs, vaccines, and other medical treatments. There is also a lot that can be done to enable someone with dementia to live well with the condition. Care and support should be "person-centered," valuing the person as a unique individual. Therefore, the two types of treatments are:

Non-drug Treatments and Support

A range of support, therapies, and activities that do not require medication can help someone to live well with dementia. Talking therapies such as counseling can help someone come to terms with their diagnosis. Another treatment called cognitive behavioral therapy (CBT) may be offered to help with depression or anxiety. There is also a lot that can be done at home to help someone with dementia, remain independent, and live well with the memory loss. Support ranges from devices such as pill boxes or calendar clocks to practical tips on how to develop routines or break tasks into simpler steps.

Activities that help to keep the mind active, such as cognitive stimulation, are popular. As the condition progresses, many people with dementia enjoy life story work (in which the person is encouraged to share their life experiences and memories). Such activities may help improve someone's mental abilities, mood, and well-being.

It is vital that people with dementia stay as active as they can – physically, mentally, and socially. Everyone needs meaningful activities that they enjoy doing and which can develop confidence and self-esteem.

Drug Treatments

There are drugs that can help to improve the symptoms of dementia or that, in some cases, may stop them progressing for a while. A person with mild to moderate

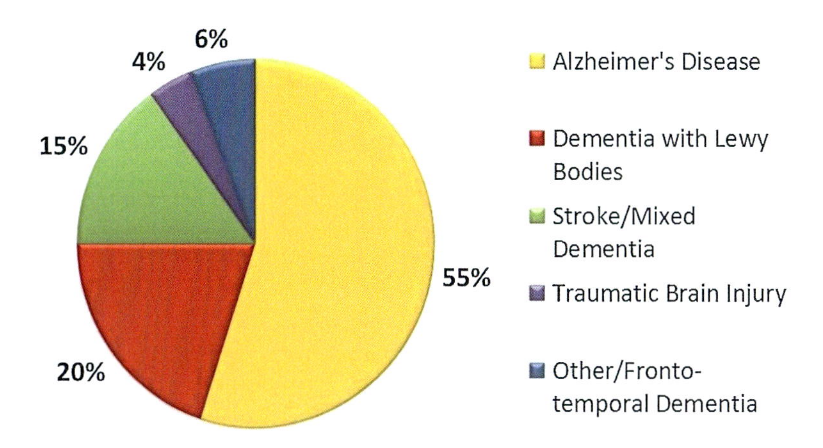

Fig. 3 Estimated figure showing causes of dementia

Alzheimer's disease or mixed dementia may be prescribed donepezil (often known by the brand name Aricept), rivastigmine (e.g., Exelon), or galantamine (e.g., Reminyl). These may temporarily relieve memory problems and improve alertness. Donepezil, rivastigmine, and galantamine can be helpful for someone with dementia with Lewy bodies who has distressing hallucinations or delusions or who has behaviors that challenge. For a person with vascular dementia, drugs will be offered to treat the underlying conditions. These conditions often include high blood pressure, high cholesterol, diabetes, or heart problems. Controlling these may help slow the progression of dementia.

A person suffering from dementia needs a lot of care to deal with the disease, as the person suffers from many problems and that leads to person's deteriorating stage physically and mentally, so to help the person struggling from physical and mental problems, the patient needs the care to be provided by the caregivers and families and to avail of the available services and awareness provided to them. Thus, there are certain caretaking services in India which can serve the patients to live a better life.

https://www.alzheimers.org.uk/info/20007/types_of_dementia/1/what_is_dementia/7

The following snapshots (Figs. 4, 5, and 6) from the resource book talk about the details of the doctors available for dementia patients and also some tips on nutrition needed for dementia patients along with the prevention measures.

61.1.2 Which Dementia Care Services Are Available in India?

Alzheimer's and Related Disorders Society of India (ARDSI) is actively involved in developing services for dementia across the country. Healthcare organizations like HelpAge India, Dignity Foundation, Nightingales Trust, the Dementia Society of Goa, Sangath, Silver Innings Foundation, Christian Medical College, Vellore,

Name of Doctors for Dementia Patients

- **Dr. Gautam S Amin M.D, Psychiatrist**

  Phone/Mobile : 0265 – 2395997, 0265 - 2426278, 9825027455
  Address : Ellora Park, Race Course, Main Road,Vadodara -01
  LandMark : Near Indraprasth
  Contact Person : Gautam S Amin
  Working Hours : Monday - Saturday: 9.30 AM - 10 PM
  Categories : Psychiatrists

- **Dr. Himanshu Chauhan, Psychiatrist**

  Address : Sanstha Vasahat, Opposite Shree Raj Medical Store, Pratap Road, Raopura, Vadodara, Gujarat – 390001
  Enquiries : (0265) 2416217, (+91) 9824266855
  Appointment : (0265) 2416217, (+91) 9824266855

- **Dr Venkat Iyer, Psychiatrist**

  Address : 305 B, Sarjan Complex, Pratap Road, Dandia Bazar, Near Anyonya Bank, Vadodara - 01
  Phone/Mobile : +(91)-9824034476 +(91)-265-2439284, 6538062

Also Listed In:

Psychiatrists

Psychotherapy Doctors

Doctors For Mentally Challenged

Psychiatrist Doctors For Children

20

Fig. 4 Names of doctors for dementia patients

Nutritional Tips for Dementia Patients

People with Alzheimer's or dementia do not need a special diet. As with anyone, eating a well-balanced, nutritious diet is important for overall health.

Proper nutrition is important to keep the body strong and healthy. For a person with Alzheimer's or dementia, poor nutrition may increase behavioral symptoms and cause weight loss.

The basic nutrition tips below can help boost the person with dementia's health and your health as a caregiver, too.

- Provide a balanced diet with a variety of foods. Offer vegetables, fruits, whole grains, low-fat dairy products and lean protein foods.
- Limit foods with high saturated fat and cholesterol. Some fat is essential for health — but not all fats are equal. Go light on fats that are bad for heart health, such as butter, solid shortening, lard and fatty cuts of meats.
- Limit foods with high sodium and use less salt. As the disease progresses, loss of appetite and weight loss may become concerns. In such cases, the doctor may suggest supplements between meals to add calories.

  Staying hydrated may be a problem as well. Encourage fluids by offering small cups of water or other liquids throughout the day or foods with high water content, such as fruit, soups, milkshakes etc

Fig. 5 Nutritional tips for dementia patients

Prevention of Dementia

Stay with a Healthy Weight

Most of us are eating too much and not being active enough. The result is often weight gain. Regular physical exercise (e.g. cycling, brisk walking), maintains healthy weight.

Take a Balanced Diet

Try to stay active and eat a healthy diet. A balanced diet is one which is low in saturated fat, does not have too much salt, dairy or meat, and includes plenty of fish and fresh fruit and vegetables

Keep fit and active

Regularly doing something active that you enjoy, like walking or gardening will help you stay fit and well. While being socially active could include visiting friends or going to a place of worship, while being mentally active could include doing puzzles or reading

Have an NHS(National Health Service) Health Check

An NHS Health Check can help spot problems before you can tell that something is wrong, for example, having high blood pressure or high cholesterol.

Fig. 6 Prevention of dementia

St. John's Medical College, Bangalore, and Voluntary Health Services, Chennai are also providing care services either alone or in collaboration.

To overcome the limited information about dementia services in the country, ARDSI made an effort to map the dementia services available in India by contacting all chapters and other partners in the country. The services available in the country that cater exclusively to the people with dementia are provided in Table 1. There are half a dozen residential care facilities exclusively for people with dementia. The available services are grossly inadequate to meet the needs of the over 3.7 million PwD (people with dementia) in India.

Dementia remains a largely hidden problem in India, especially in those parts of India where poverty and illiteracy levels are high. There is a growing realization that the care of older people with disabilities makes enormous demands on their caregivers. Seeing the increased prevalence of dementia in India among elderly and lack of facilities for the treatment of the diseases, terms like dementia and Alzheimer's

Table 1 Services exclusively for people with dementia in India

Name	Type of service	Approximate number of such facilities in India
Residential care facilities	This facility is suited for those families who find it difficult to manage the basic day-to-day activities of the PwD. They could opt for long-term care in a nursing home facility. This also includes respite services where the PwD is looked after for a short period to give the caretaker a break	6
Day care Centers	Dementia day care facilities are designed for those PwD who have a need for medical attention and supervision but who do not require institutionalization in a nursing home. In these facilities, the PwD are looked after during the day after which they return home	10
Domiciliary care services	The services are provided to the PwD at their residence. Services could range from education to providing tips on caregiving to actual formal care. Could be provided by geriatric home nurses or other trained personnel. There is evidence that nonspecialist community workers can also be engaged in providing this service	
Support groups	Support groups are groups of people who have lived through the same difficult experiences and try to help themselves and others by sharing coping strategies. Recommended for caretakers of PwD	Exact data not available
Memory clinics	Memory clinics are specialized clinics that offer assessment, support, information, and advice to those with memory problems and their caretakers	100
Dementia help lines	These are special phone numbers dedicated to addressing queries on dementia. The calls are handled by trained personnel	10

Alzheimer's and Related Disorders Society of India 2010 report should be more emphasized by creating awareness among the families, caregivers, policymakers, and civil society in general. It is forecasted that in years to come, there will be a number of seniors who will be affected and hence the society should be aware of dealing with such issues. Level of awareness varies enormously across countries and even within countries. The public awareness about dementia in India is low. Primary care doctors do not encounter many cases in their practice, and there is no special emphasis on dementia diagnosis and management in the training of healthcare professionals.

61.1.3 How Can Designing a Resource Book on Dementia Be Helpful?

Lack of awareness has serious consequences for PwD (people with dementia). Their families do not understand dementia as a health problem. Hence, dementia is not recognized at its onset, and the diagnosis is often delayed due to lack of awareness of the disease. There is no structured training on the recognition and management of dementia at any level of health service. Healthcare services remain insensitive too and do not provide the much-needed information and support for caretakers and family members. While family members are the main caretakers, they must do so with little support or understanding from other individuals/agencies or information sources available at their doorstep.

Media interest in dementia and related healthcare issues also remain low. The information covered by mass media is not enough quantity- or quality-wise. For example, an article in the newspaper may not cover the type of information a reader is looking for or one may not read the newspaper on the day article/news is published or the reader may not understand the language used in the article. Similarly, TV or radio ads are being played in few seconds and social messages are telecasted/broadcasted, which may not be enough to cover the important information on the subject. Interviews or other formats are covered rarely in such electronic media. Online free resources are available in abundance for its users. They are in variety of formats which helps the Internet users to be aware of updated information through its web sources. WHO and other professional agencies and organizations working in this area keep sharing and uploading information on dementia. Majority of these websites are hosted by the developed countries which is not completely adoptable by Indian citizens. India is a country of diversity, and people in India are in a transition phase, moving from traditionalism to modernization. Therefore, their thinking, actions, and behaviors are a mixed of traditionalism and modernism; likewise their media accessibility and usage pattern include both traditionalism and modernism and are not transformed completely to modernism yet. The researches in media studies, Media Yearbook, and other reports on communication for social change reveal about multimedia strategy as an effective form of imparting and disseminating information to masses. Hence, looking to the advantages of a printed book, availability of the updated online resources, and the intended target group (mainly adults/elderly), a resource book will be of great help which includes a number of links to the online sources.

For augmenting the capacities of caretakers/family members of people with dementia, this resource book is brought into light. It helped the targeted users to get the updated information which is there on videos, podcasts, films, books, and mobile apps. From this one resource book, they were given all the details about doctors whom they can consult with and list of helpline numbers and care centers and also organizations for dementia patients.

61.2 How Was the Resource Book Designed, Distributed, and Checked for Its Effectiveness?

The designed resource book is targeted to create awareness regarding dementia (which included causes, symptoms, treatment, food, daily activities, etc.), among dementia patients, caregivers, and their family members. The language used in the book to explain about this complex disease is simple and with lots of illustrations. It has colorful visuals to sustain reader's interest in the subject and information provided. When family members of the person with dementia refer and consult medical doctors/general practitioners for issues they face in caring and handling, medical doctors/general practitioners often fail to make them understand their level of comprehension. They often fail to give effective guidance related to the type of food the patients should be receiving. The families of people with dementia are often with half/incomplete or less information about the disease which results in providing improper care services. The condition of the patient in such cases turns bad to worse, which is a cause of great concern for all in the health profession.

61.2.1 Approach Used in Designing the Resource Book

After an extensive reference work to interpolate all the details about dementia which a caretaker/family members would look forward to a resource book, the team referred books, viewed videos, read real stories of dementia sufferers, and surfed games and mobile apps useful for them and their caretakers. The interactions with the affected families revealed that the families with dementia patients look for organizations, doctors, and helpline numbers to deal in emergency situations. The team collected a number of such need-based information for compiling it in a systematic manner. Informative cartoons, quotes, and poems on dementia were also collected to make the resource book attractive, interesting, and informative for a reader from nonmedical background to understand the complexities which dementia patient faces. This approach is a big hit as it is always a challenge for health communicators to explain the medical issues to nonmedical personnel in a way which is comprehensible.

Discussions with practicing doctors, Medical College professors, social workers, researchers, and development/extension communicators helped immensely in gathering the details about dementia and the topics to be covered for the resource book. The team referred other resource books on dementia compiled in other countries other than India. The following topics were selected after few months of brainstorming with experts, families with dementia patients, and observations by team:

1. **Introduction about dementia** – About dementia, signs and symptoms, prevention, treatment, and causes
2. **Self-check test for dementia**
3. **Facilities and services for dementia patients** – List of doctors treating dementia patients, organizations working for dementia patients, and helpline numbers for emergencies
4. **Role of a caretaker** – Roles and responsibilities while taking care of dementia patients
5. **Tips for family members and caretaker** – includes the type of behavior they should maintain with the patient and the type of activities which can be given to dementia patient for cognitive stimulation
6. **Nutritional tips** highlighting the importance of nutrition in the resource book to make the readers aware of the kind of food to be given to dementia patients
7. **Gears for dementia patients** talks about the variety of equipment's which can be used for patients with dementia
8. **Other resources** which can be very useful for dementia patients like games, mobile apps, books, films, videos, and a podcast which may be helpful to cope with symptoms of dementia and how to deal with it in a correct way. The team identified the most relevant videos available on YouTube and other web portals, and their web links were shared in the resource book. Likewise, similar approach was adopted for podcasts, films, and mobile apps.

The team progressed to the next level to organize the collected/identified details under the relevant topic heads as separate chapters by creating an index. The textual material went hand-in-hand with an illustration for better visual learning. The content in its simplest form with creative animated pictures and quotes on old age mentioned at the bottom of the page gave additional information to its readers to learn about the old age and dementia, thereby making the resource book interesting and attractive.

61.2.1.1 Designing of a Leaflet

With a purpose to publicize the resource book, a leaflet was designed. The text of the leaflet highlighted the resource book and its objective behind designing. It also mentioned briefly about the content of the resource book and the email IDs of team members who can be approached for availing the copy of it.

61.2.1.2 Validation of the Resource Book

The designed resource book underwent a thorough review with five experts to check the content, designing aspects, the color of the fonts, suitability of subtitles, and coordination of pictures along with the text, also the logical validity of the content and clarity of the language. The experts were professors in Medical College, practicing doctors, philanthropist working for elderly, and development communicators/media designers.

The team incorporated the suggestions received from the experts for improving the content of the resource book. The resource book received high appreciations with very positive and supportive words as this is a much-needed work in this field. The style/approach of presenting the content, compiling the information and the usage of visuals and images in the resource book, impressed the reviewers highly. Post validation, the resource book was all set for printing and distribution.

61.2.1.3 Budget for Printing the Resource Book

The budget/sponsorship basically was needed for the cost that incurs in designing, printing, and binding of the 50 hard copies of the resource book. The team approached a few philanthropists from Vadodara City, CSR-based NGOs working in the health sector, with the resource book as a project proposal for sponsorship. "Bhumi Procon Private Ltd" Vadodara sponsored for the printing budget of this project with an amount of Rs. 35,000. The team planned to distribute 50 hard copies to those who have genuine issues in accessing it in the form of soft copies; otherwise soft copies through email would be distributed. This strategy worked well with a limited budget; however, if more copies are printed, then a large number of the population who is more comfortable reading the print form would be benefitted.

61.2.2 The Resource Book Distribution Approach

Once the resource book is compiled, the intended target group should be made aware about its availability and the benefits. To achieve this, the team publicized it in one of the regional newspapers *Gujarat Samachar* which has wide circulation in the city of Vadodara as well as in the State of Gujarat, India. When the editor of the newspaper was approached, he showed interest in publishing about the resource book. The article described the widespread prevalence of dementia, the problems the dementia patients go through, and how to cope and deal with such patients. The article mentioned that the resource book is designed for such sufferers and their caretakers/family members. Topics covered in the resource book were mentioned in the article, and those interested and in need of the book will be given the resource book free of charge. The email IDs of the team members were mentioned in the article to avail the soft copies.

Leaflet was an another source for publicizing about the resource book, which was distributed to the members of senior citizen clubs – as it houses a large number of senior citizens as life members. The team conducted a presentation to create awareness about dementia in those clubs by announcing the dates of the presentation well in advance. It was to ensure that all the members participate and benefit from the presentation. A large number of members were addressed, and at the end of the session, leaflets were distributed to each one of them.

The published article in the newspaper had a surprising impact; in other words, an overwhelming response was received on the email IDs of the team members mentioned in the newspaper and the leaflet. Also, some really interesting incidents happened as follows. On the day the article was published in the newspaper, a person visited the office of the team members to get the resource book telling how keen he was to own a copy. Fifteen phone calls were received demanding for the copy, though phone contact details were not shared either in the newspaper article or in the leaflet. A letter demanding for a copy was received for the same. It revealed their keenness to read the book. The team announced a cutoff date for sending in the requests for the copies, and they received 146 emails till the cutoff date, out of which large number had asked for hard copies. Reply to 50 of them was sent mentioning the dates, time, and venue to collect their copy. Others were sent with the soft copy. Along with copies (hard and soft), the team also distributed feedback form. A record of the people who came to collect the resource book was maintained wherein their name, address, contact number, email ID, numbers of copies taken by them, and reason for taking the resource book were jotted down for future reference.

61.2.3 Reactions to the Resource Book from Its Readers

There are a number of models used in designing a communication strategy for the development projects and programs to achieve the objective of advocacy, social mobilization, or behavior change among the target group. The three commonly used models are ACADA (Assessment, Communication Analysis, Design, and Action) which uses systematically gathered data to link a communication strategy to the development problem, the P-Process for planning strategic, evidence-based communication programs, and COMBI model that emphasizes on ten steps which may not be followed in a linear fashion; steps are often repeated. All three models considered evaluation, evaluation and replanting, and impact assessment as an integral components of the model.

(https://www.unicef.org/cbsc/files/Writing_a_Comm_Strategy_for_Dev_Progs.pdf)

It is indeed very important to know if the objective of designing this resource book was achieved or not, was it interesting for the readers, did it give them enough information about dementia, and did the content help them in dealing with the issues. To achieve this, the team members conducted evaluation to get the reaction/feedback from the target group on the resource book.

61.2.3.1 Tool for Collecting Reactions

The reaction scale had series of questions with multiple-choice response system, few are open-ended, and few are checklist types. It consists of three sections, and questions are asked as per the title of the sections which is as follows:

Section 1: Personal Details – This section consists of questions related to personal details of the respondents like their age; contact details; type of respondent they were like patient, caretaker, or a family member; sources of information about dementia; sources to know about the present resource book; type of problems faced by the respondents; and few other questions to get the basic information about the respondents. The section includes open-ended and multiple-choice questions.

Section 2: Feedback About the Resource Book – This section housed questions related to the feedback for the resource book to understand the usefulness of the content in the situation they are in. Did they change their diet pattern and view videos and films to know further details about dementia? Was the presentation of the information with visuals and colors interesting and easy to comprehend? All the questions were to know the worthiness of designing such a resource book. The section included both checklist and multiple-choice questions.

Section 3: Suggestions – This section contains questions to gather their suggestions for creating awareness about dementia, suggestions of adding more information to the resource book, and whether the information provided was latest and updated. Open-ended- and multiple-choice-based questions were framed to seek the responses.

61.2.3.2 Procedure of Data Collection

The team distributed the reaction scale along with the hard and soft copy, and the returning date of the reaction scale was mentioned on the covering letter. Beneficiaries were given enough time to read the book and follow the links for videos and films to watch. Twenty of them sent the filled-in reaction scale on the date mentioned. The team analyzed the data based on the responses provided by them. The contents of the emails received made it very clear that the resource book was referred by dementia patients, family members/caregivers, senior citizens, and young people who wanted to be aware of the disease. The senior citizen clubs went ahead one step and requested for a copy to keep it in their library for its members for future reference.

61.2.3.3 Reactions/Responses of the Readers

The team noted that the 20 respondents who filled in the reaction scale were among these three: the family members of the patients, the dementia patients themselves, and their caretakers; however, majority of them were the family members of the dementia patients. The team went through the reactions/feedbacks thoroughly and charted out the below observations.

- Majority (60%) of the patients were suffering from dementia since last 1–2 years, little more than one-tenth percent (13.3%) of the patients were suffering from dementia since last 3–4 years, whereas, very less (6.6%) of the patients were suffering from dementia since last 10 years
- Little less than a half percent of the respondents got to know about the disease dementia through newspaper/magazine, while 35% of the respondents got the information from doctors and one-tenth (10%) of the respondents got the information through neighbors and friends.
- A very high majority (90%) of the respondents got to know about the "resource book on dementia" through the newspaper and one-tenth of the respondents by word of mouth.
- The majority (65%) of the respondents were interested in having the resource book for their own self/family member/friend/relative who had the symptoms of

forgetfulness, and also they were interested in acquiring additional knowledge to whatever they already knew about dementia. While many of them were interested in having the resource book since they were aging and felt the necessity to self-educate themselves about it, 40% of the respondents were interested in having the resource book because they can be more skilled and effective in treating the patient. Few reported that the cause is widespread; hence, they want to help the society by being a well-informed citizen.

- The following are what the respondents say about the type of problems they face while dealing with the dementia patients:
  – Unable to understand what she is going through and does not express anything.
  – Does not want to talk, sulks down when walking, forgets taking medicine, and forgets that he had meals.
  – Trying to make the patient remember one thing with many trials and lots of efforts and then still when the result is not fruitful, it is the main problem
  – Forgets road directions at night, date, days, names, things, the name of people, childhood events, etc.
  – She keeps on repeating same sentences again and again and forgets what she spoke for 5 min.
  – She does not understand the daily routine.
  – She remains absent-minded, doesn't do any work (we force her to eat), does not remember most of the things from past, keeps crying, cannot control her urine, etc. She was missing from home for 24 h; we had to find her through the newspaper.
  – I have trouble in remembering events that happened in the past.
  – My mother is very hyper at times, to the event that she hits the caretaker.
- A high majority (70%) of the respondents found that the information about dementia (signs and symptoms, causes, prevention, treatment), the chapter on the role of caretaker, were the most useful details for them in the resource book.
- A very high majority (75%) of the respondents found the list of books and poem useful, while 70% of the respondents found information related to podcast useful.
- The majority (65%) of the respondents found self-check test most useful, and equal percentage found information about organizations useful, whereas 60% of the respondents found games for dementia patients and mobile apps useful. More than half percent (55%) of the respondents found the name of doctors and helpline numbers most useful, whereas the same percentage of the respondents found nutritional tips useful.
- Less than half percent (45%) of the respondents made a change in the diet of patient's food, whereas thirty percent (30%) didn't make any change in the diet of patient's food after reading the resource book.
- Half of the respondents had viewed the video (suggested links in the resource book).

- A high majority (80%) of the respondents could relate their own experiences with real-life stories and experiences mentioned in the resource book
- A very high majority (90%) of the respondents liked the language used (simple to understand for amateur); high majority (80%) of the respondents liked the colorful presentation and list (doctors, organizations, helpline numbers) provided in the resource book, whereas majority (75%) of the respondents liked the written matter.
- All the respondents agreed that the pictures were relevant to the content and size of the fonts was appropriate. And all the respondents were interested in reading the resource book on dementia
- A high majority of the respondents felt that the training program should be organized for caretakers and family members of dementia patients. More than half percent of the respondents felt that lectures should be organized, helpline number on 24-h basis should be started, and 35% of the respondents felt that special geriatric wards should be opened.
- All the respondents felt updated about dementia-related information after going through this resource book.

Details on some of the topics which were highly appreciated by the readers such as tips for family members and the caretakers, videos for better understanding and caretaking, and podcasts for the engagement of dementia patients are depicted in the below snapshots (Figs. 7, 8, and 9) from the resource book.

61.2.3.4 What Did the Readers Suggest?

- Interviews with the patients' relatives should be recorded.
- The resource book should be put on a website (If no website exists, create one).
- Finding out primary causes of dementia, e.g., due to tension, and how can it be detected and cured at an early stage.
- A separate chapter on yoga and light exercise for keeping fit physically should be added to the resource book.

61.2.3.5 Suggestions for Future Work in Promoting Awareness About Dementia

- Seminars should be conducted on dementia for family members, caregivers, and patients suffering at an early stage.
- Helpline numbers of dementia care centers should be publicized through media.
- Lectures should be organized along with the resource book and other materials helpful in coping with different situations that occur in dealing with the patients.
- More resource books with different ideas or any other references related to dementia should be published to help the patients and their caretakers.
- The resource book or any other reference made should be shared on the website. If no website is available, then one should be created, and required details on dementia should be housed on that website.

Tips for Family and Caretaker

It is helpful to talk to the person with dementia about what they enjoy. Take clues from them and try to find creative ways to adapt activities, focusing on what can be done.

Try not to worry about getting things wrong first time; this can lead to finding the right activity. The focus of the activity should be on whether someone is enjoying it and that it has meaning for them, not the 'result' of the activity itself. The following suggestions may be helpful

Conversation

Conversation is a good example of a simple activity that is meaningful and beneficial for a person with dementia. It can be a good way for younger family members to engage with the person with dementia.

This type of activity can have a positive impact on the wellbeing of the person with dementia. Even if the person with is having difficulties with verbal communication, non-verbal communication (eye contact, gestures and touching) can be just as meaningful. The important thing is to have a connection through the social interaction.

It is important to involve the person with dementia in the conversation, not cutting them off or talking to others as if the person is not there. Do not assume that someone cannot contribute to a conversation just because they have dementia.

Some ideas for aiding conversation can include using different prompts for conversation such as a past job or a favorite sports team, reading a newspaper or magazine together, or using technology such as online videos of old TV shows or events.

Fig. 7 Tips for family and caretaker

VIDEOS

Video in Gujarati - Prabhakant and Kanta Patel – Dementia

This is a story of Chanchalben Pate, she was diagnosed with Alzheimer's disease in 1997. Prabhakant, her son, and his wife became her sole careers.

Some family members felt ashamed and would not acknowledge that she was sick.

After she died the couple were ostracised by their community for being so open about the diagnosis. She died in 2004.

http://www.youtube.com/watch?v=qksjqrdllLQ

Video in Gujarati - Stan and Denise Lintern – Dementia

This is a story in Gujurati of a Family man Stan Lintern was diagnosed with Alzheimer's disease in 1996 after a long career in the NHS.

Stan was also a talented premier league table tennis player but, as the dementia took hold, his life changed dramatically.

Now Stan is unable to walk and doesn't speak. However, he is much loved and surrounded by a large and loving family.

http://www.youtube.com/watch?v=FO5_eFMJ4hQ

Video in English-National Dementia helpline

This film is to hear about Alzheimer's Society's National Dementia Helpline, including how to access it, what we can help you with and what to expect when you call.

http://www.youtube.com/watch?v=ZPslzrlk0wM.

Fig. 8 Videos

Podcasts

Meaning of Podcast

A digital audio or video file or recording, usually part of a themed series, that can be downloaded from a website to a media player or computer

Podcast in Gujarati-

Bilingualism a boon and a curse in Dementia

This podcast suggest speaking a second language delays Alzheimer's disease, but what happens when an aging Dementia patient forgets English and reverts back to their mother tongue.

http://www.podcasts.com/gujarati/episode/bilingualism_a_boon_and_a_curse_indementia

Podcast in English-

The progression of dementia - Alzheimer's Society podcast December 2014

This podcast describes about the word dementia describes a set of symptoms that may include memory loss and difficulties with thinking, problem-solving or language. Dementia is caused when the brain is damaged by diseases, such as Alzheimer's disease or a series of strokes.

https://www.youtube.com/watch?v=CuXY33nkes&list=UU1miDO27ShatLOE4nWARYCg

Podcast in English-

Dementia and personal care: Incontinence - Alzheimer's Society Podcast August 2014

Difficulties with using the toilet, accidents and incontinence can all be problems for people with dementia, particularly as the condition progresses. These problems can be upsetting for the person and for

Fig. 9 Podcast

61.3 Conclusion

The idea of this kind of initiatives is to augment the capacities of caretakers/family members of dementia patients by designing and distributing the resource book among them so that they understand the patients better with utmost care and concern. The resource book covers a large amount of information on dementia as a disease, facilities and services available for them in the city as well as at the national level for accessing it during critical situations, list of games for cognitive stimulation, and mobile apps for ease in accessing information and help anytime. Details such as list of books, films, videos, podcasts, and real stories and experiences of dementia patients were also covered in this book for its readers. As the team adopted the strategy of analyzing the needs and characteristics of the targeted group, it turned out in designing the content appealing to the nonmedical readers. Simple and effective messages through quotes and informative cartoons helped in passing on the thoughts effectively.

While designing the resource material for a targeted group, it is always a best idea to give personal visits to the families and the caretakers to observe and understand what they really need and how can they be helped and guided accordingly with the much needed information through the resource material and the team working on the above-discussed project followed the same strategy, this strategy helped tremendously to design an apt resource book for the family members and the caretakers.

The team used a varied set of communication media to achieve the objectives of reaching out to the intended target group and to pass on the information effectively. The newspaper article and the leaflets publicized the resource book among the residents of Vadodara City and the State of Gujarat. The medium of Internet delivered the soft copies of the resource book to the family members and the caretakers irrespective of how far they resided. The team shared the email IDs through the media for effective and easy reception of the soft copies or the reaction scale for the interested parties.

The sponsorship received from the philanthropists for printing of the resource book was of great help as it helped in reaching out to the targeted group with the desired content in the right form. The CSR-based NGOs or voluntary organizations working in the sector always financially support for these kinds of projects which involve creating of awareness among the masses for the larger benefit of the society.

To cover it in a nutshell, this resource book is a big win and is highly appreciated among the researchers from gerontology discipline and development communicators working for bringing social and behavioral change in the society – for the approach adopted in designing the content, the communication strategy for distribution of the book, and on the whole as the end product as it is very feasible and completely apt for the targeted group to refer to it anytime and anywhere with ease.

Apart from the researchers, the team also received high appreciations from the readers of the targeted group through emails (inbox was flooded) for designing such a tremendously informative book. The feedbacks clearly indicated the effectiveness and worthiness of the resource book. The resource book gave a broader picture to many needy senior citizens, family members, and also to the youngsters as they got to learn about the issues and problems faced by dementia-affected patients and their family members/caretakers, which indeed is a positive sign for prevention of this disease in future. It would be really great to adopt such approaches and is highly recommended to design such communication strategies for disseminating information to masses about newly identified diseases like Ebola, swine flu, dengue, and chikungunya for creating awareness in them and capacity building among family members and caretakers for a healthy world.

61.4 Cross-References

▶ A Threefold Approach for Enabling Social Change: Communication as Context for Interaction, Uneven Development, and Recognition
▶ Empowerment as Development: An Outline of an Analytical Concept for the Study of ICTs in the Global South
▶ Family and Communities in Guatemala Participate to Achieve Educational Quality
▶ Health Communication: A Discussion of North American and European Views on Sustainable Health in the Digital Age
▶ Health Communication: Approaches, Strategies, and Ways to Sustainability on Health or Health for All

References

Brock W (2014) Social security: the qualitative dimension. http://hemispherespublishing.com/social-security-qualitative/. Accessed 8 Oct 2014

Das et al (2012) Dementia: Indian scenario. Neurol India 00283886. http://www.neurologyindia.com/article.asp?issn=0028-3886;year=2012;volume=60;issue=6;spage=618;epage=624;aulast=Das

Further Reading

Clasper et al (2013) What can social media offer dementia care practitioners? J Dement Care. http://www.careinfo.org/wp-content/uploads/2013/09/21.1.10-11.pdf

Gorman (2000) Defining old age. World Health Organization. http://www.who.int/healthinfo/survey/ageingdefnolder/en/

Kumar et al (2015) Dementia friendly Kerala-the way forward. Kerala J Psychiatry 2395–1486. http://kjponline.com/index.php/kjp/article/view/8/pdf_4

Raina SK, Raina S, Chander V, Grover A, Singh S, Bhardwaj A (2013) Identifying risk for dementia across populations: a study on the prevalence of dementia in tribal elderly population of Himalayan region in Northern India. Ann Indian Acad Neurol 16:640–644. [serial online] 2013 [cited 2017 Dec 31]. Available from: http://www.annalsofian.org/text.asp?2013/16/4/640/120494

Shaji et al (2010) The Dementia India report – prevalence, impact, cost and services for Dementia, Alzheimer's and related disorders society of India. http://ardsi.org/downloads/main%20report.pdf

Vas CJ et al (2001) Prevalence of dementia in an urban Indian population. 13(4):43950. http://www.ncbi.nlm.nih.gov/pubmed/12003250

Webliography

http://ardsi.org/downloads/main%20report.pdf
http://en.wikipedia.org/wiki/Old_age
http://www.apa.org/pi/aging/resources/guides/older.aspx
http://www.careinfo.org/wp-content/uploads/2013/09/21.1.10-11.pdf
http://www.helpguide.org/articles/alzheimers-dementia/alzheimers-and-dementia-prevention.htm
http://www.neurologyindia.com/article.asp?issn=0028-3886;year=2012;volume=60;issue=6;spage=618;epage=624;aulast=Das
http://www.webmd.com/healthy-aging/tc/healthy-aging-normal-aging
http://www.who.int/healthinfo/survey/ageingdefnolder/en/
http://www.who.int/mediacentre/factsheets/fs362/en/#Factsheets
https://dementiacarenotes.in/resources/india
https://goo.gl/329PY1
https://sbccimplementationkits.org/courses/designing-a-social-and-behavior-change-communication-strategy/
https://www.alzheimers.org.uk/info/20007/types_of_dementia/1/what_is_dementia/3
https://www.scribd.com/document/190077600/WHO-Definition-of-an-Older-or-Elderly-Person
https://www.thehealthcompass.org/sites/default/files/strengthening_tools/Field-Guide-to-Designing-Health-Comm-Strategy.pdf
https://www.unicef.org/cbsc/files/Writing_a_Comm_Strategy_for_Dev_Progs.pdf
www.hamlet-trust.org.uk/articles/dementia.html

Strategic Communication to Counter Sexual Harassment in Bangladesh

62

Nova Ahmed

Contents

Abstract

Sexual harassment is a major problem that is not openly discussed in countries like Bangladesh. Women are the major victims of this problem who hesitate to share the problem. Strategies to discuss about sexual harassment are explored here. The initial conversation was over online survey where silence was observed. It followed a series of methods using focus group discussions, individual discussions, and sharing experience over writing or over phone calls. Indirection during the communication method helped the participants (female) to discuss with certain comfort level. Once there is a level of trust, indirection is no longer required. The journey of working on sexual harassment and how various communication modes helped in generating a productive conversation is presented in this chapter.

Keywords

Eve-teasing · Sexual harassment

N. Ahmed (✉)
North South University, Dhaka, Bangladesh
e-mail: nova.ahmed@northsouth.edu

J. Servaes (ed.), *Handbook of Communication for Development and Social Change*,
https://doi.org/10.1007/978-981-15-2014-3_2

62.1 Introduction

Sexual harassment is a serious concern in Bangladesh (Ahmed et al. 2014; Ahmed 2016). The problem exists in regions across geographic and economic boundaries (The Financial Express 2012; Dimond et al. 2011; Lahsaeizadeh and Yousefinejad 2012; Li and Lee-Wong 2005; Wikipedia 2015; Action Aid 2015; United Nations Committee 2016). However, there is an ongoing *silence* when efforts are initiated to discuss about this problem (Ahmed 2016).

Currently, the third wave of feminism (Bardzell 2010) is ongoing, where gender inclusion is considered. Problems such as sexual harassment, where women are historically the major victims, take us decades back where basic rights are questioned. The problems exacerbate when it takes place in developing regions where gender disparities are widely visible.

Even in this modern world, there are victims like the medical student at Delhi (BBC 2015) who was brutally sexually abused which resulted in her death; there are girls like Sheemi, Fahima, Mahima, and Pinky (Wikipedia 2015) who committed suicide after being harassed and publicly insulted. It is a worldwide phenomenon as we can see it occurring in Iran (Lahsaeizadeh and Yousefinejad 2012), Singapore (Li and Lee-Wong 2005), the USA (Dimond et al. 2011), Canada (Macmillan et al. 2000), and many other countries. The problem takes a severe form in Indian subcontinent (Wikipedia 2015). The form of street harassment is referred as "eve teasing" which is banned currently after High Court ruling under the protest from aware citizen in Bangladesh.

The current discussion focuses on Bangladesh where females committed suicides, lost interest in going to school, and forced to early marriage in rural areas as well as in urban areas (Action Aid 2015) as a result of sexual harassment. Contemporary researchers and practitioners have addressed the problem of sexual harassment in physical domain along with cyber domain. There are global efforts to ensure social justice using local and global networks where social cultural context plays a role in adopting readily available solution approaches.

Communication regarding the existence, impact, and probable solution approach (e.g., using simple technology) was an initial goal of researchers of Bangladesh. However, the initial efforts to communicate about sexual harassment turned out to be extremely challenging as the researchers faced silence (Ahmed 2016) (Robertson et al. 2017) regarding this problem. There were series of communication patterns and strategies applied to reach out urban young adults to discuss about sexual harassment in physical world as well as cyberspace. A high-level look at the country gives us a positive picture on women empowerment and possible solution approaches. It is a developing country ruled by women since 1991. The country has major female work force in garment industry. However, women face challenges that come from patriarchal values and practices embedded in social cultural structure. On top of that, suppression is often practiced using the name of religion which is hardly contested by the Government or other authorities. Sexual harassment exists, getting more challenging over years as abusive contents are easily shared using platform of the internet. Weak law enforcement system in the country is another major problem that is prevalent. The current problem requires awareness, and discussion in the first place.

The ongoing journey of working on sexual harassment has opened up strategies, where user indirection and direction plays a role, and is explored. The entire process was based on the concept of trust which required time and patience to deal with. We share the findings from the communication strategies as follows: When the person is not visible, trust is an issue, considering online platform-based communication; in a group, certain voices dominate, Female-only focused group discussions, sharing is possible with certain level of indirection: Sharing experiences in writing and over phone call conversations; personal conversations did not work out in certain cases, one-to-one conversations or diary notes that did not turn out to have the desired outcome; and trust buildup the over years can bring one-to-one conversation, unexpected story sharing that takes place once the trust is built up on the researcher.

62.2 Related Work

The current work focuses on the complex relationship of urban girls in the context of a conservative setup. The question regarding revolves around how women perceive technology to handle sexual harassment. It is evident that this work needs to address the domain of work where our work is related to *feminist HCI*; within feminist HCI we talk about the perception of *fear* and *hope toward technology*, we discuss the *global and local status of sexual harassment* and *cross-cultural effects* in dealing with this topic from psychological perspective, and we justify our *methodology* used in this work. This discussion shows how our work relates and differs from existing research works.

Shaowen Bardzell (2010) lays down category of work referred to *Feminist HCI*, where economic, political, social, and psychological oppression of women is discussed. Currently the third wave of feminism (Bardzell 2010) is ongoing where inclusion is the main focus. However, there are primitive concerns regarding ensuring basic rights in economically backward countries. One example is presented in the works of Oyeronke Oyewumi (2002) in the context of Africa. Similarly Ishtiaque et al. (Ahmed et al. 2014) discusses issues in the context of Bangladesh.

The story of women and technology is not new which impacts the solution approach toward harassment indirectly; it has been present all along. Mary Wyer et al. (2013) look at how gender plays a role in the context of science and technology-related fields. Donna Haraway (2006) illustrates the presence of women in various work fields and its effect in the fast-changing world. Sherry Turkle (1988) goes back to possible reasons behind gender bias in technology-related fields, which is often caused by admonishment. We encounter this scenario when gender differences are coming from family as mentioned by Oyeronke Oyewumi (2002).

Recent work presented by Jill P. Dimond et al. (2011) illustrates other factors that create technology-divide among women caused by using technology to abuse vulnerability of women. Either this weakness comes from the existing differences of women spending less time with technology compared to their male counterparts or it comes from the emotional instability after some form of harassment. This scenario is present in Bangladesh as well, where we find technology used to harass college going girls who are considered to be advanced technology users of their generation. It must be mentioned that the fear is present among elderly, women, and black community living alone, compared to others (Schweitzer et al. 1999), which refers back to the social impact and the feelings of being vulnerable.

The story of women and technology is not only about fear; that would be a misinterpretation of facts. There is the other side of the story where technology is providing hope and is being used to build up connectivity and increase confidence. Technology can be a great way to start a new beginning after a major incident of harassment as we can see in the work of Michael Massimi et al. where people are slowly going back to their normal life using technology as their subtle friend (Massimi et al. 2012). A similar tone of coming back story is presented by Rachel Clarke et al. using a photo sharing application which supports victims of domestic violence (Clarke et al. 2013, April). Hollaback! (Dimond et al. 2013) is used to create a sense of togetherness as mentioned by Jill P. Dimond et al., which is an ongoing international effort and has become integral part of feminist movement. The feeling of being connected provides a sense of safety, and technology fits right in its place – it can be a website, anonymous storytelling platform, or a simple device like a mobile phone to provide connectivity. Jan Blom et al. (2010) show this situation among urban users in India using mobile phones. A wearable technology is proposed to reduce fear of crime among elderly people in an English city as mentioned by Mark A. Blythe et al. (2004) to reduce the fear of crime. The role of technology can play a role in coping after women face harassment which can be of the form of advocacy seeking or social coping as mentioned by Lilia M. Cortina et al. (2005) in describing the scenario of coping over social cultural boundaries.

One of the major challenges faced was about the communication methodology as we were not able to find a particular way in which users were comfortable to discuss such a sensitive topic. Existing work was used as reference models to communicate including the work conducted in the USA by Jill P. Dimond et al. to get familiar with users and talk to them; a similarly work discussed by Macmillan et al. (Li and Lee-Wong 2005) in the context of Singapore, researchers utilize use interview-based methods. We have used mixed method research designs to meet the user's comfort level inspired by the work presented by William E. Hanson et al. (2005). There were cases found at the beginning where participants were trying to give us input to meet the research expectations, a phenomenon mentioned by Barry Brown et al. (2011, May). There was a requirement to accommodate great flexibility in current methodology which could be altered as the research demanded.

In the context of Bangladesh: Young male adults gathering around the tea stalls and chatting for a while has been socially accepted and is present in our literature and media (Wikipedia Humayun Ahmed 2015). It became a social issue since the group of boys started to make fun of girls passing by often referred to as "eve-teasing" (Wikipedia 2015). The problem remained unnoticed until several teenage school going girls committed suicide in recent years – as a result of humiliation. Pinky committed suicide after being publicly harassed where no one around her protested. In a study conducted by actionaid, it is observed that 73% of the parents believe that it is better to have their children married rather than facing harassment problems, it is a major problem believed by 79.8% parents and they are scared of it 84.5% times, the fear factor is 97.4% for the girls themselves (Action Aid 2015). Often religious views are used to suppress and blame females as can be learned from Islamic talks by religious leaders (The Daily Star 2015b, c, d; UbConnect 2015; Daily Tribune 2015a).

The picture is similarly depressing in the Indian Subcontinent which includes girls from India and Pakistan. A medical student from Delhi was forced to brutal death after sexual harassment (BBC 2015; HRCBMDFW 2015). The global scenario is not better. Females from countries with higher economic growth like USA (Dimond et al. 2011), Canada (Macmillan et al. 2000), Singapore (Li and Lee-Wong 2005) or Iran (Lahsaeizadeh and Yousefinejad 2012) face harassment of similar magnitude.

It is surprising to find out the limited number of research work conducted on cross cultural issues regarding sexual harassment. Macmillan et al. shows cultural differences in Singapore (Li and Lee-Wong 2005) and Lengnick-Hall et al. consider various study methodologies in such cases. The current work shares discussion methods that worked in the context of urban female youth to discuss about sexual harassment of Bangladesh.

62.3 Context of Bangladesh

The existing scenario of Bangladesh regarding sexual harassment and corresponding law enforcement support will help the readers to understand the reason behind ongoing silence among urban young female adults. Urban females in various undergraduate and graduate institutes are enlightened, educated and aware about their rights. They are privileged in receiving exposure to information and support, compared to suburban or rural women.

An anonymous law enforcement officer, who is very active in the online community and connected to young adults, has shared the experience of getting many personal messages from women regarding their problems. He mentioned the system of enforcing a complaint to be placed in person is a major limitation why women do not actively seek support from law enforcement agency. It is often the case where a woman is harassed the second time when she is trying to file a complaint. In many cases our society blames the women as the cause of harassment.

A recent incident of women being harassed publicly during the Bengali New Year celebration (14 April 2015) in the premise of University of Dhaka revealed the vulnerability of women in festive celebrations. There was group of abusers who took advantage of the crowd and physically abused several women. The law enforcement officers were not around to handle the situation (Dhaka Tribune 2015b; The Daily Star 2015a; BDNews24 2015). There were evidences of the abusive behavior on CC-TV footage but not a single complaint was filed. There have been discussions on the media heavily regarding this subject where a group of people pointed fingers to the victims bringing in religion, and social cultural norms to indicate it was the responsibility of the women not the abusers.

The background illustrates that the roads are harsh even in urban Bangladesh for women and silence is used to avoid the turbulence. Our group discussion revealed that women are interested to seek help from friends rather than family members in the fear of losing their independence to move alone.

There must be major changes in the policy level to support women in all aspects in Bangladesh to break the silence.

62.4 Approaching Young Adults

The study was focused on young adults using qualitative studies where group or individuals shared detailed experiences with the researcher. Majority of our participants were students in various undergraduate and graduate institutes. Participants were contacted through emails, social media calls and poster setup on campus compounds. The author (female) was assigned as a contact person as preferred by some women who were reached out informally. Key points were drawn from various discussion forums that took place during the time period of 2012–2017. All the discussions took place in Bengali, and the participation was voluntary without any monetary incentive. The interviews were noted down, recorded only if participants agreed to audio recording. The interviews were conducted, noted, and transcribed by the researcher herself. This study considered various forms of discussion methods ranging from online survey to focus group discussions to personal interviews.

The study initiated with an *online survey*. The online interview was placed for a week receiving 121 responses (79 female).

There were *focus group discussions* in a semi structure format which took place in various locations suggested by participants. There were less than 10 participants in each focus group who were known to each other. It was often a circle of friends who scheduled the appointment with the researcher. We had total seven focus group discussions – four in the 2012 and the rest took place later in time.

The *one to one discussions* continued as individuals would set up schedule at their convenient time and place to meet the researcher. Individual discussions continued to be easier as the researcher initiated discussions, awareness campaigns regarding the problem of sexual harassment over time.

The variation in interview time frame requires clarification. Our work on sexual harassment was an incremental work initiated in 2012. There were two major sexual harassment events in the country within the specified timeframe. Each of the incidents initiated discussions and rethinking over the problem and probable solution approach.

62.5 Experiences Regarding Communication Strategies

Discussion about sexual harassment was challenging in the context of Bangladesh. There was various communication methods tried out to find out the actual picture of the problem in urban setup.

Missing Trust on Online Platform: The initial online survey was filled up fast by many participants (total 121) within a short time. However, many of the participants left the questions relating to sexual harassment *blank*. The questionnaire allowed partial responses considering the unconventional topic of sexual harassment. Majority of online participants mentioned about not facing sexual harassment at all in the survey responses. The researchers were convinced that sexual harassment was not a serious problem in urban setup from the survey data analysis.

Later, in informal conversations, participants, who followed up with the researcher, shared their fear of being exposed to the wrong set of people. It must be noted that the survey was anonymous. Some of the participants also suggested a female-only discussion group as a preferred way to communicate. This requirement shows how the audience is important, female personnel is preferred to discuss about sexual harassment.

Emotion in Focus Groups: The female-only focus group worked best in terms of sharing personal experiences, feelings and more. The focus group setup initiates with some reservation which takes spontaneous form as the conversation continues. The initial 1 h long focus group discussion (we conducted 4 of them), ended up having couple of hour long discussions. At the end of the discussion, participants offered the researcher food, snacks showing genuine feelings of bonding. Many of the participants are still connected with the researcher.

The group format and spontaneous discussion were a great strategy while there are two notable observations. *First*, the shy and introvert participants are often not addressed in a group setting. In many cases, these participants would agree or disagree with others without providing clear personal opinion. *Second*, emotional outburst of one or more can influence the entire group. A group of participants joined one discussion that had a close friend, sexually harassed by a male teacher in a girls school. The incident was notified to the school authority who denied all the allegations. It took the protest of students and parents which was strongly supported by media and general people to get rid of the teacher. The participants were actively involved in the protest against this teacher.

The discussion turned out to be negative as some of the participants were angry and emotional about the incident.

It was evident that the focus groups are greatly influenced by the personalities and emotional states of participants. It worked well when the participants were content and calm.

Limitations of Direct Communication: One-to-one communication appeared as a next possible way to reach out individuals. However, the reality was different. Participants met the researcher in person to mention that they were interested in a more indirect way of communication such as in written form. There was interest and eagerness to share the experiences regarding sexual harassment but there was a challenge of making a direct contact to share it.

Direct communication was not the preferred way of discussing about sexual harassment in the current context.

Comfort in Indirection: Indirect communication worked great as a way to share individual experiences. There were experiences sharing in written form as well as through phone calls. The indirect way of communication was not originally planned as a communication method during our study design.

It must be noted that, in most of the cases, the indirect method contained extremely personal experiences, personal abuse where the indirection helped. The write-ups were dropped at the researcher's office or personal space when she was away. Similarly, the phone calls were made at late nights in many cases. The timing also played an important role in the indirection.

Comfort in Direct Communication: There were voluntary participants who would come forward to share their experiences knowing about the author's work on sexual harassment. This group approached the researcher over time as they came to know about her work and had built trust in her.

The individual volunteers were from diverse backgrounds and age. In this case, one-to-one conversations regarding intimate abusive experiences were shared. This shows the importance and role of a trusted companion.

Bonding Through Online Platform: The incident of harassment during Bengali New Year was greatly discussed in media and generated public outrage. A secret Facebook group was formed to provide support for women in Bangladesh. The team contained male and female across the globe who are originally from Bangladesh. The team entire initiative to generate a tech solution along with generating awareness was conducted by the author; this group became dormant after the initial emotion subsided. The initiative showed how social media platform can play a positive role in contributing toward social justice.

Online Support Groups: An online platform was observed over a period of 2 months that provides support against crime. Interestingly, there are continuous open posts regarding online abuse of women. The posts are provided by male participants mentioning about the problem. The voluntary organization has successfully supported many cases and has received the trust of a certain community (that has access to online resources).

Diary Study that was Misunderstood: A diary study was conducted which was viewed as a viable approach intuitively. We asked individual female university students to keep a diary for 3 weeks and note down any experience she has faced during this time period. We asked three students initially who were contacted through other student volunteers. Two of the participants returned the diary before the scheduled time frame while one continued to use the diary. However, one particular participant wrote down imaginary harassment stories (probably inspired by movies) where a friend (male) came when 6/7 attackers were trying to harass her. She later mentioned, in a follow up discussion, that she was thinking of an imaginary scenario what could possibly take place. This study was not continued further considering the misunderstanding it initiated.

The exploration of different communication methods gave us insights on how various mechanisms are perceived and trusted to discuss a topic such as sexual harassment.

62.6 Lessons Learned

Current experiences of working on the problem of sexual harassment have revealed the fact that discussion is challenging in the context of Bangladesh considering the social, cultural and personal boundaries. Initially, silence was a major barrier in reaching out to urban young adults; which slowly mitigated over time. The participants provided us with valuable insights regarding their preferred way and pace of communication. Trust is the key element in building up the conversation regardless of the communication mode.

Indirection during the communication played a vital role in initiating a conversation. *Absolute indirection*, noticed during online communication was not preferred among many participants. However, indirection where communication is taking place with a known person provides great ways for communication. Female-only focus group discussions were spontaneous where the group participants created a level of indirection where *individuals are identified as a group* rather than a person. Similar responses were received during written note sharing or phone calls where the participant is aware of whom the conversation is taking place with, avoiding a personal contact during sharing of experiences. Individual personal contacts and diary studies did not work at all or did not work as expected as the participants were not very comfortable of sharing the idea in a face-to-face setup.

The social-media based awareness along with generic online and print media has played a great role in opening up fluent communication and awareness regarding sexual harassment. Trust is built up through the sincere voluntary effort and in ways to reach out for further support. Silence is broken among the aware community.

62.7 Conclusion

There is no silver bullet communication method when it comes to the topic of sexual harassment in the context of Bangladesh. The notion of building up trust over years has been the key used to break the silence regarding this topic. The indirection during communication often helped to build up trust. Indirection came in the form of communication medium, time, presence of others and distance. Once the sense of trust was created, there was a spontaneous process in place.

References

ActionAid (2015) Lingo bhittik shohingshota, Shuchi Karim, actionaid report

Ahmed N (2016, May) Discussing about sexual harassment (breaking silence): the role of technology. In: Proceedings of the 2016 CHI \conference extended abstracts on human factors in computing systems, ACM, pp 459–472

Ahmed SI, Jackson SJ, Ahmed N, Ferdous HS, Rifat MR, Rizvi ASM, ... Mansur RS (2014, April) Protibadi: a platform for fighting sexual harassment in urban bangladesh. In: Proceedings of the 32nd annual ACM conference on human factors in computing systems, ACM, pp 2695–2704

Bardzell S (2010) Feminist HCI: taking stock and outlining an agenda for design. In: Proceedings of the SIGCHI conference on human factors in computing systems, ACM, pp 1301–1310

BBC (2015) Four convicted for Delhi gang rape. http://www.bbc.co.uk/news/world-asia-india-24028767

BDNews24 (2015) http://bdnews24.com/bangladesh/2015/04/16/womens-platform-protests-harassment-at-dhaka-university-during-new-year-celebrations. Last accessed on May 2015

Blom J, Viswanathan D, Spasojevic M, Go J, Acharya K, Ahonius, R (2010, April) Fear and the city: role of mobile services in harnessing safety and security in urban use contexts. In:

Proceedings of the SIGCHI conference on human factors in computing systems, ACM, pp 1841–1850
Blythe MA, Wright PC, Monk AF (2004) Little brother: could and should wearable computing technologies be applied to reducing older people's fear of crime? Pers Ubiquit Comput 8(6):402–415
Brown B, Reeves S, Sherwood S (2011, May) Into the wild: challenges and opportunities for field trial methods. In: Proceedings of the SIGCHI conference on human factors in computing systems, ACM, pp 1657–1666
Clarke R, Wright P, Balaam M, McCarthy J (2013, April) Digital portraits: photo-sharing after domestic violence. In: Proceedings of the SIGCHI conference on human factors in computing systems, ACM, pp 2517–2526
Cortina LM, Wasti SA (2005) Profiles in coping: responses to sexual harassment across persons, organizations, and cultures. J Appl Psychol 90(1):182
Dhaka Tribune (2015a) Online edition, women's right activists denounce Shafi's statement. http://www.dhakatribune.com/bangladesh/2013/jul/12/women%E2%80%99s-right-activists-denounce-shafi%E2%80%99s-statement
Dhaka Tribune (2015b) http://www.dhakatribune.com/bangladesh/2015/apr/15/women-sexually-harassed-du-campus. Last accessed on May 2015
Dimond JP, Fiesler C, Bruckman AS (2011) Domestic violence and information communication technologies. Interact Comput 23(5):413–421
Dimond JP, Dye M, LaRose D, Bruckman AS (2013, February) Hollaback!: the role of storytelling online in a social movement organization. In: Proceedings of the 2013 conference on computer supported cooperative work, ACM, pp 477–490
Hanson WE, Creswell JW, Clark VLP, Petska KS, Creswell JD (2005) Mixed methods research designs in counseling psychology. J Couns Psychol 52(2):224
Haraway D (2006) A cyborg manifesto: science, technology, and socialist-feminism in the late 20th century. The international handbook of virtual learning environments, Springer, Dordrecht, pp 117–158
HRCBMDFW (2015) Mahima rapists remanded, autopsy report confirms suicide after gangrape. http://hrcbmdfw.org/forums/post/163.aspx
Lahsaeizadeh A, Yousefinejad E (2012) Social aspects of women's experiences of sexual harassment in public places in Iran. Sexuality & Culture 16(1):17–37
Li S, Lee-Wong SM (2005) A study on Singaporeans' perceptions of sexual harassment from a cross-cultural perspective. J Appl Soc Psychol 35(4):699–717
Macmillan R, Nierobisz A, Welsh S (2000) Experiencing the streets: harassment and perceptions of safety among women. J Res Crime Delinq 37:306–322
Massimi M, Dimond JP, Le Dantec CA (2012, February) Finding a new normal: the role of technology in life disruptions. In: Proceedings of the acm 2012 conference on computer supported cooperative work, ACM, pp 719–728
Oyewumi O (2002) Conceptualizing gender: the eurocentric foundations of feminist concepts and the challenge of African epistemologies. Jenda: A Journal of Culture and African Women Studies 2(1):1–9
Robertson SP, Andalibi N, Diakopoulos N, Forte A, Maruyama M, Ahmed N, ... Hartmann B (2017, May) CHI 2017 Stories Overview. In: Proceedings of the 2017 CHI conference extended abstracts on human factors in computing systems, ACM, pp 14–18
Schweitzer JH, June WK, Juliette RM (1999) The impact of the built environment on crime and fear of crime in urban neighborhoods. Journal of urban technology 6(3):59–73
The Daily Star (2015a) http://www.thedailystar.net/frontpage/outrage-over-sex-assault-77496. Last accessed on May 2015
The Daily Star (2015b) Online edition, PM blasts Hefajat chief. http://www.thedailystar.net/beta2/news/125852/
The Daily Star (2015c) Online edition, PM blasts Hefajat chief. http://www.thedailystar.net/beta2/news/125852/

The Daily Star (2015d) Online edition, Sermon shafi style. http://www.thedailystar.net/beta2/news/sermon-shafi-style/

The Financial Express (2012) Prevent eve-teasing and suicide, Rumana Sharmeen, University of Dhaka, published: 28 December, 2012, Link: http://www.thefinancialexpress-bd.com/index.php?ref=MjBfMTJfMjhfMTJfMV83XzE1NDY2MQ==

Turkle S (1988) Computational reticence: why women fear the intimate machine. Technology and women's voices: keeping in touch, Routledge, 2004, pp 41–61

UbConnect (2015) PM swipes at Allama Shafi for his misogynist remarks http://unbconnect.com/pm-allama-shafi/#&panel1-2 http://unbconnect.com/pm-allama-shafi/#&panel1-1

United Nations Committee (2016) A commentary on Bangladesh's combined third and fourth periodic report

Wikipedia (2015) Eve- teasing. http://en.wikipedia.org/wiki/Eve_teasing

Wikipedia Humayun Ahmed (2015) Humayun Ahmed – Wikipedia, the free encyclopedia. http://en.wikipedia.org/wiki/Humayun_Ahmed

Wyer M, Barbercheck M, Cookmeyer D, Ozturk H, Wayne M (eds) (2013) Women, science, and technology: a reader in feminist science studies. Routledge, New York

Multiplicity Approach in Participatory Communication: A Case Study of the Global Polio Eradication Initiative in Pakistan

63

Hina Ayaz

Contents

Abstract

This chapter explores the usefulness of the social mobilization and community engagement in participatory communication approach in a development program, i.e., the Global Polio Eradication Initiative in Pakistan for its UNICEF's communication strategy. The Global Polio Eradication Initiative in Pakistan was highly affected by negative perceptions about the polio vaccination. After the unfortunate incident of May 2011, a number of Islamic Ullema gave fatwa against the polio vaccination to ban polio vaccination campaigns in Pakistan during 2012–2013. In these extremely challenging circumstances, the UNICEF's broader shift for its communication strategy and the use of the "social mobilization and community-involvement" guided by the multiplicity model of participatory communication was highly useful. This study demonstrates the significance of the multiplicity approach ("social mobilization and community-involvement") to increase the effectiveness of the Global Polio Eradication Initiative in Pakistan.

H. Ayaz (✉)
Berlin, Germany
e-mail: hina.ayaz5@gmail.com

J. Servaes (ed.), *Handbook of Communication for Development and Social Change*,
https://doi.org/10.1007/978-981-15-2014-3_91

Keywords

Development · UNICEF · Communication for development · Modernization · Participation · Participatory communication approach · Community involvement · Social Mobilization · Multiplicity approach · The global polio eradication initiative

63.1 Introduction

If development can be seen as a fabric woven out of the activities of millions of people, communication represents the most essential thread that binds them together. (Colin Fraser and Jonathan Villet 1994)

Communication plays a significant role in promoting human development. Communication is central to development in many ways. Communication holds interesting possibilities to divulge traditional wisdom and people's underlying attitude. Communicating enables policy makers and practitioners to take into account peoples' views, attitudes, needs, and traditional knowledge, when identifying and formulating development programs. Communicating with people at all levels, on the one hand, empowers them to recognize important issues. On the other hand, it helps people to find common ground for action. In this way, communication builds a sense of participation and ownership in order to implement decisions. Communicating with people extensively at all stages of a development program invigorates the development program and helps to sustain the development process. In this way, communication is a powerful tool and the strategic use of communication can bring social change in a society.

The first and dominant paradigm of modernization during 1940s–1960s envisioned the strategic use of communication to bring a social change in a society through mass communication. The linear model of this paradigm was mainly based upon the Lasswell's model of communication: – "Who says What through Which channel to Whom with What effect?" This model was top-to-down in nature and believed in linear one-way flow of information. Guided by Lasswell's model of linear communication, the work of two communication theorists, Daniel Lerner and Evert Roger, was seminal in establishing the very powerful role of the mass communications in forming public opinion, as well as changing people's attitude and behaviors by putting communication at the heart of the modernization paradigm (Lerner 1958; Roger 1982). Wilbur Schramm (1964) further built on Lerner's theory. Wilbur Schramm's work was seminal in shaping and defining the role of the modernization paradigm at the heart of national development.

The assumption of this approach is that the causality (cause-and-effect) is a linear activity. The criteria of rationality can be determined externally. Moreover, the causality *(the cause-and effect)* can be understood and explained in *independent and isolated two factors.* In this way, the process of social change is assumed to be a linear activity (Servaes 1999, p. 28). Thus, in this reductionist approach the behavior and social change is predetermined and the future outcome is predicted accurately.

Over the subsequent years, the theoretical premises of the modernization approach has faced enormous criticism. For instance, it has been widely criticized for not taking into account the local context and culture in which communication process occurs (Servaes 1999, p. 29), perceiving the "complex systems into the linear one" (Joseph 2011), "associating with things rather than people" (Chambers and Pettit 2004) and ignoring the process and putting people into categories like things (Earle 2003).

Unlike the modernization approach, a participatory communication approach based upon the multiplicity model has quite a potential to stimulate people's awareness, their involvement, and capabilities regarding an issue. According to Servaes (1999), the central idea of the holistic approach of multiplicity paradigm is that *"there is no universal development model, that development is an integral, multidimensional, and dialectic process that can differ from society to society."* The multidimensional and dialectic nature of development calls for a delineate analysis of *the many faces of power* and the interrelationships of *local, national, and extern-national levels, social* groups, and *actors. "The power concerns the possible effectuated and asymmetrically divided ability of one actor (powerholder) to put into order, inside a specific interaction system, the alternatives of actions of one or more actors (power subjects). Power centers around the capability to regulate and structure the actions inside asymmetrical relations."* Hence, understanding the concept of power and its multiple faces is a crucial aspect of the development due to the fact that the power and its dynamics in a society are *essential in understanding the social reality* of development process (Servaes 1999, pp. 53–59). When it comes to how the power is exercised, the role of communication and participation is very important. Paolo Freire (1970), who is known to be the pioneer in the field of the participatory communication, argues that *"Communication must be dialogic. Dialogue is at the heart of participation, communication and empowerment."* Freire believes that the involving people in the dialogue and discussion as a communication process is the only way to liberate the mankind. The liberation is the only possible way to empower the oppressed and it is the only way to bring the real social change in a society. Freire further stresses that in the process of social change, we should focus on the process not on the outcome.

In a participatory communication activity, people are involved and engaged in dialogue and debate. It increases freedom of expression and provides people opportunities to share information and exchange of ideas in a more productive manner. The use of communication to bring social change in a society, based on the philosophy of "two-way communication process" leads to a productive outcome. In a communication activity, a blend of communication approaches, i.e., use of skilled interpersonal communication personal and the use of range of communication tools (such as audio-cassettes, videos) can be employed to enrich people's understanding of the social and moral repercussions of a variety of development issues such as illiteracy, gender discrimination, violence, the exploitation and abuse of children, undernourishment, HIV/AIDS, nonimmunization, adolescent fertility, poor sanitation, unhygienic way of living, drug abuse, and environmental degradation. On the one hand, people's participation in a meaningful discussion enables them to identify their problems. On the other hand, it also helps them to recognize where their capabilities lie. Participation also opens up avenues for people to reach consensus. And their involvement in decision-making processes develops a sense of ownership.

Servaes and Malikhao (2005) argue that in the participatory communication approach *"the point of departure must be the community. It is at the community level that the problems of living conditions are discussed, and interactions with other communities are elicited. . ..* However, *participation doesn't imply that there is no longer a role for the development specialists, planners and institutional leaders."*

In a participatory communication process, involving the community is very important in order to know their concerns. Engaging the community at all levels of participation during all stages of a development program ensures the real behavior change in society. Below is a case study of the global polio eradication initiative in Pakistan which highlights the community-involvement into the participatory communication process and its usefulness in bringing the behavior and social change in the society.

63.2 A Case Study of the Global Polio Eradication Initiative in Pakistan

The Global Polio Eradication Initiative (GPEI) is one of the world's largest public and private health initiatives. This development program aims to eradicate poliomyelitis from the globe. Polio (poliomyelitis) is a disease caused by the poliovirus (mainly in children under 5 years). It leads to paralysis in the lower leg muscles (WHO 2017). The development program of the Global Polio Eradication Initiative (GPEI) is run under the national governments in collaboration with spearheading partners. The Global Polio Eradication Initiative has come a long way in its efforts to eliminate poliomyelitis (polio virus) from the various regions of the world. Since the launch of GPEI (the Global Polio Eradication Initiative) in 1988, there has been a significant decrease in the number of polio cases and subsequently of endemic countries. In 1988, there were an estimated 35,000 cases in 125 countries in the world. According to an estimate, during the period (1988–2013), the overall numbers of polio cases fell by over 99% and over 10 million people were saved from being paralyzed. The number of polio cases was reduced from 35,000 to 416 by 2013. By 2014, four out of a possible six WHO regions (of the world) were certified as polio-free (Americas in 1994, followed by the Western Pacific Region in 2000, the European Region in 2002, and the South-East Asian Region in March 2014). By 2014, there were only three red zones, which meant that the polio virus had already been eliminated from most of the world except for three countries: Afghanistan, Pakistan, and Nigeria. Today the polio virus only exists in two countries: Pakistan and Afghanistan (Matlin et al. 2017).

The Global Polio Eradication Initiative in Pakistan has made significant progress since its official start in 1994. And the Polio Program in Pakistan has come a long way in its efforts to accomplish the goal of eliminating polio from its territory. The program continued to make progress from year to year. The number of polio cases decreased each year. According to a report of the Rotary International, it was estimated that Pakistan could achieve the status of being "*polio-free*" by the end of 2001 to 2002, when about 27 million children were vaccinated in March, 2001 (Silver 2001). The last follow-up polio round was supposed to be held and completed in October, 2001. However, the incident of the 9/11 (2001) happened. The incident of the 9/11 changed the destiny of the Global Polio Eradication Initiative in Pakistan. It made Pakistan a frontline ally in the war on terror. Thus, a country with a lack of resources had to shift its focus from various development activities to defense activities, diverting many funds from one department to the other. Pakistan has faced many challenges in achieving the goal of eliminating the polio virus from its territory since 2001. Along with the financial and institutional constraints, one of the major hurdles was misconceptions about the polio vaccination. The issue of addressing the misconceptions became especially challenging after the May, 2011 incident. In May 2011, the CIA launched a fake hepatitis-B door-to-door campaign to collect DNA samples from Osama bin Laden and his family in the city of Abbottabad in Pakistan (Johns Hopkins Bloomberg School of Public Health 2014). This incident had serious repercussions for the Global Polio Eradication Initiative in Pakistan as parents started refusing to allow their children to be vaccinated with polio drops (GPEI 2013, 2014). There are, of course, also other countries where fractions of the population have refused to vaccination for polio due to sociocultural and religious reasons such as Nigeria and India (Obregón et al. 2009). However, in Pakistan, the May 2011 incident brought severe repercussions for the polio program because the CIA used the fake hepatitis-B door-to-door vaccination campaign and the polio vaccination campaign was/is done in the same way, i.e., going door-to-door to give children polio drops. The incident of this fake hepatitis-B door-to-door campaign created doubts in the population and mistrust towards the polio vaccination campaigns.

The parents' refusal to allow their children to be vaccinated with polio drops, consequently, started affecting the number of polio cases. There were 58 polio cases in 2012. The polio cases increased from 58 in 2012 to 93 in 2013 and 306 cases in 2014. These 306 cases in 2014 was the highest number of polio cases reported in the last 10 years (GPEI 2014). Furthermore, the misconceptions about the polio vaccination also made the position of frontline workers (who go door-to-door for the polio vaccine) very difficult. Indeed, there was a first attack on the staff of the Global Polio Eradication Initiative in Pakistan in July 2012 (GPEI 2013). Moreover, some religious scholars also gave fatwa against polio vaccination. In this critical situation, the Global Polio Eradication Initiative in Pakistan had to occasionally stop work. Therefore, it became extremely important to clear up these misconceptions about the polio vaccination.

In this complex situation, in order to clear up the misconceptions about the polio vaccination and to raise awareness about its importance for children, the UNICEF took a strategic shift for its large-scale mass communication campaign called "*communication and social mobilization interventions*" which was launched throughout the country in January 2012. The UNICEF broadened the focus of its communication strategy from national-level mass communication activities to incorporating extensive efforts strongly focused on community-level involvements. The campaign utilizes various modes of mass communication, for instance, newspaper, radio, television, cable TV channels, wall paintings, billboards, banners, streamers, bus branding, and airport signage all over the country and in particularly in the high-risk areas. According to an estimate, this mass media campaign reached more than 100 million people throughout the country. Along with the use of mass mediums, many other social and cultural activities were also organized. Some of them were very unique in their approach. For instance, UNICEF organized a theater play named *"Rescuing Oppressed and Destitute (ROAD), the Nomad Children's Festival for Polio Awareness"* for Punjab's nomadic tribes in Allhamra Cultural Complex, in Lahore, in March 2012. It was a one-day event which included a variety of performances such as magic shows, songs, and dances along with other entertainment activities for the children such as rides, roving clowns, and cartoon characters. A polio awareness stage drama was also performed. During the entire event, polio vaccination teams were circulating with fix vaccination points. More than 1,000 children were present at the festival and received OPV (Oral Polio Vaccine). Other social and cultural activities include "Clean-up Gadap"135, "Punjab Polio Photo Exhibition and Awards Ceremony," a literary festival "*Touching sensitive souls: World Pashto Literary Convention supports polio eradication."* Under this large-scale mass communication campaign, various advocacy and social mobilization activities took place all over the country (UNICEF 2012).

Involving religious leaders was one of the main focuses of the UNICEF's social mobilization communication strategy for the Pakistan Polio Program. The program reached out to renowned religious scholars of the country and convinced them to support the polio campaign in Pakistan to eliminate polio from Pakistan. Engaging the religious scholars enabled the program to respond to negative perceptions about the polio vaccine based upon religion. Initially, the program reached out to individual religious scholars and gradually in order to scale up the support of religious leaders, a press conference was also organized (in the high-risk areas, i.e., KPK and FATA) in which more than 50 renowned Ullema participated from Pakistan (End polio Pakistan 2014, 2015a).

Moreover, regular coordination and communication among all stakeholders was ensured by developing an online (web-based) dashboard which was accessible to all stakeholders. This dashboard is a compilation of all support tools being used by Pakistan Polio Program for its implementation, including monitoring indicators. It would maintain the historical data and generate profiles of all districts and union councils. It would also generate comparative analysis, which would help adaptive planning for high-risk areas. Since dashboard is web-based, it is regularly updated and shares all information of on-going program. In this way, the dashboard helps evidence-based planning and facilitate EOCs (Emergency Operation Centers). Dashboard also sends feedback alerts through E-mails and SMS (End polio Pakistan 2015b).

Involving all stakeholders and the participatory communication strategy of involving the Ullema community in the dialogue and discussion proved to be highly successful. One of the biggest achievements of the participatory communication strategy of involving the community in dialogues and discussions was that those Islamic Ullema who had given fatwa against polio vaccination after May 2011 incident were convinced that the polio vaccination was not part of the fake hepatitis-B vaccination campaign of the CIA. Eventually, at the end of 2014 and in 2015, these Islamic Ullema revoked their earlier fatwa and gave fatwa in favor of the polio vaccination for the health benefits of children. The fatwa in favor of the polio vaccination for the health benefits of children by a number of Islamic Ullema brought significant decrease in the number of refusals by parents to allow their children to be vaccinated by the polio drops. According to the End Polio Pakistan, the number of polio cases dropped from 306 in 2014 to 54 in January 2016 (End Polio Pakistan 2016a, b).

Therefore, it can be safely analyzed that the participatory communication approach has proven to be highly effective for the context-relevant communication in this highly context-dependent nature of the situation. Moreover, this major breakthrough, i.e., the success of convincing Islamic scholars' community is surely attributed to the persistent efforts of participatory communication activities for involving the community in dialogue and discussion at all levels and likely reveals the sincere intentions and the best efforts of all the hard-working and dedicated people involved in this noble cause of eliminating polio from Pakistan.

63.3 Conclusion

The above case study of the global polio eradication initiative in Pakistan for its UNICEF's communication strategy has demonstrated the strength of the participatory community-involvement approach underpinned by the multiplicity model of the participatory communication. This model has helped overcome barriers and moved the global polio eradication initiative in Pakistan forward.

There is no better conclusion to this story than what Servaes (1999, 89) has argued for the participatory communication approach in his multiplicity model:

> *In the participatory model, the focus moves from a "communicator" to a more "receiver-centric" orientation, with the resultant emphasis on meaning sought and ascribed rather than information transmitted. With this shift in focus, one is no longer attempting to create a need for the information one is disseminating, but one is rather disseminating information for what is a need. Experts and development workers respond rather than dictate; they choose what is relevant to the context in which they are working. The emphasis is on information exchange rather than on persuasion as in the diffusion model. Listening to what the other say, respecting the counterpart's attitude, and having mutual trust are needed. Participation supporters do not underestimate the ability of the masses to develop themselves and their environment.*

References

Chambers R, Pettit J (2004) Shifting power to make a difference. In: Groves L, Hinton R (eds) Inclusive aid: changing power and relationships in international development. Eart fiscal, London

Earle L (2003) Lost in the matrix: the logframe and the local picture. Paper presented at the INTRAC's 5th evaluation conference: measurement, management and accountability? The Netherlands, 31 March–4 April, 2003

End polio Pakistan (2014) EOC Pakistan: communication update Dec 19, 2014. http://www.endpolio.com.pk/knowledge-centre/communication-update. Accessed 5 Jan 2015

End polio Pakistan (2015a) EOC Pakistan: communication update Jan 2, 2015. http://www.endpolio.com.pk/images/communication_update/EOC-Comms-Update-Jan-2-2015.pdf. Accessed 10 Feb 2015

End polio Pakistan (2015b) EOC Pakistan: communication update Jan 9, 2015. http://www.endpolio.com.pk/images/communication_update/EOC-Comms-Update-Jan-9-2015.pdf. Accessed 10 Feb 2015

End polio Pakistan (2016a) EOC Pakistan: communication update Jan 31, 2016. http://www.endpolio.com.pk/images/communication_update/EOC-Comms-update-January-31.pdf. Accessed 2 Feb 2016

End polio Pakistan (2016b) EOC Pakistan: communication update Jan 1, 2016. http://www.endpolio.com.pk/images/communication_update/EOC-51-comms-update-final.pdf. Accessed 2 Feb 2016

Fraser C, Villett J (1994) Communication: a key to human development. FAO, Rome. http://www.fao.org/docrep/tl815e/tl815e00.htm. Accessed 13 June 2015

Freire P (1970) Pedagogy of the oppressed (trans: Bergman Ramos M). Herder and Herder, New York

GPEI (2013) Annual report 2012. WHO, Geneva. http://www.bitly/GPEI_AR2012. Accessed 10 May 2013

GPEI (2014) Status report July–December 2014: progress against the Polio Eradication and Endgame Strategic Plan 2013–2018. Geneva. http://bit.ly/GPEI_Status2014. Accessed 2 Feb 2015

Johns Hopkins Bloomberg School of Public Health (2014) White house responds to public health deans: the central intelligence agency makes no use of operational vaccination programs https://www.jhsph.edu/news/news-releases/2014/white-house-responds-to-public-health-deans-the-central-intelligence-agency-makes-no-use-of-operational-vaccination-programs.html. Accessed 12 July 2014

Joseph S (2011) Changing attitudes and behavior. In: Cornwall A, Scoones I (eds) Revolutionizing development: reflections on the work of Robert Chambers. Earthscan, London

Lerner D (1958) The passing of traditional society: modernizing Middle East. Free Press of Glencoe, New York

Matlin SA, Haslegrave M, Told M, Piper J (2017) The Global Polio Eradication Initiative: achievements, challenges and lessons learned from 1988–2016. Global Health Centre, the Graduate Institute of International and Development Studies, Geneva

Obregón R, Chitnis K, Morry C, Feek W, Bates J, Galway M, Ogden E (2009) Achieving polio eradication: a review of health communication evidence and lessons learned in India and Pakistan. Bull World Health Organ 87(8):624–630

Rogers EM (1982) Diffusion of innovations, 3rd edn. Free Press, New York

Schramm W (1964) Mass media and national development: the role of information in developing countries. Stanford University Press/UNESCO Press, Palo Alto

Servaes J (1999) Communication for development: one world, multiple cultures. Hampton Press Inc., Cresskill, pp 84–101

Servaes J, Malikhao P (2005) Participatory communication: the new paradigm? In: Hemer O, Tufte T (eds) Media and glocal change: rethinking communication for development. Sweden/Argentina, NORDICOM/CLASCO, pp 91–103

Silver C (ed) (2001) Pakistani Rotarians help immunize 27 million children in NIDs. Rotarian (Rotarian Int) 54. ISBN: S5LQ-BE8-DD3Q

UNICEF, Pakistan (2012) Pakistan polio communication review. Issue no.: 1

WHO (2017) (a) Poliomyelitis (polio). http://www.who.int/topics/poliomyelitis/en. Accessed 23 Jan 2017

Part X

Participatory Communication

Participatory Development Communication and Natural Resources Management 64

Community Participation and Communication in Managing Land and Water

Guy Bessette

Contents

Abstract

Globally, governments, civil society, and the private sector now recognize that the meaningful participation of local people is essential for sustainable natural resource management, agricultural productivity, and food security. However, traditionally, in the context of agriculture, forestry, and natural resources management, many communication efforts have focused on the dissemination of technical packages toward end users who were expected to adopt them or on the promotion of behavioral change. Not only did these practices have had little impact, but they also ignored the need to involve local people in decision-making and to address conflicts or policy gaps. Other more participatory approaches such as innovation

G. Bessette (✉)
Gatineau, Canada
e-mail: gbessette3@gmail.com

J. Servaes (ed.), *Handbook of Communication for Development and Social Change*,
https://doi.org/10.1007/978-981-15-2014-3_71

circles or farmer's research have also met with limited results, mostly because they have been driven from the outside. This chapter discusses this situation and presents different cases from Asia and Africa illustrating how participatory development communication can enable local communities to identify their development needs and the specific actions that could help to fulfil those needs, while establishing an ongoing dialogue with the other stakeholders involved.

Keywords

Agricultural practices · Burkina Faso · Community-based natural resource management (CBNRM) · Food security · Forum for Agricultural Research in Africa (FARA) · Integrated Agricultural Research for Development (IAR4D) · *Majlis* · Natural resource management · Top-down dissemination approach

64.1 Introduction

Somewhere in a Sahel country, a small group of researchers and visitors are visiting a demonstration plot showing how live fences – an innovation that seeks to protect fields from the encroachment of desert sand – can prevent soil degradation and increase agricultural productivity for smallholders. Two farmers, who maintain the plot, explain the technology: what is it about, why is it useful, and what kind of vegetation can be used. Researchers and visitors ask a few questions, then the visit ends, and the group, enthusiastic about the experimentation and its potential, goes back to the cars. Lagging behind, I ask the two farmers if they use this technology in their own plots. "Oh no! "says one, "this is too much work!".

Somewhere else, a group of researchers are on their way to one of their field experimental sites with the intention of visiting farmers that manage them and monitoring the performance of soil productivity-improving technologies that were introduced earlier in the year.

After having enjoyed the traditional hospitality of being offered some water during a visit to community elders, they were told: "Come with us now, we will lead you to your demonstration plots." Of course, outside of these plots, which were not *theirs*, none of the innovative technologies were being used.

What is the problem? What do we need to do to facilitate the sharing and adoption of knowledge that can make a difference? What can be the role of development communication?

64.2 Conceptual Models

One of the conceptual models that had a large influence on agricultural communication practices has been the innovation dissemination model. Formulated more than 50 years ago (Rogers 1962, 1976), and originating from the extension of agricultural practices exported from the North to developing countries, it involves the transmission of information to farmers by a resource person (the researcher or extension agent) and rests on three main elements: the target population, the innovation to be transmitted, and the sources and communication channels. Of course, this model has been rightly criticized for its reductionism, but in spite of all the evidence, there are still practitioners today who regards innovation as coming from the outside and which is delivered through information and demonstration with the aim of convincing farmers to adopt it in order to increase their productivity.

One model that challenged this approach was formulated by the Brazilian adult educator Paolo Freire (Freire 1970). In his model, he insisted that the mere transfer of knowledge from an authority source to a passive receiver did nothing to help promote growth in the latter as a human being with an independent and critical conscience capable of influencing and changing society. For communication to be effective, it has to be linked not only to the process of acquiring technical knowledge and skills but also to awareness-raising, politicization, and organizational processes.

In the 1970s, another communication model, developed around the emergence of community media, and championed by UNESCO, insisted on community participation:

> Whenever carefully developed programs have failed, this approach, which consists in helping people to formulate their problems or to acquire an awareness on new options, instead of imposing on them a plan that was formulated elsewhere, makes it possible to intervene more effectively in the real space of the individual or the group. (Berrigan 1981)

Through the use of these conceptual models and the learning acquired through experience, we began developing new ways of looking at the adoption of agricultural innovation. Participation, by putting the emphasis on the needs, perspectives, and knowledge of farmers and on inclusive decision-making processes, became the key concept of communication for the adoption of innovations. Many scholars engaged in agricultural research and extension, such as Röling (2004), have recognized that it is not useful to consider innovation as purely the outcome of transfer or delivery of results of scientific research to "ultimate users" or farmers and that participation needs to play a major role.

Over the years, there have been many efforts in trying to articulate this participation into concrete models aiming to share agricultural knowledge. In Africa, the Forum for Agricultural Research in Africa (FARA) (Together with the West and Central Africa Council for Agricultural Research and Development (CORAF-WECARD), the Association for Strengthening Agricultural Research in Eastern and Central Africa (ASARECA), the Centre for Coordination of Agricultural Research and Development for Southern Africa (CCARDESA), and National Agricultural Research Systems (NARS)) has been championing Integrated Agricultural Research for Development (IAR4D) multi-stakeholder processes using innovation platform (IP) as a key tool to facilitate livelihood and development initiatives' impact. The principles of this approach integrate:

- The perspectives, knowledge, and actions of different stakeholders around a common theme
- The learning that stakeholders achieve through working together

- Analysis, action, and change across the economic, social, environmental, and livelihoods and welfare of end users and consumers
- Analysis, action, and change at different levels of spatial, economic, and social organization (CORAF-WECARD 2011)

IAR4D's conceptual framework involves all stakeholders (communities, R&D organizations, and private sector) in the different phases of the process aiming to the adoption and setting in place of innovations. This approach "seeks to transform the organizational architecture of R&D actors from a linear configuration (research➔dissemination➔adoption) to a network configuration, comprising all actors in the agricultural **Innovation Sphere**" (Adekunle et al. 2013, p. 8).

This model is encouraging and has been shown to deliver promising results (Id., p. 27 ss). However, there are still a number of issues linked to the adoption of improved agricultural technologies that deserve attention. The nature of the relationships between researchers and communities, the communication methodology, and the costs of technology adoption are some of the main ones.

64.2.1 Building Relationships Between Farmers and Researchers

First and foremost, relationships between farmers and researchers need to be built and nurtured, and this usually takes time. Some researchers have had to work for many years with the same communities before mutual trust developed.

In other situations, researchers work with the same community members for many years, without engaging with the community as a whole.

Recently, in the context of needs analysis with a farming community, farmers told us "You need to respect us, even if we are illiterate, and do not have nice clothes. You need also to listen to us. Often, you come here and discuss with our sons that have received some education, you don't come to us."

We can add to this that too often, women as well as the poorest farmers are not involved in the relationship a researcher establishes with a community. Gender and equity are essential to take into consideration.

Often, researchers are perceived by communities as very important and smart people who come from the city in big 4x4, do all the talking, and sometimes provide some sort of monetary or in-kind benefits. If this perception is not challenged, it may be difficult to see a collaborative relationship develop between researchers and farmers. Moreover, if farmers see researchers only as monetary and in-kind benefits providers, (regretfully, this has been one of the side effects of many development projects), the new relationship might start on a wrong footing.

Some work certainly has to be done to establish and nurture relationships with the communities before establishing new field sites, and the strategy used to approach community members is critical for success.

I once met a researcher who was working by himself in the desert. I asked him what was going on. He explained that he was studying rainfall pattern and that initially, he was part of a multidisciplinary team. Social scientists in the team started off by asking what the community members perceived as intrusive questions to which they were reluctant to provide answers, such as "How many heads of cattle do you possess?", or "How many wives do you have?", They were therefore asked to leave. "And what about you?" I asked?

"Oh, me" he said, *"I'm just a specialist in natural sciences. I install my equipment to record data on rain and humidity and when someone would come and ask me what I was doing, I would explain. They would then invite me for tea and I would go. Then, once inside, I would look at the tea and at the food, and tell myself 'If I drink this water and eat this food, I'll be sick'. I would then drink the tea, eat the food, get sick, and do it again and again, and I've been working here now for ten years."*

64.2.2 Using a Participatory Communication Approach and Methodology

The concept of dissemination is a tricky one. It is a bit like traditional class teaching methods used in formal education. It is not because a teacher explains a subject in a clear and logical way that students will automatically understand. Teaching and learning are two different complementary processes. And it is not because a researcher presents and explains a given technology to farmers that they will adopt and experiment with it. Some farmers, often the one that are more successful and who have had some contact with researchers in the past (not unlike a few bright students in a class), will take advantage of the information and experiment with the improved technology. But in most cases, the majority will not, because it is an answer to a question they did not ask.

This is why it is useful to move from information dissemination, to engaging farmers and researchers in a joint dialogue. A discussion on how climate change is perceived and how it affects productivity in the field for example, can lead to a decision to test more resilient plants and adopt improved agricultural technologies.

Traditionally, researchers tend to identify a particular problem and try out different solutions (including improved technologies) with the collaboration of local farmers. Then they try to disseminate their findings and have other farmers to adopt these solutions. It seldom works. But when communities are engaged in looking for a solution to a problem that affects them, and after having jointly found the solution, farmers then share their own experiences with other farmers; the situation evolves in a very different manner.

This approach demands a change in attitude and practices. Many researchers still perceive community members as beneficiaries and end users of research results. Even though it has been well documented that the one-way delivery of technologies to end users has had little impact, the required attitudinal shift is apparently not easy.

One methodology that can support and accelerate this process of change in attitude and practice is participatory development communication (See Bessette 2004, 2006). It is usually represented as having four main interlinked phases, diagnosis, planning, intervention, and assessment, and comprises the following 18 steps:

Phase 1: Understanding, relating, researching

1. Clarifying the mandate and the intervention
2. Developing a prior understanding of the local setting (situational analysis or preliminary research)
3. Establishing a relationship and negotiating a mandate with a local community
4. Setting the goal: involving the community in the identification of a problem, its potential solutions, and the decision to carry out a concrete initiative;
5. Identifying the key stakeholders concerned by the identified problem and initiative and learning from them (participatory communication appraisal; stakeholder analysis; SAGA, social and gender analysis; KSAP, knowledge, skills, attitudes, practices analysis; analysis of communication resources; social network analysis)
6. Identifying other stakeholders and potential partners concerned by the identified problem and initiative

Phase 2: Formulating and developing the strategy

7. Identifying and formulating communication objectives
8. Identifying key messages, content, and topics
9. Selecting appropriate communication tools and media
10. Facilitating partnerships and establishing agreements
11. Identifying the communication materials and activities to develop
12. Planning the pretesting of communication content and materials
13. Planning participatory monitoring and evaluation
14. Planning documentation
15. Defining participation modalities at each step of implementation, monitoring, and evaluation
16. Planning the sharing and utilization of results

Phase 3: Validating and organizing

17. Validating the whole strategy with the community
18. Producing a communication plan to implement the strategy

Other methodologies (vg. Acunzo et al. 2014) suggest different variants of this process, but they all have one thing in common: the emphasis on a systematic planning process that enables and encourages the active participation of stakeholders and the implementation of two-way communication between farmers and researchers.

64.2.3 Discussing the Issue of Costs

As we saw earlier, in order to promote the adoption and use of an improved agricultural technology, the innovation must go hand in hand with the building of a two-way relationship between farmers and researchers and with using a systematic methodological approach to engage communities in a process of experimentation and adoption. There is also a third element at play: costs.

Any technology has costs associated with it. It could be extra labor costs or additional input costs. This complicates things for poor farmers who struggle to survive and who do not have the means to buy what is needed. Moreover, in many settings, labor-associated costs often have to be assumed by women, on top of their heavy responsibilities.

Usually, researchers will demonstrate an improved technology, but without considering the associated costs. This is why, in many rural areas, in spite of the improved yield related to the use of certified seeds of improved varieties, a majority of farmers still use traditional seeds. This is also why, in the example mentioned above, farmers did not practice the new technology of live fences, although they were highly efficient.

There is no easy solution. The issue of costs has to be discussed between researchers and communities. In some cases, solutions can be found, if farmers decide they want to adopt an innovation. In other cases, it would be necessary to discuss the issue with other stakeholders such as the state decentralized services and policy-makers.

64.3 Engaging with Government Stakeholders

Engaging with government stakeholders is not something researchers are usually trained to do. But in their new roles as facilitators of change, it is an important dimension of the work of promoting innovative technologies for improving food security.

Sometimes, solutions to the cost issue can be found through in-kind support from the state decentralized services, including extension services of the Ministry of Agriculture. Such in-kind support may come in the form of material or farm inputs, such as seeds or fertilizers, or advices. But this, of course, remains at a small scale and is sporadic.

What is often needed is to engage government actors and civil society organizations in policy dialogue, in order to advocate for in-kind loans to farmers, subsidies for farm inputs, certified seeds or for conserving traditional seeds, extension services for improving organic soil management techniques, access to microcredit, improved distribution network for seeds, secure land tenure, etc.

Researchers especially those working for the State could also play a role by sensitizing administrative authorities of their departments, who can then approach parliamentarians (especially those active in agriculture or environment commissions) to discuss the issue. They can also link up with and sensitize civil society organizations such as farmers' federations or NGOs who already support advocacy efforts on the matter or who can start work in this area.

Promoting the adoption of an improved technology that would enhance agricultural productivity cannot be done in a policy vacuum. The costs and conditions (such as tenure security for women or for migrant farmers, farmers' rights, distribution networks, etc.) linked to its use must be brought to the attention of civil society organizations and policy-makers, in order to open a policy dialogue on the subject.

Sometimes new laws and regulations may be required (e.g., for the implementation of the International Treaty on Plant Genetic Resources for Food and Agriculture); and such issues may require time and effective policy dialogue for a resolution.

64.4 Community Participation and Innovation Adoption

Salimata is sitting at a local radio station and is singing a few traditional and local songs. Then she reminisces about when she was trained to make organic fertilizers but had to stop doing it. This year she harvested only two bushels of peanuts. She had to stop using the technology because she was afraid that if her plot of land produced too much, her field, which was granted to her by her village chief, would be taken from her.

Not far from there, other women at a radio club are listening to the broadcast and start sharing their own experiences. The issue of land tenure and poor access to land and the consequences on food security are being addressed. Then back home at their village, they bring up the issue, and finally a community meeting with the village elders and men is organized. Pledges are made that women can now improve their plot of land and put more food on the table without fear of losing their field.

This is a real story. The outcome might not be so positive every time, but the participatory communication process, where farmers and community members take the lead, nurtures change.

Improved agricultural technologies, whether modern or traditional, need to be shared widely to increase agricultural productivity and food security. For this to happen, there is a need to move from practices of information dissemination and demonstration to engaging community participation, fostering the appropriation of development initiatives by the communities, and supporting dialogue between all stakeholders.

64.5 Learning About Participatory Development Communication

In Southeast Asia, a project led by a group of regional organizations (The College of Development Communication of the University of the Philippines Los Banos, (CDC-UPLB), the Regional Community Forestry Training Center for Asia and the Pacific (RECOFTC), the International Institute of Rural Reconstruction (IIRR), the Community-based Natural Resource Management Learning Center (CBNRM LC) and the International Potato Center-Users' Perspectives With Agricultural Research and Development (CIP-UPWARD)) promoted an innovation in learning among multiple groups of learners, involved at different degrees in community-based natural resource management (CBNRM), as community organizations, NGOs, researchers, and government members and located in different countries of Southeast Asia. As a capacity-building and networking program, ALL in CBNRM (adaptive learning and linkages in CBNRM) focused on the social, communication, and collective problem-solving processes enabling communities to manage their environment (Adaptive learning and linkages in community-based natural resource management, CDC-UPLB 2007).

The program was a continuation and an evolution from the Isang Bagsak program, aiming to build capacities in participatory development communication in Southeast Asia and in East and Southern Africa. In Southeast Asia, Isang Bagsak involved more than 50 NRM researchers and practitioners in government, non-government, research and academic institutions, and community-based organizations in Cambodia, the Philippines, and Vietnam. The African program was active in Uganda, Malawi, and Zimbabwe. ALL in CBNRM built on this experience by integrating the participatory development communication approach more closely with knowledge related to community-based management in wetlands, forests, and coastal ecosystems.

The learning modalities were the following: (1) an introductory workshop; (2) face-to-face discussion of learning themes on participatory development; (3) exchange of experiences and reactions between the different learning groups through a regional online thematic discussion, with technical inputs form institutions and resource persons; (4) research support and field mentoring; (5) development of learning resources designed for community members and local stakeholders; and (6) two regional workshops at midterm and at the end of the learning cycle.

As for the learning themes, they integrated participatory development communication with the different approaches used in community-based natural resource management. There were 13 themes: (1) establishing a relationship with the local community; (2) involving the community in the identification of a problem, potential solutions, and the decision to carry out a concrete initiative; (3) identifying the different community groups and other stakeholders concerned with the identified problem(or goal) and initiative; (4) emerging understanding of roles and concepts in CBNRM; (5) arriving at a collective understanding of the local community and the CBNRM context; (6) enhancing the community's capacity for identifying problems and setting goals; (7) understanding stakeholders relationships in a CBNRM setting; (8) developing and implementing a participatory development communication plan; (9) enhancing processes for collective action; (10) developing partnerships; (11) monitoring, documenting, and evaluating CBNRM experiences; (12) encouraging CBNRM adaptation and innovation; and (13) participatory processes for policy change.

Such efforts enabled the sharing of knowledge and the building of networks of practitioners, community organizations, researchers, and government stakeholders during 8 years (2001–2009). The lessons from this initiative are still available. Successfully facilitated participatory approaches depend on the sharing of knowledge and the creation of networks between community organizations, practitioners, researchers, and other stakeholders.

64.6 Empowering Community Members in Taking an Active Part in Their Own Development

There are many examples demonstrating the use of participatory development communication approaches for empowering communities so that they can take action. Here are some recent examples (Moving from information dissemination to community participation in forest landscapes 2017).

A project from People and Nature Reconciliation in Vietnam (Nguyen 2017) worked with two women's unions in ethnic communities in order to enhance women's capacities and knowledge to enable them to influence and take a lead in exploring alternative livelihood opportunities for their families and communities. The PDC approach led the project team to begin with a self-needs assessment and consultation with the women before planning training and activities. Those were identified by the women. The training courses helped them to identify potential forest products for market development. The participants chose to focus and explore the potential of three products. Through the process, the women also expressed their desire to be more involved in forest management in order to negotiate their rights and access to forest resources. The project supported them in identifying all the stakeholders involved in the management of the forest, their needs, and situation so that women could negotiate with them. Cooking competitions and cultural performances helped them to gain attention and interest.

Another initiative, in the Philippines (Diaz 2017), incorporated participatory development communication approaches to build the capacities of indigenous youth in participating in the production of communication tools to enable the sharing and expression of their cultural heritage. In this case, two-way communication among indigenous youth was supported in order to discuss priorities in terms of community development. At the end of the process, a youth festival including communication workshops, a newsletter, a video, and a declamation piece about youth's aspirations and dreams were produced by the participants, and all contributed to the promotion of indigenous people's identity, rights, and culture.

Participatory development communication was also used in the Philippines in the context of promoting and preserving indigenous people's sustainable food systems (Dagli et al. 2017). The project was able to facilitate a process where indigenous people identified innovative ideas and practices that could be further developed into community innovation projects to address specific food systems-related problems.

In Fiji, a similar approach was used in the context of adapting to climate change and ensuring food security (Elder 2017). At the outset of the project, PDC was used to develop relationships and trust between the communities involved in the project, the project team, and the government. It then led stakeholders in identifying problems and initiatives to address them in the context of rising sea levels. They identified a set of actions that they, together with eternal specialists, would carry on to address the identified issue of food security. In Ghana, Radio Ada also used participatory development communication to facilitate consensus-building and decision-making at the community level in order to restore a waterway which used to be at the heart of the lifestyle of the communities bordering it (Larweh 2006).

64.7 Involving Communities in Their Own Development

In the context of natural resource management and agriculture, many practices of researchers and extension agents still rely on top-down dissemination methodologies. In such a context, farmers are not involved in the decision-making process regarding technologies and practices to use in their field. Despite the hardship and difficulties associated with their traditional activities, they tend to overlook the technologies or practices shared by researchers or extension agents or use them only in the context of a given project that enlist and retribute them for a definite span of time.

Participatory development communication and its two-way communication model helps in improving the exchanges between researchers, extension agents, and farming communities, enhancing farmers' participation in experimenting different approaches and technologies, and fostering farmer to farmer training.

In Uganda, a project led by the Banana Program of the National Agricultural Research Organization used such an approach to foster active participation of local communities in identifying problems in their banana gardens, as well as their causes and possible solutions (Odoi 2006). The research team found that certain community members had extensive indigenous knowledge related to the concerns identified collectively, but it needed validation and complement of information, as well as a mechanism to share such knowledge among farmers and communities.

Farmers got involved not only in the experimentation of the techniques they had selected but also in producing and using communication tools to share their knowledge among their community and in a second step with neighboring communities. They made posters with photographs they took in their field before, during, and after experimentation and also produced a video with the help of the project team (This had not been planned. The team produced a video but the farmers were not happy with it and asked to be shown how to do one.) that they used during diversity fairs they organized or in which they participated. It is interesting to note that the community members involved in this project later organized themselves into a farmer association that became quite influent in the country.

In Vietnam, reversing the top-down dissemination approach to a participatory one helped engaging disadvantaged groups in local development. Following a policy change in the central mountains, new regulations aimed to protect the forest drove ethnic minorities used to a slash-and-burn model of agriculture in search for alternative livelihood solutions. A team of the Hue University of Agriculture and Forestry (Le Van An 2006) developed an initiative to engage these community members, and especially the most disadvantaged groups of the community, in participating in a collective journey to develop new livelihood practices. Community members regrouped according to their farming activities and interests and shared ideas and opinions on the problems they now were confronted with. With the help of the researchers, they then started looking for solutions and experimenting them. For example, in the case of rice production, farmers decided to test new varieties and to try out the application of a fertilizer, together with transplanting and direct sowing methods. Training was organized to support the different initiatives. Role plays, videos, posters, and leaflets were used as needed.

64.8 Managing Local Conflicts

Conflicts and opposed interests are linked with the management of natural resources. Finding solutions is never easy, and sometimes these are out of reach of the communities themselves. The case of land reform is a good example.

But there are many conflicts at the local level on which people can act and where a participatory development communication approach can make a difference.

In Lebanon, conflicts over land use among animal herders, fruit growers, and quarry owners in a semiarid area were addressed by the intervention of a research team from the American University of Beirut who set up a user's network between different stakeholders, using a wide range of tools, including traditional communication tools such as the *majlis*, tribal get together during which issues are brought up in the community, short video documentaries on different issues which were used as powerful participatory tools, a series of workshops related to natural resource management and community development, etc. The network provided an environment in which conflict resolution could take place among different land users since the needs of conflicting parties could be voiced and compromises explored (Hamadeh et al. (2006) in Bessette G (ed), op. cit.).

In Burkina Faso, a local conflict over land use between men and women was addressed through a process of involving the women, who were victims of a local usage of taking good land from them, into addressing the situation (Thiamobiga 2006, in Bessette G, op. cit.). Five villages participated to this initiative. The issue could not be discussed publicly in regular community meetings because women of this area did not have the right to speak in public.

They used a traditional custom: once a year, women would dress themselves as men and could do as they wished without retaliation of any sort. With the help of a local NGO, CESAO, and the Theâtre de la Fraternité, a group of women from the five villages developed and organized a play in which, dressed as men, they addressed a number of issues linked to property rights but also soil productivity. A debate followed the theater play and served as a learning experience for everyone. Farmwives giving advice on soil fertility to men was a first to the region. At the end, the village elders took a pledge of respective women's land rights. If their plots produced more than others, they would not be taken from them anymore.

64.9 Conclusion

There are many examples of applying PDC in natural resource management. In all of them, the common tread is the use of communication to empower people to identify and understand the problems they face (it could also be the case of a collective vision they wish to contribute to) and identify a set of actions to address them and then to support such initiatives. In a review of similar cases from Africa, Indonesia, Lebanon, and others, it was suggested that the processes that PDC practitioners concentrate on can be grouped in four categories (Saik Yoon 2006), communicating effectively, creating knowledge, building communities, and enabling action.

As the author of this review puts it, "Conventional communication for development efforts usually concentrate on the first and last cluster (communicating effectively and enabling action. PDC covers two additional clusters (creating knowledge and building communities) that aim to self-empower people through augmenting and validating their knowledge of critical issues and subjects that affect their lives, and through forging strong alliances between people, groups and communities so that they can consult and act effectively together in order to address problems and realize aspirations" (Idem. p. 275).

This being said, as Nora Cruz Quebral, one of the main founders of our discipline puts it, "New models of communication do not necessarily replace older ones. They just co-exist" (Quebral 2006). What is important for practitioners is to be clear about the kind of communication and what kind of development they advocate. Participatory development communication takes a stand on empowerment as the key to effective community-based natural resource management.

It is also is a communication of proximity. As such, it is most effective in working with local communities. It is also a useful complement to working at a policy level on issues such as climate change, agricultural policies, land tenure, etc., because at the end, such policies can only be effective if people see them as an answer to their needs and implement them.

References

Acunzo M, Pafumi M, Torres C, Tirol MS (2014) Communication for development sourcebook. FAO, Rome

Adaptive learning and linkages in Community-based natural resource management (2007) Los Banos: CDC-UPLB

Adekunle AA, Ayanwale AB, Fatunbi AO, Agumya A, Kwesiga F, Jones MP (2013). Maximizing impact from agricultural research: potential of the IAR4D concept. Forum for Agricultural Research in Africa (FARA), Accra

Berrigan FJ (1981) Community media and development. UNESCO, Paris

Bessette G (2004) Involving the community. A guide to participatory development communication. IDRC/Southbound, Ottawa/Penang

Bessette G (ed) (2006) People, land and water, participatory development communication for natural resource management. IDRC/Earthscan, Ottawa/London

Dagli W, Roquino EF, Guzman Diaz VP (2017) Establishing relationships in co-creating local innovations in indigenous people's sustainable food systems in the Philippines. In: Moving from information dissemination to community participation in forest landscapes, op. cit.

Diaz EP, Non-Timber Forest Products-Exchange Programme for Asia (2017) Using participatory development approaches to engage indigenous youth to protect their cultural heritage in the Philippines. In: Moving from information dissemination to community participation in forest landscapes, op. cit.

Elder M (2017) Adapting to climate change and ensuring food security in Narikoso village as communities prepare for rising sea levels in Fiji. In: Moving from information dissemination to community participation in forest landscapes, op. cit.

Freire P (1970) The pedagogy of the oppressed. (ed) Herder and Herder (first translation in English), Continuum, New York

Hamadeh S, Haidar M, Zurayk R, Obeid M, Dick C (2006) Goats, cherry trees and participatory development communication for natural resource management in semi-arid Lebanon. In: Bessette G (ed) op. cit.

Integrated Agricultural Research for Development (IAR4D) Multi-stakeholder Innovation Platform (IP) Processes Dakar: CORAF-WECARD, 2011

Larweh K (2006) And our perk was a crocodile: Radio Ada and participatory natural resource management in Obane, Ghana. In: Bessette G (ed) op. cit.

Le Van A. (2006) Engaging the most disadvantaged groups in local development. A case from Viet Nam. In: Bessette G (ed) op. cit.
Moving from information dissemination to community participation in forest landscapes (2017) Bangkok : FAO/RECOFTC
Nguyen D .T. L. (2017) Using participatory development communication approaches for empowering ethnic women in sustainable forest management in Viet Nam. In: Moving from information dissemination to community participation in forest landscapes, op. cit.
Odoi NN (2006) Growing bananas in Uganda: reaping the fruit of participatory development communication. In: Bessette G (ed) op. cit.
Quebral N (2006) Participatory development communication: an Asian perspective. In: Bessette G (ed) op. cit.
Rogers E (1962) Diffusion of innovations (1st ed.). 1962 New York: Free Press of Glencoe.
Rogers E (1976) Communication and development, critical perspectives. Sage Publications, Beverly Hills/London/Delhi
Röling N (2004) *Communication for development in research, extension and education*. In: Selected papers from the 9th UN roundtable on communication for development. FAO, Rome, pp 57–77
Saik Yoon C (2006) Facilitating participatory group processes: reflections on the participatory development communication experiments. In: Bessette G (ed) op. cit.
Thiamobiga JD (2006) When farm wives take to the stage. In: Bessette G, op. cit.

Participatory Communication in Practice: The Nexus to Conflict and Power

65

Saik Yoon Chin

Contents

Abstract

Practitioners of participatory communication commonly encounter episodes of conflict in their work with communities. The first type of conflict is among peers and members of the community which occurs when people with diverging priorities attempt to address shared problems. Conflict may persist due to inability of processes to resolve disagreements and competition. The second type of conflict is between the community and influential elites who possess powers which they wish to continue asserting in a top-down fashion to shape events and influence outcomes within the community. These experiences suggest a necessity for practitioners of participatory communication to build interdisciplinary linkages to the area of conflict management and theories of power in their continuing efforts to develop approaches which are capable of processing conflict involving communities they serve. This contribution explores the nexus between participatory communication and the closely related theories and realities posed by conflict and unevenly distributed power within communities.

S. Y. Chin (✉)
Southbound, George Town, Penang, Malaysia
e-mail: chin@southbound.my; chinsaikyoon@gmail.com

J. Servaes (ed.), *Handbook of Communication for Development and Social Change*,
https://doi.org/10.1007/978-981-15-2014-3_19

Keywords

Development · Participation · Participatory communication · Conflict · Power · Empowerment of people and groups

65.1 Introduction

65.1.1 Why Practitioners Adopt Participatory Approaches

The first practitioners of participatory communication (practitioners) began their careers working with conventional mass communication methods. They discovered and then adopted participatory processes after experiencing setbacks using top-down approaches with the communities they worked. Many of these practitioners started their work in offices far away from farms and villages, designing extension-support print material, recording radio programs, and writing agriculture stories for provincial newspapers. Many were employed by organizations and government agencies set up to encourage villagers to adopt new ways of farming to increase yields and income.

The Training and Visit (T&V) system of agricultural extension was one such initiative promoted by the World Bank (Benor et al. 1984) where a continuous series of regimented training sessions is conducted by "subject-matter specialists" about scientifically prescribed farming methods. Such training originated at the "zone" headquarters of the extension services. It was then repeated at the "district" and afterward the "subdivisional" levels before being finally taught to the village extension worker. This worker, at the end of the pipeline of specialist prescriptions, then mounted a bicycle or motorcycle to visit his or her circle of about eight farming groups. At each stop the extension worker taught the methods which he (they were mostly men) had just learned to about 10 "contact farmers." These farmers are provided with communication and learning support materials and "inputs and supplies" – these may be seeds, fertilizers, pesticides, or tools and sometimes even lent money – so that the farmers could carry out what they have been taught.

The role of communication then was in the classic mode of "development support communication." Communicators were there to produce media materials specified by the technical staff of the program. The work was essentially "mass" communication – to take the centrally designed messages from the "expert" few to the groups spread across a farming district. As the T&V system and other similar programs proved over time to be less effective than conceived, everyone involved scrambled to review and pin down inadequacies in their work. A fairly common complaint was that the "communication shop" was issuing boring and unclear training support material which the contact farmers either did not understand, will not use, or both.

Some communicators responded to the criticism by pretesting the media and training support materials they were working on with the intended audiences. For many, it was the first time they had travelled to the villages and farms and showed their prototype print designs, played back recorded radio programs, and screened training films to the people who comprised the readers and audiences for these material to determine scientifically if they understood the messages being mediated.

The efforts invested in pretesting did lead to the production of more effective development support materials that better served the purposes of top-down initiatives. However, it did not solve all the problems. Communicators began probing deeper during their fieldwork and sometimes found while the materials they had produced were effective in communicating intended messages, they failed in changing people's behavior and practices. The simple answer they sometimes discovered was that targeted audiences were not interested in the issues and topics being promoted. The top-down objectives defined by the experts were not the ones the targeted "audiences" wanted to work on. Technologies and solutions were being promoted that did not meet an urgent need nor solve a priority problem of the people.

Some communicators took liberty in altering their own work-briefs, and instead of just broadcasting recommendations of the experts to the people, they also attempted to open feedback channels to the experts and subject-matter specialists in attempts to reorientate program and project objectives to match the people's needs and wants. Many of these initiatives by communicators to make available feedback from the field were not appreciated by the hierarchically more senior subject-matter specialists who found the intervention to be insolent of their "juniors" and promptly snubbed their efforts.

While this was happening, communicators made contact on their field trips with social scientists immersed in field projects aimed at doing just what they had attempted with opening feedback channels. Some of the social scientists had adopted participatory research methods (Tandon 2005) which encouraged people to take an active part in the research process and shape the directions of their work. Paulo Freire's (1970) critical thoughts on pedagogy – or more appropriately "andragogy" – would at about the same time strike a chord with some of the communicators who found resonance of their work experiences with parts of his book *Pedagogy of the Oppressed*. It proposed a new relationship between the teacher, student, and society. Freire criticized "traditional pedagogy" – which was very much the way in which development support communication engaged with target audiences – as the "banking model of education." In this model learners were approached as "empty vessels" to be filled with knowledge. He proposed that students be treated instead as co-creators of knowledge with their own contributions to what should be taught and shared. This was the new pedagogy to free the "oppressed." The early practitioners' adoption of this Freirian concept may have marked the beginnings of participatory communication.

The significance of this approach changed the focus of the work of practitioners from media materials production to the facilitation of communication and social processes for social change. The emphasis moved from the mechanics of mass communication – the printing presses and recording studios – to the people, to facilitating the expression of their ideas and group-processing these ideas to trigger collective action to solve community problems.

Such an approach was very difficult for early practitioners to reconcile with the exigencies of their organizations which had a predetermined agenda and a clearly defined mandate that often did not match the interests of the people they tried to engage with. Very few communicators working in development enjoyed the freedom of engaging with a community in a completely open-ended manner in which the people were free to address only the issues and problems that were important to them. One of the earliest groups which did enjoy this freedom was Colin Low, a filmmaker, who worked with Fred Earle of the Extension Department of Memorial University in making a number of documentary films about the 5000 residents from 10 fishing villages on Fogo Island in Canada. The fishing folks on the island were in dire straits. Catches had dwindled to uneconomic levels, and the government was seriously mulling the options of making welfare payments to the islanders or relocating them to the mainland where they could seek other forms of employment. Low had picked the island because of three factors: (1) the residents were experiencing many social and economic problems; (2) government policy for the region was being formulated, and hence ideas and options from the people were being sought; and (3) the island was large enough to experience "inter-community communication" problems which the team could attempt to develop methods to overcome (Nemtin and Loh 1968).

What emerged from the innovative approaches and processes pioneered by the team as they filmed from 1964–1967 became known as the "Fogo Process" – often described by practitioners as the first participatory film/video method. The National Film Board of Canada which funded the work called it "The Newfoundland Project," while others refer to it as "Fogo Island Films." This was before portable video cameras were available and film was shot on celluloid. The participatory element did not come in the islanders handling the camera but in their close involvement in deciding what to film, in agreeing to be filmed, in stating their views on record, in watching the "rushes" (raw, unedited footages) and deciding how sequences and footages should be edited to make up a complete film, and in watching the finished film and reflecting on the issues and views it presented.

A challenge by some of the Fogo islanders that the filmmakers "would not dare to run the statements (which the islanders had made on the films) in St John's or Ottawa" (where the provincial and national governments are based) encouraged the filmmakers to screen the films to government officials. The officials' response was "critical but supportive," and the filmmakers filmed one of the officials responding to the statements made on film by the fishing folk in Fogo Island and then took the reel back to the island and screened it for the islanders, thereby creating a two-way communication opportunity where people in different locations could see and hear each other's views, decades before the advent of videoconferencing and Skype.

In the Fogo Process, people who would otherwise be treated as the "target audience" played an active part in focusing the communication medium on people, scenes, things, and issues that were important to them. The finished product was presented back to themselves and other people of their community. It served as a mirror for the community. The reflected reality of their problems helped define these problems and mobilized community to act on the issues identified. Fogo Island communities were ultimately not disbanded and relocated. They continue to thrive today.

65.1.2 Participatory Communication in Practice

Early practitioners of participatory communication had to be creative and developed new ways of working as there were no manuals and not many cases of tried and tested methods for them to follow. Some of these new methods were borrowed and adapted from others working in development following participatory approaches.

Putting people of the community in the "director's chair," as we saw in Fogo, proved to be a sound approach to participatory communication. Community radio soon evolved (UNESCO undated). The practitioner's main role in this case was installing the hardware – recording equipment, transmitter, and broadcast tower – finding the money to pay for electricity bills and maintenance, and training members of the community in using the equipment and managing a broadcast station. The differentiation between the broadcaster and listener blurred as people took turns in "appearing" on broadcasts and programs got recorded in different homes, workplaces, markets, and community centers. Program producers did much of their recording "on the go," travelling to where the stories were breaking and visiting the homes of people who had news and stories to share. Many adopted the position that production values were not as important as having the people of the community – the listeners – in charge. Apart from in-depth programs on community issues, some of the more popular programs involved entertainment where different families and groups took turns in performing music or oral presentations which were often broadcasted live to save the cost of magnetic tape and the time required in postproduction. Personal and family information were also closely listened to on most community stations. The author recalls visiting a remote radio station which served communities living in an isolated valley dispersed on the slopes of two high mountains, where it sometimes took a day to hike from one side of the valley to the next. These quick and dry interjections over the air of family information ranged from calling family members to return home due to sickness or deaths to happier announcements of births and marriages. Besides serving a telegraphic service, such messaging knitted the communities closer in the sharing of sad and happy milestones of their members.

The print medium followed suit. Village wall newspapers were "published" on large bulletin boards hung in central locations such as community store, tea shop, meeting hall, or school (Practical Action 2007). People could pin stories and drawings on to the boards for all to read. An editorial committee would keep postings going in an orderly manner and, when the occasion called for it, write editorial pieces on community issues. The illiterate members of the community or the elderly with poor eyesight would get help from people present around the boards to read the postings to them.

Practitioners also introduced visualization methods into community gatherings and meetings. The most commonly adopted method involved writing answers and comments about issues raised during meetings on cards and dropping them anonymously into a heap (Salas et al. 2010). Participants would then work together to sort and group the cards according to themes and pin them on a wall in clusters. Discussions which follow tend to be more focused and productive when guided by the thematic clusters of ideas and comments visualized on the cards. As participants wrote the cards at the same time, rather than take turns at speaking, more ideas were collected, and everyone got an equal chance at contributing points to the discussion, and the risk of the articulate few dominating discussions was minimized. The other advantage of cards was that ideas and comments were not easily identified with the people making them and therefore got sorted and discussed less emotionally, especially when issue addresses proved to be personal and divisive. The process recommends pasting the cards on large pieces of paper after a meeting to serve as a ready record of points discussed which could be conveniently displayed in follow-up meetings so that the group did not go around in circles tackling the same issues at subsequent meetings. In communities with large numbers of illiterate members, agreed symbols are drawn on to cards instead of written words.

When done effectively, visualization processes will help groups arrive at a unanimous decision on action to be taken by the community. Often, unanimity may not be the outcome of visualization in which case a short list of options and alternatives proposed by the members will be the outcome. If not all options on the short list can be acted on at once, conducting a poll among all members to pick the priority action is often the most commonly accepted way forward. Practitioners could play the role of the neutral pollster in such a case and design a more engaging process of polling rather than a simple vote for the preferred option. A municipality in the Malaysian state of Penang involved all members of the community above 10 years of age in a gender-responsive and participatory budgeting process to decide how municipal funds budgeted for their neighborhood should be spent (Shariza 2016). After a participatory consultation process which identified a short list of neighborhood facilities which could be built using municipal funds, a poll was called. Every resident above the age of 10 years was given five ballots, each ballot representing RM (Ringgit, the Malaysian currency) 100. The residents had to decide for themselves how to spend the RM 500 they held and drop the five ballots into the respective boxes for the various options. The children in the community had mobilized themselves to lobby their parents, relatives, and friends to allocate their money votes to a playground. They won convincingly and a playground has been built for the community. While the playground is the concrete result, the more profound achievement by the people is their meaningful role in deciding how public resources allocated for their benefit are used.

Practitioners make use of different combinations of methods and strategies, such as the examples mentioned above, in designing processes for the communities they work with. Field experiences of the practitioners and advice sought from members of the community and people who know the communities intimately should guide the design of processes. Practitioners should keep an open mind about their designs and be alert to the responses of members of the communities once these processes are launched and be ready to tweak and even make major changes to processes to improve them once members of the communities engage with these processes and test their acceptability and efficacy.

The unsung hero of designs and processes that do work is often the "facilitator." This is the person who convenes the meetings and chairs them. He or she also acts as the liaison between members of the community, practitioners, and development workers who provide support to the community's efforts. The facilitators are often overlooked because they tend to work in a low-key manner and many of them are not trained media practitioners. In the case of the Fogo Process, most scholars who have written about the experiment came from specialized media and communication backgrounds and had researched the impact of the Fogo Process from the perspective of filmmaking. What was often overlooked or given less prominence in their analysis was the quiet work behind the scenes rendered by Fred Earle, the extension agent.

According to Susan Newhook (2009) and sociologist Robert DeWitt, who undertook sociological research on Fogo Island in the autumn of 1966, Fred Earle's arrival as the new extension field officer in 1964 had given "new hope to all those who claimed an interest in improvement. Here was the man, they felt, who could provide a well-needed link with the remote government in St. John's. It is hard to imagine any of the developments that followed Earle's appointment happening without him and his employers at the Extension Service. In the Fogo Island films, he appears to be an interested but uninvolved observer of events, but nothing could have been further from the case. Earle was no outsider: he was born on nearby Change Islands and moved to Fogo Island in his teens . . . had once worked . . . as an errand boy and later as a bookkeeper" on the island. It is very likely that Colin Low, the filmmaker, had selected Fogo Island for his experiment from a short list of several potential sites, because of the availability of Earle as a facilitator for the project.

Experienced practitioners of participatory communication have long recognized that an effective facilitator is the key element in participatory communication that is the hardest to replicate when a project proves successful in one community and it attempts to scale up or expand its processes to serve other communities. This chapter will attempt to understand what is it about good facilitators that remain elusive to upscaling and replication. The above analysis of Fogo Island suggests that facilitators must know the community as intimately as an "insider." In addition to inside knowledge and an empathy for the place and people, good facilitators must also be credible individuals who engender trust and goodwill among most, if not all, possible fractions or interest groups within a community. A facilitator who possesses these attributes enjoys a better chance of engaging a community in activities and processes launched. Just as importantly, if not more so, such a facilitator will play a critical role in helping the team from the outside understand sensitivities affecting members and groups within the community and disagreements on divisive issues. An effective facilitator is the frontline member of the project team who manages conflict when they happen and transform such conflicts into opportunities for solution and change.

65.1.3 Transforming Conflict and Competition

Practitioners sometimes begin their work in communities believing so deeply in the intrinsic good of participatory communication that they are sometimes caught unprepared when processes lead to conflicts among the people taking part and the mood in the community sours. An experienced facilitator is indispensable at the outbreaks of conflicts to tap the opportunity it provides to define disagreements and find ways to resolve them and move the community pass the conflicts, hopefully stronger for their occurrence.

Colin Low and his colleagues filming on Fogo Island saw an important role for conflicts. This is what they wrote in their report reflecting on the processes they innovated:

> We think the intensity of discussion should be related to the effect discussion has on direct development. In other words, if, as we were doing, the purpose of the project is inter-community communication and creation of consensus, the degree of conflict then can be constructive; will be less than with other projects. A project that channels the responses into direct action can handle more. The reason for this is obvious. A small isolated community can be torn apart if its people and problems are left exposed and unresolved. The motivation to reconcile problems is greater when direct action is being considered. In a discussion the influence to reconcile conflict comes from the discussion leader. His is a difficult role, for if no conflict or tension is elicited, it usually means that little communication has gone on. The position between definition and recognition, and division, is quite precarious.
>
> Perhaps our films could have catered to the defining of conflicting opinion, more than they did. Although we presented views supporting the adoption of longliners [large ocean-going fishing boats,] and a centralized school, and also opinions against centralization off the island and the effects of welfare, we never heard opinions counter to these. The closest we came was with the *Billy Crane Leaves His Island* (NFLD Archive 2015) reel, and it was the most successful one.

Colin Low felt that the film of fisherman Billy Crane quietly but very thoughtfully and forcefully explains why circumstances on Fogo Island had deteriorated to the extent that he had lost hope for its future and had decided to move to the mainland, as his most successful reel. Crane's repeated mention in the film of "long-liners" – larger fishing boats that could sail further from shore to richer fishing grounds – as a solution and his rejection of government "welfare" payments to islanders proved eventually to be the solution that saved the fishing industry in Fogo.

Given that conflicts are unavoidable and even "good" when effectively processed, how should facilitators and practitioners prepare themselves for their occurrences? Referencing conflict at the workplace, Madalina (2016) identified four types of conflict classified according to "who" are the people involved:

- **Interpersonal conflict** refers to a conflict between two individuals. This occurs typically due to how people are different from one another.
- **Intrapersonal conflict** occurs within an individual. The experience takes place in the person's mind. Hence, it is a type of conflict that is psychological involving the individual's thoughts, values, principles, and emotions.
- **Intragroup conflict** is a type of conflict that happens among individuals within a team. The incompatibilities and misunderstandings among these individuals lead to an intragroup conflict.
- **Intergroup conflict** takes place when a misunderstanding arises among different teams within an organization. In addition, competition also contributes to the rise of intergroup conflict.

Another way of classifying conflict is according to "what" is causing it. Jehn and Mannix (2001) proposed that conflict in work groups may be categorized into three types:

- **Relationship conflict** where there is an awareness of interpersonal incompatibilities involving personal issues such as dislike among group members and feelings such as annoyance, frustration, and irritation.
- **Task conflict** is an awareness of differences in viewpoints and opinions pertaining to the group's task which may coincide with animated discussions and personal excitement but, by definition, are void of intense interpersonal negative emotions that are more commonly associated with relationship conflict.
- **Process conflict** is where controversies arise about aspects of how a task is accomplished, specifically regarding work and how resources are shared – such as who should do what or how much should one receive to get the job done.

A third and potentially more disruptive form of conflict comes from the much appreciated "empowering" nature of participation. When the processes adopted by the people begin to take effect, a couple of things happen – the first may be that participants have been motivated to set aside apathy about an immediate need and begin to work together on addressing that need. The other possibility is that people want to tackle root causes of their problems so as to regain control of their lives and do things their way rather than listen to powerful elites who hitherto have led the community in a top-down manner. Participation is therefore empowering because people are reclaiming power to shape their own lives. Conflict may then follow as previously influential elites attempt to resist and retain the power the people are cooperating toward seizing back. Experienced facilitators are sensitive to preexisting power distributions and of the elites who hold power over the community they are working with. Most facilitators will prefer not to threaten existing arrangements until such time the community is ready to negotiate for new power-sharing arrangements to be made.

Just as many practitioners of participatory communication are not trained in processing conflict, many are also not familiar with the concept of power and how it is wielded.

65.1.4 Issues of Power in the Practice of Participatory Communication

Power has been much studied by both civilians and the military because it rests at the core of administering and protecting communities, kingdoms, and nation-states. Civilians study power as it is the grand prize in politics, business, and religion. The military was the earliest to master power and has been projecting and wielding it over the past millenniums. Practitioners of participatory communication should study power to become sensitive to the potentials of processes as well as the unpredictable dynamics we expose members of the community to when participatory processes begin to adjust power distribution within the community.

Practitioners from a media background have always worked against a backdrop of power issues. In fact, the news media has been intimate enough with power to be referred to as the "Fourth Estate" across the landscape of power found in a nation-state. The First Estate in the realm of a kingdom is the clergy. The Second comprises the nobility and the Third is made up of the people or commoners, hence "The Commons" in the British Parliament, where the "Lords Spiritual" make up the First Estate and the "Lords Temporal" form the Second. It is useful for facilitators and practitioners to keep these "four estates" in mind when engaging with a community.

One of the earliest authors on power was Niccolò di Bernardo dei Machiavelli, a famous writer of the Renaissance period who was variously a humanist, diplomat, politician, historian, and philosopher. His book *The Prince* (translated from Italian) (Machiavelli 1532) published 5 years after his death is the oldest source on thinking about power – not all of it flattering. His name spawned the term "Machiavellian" that refers to the ruthless use of cunning and duplicity in statecraft to achieve one's goals, revealing the bare-knuckle approaches of sixteenth-century, feudal Europe which many practitioners will tell you are still very much the current approaches in many places where they work. "Machiavellianism" has also become a term personality psychologists use to describe a person' who is unemotional, able to detach himself/herself from conventional morality so as to deceive and manipulate others.

A vast literature on power has been published in the six centuries after Machiavelli's book. Michel Foucault (2001) who wrote extensively on the subject from 1954 to 1984 pointed out the limited power of the Fourth Estate. He wrote about the necessity to distinguish power relations from "relationships of communication that transmit information by means of a language, a system of signs, or any symbolic medium. No doubt, communicating is always a certain way of acting upon another person or persons. But the production and circulation of elements of meaning can have as their objective or as their consequence certain results in the realm of power; the latter are not simply an aspect of the former." In other words communication is a necessary but not a core element of power. To Foucault power refers to relations between individuals or between groups where certain persons are able to exercise influence over others. The media is a mechanism in this exercise.

John French's and Bertram Raven's chapter on "The Bases of Power" (French and Raven 1959) is a useful reading to grasp the types of power that the community need to deal with when interacting with elites who wield it. There are five bases or fundamental elements of power:

- **Reward power**: the ability of a person or social agent to reward another person. For example, agricultural-extension officers who can hand out to farmers free seeds and fertilizer.
- **Coercive power**: similar to reward power but the social agent also possesses the ability to manipulate conditions to obtain the rewards. For example, a minister in the government who can approve the above farming subsidies as well as set new policies to increase such subsidies or waive regulations which disallow them.
- **Legitimate power**: created by the beliefs and values of a group which accord individuals the legitimate right to influence members of the group who in turn have an obligation to accept this influence. For example, a traditional ruler, or a religious leader, who enjoys the faithful support of either a community that reveres the royal institution or a pious community that embraces a set of religious doctrines.

- **Referent power**: an attractive person who has power over others because they in turn may derive prestige from being closely associated with this attractive person. A celebrity, such as a famous actress, wields such power. The greater the attractive attributes of this actress, the greater is her public identification, and consequently the greater the referent power that she possesses and wields. This is the reason why development agencies appoint popular celebrities as "ambassadors" of the causes they promote.
- **Expert power**: people with more knowledge and more information have power over others who don't have the knowledge or information. The strong influence doctors exert over their patients who are ill is a clear example of such power. External development "experts," participatory communication practitioners included, may also possess such power.

Practitioners when faced with challenges posed by the strong influence that the powerful have over a community that resists change are often puzzled by such apparent apathy. Gaventa (1980) suggests that such powerlessness or "quiescence" or motionlessness is due to the mechanisms of power being wielded to curb independent decision-making by the people. He established three "dimensions" within such mechanisms:

- **First dimension**: the powerful is usually the one who always prevail in bargaining over the resolution of key issues – the community or group also frequently surrenders decision-making to this powerful person.
- **Second dimension**: the reason why people surrender their independence in making decisions is because the powerful individual is able to mobilize bias and support through various means including the use of force, sanctions, manipulation, and invocation of existing biases.
- **Third dimension**: the way the powerful individual influences, shapes, and determines issues includes his/her mobilization of communication and information. The Fourth Estate hard at work for the Lords Temporal instead of the commoners, for example! In the same way, this role can be reversed – participatory communication processes can be adopted by the people to mute such influence, to clarify issues, and to shape solutions that better serve the community's interests.

65.1.5 Empowering the People

Practitioners often find that one of the unspoken, first goals of a participatory communication brief is to design processes and messages that enable people to discover that they are capable of making independent efforts to improve their lives. This may in turn motivate the people to work together to reclaim power and opportunities that they had surrendered to others so as to improve their lives and environment, in short to "empower" the people.

"Empowerment" interestingly emerged as a professional term in the 1980 presidential address by Julian Rappaport to the American Psychological Association where he was referring to the work of psychologists and the expert power they held over patients. He explained to his colleagues: "By empowerment I mean our aim should be to enhance the possibilities for people to control their own lives." Two decades would pass before his colleague Marc Zimmerman (2000) would propose the empowerment theory in a handbook on community psychology edited by Rappaport himself. The theory calls for a "value orientation" which it succinctly describes:

> An empowerment orientation also suggests that community participants have an active role in the change process, not only for implementing a project, but also in setting the agenda. The professional works hard to include members of a setting neighbourhood, or organization so they have a central role in the process. Participants can help identify measurement issues and help collect assessment and evaluation data, but the results are also shared. Feeding back information to the community and helping to use it for policy decisions is a primary goal. An empowerment approach to evaluation focuses as much attention on how goals are achieved as on outcomes.
>
> The theory emphasizes empowerment as context and population specific. Empowerment "takes on different forms for different people in different contexts. A distinction between processes and outcomes is critical." Empowering processes attempt to establish control, secure needed resources and critically understand one's environment. Empowered outcomes refer to how these processes were implemented and their impact; not only in material terms but more importantly at the social or community level. Outcomes are determined through studying consequences of the people's attempts to gain more control of how their lives are lived within their community. In carrying out such study each level of analysis may have to be undertaken separately but it is crucial that the people undertaking the analyses recognize that all these different levels are inherently connected to each other. Such study has to determine how individuals, organizations, and community empowerment are mutually interdependent and that they are all at once a cause and a consequent of each other. "Efforts to understand empowering processes and outcomes are not complete unless multiple levels of analysis are studied and integrated."

The preceding discussion of how conflict is an inevitable dynamic within participatory processes and its connection to the bases of power is an extremely brief one. It barely skims the surfaces of the disciplines involved and serves to introduce practitioners to these areas for further study and interdisciplinary research by practitioners and their colleagues who work on these issues. Elisheva Sadan (2004) who teaches social work researched the rather wide field of power and concluded that the literature on it diverges. She found that the divergence made it difficult even for us to define power. She proceeded to develop a theory of empowerment in the "shadow" of those disagreements. Her theory is shared here because it had been proposed in the context of community development that most practitioners of participatory communication do their work. Her starting point echoed that of the psychologists – the process of empowerment means a transition by people "from a state of powerlessness to a state of more control over one's life, fate, and environment." She suggests that the process is aimed at achieving three conditions:

- Changing feelings and capacities of individuals
- Changing the life of groups and communities

- Changing the professionals (such as practitioners of participatory communication who are involved in projects aimed at bringing about social change) as they learn from their collaboration with people and improve upon their skills in further practice within communities

It is an interesting theory that situates the outsiders – development workers and practitioners – within the process. She sees the work of outside professionals as "methodical intervention aimed at encouraging processes of individual and community empowerment." Her perspective is encouraging to practitioners who are often reminded by critics who hold the view that true empowerment cannot come from the outside. Their argument is that facilitators and professionals are creating pseudo-empowerment at best or introducing new kinds of dependency among the people they work with if the outsiders became overly active in the processes they are facilitating within communities.

The starting point, for Sadan, is the empowerment of the individual. It can happen in "an immense variety of circumstances and conditions" quite independently of the presence of outside facilitators or the empowerment of the groups that the individual is a part of. However, individual empowerment carries greater value when it occurs as a result of the individual's participation in social change processes and activities that are connected to groups and organizations as both the individual and groups derive "special value" from this communal experience.

Next is empowerment of a community which involves organizing and creating a community which shares a "common critical characteristic," for example, suffering from social stigmas and discrimination – this may range from small communities of people living with HIV to population-wide groups such as women who are segregated by glass ceilings within their communities. Sadan says such community organizing and building help strengthen the ability of people to "control its relevant environment better and to influence its future... develop a sense of responsibility, commitment, and ability to care for collective survival, as well as skills in problem solving, and political efficacy to influence changes in environments relevant to their quality of life."

Among the many cases of community empowerment, some of the most effective and inspiring have been led by women who rally their own strength to improve their lives and conditions. It was perhaps Caroline Moser's (1993) book on gender planning and development that made the term empowerment familiar to many professionals working in development. "The Empowerment Approach" proposed by Moser, while acknowledging inequalities between men and women and the subordination of women in the family, questions assumptions between "the interrelationship between power and development." The approach identified power "less in terms of domination over others... and more in terms of the capacity of women to increase their own self-reliance and internal strength." It was an approach which did not frame power as a zero-sum equation – the power gained by women did not mean a corresponding loss by the men. It was an approach with a conflict-transforming ethos embedded within. The other notable conflict-transforming feature of this approach is how it "utilizes practical gender needs as the basis on which to build a secure support base, and a means through which strategic needs may be reached." Moser's differentiation of these two gender needs was an elaboration of Maxine Molyneux's (1985) conceptualization of the practical and strategic interests of women in the following manner:

- **Strategic interests** are the "real" interests identified logically via analysis of the root causes of problems faced by the people – in the case of women, their subordination and discrimination – and from the visualization and articulation of a happier and more acceptable way of living and working for the hitherto disempowered people.
- **Practical interests** are about immediate needs which even if met do not help to advance the people's strategic interests. In other words practical interests don't help to overcome root causes of problems, for example, government handouts to poor farmers which while meeting the immediate practical needs of the farmers to put food on the table fail to overcome the conditions causing them to be poor.

65.1.6 The Approach to Practicing Participatory Communication

65.1.6.1 Where to Start?

Moser's and Molyneux's advice is to begin with practical issues even though the overriding imperative in development is often to tackle root causes. Beginning directly with root causes, particularly if you are new in practice, is designing difficulties – and even failure – into your project. New practitioners should not only consider practical issues but also the less complex practical issues. For example, work initially on water shortage for farming in a small neighborhood rather than attempting to reform the entire irrigation system of a district. Starting with little practical things allows everyone to get acquainted, develop robust processes, and build trust among those participating. Trust is essential for sustaining members of a group through difficulties they will face when tackling the bigger issues. Keep people who hold reward, coercive, and legitimate powers informed. Involve them in a support role if their involvement does not entail risks of curtailing what is planned. Beginning with small issues is unlikely to threaten them into undermining your work.

65.1.6.2 Begin with Individuals

The urge is to work with groups given that participatory communication is the process. However, groups are only as vibrant as the individuals who form them.

Paulo Freire (2007) in his influential book about critical pedagogy – *Pedagogy of the Oppressed* – highlighted the relationship of nonparticipation to nonconsciousness of deprived groups trapped in highly unequal power relationships. The people in such groups live in "closed societies" that renders them powerless and highly dependent upon the powerful. Their isolation paralyzes people from initiating independent action. They are unable to reflect on their actions and understand outcomes of their quiescence. Freire's solution to these two conditions is "conscientization" – from the Portuguese term *conscientização* or consciousness-raising (Freire Institute undated). The concept of conscientization is often mentioned by facilitators as one of the processes that help to "empower" passive individuals and groups. This is best achieved by first focusing on the needs of individuals participating in group activities.

A good way to start is with a small group of about half a dozen to a dozen people in a community who show an interest of working with the participatory communication team. This may take the form of a nonformal education activity which will be the least threatening to everyone within and outside the community. This small group can together select a topic or theme that they all share an interest in. This could be learning to read and write or how to get online and do e-commerce. The manner in which the activities are conducted is probably more important than the topics selected. Adopt participatory learning approaches where the participants play an active role in planning the program which could include fun, social items if the participants opt to do so. The facilitator is the resource person who responds to the queries posed by individuals.

The aim is twofold to Rowlands (1997):

- Developing a sense of self, individual confidence, and capacity and undoing the effects of internalized oppression
- Developing the ability to negotiate and influence the nature of relationships and decisions made with others whom an individual works with or relates to

The "internalized oppression" that needs undoing is succinctly described in a case study (Chan 1997) of young women recruited from villages to work on the high-pressured assembly lines of high-tech factories in an export-processing zone: "Daily experiences of emotional subordination generate feelings of shame, guilt, inadequacy, self-doubt and inferiority. The cumulative effect of these experiences is to ingrain a deep sense of helplessness, fear, ineptitude and incapacity. Minds become blank and dulled over time ... various approaches have been experimented with, to facilitate the healing and recovery... Story-telling-sharing has been used extensively and effectively as a tool for consciousness-raising and mobilization..." This overcoming of internalized oppression is at the same time also the resolution of intrapersonal conflict discussed previously.

65.1.6.3 Iterate with a Community

Storytelling and sharing is also an effective process for groups. Stories told or presented by actors may be used to discover causes and define issues; these are the "stories without a beginning" where the actors or storyteller presents a story which mirrors the current plight of the group or community. After the presentation the facilitator invites the audience to suggest reasons for the problems faced by the characters in the short play. In "stories without the middle," actors present the recent past of the community followed by a portrayal of its preferred future. The audience is then invited to essentially discuss what happened in the present that helped the community overcome the past and achieve its future. The third variation is "stories without the end" where the past and present are acted out and then the audience visualizes how it all ends up. These "stories withouts" process is potentially non-threatening and conflict-managing if the performances, while mirroring the situation of the community, are presented as entertaining street theater about an imaginary place and people so that members of the community in the audience may be free to analyze causes, make criticism of their current plight, and visualize their desired future, as if they were addressing it to the supposedly entertaining play which serves as a proxy punching bag. If the actors or storyteller are/is skilled and experienced in the process, the sessions can end with the missing part of the stories presented spontaneously based on the story line which emerged in the discussions. The process can make a profound impact on the actors if they happen to be members of the community itself. The preparatory processes involving scripting of the stories, rehearsing the lines, and presenting it before the community can leave deep conscientizing and empowering impressions on those actively involved.

Participatory communication teams may engage with a group or community directly, without the prior step of conscientization of individuals within the group, if they are able to engage with organized communities, such as those on Fogo Island, who have already established a working relationship with development agencies that share similar interests and objectives of the participatory communication team. In other communities, individual-focused programs such as the ones described above may naturally lead to the formation of groups. For example, individuals learning together how to start an e-commerce business may decide it is advantageous to group together and run a community website and shared order-fulfillment facilities to enjoy some benefits of scale and at the same time take advantage of the possibilities of cross-selling to each other's customers. Efforts which began with "training" the individuals can now progress to facilitating the individuals consulting with each other and exploring how they can work together to tap synergies and powers offered by collaborating within a group.

In the past practitioners had thought of communities as groups of individuals who live and work in locations close enough for them to share the same problems and environment. Their proximity gave them the opportunity to meet face-to-face to share their problems and work together on possible solutions. However, with the advent of the Internet, online communities comprising people who live in locations distant from each other have been formed. They are also known as "virtual" or online communities. They meet remotely via email threads, chat groups, or gaming sites, and these exchanges can be either "public" (where anyone can chip in) or "private" where a participant needs to be preregistered. The remote and virtual nature of these online gatherings limits what these communities can do to address local development issues. Online communities have proven to be very effective in mobilizing support and conducting advocacy about issues at both the national and international levels. Such communities have also been effective in raising funds and asserting political support for disadvantaged and marginalized individuals and groups trapped within harsh regimes.

Communities of neighbors who live in the same village or neighborhood also frequently make use of online networking facilities to keep in touch with each other. This fusion of actual and virtual networking processes creates some of the most powerful linkages for keeping a community informed, organized, and mobilized for coordinated action. The advent of web-based social media has offered practitioners with a low-cost but highly potent channels for participatory communication.

In communities where Internet access is affordable and universal among all members of the group, set up a chat facility or a simple group email account where members can keep everyone in the group updated on what has been done, and is being done, by individuals on furthering the group's work plan. The success or failure of the individuals and subgroups can be shared. The facilitator can take the lead in reflecting on the action taken and encourage everyone to share their analysis on why efforts worked or failed and to suggest the next course of action. This process is what Freire described as "praxis" – people acting together to address issues and afterward reflecting on what they had done (Freire Institute undated). Such action and reflection, which is usually followed up by further action and reflection, helps members of the group to gain deeper understanding of their lives. It also strengthens senses of affinity and confidence among members of the group to mount increasingly complex action which addresses the bigger challenges or strategic interests of the community.

The aim here is to discover the attributes of the collective. To learn that individuals working together will make more extensive impact than each could have achieved on their own. This includes appreciating the efficacy of collective action based on cooperation rather than competition (Rowlands 1997).

65.1.6.4 Everyone Is Unique

Facilitators should always remind themselves that all individuals are unique. When they get together and form groups, these individuals also make their communities unique. As such the practice of participatory communication is rarely always done according to the same methods and designs and at the same pace. Always get the individuals and groups involved in deciding what to do next and how to get it done. In this way their unique attributes and considerations will drive their unique processes forward at a pace and level of risk they are comfortable with.

65.1.6.5 Participation Costs Valuable Time

Practitioners and outside development workers often do not realize that members of a community sometimes decide not to participate in an initiative because they cannot afford the time to do so. Members of the project and participatory communication teams are often paid a salary to do their work; they assume that the people in the community will be happy to participate in activities they initiate because it aims to benefit the people.

However, time is more precious to people in a community than what outsiders can sometimes sense. People in villages often quietly work longer days than people in "9-to-5 jobs." Some of the work may be done at home – men busying with postharvest processing of crops they have grown, women sewing and weaving, and children often taking small animals out to graze after school. Find out how much time people can spare for activities and what times of the day, week, and year are most convenient for them. Prepare the project team to work according to the convenience of the people rather than your own "office hours."

A fair way of compensating the participants in project activities is to host a simple meal before or after a gathering or work session. The meal also offers members time to socialize and bond. Allow parents to bring their children with them – many mothers have no one at home to look after their youngest children and therefore may choose not to leave them at home alone and uncared for. Make arrangements for a member of the team to plan activities for the children so that their parents may focus on project activities.

65.1.6.6 Value Local Contributions and Risk-Taking

Practitioners need to be mindful that seemingly modest contributions by the community may be more significant than they appear to the better-resourced project team. Villagers who rarely treat themselves to snacks and hot beverages are demonstrating generosity when they serve such snacks and drinks at gatherings. They probably had to make use of laboriously gathered fuel to boil water and contribute items from their family's sometimes sparse pantry to make the snacks.

Subsistence farmers who agree to take part in trial planting of a new seed are taking a huge risk on behalf of their families who will face food shortages for a long time if the trials prove unsuccessful. For the same reason, allow the people to do their own risk assessment when opting for a solution that makes use of unproven solutions. This is where the strength of the group and participation may be demonstrated and learned. The community may opt to share risks and carry out the field trials on a shared plot and apportion the cost of failure or taste the harvest of a successful crop. Success is not measured by how many farmers and plots get planted with the new seeds but by how risks are appraised in a participatory way and how the risks of trials are shared.

65.1.6.7 Be Prepared to Transform Conflict into Praxis

Make sure everyone in the project team is trained to recognize conflict and how to respond when it occurs. A useful initial step when convening a group for the first time is to develop with the help of all members the ground rules for working together, including how to process differences of opinion and approaches. Disagreements need to be recognized from the start as a normal occurrence in group participatory efforts. It should also be agreed that everyone will work on such differences in an unemotional way. At the same time, it is useful to introduce Friere's concept of "praxis" which values group efforts at solving disagreements and that talking and working through a disagreement is a process of strengthening the community. Give quality time to solving all disagreements and problems. A good device to establish right from the beginning is a section of the wall where the group meets as an "issue park." When the group meets an impasse, summarize the issue on a card or a piece of paper, and post it on the wall, and invite everyone to help find a solution by posting their ideas for the solution on cards under the issue parked. Then move on to other issues, and give members time to reflect and form ideas about how to resolve the parked issue. The facilitator should use her/his good judgment to chat about the issue informally with the members most affected by the parked issue and explore acceptable solutions. Consult members informally for their readiness to return to the parked issue before facilitating another group discussion about it.

When a commonly accepted solution is found, implement it carefully following the way the group has worked out the solution. After the solution has been implemented and its efficacy proven (or not), give the group time to appraise the results; make sure everyone has their emotions in check before the group is invited to reflect on the issue. Recognize the achievement of the group in resolving the disagreement and developing a solution for it. Celebrate the advancement of the group if members choose to do so. Acknowledge the contributions and graces of members who made it possible.

Opt to leave deeply divisive issues in "long-term parking mode" if members of the community and the facilitator sense the group is not ready to tackle them productively. Work on other practical issues first.

65.1.6.8 Listen for Silences and Watch Body Language

Most facilitators find a silent group one of the hardest to work with. This is common in newly formed groups where members need time to get comfortable with each other and unfamiliar processes introduced by the facilitator. Starting with practical issues and working on "simple" problems that don't involve tricky power considerations will help members develop confidence in processes and each other.

More challenging silences will occur when the group progresses to issues that touch on strategic considerations. One of the reasons may be cultural. Keeping quiet, not responding or reacting, is often considered the polite thing to do when confronted with a delicate issue that one does not wish to deal with as it may provoke or embarrass others. There are deep meanings in silences. A facilitator has to be culturally sensitive to decipher such meanings.

A facilitator may also read what the silent participants are saying via their body language. This "includes facial expressions, eye contact, voice modulation, posture and gestures, attire, appearance, handshake, space, timing, behavior and smile" (Sharma 2011). Body language is complex and requires a comprehensive understanding of the cultural norms of the "speaker." For example, in many Asian communities, maintaining constant eye contact is impolite and under some circumstances may be decoded as being hostile, while in most Western cultures, avoiding eye contact may be considered impolite or suggest the telling of a lie or hiding a truth.

Understanding body or nonverbal language can help facilitators and practitioners better appreciate the emotional elements of what is being said. This enables the facilitator to determine the emotional subtext being transmitted along with the spoken words.

65.1.6.9 Practitioners to Learn and Share with Community Lessons Learned

Facilitators usually adopt quick self-tabulating methods of evaluation which routinely wraps up an activity. This maybe a simple "mood-meter" where participants paste colored sticky dots under emoji symbols sketched on a piece of flip-chart paper. The paper is usually hung on a board that faces away from the group so that participants can paste their dots anonymously. Once everyone has pasted their dots, the board is turned around, and the mood of the group is immediately apparent from this self-visualizing straw poll.

The facilitator and practitioners present may then do a quick reflection of what she/he has learned from the group during the activity that just concluded. The members of the group may then share their reflections and at the same time suggest how future activities may be tweaked to make them better. The process can be alternated with members of the group taking the lead in reflection before the facilitator and practitioners. This will be in line with the action-reflection process promoted by Friere.

65.1.6.10 Plan for Gradual Withdrawal of the Outsiders and Handover to the Community

Members of the group and community should be encouraged to play an increasingly central role in planning and facilitating activities. Correspondingly, outside facilitators and practitioners should scale back their roles but remain present at activities. When the time is right – which may be weeks or years – depending on the people, the issues, and the location, the outside team could ask the community for help in planning for the withdrawal of the outsiders and the handover of project management to the community. This may be a gradual process with a long-term agreement that the people can contact the outsiders for "tech support" especially in projects that involve new technologies or outside government and development agencies whenever that is required.

The outside project team may also offer to stay in touch with the communities via social media spaces set up for the project or even via remote participation such as a Skype hookup if the community has affordable Internet access.

65.1.6.11 Community Sharing Experiences with Other Communities and Developing Ways of Upscaling

An element of the above withdrawal and handover plan may include participation of members of the community with the outside team in similar projects involving other communities. A significant limitation of participatory communication initiatives has always been the difficulties of "scaling up" approaches which have been found to work. Part of this limitation stems from the people-embodied nature of participatory methods and approaches which require the training of members of the community as well as the project team itself. Given the usually limited number of practitioners of participatory communication teams, they can reach only a limited number of communities during their deployment in the field.

The ultimate solution may reside in the pure spirit of participation and empowerment where people in the community expand and extend processes among themselves. In this way they will make the processes better match goals they want to achieve and nurture cultures and values they appreciate intimately.

References

Benor D, Harrison JQ, Baxter M (1984) Agricultural extension: the training and visit system. The World Bank, Washington, DC

Chan LH (1997) Women on the global assembly line. In: Walters S (ed) Globalization, adult education and training: impact and issues. Zed Books, London, pp 79–86

Foucault M (2001) Power: the essential works of Foucault, 1954–1984 (Faubion JD [ed], trans: Hurley R), vol 3. The New Press, New York
Freire P (2007) Pedagogy of the oppressed. Continuum, New York
Freire Institute (undated) Concepts used by Paulo Freire. Available via http://www.freire.org/component/easytagcloud/118-module/conscientization/. Accessed 18 Dec 2017
French JRP, Raven B (1959) The bases of social power. In: Cartwright D (ed) Studies in social power. Research Center for Group Dynamics/Institute for Social Research/University of Michigan, Ann Arbor, pp 150–165. Available via: http://web.mit.edu/.../Power/French_&_Raven_Studies_Social_Power_ch9_pp150-167.pdf. Accessed 15 Dec 1017
Gaventa J (1980) Power and powerlessness: Quiescence and rebellion in an Appalachian valley. University of Illinois Press, Urbana
Jehn KA, Mannix EA (2001) The dynamic nature of conflict: a longitudinal study of intragroup conflict and group performance. Acad Manag J 44(2):238–251
Machiavelli N (1532) De Principatibus/Il Principe. Antonio Blado d'Asola
Madalina O (2016) Conflict management, a new challenge. Procedia Econ Financ 39:807–814. Available via http://www.sciencedirect.com/science/article/pii/S2212567116302556. Accessed 12 Dec 2017
Molyneux M (1985) Mobilization without emancipation? Women's interests, the state, and revolution in Nicaragua. Fem Stud 11(2):227–254
Moser C (1993) Gender planning and development: theory, practice and training. Routledge, London
Nemtin B, Loh C (1968) Fogo Island film and community development project. National Film Board of Canada, Ottawa. Available via http://onf-nfb.gc.ca/medias/download/.../pdf/1968-Fogo-Island-Project-Low-Nemtin.pdf Accessed 5 Dec 2017
Newhook S (2009) The Godfathers of Fogo: Donald Snowden, Fred Earle and the roots of the Fogo Island Films, 1964–1967. Newfoundland and Labrador Studies, [S.l.], June 2009. Available via https://journals.lib.unb.ca/index.php/NFLDS/article/view/14408/15473. Accessed 7 Dec 2017
NFLD Archive (2015) Fogo Project Billy Crane Moves Away. Available via https://www.youtube.com/watch?v=e3NKnUpj22I. Accessed 12 Dec 2017
Practical Action (2007) Wall Newspapers. Practical Action, Rugby, Warwickshire. Available via https://www.sswm.info/library/2526. Accessed 9 Dec 2017
Rappaport J (1981) Praise of paradox: a social policy of empowerment over prevention. Am J Community Psychol 9(1):1–25
Rowlands J (1997) Questioning empowerment: working with women in Honduras. Oxfam, Oxford
Sadan E (2004) Empowerment and community practice (trans: Flantz R). Available via: http://www.mpow.org. Accessed 18 Dec 2017
Salas MA et al (2010) Visualisation in participatory programmes: how to facilitate and visualise participatory group processes. Southbound, Penang
Schellenberg JA (1996) Conflict resolution: theory, research, and practice. State University of New York Press, Albany
Shariza K (2016) Gender responsive and participatory budgeting in Penang: the people-oriented model. In: Ng C (ed) Gender responsive and participatory budgeting: imperatives for equitable public expenditure. Springer/SIRD, Switzerland/Malaysia
Sharma V (2011) Decoding non-verbal communication. Available via https://www.researchgate.net/publication/283794137
Tandon R (ed) (2005) Participatory research: revisiting the roots. Mosaic Books, New Delhi
UNESCO (undated) Defining community broadcasting. UNESCO, Paris. Available via https://en.unesco.org/community-media-sustainability/policy-series/defining. Accessed 9 Dec 2017
Zimmerman M (2000) Empowerment Theory. In Rappaport J and SeidmanHandbook E (eds) Handbook of Community Psychology. Springer US

Capacity Building and People's Participation in e-Governance: Challenges and Prospects for Digital India

66

Kiran Prasad

Contents

Abstract

Access to ICTs alone does not make for successful national e-governance projects in developing countries but requires participatory efforts to promote democratic practices. India's ambitious Digital India project, key to its administrative reform agenda, proposes to extend the Internet to the remotest of villages by 2017. The foundation of this initiative is a program of e-literacy, capacity building, and installation of ubiquitous broadband-enabled computer kiosks based on entrepreneurial public-private partnerships. The best example of this is the Akshaya

K. Prasad (✉)
Sri Padmavati Mahila University, Tirupati, India
e-mail: kiranrn.prasad@gmail.com

J. Servaes (ed.), *Handbook of Communication for Development and Social Change*,
https://doi.org/10.1007/978-981-15-2014-3_12

centers project in Kerala, a potential model for the rest of India and other developing nations interested in e-governance initiatives to bring innovative administrative reform. The capacity-building Akshaya e-literacy project of Kerala Government was implemented in 2002 jointly by the Kerala IT Mission and Department of Science and Technology, with tie-ups with local bodies and voluntary agencies. This project aimed at making Kerala the first completely e-literate state in India. It is expected that the IT infrastructure expanded through Digital India project will create and expand economic opportunities, empower individuals and communities through enhanced access to information, modernize and upgrade skill sets, integrate communities through creation of e-network, create awareness of ICT tools and usage, generate locally relevant content, and lead to an environment for digital democracy. The chapter will also critically examine the challenges and prospects of Digital India for inclusive governance and citizen participation through education and training that could change priorities, save money, and deliver better results through digital empowerment of people.

Keywords

e-governance · Capacity building · People's participation · Digital India · Inclusive governance · Digital democracy

66.1 Introduction

The UN e-government rankings assess the capacity of a country to fully leverage information and communication technologies (ICTs) in daily activities and production processes with efficiency and competitiveness to create an e-participation index to track the extent to which citizens participate in e-governance (UN 2010). While political processes such as decentralization and participatory governance are high on the international development agenda, it must be recognized that the motivations and imperatives for adopting e-governance in a developing country like India are vastly different from those in the developed countries. India is the second largest nation with a population of over one billion, with the largest rural population (857 million) (UN 2014) in 640,867 villages (Census of India, 2011). India is ranked 135 out of 185 countries in human development with 55% of its population experiencing multidimensional poverty (UN 2014). Hence e-governance initiatives need to be planned with reference to these ground realities in the country.

e-governance emerged out of a realization that ICTs can be utilized to effectively provide services to a population of over one billion people (Prasad 2004, 2009). e-governance is now seen as a key element of the country's governance and administrative reform agenda. Taking note of the potential of e-governance to improve the quality of life of the vast population of the country, the Government of India has formulated a national program – the National e-Governance Plan (NeGP). The NeGP has been scaled up into the Digital India project with an investment of Rs 1.13-lakh crore ($ 2.1 billion) to provide thrust to nine pillars which include broadband highways, everywhere mobile connectivity, Public Internet Access Programme, e-governance, e-Kranti (which aims to give electronic delivery of services), information for all, electronics manufacturing, IT for Jobs, and early harvest programmes (http://digitalindiaproject.com/). The Prime Minister is the Chairman of the Monitoring Committee of the Digital India project under which all central government ministries and departments will come up with their individual projects that can be delivered to public using ICT like health services, education, judicial services, etc. The Digital India project aims to offer a one-stop shop for government services using the mobile phone as the backbone of its delivery mechanism. Government prefers to adopt public-private partnerships (PPP) wherever feasible for rolling out Digital India programme. The best example of this is the Akshaya project in Kerala, a potential model for the rest of India and other developing nations interested in participatory e-governance initiatives. What can one learn about the Akshaya e-literacy project in India about capacity building and inclusive governance for designing the Digital India project? Access to ICTs alone does not make for successful national e-governance projects in developing countries but requires participatory efforts to promote democratic practices.

66.2 Digital India Project

India's ambitious Digital India project, key to its administrative reform agenda, proposes to extend the Internet to the remotest of villages by 2017. Digital India has three core components. These include creation of digital infrastructure, delivering services digitally, and digital literacy. Digital India proposes to extend the Internet to the remotest of villages, but making this relevant at the local level requires participatory efforts to promote democratic practices. The foundation of this initiative should be e-literacy, capacity building, and installation of ubiquitous broadband-enabled computer kiosks based on entrepreneurial public-private partnerships (Prasad 2012: 183). This chapter seeks to situate and highlight the Akshaya e-literacy project in the specific context of Kerala. The only initiative to make ordinary citizens e-literate in India is in the state of Kerala, the study of which can assist in determining how e-literacy can impact e-governance. The findings of this study have implications for planners and policy makers in India seeking to advance people's participation in e-governance and apply to India as a whole – with its strategic geopolitical position as the largest country in South Asia and the largest stable functioning democracy in the world – and have broader relevance in the context of other developing countries in Asia and Africa that have similar demographic, socioeconomic, and cultural characteristics.

The e-governance initiatives in Kerala have been commended by international agencies and have also won admirers from outside India. The World Bank delegation found the Kerala State IT Mission (KSITM) competent to perform the role of an international consultant, especially to developing countries in Asia and Africa (Praveen 2011). The KSITM had the added advantage of practical experience in rolling out e-projects and is a pioneer in the use of free and open software in e-governance. A delegation from Zimbabwe visited Kerala in December 2010 to study how to revive its economy through the application of ICT solutions. A delegation from Bangladesh visited Kerala in May 2011 to learn from its experience in mobile governance applications and the citizen-centric delivery of e-governance services undertaken by the state through the KSITM. Bangladesh is in the process of setting up 4500 net-enabled information centers similar to Kerala's Akshaya Common Service Centres (CSCs) and wants an integrated application of ICTs in the delivery of e-governance services which are presently scattered across multiple networks and servers.

The subsidiary research questions:

1. Are the Akshaya centers being implemented in Kerala in line with the objective of increasing citizen participation in e-governance?
2. Are the Akshaya centers designed within a public-private partnership framework which would augment the financial viability of the centers?
3. Do these centers serve as anchor institutions or demand aggregators and which provide digital literacy instruction, continuing education, job training, and entrepreneurship classes for capacity building?

66.3 e-Governance and People's Participation

The role and importance of information and communication technologies (ICTs) attracted the attention of the Indian government, and the deployment of ICTs began as early as the 1970s. In 1985, the Indian government decided to increase the pace of ICT use in the 1990s. The National Informatics Centre Network (NICNET) connected district-level and rural-level government offices to government secretariats in the state capitals and was in turn connected to the national network in New Delhi. e-governance in India steadily evolved from computerization of government departments to fragmented initiatives aimed at speeding up e-governance implementation across the various arms of the government at the national, state, and local levels (Arul Aram 2004). These fragmented initiatives were unified into a common vision and strategy provided by the National e-Governance Plan (NeGP) in 2006 (NeGP 2006).

The NeGP takes a holistic view of e-governance initiatives across the country and envisages a model for delivery of web-enabled anytime and anywhere access to information and services in rural India. Around this idea, a massive countrywide infrastructure reaching down to the remotest of villages is evolving, and large-scale digitization of records is taking place to enable easy and reliable access over the Internet. The vision of the NeGP is to "make all Government services accessible to the common man in his locality" (http://arc.gov.in/11threp/ARC_11thReport_Ch7.pdf). The integrated projects include e-business, Common Service Centres, India portal, e-procurement, and e-courts.

According to Kalam (2005: 37), "e-governance should enable seamless access to information and a seamless flow of information across the state and central government in the federal set-up." Research has suggested that in developing countries, ICTs have largely been employed in efforts to streamline labor-intensive bureaucratic transactions rather than in participatory or consultative efforts to promote democratic practices (Bekkers and Homburg 2007) or citizen engagement (Mosse and Whitley 2009). Many e-government initiatives focused on adapting e-commerce models to increase efficiency, for example, tax administration (Jamaica, Guatemala); better services to customers, businesses, and stakeholders in general (Brazil, India); and government for transparency and business efficiency (the Philippines, India, Chile) (Raman 2013). Colle (2000) emphasized the need for participation by local communities in telecentres.

The Common Service Centre (CSC) is a strategic cornerstone of the National e-Governance Plan (NeGP), approved by the government in May 2006, as part of its commitment to introduce e-governance on a massive scale. CSCs or broadband-enabled computer kiosks offer a range of government to citizen and business-to-customer services, besides promoting sheer access to the Internet. Information management systems are focused to ensure that relevant information is available anywhere anytime and in any way for interactions between government to government (G2G), government to citizens (G2C), and government to businesses (G2B). The scheme creates a conducive environment for the private sector and NGOs to play an active role in implementation of the CSCs and become partners of the government in the development of rural India.

The PPP model of the CSC scheme envisages a three-tier structure consisting of the CSC operator (called Village Level Entrepreneur or VLE), the Service Centre Agency (SCA) that will be responsible for a division of 500–1000 CSCs, and a State Designated Agency (SDA) identified by the State Government responsible for managing the implementation over the entire state. The CSCs are aimed at providing high quality and cost-effective video and voice and data content and services, in the areas of e-governance, education, health, telemedicine, entertainment, as well as other private services. CSCs also offer web-enabled e-governance services in rural areas, including application forms, certificates, and utility payments such as electricity, telephone, and water bills (http://www.csc-india.org/). The total number of CSCs in India as of 31st March 2016 is close to 200,000 that are operational in the country (http://www.csc.gov.in/).

The Digital India project launched in August 2014 augments the services of CSCs to offer a one-stop shop for government services using the mobile phones as the backbone of its delivery mechanism. The Telecom Regulatory Authority of India (TRAI) pegged the number of Internet users in India at 278 million as of October 2014 (www.trai.com). It is expected that India's web user base will grow to 354 million by June 2015, and it is largely driven by increased Internet use on mobile phones with 173 million mobile Internet users (128 million urban users and 45 million rural users) in December 2014 (IAMAI and IMRB, 2015). India has the world's third largest Internet users after China and the United States with three-fourths of its online population under the age of 35. Access to the Internet media is largely concentrated in the urban and semi-urban areas. One would think the number would be much higher, given the country's fairly advanced capabilities in the software field, but this is typical of India's political economy paradox, large swathes of backwardness amid high economic growth rates (Ram 2011).

Communities with greater capacity for learning will achieve more desirable outcomes from their telecentres than communities with low capacity for learning; this is identified as a success factor for telecentres (Harris 2001). The capacity building among citizens began with awareness to the right to information created by all media like Internet, cable TV, community/FM radio, and the vernacular press. Combined with appropriate content, connectivity, and capacity-building measures, the media has helped in ushering in higher awareness about governance and motivating people's participation in political processes. Since India has opted for a model of assisted access, particularly in rural areas, building capacity among the service center operators is a key area of attention (Das and Chandrashekhar 2007).

The availability of the new media including mobile communications, social networking sites, and the Internet supported people's participation in movements for the right to information, anti-corruption, and environmental conservation in India. The anti-corruption national people's campaign led by veteran social activist Anna Hazare who fasted for 13 days from August 16, 2011, to August 28, 2011, ended only after both houses of the Indian Parliament agreed to consider appointing an ombudsman with legal powers to act against corruption. The protest movement drew hundreds of volunteers who managed the telephone helplines; gave bytes to newspapers, radio, and television; went online; sent emails; tweeted; formed online forums; and sent mobile clips to media on the local protests organized in the country and even abroad (Saxena 2011: 46–48). The anti-corruption movement across India saw a convergence of social movements, new media, and civic engagement never witnessed before in post-independent India (Prasad 2012, 2013).

There were more than 58 million tweets in the 2014 elections in India, a country which accounts for less than 5% of the world's Internet users. Far from the days of using just posters and street banners, political parties now use digital, SMS, MMS, and online media quite effectively in their campaigns. News organizations have made social media as the "second screen" by reading tweets and comments on TV, inviting the audience to raise issues and questions and even hosting debates. The social media networks have also been influential in spreading protest movements such as the Anti-Corruption Campaign in India and environmental movements.

The Twitter Samvad, a service that allows people to receive tweets by government leaders and agencies as text messages over mobile phones, was launched on March 24, 2015, the same day on which the Supreme Court struck down the controversial Section 66A of the Information Technology Act, 2000, enacted through an amendment in 2008 which could be used by the government to regulate and curb the use of social media. The adoption of information and communications technologies is having a significant, yet often unacknowledged, impact on the way that governments administer policies and conceive of citizens (Henman 2010). India's experience in e-governance/ICT initiatives has demonstrated significant success in improving accessibility, cutting down costs, reducing corruption

(Bhatnagar 2012), and increasing access to unserved groups (Das and Chandrashekhar 2007). It is clear that the Internet and new media in its myriad forms are gradually moving center stage to contour a digital culture even in societies like India which are sharply marked by the digital divide.

66.4 Capacity Building for e-Governance

The state of Kerala presents an interesting case in the study of e-governance in a region with high literacy and educational status, access to ICTs, civic engagement, and political participation as compared to other states in India. Kerala occupies the first position among all states in India with its high human development status. It is in the forefront of implementing e-governance and m-governance. It is also the only state to implement the Akshaya e-literacy project toward facilitating capacity building for citizens to participate in modernizing governance and implementing an effective plan to bridge the digital divide (Prasad 2011). These factors make Kerala an ideal setting to study people's participation in governance. Pathanamthitta in Kerala was purposively selected for study as it is a developed district. The overall literacy rate in the district is 95.09%. The male literacy rate is 96.62%, while the female literacy rate is 93.71%.

Pathanamthitta district has its headquarters at Pathanamthitta. Tiruvalla and Adoor are the revenue subdivisions. The district is divided into nine community development blocks. Subsequently three community development blocks from Pathanamthitta district, Mallappally, Parakkode, and Koipuram, were selected for study. From the selected community development blocks, Kaviyoor, Mallappally, and Kallooppara from Mallappally, Kalanjoor from Parakkode, and Eraviperoor from Koipuram were selected for data collection. A random sample of both men and women were selected from each block to study people's participation in e-governance. The sample consists of 345 males and 355 females. The total sample from the three community development blocks is 700. The study used an analytical survey to examine the factors influencing the participation of people in e-governance through a structured questionnaire framed on the basis of the objectives of the study.

Apart from using a quantitative survey, a qualitative method was also considered appropriate because the study intended to find from the Akshaya entrepreneurs the services provided by the Akshaya centers to the people in Pathanamthitta district of Kerala and to identify the e-governance interactions between the government, citizens, and businesses. Case studies were carried out using in-depth interviewing to gather information from ten entrepreneurs who run the Akshaya centers that is the mainstay of e-governance initiatives aimed at bridging the digital divide in Kerala. The officials responsible for the functioning of the Akshaya project were also interviewed to generate primary data for the study.

The limitations of the study lie in the fact that Kerala is a highly literate state with the highest human development indicators and better gender equality; the findings of the study may not apply to backward regions of India. It is also the only state with the Akshaya e-literacy policy initiative in India which makes comparisons with other states rather limited. The study was limited to a developed district in Kerala, and the gender-related findings may also differ in other states of India with a wide gap in education and status of women.

66.5 Socioeconomic Characteristics of Respondents

The socioeconomic characteristics of the respondents of the study reveal that while over one-third of the respondents (35.4%) live in the rural areas, around two-thirds of them live in semi-urban areas. This is due to the characteristic nature of Kerala where rural areas and urban areas are contiguous. Over one-third of them (40.3%) are aged between 18 and 35 years, whereas 46.1% are 36 to 50 years old, while 13.6% of them are above 50 years. While 49% of the respondents are male, 51% of them are female. This is in line with the sex ratio of the state of Kerala where women outnumber males. Around two-thirds (73.4%) of the respondents are married, and around one-third (26.3%) of them are unmarried. While half of the respondents (50.9%) are Hindus, 44.4% are Christians, and 4.7% are Muslims.

It is found that 40.9% of the respondents hold a degree or have higher educational levels, while 38.9% have secondary or higher secondary education, while one-tenth of the respondents hold diploma or certificate in technical education. Only 7.9% have primary education especially those who are working in the agricultural sector. More than one-tenth (12%) of the respondents are employed in government or public sector, and 15.3% are employed in private sector. About 14.6% are self-employed or have their own business, and 13.1% are employed in agriculture sector. Interestingly nearly half of the respondents (45%) are unemployed, women account for a majority of the unemployed in the district. While over a third of the respondents have a family member working abroad, one-fifth (20.4%) of the respondents had a family member employed in government service. Over half of the respondents (52.4%) belong to the middle-income group followed by nearly one-third of them (29.7%) who belong to the low-income group and less than one-fifth of the respondents (17.9%) in the high-income group.

66.6 Exposure to Mass Media and e-Governance Information

An overwhelming majority (92.3%) of the respondents both men and women read newspapers daily. While about two-third women (73%) have expressed their interest in reading political news, an overwhelming majority of men (92.5%) expressed their interest in reading political news. Men and women were equally interested in reading social and development news. While men and women were equally interested in reading news on e-governance programs and services, men were more interested in reading business news and economic matters compared to women. Over half of the respondents (54.7%) were interested in reading news related to agriculture since many of the respondents reside in the rural areas where agriculture is a dominant occupation. While 46.7% had read governance-related news in newspapers, one-third of the respondents (33.1%) made use of governance programs that they read in the newspapers. Half of the respondents (50.1%) recall seeing posters which convey e-governance-related news, advertisements, and messages. About three-fourths of the respondents (71.1%) watch e-governance-related programs on TV.

It is widely recognized that timely availability of information enables good decisions, increases productivity, and improves governance. However, to ensure that information is available in a timely manner, systems for collating data and converting them into information inputs are essential. It is essential to have systems, which ensure that relevant information is available anywhere anytime and in any way for interactions between government to government (G2G), government to citizens (G2C), and government to businesses (G2B). The emphasis of citizen relevant data has been on reducing the transaction costs incurred by the individual citizen as on date. The transaction cost here does not refer to the pricing of services (user charges) but the cost involved in getting information, like cost of repeated travel, wastage of time and opportunity cost, etc. It is critical to have decentralized information access centers that cater to a range of citizen needs and has an inbuilt integrated front-end system.

66.6.1 Are the Akshaya Centers Being Implemented in Kerala in Line with the Objective of Increasing Citizen Participation in e-Governance?

The Akshaya e-literacy project of Kerala Government was implemented in 2002 jointly by the Kerala IT Mission and Department of Science and Technology, with tie-ups with local bodies and voluntary agencies. This project aims at making Kerala the first completely e-literate state in India. Under this project, state and local self-government bodies are connected via the Internet and mailing facility in Malayalam; the local language is also provided. Akshaya centers were set up within a distance of 2–3 km to serve as an ICT access points, 1 for every 1000 families. There are 2328 Akshaya centers out of which 2662 (87.50%) are in the rural areas. It is to be enhanced to 3180 centers in the near future, thus covering every part of the state, even the remotest villages (http://akshaya.kerala.gov.in/). All these centers are recognized by the central government as CSCs. Thus the access to Akshaya centers is the first step toward bridging the digital divide.

The Akshaya centers being implemented in Kerala are in line with the objective of increasing citizen participation in e-governance. They are access centers for G2C information interaction as well as for a substantial range of G2B information interaction. Apart from G2C information flows, G2G information flows especially the data from field-level offices to higher tiers are a critical information management issue. Given the fact that all the government departments and individual offices do not have automated systems or computerized backend information systems, currently such data flows are in the manual mode using defined proformas in specified periodical statements and have issues of delay and retrievability for analysis attached to it. The capturing of data in proformas and forwarding the same in electronic format to central repositories through the infrastructure available at Akshaya centers are crucial in strengthening e-governance until the capabilities are acquired in all the government offices concerned.

Apart from G2G, G2C, and G2B information interactions, it is also critical that some of the transactions in these categories are also brought to decentralized and integrated front-ends. This would essentially mean that government can then concentrate more on the critical core backend operations that it is mandated to do and would channelize its information and transactions through a widespread network of access points. A key component of such a decentralized system wherein the information and if possible transaction services happen outside the premises of government offices/institutions would be the presence of electronic access points which could serve as information/transaction dissemination points as well as data collection and capture points. Akshaya seeks to provide such access points. The Akshaya centers also offer over 23 government services on continued e-learning program; data entry under e-governance program; DTP and job-typing work; computer training for public; design of invitation cards, visiting cards, banners, posters, and paper bags; screen printing; data bank services; and telemedicine applications.

66.6.2 Are the Akshaya Centers Designed Within a Public-Private Partnership Framework Which Would Augment the Financial Viability of the Centers?

There are several community telecentres in different countries initiated by the public and private sector funded by individual entrepreneurs, local bodies, government, and international institutions. In several developing countries, the telecentres showed no systematic participation of people in learning, information, and training except for providing incidental technical assistance as in Nepal (Tulachan 2010). Similarly, in Vietnam the telecentres have yet to make people part of the information society (Colle and Van Dien 2013). In a study of Akshaya and the two UNESCO ICT initiatives Nabanna in West Bengal and Namma Dhwani in Karnataka, the most striking difference was one of scale (Nair et al. 2006). While Nabanna and Namma Dhwani both operated in a maximum of five centers, Akshaya had state-wide operations. The Akshaya project was on a sound financial footing due to its backing by the State Government and municipality which together invested 13.2 crores (132 million in 2006), and it was amply supported by the technical teams at KSITM and TULIP IT services and given entrepreneurial training by STED (Science and Technology Entrepreneurship Development). When compared to the UNESCO initiatives, the Akshaya project has achieved an impressive level of success in putting into place various administrative, technical, and training responsibilities. The critical difference between Akshaya and UNESCO's ICT centers lies in Akshaya's strategy of involving and building ICT entrepreneurs from the community, while both Nabanna and Namma Dhwani are projects that are working with the community, but nonetheless have largely been conceived and executed by the NGO partners (Nair et al. 2006).

Akshaya has been modelled as ICT access points that are not state-run institutions but entrepreneurial ventures. The Akshaya centers as key stakeholders in information and transaction dissemination are designed within a public-private partnership framework which would augment the financial viability of the centers. The investment for setting up the e-centers is made by the entrepreneurs. It costs around Rs.3.00 lakhs ($4800 approx.) for setting up an Akshaya e-center with five to ten computers, printers, scanners, webcam, other peripherals and necessary software. The entire recurring expenditure for running the e-center is also borne by the entrepreneur. KSITM provides the support required to sustain the project by way of e-literacy, training fund, connectivity, advanced courses, content CDs software, e-governance, and various other services. KSITM focuses on creating effective market mechanisms for demand-driven delivery of services in a public-private partnership (PPP) framework.

Akshaya's strategy builds ICT entrepreneurs from the community supported by the technical teams at KSITM and TULIP and given entrepreneurial training by STED (Science and Technology Entrepreneurship Development). The project has achieved an impressive level of success in putting into place various administrative, technical, and training responsibilities (Nair et al. 2006). Akshaya operates within an economic model linked to the management skills and social commitment of the entrepreneur in terms of its social and financial sustainability. The service delivery phase opens up ample opportunities for entrepreneurs to earn a steady income by offering a variety of services such as e-learning, e-commerce, advanced IT training, e-governance, communications, and specific community-based services.

The planned interventions that can come from government initiatives, especially in the developing countries, are important and inevitable. The government usually has the resources, the infrastructure, and the authority to implement programs aimed at reducing the digital divide (Joshi 2004, Jayakar 2012). Government policies will influence the ability of telecentres to induce desirable development outcomes (Harris 2001). Market forces can play an important part only if the government policies are favorable. The private sector is primarily responsible for providing access, and competitive private sector-led markets go a long way toward making these services widely available. The public sector's main role is to provide a sound policy framework, regulate markets where they do not work well enough on their own, and support additional service provision where markets do not achieve economic and social objectives. The Government of India has undertaken an initiative, namely, BharatNet – a high-speed digital highway to connect all 2.5 lakh Gram Panchayats (village units) of the country. This would be the world's largest rural broadband connectivity project using optical fiber. The public telecom service provider BSNL has undertaken large-scale deployment of Wi-Fi hotspots throughout the country which will render affordable access to the Internet.

66.6.3 Do these Centers Serve as Anchor Institutions or Demand Aggregators and Which Provide Digital Literacy Instruction, Continuing Education, Job Training, and Entrepreneurship Classes for Capacity Building?

The Akshaya project was aimed to make at least 1 person in each of the 65 lakh families in the state e-literate through a 15-h course in the first phase of Akshaya project from 2002 to 2006. A unique method was adopted to spread e-literacy in Ernakulam after it was found that the 14 Akshaya centers initially allotted within its limits were insufficient to achieve the stated objective. To overcome this, 30 efficient Akshaya entrepreneurs from the Kochi city outskirts were roped in to cover the corporation area with the approval of the Corporation Council. Each entrepreneur was allotted two divisions each with subcenters in at least six different locations within the division. Thus, Kudumbashree units (women's self-help groups), anganwadis (child care centers), political party offices, vacant buildings of the corporation, individual households, and even police stations were turned into centers for computer learning (Praveen 2009). Of the 5,12,270 e-literate people in the Ernakulam district, 3,43,753 were trained in basic computer literacy by Akshaya (Praveen 2009). Near universal e-literacy was achieved in eight districts in Phase 1. The pilot project in Malappuram district had multipurpose community training centers to train people to handle computers, data entry, desktop publishing, and Internet browsing. Malappuram was declared as the first e-literate district in India. Malappuram and Kannur districts were declared as 100% e-literate. Kollam, Kozhikode, Thrissur, and Kasaragod districts achieved e-literacy above 90%. So far around 33 lakh beneficiaries have been trained under the Akshaya e-literacy project.

In 2007 Akshaya moved into the second phase two of the project, covering the balance six districts and also rendering new G2C and B2C services. Akshaya has to its credit over 200 crore rupees worth transactions besides providing multitude of services through its 2000+ Akshaya centers. The Akshaya project has an additional objective of enhancing the quality of available IT infrastructure in the state to bridge the rural-urban digital divide. It is expected that the IT infrastructure will be expanded to the rural areas to create and expand economic opportunities in the knowledge economy, empower individuals and communities through enhanced access to information, modernize and upgrade skill sets, integrate communities through creation of e-networks, create awareness of ICT tools and usage, generate locally relevant content, and generate direct employment opportunities (http://www.akshaya.kerala.gov.in).

The Akshaya project is an e-governance initiative that has succeeded in drawing people to use technology in socially deterministic ways to satisfy local needs. In an evaluation of the Akshaya project, it was found that it was successful in generating employment, providing IT literacy, enhancing communication, and providing e-services (Ghatak 2006). The Akshaya program generated employment for the youth, particularly women, for work like DTP, typing, etc. Trainees particularly women could search for better employment opportunities at the end of their course. Akshaya program provides cheaper e-literacy courses to the people. The courses offered ranges from easier ones (like MS Office, DTP) to harder ones (like Diploma courses). E-learning programs like Intel learning, Learn English, Arabic Tutor, Internet for Mass, Medical Transcription, e-Vidhya, etc. are already introduced and are implemented by a number of e-centers. Internet-enabled kiosks are used by people to contact their relatives/friends who are staying abroad (such as the Gulf countries) or other states in India. Communication is also done for marketing of products. Akshaya kiosks are providing a range of services like registration of births and deaths and collection and feeding of health-related data (in a way acting as databanks) of the local population (by tying up with village administrative units) (Ghatak 2006).

In the third phase from 2011, the Akshaya centers were recognized as CSCs by the central government to provide a range of 23 citizen services. A flagship project funded by UNESCO in association with Akshaya called *Entegramam* (My Village) was launched which is an online community portal in Malayalam and maintained by the citizens of each village. The project aimed at bringing forth web portals that cater to the needs of the citizens locally. Each village has its own space in the portal, and the information ranges from a catalog of "useful services" from coconut tree climbers, carpenters, and more along with their phone numbers and location where they live. The portal also has the history of the land, governance, information on public services, locally relevant news, and announcements. This project has been implemented in Kannur district. In the course of time, the portals will be used for local transaction, enabling in it with more business features.

The access to information, backed with relevant infrastructure and services, not only allows rural populace to improve its quality of life but also support and supplement its existing incomes in a sustainable way. Access to information and services, like e-governance, micro-credit, literacy, education, health, etc., can provide a solid foundation for the economic prosperity of rural villages. Rural consumers were willing to pay for products and services that meet their needs and are offered at affordable prices. The potential of extending Akshaya ICT facilities would not only facilitate better communication, with a direct improvement in the quality of family lives and positive impacts on savings, but it would also open up unlimited opportunities to develop training resources, upgrade human resources, and improve trade and job opportunities (Nair et al. 2006).

66.7 Akshaya Centers, e-Literacy, and Participation in e-Governance

A great majority of the respondents (84.3%) reported that Akshaya center is available near their residence, while less than one-fifth of the (15.7%) respondents reported that there was no Akshaya center near their home. An attempt is made to find how the respondents came to know about Akshaya centers. Nearly half of the respondents (43.6) know about Akshaya through their friends and neighbors followed by one-fifth of them (22.6%) through mass media and advertisements. Less than one-fifth of them (17.3%) came to know Akshaya through Akshaya entrepreneurs, while less than one-tenth of them (7%) came to know through local government functionaries. About one-tenth of the respondents (9.6%) had never heard about Akshaya centers. The results about awareness of Akshaya centers among men and women are highly significant.

About two-thirds (64%) of the respondents were aware of the various services available at Akshaya centers, while one-third of them (36%) did not know about them. Irrespective of gender, people's awareness about various services is high. The services offered by the Akshaya centers include payment of utility bills, various certificates, insurance, application for ration cards, web browsing, and e-learning. Majority of the respondents (72.7%) were aware that Akshaya centers were conducting computer training, while about one-fourth of them were not aware of it. It is also found that majority of the respondents (71.7%) were aware that Akshaya centers has Internet browsing facility, while one-fourth (28.3%) of them were not aware about it.

Over two-thirds of the respondents (69.9%) favor utilizing the services available at Akshaya instead of going to various offices. About one-third of them (30.1%) reported that they prefer to go to various offices as Akshaya centers were more distant than the government offices. During the course of the study, interviews with people revealed that those who live in the vicinity of the government offices often go there rather than to the Akshaya centers which are further away. Kerala has a sizeable population employed outside the state and the country. They would like to make payments for property tax, electricity, telephone, and other services on an annual basis which is normally facilitated at government offices after checking their account balances for such payments.

More than two-thirds of the respondents, one of their family members, neighbors, or friends (71.9%), had received e-literacy training at the Akshaya centers. More than one-fifth of the women respondents (22%) underwent computer training at Akshaya, while more than one-tenth of men (13.6%) underwent computer training. About one-fourth (24.3%) of the respondent's family members were trained in the use of computers at Akshaya. Over a quarter of them (29.7%) reported that either their neighbors or friends were trained in the use of computers at the Akshaya center. Over a quarter of them (28%) had not received e-literacy training a majority of them being agricultural workers and mothers with small children. A great majority of women (83.4%) think that they can work more efficiently with computer skills, while three-fourth of the men (76.2%) reported the same. More women thought they could work efficiently with computer skills.

Nearly half of men (44.9%) and less than one-third of women (29.9%) use Internet for e-mail, chat, and some of the social networking groups to communicate with their friends and colleagues. The empirical findings establish that more men use Internet facility to communicate with friends and colleagues than women. Capacity-building programs can bridge the gender divide in the digital world. While a great majority of the respondents (93.3%) did not use the Internet to access market networks, it is heartening to note that 6.7% of them used the Internet to access market networks. Around one-tenth of the men had accessed market networks when compared to women who account for a majority of the unemployed in the district.

A majority of the respondents (85.9%) did not receive any information about e-governance programs and facilities through digital/social media which includes e-mail, Facebook, and Twitter. More than one-tenth of them (14.1%) received such information through social media. It was found that messages related to e-governance would be delivered only to those who registered their e-mail addresses or mobile numbers to the concerned authority. The Digital India project proposes to use mobile platforms to reach people who were hardly in touch with government departments.

An attempt was made to evaluate whether the information they received from the Internet was beneficial to the respondents. More than half (59.4%) of them reported that they benefited from the educational opportunity information through the Internet followed by around half of them (43.9%) who reported that they benefited from information about job openings/opportunities. Less than one-fourth of them were benefited from health information followed by about one-tenth of them who reported that they were benefited for income generation and business opportunity from the Internet.

An overwhelming majority of the respondents (87.1%) opined that e-literacy is needed to get more access to government services and information online compared to about one-tenth of them (12.9%) who do not think so. This revelation has an important bearing on policy making. The government needs to continue capacity-building efforts to offer free e-literacy classes to all those who do not have e-literacy. The Kerala Government's initiative of providing e-literacy through Akshaya centers has been successful in enabling people to communicate and interact online and access online services with the help of the Akshaya staff, but they need more digital skills to engage in e-commerce and e-governance.

Kerala is also in the forefront of setting up mobile governance. M-governance is defined as the strategy and implementation involving the utilization of wireless and mobile technology services, applications, and devices for improving benefits for citizens, business, and all government units. The rapid diffusion of mobile ICT gadgets such as laptops, mobile phones, and personal digital assistants (PDAs), along with emails, instant messaging, and other networking services has rapidly fuelled the mobile interaction. In order to take advantage of mobile and wireless ICT technologies as well as dealing with the fluidity of the interaction with the mobile society and booming mobile usage rates, the Kerala State Government has initiated action to set up about 20m-government services offered by 8 departments identified for pilot-level implementation and to deliver services through mobile phones accessible to the citizens in the field, in the street, at home, or other convenient locations on a 24 × 7 basis, rather than the users having to visit government offices or log on to the Internet portals to access services.

The case studies of the Akshaya centers and personal interviews with entrepreneurs revealed that there are some factors that contribute to the successful running of the centers. These factors are the knowledge of computers, IT skills, ownership of premises, an additional occupation to beset the operational costs of running the center, entrepreneurial skills, and the ability to build a rapport with the community. Entrepreneurs who understood the nature of the program and the need to provide e-literacy to bridge the digital divide and felt that running Akshaya center was a social responsibility were successful in running the CSC than those who viewed it purely from a business perspective (Prasad 2011).

66.8 Conclusion

Evidence from both theoretical and empirical studies reveals that ICTs and new media technologies have become inevitable in e-governance. But the motivation for adopting e-governance in developing countries like India is quite different from that in developed countries. The findings of this study suggest that the Indian government's e-governance program, with the CSCs promoting sheer access to the Internet, wants to promote citizen access to ICTs for encouraging their participation in e-governance. Providing access to the Internet alone is not enough – people must be enabled to use ICTs for citizen-government interaction.

It is also clear that literacy skills, greater awareness, education, and capacity-building efforts such as the Akshaya e-literacy project in Kerala are important factors that will enable greater civic engagement and citizen participation in e-governance. The Akshaya project is a unique partnership involving the government, private entrepreneurs, community volunteers, and citizens in improving the e-literacy skills of the community. Akshaya project is a bottom-up model for imparting e-literacy training, delivery of content, services, information, and knowledge, that integrates public and private enterprises – through a collaborative framework – to integrate their goals of profit as well as social objectives, into a sustainable business model for achieving rapid socioeconomic change in the state. It is unique as local ownership of Akshaya centers has fostered its sustainability even in difficult times. Many projects that were externally owned and donor-funded and had limited timeframes often resulted in wastage of resources and had to close down as the necessary skills to sustain them were not honed through entrepreneurial capacity building.

Future research projects can study various capacity-building measures to determine the efficiency of e-literacy of communities and frame suitable policies for digital inclusion of people of developing countries. ICT initiatives can be studied for UCD (user-centered design) that gives attention to the target audience's perspective. Capacity-building initiatives like the Akshaya project can bridge the digital divide and advance digital democracy by including women who have often been restricted and received limited benefits of access to ICTs (Prasad 2007, 2008). The study demonstrates that Akshaya has provided equitable access to women in e-literacy training, ICT access, availing various services, and even becoming ICT entrepreneurs themselves. Kerala had achieved almost total literacy before embarking on providing e-literacy to the people of the state. Though the 29 states of India are at various stages of development, the project attempts to highlight the possibilities for other states that are similar to Kerala in levels of development. It can be regarded as a model for emulation in other states of India with the implementation of Digital India project and has also generated considerable interest throughout South Asia. Nevertheless, financing affordable Internet access and ICT competence – including investment and training to create, maintain, and expand computer networks – may challenge the sustainability of e-governance in developing countries like India as they continue to grapple with the many complexities of development (Prasad 2011).

References

Arul Aram I (2004) e-governance: ushering in an era of e-democracy. In: Prasad K (ed) Information and communication technology: recasting development. B. R. Publishing Corporation, New Delhi, pp 355–372

Bekkers V, Homburg V (2007) The myths of e-government: looking beyond the assumptions of a new and better government. Inf Soc 23(5):373–382

Bhatnagar S (2012) e-government and access to information. In: Transparency international: global corruption report 2003. Accessed 26 Apr 2012. http://www.transparency.org/publications/gcr/gcr_2003#download. pp 24–32
Colle R (2000) Communication shops and telecentres in developing countries. In: Gurstein M (ed) Community informatics: enabling communities with information and communications technologies. Idea Group Publishing, Hershey
Colle R, Van Dien T (2013) A role for universities in ICT for development interventions. In: Servaes J (ed) Sustainable development and green communication: African and Asian perspectives. Palgrave Macmillan, New York
Das SR, Chandrashekhar R (2007) Capacity building for e-governance in India. Available at www.apdip.net/projects/e-government/capblg/. Accessed on 16 Aug 2011
Ghatak S (2006) Gender facts in Malappuram, Kerala: evaluating the Akshaya Programme. Retrieved on 9 Jan 2009 from http://www.i4donline.net/news/top_news.asp?catid=3&newsid=8704
Harris R (2001) Telecentres in rural Asia: towards a successful model. Conference proceedings of international conference on information technology, communications and development (ITCD 2001), 29–30 Nov 2001, Kathmandu. www.itcd.net. Accessed 3 July 2015
Henman P (2010) Governing electronically: e-government and the reconfiguration of public administration, policy and power. Palgrave Macmillan, London
Jayakar K (2012) Promoting universal broadband through middle mile institutions: a legislative agenda. J Inf Pol 1(2011):102–124
Joshi SR (2004) Bridging the digital divide in India. In: Prasad K (ed) Information and communication technology: recasting development. B. R. Publishing Corporation, New Delhi, pp 415–449
Kalam APJA (2005) A vision of citizen-centric e-governance for India. In: Bagga RK, Keniston K, Mathur RR (eds) The State, IT and development. Sage, New Delhi
Mosse B, Whitley EA (2009) Critically classifying: UK e-government website benchmarking and the recasting of the citizen as customer. Inf Syst J 19(2):142–173
Nair S, Jennaway M, Skuse A (2006) Local information networks: social and technological considerations. UNESCO, New Delhi
NeGP (2006) National e-governance plan. Retrieved on 2 Feb 2011 from http://arc.gov.in/11threp/ARC_11thReport_Ch7.pdf
Prasad K (2004) Information and communication technology for development in India: rethinking media policy and research. In: Prasad K (ed) Information and communication technology: recasting development. BRPC, New Delhi, pp 3–48
Prasad K (2007) From digital divide to digital opportunities: issues and challenges for ICT policies in South Asia. In: Global Media Journal, Indian Edition, July 2007. http://www.manipal.edu/gmj/issues/jul07/prasad.php
Prasad K (2008) The digital divides: implications of ICTs for development in South Asia. Papers in international and global communication, no 3/08. Centre for International Communication Research, Institute of Communication Studies, University of Leeds, available at http://ics.leeds.ac.uk/papers/cicr/exhibits/56/Leeds-CICRworkingpaper-Kiran.pdf. ISSN 1752-1793
Prasad K (2009) Communication for development: reinventing theory and action, vol 1 & 2. BRPC, New Delhi
Prasad K (2011) ICTs and e-governance: a study of people's participation in Kerala, UGC Major Project Report. University Grants Commission, New Delhi
Prasad K (2012) e-governance policy for modernizing government through digital democracy in India. J Inf Pol 2:183–203
Prasad K (2013) New media and pathways to social change: shifting development discourses. B.R. Publishing Corporation, New Delhi
Praveen MP (2009) Ernakulam district to become e-literate. The Hindu, Feb 24, 2009. Retrieved on 24 Feb 2009 from http://www.hindu.com/2009/02/24/stories/2009022458320200.htm
Praveen MP (2011) e-governance scheme gets global acceptance. The Hindu, May 22, 2011. Accessed 26 Apr 2012, http://www.hindu.com/2011/05/22/stories/2011052258340400.htm
Ram N (2011) The changing role of the news media in contemporary India, Indian History Congress, 72nd Session. Punjabi University, Patiala, December 11–13

Raman V (2013) Technologies for governance reconsidered: a capabilities-based model from developing countries. In: Prasad K (ed) New media and pathways to social change: shifting development discourses. B.R. Publishing Corporation, New Delhi

Saxena P (2011) The protest party. The Week, 46–48

Tulachan A (2010) Telecentres as local institutions: present landscape in Nepal. J Dev Commun 21(1):62–68

United Nations (2010) United Nations e-government survey: world e-government rankings. United Nations, New York

United Nations (2014) World urbanization prospects: the 2014 revision, highlights. UN, Department of Economic and Social Affairs, Population Division, New York

Fifty Years of Practice and Innovation Participatory Video (PV)

67

Tony Roberts and Soledad Muñiz

Contents

Abstract

Benefiting from more than 50 years of practice and innovation, participatory video (PV) is a firmly established approach in the field of communications for development. The term “participatory video” is used to refer to a very wide range of practices that involve nonprofessionals in making their own films as a means to engage communities, develop critical awareness, and amplify citizens’ voices to

T. Roberts
Institute of Development Studies, University of Sussex, Sussex, UK
e-mail: T.Roberts@ids.ac.uk

S. Muñiz (✉)
InsightShare, London, UK
e-mail: smuniz@insightshare.org

J. Servaes (ed.), *Handbook of Communication for Development and Social Change*,
https://doi.org/10.1007/978-981-15-2014-3_39

discuss social problems that they prioritize. The canonical texts on participatory video all make reference to PV's grounding in the praxis of Brazilian popular educator Paulo Freire, and the influence of feminist practice is often also noted in the literature. The authors also draw on affordance theory as a way of clarifying the possibilities for social action enabled by participatory video. In recent years, a number of important critiques have been leveled at PV which have reopened a normative debate about what practices, values, and objectives should constitute participatory video. Rapid recent advances in digital filmmaking technologies coupled with falling costs of mobile devices are opening up exciting new future possibilities and challenges for PV. This chapter reviews a range of PV practices, examines key critiques, and assesses potential future directions for participatory video in communication for development.

Keywords

Affordance theory · Critical theories · Degrees of participation · Fogo process · Paper edit · Participatory video (PV) · Practitioner network (PV-NET) · *Tyranny of Participation* · Visual storyboard

67.1 Introduction

Participatory video (PV) is a process of involving people without filmmaking expertise in making short films about social issues affecting them, using their own words and their own images. Unlike other kinds of filmmaking, participatory video hands over control of the filmmaking equipment and of the editorial process to inexpert users. An external practitioner or "facilitator" is often responsible for structuring and guiding the PV process in which participants first reflect critically on the social issues that are affecting them and the change they wish to see, before producing their own film to represent their priority issues from their perspective.

In one of the foundational texts on the subject, Shaw and Robertson (1997: 26) describe participatory video as: "a process of media production to empower people with the confidence, skills and information they need to tackle their own issues." However, there is no commonly agreed definition of participatory video. What the term PV *should* refer to and what the focus of PV *should* be are contested normative debates. In practice the term participatory video is used to describe a wide range of distinct community-led filmmaking practices, and there are many processes that closely resemble participatory video that are not referred to as PV by their practitioners (High et al. 2012). For the purposes of this chapter, we take an inclusive approach as to what is considered to be participatory video. This inclusive perspective is perhaps best captured in the definition of PV developed collectively by the practitioner network "PV-NET," which described participatory video as: "a collaborative approach to working with a group or community in shaping and creating their own film, in order to open spaces for learning and communication and to enable positive change and transformation" (PV-NET 2008).

67.1.1 Description

There is also no single agreed correct way to do participatory video. However, it is possible to identify many elements common to different participatory video processes from practical "how-to" guides, including those authored by Shaw and Robertson (1997) and Lunch and Lunch (2006). In most PV processes, an experienced facilitator or team arrives with cameras, tripods, and any other equipment necessary to make the film. A group of participants take part in practical exercises to familiarize themselves with the functioning of cameras and related equipment. Participants are engaged in group dialogue about the social issues most important to them, and they decide what kind of film they would like to make. Participants then collaborate in the group production of a script or visual "storyboard" which is sketched on paper. This "paper edit" then becomes the plan which participants use to guide them as they take up the camera to begin filming the scenes depicted in each frame of the storyboard in order to produce their own film.

Periodically during the filming and editing process, participants pause to review their progress and to critically reflect on both the technical process and the social issues that they are trying to articulate on film. For some practitioners this group process of dialogue about the social issues is the most important aspect of participatory video. From this perspective, participatory video is designed to create a space for participants to critically reflect and learn together about the injustice that they experience and its root causes and to discuss how they can best communicate their views and act together to produce the social change that they value. From this perspective the process of participatory video has an intrinsic value in creating new spaces for group reflection and dialogue about issues that may be socially taboo or are not prioritized. The films produced fulfill an epistemological role as an iterative process of reflection, filming, and editing, through which participants come to a deeper understanding of their own situation and produce a clearer articulation of their priorities for change. As such participatory video is a means for collective self-inquiry and reflection (Freire 1970) that can enhance participants' ability to "experience their capability and power to produce knowledge autonomously" (Fals-Borda and Rahman 1991: 17).

In some processes it may be only the immediate group of filmmakers themselves who reflect critically on the issues. In other processes wider community screenings are central to the iterative processes. Early screenings of the rough footage may be held with the wider community in order to stimulate broader debate and inclusion and input about the issues and to generate agreement about the core message that they wish to communicate in the final edit (Braden and Huong 1998; White 2003; Lunch and Lunch 2006).

For other practitioners it is important that participatory video processes go beyond consciousness-raising of the participants and there is an extrinsic goal to communicate to external audiences a call to action for social change. This often involves constructing films that contain an advocacy message targeted at distant decision-makers. This participant-authored representation format of PV (Shaw 2012) is perhaps the classic form of participatory video and echoes back to the original 1957 Fogo Process (see below).

67.1.2 Product Versus Process

Put another way, some participatory video initiatives are primarily process-focused, while others are primarily product-focused. In some PV initiatives, the primary objective is to create a space for group dialogue and learning, and the production quality of the film is of secondary concern (Nemes et al. 2007). It may be the case that the subject matter of the film is so personal or political that disclosing the content to a wider audience would raise serious ethical or personal safety risks (Thomas and Britton 2012; Teitelbaum 2012). However, in other PV initiatives, the primary objective may be to communicate an advocacy message in pursuit of social change. If the intended audience is the government or other high-level policy makers, a decision may be taken to invest a larger share of the project's finite resources to achieving high production values in the film. This process-product distinction is not binary. Many PV initiatives value process and product to varying degrees (Kyung-Hwa Yang 2012). Some participatory video processes begin life as internally focused, but over time participants develop an aspiration to project their perspective to external audiences (Shaw and Robertson 1997). In an interview for this chapter, Namita Singh, an experienced PV practitioner at Digital Green in India, said that "[over time] *I started to focus more on the* technical quality of the video. Earlier I used to think that technical quality is not important, the content is what actually matters. Now I place a lot of emphasis is on being able to share the info clearly and being entertaining. It has to compete with highly entertaining content like Bollywood." Given finite resources, a trade-off is necessary between dedicating time and resources on the group consciousness process and taking time to perfect the technical quality of the sound recordings and editing effects. There is no single agreed correct way to practice participatory video, but being aware of and transparent about these trade-offs and opportunity costs is important.

This rest of this chapter will first locate participatory video in a historical and theoretical context before drawing on practitioner interviews and desk research to reflect on the perceived benefits, limitations, and future directions for participatory video practice.

67.1.3 History

The earliest cited example of participatory filmmaking is the 1967 work by the people of Fogo Island, Newfoundland, facilitated by a team headed by Donald Snowden (Snowden 1984). The "Fogo Process" began by filming community members' views on pressing social issues and then organizing 35 separate community screenings to a combined audience of 3000 islanders (60% of the total population). Post-screening discussions were used to animate social dialogue and to identify key issues of common concern to Fogo Islanders. The islanders then co-authored and appeared in a film to represent their concerns. Snowden and his team filmed and edited the film, and it was shown to government officials. The Minister of Fisheries then recorded a filmed response to play back to the communities. This film-mediated dialogue improved communication and dialogue, altered government policy, and led to Fogo Islanders forming a fishing cooperative that increased employment and improved livelihoods (Crocker 2003; Corneil 2012). The Fogo Process became an exemplar of communication for social change and, in modified formats, has been used in numerous locations around the world.

Participatory video practices have diverse roots including the community arts movement, feminist documentary filmmaking, action research, and video for change activism. Social workers and community arts activists began harnessing video's potential from the 1970s onward (Shaw and Robertson 1997). Feminist documentary filmmakers emphasized the need to express women's knowledge in their own words (Lesage 1978). In action research PV has long been used as a process for citizen groups to analyze the unequal power relationships that they experience (Kindon 2003), and in the video for change movement, Gregory (2005) reminds us of indigenous media initiatives in Chiapas, Mexico, and the Drishti Media Collective in India from the 1990s onward.

67.2 Theoretical Framework

The following sections provide a theoretical framework for the chapter. Foundational texts by practitioners of participatory video all ground their work in the praxis of Brazilian popular educator Paulo Freire. Contemporary authors increasingly incorporate critical feminist perspectives in their approaches. To these framings here we add the concept of "affordances" as a way of illuminating the unique action possibilities made available to practitioners by participatory video, when compared to other participatory methodologies.

67.2.1 Critical Theories

Although diverse practice traditions contribute to participatory video processes, in terms of theoretical grounding, all of the canonical texts on PV all refer to Freire's (1970) critical pedagogy as having influenced their participatory video practice (Shaw and Robertson 1997; Braden and Huong 1998; White 2003; Lunch and Lunch 2006). These core PV texts all refer to Freire's most well-known texts (1970, 1974) in which he introduced his critical pedagogy of group reflection on generative images (sketches) in order to enhance participant's "critical consciousness" about their own experience of social injustice and its causes. Freire described this process as one of learning to "read the world" more critically in order to "act in the world" for social change. It is interesting to note that Freire does not mention video in these (1970, 1974) texts but 20 years later he did write specifically about the value of using video images to generate critical consciousness (Horton and Freire 1990). Given the centrality of Freirean theory to much of participatory video and Freire's specific advocacy for the use of video in this publication, it is perhaps surprising that it is not more widely read and cited (Roberts 2015). While Freire did not himself use video, he was explicit in this text that video could be used, in pace of the original sketches, as generative images used to stimulate critical dialogue about the often hidden structures shaping experienced inequality (Horton and Friere 1990: 88):

> Give a camera to several people and say: 'Record what you want, and next week we meet together [to discuss your videos]... They [will be] reading reality through the camera. . . now it is necessary to deepen the reading, and to discuss with the group lots of issues that are behind and sometimes hidden. . . trying to understand the concrete reality that you are in (Freire in: Horton and Friere 1990: 88)

Despite participatory video's roots in emancipatory social movements of the 1970s and 1980s, participatory video practice was arguably diluted in the 1990s and 2000s when, it has been argued, participatory methods in general were "co-opted for a range of agendas other than those with the needs of the poor and oppressed at the their heart" (Cooke 2001: 120). By the late 1990s, participation had become institutionalized as a condition of receiving funding for research projects, establishing what Braden and Huong (1998: 94) referred to as the "new participatory orthodoxy." This practically compelled all those seeking grant-funding to claim that their initiatives were "participatory" in some way. All participatory methods, including participatory video, were in danger of being co-opted. Cooke and Kothari (2001) famously called this the "Tyranny of Participation." This "compulsory participation" resulted in both a dilution of the original, radical meaning of participation and attempts to manipulate "participatory" processes in order to legitimize the top-down agendas of governments and funders (Shaw 2007, 2012).

In recent years there has been a concerted effort to reclaim and reaffirm an emancipatory participatory video practice that enhances critical consciousness (Benest 2010, Roberts and Lunch 2015) and political capabilities (Williams 2004). This movement to reorientate participation back "From Tyranny to Transformation" (Hickey and Mohan 2004) does not deny that forms of fake participation continue to exist. What it claims is that the existence of fake participation does not prevent other practitioners from practicing a participatory video process in which they create space for participants "to perceive social, political and economic contradictions, and to take action against the oppressive elements of reality" (Freire 1970: 17).

In the last 15 years, progress has also been made in developing a consciously feminist practice of participatory video (Kindon 2003). It has been shown that PV can be an effective means to hear women's voices, validate women's experiences, and project women's representations of issues that they prioritize (Waite and Conn 2012). Feminist practitioners seek to bear witness to issues of gender injustice drawing on the knowledge and experience of those most directly and faithfully articulating their standpoint (Harding 2004). Participatory video can also be used to create a space for participants to effect their own critical reading of what Naila Kabeer (2013: 2) calls "the dense root-structure of gender injustices experienced in daily lives" and to articulate the intersectionality of the disadvantage and discrimination women experience (Poveda and Roberts 2017).

67.2.2 Affordance Theory

The field of communication for development provides a range of approaches that participants can draw upon to produce social change. In order to produce the development outcomes associated with participatory video, practitioners might alternatively use, for example, digital storytelling, participatory photography, or participatory theater. Therefore, the question arises why should a practitioner choose participatory video? One way of answering this question is with reference to the idea of "affordances."

The concept of affordances was originated by the visual psychologist Gibson (1977) to refer to actionable properties suggested to a user by something's visual appearance. Later, in the field of technology design, Norman (1988) appropriated the term in order to signify elements of a technology's design that *invite*, *allow*, *or enable* a person to act in a particular way; a cup, for example, signifies the affordance of "liquid conveyance." The technology of a cup, from this perspective, *invites*, *allows*, *or enables* a user to convey liquid. In other words the cup affords the user the *action possibility* of conveying liquid – in a way that a fork does not.

The concept of affordances is relevant to the field of communication for development as it provides us with a basis for determining which participatory methodology to use; it also provides us with an analytical tool for distinguishing which actions are made possible by each participatory methodology (Roberts 2017). For example, from this perspective, participatory video can be argued to have specific affordances for the co-construction of a group advocacy position, compared with digital storytelling which has affordances for eliciting individual participant narratives. As a digital medium, participatory video has the affordance of transmitting its product instantly worldwide, which is not an inherent affordance of participatory theater.

67.3 Benefits and Limitations

As background research for this chapter, in addition to a desk review of the existing literature, the authors conducted semi-structured interviews with a range of participatory video practitioners, each having over a decade of experience in diverse settings. They were asked among other things to reflect on the perceived benefits of participatory video, its limitations, and potential future directions. In the following sections, these practitioner perspectives are used to both ground and illuminate the academic literature on PV.

67.3.1 Perceived Benefits

The increasing popularity of participatory video within the field of communication for development and social change is related to a range of perceived benefits of the approach. Different elements in the participatory video process are perceived to contribute to different development outcomes at three levels: individual, group, and societal. This section will present and examine these perceived benefits.

In the participatory video process, individuals may learn to interview others, to be interviewed themselves on camera, to hear themselves speaking authoritatively on issues of importance, and to see themselves projected on-screen (Roberts 2016). These experiences can develop an individual's sense of having something worth saying and of being heard. Interviewed for this research about her work using participatory video for peace-building, Valentina Bau explained how participants expressed a sense of agency and freedom at being able to speak their truth: "Those who told their stories had the perception of finally being given the opportunity to speak and to tell their own story, and to contribute towards peace" (see also Bau 2014). By taking on the roles of film producers, videographers, interviewers, presenters, editors, and sound engineers and incrementally building their skills and capabilities, participants increase their sense of self-efficacy (Bandura 1995) and self-esteem (Bery 2003). By constructing documentary narratives and expressing themselves through film, participants can increase their confidence to voice their opinion and their sense of authority to speak out on subjects of concern to themselves (Underwood and Jabre 2003).

The group reflection and critical dialogue that are at the heart of much participatory video practice are perceived to contribute additional benefits. In an interview conducted for this chapter, Nick Lunch, Director of InsightShare, recounted that, in his early experience with participatory video working with young people in Nepal, he found that participants used the process to "build consciousness within the group." This resonates with other practitioner experiences that the process of group dialogue enables participants to build their critical consciousness as actors for social change (White 2003; Benest 2010; Walsh 2012). What distinguishes participatory video from other filmmaking is the emphasis that it places on the collective process rather than the film product; it is by working as a group that participants are allowed the time and space to build the political capabilities of collective prioritization and decision-making. It is this iterative process of moving from "reflection on action" to "action on reflection" (Freire 1970) that builds the group's critical consciousness about themselves and their social situation (Thomas and Britton 2012).

To have social change impacts at the societal level, some practitioners argue that participatory video must go beyond engaging with participants themselves and the immediate community (Muniz 2008, 2010). From this perspective, "voice is not enough"; the objective should be social action to change the unjust circumstances that the video highlights. One means of seeking to effect social change is by structuring participatory video initiatives so that the resulting film targets distant decision-makers or power-holders with advocacy messages calling for action. This normative approach also connects with ideas of citizenship; by communicating their messages to those in positions of power and authority, participants are able to develop their sense of themselves as citizens with rights who are able to hold power-holders to account (Wheeler 2012). This form of citizen-representation PV can also be seen as one logical outcome of the Freirean approach of critical reflection to inform critical action; having used group dialogue to develop their enhanced "reading of the world," projecting their video to a wider audience as a call to action can be seen as "acting in the world" to produce social change (Freire 1970).

We can use the concept of affordances to analyze the contribution of participatory video at each of the three levels considered above. Participatory video has some affordances that it shares in common with other participatory visual methodologies, but arguably it enables some distinct action possibilities. Like other visual methods, participatory video affords the possibility to work with low-literacy groups or with groups with mixed literacies (Braden and Huong 1998), and, as an audiovisual medium, PV affords its advocacy message greater "recallability" when compared to text-only content. Research has established that we more readily recall what we hear and see compared to what we read (Fraser and Villet 1994). As PV practitioner Fernanda Baumhardt from Pro Planeta commented when we interviewed her for this chapter: "PV is a particularly powerful medium because it projects the voice, and combines video and sound."

Participatory video also has particular affordances of "reflexivity" due to the technologies' functionalities of rewind, replay, and projection on-screen. This provides participants with new action possibilities for reflecting on their ability to act in the world and to revise and rehearse their performance in a "safe space" (Miller and Smith 2012; Waite and Conn 2012). Lunch and Lunch (2006) have likened participatory video's "reflexivity" affordance to Lacan's mirror stage in enhancing a person's sense of selfhood.

Users of participatory video often discover that a tripod and digital camera offer them a reason and excuse to approach and question higher-status people (Shaw and Robertson 1997). In the first author's own research, young women users in Zambia found that holding a digital camera with the recording light flashing enabled them to elicit a respectful and considered response from older men, which they had not been afforded previously (Roberts 2015). This affordance of "respect-ability" can have the effect of raising users' self-efficacy and sense of agency.

It has also been argued that participatory video has distinct affordances for participatory monitoring and evaluation (PM&E). PV's affordance of "accessibility," removing literacy barriers to participation, provides the action possibility to incorporate PM&E feedback that is more widely representative of the whole population. Lunch (2006) has argued that the iterative process of PV (taking participants through several cycles of action-orientated filmmaking alongside sessions of critical reflection and analysis) lends itself well to the task of monitoring and evaluation. Lunch and colleagues at InsightShare drew on the "mobility" and "accessibility" affordances of video to encourage participants to take the cameras out into the villages and interview people of all backgrounds, asking them evaluative questions concerning a particular initiative. It was also their experience that hearing evaluation feedback directly from villagers themselves, through the medium of film, helped managers and funders to better ground evaluation findings contextually. They also found that evaluation feedback in the form of film can connect at affectual and emotive levels not normally achieved by textual or statistical reports. In order to add rigor and depth to their M&E process,

Lunch and colleagues then incorporated the widely used "most significant change" (MSC) approach (Lemaire and Lunch 2012; Asadullah and Muñiz 2015). The MSC approach provides multiple stakeholders with open-ended opportunities to identify the most significant change that they associate with a given intervention and a structured means to interpret the arising data for evaluation purposes (Davies and Dart 2005).

The participatory video process also has affordances that make it a popular choice for action research processes (Plush 2012). Like PV, action research seeks to bring together action and reflection, theory, and practice, in participation with others in the pursuit of practical solutions to issues of pressing concern to people (Reason and Bradbury 2006; 2). Given this shared objective, what are the specific affordances that participatory video contributes to action research? Participatory video brings together marginalized groups to co-produce knowledge on social issues; this affordance of "co-produce-ability" of marginalized groups' own audiovisual representations on key social issues is a unique form of knowledge input to social change processes, which is not afforded by some other means. PV also has the advantage of producing a digital product which has the affordance of "send-ability," meaning that it can be immediately attached to an email or uploaded to social media in order that participants can inexpensively communicate their words and images directly to a single recipient, or to any number of people or groups, whom they may wish to influence or with whom they may wish to make common cause.

67.3.2 Degrees of Participation

The use of participatory video is not unproblematic. Not all PV is equally participatory. Power relationships exist between participants, facilitators, intermediary organizations, and funders. The early literature on participatory video understandably focused primarily on establishing the benefits of the approach and was largely uncritical (Shaw and Robertson 1997; White 2003). More recently, and especially in the wake of the "Tyranny of Participation" critique (Cooke and Kathari 2001), the participatory video literature has become less celebratory and more critical, questioning the extent to which control is handed over (and to whom) and scrutinizing claims that PV produces social change (Muñiz 2010; Shaw 2012; Roberts and Lunch 2015).

The basic theory of change underlying all participatory video is that marginalized individuals and groups who take part in participatory video processes can gain a range of benefits which contribute to development or social change. This theory of change has a number of embedded assumptions that give rise to questions including who participates? In which elements of the project cycle? Who is defining development and social change? As well as who benefits? And according to whose evaluation/criteria? While all participatory video involves a group of people making their own film, significant differences exist regarding who participates in which stages of the film's conception, planning, filming, editing, evaluation, and distribution. The next section offers a highly schematic chronology for understanding how participatory control over elements of the PV film production cycle is expanding over time.

The groundbreaking innovation of the 1967 Fogo Process was to extend to communities participatory control over the content of a film about them, enabling them to voice their own concerns in their own words. They did not operate the cameras or any of the editing equipment. These tasks were reserved for external experts. Editing took place away from the islands (Quarry 1994). There was no editing suite on the Fogo Islands, and it could be argued that professional training was necessary to operate the filming and editing equipment available at that time. In this "PV version 1.0," participants were able to take participatory control of determining the content of the film and had the ability to require edits and re-edits, but the people with their hands on the camera and on the editing equipment were external professionals. This reflected existing power relationships about ownership of the means of film production.

By the 1990s the cost of video cameras was falling dramatically, and their ease of use and resolution quality were increasing rapidly. This allowed the innovation of handing over control of operating the camera equipment to participants themselves. In this "PV version 2.0," participation was no longer limited to co-creating a script for external experts to film and edit; participants were now able to take control of the filmmaking equipment and to carry out their own filming. A typical PV process now involved participants taking rotational turns to play roles including camera operator, sound engineer, director, and interviewer. However, they did not yet do the editing themselves. It still remained common practice for all of the film media and equipment to be collected up by the facilitating team and removed to an editing suite, often in a different location, with the final film being returned to the participants at a later date for their approval or suggested further edits (Mak 2012). In the majority of projects, it also remained the case that the cameras would leave when the facilitating organization left at the end of the funded period, which had negative implications for ongoing dependency and sustainability of benefits (Colom 2009). Despite the emancipatory intent of external facilitators, power relationships remained, and these unequal power relationships continued to reflect unequal ownership of the means of film production.

After 2010 the availability of cheap and powerful laptops, free and open-source film editing software, and low-cost and high-definition video cameras made it possible for participants to carry out all of their own editing as well as the filming. In this "PV version 3.0," participants are able to take not only full editorial control over the film product but also full control over both the filming and editing process. Editing could now be carried out on standard laptops, by participants themselves, on the same day and at the same location as the filming. This makes it possible for external PV facilitators to operate a "hands-off" approach where they decline to touch either the camera or the editing equipment at any stage in the process, ensuring that the film is entirely the participants' work and that their sense of agency and ownership of both the process and the product is fully optimized (Poveda and Roberts 2017). This has the advantage of accelerating participants' hands-on skill development, which can boost their self-efficacy and autonomy. The relatively low cost of high-definition cameras has also made it possible for the cameras, tripods, editing software, and laptops to remain the property of participants beyond the initial funded project engagement. The transfer of ownership of the means of film production from external facilitators to local experts is a substantial advance that arguably shifts power relationships by removing dependencies and handing independent control for determining all aspects of film conception, production, and dissemination.

This progressive extension of participatory control over elements of the participatory video process should not be read as a simple linear unfolding logic. There have always been, and continue to be, exceptions to and contradictions within the schematic model presented above. Even in cases where participants have control over film content, filming equipment, and editing equipment, there are always people who are excluded from the process. Inequalities of power between funders, facilitators, and participants mean that initial framings or dissemination processes may continue to lack the participation of all players.

However, some progress continues to be made. Recent years have also seen a shift away from short interventions led by externally funded facilitators toward long-term participatory video processes and processes designed to build the capacity of independent and sustainable participatory video capacity in grassroots organizations. Networks of such independent PV organizations are forming, and south-to-south exchanges have happened. In this "PV version 4.0," independent participatory video organizations have full control of the entire project cycle and are free to conceptualize their own PV initiatives from the outset and have the freedom to film, edit, and disseminate their work without the historical dependence on external experts. Examples include Drishti, Video SEWA, and Digital Green in India (Singh 2014) and the community video hubs developed by InsightShare in South Africa, Peru, and London (Muñiz 2011). This is not to claim that these entities operate free from structural constraints and power relations; it is only to say that the "PV version 4.0" innovation of building independent long-term capacity is preferable to parachute interventions, which may raise agency in the short term only to accentuate a sense of powerlessness in the medium term (Milne et al. 2012). Longer-term participatory video engagements that build permanent local filmmaking capacity are arguably more effective at sustaining the political spaces that groups are able to open up through the use of participatory video and other means (Colom 2009).

67.3.3 Conflicting Objectives

One other way that power relations often play out in participatory video initiatives is in the form of multiple and sometimes conflicting objectives. PV practitioners are motivated by a wide range of objectives, and their initiatives are informed by a wide range of explicit, or implicit, theories of change. Within a single participatory video process, there may be multiple and contradictory theories of change in operation. The motivations of the funder, the practitioner, and the participants, for example, may not be the same and may not be mutually compatible. The contradictory nature of these objectives may not be known, or may not be made explicit at the outset, and may only become apparent when conflict arises. This reality is not specific to participatory video and occurs in a wide range of development initiatives.

Several practitioners have written about the challenges of reconciling the contradictory demand of stakeholders including funders, practitioners, and participant groups themselves. Shaw (2012) has reflected on what happens when it transpires that the commissioning organization's objective was not to elicit community input into a decision-making process but rather to secure community validation for decisions that they had already made; Milne et al. (2012) provides an example of a community doubting the funder and research objectives and refusing en masse to participate; and Dougherty and Sawhney (2012: 447) point out that based on their extensive experience: "Quite often, community-produced video projects have obligations to satisfy funders or non-governmental organisation's agendas; the degree to which the community owns and controls the process can be in question as the project advances." Clearly it would be optimal to co-develop the objectives of a participatory video process jointly with all stakeholders and to agree in advance what happens if participants wish to take the PV process in an unforeseen direction midway through the process. But achieving this is not always practical due to time and resource constraints and the power relationships that always exist between funders, intermediary organization, practitioners, and participants (Wheeler 2012).

In clarifying objectives, it may be useful to think through the stages of the participatory video process and the kind of outcomes that it is possible to generate at each stage. This can then inform a discussion about which elements participants wish to focus their finite time and energy on. The first operational phase of a participatory video process often involves setting shared expectations and developing ground rules. This preparatory phase may involve some icebreaking exercises to establish trust and familiarity. If there has been success in bringing together people from the margins who are not normally heard, this group-building phase can be key in breaking down inhibitions and building a sense of shared enterprise. Often the first phase also includes some games or demystifying exercises with the camera equipment (Lunch and Lunch 2006).

There is more than one way to break down the stages of the project cycle of participatory video, but Shaw (2017) breaks her extended video process into four key phases: (1) ensure inclusive engagement during group forming and building, (2) develop shared purpose and group agency through video exploration and sensemaking, (3) enable horizontal scaling through community-level videoing action, and (4) support the performance of vertical influence through video-mediated communication. Inclusive participation is a key outcome in phase 1. Unless all stakeholders, from commissioning funders to low-caste women, are in the room at this stage, there is potential for later conflict when the group discovers that conflicting objectives disrupt progress. In phase 2 participants begin to develop shared intent and a collective voice as they critically reflect on their priority issues. In Shaw's third phase, the active engagement and input of the wider community are secured, and it is not until phase 4 that vertical influence is attempted through video-screening with decision-makers. This staged approach reflects Shaw's (2012) concern that having voice is not the same as being heard and being heard is not the same as having influence. According to this logic, producing external social change in a context of complex power relations requires a staged strategic approach that connects vertically with relevant power-holders. This resonates with the work of PV practitioner Tamara Plush, which analyzes the capacity of PV to transform power relations. When we interviewed her as part of the background research to produce this chapter, she pointed out that: "Just because a voice is amplified it doesn't mean that you are shifting power relations." Like Shaw her concern was whether anyone was listening and being influenced, and also like Shaw, her conclusions include that: "PV needs to be included in wider projects of citizen engagement and mobilisation that connect it to local and national advocacy campaigns" (see also Plush 2015).

Ultimately it can be argued that no PV process is "maximally participatory" as not everyone wants to participate in every element of a film's conception, production, and dissemination. Realistically external funders, government officials, and even facilitators will exercise unequal influence over some element of project framing, partner choice, process, or product in most participatory video. It may be that being as open and transparent and as reflexive as possible about these realities and aiming for PV that is "optimally participatory" rather than "maximally participatory" is a more realistic objective.

67.4 Future Directions

Continuing rapid technological change, practitioner adaptations, and theoretical developments suggest that innovation in participatory video practice will accelerate in the years ahead. The incorporation of microprojectors into video cameras; the advent of high-specification, low-cost tablet computers; and the increasing video capabilities of mobile phones are creating exciting new opportunities and challenges for participatory video practitioners. In an interview for this research, experienced PV practitioner and academic Jay Mistry commented that "in our current projects we are already using tablets because they are cheaper and because we can edit directly on them" and also noted that "Community researchers bought smart phones and in their spare time have been training a lot of people and doing conservation projects with other organisations. They are becoming independent facilitators (see also Mistry 2014)." Outside of "projects" people are independently acquiring the equipment and the skills to produce their own films and to incorporate them into their communication and social change work.

The first book on mobile phone-based participatory video appeared in 2016 following the first conference on incorporating mobile phone films into visual research and activism (MacEntee et al. 2016). Perhaps we will soon be incorporating 360-degree virtual reality cameras into PV. The digital nature of video media certainly has the potential affordance of "combine-ability" with other digital visual methods to create new participatory methods. Participatory video practitioner Namita Singh, in an interview for this chapter, stated her experience that: "People want to transfer videos between phones using WhatsApp. The video format will change to create shorter videos that don't make people spend too much data to download." So the question arises whether in the future PV content will be compromised in order to accommodate new communication practices and digital formats. We should be asking ourselves whether moving away from cameras to tablets and mobile phone-based PV dilutes group dialogue in favor of more individualized processes. Is there a danger in the rapidly changing technological landscape of losing sight of the original radical purpose and emancipatory objectives of participatory video?

We have seen in this chapter that participatory video practice is always in flux over time and that, like any approach, PV can be used for repressive or for emancipatory ends. Participatory video can be used to legitimate top-down decision-making or to create critical consciousness and shift power relationships. Given the rapid pace of technological change, practitioners will have to resist the glittering temptation of the new for its own sake and instead be mindful of the specific affordances each new video technology has for stimulating critical consciousness and collective action. What came over in the interviews that we conducted with practitioners in advance of writing this chapter was a desire to remain focused on challenging power relations and supporting people to create the change that they want in their lives. Achieving this in practice means not expecting PV to achieve this in isolation but rather to integrate PV practice vertically and horizontally within wider programs of communication and action for transformative social change.

References

Asadullah S, Muñiz S (2015) Participatory video and the most significant change: a guide for facilitators. Insightshare, Oxford

Bandura A (1995) Self-efficacy in changing societies. CUP, Cambridge

Bau V (2014) Building peace through social change communication: participatory video in conflict-affected communities. Community Development Journal 50(1):121–137

Benest G (2010) A rights-based approach to participatory video. Insightshare, Oxford

Bery R (2003) Participatory video that empowers. In: White S (ed) Participatory video: images that transform and empower. Sage, London

Braden S, Huong T (1998) Video for Development. Oxfam, Oxford

Colom A (2009) Participatory Video and Empowerment: The role of Participatory Video in enhancing the political capability of grass-roots communities in participatory development. Unpublished masters dissertation, SOAS, London

Cooke B, Kothari U (2001) Participation: the new tyranny. Zed, London

Corneil M (2012) Citizenship and participatory video. In: Milne EJ et al (eds) Handbook of participatory video. AltaMira, Plymouth

Cornwall A (2004) Spaces for transformation? Reflections on issues of power and difference in participation in development. In: Hickey S, Mohan G (eds) Participation: from tyranny to transformation? Zed Books, London

Crocker S (2003) The Fogo process: participatory video in a globalizing world. In: White S (ed) Participatory video: images that transform and empower. Sage, London

Davies R, Dart J (2005) The most significant change technique. In: Mathison S (ed) Encyclopedia of evaluation. Sage, Thousand Oaks, pp 261–263

Dougherty A, Sawhney N (2012) Emerging digital technologies and practices. In: Milne EJ et al (eds) Handbook of participatory video. AltaMira, Plymouth

Fals-Borda O, Rahman M (1991) Action and knowledge: breaking the monopoly with participatory action research. Intermediate Technology Press, London
Fraser C, Villet J (1994) Communication; a key to human development. FAO, Rome
Freire P (1970) Pedagogy of the oppressed. Continuum, NewYork
Freire P (1974) Education for critical consciousness. Continuum, NewYork
Gibson J (1977) In: Shaw R, Bransford J (eds) The theory of affordances in perceiving, acting, and knowing. OUP, London
Gregory S (2005) Video for change: a how-to guide for activists. Pluto Press, London
Harding S (2004) The feminist standpoint theory reader: intellectual and political controversies. Routledge, London
Hickey S, Mohan G (2004) Participation: from tyranny to transformation. Zed, London
High C, Singh N, Petheram L, Nemes G (2012) Defining participatory video from practice. In: Milne E-J et al (eds) The handbook of participatory video. AltaMira Press, Lanham
Horton M, Freire P (1990) We make the road by walking. Temple University Press, Philadelphia
Kabeer N (1994) Reversed realities. Verso, London
Kindon S (2003) Participatory video in geography research: a feminist practice of looking. Area 35:142–153
Kyung-Hwa Y (2012) Reflexivity, participation and video. In: Milne EJ et al (eds) Handbook of participatory video. AltaMira, Plymouth
Lemaire I, Lunch C (2012) Using participatory video in monitoring and evaluation. In: Milne EJ et al (eds) Handbook of participatory video. AltaMira, Plymouth
Lesage J (1978) The political aesthetics of the feminist documentary film. Q Rev Film Stud 3(4):507–523
Lunch C (2006) Participatory video for monitoring and evaluation: experiences with the MSC approach. Capacity.org 29. http://insightshare.org/wp-content/uploads/2017/05/PV-for-ME-Experiences-with-the-MSC-approach-English.pdf
Lunch N, Lunch C (2006) Insights into participatory video. Insight, Oxford
MacEntee K, Burkholder C, Schwab-Cartas J (2016) What's a Cellphilm?: integrating mobile phone technology into participatory visual research and activism. Sense, Rotterdam
Mak M (2012) Visual post-production in participatory video-making processes. In: Milne et al (eds) Handbook of participatory video. AltaMira, Plymouth
Miller E, Smith M (2012) Dissemination and ownership of knowledge. In: Milne EJ et al (eds) Handbook of participatory video. AltaMira, Plymouth
Milne EJ et al (2012) Handbook of participatory video. AltaMira, Plymouth
Mistry J (2014) Why are we doing it? Exploring participant motivations within a participatory video project. Area 48(4):412–418
Muñiz S (2008) *Participatory development communication: rhetoric or reality?* Unpublished masters dissertation, University of Reading
Muñiz S (2011) Insightshare's global network of community-owned video hubs. PLA Notes 63:64–130
Muñiz S (2010) Participatory development communication: between rhetoric and reality. Glocal Times 15
Nemes G, High C, Shafer, N, Goldsmith R (2007) Using participatory video to evaluate community development. Paper for Working Group 3, XXII European Congress of Rural Sociology, Wageningen
Norman D (1988) The design of everyday things. Basic Books, New York
Plush T (2012) Fostering social change through participatory video: ac conceptual framework. In: Milne et al (eds) Handbook of participatory video. AltaMira, Plymouth
Plush T (2015) Participatory video and citizen voice – We've raised their voices: is anyone listening? MUEP, Glocal Times 21:1–16
Poveda S, Roberts T (2017) Critical agency and development: applying Freire and Sen to ICT4D in Zambia and Brazil. Info Technol Devel J 24:119–137

Quarry W (1994) The Fogo process: an experiment in participatory communications. University of Guelph
Reason P, Bradbury H (eds) (2006) Handbook of action research. Sage, London
Roberts T (2017) Participatory technologies: affordances for development. In: Choudrie J, Islam M, Wahid F, Bass J, Priyatma J (eds) Information and communication technologies for development. IFIP advances in information and communication technology, vol 504. Springer.
Roberts T (2016) Women's Use of Participatory Video Technology to Tackle Gender Inequality in Zambia's ICT Sector. In: Proceedings of the 8th International Conference on Information and Communication Technology and Development, Ann Arbor, USA
Roberts T (2015) Critical agency in ICT4D. Unpublished PhD, Royal Holloway University of London
Roberts T, Lunch C (2015) Participatory video in: international encyclopedia of digital communications and society. Wiley, London
Shaw J, Robertson C (1997) Participatory video: a practical guide to using video creatively in group development work. Routledge, London
Shaw J (2007) Including the excluded: collaborative knowledge production through participatory video. In: Dowmunt M et al (eds) Inclusion through media. Goldsmiths College, London
Shaw J (2012) *Beyond Empowerment Inspiration: Interrogating the Gap between the Ideals and Practice Reality of Participatory Video* in. In: Milne EJ, Mitchell C, de Lange N (eds) Handbook of participatory video. Altamira Press, Lanham
Shaw J (2017) Pathways to accountability from the margins: reflections on participatory video practice. Institute for Development Studies, Brighton
Singh N (2014) Participation, agency and gender: the impacts of participatory video practices on young women in India. Unpublished PhD thesis, Open University Press, Milton Keynes
Snowden D (1984) Eyes See, Ears Hear: supplement to a film under the same title. St. John's Memorial University of Newfoundland
Teitelbaum P (2012) Re-seeing participatory video practices and research. In: Milne EJ, Mitchell C, de Lange N (eds) Handbook of participatory video. Altamira Press, Lanham
Thomas V, Britton K (2012) The art of participatory video. In: Milne EJ, Mitchell C, de Lange N (eds) Handbook of participatory video. Altamira Press, Lanham
Underwood C, Jabre B (2003) Arab women speak out: self-empowerment via video. In: White S (ed) Participatory video: images that transform and empower. Sage, London
Waite L, Conn C (2012) Participatory video: a feminist way of seeing? In: Milne EJ et al (eds) Handbook of participatory video. AltaMira, Plymouth
Walsh S (2012) Challenging knowledge production with participatory video. In: Milne EJ et al (eds) Handbook of participatory video. AltaMira, Plymouth
Wheeler J (2012) Claiming citizenship in the shadow of the state. PhD thesis, Institute of Development Studies, University of Sussex
White S (2003) Participatory video: images that transform and empower. Sage, London
Williams G (2004) Evaluating participatory development: tyranny, power and (re)politicisation. Third World Q 25(3)

Reducing Air Pollution in West Africa Through Participatory Activities: Issues, Challenges, and Conditions for Citizens' Genuine Engagement

68

Stéphanie Yates, Johanne Saint-Charles, Marius N. Kêdoté, and S. Claude-Gervais Assogba

Contents

The authors received a grant from the Social Sciences and Humanities Research Council of Canada (430-2016-01029) and from the International Development Research Center (107347) for this research.

S. Yates (✉)
Département de communication sociale et publique, Université du Québec à Montréal, Montréal, QC, Canada
e-mail: yates.stephanie@uqam.ca

J. Saint-Charles
Département de communication sociale et publique, axe santé environnementale, CINBIOSE, Université du Québec à Montréal, Montréal, QC, Canada
e-mail: saint-charles.johanne@uqam.ca

M. N. Kêdoté
Institut Régional de Santé Publique, Comlan Alfred Quenum, Université d'Abomey-Calavi, Abomey Calavi, Benin
e-mail: kedmar@yahoo.fr

S. C.-G. Assogba
Faculté d'Agronomie, Université de Parakou, Abomey-Calavi, Benin
e-mail: a_claude2003@yahoo.fr

J. Servaes (ed.), *Handbook of Communication for Development and Social Change*,
https://doi.org/10.1007/978-981-15-2014-3_25

Abstract

Largely supported by international donors, participatory initiatives are multiplying in Africa, to such an extent that some observers refer to a new "tyranny of participation." The challenges associated with participatory democracy are even more acute in developing countries. In a context where the notion of civil society in these countries remains unclear, the inclusion principle is difficult to enact, particularly with regard to women who are still largely underrepresented in this type of process. Moreover, there is a risk that participatory initiatives widen the gap between "politically engaged" citizens and their more apathetic counterparts. Power relationships between participants are another important issue. If not properly tackled, they can lead to the maintenance of traditional hegemonic discourses – rather than to innovative ways of thinking. Lastly, the concrete implementation of the solutions or compromises emerging from participatory processes is paramount for their legitimacy.

This chapter examines and compares two cases of participatory processes put forward in the context of a public health project aiming at reducing air pollution in the cities of Cotonou (Benin) and Dakar (Senegal). This allows us to reflect on the issues, challenges, and conditions of success of the participatory processes orchestrated in these contexts. It also brings support to the need to address various challenges when participatory initiatives are fostered and shows that these challenges are met differently in the two countries.

Keywords

Public participation · Air pollution · Public health · Senegal · Benin · Communication · Dialogue · Inclusiveness · Power relationships

68.1 Introduction

In the last two decades, acknowledgment of the relevance of citizen participation by governmental actors has risen in Occidental countries. Given the limits of representative democracy, participatory democracy initiatives are more and more considered legitimate, if not necessary, in some instance (Bacqué and Sintomer 2011; Blondiaux and Sintomer 2002; OCDE 2001). Science communication has also adopted a more participatory framework, moving from the deficit model to a dialogic one (Brossard and Lewenstein 2010; Nisbet and Scheufele 2009). This paradigm shift is also observed in some African countries (Cissoko and Toure 2005; OCDE 2013) where there is a movement from an elitist mode of management to modes of governance that are said to be more and more participative (Fung and Wright 2003; Tenbensel 2005). Indeed, participative governance is seen as the "missing link" to fight against poverty (Carrel 2007; Cling et al. 2002; Dom 2012; Schneider 1999). The objective is, of course, to gain greater legitimacy of the decisions made by governing bodies but also to support the efficacy of those decisions through the acknowledgment of citizen or "lay" knowledge (Callon et al. 2001; Fisher 2009). The consideration of this lay knowledge is helpful in the development of these decisions and for their uptake by communities, hence favoring their implementation.

This openness to participation comes with numerous challenges. A consensus is slowly emerging around what would be good practices in terms of public participation (Bacqué and Gauthier 2011; INM 2013; Blondiaux 2008; Fung 2006; Gastil and Levine 2005). These are based on cumulative learning emerging from participatory experiences in Northern and Western Europe (Anderson and Jaeger 1999; Gourgues 2012; Koehl and Sintomer 2002), in North America (Bherer 2006; Rabouin 2009), and to a certain extent, in Latin America (Baiocchi 2003; Garibay 2015; Weyh and Streck 2003). It remains to see whether such "basic and procedural rules" to insure sound processes can be transposed to other social contexts such as those of African, Asian, or Eastern Europe societies. Indeed, the mechanical application of so-called "universal" rules on participation, which obscure local specificities, may lead to unacceptable experiences of participation or, worse, to the disinterest – or even cynicism – of the populations one seeks to mobilize.

As a step toward a better understanding of participatory processes in other social contexts, we propose to highlight the participatory challenges encountered in two cases of participatory processes put forward in the context of a public health project aiming at reducing air pollution in the cities of Cotonou (Benin) and Dakar (Senegal). Our analysis is drawn from the direct observation of participatory activities organized in these two cities, from semi-structured interviews with participants, and from conversations with the team implementing the project. A guided visit of the discussed neighborhoods has also highlighted our perspective. This empirically based approach allows us to reflect on the issues, challenges, and conditions of success of the participatory processes orchestrated in these specific contexts. The actors' mutual influence, resistance and opportunities emerging from these processes, political significance of these experiences for the participants, as well as the social effects generated by them are looked upon, thus answering the call of researchers in the field (Lavigne Delville 2011).

68.2 Context

In 2012, more than 7 million of deaths in the world were attributable to the exposure to air pollution (WHO 2014). Air pollution was thus identified as an important factor for non-transmittable diseases such as cardiovascular diseases, asthma, and *bronchopulmonary* carcinoma (Clark et al. 2013; Laumbach and Kipen 2012; Martin et al. 2013; Pascal et al. 2013; Pope and Dockery 2006). West African cities are strongly affected by this problem, notably for chaotic industrial activities (with some factories directly located within cities), weakly organized transportation systems characterized by the multiplication of motorcycles and old vehicles, and polluting sources of energies, like traditional fire oven inside households (Liousse and Galy-Lacaux 2010; Liousse et al. 2014). There are few available data allowing for the quantification and characterization of air pollution in West African cities, the evaluation of its effects on health, and the influence of lifestyle habits on the situation (Perez et al. 2013).

In this perspective, the International Development Research Center of Canada is founding a research that aims at improving the understanding of urban air pollution and its impacts on non-transmittable breathing diseases in four west African cities, Abidjan (Ivory Coast), Cotonou (Benin), Dakar (Senegal), and Ouagadougou (Burkina Faso). The project is a collaboration between four African universities and is under the auspices of the Communauté de pratique en écosanté en Afrique de l'Ouest et du Centre (CoPES-AOC). Since 2006, the CoPES-AOC has been very active in the region to promote an ecosystem approach to health (ecohealth), to foster collaboration between scholars and practitioners, and to train a new generation of African scholars. The research also aims at developing prevention strategies in partnership with community organizations in these cities. More specifically, the Air Sain project, a component of the research, has the objective to develop and evaluate, in these urban west African contexts, adaptation strategies that benefit municipalities and communities and contribute to environmental protection, population health, and social and gender equity. The involvement of the actors directly concerned by the issue is paramount for the success of the project, so they can have their say in the definition of realistic and efficient interventions. Political actors, industries, funding organizations, as well as local communities are thus targeted, knowing that the more vulnerable members on the economic and healthy plans of the latter are not always aware of the public health problems posed by air pollution. Hence, the Air Sain project goes beyond the "deficit model" in communication (Brossard and Lewenstein 2010; Nisbet and Scheufele 2009), which essentially consists in transmitting the knowledge. It is rather a project based on an ecohealth perspective, where different types of knowledge – local, traditional, scientific, and practical – must be considered so that people can relate to them and integrate them in their reality (Saint-Charles et al. 2014; Charron 2012; Webb et al. 2010). This is particularly important since the changes of behavior promoted to reduce air pollution may appear in contradiction with deeply rooted cultural or religious practices that are considered as "good" by the persons (such as using coal or fire for cooking or burning incense to remove bad smells). The recommended solutions must take into account the preoccupations and restrains of the different groups of the population in order to be socially acceptable. Social acceptability can be defined as "the result of a process through which stakeholders build together the minimal conditions under which a project, a program, or a policy will be harmoniously integrated in its natural and human environment" (Caron-Malenfant and Conraud 2009, our translation). The co-construction principle underlies the notion of social acceptability and calls for public participation (Fortin et al. 2013; Batellier 2015).

Air Sain is thus a participatory project. As shown by a growing number of studies, public participation comes with many challenges, which can be grouped in four categories (INM 2013). The first challenge consists of fighting the assumed popular apathy by convincing citizens to take part in participatory initiatives. The second challenge consists of avoiding reproducing inequalities in participation, which questions the inclusion of marginalized populations, notably women (Raibaud 2015). As such, there is a risk that participatory initiatives widen the existing gap between engaged citizens and their more apathetic counterparts (Blacksher 2013). Another challenge concerns the balance to reach between experts and citizens' voices and the difficulty to avoid the reproduction, within participatory initiatives, of the usual power relationships leading to the traditional "hegemonic discourses" (Abelson et al. 2011). Hence, an effort has to be made to make sure that "lay citizens" are not evacuated from the process to the benefit of a "new elite of participation" (Bherer 2005). A last challenge concerns the implementation of the solutions or compromises emerging from participatory initiatives. The initiatives that are not followed by concrete results will likely be considered as mere legitimation attempts of public decisions and will be associated with an instrumentalization process (Bherer 2011; Blondiaux and Fourniau 2011; Clarke 2013; Levine and Nierras 2007).

We can think that all those challenges are even more acute in West African countries. Promoted by international donors (Norad 2013), public participation in governmental decision is often seen as a "'bureaucratic populism' that has structured state interventions from the beginning of the colonisation" (Lavigne Delville and Thieba 2015, p. 214, our translation). Considered as hegemonic, public participation is even deemed, by some observers, as "a new tyranny" (Cooke and Kothari 2001). It becomes particularly difficult, in this context, to fight against the ambient cynicism that often accompanies public participation initiatives in emerging countries. Moreover, with the very notion of civil society remaining unclear (Hearn 2001; Bayart 1986; Otayek 2009), the inclusion principle is hard to attain, particularly toward women, still largely underrepresented in participatory processes. Even when the inclusion principle is respected, having one's voice heard is not always easy when the presence of other participants is deemed more legitimate, based on hierarchical and status differences. Even if one cannot conclude that these processes come under "pure instrumentalization" (Lavigne Delville 2011), the political and institutional contexts specific to each initiative must be taken into account in order to understand the dynamics that occur, hence the relevance of case studies (Lavigne Delville and Thieba 2015). Empirical approaches that encourage the consideration of citizens' daily ways of life and sense of citizenship are a good start to turn back to a strictly normative vision of participation (Robins et al. 2008) and to foster more sustainable changes.

68.3 Participation in Practice

Our empirical perspective is based on two participatory activities in each of the country studied, mostly held in local languages (Fon in Benin and Wolof in Senegal). The activities slightly varied between the two countries. In Benin, we report on a preparatory meeting with four women identified as leaders in the Cadjehoun neighborhood; the objective of this meeting was to prepare the mobilization for the larger assembly to be held in this neighborhood afterward, on which we also report. In both cases, participants had already taken part to a public awareness activity on air pollution before and, as such, had already some knowledge on this matter. A large part of the meeting in the second neighborhood, Dantokpa, aimed at discussing the information that was presented at the previous meeting in order to see what participants remembered and how this knowledge had changed their practices. A leaflet and a poster, both presenting handmade drawings, were used as a communication support during these meetings.

In Senegal, we discuss two public awareness meetings hold at 1 week of distance with about 20 community leaders from Fass-Colobane and Medina neighborhoods. The first meeting had the objective to present information on air pollution and the second, which gathered the same participants, to discuss this information in order to see what elements were remembered and what changes of practices were envisioned. A series of slides were presented during the first meeting showing, in a first part, definitions, figures, and explicative drawings and, in a second part, different sources of pollution, their impacts on health, and the measures that can be taken to reduce exposure, all of this illustrated with photos. The last slides invited participants to discuss the information just presented, for instance, by asking them to share their own experiences related to pollution. Photos showing situations of dense pollution were reused during the second meeting as a reminder and to generate discussions.

Many aspects of these four meetings support our discussion: who is talking to whom and after whom, the role of the presenter, the level of bidirectionality between her and the attendees, the thematic of the exchanges and the reactions generated by them, the level of agreement or conflict, the openness or closeness toward the subjects discussed, the diverse calls to legitimacy claimed by participants (expertise, representativeness, tacit knowledge, seniority), and the presence of leaders and their overall effects on the attendees. We also noted the various types of participants, the rules of the discussion, the spatial organization of the room, and the possible effects of the observers on the behavior of the participants. An observation grid allows for the compilation of these elements.

Interviews with activities' participants from a diversity of gender, age, and profession conducted by a locally engaged student allow for an insider view of the participating process. Interviews were conducted in the local language and then translated and transcribed. Several aspects of the participatory meetings were discussed during these interviews, such as the mobilization for these meetings, the impression on their utility, the easiness to speak during the meetings, the knowledge emerging from them, the changes of perception and behavior following them, as well as the awareness efforts toward third parties.

68.4 Learnings

By grounding our reflections in empirical concrete situations, we were able to shed light on several aforementioned challenges associated with participatory initiatives. Indeed, our results highlight risks of reinforcement of inequalities by the recruitment of already engaged citizens, the reproduction of pre-existing power relationships that poses a threat to the inclusion principle, and the difficulty to co-construct and implement solutions emerging from these participatory processes.

68.5 Preaching to the Converted?

As stressed by Blacksher (2013), participatory initiatives pose the risk to widen the gap between politically engaged citizens and those who are mostly detached from the discussions associated with collective issues, if not apathetic or bluntly cynical about them. The different strategic choices between Senegal and Benin when it comes to the recruitment of participants highlight this issue.

In Senegal, Air Sain organizers banked on community leaders who can be associated, following Bherer (2005), with a certain "elite of participation." These individuals, who attended the two public awareness meetings we observed, were selected based on their role in the Medina and Fass neighborhoods and on their capacity to disseminate public health messages to a broader population. For instance, one member of the Medina Women Association, another one from the Medina Notability Association, several representatives of diverse community organizations, one teacher in women entrepreneurship, two imams, one engineer, and several Badiene Gokh were parts of the meetings (literally, Badiene means auntie or mommy and Gokh means neighborhoods. Nowadays, their role is to conduct awareness campaigns on health-related topics. Their work schedule is dependent upon the topics elected by the Ministry of Health). The recruitment of participants was under the responsibility of an influent local public servant. The facilitator of the meeting was a young woman who has studied in Paris and was now doing her master with the head of the research project for Senegal.

Most of the interviewees asserted that the meetings were an occasion to learn about air pollution. One can think that participants were particularly open to new learning – in fact, several of them took some notes – in a context where many of the harmful practices that were discussed did not directly concern their own behavior. Organizers pointed up, for instance, pollution generated by urban transportation, the burning of waste or their tipping out in open-air canals, the smoking of fishes or the smoke coming from bakery (which usually cook their breads on wood fires), the detergent used by women whose occupation is to do the laundry, the cigarettes, or the selling of perfumes or other potentially toxic products. Hence, we can put forward that the discussions around the changes of practices associated with these sources of pollutants did not represent an immediate threat to these participants' way of living. In line with this remark, several interviewees mentioned that it would have been relevant to invite at the meetings other types of participants more directly concerned by the harmful practices identified, such as mechanics, metal welders, drivers, carpenters, shoemakers, or sellers exposed to wastes in the market.

Of course, even these "elite" participants themselves generate some types of pollutants, and as such, they may have been concerned by the discussions around some household pollutants, like incense, insecticides, or spray deodorants. Indeed, when asked about the changes in their behavior they put forward following the public awareness meetings interviewees mostly referred to a less intense use of incense. But it is probably accurate to state that, globally speaking, participants in Dakar did not feel directly concerned with the harmful practices associated with air pollution that were discussed. In this perspective, their participation to these meetings may have accentuated the gap between their perception on their "good" way of living and the noxious practices of other actors in their environment. Supporting this assumption is the fact that some interviewees mentioned, as a change in their behavior following the meetings that, from now on, they would be less reluctant to "denounce" improper behaviors of others, notably through the environmental emergency phone line put in place in Dakar. As such, one interviewee stated that, "Yes, a change in behaviour must occur because there are practices in my neighbourhood that I didn't like at all, but I had no means to remedy them. From now on, I will make a good use of the emergency number you gave us" (Dakar 4) (all of the interviews excerpt have been translated in English by the authors.). Another interviewee mentioned: "I am exposed to emanations coming from a carpenter shop in our neighbourhood, but I plan to remedy to this situation with the environmental emergency number you gave us" (Dakar 12).

In Benin, in each of the neighborhood selected for the project, Air Sain organizers rather chose to hold one preparatory meeting with a few women identified as leaders in their community and with the neighborhood delegate, assisted by a woman animator and a woman co-animator. These women had then the responsibility to recruit a total of 60 participants for the participatory assembly that would follow, which raised some tension for the difficulty to attain such a number. The assembly that we observed in Dantokpa market gathered 15 participants, mostly women merchants (although some of them also consider themselves as leaders in their community) as well as the head of the neighborhood (a man). The facilitator was a woman from the area. Interestingly, a comparative analysis of the vocabulary used in the two countries shows that the organizers tended to refer to scientific terms in Senegal in order to talk about air pollution and its effects on health. For instance, the presentation slides made references to air composition, to the respiratory system and to the effects of asthma on it, and to cardiovascular or coronary diseases, using detailed figures to illustrate these concepts. In Benin, organizers rather favored a more accessible language, for instance, they talked about "dangerous pollutants" when referring to toxic substances coming from wastes burning. This finding reinforces the idea of a certain "elite of participation" in Senegal, which would be deemed more capable to understand a scientific presentation. The choice of facilitator (a university student in Dakar, local facilitators in Benin) also nourishes this distinction between the two countries.

Contrary to what was stated earlier for Dakar, we can think that the individuals who were part of the participatory assembly observed in Cotonou felt directly concerned by the harmful practices associated with air pollution that were discussed during the meeting, since these practices are part of daily behaviors that take place inside many households. Cooking on a wood fire oven was mentioned, as well as the use of vegetal fuel (*kpèlèbè*, derived from the transformation of palm nuts into palm oil) or plastic bags to make fire, the burning of household wastes, and the use of insecticides. A large part of the discussions also turned around household cleanliness, despite that this element was not specifically mentioned on the posters and the leaflet produced for the Air Sain campaign. Besides, the link between insalubrity and air pollution remains tenuous and mostly concerns unpleasant smell issues (of course, insalubrity is the source of many public health problems, but they are not directly related to air pollution). This emphasis on insalubrity as a source of air pollution also clearly stands out from our interviews. In the following excerpt, a participant unequivocally associated the participatory assembly with hygiene measures: "I think that we were called to attend this meeting for reasons of good hygiene, for us to adopt good practices in terms of cleanliness" (Cadjehoun 4).

Admittedly, interviewees mentioned other actors who should also be targeted by the public awareness campaign, such as canteen owners who produce a lot of wastes and wastewater, or taxi-moto drivers. That being said, they generally acknowledged that they had to change their own practices in a much stronger way than what was the case for Dakar. The following excerpt gives an idea of the general state of mind after the assembly hold at the Dantokpa market: "Since we had this meeting on air pollution, I did not completely stop using the mosquito (an insecticide), but I reduced the frequency of its use. It is already a change of behaviour. It is true that I hate uncleanliness, and I reinforced hygienic measures around me. All of this is a result of the meeting" (Dantokpa 1).

Thus, we can put forward that the approach that was favored in Benin allowed the engagement of actors whose behaviors produce air pollution, who are directly affected by it, and who have the possibility to partly modify their own practices in order to reduce their exposure. Contrary to what was observed in Dakar, the campaign does not seem to be an initiative that reinforces a certain "elite of participation." Nevertheless, under this "grassroot" participation, more subtle power relationship issues appear. During the preparatory meeting observed in Cadjehoun, the head of the neighborhood, who chaired the reunion, mentioned that he "had fought" in order to have his neighborhood selected by the Air Sain project, in a context where organizers plan to distribute gas ovens to selected households for free at the end of the campaign. Moreover, although the recruitment of participants for the participatory assemblies was officially under the charge of the women identified as leaders in each of the selected neighborhoods, heads of neighborhood stepped in to convince people to participate. In Dantokpa, it seems that this was notably done on the base of past collaborations, as illustrated in the two following excerpts:

> When the (head of the neighbourhood) gives me a mission or asks me a service, I duly carry it out. This is certainly why I was invited to the meeting (Dantokpa 1).
>
> Another reason to explain my invitation is the trust and the conscientiousness that I am capable of when I carry out the missions that the head of the neighbourhood entrust me with (Dantokpa 4).

Furthermore, it seems that the head of the neighborhood in Dantokpa exerted a certain pressure over the participants to make sure they attend the meeting, as underlined in the following excerpt "I was opposed to the idea of using constraint in order to mobilize women for the meetings. The head of the neighbourhood was using this threat. I rather believe that participation must be voluntary and free" (Dantokpa 2). These elements show the difficulty to completely evacuate power relationships when it comes to selecting and recruiting attendees for participatory initiatives. As we will see in the following section, power relationships persisted during the participatory activities we observed, despite the intention to hold initiatives that were as inclusive as possible.

68.6 Persistent Power Relationships

Since they gather actors coming from diverse horizons, participatory initiatives are prone to reproduce traditional hegemonic relations, thus impeding the emergence of new ways of thinking (Abelson et al. 2011) and the expression of more marginalized groups, such as women (Raibaud 2015). Following ecosystemic approaches to health and its principle of gender equity, the participatory initiatives we observed in Senegal and Benin were both characterized by the presence of a majority of women attendees. Nevertheless, we noticed the persistence of traditional power relationships during these activities, in both countries. Hence, in all of the meetings we observed, men with a high social status took more room in the discussions than women, despite the fact they were less numerous.

In Dakar, it is not trivial to notice that during the two public awareness meetings, four men out of a total of six men attendees sat in front of the room, while women "naturally" took place behind. This configuration favored dyadic exchanges between the presenter, also in front of the room, and some men attendees who had private conversations with her at several occasions. These situations notably happened in relation with the use of incense, a delicate matter since it is associated with women's seduction practices; besides, some joking comments were made on the fact that the woman facilitator was still unmarried. These interludes, which excluded the other participants, each time gave rise to a hubbub and diverted the attention of the audience. It is also worth mentioning the propensity of participants with a high social status – heads of neighborhood or imams – to speak louder and for longer periods of time than the rest of the attendees, sometimes with an authoritative voice and often standing up while speaking. Admittedly, this situation did not totally impede the expression of women as several of them appeared very at ease to speak up; however, it created an imbalance between the number of women attendees at the meetings and the scope of their voice.

This dominating place took by men in assemblies mostly composed of women was also observed at the meeting of the Dantokpa market, in Benin, where the head of the neighborhood took the floor for long periods of time, at several occasions, many times in French and in a refined language. Some of his remarks could be deemed as paternalistic, for instance, when he stressed that the country was a patriarchal society and asked what was the place of men in the campaign, asserting that "no woman had confronted his man" on air pollution matters. It is true that using wood fire to cook fish or chicken, for instance, is deeply rooted in tradition. Changing this practice necessitates to heighten the whole family awareness of the problems that it may cause. Otherwise, a blunt change of practices will likely generate strong resistance in the household. As it was the case in Dakar, some women did speak, but overall the head of the neighborhood was a dominating presence.

When asked in the interviews if they believed that participants were able to express themselves as they wished, respondents (both in Dakar and in Cotonou) noted that there was not enough time for most people to express themselves and that, in general, those who spoke were those familiar with public speaking. Respondents expressed global satisfaction with the facilitators at the meetings.

68.7 Experts' Discourses and Changes of Practices

Surprisingly, resistance to change was rarely mentioned by participants and interviewees from both countries. If a few people did underline that some harmful practices were rooted in ancestral traditions and that, consequently, change had to be progressive, the majority of the interviewees asserted, on the contrary, that they had already modified their behavior:

> From then on, I forbad the use of plastic bags in fire as well as the use of vegetal fuel (kpèlèbè). Wastewater is filtered to remove its organic components – pasta, vegetables and so on. Household wastes are not exposed in an untidy heap anymore: they are put in bags and dropped in a specific spot where they will be collected. (Cadjehoun 1)
>
> From now on, I buy my gas. I reinforce cleanliness in my household and in the toilets. When I have to, I light the fire before starting to cook. I don't use coal and plastic bags anymore, when I leave my house I spray my bedroom so it can be in good condition when I come back in the evening. I don't rush at these necessities anymore. (Dantokpa 11)

Some changes of practices associated with the use of insecticides and incense can be done quite easily: one can simply wait outside the room by the time these products are burning or are settling down, in the case of spray insecticides. On the opposite, changes of practices associated with "clean" fuel are far more complex, for they imply more costly alternatives such as gas ovens. And yet, participants at the meetings rarely mentioned the costs increase associated with these changes, and only a few persons brought up this issue during the interviews. Most interviewees in Benin declared having changed their practices to some extent. Such a finding could be explained as a desirability bias to please the organizers – or not to displease them – by presenting views that correspond to what was taught during the meetings. But another as likely explanation is the deep trust toward experts' discourse that filtered out of the interviews. In Senegal but also in Benin, many interviewees mentioned the quality of the information given by experts during the meeting, and the fact that they trust this information. In both countries, "experts" were prominent during the activities, stepping in on many occasions to answer a question or correct a false information or perception. Interestingly, findings from the interviews clearly show that the confidence toward experts is partly attributable to the fact that they are not tied to the "political" universe, broadly taken: "At first, I thought this was a political thing, but I have been pleasantly surprised to see that is was a health affair, which motivated me more" (Dantokpa 10). The clear "nonpolitical" positioning of Air Sain could be construed as a relevant strategy to fight the cynicism related to public participation.

This acknowledgment – if not deference – toward the experts' discourse tends to limit, in return, the emergence of "lay knowledge" that would come from citizens' experiences and perceptions. This could potentially explain why resistance to changes of practices that could be considered as "good" notwithstanding their consequences on health – fire to cook fish or meat and incense to seduce – is seldom mentioned. In this perspective, the "solutions" presented to fight air pollution are mainly imposed by experts.

68.8 Preconceived Solutions

Another challenge associated with participatory initiatives relates to the concrete implementation of the solutions emerging from these endeavors (Bherer 2011; Blondiaux and Fourniau 2011; Clarke 2013). Like we just mentioned, in the two cases studied, solutions aiming at fighting air pollution were thought beforehand by Air Sain organizers, a budget having been planned from the outset in order to buy and distribute gas ovens. Admittedly, participants were involved in the discussions aiming at producing the leaflet in Cotonou, proposing, for instance, to add or change some images in order to make it more meaningful. In Dakar, participants suggested to produce a leaflet in the local language, a proposition that was welcomed by the organizers. But overall, respect, confidence, and deference toward experts' discourse did not favor the emergence of lay or citizen knowledge. The challenges posed by the concrete implementation of the identified solutions remained seldom touched on. Rather, participants proposed to organize more meetings with other participants but with the presence of experts. In this context, Air Sain activities were closer to communication models of persuasion and knowledge transmission rather than to genuine participatory models. These models would be more in phase with the traditional African palaver model, characterized by small assemblies where deliberation occurs through an informal setting, "transcending status and centered on content" (Lanmafankpotin 2015, our translation).

And yet, the type of engagement privileged by the Air Sain program is not vain. We can think that the form of citizen participation that took place through this project had a significant political meaning for participants. Indeed, McComas (2010) distinguishes three motives that justify citizen engagement in participatory initiatives (the author proposes these motives in relation to citizen engagement in crises management, but the concept can be applied to other contexts.). First, the motive can be normative, which means that it is based on the profound belief that individuals have a fundamental right to be heard in democratic regime. Second, the motive can be instrumental when it is thought that participation will likely favor a better acceptance of the decisions that have been debated and a greater confidence toward authorities. Finally, the motive for citizen engagement can be substantive, which is the case in situations where the benefits of such engagement are considered to go beyond the issue at stake. In these situations, participation is seen as a means to generate social capital, to lead to greater political efficiency and increased capacities for a community. Air Sain activities could be associated with the last two perspectives. First, taking into account the persuasion model of communication that was favored, the meetings created better conditions for uptake notably thanks to direct interactions with experts. Second, the Air Sain campaign also gave rise to participants' empowerment; this would be in phase with the participatory approach favored by international organizations, which nowadays call for "the valorisation of personal experience in participation to development projects" (Parizet 2016, p. 84, our translation). Beyond the knowledge acquired through public awareness activities, participants can get a feeling of empowerment associated with the ability to heighten third-party awareness on air pollution. As such, Air Sain has the potential to generate "social effects" (Lavigne Delville 2011) among its participants. The following excerpts speak to this idea:

> Whenever I had the chance to exchange with people and I had the occasion to share learning from the meeting with them, I did it, and I encouraged these individuals to share these learning as much as they can. (Cadjehoun 2)
>
> The meeting will be instrumental in the success of the future public awareness campaign I'll put forward, since I now possess many assets and scientific proofs concerning pollution ravages. (Dakar 1)
>
> I even apprehend people that I meet for the first time, go out of a taxi and commit gestures that are the opposite of what we learnt at the meeting. I hail them directly and draw their attention on the unfortunate consequences that could follow their behaviour. (Dantokpa 4)

This is consistent with an ecohealth perspective, according to which the emphasis must be put on knowledge-to-action, notably with regard to scientific knowledge. This bridge between experts and citizens is viewed as a way to insure projects' sustainability and better social cohesion. That being said and as stressed earlier, a strict transmission model tends to impede the emergence of lay knowledge that would allow the identification of more socially acceptable, and thus more sustainable, solutions. The recourse to incense, for example, remains strongly rooted in common habits, and one can question the sustainability of a solution promoting the reduction of its use.

68.9 Conclusion

In sum, by anchoring the challenges identified in the literature into field observations and interviews, we were able to bring support and nuances to the need to address various challenges when participatory initiatives are fostered. As we have seen, the initiatives presented may have contributed to widen the gap between informed citizens and those not involved in discussions related to collective issues. This was more so in Dakar where those invited to the meetings were community leaders that can be think of as an "elite of participation." The situation was different in Benin, even though our interviews have revealed that several participants to the Dantokpa meeting were recruited based on their pre-existing (positive) relationship with the community leader. We have also underlined how difficult it is to ignore the "usual" power relations during meetings in which, in principle, participants have an equal voice. We have seen that men, and notably those with an acknowledged social status, had a tendency to speak more often and for a longer time than women despite their minority status in the meeting. Our analysis has also revealed the central role played by experts in the conversations; even though the information they brought was highly relevant and well received by participants, their knowledge dominated the meetings leaving very few room for lay knowledge. Finally, our analysis has highlighted the complexity of the concept of "co-construction" at the heart of participatory initiatives: we were led to conclude that the initiatives we observed were close to persuasive communication and knowledge transfer models focusing on awareness and offering prepackage solutions.

This global conclusion questions the participatory nature of the observed meetings, the selected format leaving little room for the emergence of a dialogue between participants. Yet, dialogue is a foundation of communication for social change: "Communication for social change (. . .) is defined as a process of public and private dialogue through which people define who they are, what they want and how they can get it. Social change is defined as change in people's lives as they themselves define such change" (Gray-Felder and Dean 1999, p. 8. On dialogue, see also Cohen and Fung 2004, De Bussy 2010, Delli-Carpini et al. 2004, Fishkin 2009, Kent and Taylor 2002, and Sintomer 2011). In the meetings we observed, leaving more room for dialogue could have permitted the emergence of lay knowledge regarding some sources of pollutants and, more importantly maybe, could have exposed the challenges posed by the desired changes in perceptions and practices.

What has been said above does not question the relevance of the Air Sain program. As studies in communication have shown (Figueroa et al. 2002; Mertens et al. 2005; Kirk 2004), peer pressure and influence are major determinants for behavior change. Therefore, dialogue and interinfluence are crucial for these changes to go beyond individual behaviors to embrace collective practices: "For social change, a model of communication is required that is cyclical, relational and leads to an outcome of mutual change rather than one-sided, individual change" (Figueroa et al. 2002, p. iii). Indeed, even if part of the fight against air pollution can be conducted within households, it requires collective awareness and behavior changes at the society level. In this regard, our results clearly show that participants have learned and that many of them feel better equipped to convince their peers of the need to change some of their practices. Hence, one can think that dialogue and interinfluence will happen on the long run within participants' social networks. This is bound to contribute to behavior changes that could help fight air pollution and hence diminish its impact on people's health in the two cities.

68.10 Cross-References

▸ A Community-Based Participatory Mixed-Methods Approach to Multicultural Media Research
▸ A Threefold Approach for Enabling Social Change: Communication as Context for Interaction, Uneven Development, and Recognition
▸ Bottom-Up Networks in Pacific Island Countries: An Emerging Model for Participatory Environmental Communication
▸ Empowerment as Development: An Outline of an Analytical Concept for the Study of ICTs in the Global South
▸ Multidimensional Model for Change: Understanding Multiple Realities to Plan and Promote Social and Behavior Change

References

Abelson J, Gauvin FP, Martin É (2011) Mettre en pratique la théorie de la délibération publique: étude de cas du secteur de la santé en Ontario. Télescope 17(1):135–155

Anderson JE, Jaeger B (1999) Scenario workshops and consensus conferences: towards more democratic decision-making. Sci Public Policy 26(5):331–340

Bacqué MH, Gauthier M (2011) Participation, urbanisme et études urbaines. Quatre décennies de débats and d'expériences depuis *A ladder of citizen participation* de S. R. Arnstein. Participations 1(1):36–66
Bacqué MH, Sintomer Y (2011) La démocratie participative: histoire et généalogie. Éditions La Découverte, Paris
Baiocchi G (2003) Emergent public spheres: talking politics in participatory governance. Am Sociol Rev 68(1):52–74
Batellier P (2015) Acceptabilité sociale. Cartographie d'une notion et de ses usages. Cahier de recherche, UQAM, Les publications du Centr'ERE
Bayart JF (1986) African civil society. In: Chabal (ed) Political domination in Africa: reflections on the limits of power. Cambridge University Press, Cambridge, pp 109–125
Bherer L (2005) Les promesses ambiguës de la démocratie participative. Éthique publique 7 (1):82–90
Bherer L (2006) Le cheminement du projet de conseils de quartier à Québec (1965–2006): Un outil pour lutter contre l'apolitisme municipal ? Politiques and Sociétés 25(1):31–56
Bherer L (2011) Les relations ambiguës entre participation et politiques publiques. Participation 1 (1):105–133
Blacksher E (2013) Participatory and deliberative practices in health: meanings, distinctions, and implications for health equity. J Public Delib 9(1):art. 6
Blondiaux L (2008) Le nouvel esprit de la démocratie. Actualité de la démocratie participative. Éditions du Seuil, Paris
Blondiaux L, Fourniau JM (2011) Un bilan des recherches sur la participation du public en démocratie: beaucoup de bruit pour rien. Participations 1(1):8–35
Blondiaux L, Sintomer Y (2002) L'impératif délibératif. Politix 15(57):17–35
Brossard D, Lewenstein BV (2010) A critical appraisal of models of public understanding of science: using practice to inform theory. Commun Sci:12–39
Callon M, Lascoumes P, Barthe Y (2001) Agir dans un monde incertain. Essai sur la démocratie technique. Le Seuil, Paris
Caron-Malenfant J, Conraud T (2009) Guide pratique de l'acceptabilité sociale: piste de réflexion and d'action. Éditions D.P.R.M, Québec
Carrel M (2007) Pauvreté, citoyenneté and participation. Quatre positions dans le débat sur la participation des habitants. In: Neveu C (ed) Cultures and pratiques participatives: perspectives comparatives. L'Harmattan, Paris, pp 95–112
Charron DF (2012) Ecohealth research in practice innovative applications of an ecosystem approach to health. International Development Research Centre, Springer, Ottawa/New York
Cissoko K, Toure R (2005) Participation des acteurs sociaux and gouvernance d'État. Le cas du Cadre stratégique de lutte contre la pauvreté au Mali. Politique africaine 99:142–154
Clark ML, Peel JL, Balakrishnan K, Breysse PN, Chillrud SN, Naeher LP, Rodes CE, Vette AF, Balbus JM (2013) Health and household air pollution from solid fuel use: the need for improved exposure assessment. Environ Health Perspect 121(10):1120–1128
Clarke J (2013) L'enrôlement des gens ordinaires. L'évitement du politique au cœur des nouvelles stratégies gouvernementales? Participations 6(2):168–189
Cling JP, Razafindrakoto M, Roubaud F (2002) Processus participatifs et lutte contre la pauvreté: vers de nouvelles relations entre les acteurs ? L'Économie politique 16(4):32–54
Cohen J, Fung A (2004) Radical Democracy. Swiss Polit Sci 10(4):23–34
Cooke B, Kothari U (eds) (2001) Participation: the new tyranny? Zed Books, Londres/New York
De Bussy NM (2010) Dialogue as a basis for stakeholder engagement. Defining and measuring the Core competencies. In: Heath R (ed) The Sage handbook of public relations. Sage, Thousand Oaks, pp 127–144
Delli-Carpini M, Cook FL, Jacobs LR (2004) Public deliberation, discursive participation and citizen engagement: a review of the empirical literature. Annu Rev Polit Sci 7:315–344
Dom C (2012) Empowerment through local citizenship. In : Organisation de coopération et de développement économique (OCDE), poverty reduction and pro-poor growth: the role of empowerment

Figueroa ME, Kincaid DL, Rani M, Lewis G (2002) Communication for social change: an integrated model for measuring the process and its outcomes. Communication for social change working paper series, The Rockefeller Foundation 1913
Fisher F (2009) Democracy and expertise: reorienting policy inquiry. Oxford University Press, Oxford
Fishkin J (2009) When people speak. Deliberative democracy and public consultation. Oxford University Press, Oxford/Angleterre
Fortin MJ, Fournis Y, Beaudry R (2013) Acceptabilité sociale, energies and territoires: De quelques exigences fortes pour l'action publique – Mémoire soumis à la Commission sur les enjeux énergétiques. GRIDEQ/CRDT/UQAR, Rimouski
Fung A (2006) Varieties of participation in complex governance. Public Adm Rev 66(1s):66–75
Fung A, Wright EO (2003) Deepening democracy. Institutional innovations in empowered Paticipatory governance. Verso, Londres/Angleterre/New York
Garibay D (2015) Vingt-cinq ans après Porto Alegre, où en est (l'étude de) la démocratie participative en Amérique latine. Participations 11(1):7–52
Gastil J, Levine P (2005) The deliberative democracy handbook: strategies for effective civic engagement in the 21st century, Project of the deliberative democracy forum. Jossey-Bass Publishers, San Francisco
Gourgues G (2012) Des dispositifs participatifs aux politiques de la participation. L'exemple des conseils régionaux français. Participations 2(1):30–53
Gray-Felder D, Dean J (1999) Communication for social change: a position paper and conference report. New York: Rockefeller Foundation Report
Hearn J (2001) The uses and abuses of civil society in Africa. Rev Afr Polit Econ 28(87):43–53
Institut du Nouveau Monde (INM) (2013) État des lieux des mécanismes de participation publique au Québec and relevé d'expériences inspirantes de participation publique hors Québec. On-line: http://inm.qc.ca/Centre_doc/27-Etat_des_Lieux_participation.pdf. Consulted on 19 Dec 2017
Kent ML, Taylor M (2002) Toward a dialogic theory of public relations. Public Relat Rev 28:21–37
Kirk P (2004) Community leadership development. Community Dev J 39:234–251
Koehl É, Sintomer Y (2002) Les jurys de citoyens berlinois. Rapport remis à la Délégation interministérielle de la ville
Lanmafankpotin PGY (2015) Appropriation de la décision collective: évaluation environnementale comme champ d'application de la participation publique au Bénin. Thèse présentée à la Faculté des études supérieures en vue de l'obtention du grade de Philosophiæ Doctor (Ph.D.) Doctorat en géographie Géographie environnementale et de développement
Laumbach RJ, Kipen HM (2012) Respiratory health effects of air pollution: update on biomass smoke and traffic pollution. J Allergy Clin Immunol 129(1):3–11
Lavigne Delville P (2011) Du nouveau dans la participation au développement ? Populisme bureaucratique, participation cachée and impératif délibératif. In: Jul-Larsen E, Laurent P-J, Le Meur P-Y, Léonard E (eds) Une anthropologie entre pouvoirs and histoire. Conversations autour de l'œuvre de Jean-Pierre Chauveau. APAD-IRD-Karthala, Paris, pp 161–188
Lavigne Delville P, Thieba D (2015) Débat public and production des politiques publiques au Burkina Faso. La politique nationale de sécurisation foncière. Participations 11(1):213–236
Levine P, Nierras RM (2007) Acitivists' views of delibation. Journal Of Public Deliberation 3(1):1–14
Liousse C, Galy-Lacaux C (2010) Pollution urbaine en Afrique de l'Ouest. La Météorologie (71):45–49
Liousse C, Assamoi E, Criqui P, Granier C, Rosset R (2014) Explosive growth in African combustion emissions from 2005 to 2030. Environ Res Lett 9(3):1–10
Martin WJ II, Glass RI, Araj H, Balbus J, Collins FS, Curtis SN, Bruce NG (2013) Household air pollution in low- and middle-income countries: health risks and research priorities. PLoS Med 10(6)
McComas KA, Community Engagement and Risk Management (2010) In: Heath RL (ed) The SAGE handbook of public relations, 2nd edn. SAGE, Thousand Oaks, pp 461–476

Mertens F, Saint-Charles J, Mergler D, Passos CJ, Lucotte M (2005) Network approach for analyzing and promoting equity in participatory Ecohealth research. EcoHealth 2:113–126. https://doi.org/10.1007/s10393-004-0162-y

Nisbet MC, Scheufele DA (2009) What's next for science communication? Promising directions and lingering distractions. Am J Bot 96(10):1767–78

Norwegian Agency for Development Cooperation (Norad) (2013) A framework for analysing participation in development. Evaluation Department, Oxford Policy Management

Organisation de coopération and de développement économique (2001) Des citoyens partenaires. Manuel de l'OCDE sur l'information, la consultation and la participation à la formulation des politiques publiques. OCDE, Paris. Online: http://www.bourgogne.gouv.fr/assets/bourgogne/files/dvlpt_durable/OCDE_participation_des_citoyens.pdf. Consulted on 25 Jan 2016

Organisation de coopération and de développement économique (OCDE) (2013) Accountability and democratic governance: orientations and principles for development. DAC guidelines and reference peries, preliminary copy

Otayek R (2009) La problématique africaine de la société civile. In: Gazibo M, Thiriot C (eds) Le politique en Afrique. Karthala, Paris, pp 209–226

Parizet R (2016) Le pauvre d'abord. Une analyse des dynamiques circulatoires de la participation populaire au développement. Participations 14(1):61–90. https://doi.org/10.3917/parti.014.0061

Pascal M, Corso M, Chanel O, Declercq C, Badaloni C, Cesaroni G, Medina S (2013) Assessing the public health impacts of urban air pollution in 25 European cities: results of the Aphekom project. Sci Total Environ 449(0):390–400

Perez L, Declercq C, Iñiguez C, Aguilera I, Badaloni C, Ballester F, Künzli N (2013) Chronic burden of near-roadway traffic pollution in 10 European cities (APHEKOM network). *ERJ Express*

Pope CA III, Dockery DW (2006) Health effects of fine particulate air pollution: lines that connect. J Air Waste Manage Assoc 56:709–742

Rabouin L (2009) Démocratiser la ville. Le budget participatif: de Porto Allegre à Montréal. Montréal, Lux Éditeur

Raibaud Y (2015) La participation des citoyens au projet urbain: une affaire d'hommes! Participations 12(2):57–81

Robins S, Cornwall A, von Lieres B (2008) Rethinking 'citizenship' in the Postcolony. Third World Q 29(6):1069–1086

Saint-Charles J, Webb J, Sanchez A, van Wendel J, Nguyen-Viet H, Mallee H (2014) Ecohealth as a field – looking forward. EcoHealth 11(3):300–307

Schneider H (1999) Gouvernance participative: le chaînon manquant dans la lutte contre la pauvreté. Centre de développement de l'OCDE, Cahier de politique économique no 17. Online: http://www.oecd.org/fr/social/pauvrete/31649590.pdf. Consulted on 12 Jan 2016)

Sintomer Y (2011) Délibération and participation: affinité élective ou concepts en tension ? Participations 1(1):239–276

Tenbensel T (2005) Multiple modes of governance. Disentangling the alternatives to hierarchies and markets. Public Manage Rev 7(2):267–288

Webb JC, Mergler D, Saint-Charles J, Spiegel J, Woollard RF (2010) Tools for thoughtful action: the role of ecosystem approaches to health in enhancing public health. Revue canadienne de santé publique 101(6):439–441

Weyh CB, Streck DR (2003) Participatory budget in southern Brazil. A collective and democratic experience. Concept Trans 8(1):5–42

WHO (2014). Burden of disease from the joint effects of Household and Ambient Air Pollution for 2012. Online: http://www.who.int/phe/health_topics/outdoorair/databases/FINAL_HAP_AAP_BoD_24March2014.pdf?ua=1. Consulted on 31 Aug 2015)

Community Radio in Ethiopia: A Discourse of Peace and Conflict Reporting

69

Mulatu Alemayehu Moges

Contents

Abstract

In the states, like Ethiopia, where internal conflicts, mainly ethnic conflicts, have currently been appearing in many parts, a medium that deals with those issues appropriately is vital. Among the three main media branches, such as public service, commercial, and community radio, the later can be described as an appropriate medium that can, perhaps, bring not only possible timely solutions to the cases but also some social changes in the society. This is because the nature of the community radio, which is close to the societies and covers the issues immediately, can serve the people well by raising directly relevant issues in relation to conflict and peace. By taking some cases from Ethiopia, it is the purpose of this chapter to show how community radio is a best platform in dealing with internal conflicts in particular and social issues in general in the marginalized societies that are vulnerable to various social and political problems.

M. A. Moges (✉)
School of Journalism and Communication, Addis Ababa University, Addis Ababa, Ethiopia
e-mail: mulatu_alem@yahoo.com

J. Servaes (ed.), *Handbook of Communication for Development and Social Change*,
https://doi.org/10.1007/978-981-15-2014-3_37

Keywords

Community radio · Ethiopian Broadcast Authority (EBA) · Kembat community radio · Kore community radio · Public broadcasting service

69.1 Introduction

In a broad classification of broadcasting, there are three groups of electronic media in Ethiopia. The first one is public broadcasting service, which is funded by the public. In the case of Ethiopia, these media are financed and controlled by the government. This can also be named government-run media. The second is commercials, which are privately owned, and their goals are profit making. The third is community radio, which is supported by the community and local and international NGOs. And, the main goal of the community radio is to serve and benefit the target groups in some areas. As it is defined by AMARC, community radio is a "nonprofit" station, currently broadcasting, which offers a service to the community in which it is located, or to which it broadcasts, while promoting the participation of this community in the radio (AMARC-Europe 1994 cited in Carpentier et al. 2003, p. 240). Among these three types of media in the country, it is the argument of this paper that the later one (community radio) can be one of the best means of communication for social changes in particular, dealing with internal conflicts in a given country. This will be explained throughout this chapter. But, first, it is fair to note briefly about the development of community radio and the extent of conflicts in the country.

Currently, there are a number community radio stations in Ethiopia. Most of them are well established and structured. Unlike the other media in the country, such as the national radio, television, and newspaper, the community radio can be described not only as a recent phenomenon but also as a fast-growing medium in the country (Tadesse 2006; Eshetu 2007). As of 2008, the government of Ethiopia, in its office, called Ethiopian Broadcasting Authority (EBA), has begun licensing the community radio in a formal way, but most of them have started properly functioning as of 2010 and 2011 (Infoasaid 2011). Presently, there are about 26 community radio stations, which are functioning well across the country; the other 30 got license, but have not started providing service to the community (EBA 2016). Since it is being recent and comparing it with other mainstream media, the community media can be described as a media in good progress in the country.

In advocacy communication, which is an important tool to foster public policies in the audiences by producing issues or program continuously in the media, in which they could bring solution, for instance, to the conflicts (Servaes and Malikhao 2012), the community radio can be the best medium in taking such roles in the country. This can be discussed into two points in relation to internal conflicts. Firstly, there are several internal conflicts in the country. Ethiopia, one of the Horn of African countries, has been experiencing both inter- and intra-conflicts in its history due to political, socioeconomic, border, and administrative structure, ethnic and religious tensions, the geopolitics of the country, and other causes. Specifically, intra-conflicts have currently become so serious and vast in numbers. Various research and reporters indicated that there have been several internal conflicts, which have had serious consequences, which led to losing the lives of many people, destruction of proprieties, and other social problems in the country (Armed Conflict Location and Events Data Project-ACLED (2016). In fact, there were some regional states in the country which were vastly experiencing ethnic conflicts. Oromia, Southern Nations, Nationalities, and People's Region (SNNPR), and Somalia regions were among the most vulnerable regions to ethnic conflicts (iDMC 2009). But recently, the conflicts have sporadically happened in the other regions due to the political tension, and many people lost their lives. In pastoralists' areas where grazing and farming land are so scarce and where the border structures are poorly demarcated, they have been accommodating many clashes and conflicts.

Secondly, the community radio stations in the country are mostly situated in marginalized areas, in which the people are vulnerable to various conflicts, specifically ethnic conflicts. These make the importance of community radio so magnificent in alleviating conflict and in bringing the solutions to the problems.

69.2 The Community Radio as Tool to Solve the Internal Conflicts

There have been two different arguments on the importance of the media, in general, in relation to their interventions in conflicts. Some groups, for instance, Des-Forges (1999), Thompson (2007, 1999), by taking the cases of Rwanda and former Yugoslavia, argued that the media play a negative intervention in the conflict by presenting the cases in a more sensational ways that motivate the people to stand against the other perceived enemies. On the other hand, there are groups who are strongly arguing that the media are still one of the best means to approach conflict and bring sustainable solution to a given conflict (Wolfsfeld 2004; Spencer 2005; Lynch and McGoldrick 2005; Howard 2008). The best-case scenario of the positive intervention of the media in conflicts is the crisis between the Catholic and the Protestant groups in Ireland. The media in this country are appreciated in creating a conducive environment for the two-antagonist groups to settle their differences. To note here, both the negative and positive interventions have been observed in the mainstream media in the respective countries. It is the argument of this chapter that unlike the mainstream media, community radio differently serves the people in the discourse of peace and conflicts if it is properly established and managed and the volunteers' journalists are well trained. As Ethiopia is ethnically a divided country where its societies are grouped based on their ethnic lines which could help to mobilize themselves in their groups as well as to get some benefits because of their groupings (Abbink 2006; Østebø 2007; ICG 2009), a recent study on the mainstream media found that journalists in these media are always working in fear where their report may incite conflicts among the ethnic groups (Moges 2017). Due to the fear of inciting conflicts as well as fear of creating animosity among a perceived ethnic enemy, most of the mainstream media in the country do not dare to report internal conflicts, particularly ethnic conflicts (Moges 2017). The media prefer to silence the cases.

Taking this finding into account, it is the main argument of this chapter that the community radio can perform better than the mainstream media in relation to reporting conflicts and possibly bringing solutions in Ethiopia. This is because their nature, approaches, and autonomy could help them to deal with internal conflicts in the society in a way to ameliorate cases. In support of this idea, Terizs and Vassiliadou (2008, p. 26) argued that the community radio has the power to reach the target audience and enable them to participate in the decision-making process both in the content selection and content production. Carpentier et al. (2003) also explained the importance of community radio into four multi-theoretical approaches. These are in terms of serving a community, the community media as an alternative to mainstream media, linking community media to the civil society, and community media as rhizome. These multi-theoretical approaches, except the latter, are very important to elaborate the cases of community radio in Ethiopia in the case of ensuring peace and resolving conflicts. In this chapter, the community radio are present in terms of their ownership, proximity and immediacy, language and values, and volunteerism and nonprofit-making interest.

69.3 Ownership of the Media

In the political economy of the media, ownership is one of the elements that influence the mainstream media content production and dissemination in the given society. However, it is possible to argue here that the influence of ownership in community radio becomes minimal. It is because the community owns the media and they can also select, propose, and even participate in the content production in the programs aired in the stations (Fraser and Restrepo-Estrada 2002). Unlike the mainstream media that are mostly working to meet either political or economic benefits of the owners by reporting some issues, such as conflicts, perhaps sensationally to get more sells (Allen and Seaton 1999), the community radio are, nevertheless, less likely rushing to report issues in a way to exaggerate the reports so as to maximize incomes. For instance, when there is conflict in a given society, the community radio stations are believed to report the cases in a very fair manner that benefits their target communities. It is because the community radio stations are always reporting cases in a way that helps that particular society. In support of this idea, Fraser and Restrepo-Estrada (2002) emphasize that the decision on all aspects of management is transparent and democratic which enable them to serve the community. This indicates that, for example, if there is a conflict in some areas, the journalists in the community radio are mainly concerned with how the cases could be reported in a way to bring the solutions to the problems. In other terms, their (the community media) main interest is to deal with the issues in line with the interests of the community, which are witnessing the conflict. A study on internal conflicts in Ethiopia found that owners (in terms of political and economic) are highly influencing the media in both the extent of and framing of those issues

(Moges 2017). However, when it comes to the community radio, unlike the other mainstream media, the owners' influence in the community media is so little. Instead, since the communities are the owner of the radio station, they are highly privileged in participating in the content selection and production freely (Carpentier et al. 2003). They also underscored that in the community radio, the community are getting better access in participation in the content production and in the discussion in the programs (Carpentier et al. 2003). Thus, the discourse of conflict and peace in the case of Ethiopia can be easily and fairly facilitated by the community radio stations, which are mostly owned by the communities.

69.4 Proximity and Immediacy

Proximity and immediacy are some of the most valuable elements in story selection in the media. When it comes to the conflict cases, these two elements become the most important criteria of story selection in the media (Gultung and Ruge 1965). More specifically, these two elements can be most valuable and well used in the community radio. This is because the radio stations can be the most close media to the society in providing information to them as soon as possible. This further benefits that particular society in four ways. Firstly, the target audiences can get the chance to know what is going on in their surroundings since journalists in the community radio relatively report the case as soon as it happens. Secondly, unlike the mainstream media at the national and international level, the community radio stations are situated close to the target community, in which their journalists can easily identify the possible violence, perhaps, before the outbreak of the conflicts, as well as suggest possible solutions to them. As it has been argued by Lynch and McGoldrick (2005), reporting conflicts proactively can contribute to containing the outbreak of serious violence and their consequences in the community. This is because when the report identifies some problems ahead and suggests the way out, the concerned bodies can also come up with the solutions. It is also easy to get the solution to the problem from the community themselves. This can happen in the community radio, which is situated in the community. In support of this idea, Fraser and Restrepo-Estrada (2002, p. 71) state:

> *The collective perception can only be achieved through internal discussions to analyse specific problems, identify possible solutions, and mobilize the appropriate people or groups for action. Community radio provides the perfect platform for this internal discussion.*

Thirdly, since the journalists and the owners (the community) are part of the same society, the former can bring issues of conflict to the audiences so as to discuss the case in the way they want to be. The journalists know the social and cultural values of that particular community (will be discussed next), and they can present the cases that the audiences can understand by using its language and culture. In light of that, reflecting and promoting local identity, values, and culture of the community are some of the main functions of the community radio (Fraser and Restrepo-Estrada

2002). The society is also considering the community media as a safeguard for their culture (da Costa 2012). This can be further discussed in the next subtopic.

Lastly, the community can get the chance to easily participate in the program, which also helps them to promote the rights of the societies (Fraser and Restrepo-Estrada 2002, p. 71).

> *(The community radio help) to promote good governance and civil society by playing a community watchdog role that makes local authorities and politicians more conscious of their public responsibilities. The marginalized and the oppressed normally have no way to complain when authorities take advantage of them, but community radio gives them a voice to air their grievances and obtain their due rights.*

In a similar vein, Carpentier et al. (2003) noted that the community media has the power to provide access to the community, particularly the marginalized, in participating (getting the chance to have their voice heard) and facilitating communication in the given society. The Kore community radio in Ethiopia can be a case in point for such role. Historically, the Kore community is likely isolated both politically and geographically from the central government. The area, which is located in SNNP region, is not easily accessible for transportations. Like many other ethnic groups, this community also has a distinct language, which is not catered by the national media. Of course, as the Ethiopian populations are large in number and there are many ethnic groups with their own language, it has become difficult to provide coverage for all groups in their languages at the national level in the mainstream media. As a result, the community radio, Kore Community radio, has become the best platform to the community to use the media in expressing whatever they want. Kembata community radio has also shared similar roles. Generally, the issues of proximity and immediacy make the community media more preferable stations in the discourse of peace and conflicts. There is a strong need for the proximity of radio stations to the rural audiences and localization of program materials (Diedong 2014). This ultimately helps the community to raise more issues, and then, they can get the chance to be heard by the concerned people. In addition, as it is noted in the multi-theoretical approach of Carpentier et al. (2003), topics of the discussion in the programs in the community radio are considered for selection if and only if they are found relevant to that particular community. This has also another advantage in the community. Issues, which need immediate action by the decision-makers, could get answer shortly.

69.5 Languages and Values of the Community

As noted above, briefly, the other most important elements in the discourse of peace and conflicts in the community radio in Ethiopia are the use of local language and social values of the community in the program production. It is clear that language is one of the means that reflect values and cultures. The community radio is also using this advantage by producing and disseminating information to the audiences that they can clearly understand the cases. In addition to language, the communities have different identities, cultures, or perceptions that may bring some forms of different understanding. This can be described as the main challenges to the mainstream media at national, foreign, or international level in reporting cases in line with the cultural context of the cases. As it is argued by Terizs and Vassiliadou (2008), the international and foreign media can be highly challenged in understanding and translating words and concepts of the local community. The low level of understanding and translating the local language by these mainstream media may lead to prejudice and biases toward the cases, the context, and the society they are reporting to (Terizs and Vassiliadou 2008, p. 386). In other terms, Servaes and Malikhao (2012) argued that media, which aspire to bring social changes, particularly build peace, and bring resolution to the conflicts, shall frame the message based on the culture context of the target community. "The information should be trailered to the audiences and be in line with the understanding and expectation of the people or stakeholder" (Servaes and Malikhao 2012, p. 237). In his handbook, *Conflict-Sensitive Reporting*, Howard (2008) also noted that journalists should describe conflicts accurately, which needs understanding to the background of the case, the context, and other: ". . .the spelling of names, the facts as they happened, and the real meaning of what was said" (Howard 2008, p. 21). All the above authors suggested that considering the language and values of the given community where the media are working in is vital. In the case of community radio, the journalists do understand well not only the language and the values of the community but also the interest of the target audiences. This can help the community to get more chances to speak out their problems freely. And they can also suggest possible solutions freely that can work in line with their cultural values.

This can be described in terms of presentations of the issues in the community media. Some communities have the culture that can colorfully express certain issues. This can help the community radio to use such values of the society in presenting issues of reconciliation and arbitration of the conflict artistically, which ultimately make the discussion very soft and smooth. As one of the roles of the journalists who are aspiring to bring conflict cases and present them in a non-violent way, Lynch and McGoldrick (2005) suggest that the presentation shall be more creative and be in line with the social values of the community. The authors also advise that the story shall include the voice of the people in general and influential cultural and other social groups, in particular, who can influence the public opinion easily. Unlike other mainstream media, the community radio is a best platform to bring those values from the societies, and they can present those in the form of music, dialogue, art, poem, and others (Fraser and Restrepo-Estrada 2002). In its nature, the community radio is producing programs in the local language about local issues, music, cultures, and news based on the interests of the target listeners (Milius and Oever NY.). This makes the community radio the right medium in solving the problem and bringing peace to the community. To note here, all the community radio stations in Ethiopia are using the local languages of the people who are getting the service from the aired programs. In other terms, most of the community radio in Ethiopia is serving small ethnic groups, which have a strong sense of local identity (Infoasaid 2011, p. 18).

This can be further explained that some community radio stations are serving some marginalized communities, which are somewhat isolated in the socioeconomic benefits of the country. At the same time, these marginalized societies are most vulnerable to minor conflicts. This is one of the advantages of community radio, specifically in Ethiopia, which are producing contents and disseminate them to the societies in different languages of the communities.

Languages and values can also be discussed in relation to having clear understanding to the cases. Facts are the most important elements in the objective reporting in the media; it thus becomes so serious in writing accurate stories within the context of a given society. It is the job of the journalists to work hard to figure out facts from the trashes. There are times that journalists who are not close to the issues or particular communities are challenged in identifying "which is which." However, these problems may not be a big deal with the journalists who are working in the community radio. It is because they either can know the facts well or can easily cross-check them from various sources. Moreover, they can know the context of the conflicts, the background of the conflict, and the values that should be dealt with while reporting the cases. They can also easily anticipate the intended consequences of their reports in the political, social, and cultural matters in the target community. Hence, the presentation about a certain conflict that aims at bringing reconciliation in the community or arbitrating the groups by taking their values in the program needs to have clear knowledge about the case, which is not a challenge to the journalists who are working in the community radio.

69.6 Volunteerism and Nonprofit Making

The last point that this chapter wants to mention to strengthen the discussion in relation to the importance of the community radio in the intervention of peace and conflict reporting in Ethiopia is volunteering and nonprofit-making interest of both the journalists and the stations. Journalists who are working in the community radio are volunteers who have ardent interests to serve the society for free. Of course, they shall have some skills to write and report issues. They work in the community radio parallel with their jobs. The radio station may not pay them. As their aim is only to serve the target community, they do not have extra interest in working in the media. This is to mean that journalists in the mainstream media are highly interested in reporting conflicts mostly in a sensational manner. This can be an advantage to promote them and become famous in the media by dealing with sensitive issues. However, when it comes to the community radio, its journalists do not have an extra interest in rushing to report some issues in a sensationalized manner for monetary gain or personal promotion. In fact, this has been argued in a different way.

The flow of information is another advantage of the community radio. Unlike other mainstream media, the community radio does not highly promote the top-down flow of information. In fact, there are times that the community radio stations set some agendas from the NGOs or other funding organization. For instance, there were some local NGOs that have been financing Kembata community radio to deal with women and children issues with the aim of minimizing women-children mortality rate. This is also common in Africa in general. NGOs and development agencies consider the community radio as conduits for message that aimed at educating the community, fostering behavioral changes, and empowering them (da Costa 2012, p. 135). This indicates that the community radio not only develops volunteerism but also promotes social changes at the grassroots level in its community. This ultimately has an impact in dealing with issues of conflicts.

69.7 Summary

Generally, community radio in Ethiopia can be considered as one of the tools that can be used to intervene positively in the discourse of peace and conflicts and bring solutions. Since community radio stations are owned by the community and run by the volunteers, and they are close to the society, they become so vital not only becoming a platform for discussion on conflicts and peace as well as other sociopolitical issues but also deepening the participation of the community in many aspects which ultimately bring social changes. It also maximizes the two factors of development, such as communication and people's participation. Taking all these points into account, community radio in Ethiopia is one of the catalysts for social changes and solves ethnic conflicts mostly happening in the remote parts of the country. According to Fraser and Restrepo-Estrada (2002, p. 69), "the community radio station is a platform for identifying and analyzing problems and their solutions, thereby determining development inputs that truly meet local needs." Hence, those who want to use and advocate certain issues, including peace building, in Ethiopia, shall approach the community radio to meet their interests. This is because the community radio is the best medium that can bring social issue up front to the public discussion.

References

Abbink J (2006) Ethnicity and conflict generation in ethiopia: some problems and prospects of ethno-regional federalism. J Contemp Afr Stud 24(3):389–413

Allen T, Seaton J (1999) Introduction. In: Allen T, Seaton J (eds) The media of conflict: war reporting and representation of ethnic violence. Zed Books, London/New York, pp 1–9

Carpentier N, Lie R, Servaes J (2003) Is there a role and place for community media in the remit? In: Hujanen T, Lowe GF, Göteborg S (eds) Broadcasting & convergence: new articulations of the public service remit. Nordicom, Göteborg, pp 239–254

da Costa P (2012) The growing pains of community radio in africa: emerging lessons towards sustainability. Nordicom Rev 33(Special Issue):135–147

Des-Forges A (1999) Leave none to tell the story: genocide in Rwanda. Human Right Watch, New York

Diedong AL (2014) The relevance of integrating models of radio into development process. Int J Humanit Soc Sci 4(5):50–61

EBA (2016) Report about the community radio in 2016. Ethiopian Broadcast Authority, Addis Ababa

Eshetu A (2007) Overview of community radio development in Ethiopia. A paper prepared for the First AMARC AFRICA MENA Conference, Rabat, October
Fraser C, Restrepo-Estrada S (2002) Community radio for change and development. Soc Int Dev 45 (4):69–73
Gultung J, Ruge MH (1965) The structure of foreign news, the presentation of the Congo, Cuba and Cyprus crises in four Norwegian newspapers. J Peace Res 2(1):64–91
Howard R (2008) Conflict sensitive journalism. International Media Support, Nairobi
ICG (2009). Ethiopia: ethnic federalism and its discontents. Africa Report N°153, Working to prevent conflict worldwide. International Crisis Group.
iDMC (2009) Ethiopia: human rights violations and conflicts continue to cause displacement, A profile of the internal displacement situation. Internal Displacement Monitoring Centre, Addis Ababa
Infoasaid (2011) ETHIOPIA media and telecoms landscape guide. Addis Ababa
Lynch J, McGoldrick A (2005) Peace journalism. Hawthorn Press, Stroud
Milius A, Oever NT (n.d.) Community radio: a practical guide. The Medication Foundation, Netherlands
Moges MA (2017, August) Why silence? Reporting internal conflict in Ethiopian newspapers. PhD dissertation. University of Oslo, Oslo
Østebø T (2007) The question of becoming: Islamic reform-movements in contemporary Ethiopia. CMI, Bergen
Servaes J, Malikhao P (2012) Advocacy communication for peacebuilding. Dev Pract 22 (2):229–243
Spencer G (2005) The media and peace: from Vietnam to the “war on terror”. Palgrave Macmillan, New York
Tadesse M (2006) The challenges and prospects of community radio in Ethiopia: the case of Harar community radio. Addis Ababa University, School of Graduate Studies, Addis Ababa
Terizs G, Vassiliadou M (2008) Working with media in areas affected by ethno-politcal conflcit. In: Servaes J (ed) Communication for development and social change. SAGE, Los Angeles/London/New York, pp 374–388
Thompson M (1999) Forging war, the media in Serbia, Croatia, Bosnia, and Herzegovina. University of Luton, Luton
Thompson A (2007) Introduction. In: Thompson A (ed) The media and the Rwanda genocide. Fountain Publisher, Kampala, pp 1–11
Wolfsfeld G (2004) Media and the path to peace. Cambridge University Press, Edinburgh

Part XI

Regional Overviews

Political Economy of ICT4D and Africa

70

Tokunbo Ojo

Contents

Abstract

With hype surrounding the leapfrogging power of information and communication technologies (ICTs) in the national and international development agenda, there have been significant scholarly research interests in what ICTs "will do" *and* "can do" for Africa and its people if appropriated within national development planning agenda. Several of these studies have examined the utopia and dystopia dimensions of the ICTs from the standpoints of socioeconomic development. Against this background, this chapter examines the political economy of ICTs for development in Africa. In this context, the chapter explores ICTs as an economic sector and also as an interventionist tool in the development process.

Keywords

Digital capitalism · Flashing · Information and communication technologies for development (ICT4D) · Mobile phones and economics in Africa · TRACnet · World Health Organization (WHO)

T. Ojo (✉)
Department of Communication Studies, York University, Toronto, ON, Canada
e-mail: ojotoks@yahoo.com

J. Servaes (ed.), *Handbook of Communication for Development and Social Change*,
https://doi.org/10.1007/978-981-15-2014-3_64

70.1 Introduction

In their mapping of the published scholarly peer-reviewed journal articles on the communication and international development from 1998 to 2007, Ogan et al. (2009) found that "of all the primary approaches adopted for the study, information and communication technologies for development (ICT4D) comprised 42.3 percent (and an additional 12.5 percent of the secondary approaches) of the total and 40.8 percent used ICTs as the media focus of the research" (p. 660). As an acclaimed buzzword in the scholarship on international development and communication, ICTs are viewed as a "magic multiplier for the poorest of the poor" (Ogan et al. 2009, p. 656). Consequently, development actors such as civil society groups, corporate entities, UN agencies, and cross section of nation-states "view projects like telecenters or information kiosks as the solution to becoming more globalized and stimulating the local economy" (Loh 2015, p. 235). Several of the studies, which focused on Africa, examined the continent's digital divide and the role of international organizations in the ICT policy formation.

While the pessimistic perspectives largely dismissed the notion of ICT4D on the ground that it is another subtle imperialist project of the Western economic power, the overtly optimistic perspectives see the ICTs as a leveler in the global economy and also as a positive index of the twenty-first century human development agenda (Ojo 2016; Murphy and Carmody 2015). Outside of these polarized perspectives, there are other sporadic works that looked at the ICTs in the contexts of political institutions and governance, and social movement, following the 2007 Kenyan presidential election crisis and 2011 Arab spring. There is also an emerging wave of scholarship on the ICTs and journalism practice in Africa, with a particular focus on the citizen journalism and integration of new technologies into the operation of legacy news media outlets in a cross section of African countries. This area of research on African journalism is partly inspired by the broader questions of future sustainability of the legacy news media and "old" technologies such as newspapers and books, amid the rapid growth of digital mobile communication technologies. From an Afrocentric perspective, it is also an important area of scholarship on political economy of African media systems in the post-military rule era.

In spite of all these interesting research works, there is still a dearth of research on "what Africans *do* with ICTs through enculturation" (van Binsbergen 2004 cited in Nyamnjoh 2005, p. 205) and also practical understanding of ICTs as a set of economic practices in African contexts. While acknowledging the validity and contributions of the previous studies to the knowledge production and understanding of ICT4D in Africa, this chapter explores the political economy of ICT as an economic sector and also as an interventionist tool in the development process.

70.2 Digital Capitalism and Business Economies of ICT4D

Digital capitalism, which is the resuscitation of the liberal economic policy of the Victorian age Britain, emphasizes the social and cultural dimensions of the capitalist-based networked economy, amid increasingly consolidation of the public and private interests in digital platforms since the turn of the century (Schiller 1999). Both Schiller (1999) and Pieterse (2010) characterize the materiality and imaginaries of ICTs as an assemblage of creative capitalism, which increasingly organizes accumulation of sociocultural capital and economic capital for individuals, universities, corporate entities, social entrepreneurs, and nation-states in multiple frontiers of transnational spatial spaces and diverse economies. In a utilitarian sense, digital capitalism packages ICT4D as all-inclusive progressive initiatives for the greater good of greater number of the people, provided every region of the world is interconnected and networked into global market systems that is fostered by advancement in digital and mobile communication technologies. As such, "when they are not trumpeting the wonders of digital networks, the stewards of digital capitalism remain basically complacent about their project's human face" (Schiller 1999, p. 208). But, the market-oriented capitalistic ventures are conflated with social progress, without making "distinction between development and capitalism, or between modernization and capitalist transformation" (Lin 2006 cited in Zhao and Chakravartty 2008, p. 14). The conflation naturalizes ICTs as an integral "force that shapes, determines, constrains, or otherwise controls social development" and economic growth (Mosco 1996, p. 143). Hence, while ICT4D is ideologically rooted in the marketization of communication services and infrastructure, the ICT4D discourse operates on the terrain of "technological causation" that justifies technologies as enabler of equality, access, freedom, participation, and control (Mosco 1989; Ojo 2016).

With the metaphoric phrases of "information society" and "technological revolution" well-enshrined in the international development agenda, the deepening levels of investment on infrastructural development and ICT networks, coupled with neoliberal friendly policy frameworks of several nation-states, enabled the opening of new markets and economic of scales for transnational telecommunication and technology companies. In 2015, global trade in ICT goods and services exceeded $US 2.5 trillion (UNCTAD 2017). The World Bank's data in the *2016 World Development Report* indicated that, "the accumulation of ICT capital accounted for almost 20 percent of global growth between 1995 and 2014" (World Bank 2016, p. 55). This is further underscored by (a) the market capitalization and profitability of a cross section of transnational telecommunication and technology companies and (b) increasing growth of ICT industry in the world economy.

In 2017, 10 of the top 20 companies in the world are in the ICT sector of the global economy (PwC 2017). As Table 1 shows, with market capitalization of $US 754 billion, the Apple Inc. is the number one company in the world. It is followed by Alphabet Inc. and Microsoft, with market capitalization of US$ 579 billion and $US 509 billion, respectively (PwC 2017). Between March 2009 and March 2017, Apple Inc. had approximately eightfold increase in its market capitalization value. In short, "ICT4D may be a terrain in its own right but it is also part of a general ICT boosterism in which ICT is the latest major wave of capital accumulation" (Pieterse 2010, p. 173).

The alignment of geopolitical interest with corporate power has not only expanded the transnational corporate supply chains of transnational ICT and telecommunication companies in this context; the networked global economy

Table 1 Global top 20 companies in 2017

Company	Country of origin	Industry	March 2017 Market cap rank ($US billion)		March 2009 Market cap rank ($ US billion)	
Apple Inc.	USA	ICTs	754	1	94	33
Alphabet Inc.	USA	ICTs	579	2	110	22
Microsoft	USA	ICTs	509	3	163	6
Amazon Inc.	USA	ICTs	423	4	31	N/A
Berkshire Hathaway Inc.	USA	Financial services	411	5	134	12
Facebook Inc.	USA	ICTs	411	6	–	–
Exxon Mobil Corp	USA	Oil and gas	340	7	337	1
Johnson & Johnson	USA	Health-care	338	8	145	8
JPMorgan Chase & Co	USA	Financial services	314	9	100	28
Wells Fargo & Co	USA	Financial services	279	10	60	55
Tencent Holdings Ltd	China	ICTs/telecom	272	11	13	–
Alibaba Group Holding	China	ICTs	269	12	–	–
General Electric	USA	Industrial energy	260	13	107	24
Samsung Electronics	South Korea	ICTs	259	14	61	53
AT&T Inc.	USA	Telecomm	256	15	149	7
Ind.& Comm. Bank of China	China	Financial services	246	16	188	4
Nestle	Switzerland	Food and beverage	239	17	129	15
Bank of America	USA	Financial services	236	18	44	87
Procter & Gamble	USA	Consumer goods	230	19	138	10
China Mobile	China	Telecoms	224	20	175	5

Source: PwC (2017)

has also reconfigured economic geography by facilitating ICT capital accumulations in the periphery African cities such as Lagos, Kigali, and Nairobi where clusters of technology hubs and ICT-oriented start-up firms have emerged. There are 314 technology hubs that are operational in 93 cities in 42 African countries in 2016 – that is about 15% increase in comparison to 2014 (Du Boucher 2016). These technology hubs, which are metaphorically known as the Silicon Valleys of the Global South, are embedded in the transnational network capitalism, and are partially funded by international venture capital and philanthropic foundations such as Mark Zuckerberg's the Chan Zuckerberg Initiative.

As the specters of what Schiller (2007) termed "accelerated commodification" of communication resources, these technology hubs serve as the axis of regional ICT economies and knowledge production whereby partnership are forged among international investors, telecommunication operators, ICTs software companies, civil society, and local entrepreneurs for innovative product development and capacity building. By so doing, the temporal dynamism and spatial configuration of these technology hub cities hinge on the sociocultural variation and combined forces of accumulation logic, agglomeration of creative labor, and mass consumer culture. As such, it is no wonder that half of these technology hubs in Africa are situated in five countries – South Africa, Egypt, Kenya, Nigeria, and Morocco (GSMA 2016). While these five countries are on the margin of the interdependent global economic system, they are among the leading emerging economies in Africa (AfDB, OECD & UNDP 2017). The capital flow from the transnational ICT companies is also predominately clustered in these countries' economic core cities as well – in Nigeria, Lagos; in Kenya, Nairobi; in South Africa, Johannesburg, Port Elizabeth, and Cape Town; in Egypt, Cairo and Port Said; and in Morocco, Casablanca and Tangier (World Bank 2016). Invariably, by the geometry of entrepreneurial development intervention, technology hubs are another geostrategic expansionary tools for the ICT economy sector in the global and local contexts.

As Table 2 shows, 9 of the overall top 10 African ICT and telecommunication companies originated from South Africa, Egypt, Kenya, and Morocco. As a corollary, the fortunes of these regional big companies and those of internationally backed new local start-up technological firms might become intertwined with those of the transnational ICT/telecommunication companies such as Alphabet Inc. and Facebook that are now offering basic Internet services in the continent (Ojo 2013; Adejunmobi 2011).

Although social appropriation of ICTs and technology hubs can neither be seen as a secured foundation for Africa's economies nor be considered yet as a major driver of the continent's economic growth at the moment, the continent provides a "spatial fix" for the digital capitalism that underpins the ICT4D and transnational digital

Table 2 Top 10 African ICT/telecommunication companies as of June 2017

Rank	Company name	Country	Total market cap (US$ million)
1	Naspers[a]	South Africa	91,763.51
2	Vodacom Group	South Africa	18,224.52
3	MTN Group	South Africa	17,033.81
4	Maroc Telecom	Morocco	12,492.37
5	Safaricom	Kenya	7854.06
6	Sonatel	Senegal	3895.57
7	Telkom	South Africa	3032.19
8	Global Telecom Holdings	Egypt	1883.74
9	EOH Holding	South Africa	1440
10	Telecom Egypt	Egypt	946.01

Source: [a]African Business Magazine, 2017

economy of the twenty-first century (Unwin 2017; Harvey 2001). Approximately 56% of continent's 1.2 billion populations are of the working age, which is the age group of 15–64 (AfDB, OECD & UNDP 2016). "A young and growing population is generally seen as providing a 'demographic dividend' to GDP growth and GDP per capita growth through labour supply" (AfDB, OECD & UNDP 2016, p. 41). At the heart of this "demographic dividend" factor is the spending and consumption power of the continent's growing educated middle class, which represents 34% of the continent-wide population (Thomson 2016). With such a number, the new and old class of African bourgeoisie constitutes a critical mass for the transversal production, circulation, and consumption of information through various digital communication platforms and mobile phones. The acculturation of new technologies would lead to "a 'double movement' in which profits flow downwards to direct producers and upwards to global corporations" (Murphy and Carmody 2015, p. 20 cited in Mann 2017). Accordingly, in terms of values, African ICT and telecommunication market is a prime market for capital accumulation and new customers.

70.3 ICTs, Africa, and Global Economies

The annual contribution of ICT sector to the global gross domestic product (GDP) has been in the range of 3–6.5% since early 2000s (UNCTAD 2017; Heeks 2018). At the national level, its contribution to the GDP varies from country to country. In the Organization for Economic Cooperation and Development (OECD) countries, it accounts for about 7% of the GDP and about 20% of the overall economic growth (Fransman 2010; OECD 2017; World Bank 2016). In the USA, which is home for several of the transnational ICT companies in the world, the sector adds an average of 7% to the country's GDP annually (World Bank 2016). Whereas quantitative analysis of the linear relationship between ICTs and economic growth is generally weak and sometimes inconclusive in Africa, the ICT sector's annual contribution to the national GDP in Rwanda, Kenya, Nigeria, South Africa, Africa, Ghana, and Egypt has been fluctuating between 5% and 10% annually since 2010 (Jerven 2013; Souter 2015; Ndemo and Mureithi 2015; GSMA 2017). Combined, ICT sector's contribution to African economies was $US 102 billion in 2015 (GSMA 2015; Ford 2015).

Although the Africa's economies still largely resolve around the extractive and commodity industries, the mobile telephony subsector has been heralded as the transformational dimension of the convergent ICT and telecommunication sectors in African contexts. Relative to other forms of ICTs such as Internet, mobile telephony has grown exponentially in the continent in the last decade. With the take-up growth of over 550% in the last decade, the continent is considered to be the world's fastest-growing mobile phone market (GSMA 2017; ITU 2015). The deepening penetration of mobile phone is considered as a form of social progress and modernity. That is, a form of sociocultural mobility in which "the mobile phone has quietly provided people at the bottom of the income pyramid access to electronically mediated communication; often for the first time" (Ling and Horst 2011, p. 364). In

2016, the ecosystem of the mobile telephony added around $37 billion (or 2.6% of GDP) to the economies of the sub-Saharan African region (GSMA 2017).

While the overall contribution of the ICT sector to the broader global, regional, and national economies seems modest on the surface, what is significant is that its "capital contribution to the GDP growth has been fairly constant over the past two decades" (World Bank 2016, p. 12). The sector's spillover effects are considered enormous in other established economic sectors such as banking, retail services, health-care services, tourism and advertising, and media and entertainment where various variants of ICTs are considered integral to business transactions, supply chain, and labor productivity (Souter 2015; World Bank 2016; Albiman and Sulong 2017). Evidently, the boom in the mobile phone ownership and subsequent growth in the mobile Internet access have boasted video on demand and digital content production in African movie and entertainment industry, which has grown exponentially over the last decade (Kacou 2015). Regardless, the linkage between the ICT sector and macroeconomic economic development is complex. As such, it cannot be deduced in the linear fashion.

Unlike Asian region, Africa has not turned itself into "workshop of the world" by becoming a major producer and major exporter of the ICT goods and services (Zhao 2007; Ojo 2013). With the exception of fragmented presence of internationally backed start-ups like Nigeria-based software firm Andela and South African smartphone manufacturer Onyx Connect, the local African ICT sector is predominately dominated by few local big companies (such as Naspers, MTN Group, Vodafone, and Safaricom) and major global players (such as Huawei and Ericsson). Most of the continent's socioeconomic activities are informal or simply economies of affection (Hyden 2001).

70.4 Mobile Phones and Economics of Affection in Africa

There are adaptation and appropriation of mobile telephony in economies of affection, which is underpinned by the notion of "people as infrastructure," that cannot be quantified in the macroeconomic terms. High-profile examples include hawking of mobile phone airtime scratch cards, recirculation of disposed mobile phone handsets, and parallel market for sales of fake and stolen mobile phones. The notion of *people as infrastructure* refers to people's activities to generate concert social relations and economic actions within the nodes of cities that are entrapped in embedded structures of (a) unproductive postcolonial nation-state, (b) socioeconomic failings from truncated state-enacted grandiose national development plans, and (c) spatial and class inequalities from sociocultural and political logics of everyday life (Simone 2004; Hyden 2001). In other words, *people as infrastructure* encapsulate everyday "politics of suffering and smiling" (Chabal 2009) of people, and how they cope with their socioeconomic predicament by "using their wit in ways that fit neither neoclassical nor Marxist assumptions about human behavior" (Hyden 2001, p. 10023). In general, the enacted socioeconomic actions of survival in this context are often ad hoc, unsustainable, and operate outside the formal institutional boundaries of recognized universal economic activities that normally count in the calculation of the GDP (Chabal 2009).

In the backdrop of the increasingly digitalization of everyday life fueled by digital capitalism, the surge in mobile telephony blurs the line between livelihoods means and communicative action. At the functional social level, mobile phone provides a basic connectivity and sociability accorded at the baseline of communal exchange by increasing frequency of interaction with geographically distant friends, families, and relatives (Donner 2009; 2015). In the same vein, while it strengthens social communication networks, the surge in mobile telephony also leads to socioeconomic survival-driven entrepreneurial ventures such as street hawking of phone cards, exchanging of phone accessories for phone minutes credits, and selling of second-hand or recycled mobile phone handsets, “thereby commodifying the ‘social capital’ of poor people into multinational chains of accumulation” (Elyachar 2005 cited in Mann and Nzayisenga 2015, p. 33).

In Uganda, it is estimated that the government loses about $US 9 million annually in sales tax revenues due to the flourishing underground markets for stolen or fake mobile phone handsets in the country and its neighboring countries of Democratic Republic of Congo and South Sudan (Mahajan 2009; Bagala 2016). Whereas this is a reflection of the complex disjunctive order that characterizes the contemporary global cultural economy and technological flows (Appadurai 1990), the generative political economy of *people as infrastructure* underscores the fact that adaptation and penetration of ICTs, in particular mobile phones, “*transgress* the boundaries imposed by the state, the culture, the economy, and by the technology-capitalism complex itself” (Wasserman 2011, p. 150 – italics in the original).

Based on the transmission model of ICT4D, the view is that ICTs would revolutionize economic practices and steer path for economic development. Accordingly, the linear effects of ICTs, mobile phones in particular, on the ecosystem of African economies are often highlighted with micro-financing and e-banking initiatives such as M-Pesa in Kenya and WIZZIT in South Africa. But such macro-level perspectives on transmission effects did not fully capture the intricacy of appropriation and adaption of ICTs in everyday life of those at margin of the local economy in many African nation-states. One of the standard entrepreneurial features of the economies of affection in Africa is the open public square market where people, in particular women, sell all forms of commodity goods that range from raw food products to household commodities and mundane retail products such as street-made fried beans and roasted corns. For several of these traders, mobile phone is a necessity of life. However, the particularities of adaptation and usage extend beyond the forecasted linear patterns of economic usage highlighted in the scholarly literature (Aker 2010; Akers and Mbiti 2010) and position papers of international development agencies about African farmers and market women’s use of mobile phones for market pricing information and competitive price adjustment.

Based on the author’s nonparticipant observation of the activities in several of these street markets in Nigeria and Kenya, the petty traders’ utility of mobile phone is largely for personal use and to maintain business contact with their good suppliers and customers. In this respect, the trade-related utility of mobile phone is generally about demand-supply coordination of goods and services. As Burnell (2014) also observed, in regard to market women in Ghana, "the calls between market women and the wholesalers/suppliers they purchased goods from often involved requests to set aside quantities for later collection. . . . In the coordination work described by many market women, what is exchanged by phone is information about one's present location, predictions about time of arrival, details that help to smooth the eventual in-person exchange between buyer and seller" (pp. 585–586). Though at the periphery of the official normative economic frameworks, the traders in this context appropriated mobile phones as a constitutive part of the wider social networks of their established trading norms in an informal economic environment, which is not completely detached from the domesticated communal spaces of everyday life. Hence, rather than changing the established relationship dynamics and nature of informal market economic structure in this context, the mobile phones actually help to strengthen the social embeddedness of informal market transactions, especially among the petty traders and their suppliers (Burnell 2014; Avgerou 2010). In a nutshell, the market women and petty traders adopted "mobile phone as an enabling device – and as a symbol of their own particular form of a 'modern life-style', which enables them to escape the designation of cultural backwardness while distinguishing themselves from the 'official' secular subjectivity of the modern nation-state" (Morley 2017, p. 177).

The generative political economy that manifests through the commodification of the social capital of everyday life shows the widening gap between the ritualized official discourse of ICTs' leapfrogging power and the materiality of sociocultural appropriation of ICTs. On average, Africans spend between 10% and 17% of their monthly incomes on the mobile phone calls and SMS texts (Gillwald 2017; Ojo 2013). To save on the economic cost of mobile phone conversations, people often resort to the short messaging system (SMS) texts *and/or* "flashing." Flashing is an act of dialing a mobile phone number, allowing it to ring once or twice, and then hanging up before the call is picked up on the other end (Hahn and Kibora 2008). It is a culturally invented tactic for callers not to incur the costs of phone calls but a coded message for the call receiver to call back. Importantly, it is a subtle message to the call receiver that the caller might not have had enough "credit" to complete the call (Hahn and Kibora 2008; Smith 2006). On the other hand, it also speaks to sociocultural wit of staying connected and coping in the mediated social relation that is forged in the Africa's mobile revolution era.

Depending on the context, flashing can also be seen as a form of social-cultural communication act that is similar to *aroko* in the Yoruba traditional social communication thoughts. *Aroko* is a nonverbal form of interpersonal communication that is deployed in the transmission of an emissary message of promises, pledges, solidarity, and even warnings, in social and diplomatic occasions (Ojo 2017). As Hahn and Kibora noted: "Some people give special instructions to relatives and friends, encouraging them to flash whenever necessary, particularly if these friends have previously been asked for a favour by the owner of the mobile phone" (Hahn and Kibora 2008, p. 95). Notwithstanding the technical appropriation of nonverbal cultural communication code, flashing reveals deeply embedded micro-politics of socioeconomic inequality in everyday mobile phone usage in African contexts.

70.5 ICTs, Public Services, and Social Entrepreneurship

As a result of the increasingly ubiquity of mobile phone in public life, the government, corporate entities, and civil society groups also deploy the mobile phones as the new "talking drums" for public services announcements and development-support communication activities that address "information poverty." Being that the use of the ICTs for positive health outcomes is one of the key features of the United Nations (UN)'s Millennium Development Goals (MDGs), health sector represents one of the significant areas where national governments, in collaboration with health workers, civil society groups, and international development agencies, frequently leverage the mobile phones and Internet platforms in the public safety and health-care-related issues. In this respect, Rwanda government's TRACnet, which is both Internet- and mobile phone-based platforms for the monitoring of HIV/AIDS treatment, remains one of the most successful government-led development-support communication projects. Established in 2005, the TRACnet is wired to virtually HIV/AIDS treatment clinics across the country for data collection on HIV/AIDS patients as well as HIV/AIDS treatments of these patients. "As a result, public monitoring of HIV/AIDS transmission patterns has improved, doctors and patients have access to more reliable information, and real-time monitoring of antiretroviral drug stocks lead to quicker replenishments" (Fox 2016, p. 63). Rwanda is the digital leader in this context in Africa (Unwin 2017; Heeks 2018).

Given the long-standing issues of sustainability and continuity with several of the government and international donor-funded development-support communication projects, social enterprises are springing up in many African cities as a third-wave sector between the corporate-driven ICT4D projects and international donor cum charity-based projects. Among new wave of the ICT-enabled social entrepreneurship is the M-Pedigree, a drug-monitoring technological platform that "permits health workers and consumers to send a text code to a central hotline quickly to verify, quickly, whether a medicine is counterfeit or genuine" (Rotberg 2013, p. 148) in Rwanda, Kenya, Nigeria, Tanzania, Uganda, and Ghana. Set up as a social entrepreneurship venture by Ghanaian, Bright Simons, in 2007, M-Pedigree partners with telecommunication operators (such as Safaricom and MTN Group), technology companies (such as Hewlett-Packard) and government drug regulatory agencies to minimize the circulation of counterfeit drugs in the public domains. Based on the World Health Organization (WHO)'s recent data, over 100,000 deaths per year in Africa are due to the use of the counterfeit drug (Hirschler 2017).

By simply texting M-Pedigree or calling the M-Pedigree's 24-h operated call centers, individuals and wholesale buyers can verify the authenticity of their purchased drugs. Alternatively, the buyers with smartphones can scan the barcode on the package to verify that the drugs are not fake or counterfeit. Although public good is partly served through social entrepreneurial activities such as the M-Pedigree and counterfeit drugs, ethical questions on privacy and the use of the generated personal data from the citizens might pose a future challenge, in view of the rapid growth rate of the Internet of things and big data industry in the Western hemisphere. Whereby privacy is one ethical concern with data analytics and Internet of things, political data surveillance by the state intelligence agencies as well as corporate entities' commodification of the data are among other emerging areas of ethical concerns and future risks of data-based ICT development initiatives and ICT-enabled social enterprises.

70.6 Conclusion

This chapter provided an overview of the ICT4D in the global and African economies. In doing so, it captured the intertwined dynamics of socioeconomic construct of ICT4D and the growth of ICT sector in the global and African contexts. As an economic sector, ICTs add value to public life through job creation. The sector also generates revenues for government coffers through the mechanism of direct and indirect taxation schemes – sales taxes on ICT goods and services, income taxes of ICT companies' employees, and corporate taxes. For the entrepreneurs and investors in the private sector, the investments in the leading transnational ICT/telecommunication companies have yielded positive investment returns. This is partly due to the massive increase in the market capitalization of these companies in the last decade; with few ones such as Blackberry losing their market values and dominance. In hindsight, the political economic processes for the development of critical network infrastructures for the twenty-first century economy have transformed ICT sector and made it an important orbit for global information flows and trade.

In Africa, the uptake of mobile telephony in the continent has modified the politics of everyday life by offering socioeconomic actions on one hand; while it also normalizes social inequality gaps at the same time. In other words, the influx of mobile phone produces dialectical effects of empowerment and disempowerment in socioeconomic and political domains. To this end, the chapter notes that mobile phone as a development apparatus is not necessarily a machine for the alleviation of poverty and other socioeconomic problems; but a machine that re-embeds social and economic relations within sociocultural African spaces.

References

Adejunmobi M (2011) Nollywood, globalization, and regional media corporations in Africa. Pop Commun 9(2):67–78

AfDB, OECD & UNDP (2016) African economic outlook: sustainable cities and structural transformation. AfDB, Abidjan

AfDB, OECD & UNDP (2017) African economic outlook: entrepreneurship and industrialization. AfDB, Abidjan

Aker JC (2010) Information from markets near and far: the impact of mobile phones on grain markets in Niger. Am Econ J Appl Econ 2(1):46–59

Aker JC, Mbiti IM (2010) Mobile phones and economic development in Africa. J Econ Perspect 24 (3):207–232

Albiman M, Sulong Z (2017) The linear and non-linear impacts of ICTs on economic growth, of disaggregate income groups within SSA region. Telecommunic Policy 41(7–8):555–572

Appadurai A (1990) Disjuncture and difference in the global cultural economy. Theory Cult Soc 7 (2):295–310
Avgerou C (2010) Discourses on ICT and development. Inf Technol Int Dev 6(3):1–18
Bagala A (2016) Shs170m smartphones stolen in Kampala traffic jams. Daily Monitor, March 26 (online). http://www.monitor.co.ug/News/National/Smartphones-stolen-Kampala-traffic-jams/688334-3133680-9flad9z/index.html
Burrell J (2014) Modernity in material form? Mobile phones in the careers of Ghanaian market women. Rev Afr Polit Econ 41(142):579–593
Chabal P (2009) Africa: the politics of suffering and smiling. Zed Books, New York
Donner J (2009) Blurring livelihoods and lives: the social uses of mobile phones and socioeconomic development. Innovations 4:91–101
Donner J (2015) After access: inclusion, development, and a more internet. MIR Press, Boston
Du Boucher V (2016) A few things we learned about tech hubs in Africa and Asia. GSMA, August 5 (online). https://www.gsma.com/mobilefordevelopment/programme/ecosystem-accelerator/things-learned-tech-hubs-africa-asia
Elyachar J (2005) Markets of dispossession: NGOs, economic development and the state in Cairo. Duke University Press, Durham
Ford N (2015) A maturing sector puts pressure on operator revenue. Afr Bus Mag, November (424):28–29
Fox B (2016). How can Africa bridge the digital divide? Afr Bus Mag, April (429):61–63
Fransman M (2010) The new ICT ecosystem: implications for policy and regulation. Cambridge University Press, Cambridge
Gillwald A (2017) Beyond access: addressing digital inequality in Africa (CIGI paper series: No. 48). CIGI, Waterloo
GSMA (2015) The mobile economy: Sub-Saharan Africa 2015. GSM Association, London
GSMA (2016) *A few figures on tech hubs in Africa*. GSM Association, London
GSMA (2017) The mobile economy: sub-Saharan Africa 2017. GSM Association, London
Hahn HP, Kibora L (2008) The domestication of the mobile phone: oral society and new ICT in Burkina Faso. J Mod Afr Stud 46(1):87–109
Harvey D (2001) Globalization and the "spatial fix". Geogr Rev 3(2):23–30
Heeks R (2018) Information and communication for development (ICT4D). Routledge, London
Hirschler B (2017) Tens of thousands dying from $30 billion fake drugs trade, WHO says. Reuters, November 28 (online). https://www.reuters.com/article/us-pharmaceuticals-fakes/tens-of-thousands-dying-from-30-billion-fake-drugs-trade-who-says-idUSKBN1DS1XJ
Hyden G (2001) Moral economy and economy of affection. In: Smelser N, Baltes P (eds) International encyclopedia of the social and behavioral sciences. Pergamon Press, New York, pp 10021–10024
ITU (2015) Measuring the information society report 2015. ITU, Geneva
Jerven M (2013) Poor Numbers. How we are misled by African development statistics and what to do about it. Cornell University Press, Ithaca
Kacou E (2015) Entertainment and media. In: McNamee T, Pearson M, Boer W (eds) Africans investing in Africa: understanding business and trade, sector by sector. Palgrave Macmillan, New York, pp 203–225
Ling R, Horst HA (2011) Mobile communication in the global South. New Media Soc 13 (3):363–374
Loh Y (2015) Approaches to ICT for development (ICT4D): vulnerabilities vs. capabilities. Inf Dev 31(3):229–238
Mahajan V (2009) Africa rising. Pearson Education Inc, Upper Saddle River
Mann L (2017) Corporations left to other peoples' devices: a political economy perspective on the big data revolution in development. Dev Chang. https://doi.org/10.1111/dech.12347
Mann L, Nzayisenga E (2015) Sellers on the street: the human infrastructure of the mobile phone network in Kigali, Rwanda. Crit Afr Stud 7(1):26–46
Morley D (2017) Communications and mobility. Wiley Blackwell, Hoboken

Mosco V (1989) The pay-per society: computers and communication in the information age. Garamond, Toronto
Mosco V (1996) The political economy of communication. Sage, New Delhi
Murphy JT, Carmody P (2015) Africa's information revolution: technical regimes and production networks in South Africa and Tanzania. Wiley Blackwell, Boston
Ndemo B, Mureithi M (2015) Information and communication technologies. In: McNamee T, Pearson M, Boer W (eds) Africans investing in Africa: understanding business and trade, sector by sector. Palgrave Macmillan, New York, pp 177–202
Nyamnjoh F (2005) Africa's media, democracy and the politics of belonging. Zed Book, London.
OECD (2017) OECD digital economy outlook 2017. OECD Publishing, Paris
Ogan C, Bashir M, Camaj L, Luo Y, Gaddie B, Pennington R, Rana S, Salih M (2009) Development communication: the state of research in an era of ICTs and globalization. Int Commun Gaz 71 (8):655–670
Ojo T (2013) ICTs and mobile phones for development in sub-Saharan African region. In: Servaes J (ed) Sustainability, participation and culture in communication. Intellect, Bristol/Chicago, pp 83–100
Ojo T (2016) Neo-Gramscian approach and geopolitics of ICT4D agenda. Global Media J (Canadian Edition) 9(1):23–35
Ojo T (2017) Corporate social responsibility (CSR) activities of Huawei and ZTE in Africa. In: Batchelor K, Zhang X (eds) China-Africa relations: building Images through cultural co-operation, media representation and on the ground activities. Routledge, London, pp 218–230
Pieterse JN (2010) Development theory, 2nd edn. Sage, Thousand Oaks
PwC (2017) Global Top 100 Companies by market capitalization. PricewaterhouseCoopers, London
Rotberg RI (2013) Africa emerges. Polity, Malden
Schiller D (1999) Digital capitalism: networking global market system. MIT Press, Cambridge, MA
Schiller D (2007) How to think about information. University of Illinois Press, Urbana
Simone A (2004) People as infrastructure: intersecting fragments in Johannesburg. Publ Cult 16 (3):407–429
Smith D (2006) Cell phones, social inequality, and contemporary culture in Southeastern Nigeria. Can J Afr Stud 40(3):496–523
Souter D (2015) ICT4D and economic development. In: Mansell R, Hwa Ang P (eds) The international encyclopedia of digital communication and society (volume one: A–K). Wiley Blackwell, Malden, pp 345–352
Thomson A (2016) An introduction to African politics, 4th edn. Routledge, New York
UNCTAD (2017). Information economy report 2017: digitalization, trade and development. Geneva UNCTAD
Unwin T (2017) Reclaiming information and communication technology for development. Oxford University Press, Oxford
Wasserman H (2011) Mobile phones, popular media and everyday African democracy: transmissions and transgressions. Pop Commun 9(2):146–158
World Bank (2016) World development report 2016: digital dividends. World Bank, Washington, DC
Zhao Y (2007) After mobile phones, what? Re-embedding the social in China's 'digital revolution. Int J Commun 1(1):92–120
Zhao Y, Chakravartty P (2008) Introduction: toward a transcultural political economy of global communication. In: Zhao Y, Chakravartty P (eds) Global communications: toward a transcultural political economy. Rowman and Littlefield, Lanham, pp 1–19

Mainstreaming Gender into Media: The African Union Backstage Priority

71

Bruktawit Ejigu Kassa and Katharine Sarikakis

Contents

Abstract

The chapter examines the ways in which mainstreaming gender into the media has been overlooked in the African Union (AU) gender equality frameworks. The chapter argues that in AU policies, gender in the media is effectively treated as second and last in importance to problems such as poverty, health, security, and education for women. Given the fact that the European Union (EU) has failed to implement gender mainstreaming strategies in the field of the media, and given the significant role the EU has played in global efforts as donor and source for policy inspiration, the chapter explores the possibility of a global failure in achieving gender equality in the media. The implications of such failure on social change and development are also considered.

B. Ejigu Kassa (✉)
Department of Communication, University of Vienna, Vienna, Austria

Department of Journalism and Mass communication, Haramaya University, Dire Dawa, Ethiopia
e-mail: bruktimail@gmail.com

K. Sarikakis (✉)
Department of Communication, University of Vienna, Vienna, Austria
e-mail: katharine.sarikakis@univie.ac.at

J. Servaes (ed.), *Handbook of Communication for Development and Social Change*,
https://doi.org/10.1007/978-981-15-2014-3_103

Keywords
Gender mainstreaming · African Union · Media · Gender equality · Africa · European Union

71.1 Introduction

The media, spelled out as 1 of the 12 critical areas of concern in the Beijing Declaration and Platform for Action (BPfA) adopted in 1995 at the Fourth International Conference on Women, have been referred to as "the most important yet challenging area of work for advancing gender equality" (Lowe Morna 2002, p. 1). Gender stereotypes, women limited decision-making roles in the media, and their unequal access to ICTs have been highlighted in the Declaration as key challenges to realizing equality of men and women in and through the media (BPfA 1995).

Despite being among the prominent figures in the adoption of the Declaration, the African Union (AU) has been slow to mainstream gender perspectives into the media. More than two decades later, integrating gender dimensions into the media has not yet received the weight it deserves. The nexus between gender and the media has been considered as an "add-on" in AU gender equality frameworks in which poverty, health, security, and education have assumed key priorities. At the same time, incorporating gender notions into the media is an unaccomplished mission in the European Union (EU) as well. Throughout the years, the EU has failed to implement gender mainstreaming strategies in the field of the media. Given the significant role the EU has played in global efforts as donor and source for policy inspiration, its failure in this context might be a potential indicator of a global disappointment in achieving gender equality in the media.

The chapter first discusses realities of problems in gender and the media. It then traces the historical development of AU gender equality instruments and explores the AU and the EU failures at keeping gender equality promises in general and mainstreaming gender into the media in particular. Finally, the exclusion of gender and the media issues from AU gender architecture is analyzed.

71.2 Gender and the Media: Why Should It Be an Area of Concern?

The media have been central to the construction of gender ideologies. They have a significant impact on the ways in which gender is defined and understood. The media do not only mirror reality but also shape public opinion and culture (Ross and Padovani 2017). Largely dependent on the ways in which they are used, the media can contribute to the promotion of gender equality. For instance, by using sensitive contents and language, and non-stereotypical representation of women and men, the media can enhance the equality of men and women in society (Williams 2000).

Further, the media have huge potential in advancing and empowering women by enabling them to participate and to be heard in the process of development and transformation (Bhagwan-Rolls 2011).

However, the media have been mostly criticized for standing on the way of gender equality (Lowe Morna 2002; Sarikakis and Nguyen 2009; Byerly 2014; Ross 2014a, b). Several studies have shown the problem of media's gender-based stereotyping, women's limited roles in media decision-making, and women's unequal access to ICTs (Lowe Morna 2002; Byerly 2014; Collado 2014; Grizzle 2014; Kareithi 2014; Ross 2014a, b; GMMP 2015; Webb 2016). These studies have been documenting and analyzing the ways in which women have been systematically and systemically put in disadvantaged positions both in content and decision-making in the media industries for decades.

The Global Media Monitoring Project (GMMP), the most extensive and longest-running research advocacy initiative for gender equality in and through the news media conducted every 5 years since 1995, has been reporting the persistent invisibility of women as news subjects. According to its 2015 report, women make up only 24% of the persons read about, heard, or seen in newspaper, radio, and television news, exactly as they did in 2010 (GMMP 2015). That means that for every three males we hear or see or read about in the media, we see or hear or read about only one female. Besides, the report reveals that almost 48% of all the stories reinforce stereotypical representations of men and women. The same report reveals that in Africa women's relative presence in the news has increased in half a decade from 19% in 2010 to merely 22% in 2015 yet below the global average. Such unfair representation of women maintains the status quo, contributing to the perpetuation of gender inequality in everyday life (Ross and Padovani 2017).

Although underrepresentation and misrepresentation of women in the media is a global challenge, it is more intensified in Africa where the problem is further reinforced by patriarchal tradition and social customs (Kareithi 2014). A paper prepared by Gender Links for the AU Specialized Technical Committee on Information and Communications (STC-IC) on "Media Portrayal of Women and Media Gender Gap in Africa" stresses out that the media in Africa persist underrepresenting and misrepresenting women and systematically silencing their voices through invisible "gender censorship" (Gender Links 2017). The paper further points out that despite the intensity of the trouble with gender and the media in the continent, most African countries lack legislation, policies, and a positive environment for promoting gender equality in the media.

Furthermore, with the advancement of electronic media such as the Internet and cable channels, media images and contents transcend the local (Zubair 2016). Hence, global media would mean global content in which images of women circulate in ways that are more readily accessible. These global imageries perpetuate the persistent stereotypical portrayal of women and men (Gallagher 2008). Gallagher (2008, p. 24) has noted that "the picture that emerges from many analyses of Internet content is of a masculinist rhetoric and a set of representations that are frequently sexualized and often sexist." Given the transcending nature of these platforms and the lack of policies governing global contents, the struggle to have a fair and balanced representation of women can face even more difficult challenge.

The small number of women in media decision-making is another challenge in promoting gender equality in and through the media (Lowe Morna 2002; Byerly 2014; Ross 2014a, b; Byerly and Padovani 2017). It is assumed that an increased presence of women decision-making roles in the media is likely to make a difference to the stereotypical and portrayal of women (Byerly and Padovani 2017). If women had a greater access to decision-making in the media, they could contribute to more gender-sensitive content. Nevertheless, the number of women in media decision-making is strikingly small worldwide. The Global Report on the Status of Women in News Media in 2011 shows that women held only about a fourth of "the jobs in governance (i.e., boards of directors) globally" (Byerly 2014, p. 40). Similarly, the report reveals that women in sub-Saharan African media account for 28% of the board members, 24% of the top and 54% of the senior management positions, despite constituting 49% of the technical professional field (Gender Links 2017). Kareithi (2014, p. 335) argues that though the number of African women working in media decision-making has shown some improvement after the Beijing Platform, women in the African media "can only get as far as the men will let them."

Women's access to ICTs also raises concern as the digital divide increases and poses a problem for the development of equal information society (IT) (Collado 2014; Grizzle 2014; Webb 2016). Although the issue of women's unequal access to ICTs was acknowledged in the BPfA, the subject was paid greater attention in the Beijing +5 review in 2000 (Gallagher 2011). In addition to ICT access, issues such as "infrastructure and content as well as the role of ICTs in the development of culture, and the impact of all these on women's rights and gender equality" have become new concerns that needed to be addressed (Gallagher 2011, p. 454). Subsequently, the World Summit on the Information Society (WSIS), held in two phases in Geneva (2003) and Tunis (2005), recognizes the potential of ICT as a tool for promoting gender equality and women empowerment and the existing "gender divide." The Geneva Declaration, the outcome of the first phase, affirms a commitment to ensuring women's inclusion and full participation in the information society "on the basis on equality in all spheres of society and in all decision-making processes" and to "mainstream a gender equality perspective and use ICTs as a tool to that end" under para 12 (WSIS 2003). Again, in a 10-year review of the implementation of the WSIS principles (WSIS+10), the WSIS acknowledged the persisting "gender divide" and reaffirmed its commitment to narrow the gap (ITU 2014).

The "gender divide" has been a significant impediment to women's full participation in development process and in their enjoyment of the benefits of new electronic contents and services. Even if women's unequal access and use of ICTs is a worldwide problem, women in developing countries are considerably more affected by obstacles to the access and beneficial use of ICTs (Grizzle 2014). Gender Links in its paper for the AU-STC-IC highlights that African women, particularly rural women, consistently struggle with unequal access to ICTs due to "inadequate infrastructure, affordability and availability, language barriers, illiteracy and even discriminatory social norms capacity and skills relevant content" (Gender Links 2017, p. 9).

Overall, the relationship between gender and the media remains one of the major challenges in achieving gender equality globally. Yet, the progress to address the problem has been far too slight to make gender equality in the media a reality. Further, any considerable advancement "can even be considered relative since it varies according to the region of the world" (Lourenço 2016, p. 927).

71.3 African Union Gender Equality Frameworks

The AU, which replaced the Organization of Africa Unity (OAU) at its inaugural meeting in Durban, South Africa, in 2002, has expressed its commitment to promoting gender equality and empowering women and girls in several of its legal instruments. Its commitment to gender equality can be traced to the 1948 UN Charter and the Universal Declaration on Human Rights, which emphasized the enjoyment of rights and freedom without distinction of any sex (Martin 2013). Since its establishment, the AU has been working to narrow the social, political, and economic inequality of men and women in the continent (Joshua Omotosho 2015). Acknowledging the fact that gender equality is a fundamental human right, the organization has promised to ensure "the absence of discrimination on the basis of one's sex in the allocation of resources or benefits or in access to services" (The African Union Gender Policy 2009, p. 28).

Historically, the legal mechanisms of the AU to fostering gender equality have been informed by the UN frameworks (Martin 2013). Prominent among international legal frameworks adopted by the AU, the then OAU, is the Convention on the Elimination of All Forms of Discrimination against Women (CEDAW) which is described as the "international bill of rights for women." The AU was also among the significant institutions taking part in the UN conferences held in Mexico City (1975), Copenhagen (1980), Nairobi (1985), and Beijing (1995) and its Platform for Action, as well as conferences in the 1990s on population and development, human rights, social development and human settlements, and trade and poverty reduction strategies (Martin 2013).

Adhering to its commitment to fostering gender equality and women empowerment, the AU has adopted the Millennium Development Goals (MDGs), the primary global framework for international development intended to reduce poverty and to empower women (Goal 3) by 2015, and the Sustainable Development Goals (SDGs), another global mechanism aiming to end poverty by 2030 and to achieve gender equality and empower women and girls (Goal 5).

In line with the SDGs, the AU has set Agenda 2063, the continent's 50-year strategic plan aiming to achieve an integrated, peaceful, and prosperous Africa for all. This framework, officially adopted by the AU Assembly in 2015, provides a new collective vision and road map to build a developed and united Africa based on shared values and a common destiny (The African Union Commission 2017). Agenda 2063 reaffirms fostering gender equality and acknowledges women's contribution to Africa's structural transformation and sustainable development (African Union Commission 2015).

Furthermore, in accordance with the UN gender equality mechanisms, the AU has formulated a number of gender equality instruments to accelerating gender equality and women empowerment. The first ever African Union Gender Policy was developed in 2009 with the proclaimed goal of setting up a clear vision and making commitments to guide the process of gender mainstreaming and women empowerment to influence policies, procedures, and practices which advance the achievement of gender equality, gender justice, nondiscrimination, and fundamental human rights in Africa (African Union Gender Policy 2009). Other key gender equality frameworks set by the AU include the African Charter on Human and People's Rights (1981), a significant instrument for the protection of human rights in Africa; the Protocol on the African Charter on Human and People's Rights on the Rights of Women in Africa (2003), a legal framework; the Solemn Declaration on Gender Equality in Africa (2004), a non-binding commitment to the principles of gender equality; the Ouagadougou Action Plan to Combat Trafficking in Human Beings, especially Women and Children (2006), a tool for combating human trafficking; and the Charter on Democracy, Elections and Governance (2007), a key instrument for the promotion of gender balance and equality in the governance and development process. In addition, with the aim of accelerating the implementation of the gender equality policies, the AU has declared this decade, 2010–2020, the African Women's Decade.

More importantly, the AU has made a revolutionary decision on gender parity (50/50) taken at the Inaugural Session of the AU Assembly of Heads of State and Government in July 2002. Standing by its decision, the AU elected five female and five male Commissioners during the Second Ordinary Session of the Assembly in 2003. In addition, the appointment of Dr. Nkosazana Dlamini-Zuma as Chairperson of the AU Commission, who became the first woman to lead the Commission from 2012 to 2017, further demonstrates AU effort to advancing gender equality principles.

Expanding the gender parity principle to other AU organs, AU Executive Council's formal decision in January 2016 has pledged to ensure that the voices of both women and men are equally represented in all AU organs. Accordingly, at least one of the two representatives from the five African regions in all AU organs shall be a woman (The African Union Commission 2017).

The AU has also played a significant role in promoting gender equality by pursuing its regional economic communities (RECs) representing Africa's subregions: the Arab Maghreb Union (AMU), the Economic Community of West African States (ECOWAS), the Community of Sahel-Saharan States (CEN-SAD), the Southern African Development Community (SADC), the East African Community (EAC), Intergovernmental Authority on Development (IGAD), the Economic Community of Central African States (ECCAS), and the Common Market for Eastern and Southern Africa (COMESA) as well as key programs and instruments such as the New Partnership for Africa's Development (NEPAD) and the African Peer Review Mechanism (APRM) to adopt gender principles and declarations (Martin 2013; Joshua Omotosho 2015).

Again, the AU has made a considerable contribution to encouraging its Member States to adopt, ratify, implement, and domesticate international and regional treaties, conventions, and decisions on gender equality and has created platforms to bring a consensus on gender equality issues among the Member States (Winyi 2009).

To facilitate the implementation of the frameworks on gender equality, the AU has formed the Women and Gender Development Directorate in 2002 which is responsible to build the capacity for all AU organs, RECs, and Member States "to understand gender, develop skills for achieving gender mainstreaming targets and practices in all policy and programme processes and actions by 2020," with the hope of closing the persisting gender gaps and delivering the promise of gender equality for African people (African Union Gender Policy 2009, p. 5).

The Directorate also organizes annual AU Gender pre-Summits in collaboration with other departments of the African Union Commission and development partners. The pre-Summits are held before each AU summit to update and assess progress toward gender equality and women empowerment. The outcome of the discussion serves as a key informant to the AU Summit of Heads of State and Government deliberation on gender equality issues (González 2017). The pre-Summits, "initially conceived as civil society consultation platforms... now draw participation from the African ministers responsible for gender and women's affairs, RECs, AU organs, AUC departments, the private sector, UN agencies and development partners" (González 2017, p. 20).

Over the years, the AU has been a pivotal organ in the promotion of gender equality in Africa. It has contributed to the advancement of "the gender agenda by recognizing the importance of women's contributions to development, reducing maternal mortality and promoting basic education for girls" (González 2017, p. 12). Through its gender equality frameworks, the AU has "enabled the member states and the RECs to advance their own legal, administrative and institutional frameworks to make progress on women's rights and gender equality" (Martin 2013, p. 16). Moreover, AU gender equality instruments have put gender mainstreaming at the center of AU policy-making (González 2017). Consequently, "most instruments adopted by the AU since 2003 make provision for gender equality and women's participation" (Martin 2013, p. 16). Nonetheless, it should be noted that implementation of these instruments has been hampered due to the persisting lack of political will and shortage of resources (Martin 2013).

In the journey of the AU to gender equality, the EU has been a key development partner. It has been a major source of donor and a role model on "the institutionalization of gender equality and on strategies for gender mainstreaming" (Martin 2013, p. 18). Though "primarily for geopolitical and economic reasons," the development cooperation with Africa entered the EU agenda since its establishment in 1957 (Debusscher and van der Vleuten 2012, p. 319). Gradually, their relationship took a different shape to bilateral cooperation as former colonies gained their independence. In their transformed partnership, promoting gender equality has taken a key place. The first references to women's rights in EU development cooperation with the sub-Saharan African region are found in the Third Lomé Convention signed in

1984 with 79 African, Caribbean and Pacific countries (ACP) (Debusscher and van der Vleuten 2012, p. 319). The issue of "Integrating Gender Issues in Development Cooperation" is also reflected in subsequent EU development policies such as the 1997 Amsterdam Treaty (Article 3(2)) and the 2005 EU Consensus on Development (Article 19) (Martin 2013). Advancing gender equality has been also one of the principles enshrined in the 2007 Joint Africa–EU Strategy (JAES) (JAES 2007) and central to EU development aid programs. In the study of the programming of EU development cooperation with sub-Saharan Africa for the period 2002–2013, Debusscher and van der Vleuten (2012, p. 327) find out that "gender is increasingly present in all budgetary sectors of the EU–sub-Saharan African development aid." Notably, the focus placed on the principles of gender equality in EU financial and technical support strategies for the AU has been crucial in enhancing the continent's progress toward gender equality.

71.4 Achieving Gender Equality in the AU and the EU: Unfinished Job

Since the BPfA, the AU and the EU, at either end of the spectra of power and economy, have made considerable improvement in promoting gender equality in their respective regions. Notwithstanding the huge difference in the extent of their achievements, their progress toward the actual equality of men and women has been slower than hoped. Particularly, despite reaffirming commitments to promoting gender equality in the media, the mission appears to be an unaccomplished one in both supranational entities.

71.4.1 The AU Unmet Promises

Gender inequality is a defining challenge for Africa. Although the AU has exerted considerable effort in tackling gender inequality, the path of progress has been slow and inconsistent for many African counties(UNDP 2016). Winyi (2009) has expressed the overall progress toward gender equality in Africa since the BPfA, as "regrettably slow and with a wide gap between commitments and actual action still persistent." Significant gaps between men's and women's opportunities remain severe obstacle to structural economic and social change. According to the Africa Human Development Report 2016, gender inequality is costing sub-Saharan Africa on average 95 billion dollars a year, peaking at 105 billion dollars in 2014, 6% of the region's GDP, jeopardizing the continent's efforts for inclusive human development and economic growth (UNDP 2016).

In narrowing the persisting gender inequality, the AU tends to place greater focus on the adoption of equality instruments than ensuring the practicability of them among Member States. Notably, the AU has taken more actions in the form of policy formulation and the development of frameworks for implementation (Winyi 2009). Although ratification of policies can help "strengthen the work of national mechanisms by creating additional accountability, harmonizing approaches across countries and circumventing resistance and challenges to gender equality at the national level," more important is their domestication and implementation (Martin 2013, p. 24).

When it comes to follow-up measures, the AU lacks effective evaluation and monitoring mechanisms both at national and regional levels. The Member States, regardless of their differences in government structures and political choices, have in common inadequate tracking and monitoring systems (Martin 2013). Several Member States have failed to submit progress reports on their implementation of gender mainstreaming. For instance, although Member States were required to send in periodic reports on their implementation of rights enshrined in the Protocol to the African Charter on Human and People's Rights on the Rights of Women in Africa, only three Member States have reported their progress. Again, only 13 Member States have submitted regular annual reporting on the Solemn Declaration of Gender Equality in Africa for the tenth annual report in June 2015 (Abdulmelik 2016).

More importantly, mainstreaming gender into the media is a promise greatly overlooked in AU policies and practices. The subject appears to gain little attention in AU policies toward gender equality and women empowerment (see section "Gender and the Media in AU Gender Architecture"). In fact, the nexus between gender and the media has gained a far better consideration in the RECs protocols like the SADC Protocol on Gender and Development (2008) and the Supplementary Act Relating to Equality of Rights between Women and Men for Sustainable Development in the ECOWAS Region (2015). These two documents cover the major challenges in gender and the media, gender stereotyping in media content, professional discrimination in the media industries, and unequal access to new ICTs, and propose specific action plans (SADC 2008; ECOWAS 2015).

At large, the AU progress to promoting gender equality in media has been painfully slow. According to "the Twenty-Year Review of the Implementation of the Beijing Declaration and Platform for Action (BPfA) +20: African Regional" report, only 22 Member States, less than half of the 51 countries which submitted progress reports, have shown significant advancement in formulating legal frameworks and strategies to increase the participation and access of women to expression and decision-making in the media (United Nations Economic Commission for Africa 2014). Progress toward the implementation of the legal instruments, however, is not indicated. In addition, any effort by the AU in pursuing Member States to fostering gender equality in the media is invisible in the report. Yet, following this review, the AU has reaffirmed its commitment to addressing the problem with gender and the media in the Addis Ababa Declaration on Accelerating the Implementation of the Beijing Platform for Action (2014).

Moreover, initiatives and dialogues on gender and the media supported by the AU are far too few and focus on gender and the media with relation to peace and security. A typical example is the "Network of Reporters on Women, Peace and Security" launched by AU Commission and UN Women in 2016. This initiative, taking the media as key partners in advancing women and peace and security agenda, seeks to empower journalists to respect the dignity of women and recognize women's contributions to social cohesion and lasting peace in their communities. The network has further published a handbook on "Practicing Gender-Responsive Reporting in Conflict Affected Countries in Africa" in July 2017. Regardless of the potential contribution of the network to a fair representation of women in the media at times of conflict, its scope is limited to women's role in peace and security matters and ignores their complex challenges in a day-to-day reality. More similar initiatives on gender and the media that address women in social, political, and economic contexts are missing. This could restrict the role of the media in effecting a positive social change.

71.4.2 The EU Unmet Promises

The EU has played a pivotal part in the advancement of international agenda such as gender equality, climate change, and peace and security. To this end, it has established and funded a number of development programs in Africa and other developing regions. Given the key role the EU has played in global efforts to social change and development, its progress as well as its failure to achieving gender equality as a whole and mainstreaming gender into the media in particular could have significant impact on a global success on the matter.

In addition, in view of the growing trend toward globalization of media and communication, and considering the EU greater contribution to global media contents, the ways in which gender equality is addressed in the European media would be relevant beyond the region.

It is, therefore, with this consideration that EU unmet promises in the practicability of gender mainstreaming with specific emphasis on mainstreaming gender into the media are explored.

Although the EU has gained a much greater achievement in the advancement of gender equality and women empowerment with relation to the AU, it cannot be regarded as a perfect reference point of success, particularly in implementing gender mainstreaming in its policies and practices (Tomlinson 2011; O'Connor 2014; Weiner and MacRae 2014). Tomlinson (2011, p. 3755) argues that "despite strong statements on the need to eliminate gender inequality, the EU's progress in this area is contested." The EU has regularly drafted a number of laws and strategic documents to promoting gender equality, albeit its accomplishment is slow and limited (Weiner and MacRae 2014). Weiner and MacRae (2014) further argue that gender equality is invisible in EU policies, and a "revolutionary change" is far from attainable. Similarly, O'Connor (2014, p. 72) contends that "much of the progress in the EU and elsewhere has been strong at the discursive level in terms of gender justice commitments, but relatively weak in outcome." According to the 2017 Gender Equality Index, gender gaps persist in many areas and in the labor market where women are still overrepresented in lower paid sectors and underrepresented in decision-making positions. With an average score of 66.2 for gender equality, the EU still has a long way to go to bring a fundamental social change in terms of gender parity (Barbier et al. 2017).

Mainly, the EU has come short in ensuring the practicability of gender mainstreaming (Bretherton 2001; Tomlinson 2011; O'Connor 2014; Weiner and MacRae 2014). As Pollack and Hafner-Burton (2000, p. 434) note, the concept of gender mainstreaming is "potentially revolutionary" which calls for the incorporation of gender perspectives into all EU policies. The European Commission (EC), which adopted a formal commitment to gender mainstreaming in 1996, defines the concept as:

> The systematic integration of the respective situations, priorities and needs of women and men in all policies and with a view to promoting equality between women and men and mobilizing all general policies and measures specifically for the purpose of achieving equality by actively and openly taking into account, at the planning stage, their effects on the respective situation of women and men, implementation, monitoring and evaluation. (Commission of the European Communities 1996, p. 2)

In practice, gender mainstreaming is, however, an extremely demanding approach which needs the adoption of gender dimensions by all central actors in the policy process (Pollack and Hafner-Burton 2000; Bretherton 2001). The success of gender mainstreaming would require internalizing its principles and practices "in all aspects, and at all levels, of EU policy process" (Bretherton 2001, p. 61).

Hence, institutionalizing the principles of gender mainstreaming in all EU policies and practices has become a challenging task. Discussing the implementation of gender mainstreaming, Weiner and MacRae (2014, p. 2) note that although there are "small, often hard-won" accomplishments, they are usually "outnumbered by several instances which block out gender mainstreaming partially or completely from EU policies." To illustrate the futile struggle of the EU in integrating gender into its policies, particularly in implementing gender mainstreaming, Weiner and MacRae use the story of Sisyphus. In the story, Sisyphus, a character in *the Odyssey*, condemned by the gods, pushes vainly a massive rock uphill, yet to find the rock always rolling back to where he started. The authors argue that like Sisyphus's rock, "rolling gender mainstreaming into EU policies either ends up rolling back out of policy, or never rolls in at all" (Weiner and MacRae 2014, p. 4). To this day, excluding gender aspects from many of EU policies mainly in the European Employment Strategy (EES) remains a challenge where "the gender equality goal has been instrumentally subordinated to other agendas" (Fagan and Rubery 2017, p. 7).

Above all, the EU has not succeeded in mainstreaming gender into the media. Although the nexus between gender and the media entered the EU agenda well before the BPfA, with the adoption of Council of Europe's Recommendations on Equality of Men and Women and its Guidelines for Television Advertising in 1984 (Byerly and Padovani 2017), the overall progress to integrating gender perspectives into the media has been slow (Sarikakis and Nguyen 2009; Gallagher 2011; Byerly and Padovani 2017; Ross and Padovani 2017).

Following the BPfA, the EU has adopted a number of documents regarding gender equality in the media (Byerly and Padovani 2017), yet these policies lack gender-specific provisions on the conduct of the media (Sarikakis and Nguyen 2009; Byerly and Padovani 2017). Almost a decade ago, Sarikakis and Nguyen (2009) have highlighted the ways in which EU media policy lacked explicit provisions on fair representation of women. As such, in the most significant legal instrument for the European audiovisual (AV) sector, the Television without Frontiers Directive (TVWF), "there was a timid reference that advertisements may not infringe human dignity or contain discrimination on grounds of sex" (p. 209). The authors argue that such general allusion had limited the directive legal impact.

Similarly, Gallagher (2011) has discussed the lack of specificity in other EU media policies. She explains that in two recommendations adopted by the Council of Europe in 2007 on media pluralism, a general phrase "Due attention should also be paid to gender equality issues" is included in each case at the end of a paragraph. In addition, a similar phrase, "Gender-related issues should also be mainstreamed with regard to these services," is found in the Political Declaration adopted at the first Council of Europe Conference of Ministers responsible for Media and New Communication Services (2009). However, she argues that there is "no indication of what these issues are, how they could be 'mainstreamed' or even how they relate to the substance of the policy documents" (p. 456).

The absence of explicit provisions to gender equality in the media is also observed in the recent EU ICT strategy known as the Digital Agenda for Europe which makes a general reference to the gender implications of digital transformations (Byerly and Padovani 2017). Byerly and Padovani contend that "the framing of gender equality in the digital context seems to be very narrow" and emphasizes the economic implications of involving women in ICTs (p. 15). Regarding the limited scope of the strategy, Padovani (2016) notes "more general concerns with gender social justice, persistent discrimination, and the gender implications of the unequal power relations in an evolving media environment are missing, demonstrating that gender mainstreaming in this domain is still unfinished job" (in Byerly and Padovani 2017, p. 15).

Furthermore, EU media policy provisions for gender equality in media content has been of "a 'soft' nature, referring to the media's 'contribution' to non-sexist portrayal and not to their obligation" (Sarikakis and Nguyen 2009, p. 209). On the persistence of this trend, Byerly and Padovani (2017, p. 18) state that "the 'soft' nature of policy provisions can still be found in the reference to codes of conduct and self-regulation as the main means through which the media sector can integrate normative guidelines in their operation, rather than legislation."

Notably, as Gallagher (2011, p. 456) argues "conflict within EU institutions between gender equality policy and media policy is clear." Nonetheless, reconciling the two seems to be the EU unaccomplished mission. As a whole, the media sector has remained reluctant to change (Ross and Padovani 2017).

71.5 Gender and the Media in AU Gender Architecture

The AU has made a commitment to address obstacles to gender equality in and through the media with the adoption of the Beijing Declaration. The Platform has spelled out two strategic objectives (Section J) to tackle the problem (BPfA 1995):

- To increase the participation and access of women to expression and decision-making in and through the media and new technologies of communication
- To promote a balanced and non-stereotyped portrayal of women in the media

To this end, the Platform calls both governments and international organizations to facilitate the implementation of these strategies to the extent it is consistent with freedom of expression. In what follows, to what degree the BPfA strategic objectives and action plans are translated into AU gender architecture is discussed.

The AU gender architecture encapsulates the Union aspirations of gender equality and is comprised of six pillars (Martin 2013):

1. The Consecutive Act of the AU – the constitutional framework
2. The Protocol to the African Charter on Human and People's Rights on the Rights of Women in Africa – the legal framework
3. The Solemn Declaration on Gender Equality in Africa – the reporting framework
4. The African Union Gender Policy – the policy framework
5. The African Women's Decade – the implementation framework
6. The Fund for African Women – the financial mechanism

The Constitutive Act of the AU, adopted in 2000, sets out the framework under which the AU is to conduct itself (African Union 2000). Under Article 4(l), the Act recognizes the "promotion of gender equality" as one of its principles. Based on this reference to gender equality, Martin (2013, p. 11) argues that "creating the necessary mechanisms for the promotion of gender equality is therefore an important aim of the AU." Nevertheless, the Constitutive Act lacks specific provisions on gender mainstreaming issues and strategies as it has mentioned the principle of gender equality very briefly. In fact, the terms "gender" and "women" are each mentioned once throughout the document. Although the Act "represents a significant departure from the 1963 OAU Charter-in particular by including the 'promotion of gender equality' as one of its foundational principles," it excludes a similar provision from its list of objectives and makes use of a gender-insensitive term – "chairman" (Viljoen 2009, p. 13). This elicited the adoption of the Protocol on Amendments to the Constitutive Act of the African Union on July 2003 in which the inclusion of the objective to "ensure the effective participation of women in decision-making particularly in the political, economic and socio-economic areas" and the substitution of the word "chairman" with "chairperson" were proposed (Viljoen 2009).

As the Act has not discussed gender issues in detail, no indications are found regarding mainstreaming gender into any sectors including the media.

The second key AU gender equality framework is the 2003 Protocol to the African Charter on Human and People's Rights on the Rights of Women in Africa (the Maputo Protocol). This is the legal framework which requires Member States to integrate gender in legal and policy frameworks, strategies, programmers, and development activities. Winyi (2009) calls the protocol as "the first African instrument to explicitly articulate the rights of women in Africa."

With regard to the treatment of gender and the media in this protocol, Article 12(b) calls for "the elimination of all stereotypes in textbooks, syllabuses and the media that perpetuate all forms of discrimination against women." In addition, Article 13(m) urges States Parties to "take effective legislative and administrative measures to prevent the exploitation and abuse of women in advertising and pornography." However, this phrase does not elaborate on what actions fall under "exploitation and abuse," leaving the matter for open interpretation. In addition, the provisions address merely non-stereotypical portrayal of women yet fail to include other aspects of problems surrounding gender and the media. There is no any indication to women media decision-making roles or their unequal access to ICTs. Again, gender in the media receives insignificant consideration in the protocol compared to the focus given to gender in relation to issues like health, peace and security, marriage, justice, political and decision-making, education, economics and social welfare, food security, cultural context, and sustainable development, which are discussed in more comprehensive manner.

Talking about the little attention that women and communications issues, in general, had received during the UN Decades (1975–1985) and even until the 1990s, within both the UN system and the international women's movement, Gallagher (2011, p. 453) notes that communications issues were regarded "as secondary in importance to problems such as poverty, health, and education for women." This problem persists in this protocol where gender with relation to peace and security, poverty, health, and education appears to be key priority, whereas gender in the media is considered as an afterthought.

The subsequent instrument is the 2004 Solemn Declaration on Gender Equality in Africa. It is a non-binding reporting framework committing Member States to report annually on their progress in gender mainstreaming.

This declaration acknowledges the existing digital divide between men and women in the continent and the role of ICTs in accelerating gender equality. Yet, no action plan to address the digital divide is mentioned. Further, women's representation in media contents and decision-making roles is not indicated at all. In terms of major areas of concern where gender inequality needs to be addressed, like the previous protocol, this document spells out health, particularly HIV/AIDS, peace and security, child soldiers and abuse of girl children, gender-based violence, and property rights. Despite the declaration reference to the BPfA in informing its direction, gender inequality in and through the media is not among the declaration's list of concerns.

Another key document in AU gender architecture is the AU Gender Policy which was approved in 2009 and adopted in 2010. The policy provides "the basis for the elimination of barriers to gender equality and fosters the reorientation of existing institutions" (Martin 2013, p. 14).

The Gender Policy recognizes the role of the media in accelerating gender equality and women empowerment. It identifies women's equal access to ICTs as a key area of concern. It also puts "the participation of the media" under its rationale. The media are also identified as part of the institutional frameworks in the implementation of the policy.

However, although there is a phrase "Elimination of gender stereotypes, sexism and all forms of discrimination" under the policy rationale, no indication is made to "the media." Moreover, with regard to the role of the media as part of institutional frameworks for the implementation of the Gender Policy, they are mainly considered as facilitators for mainstreaming gender into other sectors. Nonetheless, the policy fails to regard the media as sectors which would require the integration of gender dimensions themselves. Hence, while the Gender Policy identifies ten "key issue sectors" of development where gender needs to be mainstreamed into with the view to achieving gender equality and women empowerment, the media are not among them.

It may be argued that mainstreaming gender into the media may be encapsulated in one of the ten sectors, for instance, under "Social Affairs" or under "Science and Technology." However, bearing in mind media's huge potential in challenging gender inequality, and the recognition gender and the media gained in the Beijing Platform, mainstreaming gender into the media deserves much greater consideration.

The fifth pillar of the gender architecture is the African Women's Decade which was launched in 2010 with the aim to accelerate the implementation of Dakar, Beijing, and AU Assembly decisions on gender equality and women empowerment. It is based on ten priority themes targeting to empower women across Africa through dual top-down and bottom-up approach which is inclusive of grassroots participation.

When it comes to the place of gender and the media among the ten African Women's Decade themes, the framework appears to give no considerable thought to the subject. The only indication close to the issue is the phrase "contribution of Women Scientists and Information, Communication and Technology" under theme 4 – Education, Science and Technology. However, the reference is too general and vague to clearly show what needs to be addressed regarding women and ICTs. Such lack of clarity may undermine efforts taken to face up challenges in women's unequal access to ICTs. This can be observed in the Mid-Term Status Update Report of the African Women's Decade (2010–2015) in which only one country, Rwanda, has reported measures taken to promote women and girls' access to ICTs.

Complementing the African Women's Decade Road Map, the Fund for African Women was launched in 2010. This framework aims at providing financial support for African women development programs. It benefits women and development programs in African Union Member States, RECs, and African Civil Society Organizations. To enhance the implementation of the Women's Decade, the Fund for African Women, each year, supports a minimum of 53 projects related to 1 theme from the 10 Decade themes identified by AU Ministers of Gender and Women's Affairs.

As the projects supported by the Fund are selected among the ten themes of the Decade, the issue of gender in the media, which is excluded from the key themes, may find it hard to benefit from the Fund for African Women. This, in turn, could restrict the communicative space of initiatives on promoting gender equality in and through the media.

In general, the intersection of gender and the media is not a well thought through agenda in AU gender architecture where topics like health, peace and security, education, gender-based violence, property rights, economic, and justice have assumed key places. Except for some occasional references and recommendations on gender and the media, a genuine concern and concrete strategies are invisible. These provisions have failed to reflect the strategic objectives and action plans set in the BPfA.

The invisibility of the issue of gender and the media in these key official gender equality mechanisms of the AU could affect progress toward promoting gender equality in the African media and the potential role of the media in empowering African women since addressing gender inequality in and through the media is "clearly dependent on policy determinations" (Gallagher 2011, p. 452).

71.6 Conclusion

Much is happening but much remains to be done with regard to achieving gender equality in Africa. Despite the potential that exists for the media to accelerate this process, the AU has failed to embrace mainstreaming gender into the media as part of its mission. Promoting gender equality in the media has not been taken seriously within the broader policy framework to push forward on the equality agenda. The subject is considered as an "add-on" and treated second or last in importance to matters such as health, peace and security, education, and economic. For a continent with the highest maternal mortality rate, frequent armed conflicts, and extreme poverty, putting health, peace and security, and the economy as key areas of concern is an undisputable measure. However, given the pivotal role of the media as opinion shapers, there should still be a room left for fitting the agenda with the rest of AU priorities.

At the same time, mainstreaming gender into the media has also been the EU unfinished job. Despite being the AU major donor provider and a role model for designing and implementing gender equality mechanisms, the EU has lacked a best practice to share in this area. At large, the EU has been slow in making specific provisions to make gender equality in the media a reality. Although its effort to advance gender equality in and through the media started even before the BPfA, no substantial change is yet made in this context. This may demonstrate a troubling possibility of a global failure with regard to mainstreaming gender into the media which in turn can pose a serious threat to taking any meaningful steps to social change and development. When the media with their powerful influence on the public opinion regarding gender and gender equality and ICTs with their vital roles in empowering and advancing women remain gender insensitive, the status quo would be maintained, delaying the realization of gender equality.

To face up this challenge, the AU needs to set up strategies that address the problems in gender and the media. These strategies should provide clear and specific provisions on combatting unfair and stereotypical representation of men and women, women limited decision-making roles, and the gender digital divide. Here, the SADC Protocol on Gender and Development (2008) and the Supplementary Act Relating to Equality of Rights between Women and Men for Sustainable

Development in the ECOWAS Region (2015) can serve as good reference points. More importantly, as developing policies alone could not bring a substantial change in promoting gender equality in the media, emphasis should be placed on their implementation and evaluation.

With regard to the EU, since the existing strategies and initiatives have been slow to fostering gender equality in the media, more specific policy interventions could still be needed. Further, to make gender equality in and through the media a global reality, the EU financial and technical support for the AU on promoting gender equality in the media could be elemental in advancing the progress.

References

African Union (2000) Constitutive Act of the African Union. 11 July 2000. Lome, Togo. https://au.int/sites/default/files/pages/32020-file-constitutiveact_en.pdf

African Union (2003) Protocol to the African Charter on the Rights of Women in Africa. 11 July 2003. Maputo. https://au.int/sites/default/files/treaties/7783-treaty-0027_-_protocol_to_the_african_charter_on_human_and_peoples_rights_on_the_rights_of_women_in_africa_e.pdf

African Union (2004) Solemn Declaration on Gender Equality in Africa. 6-8 July 2004. Addis Ababa, Ethiopia. http://www.un.org/en/africa/osaa/pdf/au/declaration_gender_equality_2004.pdf

African Union (2009) African Union gender policy. Addis Ababa, Ethiopia. http://www.un.org/en/africa/osaa/pdf/au/gender_policy_2009.pdf

African Union (2010) African Women's Decade. Nairobi, Kenya. http://www.un.org/en/africa/osaa/pdf/events/2018/20180315/TheAfricanWomen.pdf

African Union Commission (2015) Agenda 2063: the Africa we want. http://www.un.org/en/africa/osaa/pdf/au/agenda2063.pdf

Abdulmelik S (2016) Implementation of the Women, Peace, and Security Agenda in Africa. African Union Commission, Addis Ababa

Barbier D et al. (2017) Gender Equality Index 2017: Measuring gender equality in the European Union 2005–2015. Main Findings. European Institute for Gender Equality. http://eige.europa.eu/rdc/eige-publications/gender-equality-index-2017-measuring-gender-equality-european-union-2005-2015-main-finding

Bhagwan-Rolls S (2011) Pacific regional perspectives on women and the media: making the connection with UN Security Council Resolution 1325 (women, peace, and security) and section J of the Beijing platform for action. Signs J Women Cult Soc 36:570–578

Bretherton C (2001) Gender mainstreaming and EU enlargement: swimming against the tide? J Eur Public Policy 8:60–81. https://doi.org/10.1080/13501760010018331

Byerly C (2014) The long struggle of women in news. In: UNESCO (ed) Media and gender: a scholarly agenda for the Global Alliance on Media and Gender. UNESCO, Paris, pp 40–43

Byerly C, Padovani C (2017) Research and policy review. In: Ross K, Padovani C (eds) Gender equality and the media, 1st edn. Taylor & Francis, New York

Collado C (2014) Women's access to ICTs in the information society. In: Montiel A (ed) Media and gender: a scholarly agenda for the Global Alliance on Media and Gender. UNESCO, Paris, pp 60–65

Commission of the European Communities (1996) Communication from the Commission: Incorporating Equal Opportunities for Women and Men into All Community Policies and Activities. Brussels http://aei.pitt.edu/3991/1/3991.pdf

Debusscher P, van der Vleuten A (2012) Mainstreaming gender in European Union development cooperation with sub-Saharan Africa: promising numbers, narrow contents, telling silences. Int Dev Plan Rev 34:319–338. https://doi.org/10.3828/idpr.2012.19

Economic Community of West African States (2015) The supplementary act on equality of rights between women and men for sustainable development in the ECOWAS region. Accra. http://www.ccdg.ecowas.int/wp-content/uploads/Supplementary-Act-on-Gender-Equality.pdf

Fagan C, Rubery J (2017) Advancing gender equality through European employment policy: the impact of the UK's EU membership and the risks of Brexit. Soc Policy Soc 1–21. https://doi.org/10.1017/S1474746417000458

Gallagher M (2008) Feminist issues and the global media system. In: Sarikakis K, Shade L (eds) Feminist interventions in international communication: minding the gap. Rowman & Littlefield Publishers, Lanham, pp 17–33

Gallagher M (2011) Gender and communication policy: struggling for space. In: Mansell R, Raboy M (eds) The handbook global media communication policy. Blackwell, West Sussex, pp 451–466. https://doi.org/10.1002/9781444395433.ch28

Gender Links (2017) Media portrayal of women and media gender gap in Africa. Paper prepared by Gender Links for the African Union Specialized Technical Committee on Information and Communications (STC-IC)

González OM (2017) The role of African institutions in promoting gender equality and the Political empowerment of women. In: Regional organizations, gender equality and the political empowerment of women. International Institute for Democracy and Electoral Assistance (International IDEA). New York, pp 11–31

Grizzle A (2014) Enlisting media and informational literacy for gender equality and women's empowerment. In: Montiel A (ed) Media and gender: a scholarly agenda for the Global Alliance on Media and Gender. UNESCO, Paris, pp 93–108

International Telecommunications Union (2014) World Summit on The Information Society WSIS +10 High-Level Event: WSIS +10 outcome documents. Geneva. http://www.itu.int/net/wsis/implementation/2014/forum/inc/doc/outcome/362828V2E.pdf

Joint Africa-EU Strategy (2007) The Africa-EU Strategic Partnership: A Joint Africa-EU Strategy. Lisbon, Portugal. https://www.africa-eu-partnership.org/sites/default/files/documents/eas2007_joint_strategy_en.pdf

Joshua Omotosho B (2015) African Union and gender equality in the last ten years: some issues and prospects for consideration. J Integr Soc Sci 5:92–104. www.JISS.org

Kareithi PJ (2014) Africa: outdated representations continue to deny women autonomy of their "personhood". Fem Media Stud 14:334–337

Lourenço ME (2016) Gender equality in media content and operations: articulating academic studies and policy – a presentation. Stud High Educ 41:927–931. https://doi.org/10.1080/03075079.2016.1147726

Lowe Morna C (2002) Promoting gender equality in and through the media. A Southern African case study. In: UN round table expert 10 group meeting on "participation and access of women to the media, and its impact on and use as an instrument for the advancement and empowerment of women"

Martin O (2013) The African Union's mechanisms to foster gender mainstreaming and ensure women's political participation and representation. International Institute for Democracy and Electoral Assistance (International IDEA), Stockholm

O'Connor JS (2014) Gender mainstreaming in the European Union: broadening the possibilities for gender equality and/or an inherently constrained exercise? J Int Comp Soc Policy 30:69–78. https://doi.org/10.1080/21699763.2014.888012

Pollack MA, Hafner-Burton E (2000) Mainstreaming gender in the European Union. J Eur Public Policy 7:432–456. https://doi.org/10.1080/13501760050086116

Ross K (2014a) Women in decision-making structures in media. In: Montiel A (ed) Media and gender: a scholarly agenda for the Global Alliance on Media and Gender. UNESCO, Paris, pp 44–48

Ross K (2014b) Women in media industries in Europe: what's wrong with this picture? Fem Media Stud 14:326–330

Ross K, Padovani C (2017) Gender equality and the media. Taylor & Francis, New York

Southern African Development Community (2008) SADC protocol on gender and development. 17 January 2008. Johannesburg, South Africa. https://www.sadc.int/files/8713/5292/8364/Protocol_on_Gender_and_Development_2008.pdf

Sarikakis K, Nguyen ET (2009) The trouble with gender: media policy and gender mainstreaming in the European Union. J Eur Integr 31:201–216. https://doi.org/10.1080/07036330802642771

The African Union Commission (2017) African Union handbook 2017. African Union Commission and New Zealand Crown. Addis Ababa https://au.int/sites/default/files/pages/31829-file-african-union-handbook-2017-edited.pdf

Tomlinson J (2011) Gender equality and the state: a review of objectives, policies and progress in the European Union. Int J Hum Resour Manag 22:3755–3774. https://doi.org/10.1080/09585192.2011.622923

UNDP (2016) Africa human development report 2016 accelerating gender equality and women's empowerment in Africa. United Nations Development Programme, New York, p 192

United Nations (1995) Beijing Declaration and Platform for Action. United Nations. Beijing. http://www.un.org/womenwatch/daw/beijing/pdf/BDPfA%20E.pdf

United Nations Economic Commission for Africa (2014) Twenty-year review of the implementation of the Beijing Declaration and Platform for Action (BPfA) + 20. In: Africa regional review summary report. United Nations Economic and Social Council, Addis Ababa

Viljoen F (2009) An introduction to the protocol to the African Charter on Human and Peoples' Rights on the Rights of Women in Africa. Wash Lee J Civ Rights Soc Justice 16. https://doi.org/10.5897/JASD12.027

Webb A (2016) Information and communication technology and other. J Inf Policy 6:460–474

Weiner E, MacRae H (2014) The persistent invisibility of gender in EU policy: introduction. Eur Integr Online Pap 18:1–20. https://doi.org/10.1695/2014003

Williams T (2000) Gender, media and democracy. Round Table 89:577–583. https://doi.org/10.1080/003585300225205

Winyi NM (2009) The Beijing Platform for Action: what has it delivered to African women? Pambazuka News. Retrieved from https://www.pambazuka.org/gender-minorities/beijing-platform-action-what-has-it-delivered-african-women

World Association for Christian Communication (2015) Who Makes the News?: The Global Media Monitoring Project 2015. http://www.5050foundation.edu.au/assets/reports/documents/gmmp-global-report-en.pdf

World Summit on the Information Society (WSIS) (2003) Declaration of Principles. Building the Information Society: A Global Challenge in the New Millennium. Geneva. http://www.itu.int/net/wsis/docs/geneva/official/dop.html

Zubair S (2016) Development narratives, media and women in Pakistan: shifts and continuities. South Asian Pop Cult 14:19–32. https://doi.org/10.1080/14746689.2016.1241348

Idiosyncrasy of the European Political Discourse Toward Cooperation

72

Teresa La Porte

Contents

Abstract

Development and cooperation policies are located at the core of the European Union foreign policy. Not for nothing, the European Union, together with its Member States, is considered the world's largest aid donor. Despite multiple challenges and failures experienced along these 60 years, the European Union strives to avoid to be blocked by past setbacks or by unresolved current conflicts and envisions new solutions to keep on going: the last initiative is the New Consensus on Development which attempts to align the European Union and the development policies of Member States with the development sustainable goals defined by the United Nations in 2015. This chapter examines the narratives

T. La Porte (✉)
International Political Communication, University of Navarra, Pamplona, Navarra, Spain
e-mail: mtalfaro@unav.es

J. Servaes (ed.), *Handbook of Communication for Development and Social Change*,
https://doi.org/10.1007/978-981-15-2014-3_83

used to launch this project so that the unique of the European political discourse toward cooperation can be identified, that is, the EU's specific contribution to the global discussion in terms of concepts, approaches, and aid management.

Keywords

European development policy · Sustainability · Migration · Partnership · Securitization · National ownership

72.1 Introduction

The European Union Foreign Affairs Council adopted on May 19, 2017, the New European Consensus on Development, a document that reflects the shared opinion that the European Union and the Member States have on the topic of European development until 2030. The Consensus was officially signed on June 7, 2017, by the Presidents belonging to the European Parliament, the Council of the European Union, and the European Commission and also by the High Representative of the Union for Foreign Affairs and Security Policy, which demonstrates the unanimous commitment from the different EU institutions to promoting the established priorities on the field of development (European Union 2017). The document updates the previous 2005 Consensus on Development and aims to define an EU framework for action in line with the Sustainable Development Goals approved by the United Nations on January 2015 (United Nations 2015). The objective is thus to establish the principles for a European development policy that further develops the concept of "sustainability" by applying the measures set out in the 2030 Agenda.

The new agreement has generated high expectations within the development community because they hope it will help solve endemic problems in the European development action. The development NGOs trust that the agreement is able to address issues that have been marginalized, such as the defense of women's rights or the participation of civil society groups (CARE 2016); the development agencies of the EU Member States expect an improvement in the policy coordination of the European Union (2012); the beneficiary countries seek that the new measures support the already signed agreements (Africa, the Caribbean, and the Pacific countries (ACP)); and the think tanks encourage the European Union to regain global leadership on development policy (Gavas et al. 2016).

However, the text has aroused controversy from the moment it was released. In the context of the refugee crisis, the recently signed Valletta agreement with Turkey, and the new EU Emergency Trust Fund for Africa, the European Union has been accused of prioritizing its protection from massive migration flows over the fight against inequality and poverty.

Irrespective of whether the agreement fulfills its proposed expectations, the Development Consensus allows to identify the specific contribution that the European Union makes to global development cooperation policy. The raised controversy justifies the interest to reflect upon the pillars that sustain the

European Union perspective on development and to analyze the narratives the Union utilizes to frame it.

This chapter will first address the global and European context in which the Consensus has been approved, taking into account those features that define the European concept of development and the legal and financial structure through which cooperation policies are articulated. Secondly, an analysis of the content of the Consensus will be carried out, and, following this, the positive and negative evaluations expressed by the different development agents will be examined. The conclusion of the chapter will summarize the main European contributions to global development policy and the most significant challenges the European Union must face in order to guarantee effective aid.

72.2 Context of the New Consensus: Understanding the Relevance of the European Development Policies

In order to assess the potential impact of the new document, it is necessary to briefly consider some ideas in order to determine the influence of the European Union in the global picture of development cooperation policy and identify the values, structures, and dynamics that define the European Union contribution to development cooperation.

72.2.1 The European Union as a Global Agent for Development: External Relevance

The European Union and the Member States are collectively the world's largest aid donor. In 2015, only the EU Commission invested 10.3 billion euro in Official Development Assistance (ODA) and more than 1.4 billion euro in Emergency Aid. In addition, together with the Member States, the European Union made the commitment to ensure that 0.7% of the gross national income was allocated to this purpose, following the deadlines established in the 2030 UN Agenda. In parallel, the 11th European Development Fund has approved an aid budget of 30.5 billion euro for the 2014–2010 period.

The European Union has a long experience in cooperating and collaborating with deprived communities and has significantly contributed to the innovative vision of the current concept of development. Europe synthesizes its development cooperation in two statements: aid provider and values promoter. In fact, along with its outstanding role as a donor, the specificities of the European concept of development are perceived in the priority that its policies give to values that are part of its idiosyncrasy, such as the respect for human rights, fundamental freedoms, peace, democracy, good governance, gender equality, the rule of law, solidarity, and justice (TEU, Article 21, paragraph 1 and "Agenda for Change" of the Union approved by the Council of the Union 2012).

Other international organizations also recognize the significant European influence on this matter. According to the last OECD evaluation, European development policies maintain a positive improving trend, despite the fact that they retain chronic weaknesses and despite the slight aid budget reduction during the last 3 years: "The DAC's Review of the Development Co-operation Policies and Programs of the European Union notes that, since the last review 5 years ago, the European Union has taken steps to make its aid more effective and give it more impact. These steps included organizational restructuring, streamlining the financial process, improving co-ordination, and working more with civil society" (OECD 2016).

The EU action contrasts with that of other major powers such as China and Russia, whose investments and interests in Africa are being criticized for the dubious effectiveness of their actions and the absence of human rights promotion. By contrast, as argued by some academic papers on the field, the recent inclusion of development objectives in the diplomatic actions of the European External Action Service allow to expect a more multidimensional and coherent collaboration of Europe in Africa (Constantinou and Opondo 2016).

To conclude, it is worth mentioning the appreciation made by the European Think Tank Group which considers that the integration of the development policy within the framework of the Global Strategy on Foreign and Security Policy is a unique opportunity for the European Union to position itself as a global development leader, since this context provides the opportunity to jointly address the diverse threats to security along with the challenges to sustainable development (Gavas et al. 2016).

72.2.2 Uniqueness of the European Notion on Development

The features that define the European perspective on development do not belong to it exclusively, but it can be asserted that the European principles on the matter were adopted earlier than in other institutions and organizations and that their complementarity and interaction give them a unique nature. The perspective adopted in this section is the one contained in the official legal documents and in political discourse, unfortunately not always present in the specific actions that have taken place on the ground.

Development in Europe is understood in an all-encompassing way, that is, recognizing the interdependence between the various aspects that give rise to poverty traps. Although the growing emphasis on "sustainability" has favored the deepening of this aspect, the European Union has always comprehensively considered the economic, human, social, and political aspects of development. Good governance achievements have always been a constant in the European Union's policies (Agenda for Change 2012).

Among the aspects that form the comprehensive European perspective on development, the European Union has shown a special sensitivity to the human dimension of the concept over other more structural or macroeconomic facets (European Union 2012, 2016, 2017). In fact, the European Union has not hesitated to focus its actions on the most vulnerable groups. The relevance granted to education, health, or food security, especially present in the programs of the 1990s, has been reinforced in the new century by its implementation on areas of a more personal nature, such as job satisfaction, the promotion of artistic expression, the education of girls, or the preservation of indigenous culture.

As a consequence, the concept of European development is essentially inclusive. Not only does it consider together the diverse causes of poverty, but it also attempts to involve all the agents that, in one way or another, are implicated in cooperation processes: European organisms, states, civil society, donors, and beneficiaries. The very structure of the European Union, articulated through various institutions, although sometimes complicates the processes and generates dysfunctions, facilitates the diversity of the participants.

The inclusive nature of development objectives implies a constant effort to advance in the coordination among agents and the coherence among policies. This effort can be seen in the recent texts on security and defense (Capacities Development in support of security and development 2015), actions in favor of good governance, the rule of law, and the defense of human rights that are initiatives of the Council (Strategic Framework on Human Rights and Democracy 2012) or of the European External Action Service (new Action Plan for Human Rights and Democracy for the period 2015–2019), in measures to address climate change (Paris Agreement 2015) or in the cross-cutting gender equality actions that focus on the stability and growth of the associated countries. Also, this attempting to improve coordination and coherence, can be verified in the policy coherence commitments that the European Union has regularly adopted and that are included in the next section of this chapter.

Participation is another defining element of European development policy (European Union 2016). In line with this inclusive perspective on development that drives its coordinated and coherent actions, the European Union seeks to determine the mechanisms that favor the real participation of all the actors involved. Participation is firstly considered in the relationship with the beneficiary partners. The European Union has been increasingly insisting on the fact that national governments are primarily responsible for the development of their countries and that they should partner with their local institutions and with their civil society since they are considered closest collaborators to face the necessary measures to contribute to the development of the nation. They are also deemed to be the ones charged with identifying their needs and the means to satisfy them (principle of "joint programming"). In this context, the beneficiary country should take the lead as far as possible, and international aid should solely act subsidiarily. It is also presupposed that local entities should also assume the obligation to evaluate the impact of their efforts and should be held accountable for the investment received.

Participation is also contemplated in relation to European agents. In recent years, the opinion of civil society groups has been incorporated into the institutional discussion, represented through trade unions, research centers, business organizations, and the private sector (A Stronger Role of the Private Sector in Achieving Inclusive and Sustainable Growth in Developing countries, COM (2014) 263 final) and, above all, NGOs specialized in development and in environmental defense. The intervention of the private sector is increasingly relevant, not only as a source of funding but also as a source of knowledge and experience, transferring know-how and "on the ground" learning processes.

Lastly, and in line with the global perspective on the topic, the European concept of development is essentially sustainable. The European Union understands sustainable development as "development that meets the needs of the present without compromising the possibility of future generations to fulfill theirs" (EU Strategy for Sustainable Development; European Commission 2007, p.6–7). The Union has added to the basic objectives of sustainability (those related to the preservation of the environment) those referring to trade, promoting projects such as "marketing with cause" or "ethical banking." Europe has also underlined the type of sustainability that focuses on human resources: investment in education; professional training; openness and tolerance; interreligious coexistence, which, in the medium or long term, can favor the peaceful resolution of latent conflicts; and the empowerment of the population so they are able to manage their development creatively and autonomously in the future.

72.2.3 The Articulation and the Dynamics of European Development Policies

In summary, it can be said that the priority objectives of the European Union in terms of development have been and are the eradication of poverty (in a sustainable and integrated way, encompassing food security, the fight against HIV, the management of migration flows, and security concerns), the defense of democratic values (which includes the protection of all human rights), and the national management of aid (which entails the responsibility and leadership of the partner countries in the management and implementation of it).

The progressive evolution in the achievement of these objectives has mainly followed three criteria: first, the improvement of the coordination and coherence between the actions of the EU institutions and those of the Member States; second, the incorporation of local partners in the participation in their own development; and third, the integration of the various facets that are involved in situations of poverty.

This evolution can be seen in the legislative documents regulating the action of the European Union. All development policies are based on the same framework: Article 21, Section 1, of the Treaty on the European Union, which establishes the guiding principles of the development cooperation of the Union; Article 4, Section 4, and Articles 208 to 211 of the Treaty on the Functioning of the European Union; contained in the same document, Articles 312 to 316 cover budgetary matters. Following this, each region has its own specific cooperation agreements: the Cotonou Agreement for Africa, the Caribbean, and the Pacific countries and other bilateral agreements with partner countries from other regions, such as the Mediterranean.

The immediate antecedents of the New Consensus on Development appear in the previous Consensus (December 20, 2015) and in other documents that are enumerated next and that accurately reflect the lines of improvement of the European action:

- "Policy Coherence for Development" (2005)
- "EU Code of Conduct on Complementarity and the Division of Labor in Development Policy" (2007) and the "EU Operational Framework on Aid Effectiveness" (2011) adopt measures suggested by international organizations to improve the effectiveness of cooperation policies: OECD Paris Declaration (2005), the Accra Agenda for Action (2008), and the Busan Partnership for Effective Development Cooperation (2011).
- The "Agenda for Change" (May 2012) identifies the basic pillars of European cooperation, establishes measures to enhance impact, identifies the group of least developed countries as a priority aid objective, and introduces the principle of differentiation to adapt cooperation to the specific and real needs of each country.
- The Addis Ababa Action Agenda (AAAA 2015) is an integral part of the 2030 Agenda, dealing with an effective use of financial and nonfinancial means, complemented by the Sendai Framework on Disaster Risk Reduction (United Nations 2015–2030) and the Paris Agreement on Climate Change (United Nations 2015), which provides a multilateral rules-based global order supported by the United Nations (2016).
- The 2030 Agenda for Sustainable Development (July 2015 and November 2016) incorporates the sustainable development goals set by the United Nations and the commitment to allocate 0.7% of GNI to ODA.

The financial cooperation provided by the Union is articulated through three channels: a) projects, subsidies, and contracts granted under different conditions to organizations that propose specific actions to be developed within a certain time frame and that are awarded through public calls; b) budgetary support through transfers to the national budget of the beneficiary country, according to the needs expressed by them, and which assume the responsibility to manage and evaluate its effectiveness; c) sectoral support, which consists of financing specific industry fields (energy, education, environment, etc.) managed by each partner country and channeled through different modalities.

The aid is managed through a double approach: either through a transversal thematic approach, giving priority to the objectives proposed by the project (gender equality or access to drinking water), or through a geographical criterion, giving priority this time to the collective needs of a given area.

The new European Consensus on Development intends not only to integrate the objectives set by the UN 2030 Agenda (taking into account the significant EU contribution to its definition) but also to continue and consolidate the improving trends initiated in the new millennium and reflected in the aforementioned legislative documents. In relation to financing, it also contemplates new mechanisms to guarantee the effectiveness of the aid and the involvement of all the implicated agents. The following analysis aims to assess the specific contribution of the new European Consensus on Development.

72.3 An Analysis of the New European Consensus on Development: "Our World, Our Dignity, Our Future"

The 21-page document is divided into five different parts: The EU's response to the 2030 Agenda; A framework for action; Partnership: the EU as a force for the implementation of the 2030 Agenda; Strengthening approaches to improve EU impact; Following up with our commitments; and its content is organized through the five pillars indicated in the 2030 Agenda: people, planet, prosperity, peace, and partnership.

At the beginning, the Consensus includes the values that articulate the proposal and that reflect those that define the principles of the EU: defense of democracy, the rule of law, the universality and indivisibility of human rights and fundamental freedoms, human dignity, the principles of equality, and solidarity and respect for the norms contained in the Charter of the United Nations and international law.

Reflecting the philosophy of the new sustainable objectives, the document addresses cross-cutting elements that are cause and consequence of poverty and undertakes to give them the utmost consideration in all policies: youth; gender equality; mobility and migration; sustainable energy; good governance, democracy, and the rule of law and human rights; innovative engagement with more advanced developing countries; and mobilizing and using domestic resources.

These transversal objectives are coherent with the priorities traditionally defended by the European Union in terms of development, such as its interest for the most vulnerable population groups, the defense of the rule of law, and the importance of policies responding to the idiosyncrasy of local partners. In relation to the defense of the rights of the youth, women, and girls, although it has been a recurring theme in the EU's discourse, especially regarding health and education issues (Agenda for Change 2012), the need to ensure a prosperous education and future for them is now incorporated, so that the youth are dissuaded from radical and violent options and are able to contribute to the future of their own nation.

In the chapter on "people," the topic of migration deserves special attention. The Consensus encourages all necessary measures be taken to address the seriousness of the situation, but always respecting the Member States' right to control the entry of migrants: ... [those policies should] "not affect the right of Member States under Article 79(5) TFEU to determine volumes of admission of third-country nationals coming from third countries to their territory in order to seek work" (art 39). Although European collaboration is contemplated in the measures that can prevent forced displacement from occurring (the promotion of employment or trade and local innovation), the emphasis is placed on those actions that involve control and border management and the return of illegal migrants (Articles 40 and 41), in accordance with the provisions of the Valletta agreement signed with Turkey. Actions like this are the main reasons as to why the European Union is being criticized: it seems that the European Union's security concerns take precedence over the needs of the affected populations.

Besides being treated under the specific heading of "planet," environmental sustainability is present as a guiding principle in the provision of basic needs: the guarantee of food security and a proper water management requires a regional solution taking into account local natural resources and the interlinkages between land, food, water, and energy.

The private sector is considered as a particularly effective collaborator in the chapter on "prosperity," where the objective of generating stable employment with future prospects, especially for the young population, is valued as a priority for achieving sustainable growth. Although it is discussed in more detail below (in Article 82), it is in this section where, for the first time, consideration is given to the possibility of relying on public and private funds jointly ("blending grants and loans"). This form of financing, increasingly widespread in the framework of the United Nations, has generated suspicion on the part of those who distrust the solidarity intentions of the private sector.

The achievement of political stability and peace weights the possibility of collaborating with the security sector, following the principle of respect for the competences of the Member States and the involvement of local actors to ensure the security of their own country. Although the final document corrected a statement previously contained in the draft ("including military actors under exceptional circumstances"), this measure has also aroused protests due to the lack of reliability of some "security actors" and the risk of relying on practices that may end up violating human rights.

In accordance with the EU tradition in this respect, political stability and good governance constitute inescapable pre-conditions of any development process and demand the creation of a state based on the rule of law that guarantees human rights and that, firmly grounded on democratic principles, promotes equal and plural participation by all individuals.

The current Consensus emphasizes the participation of civil society, with which it is intended to establish a fluid dialogue and which is meant to be integrated in the design and evaluation processes of planned actions. The European Union recognizes the right to decide on its future, but also the need to respond to the commitments it entails. Therefore, together with the Member States' governments and public funds from local partners, financial and human resources from private companies, NGOs, or other citizen groups will also be employed, making them responsible from the beginning in the design of the actions and for the accountability of their results.

In the context of sustainability, the Consensus underlines the importance of using domestic resources, both human and natural, adapting actions to local possibilities. This aspect was highlighted by Mogherini in the presentation of the document:

> "So, we move from a traditional approach of donor recipient to a partnership approach in which we do things with our partners to cover all different set of fields. It is still mainly about poverty eradication and reduction but it is also covering many other fields that affect directly the living conditions of the people on the ground." Mogherini (19 May 2017)

As far as the action of the European Union is concerned, the institution utilizes the document to improve the internal management of these policies. As stated in Article 6 of the Consensus, "the purpose of this Consensus is to provide the framework for a common approach to development policy that will be applied by the EU institutions and the Member States while fully respecting each other's distinct roles and competences." Although it constitutes a EU longstanding objective, the need to avoid overlaps or repetitions takes on in the document renewed emphasis as a condition for a more effective policy and for the reduction of unnecessary costs. This coherence is also enhanced when considering the coordination with the EU external action. In this regard, it is significant that the signature of the High Representative of the Union for Foreign Affairs and Security, Federica Mogherini, is next to that of the Presidents of the Council, Commission, and Parliament.

The New Consensus on Development follows on the EU's improving trend and replicates the principles of action already included in the Agenda for Change of 2012: "differentiation," paying special attention to those countries that are more fragile and have a higher level of poverty; "concentration," focusing the action only on three priorities considered key for each local partner; "coordination" between the European Union and the Member States, both in the programming of actions and in the information on the results obtained to guarantee a complementary and mutually reinforcing action; "coherence" in the principles that shape the different development actions.

Lastly, another significant, yet hardly new, aspect should be highlighted: the European Union, as part of the principle of policy coherence, also promotes the value of sustainability in the domestic policies of Member States, understanding it as a new mentality that should permeate all social activities. An example of this included in the new Consensus is the purpose of implementing the New Urban Agenda in the legislation of European regions and municipalities.

72.4 Discussions Concerning the New Consensus

The discussion generated by the publication of the new document on European Development constitutes a good synthesis of the state of play in the European space: what are the achievements and main objectives attained, what aspects remain unresolved, and what should be the path for future action. The positive and negative criticisms made to the Consensus well describe the framework of the current debate. The criticisms and controversies raised by the document are more numerous than the compliments. To a certain extent, they show the vitality enjoyed by the topic within the European Union, the significant self-criticism of the various agents involved, and the benefit of a healthy confrontation between the public and the private sector, which, despite the risk of stoppage of the actions, usually reverts in their improvement. All of this constitutes an outstanding feature of the communitarian policy.

72.4.1 The Consensus Contributions to Development Policy

The major contributions, even highlighted by the detractors of the document, are three: the reinforced vision on a sustainable development, the progress to a more holistic perspective that integrates security, and, finally, the role given to civil society in the design and monitoring of cooperation policies.

Sustainability as a value is present both in the actions carried out and in the measures adopted in the different European societies and by the agents that promote development. In the beneficiary countries of the aid, and in accordance with the 2030 Agenda, the proposal of the new Consensus reinforces the interest in areas that guarantee a lasting and persistent growth, such as, among others, the defense of the environment, growth based on own natural resources, the creation of stable employment for young people, the education of women and girls, the advancement of democratic participation, and the promotion of conditions that favor responsibility and national autonomy in development processes. These topics are considered cross-cutting issues that should be present in the different cooperation agreements. However, as already mentioned, the aim is also to achieve a coordinated and coherent action between the European Union and the Member States in order to ensure efficiency, effective spending, and the attainment of complementary and mutually reinforcing initiatives.

The second relevant contribution is the close link established between development and peace, which reinforces actions of conflict prevention and peacebuilding. The Consensus recognizes political instability and violence as fundamental causes of poverty: it is a transversal theme to which a specific section has been devoted. It recognizes new action scenarios, points out the need to coordinate humanitarian aid with development policies, and contemplates intervention in the entire cycle of violence, from conflict prevention measures to human and material reconstruction. According to the German Development Institute, "a holistic and sustainable security concept is however equally important for this overarching strategy so as to connect development policy to issues of stability, democracy and security in the EU foreign policy model (cf. combating the causes of forced migration)" (Henökl and Keijzer 2016).

NGOs specialized in development praise the predominance given to civil society as a crucial element to consolidate democracy. At least in terms of "ought to be," the Consensus considers it is necessary for local civil society to be present in the origin, development, and evaluation of projects that are implemented in the area, combining the citizens' control over the public body with the responsibility to collaborate in the application of measures within family and personal settings (Trimmel 2017). Lastly, also in relation to this aspect, it is worth mentioning the value given to the promotion of resilience and the desire to collaborate with civil society groups without limiting the process of pacification to governments and armies (Angelini 2017).

The most political sector praises the proposals for structural reform of the European development policies that are initiated in the document. The interdepartmental actions, the commitment for an improvement of the coordination, and the inclusion of the Consensus in the framework of the Global Strategy favor whole-of-government approaches to encompass all dimensions of the EU external policy (Gavas et al. 2016).

72.4.2 Limitations and Deficiencies of the Consensus

It seems necessary to begin by referring to the inevitable risk of Eurocentrism that the policies of the European Union entail, despite the fact that this concern is not present in any of the criticisms raised against the document. This Eurocentrism is evident in both the priorities set for development policies, in the way of understanding development and in the values exported. In this context, the first criticism that the document has received is that of the "securitization" of the development agenda, that is, the priority given to matters related to security. This issue has a positive aspect already examined, in that it involves an integration of all the variables that impact and hinder development. Yet the most controversial point is that this "securitization" may stem from an unjustified assessment of the risk that economic migrants and political refugees pose to Europe.

Reality is that the criticism of the "securitization" of the agenda is linked to that of an excessively Eurocentric analysis of the situation by placing the accent on measures aimed at avoiding and controlling immigrants who can pose a potential threat to Europe, both in terms of security (terrorist attacks) and in terms of socioeconomic imbalance. As Concorde points out, "(we) are alarmed by the instrumentalization of development cooperation toward security, commercial and migration objectives" (European Confederation of Development NGOs 2017). The European Union often takes a particularly close interest in matters related to conflict prevention and in fighting against the causes that trigger migration, which shifts the focus from the eradication of poverty and thus raises suspicions about what type of security is sought, whether that of the affected population or that of the European citizens. "EU governments (...) are trying to stop people from reaching Europe in search of safety and dignity, instead of fighting inequality and poverty," denounces Natalia Alonso, Oxfam International Deputy Director for Advocacy and Campaigns (Oxfam 2017).

In particular, there is some reticence with regard to the fact that new items concerning border control or migration management are added to the contribution that the European Union makes to the OECD Official Development Assistance without proportionally increasing the final amount destined to it. It also disconcerts the lack of rigor in the analysis of the situation: for instance, the differences that exist between migration and forced displacement are not taken into account, which would otherwise require contemplating different measures aimed at these two population groups.

Closely linked to this first criticism, there is a complaint regarding the lack of coherence with the fundamental principles of the European Union since collaboration with governments that do not guarantee respect for human rights is considered as a way of halting migration flows. Although the final version of the document corrected a controversial statement included in the first draft ("[the EU] will work with security sector actors, including military actors under exceptional circumstances..."), the Consensus still considers necessary the coordinated action with local authorities without designing any guarantee scheme that allows for the identification of unreliable partners on the ground. This proposal is consistent with other agreements parallelly signed with the aim to stop migration. The recent Valletta agreement (2015) signed with Turkey is a good example of this: it contemplates the detention of emigrants in exchange for economic aid and the granting of visas for Turkish citizens and also sets the terms for the return of citizens from Eritrea and Somalia, both to their countries and to Ethiopia, without guaranteeing their human rights protection during border management and upon landing.

The third element of criticism is the excessive recourse to the private sector and to blended finance. A general feeling among the development agents is that the Consensus has a great deal of confidence in the private sector by not articulating simultaneously control mechanisms that guarantee the priority of the service that the development cooperation requires over the benefit that companies seek to gain in any operation.

The imbalance between the ambition of the objectives and the lack of specific policies constitutes the fourth reason for controversy. Although in the presentation of the document, Neven Mimica, Development Commissioner, affirmed that the Consensus was only a framework for action and that the Commission would adopt more specific policies throughout 2018, a sceptical reaction has inevitably followed. The opinion of Ester Asin, Director of Save the Children's EU Office, is certainly widespread: "We see some good rhetorical commitments in the Consensus, but we need to see how these will be translated into reality."

Along the same lines, the concept of "efficacy" raises concerns over its meaning: it seems that it implies "to do more with less," and the risk of simply ending up "doing less" is clearly perceived. With regard to "doing more," the Consensus is very ambitious in the type of objectives that aims to achieve and in the areas in which it intends to act. However, it openly restricts resources without, on the other hand, specifying what kinds of actions will balance that reduction. The decrease of resources is justified if better coordination avoids overlaps, both with a more rational planning and with a progressive evaluation of impact. Yet the document does not determine what concrete actions are to be adopted or what type of collaboration is to be established with stakeholders or local governments (Boesman 2017).

On the other hand, there are important shortcomings in the interest shown in working with civil society groups. In particular, participatory processes that allow for the intervention of local civil society in an effective manner are not determined: coordination mechanisms are not specified, there is no information about direct financing, and there are not channels to share information regularly with local civil society groups that would allow them to be involved in policy control and evaluation (Angelini 2017). There were also other considerable objections raised from some political representatives of African countries during the Meeting of ACP-EU Economic and Social Interest groups held in Brussels on May 15–17, 2017: the talks surrounding the establishment of the appropriate channels for an effective participation of civil society should involve political parties and should not be exclusively targeted at governments.

Another objection refers to the expectations raised by the Consensus among the countries belonging to the African, Caribbean, and Pacific regions. A full revision to the current agreements that reduces the still existing imbalance in favor of European interests was expected: "The Partnership Framework, the EU Emergency Trust Fund for Africa and the External Investment Plan are primarily serving the EU's own agenda instead of helping people lift themselves out of poverty" (Tempest 2016). In order to correct these imbalances, it seems necessary to identify new parameters that define development policies for the APC bloc. The German Development Institute advocates for a modernization of the relationship with these countries that implies the disappearance of the geographical parameters that have guided development policies so far, and an evolution toward more economic and transversal criteria, such as income level. The German Institute suggests a policy redefinition toward middle-income countries, regardless of their geographical location, as well as toward those countries that begin to be considered as emerging economies (Henökl and Keijzer 2016).

Finally, the absence of references to the impact that Brexit has on development policies has given the impression of a lack of realism and an ill-founded rhetoric on tangible commitments.

In summary, it could be concluded that there are two main complaints that experts from various development cooperation areas make to the EU document: it holds a short-sight perspective that prioritizes short-term goals on security over the long-term views that development requires and lacks concrete guidelines to operationalize that ambitious agenda.

72.5 Conclusion: The EU Contribution to Global Development Policy

As to the conclusion that can be drawn from the previous analysis, it could be said that the European Union is in a position to effectively contribute to the sustainable development goals of the 2030 Agenda set by the United Nations (United Nations 2015), as credited by the concept of European development, the long experience accumulated and the policies and dynamics applied since its foundation. On the other hand, the lines of improvement in the effectiveness of development cooperation contemplated by the European Union coincide with the trends adopted globally.

The new European Consensus on Development ensures that support effectively. In full coherence with the European tradition in this area, it represents an advance in sustainability by integrating the security aspects in the analysis and in the necessary measures taken to eradicate poverty, considering the cross-cutting issues present in all facets of underdevelopment, and extending that sustainability to the improvement in the coordination of the work of the different development agents in Europe, as well as in the integration of the different policies that affect cooperation.

However, the implementation of the New Consensus requires a reconsideration of the balance between European interests and the real needs of underdevelopment. There is a sense of urgency to solve short-term problems – such as the refugee crisis – to the detriment of measures aimed at an effective poverty reduction. The focus on impeding migration flows over investments in places of origin that prevent displacement seems disproportionate, especially when restoring to measures that put at risk the coherence of European principles.

The new Consensus should also further develop many of the proposals mentioned. It is necessary to determine with precision the control mechanisms for the private sector, develop appropriate channels for the effective participation of civil society groups, specify the coordinated action of the institutions that in principle justifies budget reductions, and redefine the parameters of the development agreements signed so far. It would be irresponsible not to recognize that only a development policy that balances the positive impact on poverty reduction with reasonable economic and social spending will make European citizens willing to make the financial and human sacrifices necessary to collaborate on this field. The first objective of the European Union is to protect its Member States. Yet it is also obvious that a reasonable balance between protection and aid is needed and that there is a constant risk of discrimination in favor of European interests. This is the main challenge that the implementation of the Consensus must face.

References

Angelini, Lorenzo (2017) The new European consensus on development. EPLOBlog https://eploblog.wordpress.com/2017/08/28/the-new-european-consensus-on-development/

Boesman Wouter (2017) 'Is 'less' really 'more' in the new European Consensus on Development?' Euroactiv, 23 May. https://www.euractiv.com/section/development-policy/opinion/is-less-really-more-in-the-new-european-consensus-on-development/

Busan Partnership for Effective Development Co-operation (2011) 4th High Level Forum on Aid Effectiveness. Busan, Republic of Korea, 1 December. www.ocde.org/dac/effectiveness/49650173.pdf

CARE International (2016, September) The European Consensus on Development. file:///D:/2.%20Desarrollo/CARE_the_European_Consensus_on_Development.pdf

Constantinou CM, Opondo SO (2016) Engaging the 'ungoverned': the merging of diplomacy, defence and development. Coop Confl 51(3):307–324

Desarrollo de capacidades en apoyo de la seguridad y el desarrollo (2015) Alta Representante de la Unión Europea para Asuntos Exteriores y de Seguridad, (JOIN (2015) 17 final)

European Confederation of Development NGOs 2017, May 19. https://concordeurope.org/2017/05/19/eu-adopts-new-consensus-development/

European Economic and Social Committee (2017) ACP-EU – 28th meeting of economic and social interest groups http://www.eesc.europa.eu/en/news-media/press-releases/acp-eu-28th-meeting-economic-and-social-interest-groups

European Union (2012) Agenda for Change. European Commission

European Union (2016) Shared vision, common action: a stronger Europe' a global strategy for the European Union's foreign and security policy. https://europa.eu/globalstrategy/sites/globalstrategy/files/pages/files/eugs_review_web_13.pdf

European Union. Nuevo Plan de Acción para los Derechos Humanos y la Democracia para el período 2015-2019. https://eeas.europa.eu/human_rights/docs/eu_action_plan_on_human_rights_and_democracy_en.pdf

European Union (2017) The new European consensus on development: our World, our dignity, our future. A joint statement by the council and the representatives of the governments of the member states meeting within the council, the European Parliament, and the European Commission. https://ec.europa.eu/europeaid/sites/devco/files/european-consensus-on-development-final-20170626_en.pdf

European Union. Artículo 21, apartado 1, del TUE

Gavas M, Hackenesch C, Koch S, Mackie J, Maxwell S (2016) The European Union's global strategy: putting sustainable development at the heart of EU external action (ETTG policy brief (Jan.)). The European Think Tanks Group (ETTG), Bonn Retrieved from http://ecdpm.org/publications/eu-global-strategy-global-goals-ettg-2016/

Guía para la Estrategia Europea de Desarrollo Sostenible (2007) Un futuro sostenible a nuestro alcance. Comisión Europea. pp 6–7

Thomas Henökl, Niels Keijzer (2016) The future of the "European consensus on development." In: German development institute briefing paper 5/2016

Marco Estratégico sobre Derechos Humanos y Democracia (2012) http://www.consilium.europa.eu/uedocs/cms_data/docs/pressdata/EN/foraff/131181.pdf

OECD (2016) DAC member profile: European Union. http://www.oecd.org/dac/europeanunion.htm

OXFAM (2017) New EU development framework: self-interest trumps solidarity, 18 May 2017, https://www.oxfam.org/en/pressroom/reactions/new-eu-development-framework-self-interest-trumps-solidarity

Reforzar el papel del sector privado para lograr un crecimiento inclusivo y sostenible en los países en desarrollo, (COM (2014) 263 final)

Tempest, Mikel (2016) Cracks appear in EU-ACP unity at Cotonou meeting in Dakar. EUROActive.com, 4 May 2016

The Paris Declaration on Aid Effectiveness (2005) 2nd High Level Forum on Aid Effectiveness. Paris. www.oecd.org/dac/effectiveness/34428351.pdf

Johannes Trimmel (2017) CONCORD Europe President, May 19, 2017. https://concordeurope.org/2017/05/19/eu-adopts-new-consensus-development/

United Nations (2015) Paris agreement on climate change, FCCC/CP/2015/L.9/REV, http://unfccc.int/paris_agreement/items/9485.php

United Nations (2016) New urban agenda-Habitat III, A/RES/71/256, http://habitat3.org/the-new-urban-agenda

United Nations (2015) Transforming our world: the 2030 agenda for sustainable development'. Resolution adopted by the General Assembly, 25 (2015, September) A/RES/70/1 http://www.un.org/ga/search/view_doc.asp?symbol=A/RES/70/1&Lang=E

United Nations (2015–2030) Sendai Framework on Disaster Risk Reduction. A/RES/69/283, http://www.unisdr.org/we/coordinate/sendai-framework

Valetta Action Plan (2015) Valetta Summit on Migration. Valetta, Malta, 11–12 November. www.consilium.europe.eu/media/21839/action_plan_en.pdf

Wouter Boesman is director of policy at PLATFORMA (2017) Is 'less' really 'more' in the new European Consensus on Development? Euroactiv, 23 May 2017. https://www.euractiv.com/section/development-policy/opinion/is-less-really-more-in-the-new-european-consensus-on-development/

The Challenge of Promoting Diversity in Western Journalism Education: An Exploration of Existing Strategies and a Reflection on Its Future Development

73

Rozane De Cock and Stefan Mertens

Contents

Abstract

Western journalism is guided by a traditional occupational ideology that includes values as truth and accuracy, independence, fairness and impartiality, humanity, and accountability (http://www.ethicaljournalism.org). Nowadays, this ideology faces the pressure of two ongoing developments, i.e., multimediality and multi-culturalism (Deuze et al. 2002). Furthermore, mass audiences lose their unified and monolithic characteristics due to the demographic diversity in society. Media use becomes increasingly personalized and digital platforms gain importance, resulting in a society that is diverse, but displaying a media consumption pattern increasingly based on digital platforms, that functions as an echo chamber of one's own opinions, inspired by one's own identity. This evolution entangles the necessity of developing an adapted code of praxis for journalism education that

R. De Cock · S. Mertens (✉)
University of Leuven, Leuven, Belgium
e-mail: rozane.decock@kuleuven.be; stefan.mertens@kuleuven.be

J. Servaes (ed.), *Handbook of Communication for Development and Social Change*,
https://doi.org/10.1007/978-981-15-2014-3_125

fosters diversity as a central concept (Deuze et al. 2002). Following Botma (2016) journalism education needs to teach cultural citizenship in the digital future. Also O'Donnel states that media diversity is an important public interest policy objective and that "audience access to a wide variety of news and opinions sources enhances democracy" (2017, p. 20). This implies that journalism students need to acquire digital literacy, alongside with cultural knowledge on diversity. The DIAMOND "Diversity and Information Media: new tools for a multifaceted public debate" project in Flanders aspires to develop educational tools to face this challenge. To reach this goal, the project explores existing initiatives on diversity education in journalism schools in Flanders by conducting a survey among the staff of all the bachelor and master programs, alongside interviews with ten selected international journalism education experts, known for their trendsetting role in diversity education in journalism. Different strategies of educational dealing with differences in society regarding age, gender, ethnicity, class, disability, and sexual identity will be discussed throughout the data analysis, and suggestions for the development of future teaching material will be presented. The focus is on Western countries dealing with diversity in journalism education, with a special case study on Flanders.

Keywords

Accrediting Council on Education in Journalism and Mass Communications (ACEJMC) · Constructive journalism · Ethnicity · Flemish journalism programs · MEDIVA project · Public journalism · The media blind spot

73.1 Diversity in Journalism Education

Already in 1992, Theodore Glasser, in that time director of the graduate program in journalism at Stanford University, USA, pleas in favor of a detrivialization of diversity in journalism educational programs and a genuine recognition and operationalization of diversifying the curricula. In his exploration of what diversity means as an epistemological claim and the contradictions between (so-called value-free, neutral, and objective) professionalism and diversity, Glasser states that professionals "typically operate with assumptions and attitudes that can be (...) described as ethnocentric" (p. 131) and that it is utterly important to "combat the indifference to difference." Professionalism concentrates on what journalists have in common, it promotes standardization and homogeneity, and in general filters out personal and cultural differences to live up to the rules, journalistic routines, and beat demands. Knowledge exists in and through experience which makes facts to be seen as social phenomena. This proposition leads to the idea of multiculturalism, cultural pluralism, or what Glasser describes as "simply diversity" (p. 131). This approach to knowledge as a social construction dependent on social and cultural viewpoints conflicts directly with a belief that considers the (re)presentation of the world as it really is, a permanent, universal reality (objectivism). Glasser's viewpoint emphasizes that standards, also journalism standards, "endure in history and not in nature"

(1992, p. 132) and that, therefore, it is important to reflect upon how knowledge, its relationship with experience, and journalism standards shape journalism courses and curricula. That is, for Glasser, more important than trivializing diversity by only quantifying and reducing it to "physiographic criteria for admission and employment" as there is a more profound, qualitative, and cultural layer underneath. The quantitative aspects are necessary conditions "but seldom sufficient for the differences in experience that yield differences in thought and action" (p. 133). The work of Glasser invites us, even a quarter of a century later, to rethink the role and importance of diversity in the education of journalism professionals and to reflect upon the division between "conceptual, theory courses, or the non-skills side of the curriculum" and the practical courses in relation to the hierarchy of knowledge (theory versus knowledge-in-practice) (p. 137). Also Deuze (2006) emphasizes the importance to study journalism education more rigorously and multicultural journalism education in particular.

When reviewing the literature on diversity and journalism education, it becomes clear that the majority of these studies are conducted in the USA, followed by Australian studies. Far less European studies published in peer-reviewed journals can be retrieved on the topic, and also the educational approach in other continents such as Africa is only sporadically reported on (see, e.g., Botma 2016). When it comes to the Low Countries, information on diversity aspects taught in the Netherlands, "a typical multicultural society," has received scholarly attention (Deuze et al. 2002; Deuze 2006, p. 390), whereas a systematic overview of the approach in Flanders (Dutch-speaking part of Belgium) is lacking. The Australian studies show a remarkably strong focus on useful development of practical material and course content that can be used in journalism education to teach students "to reflect, assess, and challenge (if necessary) ingrained inequity" (North 2015, p. 182). The work of North (2015, p. 177) describes the quite recent development of the first gender unit (course) in an Australian journalism program in response to the United Nations Educational, Scientific and Cultural Organization's (UNESCO) call for a teaching model that "gives students at least an introductory understanding of (...) 'gender, cultural diversity, religion, social class, conflict, poverty, development issues and public health issues with training in (...) techniques to cover these issues'." Another Australian example is the pedagogical practice of direct personal contact used by Burns (2015) in order to teach his students how to report in an inclusive and respectful manner on people with a disability. Also the Reporting Diversity Project (Hess and Waller 2011), a partnership of universities and news media organizations supported by the Australian Federal Department of Immigration and Citizenship, has resulted in a multitude of practical educational material: cases, a website, and a bibliographical database on multicultural topics. Although diversity training "in one form or another is widely promoted across Europe" according to O'Boyle et al. (2013, p. 301) and supported by the Council of Europe (e.g., Toolbox: Journalism Training, Discrimination and Diversity), the scholarly study of the approaches used in journalism education throughout Europe remains limited. The results of the MEDIVA project (Media for Diversity and Migrant Integration across the EU) show that diversity training is not prioritized in journalistic education nor in professional development and is "commonly viewed as extra-curricular"

(O'Boyle et al. 2013, p. 305). The UK and the Netherlands are exceptions which make the authors conclude that European states with longer immigration histories generally have more developed diversity policies and binding or advisory guidelines for journalists (O'Boyle et al. 2013, p. 301–305).

According to the study of Jones Ross and Patton (2000) on the nature of journalism courses devoted to diversity, the relative newness of diversity journalism education makes many courses focus primarily on rising the awareness of journalism students for diversity topics. Climbing up the ladder of changing people's behavior, rising awareness is the first step in the direction of aiming for social change but is still many rungs away from the top of the ladder, that is, demonstrating fixed and obtained diversity inclusive behavior (Bambust 2015). Designing enduring change urges for paying attention to the stepping stones in between on the road to transitions and thinking about steps as concern, insight and understanding, intention, and trial or testing behavior before resulting in fixed behavior. This remains a challenge in Western journalism education although Jones Ross and Patton (2000) clearly stress the substantial progress booked by journalism schools in the USA, as three quarter of the surveyed schools offer courses devoted to media and diversity compared to only a quarter of the schools in previous studies throughout the 1970s until the late 1990s. An important stimulating factor in this evolution to activate journalism educators in the USA was a requirement imposed by the Accrediting Council on Education in Journalism and Mass Communications (ACEJMC) in 1982. The ACEJMC required journalism schools to make "effective efforts to recruit, advise and retain minority students, staff, and women faculty members" (Jones Ross and Patton 2000, p. 24). Ten years later, in 1992, the requirement was expanded by what came to be known as Standard 12 that demanded journalism curricula to incorporate courses that made future journalists understand, cover, and communicate with and within a multicultural, multiracial, and otherwise diverse society. By 2013, Standard 3 (Biswas et al. 2017) appeared on the stage giving priority to diversity and inclusiveness and stressing the need for a diversity plan in journalism schools in order to achieve an inclusive curriculum and a diverse faculty and student population. The plan has to be expressed in a written document and has to state clearly the school's definition of diversity.

Besides the precise position of the focal point in journalism education when it comes to diversity (rising awareness), also the thematic focus is mostly limited in scope as the majority of the journalism programs are entirely devoted to race and gender issues (Jones Ross and Patton 2000). Within ethnic minorities and gender-related topics, the courses offer much more attention to more easily available information in handbooks on African-Americans on the one hand and on female stereotypes on the other, neglecting the coverage of other ethnic groups such as Latino's or ignoring stereotypical representations of males. In addition, the courses often limit themselves to a historical approach of the lack of diversity, which might suggest the false idea of diversity issues being solely "a thing of the past," a description that no longer holds true. The study of Biswas and Izard (2010) indicated that media instructors already included more diversity topics in their courses such as class and diversity in general and that the majority of separate courses on diversity dedicated attention to both historical and contemporary diversity issues.

Nevertheless, the top three themes that received the most attention in these courses remain very typical: gender, diversity in general, and race/ethnicity (Biswas et al. 2017). It is, without any doubt, a challenge to devote sufficient attention to news media's relationship with other groups or minorities who experience discrimination and are often mis- or underrepresented in the news. In this respect, next to ethnic/racial aspects and gender, future journalists need to be aware of how to report on sexual orientation, disability, class, religion, and age in a fair way (Bodinger-de Uriarte and Valgeirsson 2015; Biswas et al. 2017; Jones Ross and Patton 2000; Martindale 1991). In their most recent update on diversity teaching in the USA, Biswas et al. (2017) added a new category, "international," to stress the importance of international journalism, global culture, and diversity and to meet ACEJMC Standard 3 demands on instruction related to "mass communications across diverse cultures in a global society" and international relationships (p. 11).

Within the field of journalism education programs, different approaches come to the fore when the precise position of diversity education is concerned. Three pathways appear in the studies of Jones Ross and Patton (2000), Biswas and Izard (2010), and Biswas et al. (2017): (a) offering a separate diversity course in the program, (b) integrating diversity issues solely into other general journalism classes, and finally (c) a combination of a and b thus devoting a course entirely to diversity matters while at the same time dealing with it in other journalism classes as well. Jones Ross and Patton (2000) state that the majority (57.5%) of the journalism schools opted for the combination, while nearly a third of the educational institutions dedicated one or more separate courses to diversity. A minority of the programs reported to only incorporate diversity topics into their general journalism courses. The most recent study on the preferred diversity teaching approach in journalism programs shows that integrating diversity content across the curriculum was given top priority and that teaching diversity separately was the most unpopular path (Biswas et al. 2017). Martindale (1991, p. 34) writes about the "infusing approach" to name the integrating of diversity issues in the range of general courses in journalism educational programs. She is in favor of the approach as she sees multicultural knowledge and sensitivity as a "standard part of a journalist's job, not an extra skill" (1991 p. 34) and something that is acquired only by repetition. Besides these advantages, the approach also reaches more students than when programs only include diversity into optional courses. Proponents of this infusing approach stress the benefits of the integration into the whole program because of the potential danger of ghettoizing the information by separating diversity courses from the rest of the journalism training (Jones Ross and Patton 2000). This is the same point of view that Glasser (1992) calls "neatly packaged" diversity and the "risk segregating it from the rest of the curriculum" (p. 132). At the same time, proponents recognize the challenge that comes along with the infusing approach, that is, matching the diversity theme with the main contextual focus of the more general journalism course. Next to the three approaches, diversity courses tend to be optional classes instead of compulsory courses (Biswas and Izard 2010). This has an impact on the total number of journalism students that can be reached in order to foster diversity in news reporting.

Strengths of diversity classes in journalism education reported in previous studies are the variety of guest speakers that are invited to interact with students and actual reporting tasks to experience a diver's viewpoint by hands-on training (Jones Ross and Patton 2000; Martindale 1991; De Uriarte 1988). De Uriarte prepares her students to address and wipe out "the media blind spot" (1988, p. 78) and states that this has to be done "as vigorously in the classroom as around the editorial table." She does so by using her former experience as a newspaper editor and personal background to assign the Latino community in Austin as beat to her journalism students. The course starts with offering an overview of US Latino cultural and history (theory) and then takes the students literally to places and city quarters where they have never been before. There, they have to look for diverse news stories giving voice to people who often remain ignored in non-diverse news media stories. Martindale describes the "urgent need" for journalists "who are sensitive to multicultural issues and have some knowledge of the (. . .) groups they will cover" already in 1991 (p. 34). Just like De Uriarte, she not only stresses the importance of knowledge transfer in a more theoretical way, but she especially emphasizes hands-on training. The courses she proposes are not only pure news and feature writing pieces but also opinion writing classes, journalism history courses, media ethics incorporating diversity topics, and courses in photojournalism. In her plea, she aims to widen up the diversity debate to go wider ("The students should be taught that they have a responsibility to reflect in their stories all parts of society, not just the white middle and upper classes") and deeper ("news beyond coverage of isolated events in order to cover long-standing situations that affect the lives of their audience.(. . .),to dig below the surface rhetoric (. . .) to cover the root causes of the controversies") at the same time (Martindale 1991, p. 35). The willingness to invest time and effort in diversity teaching of faculty members is crucial to make diversity education work. The research of Ng et al. (2013) shows that female staff members, younger staff members, and people of color hold more positive attitudes toward diversity goals in education than male and white colleagues and faculty members with more than 15 years of institutional experience.

73.2 Researching Diversity in Journalism Programs

During February and March 2018, an online survey questionnaire was developed, tested, and spread among all staff members of all bachelor (six) and master programs (three) in journalism education throughout Flanders and Brussels (the Dutch-speaking journalism education programs in Belgium). A total of 67 respondents filled out the questionnaire, resulting in a response rate of 29% after two reminders. All responses of the survey were collected anonymously in line with the ethical guidelines of the Social and Societal Ethics Committee (dossier G 2017 07 859). Respondents answered to as well open-ended as to closed questions. The questionnaire consisted of a socio-demographic part, a professional part (type of journalism education program they are working for, type of course they are teaching), a diversity aspects part (teaching approach, sub-dimensions of diversity, importance of diversity courses), and finally, a selection of statements on diversity teaching in journalism education.

In the same time period, ten in-depth interviews based on an open-ended topic guide were conducted with staff members of ten different international journalism schools known for their trendsetting role in diversity education in journalism. Interviews were audio-recorded and analyzed using open coding and case study analysis in order to examine and identify collective factors underlying successful diversity in journalism education examples. The ten experts were selected in nine countries (The Netherlands, Germany, the UK (two experts), the USA, Australia, Rumania, Canada, France, and Sweden); seven participants were female and three were male.

Based on the literature review and the research objectives, the following hypotheses and research questions were formulated:

RQ1: How important is diversity in journalism education according to faculty staff in journalism programs in Flanders?
RQ2: Which thematic sub-dimensions of diversity are the most important according to faculty staff in journalism programs in Flanders?
RQ3: To what extent are thematic sub-dimensions of journalism diversity already being taught at journalism education programs in Flanders?
RQ4: To what extent are faculty members happy with the present attention for diversity in their educational journalism program?
RQ5: What collective factors are underlying successful diversity in journalism education initiatives when looking at known international examples?
RQ6: Which diversity aspects need extra attention in future journalism education programs according to lecturers in Flanders and international journalism schools known for their diversity initiatives?
H1: Lecturers in journalism in Flanders prefer integrating diversity as an aspect in general journalism classes instead of offering specialized, stand-alone courses on diversity in journalism education.
H2: Staff teaching theoretical journalism courses believes attention for diversity in journalism education is more important than staff members teaching practical courses.
H3: Younger staff members believe attention for diversity in journalism education is more important than older staff members.
H4: Staff members that are part of a diversity group themselves believe attention for diversity in journalism education is more important than staff members not belonging to a diversity group.
H5: Female lecturers believe attention for diversity in journalism education is more important than male lecturers.

73.3 Journalism Education and Different Diversity Approaches

Of all respondents in the survey, 65% were teaching in a bachelor program in journalism education, 29% were staff members of a master program, and 6% of the respondents were both active within a bachelor as a master training program.

This proportion reflects the number of bachelor versus master programs in journalism in Flanders. Seventeen percent of the respondents report teaching mainly theoretical courses, 28% says offering practical training to students in essence, but the majority of surveyed staff members reports using a combined approach of teaching theory and practice. Slightly more male (52%) than female (48%) lecturers took part in the survey. On average, participants were 48 years old (range, 30–63), the majority of the sample had obtained a university degree (70%), 19% holds a doctoral degree, 9% obtained a higher education degree (non-university profession training), and the remaining 3% did not obtain a higher education certificate.

73.3.1 Importance of Diversity in Journalism Education

Respondents were asked how important attention for diversity is in a journalism education program that is training future journalists. On a scale from 0 to 10, on average, the faculty staff in journalism programs in Flanders awards the importance of diversity 7,77 out of 10 (SD = 1,67). Only 6% of the staff members gives this aspect a score of 5 out of 10 or lower. The largest group appoints a score of 8 to the importance of diversity (39,4%), and a quarter of the sample chooses to give diversity an importance score of 9 or 10 out of 10. It is therefore clear that diversity is widely acknowledged as a crucially relevant topic that is essential in journalism education in Flanders.

Based on previous studies, it is expected that staff members teaching theoretical journalism courses would consider attention for diversity in journalism education as more important than staff members teaching practical courses. The data indicate that this is the case as staff members teaching primarily theoretical courses give the importance of diversity a mean score of 8,17 out of 10 (SD = 1,19), whereas staff teaching mainly practical classes award a lower mean score of 7 out of 10 (SD = 2,53). The staff group that reports teaching a combination of theoretical and practical courses finds itself in between both other groups (M = 8,06, SD = 0,99). The ANOVA test is only marginally significant as the p-value is precisely 0.05 (F = 3,05; df = 2, $p = 0.05$).

Also the relationship between staff age and their perception of the importance of diversity in journalism educations shows a marginally significant result. Younger staff members consider attention for diversity in journalism education as more important than older staff members ($r = -0.20$, $p = 0.06$). When the focus shifts toward gender, results show that female lecturers do believe attention for diversity in journalism education is more important than their male counterparts ($r = 0.26$, $p<0.05$). Hereby, hypothesis 5 is confirmed whereas hypothesis 4 has to be rejected. Staff members that are part of a diversity group themselves do not believe that attention for diversity in journalism education is more important than staff members not belonging to a diversity group ($r = -0.13, p>0.05$). In the sample, 2,74% of the teaching staff in Flemish journalism programs reported that they consider themselves as being part of a minority group when it comes to ethnicity, sexual orientation, physical disability, a not-physical disability, and social class.

73.3.2 Teaching Approach of Diversity and Course Content

The lecturers in the sample were asked in what way their journalism program pays attention to diversity. The majority of the sample (75,4%) indicates that their journalism school prefers integrating diversity as an aspect in general journalism classes instead of offering specialized, stand-alone courses on diversity (3,5%). One out of ten staff members report that their program combines as well an infusing approach along with a specialized course on diversity throughout the journalism training. Surprisingly, 10,5% of the surveyed staff members say their journalism school does not pay attention to diversity at all.

These results are in line with the answers on the following question: "How important is it for you that a journalism education program dedicates a complete course to diversity"? The mean score on this question (M = 5,31;SD = 2,65) and the distribution of the answers (55% gives a score of 5 out of 10 or lower) indicates that opinions are mixed on this matter.

Respondents were asked which thematic sub-dimensions of diversity are the most important to teach in a journalism curriculum. Therefore, staff members ranked the sub-dimensions from most important (1 out of 8) to least important (position 8 out of 8). The lower the number, the more important the sub-dimension is according to faculty staff in journalism programs in Flanders (see Table 1.).

With a mean score of 1,94, the sample ranked ethnicity as the most important sub-dimension of diversity to teach throughout journalism education. Gender appears as the second most important aspect of diversity according to the staff members (M = 3,29). Class holds the third position, before sexual orientation and international journalism who both score an average of 4,42 on 8. Disability ends on the sixth position, followed by age. The "other" sub-dimension ends on the final position, and the open answering field accompanying this answering option makes clear that several respondents added religion as this final eighth sub-dimension that deserves extra attention in journalism programs.

Next to the ranking of sub-dimensions, respondents were asked to what extent the different thematic sub-dimensions were already being taught sufficiently at their

Table 1 Mean ranking position of sub-dimensions of diversity (1 = most important, 8 = least important) according to staff members of journalism schools in Flanders

Sub-dimension	Position	Mean/out of 8 (standard deviation)
Ethnicity	1.	M = 1,94/8 (SD = 1,02)
Gender	2.	M = 3,29/8 (SD = 1,88)
Class	3.	M = 3,77/8 (SD = 2,15)
Sexual orientation	4./5.	M = 4,42/8 (SD = 1,38)
International journalism	4./5.	M = 4,42/8 (SD = 2,33)
Disability	6.	M = 4,83/8 (SD = 1,39)
Age	7.	M = 5,88/8 (SD = 1,51)
Other (primarily religion)	8.	M = 7,46/8 (SD = 1,56)

Table 2 Which diversity sub-dimension is already being taught sufficiently in your journalism education program according to you? (% of sample)

Diversity sub-dimension	%
Gender	24,49
Sexual orientation	21,43
Ethnicity	15,31
Disability	13,27
International journalism	13,27
Class	5,1
Other	4,08
Age	3,06

journalism education program in Flanders (see Table 2). The sub-dimension gender was the most often (a quarter of the sample) indicated as receiving enough educational attention. Sexual orientation follows closely by 21% of the staff members. Ethnicity is indicated by only 15% of the sample as receiving enough attention. Disability as well as international journalism share the fourth position in the ranking as both are being reported as sufficiently incorporated into the program by 13,27% of the sample. Class (5%) and age (3%) are clearly not seen as topics sufficiently being taught to future journalists.

73.3.3 Future Plans Related to Diversity Education

Staff members were asked about future plans and attention aspects in their education programs for journalism training. Respondents were requested to indicate which diversity aspects needed extra attention in future journalism education programs in Flanders. Ethnicity appears to be the top priority of faculty members as 23,96% puts this sub-dimension of diversity in the picture. Class differences are reported as the second most important future aspect by 20,83% of the sample. International journalism closes the top three list (14,58%). Age discrimination holds the forth position (11,46%) and is followed by disability (9,38%), sexual orientation (8,33%), other aspects (6,25%), and finally gender (5,21%).

Faculty members were asked how happy they were with the present attention for diversity and whether they were planning to pay more attention to it in the future. 45,83% of the respondents say they are not planning to pay more attention to diversity in their future journalism program as they believe their attention for it is already alright now. 14,58% says no, and 39,58% reports to plan more attention for diversity in the future.

Of all respondents, a big majority (73,47%) believes it would be a good idea if journalism programs in Flanders voluntarily would design a written diversity plan that indicates how they want to work on diversity and inclusion. In an open text field, they could explain why they are in favor of such a diversity plan: “Without a plan, too less initiatives will arise”; “Such a plan helps carrying out ideas”; “When you design a plan on paper and you sketch how you will include diversity into the different courses throughout the program and how you see the practical coaching, you keep staff members alert"; "A plan incites to action and to reflection on the own journalism education program." Another respondent emphasizes that "structured attention for this important topic is necessary." A remaining quarter (26,53%) is opposed against this idea. The main reason there is the already high working and planning load for staff members.

73.3.4 Collective Factors Underlying Successful Western Diversity Initiatives in Journalism Education

73.3.4.1 Embedding Diversity in the Whole Program

A first very important factor that came to the fore throughout the in-depth interviews with diversity experts is that journalism education activities need to be embedded within an entire program. If students can follow a journalism education program without being obliged to follow a diversity course, it means that it is not a core issue. Diversity education, though, does not necessarily have to be concentrated into one specific course. On the contrary, it is better for diversity education to be integrated in all courses, especially the ethics course and the practical training courses. The infusion approach is the most popular, according to the interviewees.

> The attention for diversity should not be a stand-alone topic. It needs to be integrated in a way of practicing journalism, that holds the future for me. It is a way of practicing journalism that is a bit more slow, focused on inclusion and sustainability. If this form of journalism arrives, specific attention for diversity is hopefully not necessary anymore. (Interviewee 1, The Netherlands)

> Too often diversity is an afterthought. Or something we add on at the end. Or we put it in fifteen minutes in week 7. Because journalists are writing about people and issues, it should be infused in the course all semester long all time. (Interviewee 2, USA)

73.3.4.2 Accreditation

A link with accreditation for journalism programs may be a good idea according to interviewees. The link between a journalism program getting funding and its apt treatment of diversity in its journalism education program may be a particular enforcing way to promote diversity. Diversity is too often considered an afterthought and that has to be avoided.

> You have to say "these are the requirements that any course that wants to be accredited by us must reach." That buy-in from industry means that most courses want to be accredited. (Interviewee 5, UK)

73.3.4.3 Producing Content in Practice

Producing actually diverse content is a very important condition. Students can most effectively see the benefit of producing diverse content if they talk with others than their own friends and family and produce news items and articles themselves using a diverse input (i.e., ethnic differences, social class differences, age differences, etc.). Many interviewees also report award-winning initiatives within their journalism programs as a result of student's hands-on work. Students publishing their own book are another concrete example.

> The students are very excited: we are publishing a book in our class. And when they get the books, they are very proud, and they say:' I told people that I was publishing a book and they didn't believe me'. (Interviewee 2, USA)

> We have done some practical work where we created three radio documentaries and they were 30 minutes programs. They were broadcast on local public radio. (Interviewee 9, Australia)

73.3.4.4 Getting Out of the Comfort Zone

Students can be believed to live in a comfort zone. Young journalists are especially prone to being locked in in their own comfort zone, because the young generation is increasingly news dependent upon social media. This entails the risk of young people being entrapped in a so-called filter bubble, whereby they only consume news that echoes their own ideological preferences. Raising awareness that a society is diverse with diverse opinions and diverse identities may therefore be a crucial step toward educating diversity-oriented journalists.

Most big cities have a diverse cultural, ethnic, and class-related landscape, often only on walking or public transport distance from the journalism schools' main campus buildings. This can form the ultimate, unique, and non-expensive opportunity to "practice what we preach" throughout diversity classes in journalism education. This is a plea to go out of the classroom, to incorporate service-learning projects with a diversity of societal diverse and inclusive organizations into the journalism curriculum, and to install a win-win learning environment. Societal organizations win public attention and recognition; the journalists of the future are immersed into a close but up till then unknown world that deserves being the input of a wide variety of news stories.

> A very basic exercise that we have is the 'my neighborhood exercise'. They get a postal number that they are supposed to use to track people. They don't get to choose the neighborhood and after six months they get another postal code and that exercise was created because a lot of students tend to hang out in the same area with the same people. (Interviewee 4, Sweden)

> Journalism students should know that diversity exists. On many points of view society is a bit conservative. Students of journalism should learn how to deal with such issues and know that such issues exist. (Interviewee 6, Rumania)

73.3.4.5 Cross-Perspective Practice

Students should also be encouraged to work with other people who have different perspectives. This is especially easy to achieve if the students in the journalism education have diverse backgrounds themselves. If someone with, for instance, a listening disability is participating in a journalism course, this sensitizes the other participants especially to being attentive toward this particular form of diversity. Another example are classes with international students. In these classes, students coming from different countries may participate, resulting in a confrontation between the ideas of these students formed by their own national contexts of journalism production.

> For instance one student comes from Germany and the other comes from France. By making them work together and crossing their perspectives, you can open their mind on how they cover the news and how they need to include people. (Interviewee 3, France)

73.3.4.6 Working Together with NGO's

In other circumstances confrontation of students with diverse voices may not be as easy. A special good practice in this case may be to work together with NGO's who especially target diverse journalists or diverse audiences. An example may be the cooperation of a journalism education program with community media representing indigenous people.

> Interviewer: What is the initiative in journalism education that you're most proud of?
>
> Answer: We work with an NGO. They have been training editors and journalists across the world in countries where there is conflict and people cannot speak freely about issues in their society. They bring their own expertise. (Interviewee 8, UK)

73.3.4.7 Applying the Right Interview Techniques

Journalists who are currently being trained for interviewing people should also be aware of the fact that freely and openly asking questions to people is not that easy and accepted in every culture. The directness of Western interviewing might be less appreciated in non-Western cultures. Students should learn how to interview people from different cultural backgrounds. This may be of wider importance than just ethnical background, because not feeling at ease when interviewing people with a disability is another example.

> They did a workshop on how to report sensitively on indigenous people and indigenous issues. So the class was on paper a digital journalism class but the project that we did and the publication we worked on was on telling indigenous stories. (Interviewee 10, Canada)

73.3.4.8 Focusing on Alternative Diversity Dimensions

A special topic to be treated is a focus on social class differences. Many differences regarding diversity are on the agenda, such as age, ethnicity, religion, gender, disability, and sexual orientation. But many of the treated diversity topics tend to overlap with what was earlier called "social class differences." Nowadays Marxism is not as prominent on the agenda of education as it used to be once before, but the underlying phenomenon of class differences remains important. For instance, class differences and ethnic differences tend to overlap very often, while working journalists and students tend to stress the importance of culture and ethnicity too heavily.

> In the discussion, the gender and ethnicity focus has been emphasized, but the socio-economic dimension is at least equally important. The socio-economic diversity implies very often other forms of diversity. Class differences, as we would have called it in a Marxist framework, correlate very often with for instance ethnicity. (Interviewee 1, The Netherlands)

> One of the areas that is less focused on is class and social mobility, and that tends to cut across lots of other diversities. You might get women or people with different ethnic backgrounds coming forward, but if they are all middle class, we are ruling out a whole heap of people. (Interviewee 5, UK)

73.3.4.9 Raise the Quality of Public Discourse

Many can be used to promote diversity, including quota for people coming from a diversity background. The respondents have mixed feelings about the benefits of quota. They might be an interesting intermediary goal to enhance diversity, but in an ideal world quota would not have to exist. A better strategy is to work on public discourse, so that eventually the barriers for people with a diversity background would fade out, such as the barrier to participate in key sectors of society such as journalism and journalism education. Minorities are still underrepresented in newsrooms and journalism education.

> Sometimes I have to say quota might help, but the better solution is lively public discourse. Sometimes quota present a reference point, but nobody can say what the best or the correct reference point would be. (Interviewee 7, Germany)

> Quota may lead to a point that quality disappears, and we do things only because there are quota. (Interviewee 6, Rumania)

73.3.4.10 Produce a Mission Statement

Although words may not be enough, as has been said in the point on accreditation, it may be useful to integrate intentions in a mission statement of the journalism school. Another option could be a diversity charter, being a more explicit document on diversity goals of a journalism education program, next to a general mission statement with a broader scope.

> In any mission statement of a journalism program, as in any mission statement of a journalists union, a statement with how to deal with the diversity of the world is obviously a key issue. (Interviewee 7, Germany)

73.3.5 Diversity Issues that Need Extra Attention in Future Education Initiatives

73.3.5.1 The Speed of Journalism vs Slow News

Journalism needs to produce quick reporting. This forms a perennial part of journalism culture, but the development of the digital era has speeded up the process. Due to the availability of free online news, news that is paid for is always suffering from sharper deadlines. This can also be an opportunity because "slow news" can be said to be the one form of news that still merits being paid for, but in general the time pressure on journalism may still be first and foremost a constraining factor, because reporting facts in a diverse way is most often reporting beyond what is immediately available. Nevertheless audiences become more diverse as well, and this may provide a new business model.

73.3.5.2 Market Orientation for Journalism Students

Journalism programs should promote diverse reporting and ideally get accreditation for doing so. They might promote an alternative form of journalism, but in the end journalism schools need to prepare journalism students for a job in journalism. These jobs will require more traditionally educated journalists adapted to the standard routines of journalism. Journalism schools might be judged according to the extent to which they lead young people to a job.

> This job allocation is really important, therefore a lot of courses focus on training, but what is lost, is a much more general understanding of the principles and the ethics of journalism. (Interviewee 9, Australia)

73.3.5.3 The Infeasibility of Acquired Diversity Reporting Skills

It is good that journalists learn how to report diversely, but you also need to focus on skills that are feasible within a traditional journalistic job. It may be very interesting to teach a form of journalism that transcends the diversity limits of current journalistic practice, but this has no use when once employed, young journalists discover that the skills demand time and resources that are simply not available on a day to day basis.

> We can teach anything we like to students, if they can't apply it in the workforce, what is the point of that knowledge? (Interviewee 8, UK)

73.3.5.4 Avoiding Focusing on Negativity

Journalism in the Western tradition has always overemphasized negative news. This risk is especially outspoken for the reporting of ethnic minorities, who tend to appear most often in the news "in trouble" or "as trouble." The highlighting of people of ethnic minority background in other than criminal situations can be an example of a solution-based journalism. Labels such as "constructive journalism" or "public journalism" can also be used to describe this development. It is important that journalism education programs develop more progressive and positive forms of journalism, because the ideas of young beginning journalists might influence the next generation. Seeing through the classic overstressing of negativity can also be part of a media literacy approach. Being "media literate" is also being sensitive toward the bias of traditional journalism. It makes sense to teach journalists the paradigm of "constructionism" in journalism as well, because having insight in the social construction of all realities, including media reality, might lead to a sharpened feeling for pitfalls to be avoided when reporting issues concerning minorities.

> What I tend to observe today is that the debate on diversity tends to be less recognized or not considered as an important issue as it was before, due to many different reasons, but specifically due to the terrorist attacks and the migrant crisis and the debate in European societies. (Interviewee 3, France)

> We had a very long conversation in class about when news organizations use the word terrorism. A gunman murdered 9 man in a mosque and that was not called terrorism, but the guy in the car in London who hit the pedestrians on the London Bridge was called a terrorist. (Interviewee 10, Canada)

73.4 From Resemblance in Diversity Toward a More Diverse Approach

This exploratory survey research demonstrated that the importance of diversity issues for journalism education is considered high in Flanders. Although further survey research would be needed to test the hypotheses of the survey in other Western countries, the survey showed that especially journalism educators with a theoretical background (rather than a vocational background) and especially female journalism educators (rather than male journalism educators) want to see more diversity-oriented journalism education initiatives.

In-depth interviews with international journalism education specialists confirm the data from the survey inasmuch as a stress on the importance of diversity issues is reiterated. Especially the infusion approach, i.e., the integration of diversity education throughout the curriculum, is highlighted. Means supporting the infusion approach such as accreditation, the editing of a diversity plan (e.g., as part of a mission statement), and the importance of improving public discourse are emphasized throughout the interviews.

Another important focus in the in-depth interviews is on the necessity to improve journalism education through practical exercises. Journalists should undertake practical production exercises, work with other students from various backgrounds, work with societal organizations, and apply the interview techniques adapted to the people involved and the context.

The survey indicated that, at least with the Flemish case as an entry point, gender and ethnicity are the most treated diversity topics. Other cases might merit attention as well, and a particular point of interest is social class. It remains important, but is somewhat lower on the agenda as it used to be in the age of Marxism.

A major constraining factor for the development of diversity through journalism education is the features of contemporary journalism itself. Journalism needs to be fast and emphasizes negative news, and it is precisely in this context that the young journalists need to find a job; all these features may be a challenge for and in contradiction to their educationally acquired diversity reporting skills.

In all, this contribution shows the questions and issues that are on the agenda of journalism diversity education. Further empirical inquiries could show which aspects of the hereby demonstrated agenda are important in other Western but also non-Western countries.

References

Bambust F (2015) Effectief gedrag veranderen met het 7E-model. Sociale marketing. Politeia, Brussels. [Changing behavior effectively with the 7^E-model. Social marketing]

Biswas M, Izard R (2010) 2009 assessment of the status of diversity education in journalism and mass communication programs. Journal Mass Commun Educ 64(4):378–394

Biswas M, Izard R, Roshan S (2017) What is taught about diversity and how is it taught? A 2015 update of diversity teaching at U.S. journalism and mass communication programs. Teach Journal Mass Commun 7(11):1–13

Bodinger-de Uriarte C, Valgeirsson G (2015) Institutional disconnects as obstacles to diversity in journalism in the United States. Journal Pract 9(3):399–417

Botma G (2016) Cultural citizenship in the digital future(s): in search of a new code of praxis for South African journalism education and training. Critical Arts 30(1):102–116

Burns S (2015) Diversity and journalism pedagogy: exploring news media representation of disability. Journal Mass Commun Educ 7(2):220–230

De Uriarte ML (1988) Texas course features barrio as story source. Journal Educ 43(2):78–79

Deuze M (2006) Multicultural journalism education in the Netherlands: a case study. Journal Mass Commun Educ 60(4):390–401

Deuze M, Boyd-Barrett JO, Claassen G, Diederichs P, Tyler Eastman S, Jordaan D, Louw PE, Newsom D, Quinn S, Rabe L, Steenveld L, Stevenson RL, van Rooyen G, Wasserman H, Williams J (2002) Comments on the Sanef media audit: a new news culture is facing the media and journalism educators: the time to act is now! Ecquid Novi 23(1):87–151

Glasser TL (1992) Professionalism and the derision of diversity: the case of the education of journalists. J Commun 42(2):131–140

Hess K, Waller L (2011) Packaged good: responses from Australian journalism educators on the Reporting Diversity Project. Pac Journal Rev 17(2):11–26

Jones Ross F, Patton JP (2000) The nature of journalism courses devoted to diversity. Journal Mass Commun Educ 55(1):24–39

Martindale C (1991) Infusing cultural diversity into communication courses. Journal Educ 45(4):34–38

Ng J, Skorupski W, Frey B, Wolf-Wendel L (2013) ACES: the development of a reliable and valid instrument to assess faculty support of diversity goals in the United States. Res Pract Assess 8:29–41

North L (2015) The currency of gender: student and institutional responses to the first gender unit in an Australian journalism program. Journal Mass Commun Educ 70(2):174–&86

O'Boyle N, Fehr F, Preston P, Rogers J (2013) Who needs or delivers diversity training? The views of European journalists. Journal Pract 7(3):300–313. https://doi.org/10.1080/17512786.2012.740242

O'Donnel P (2017) Journalism education in Australia: educating journalists for convergent, cosmopolitan and uncertain news environments. In: Goodman RS, Steyn E (eds) Global journalism education in the 21st century: challenges and innovations. Regent Press, Berkeley

Institutionalization and Implosion of Communication for Development and Social Change in Spain: A Case Study

74

Víctor Manuel Marí Sáez

Contents

Abstract

This chapter offers an overview of the institutionalization of the academic field of communication for development and social change (CDSC) in Spain in the twenty-first century, following a period of neglect and marginalization. The ongoing expansion of this field in the Spanish context is understood as a process of implosion, i.e., an inward collapse deriving from the inconsistencies and weaknesses of its tardy and hasty institutionalization.

A triple methodological approach has been implemented in this case study: (a) an historiographical analysis of the field of the CDSC in Spain, (b) a bibliometric analysis of the papers published on this subject in the main Spanish journals between 2000 and 2015, and (c) four significant and relevant Spanish case studies in this regard.

The results of this research basically point to (1) a late consolidation of the field in Spain, which has led to its implosion during its expansion and

V. M. Marí Sáez (✉)
Faculty of Communication and Social Sciences, Universidad de Cádiz, Jerez de la Frontera, Spain
e-mail: victor.mari@uca.es

J. Servaes (ed.), *Handbook of Communication for Development and Social Change*,
https://doi.org/10.1007/978-981-15-2014-3_65

institutionalization stage (2011–to date); (2) a marginal production of scientific papers in comparison with the general academic production in the field of communication in Spain; (3) a certain subsidiarity in other more closely related fields (e.g., education for development) which has covered existing gaps and lacunas; (4) the fundamental role played by Spanish community radio; and (5) the potential of virtual communication spaces such as those deriving from the experience of the "Indignados" or 15-M movement (2011).

Keywords

Bibliometric analysis · Bibliometric research · Communication for development and social change (CDSC) · Epistemological framework · Institutionalization process

74.1 Introduction and State of the Question

The strategic importance of communication in solidarity organizations has gradually gathered steam over the past three decades. Diverse technological, communication, social, and political processes, which currently enable us to understand communication as an essential dimension of social and solidarity organizations, have contributed to this situation. This progressive relevance of communication in the field of social action and development has been accompanied by a wide array of theoretical-practical approaches from which communicative action is conceived. The majority of these approaches are dominated by imaginaries (Castoriadis 1987; de Certeau 1984) in which communication is reduced to a mere transmission of information or modification of the conduct of citizens which, generally speaking, is limited to an audience and consumer dimension.

However, communication for development and social change (hereinafter CDSC) is apt to frame the communicative reflection and action of solidarity organizations in contexts more coherent with the social goals championed by them and which allow us to identify the strategic role of communication in the process of transforming reality. In line with Enghel (2011), we believe that communication for development (Servaes 2002, 2008) and communication for social change (Gumucio 2001; Gumucio-Dagron and Tufte 2006; Tufte 2017) constitute different approaches that can serve to designate and characterize a field relating to the role of communication in the strategic efforts to overcome collective social challenges and to advance toward greater social justice.

Based on these premises, this chapter aims to describe the institutionalization process of CDSC in Spain after a long period of neglect and marginalization. As will be seen in greater detail below, in the past 25 years CDSC has gone through three major stages. After an initial stage of neglect and marginalization (during the 1980s and at the beginning of the 1990s), it entered a stage of emergence (from the mid-1990s to 2002), followed by a stage of institutionalization and implosion (from 2003 to date). The initial hypothesis is that CDSC' current expansion or boom in Spain can be understood as an implosion, namely, as an internal rupture, due to its inconsistent and weak development during the institutionalization stage.

Some of the questions underlying this research include the following: If CDSC emerged in Spain in the twenty-first century, what is the point at issue? Or, in other words, to what questions – which still have not been answered sufficiently or consistently – does the institutionalization of CDSC respond in the Spanish context? What continuities or discontinuities are there between the trends observed in the Spanish case and those inherent to the field at an international level?

74.2 Methodological Strategy for Studying CDSC in Spain

In order to study and analyze CDSC in Spain, we have implemented a triple methodological strategy. This responds to the criterion of methodological triangulation, aimed at making research more accurate and consistent by integrating qualitative and quantitative methods with a view to covering the different dimensions of the reality under study more comprehensively. We will now briefly describe each one of the methodologies used in our study.

Firstly, we have reviewed the literature in the field, following the keys proposed by authors such as Erlandson (1993), McDonald and Tipton (1993), for whom "documents constitute a third source of evidence" (Erlandson 1993:99). The historiographical perspective of our work assumes that the social sciences are historical disciplines per se (Mills 1959).

In social research, the use of multiple sources of evidence (Stake 1995; Yin 2009) allows researchers to reflect the complexity of a specific phenomenon or case study. The three most habitual sources of evidence (Erlandson 1993; Ruiz de Olabuénaga e Ispizua 1989; Vallés 1999) are observation (first source of evidence), interviews (second source), and documentation (third source). In this respect, Vallés (1999) regards documentation as a methodological strategy for obtaining information to supplement that of the other two sources of evidence, inasmuch as this enables us to perform research that addresses historical reconstructions. All these elements have been taken into consideration in our study when using documentary analysis to conduct a historiographical study of CDSC in Spain.

To this end, we have reviewed the academic literature and documents describing the advent and institutionalization of CDSC in Spain. This review has been supplemented by secondary data produced by social organizations such as the Coordinating Committee of NGDOs, the *Red Estatal de Medios Comunitarios* (State Network of Community Media – ReMC), and the *Red de emisoras locales, públicas, alternativas y ciudadanas de radio y televisión de Andalucía* (Network of local, public, alternative and citizen radio and television broadcasters of Andalusia – EMA-rtv). The purpose of breaking the development of CDSC down into stages has been to identify those institutions, research groups, researchers, and social organizations that have been capable of using the approach to CDSC in Spain to stimulate research, formation, and/or action processes.

Secondly, we have implemented a strategic bibliometric methodology for reviewing papers on "communication, development and social change" published in the top 10 communication journals in Spain. The timeline of the bibliometric study (2000–2015) is justified by the inclusion in this period of the two main stages of CDSC in Spain (*emergence* and *institutionalization and implosion*) in order to be able to determine and compare their influence on bibliographical production. The sample employed in this research comprises 3782 papers to which has been applied an evaluation sheet that uses preliminary studies, especially those performed by Martínez Nicolás and Saperas (2016), as benchmarks. A series of concepts that serve as filters to identify the link between the papers on CDSC have been established, based on the presence of the following terms in their keywords and main texts: (1) communication for development, (2) communication for social change, (3) solidarity communication, (4) NGO/NGDO and communication, (5) third sector/audiovisual third sector; 6) community/citizen/alternative media, (7) social movement/ICT/society of information, (8) citizen movement/ICT/society of information, and (9) social/citizen/ICT participation.

Lastly, we have conducted case studies of four particularly representative CDSC communication experiences in Spain, chosen on the basis of the field's

Table 1 Case studies

Case	Type of organization	Stage of CDCS in Spain	Characteristics
IEPALA (Latin America and Africa Political Studies Institute)	Development research institute	Pioneer stage (1980–1994)	Institution playing subsidiary role in development research in the Spanish case
Communication, Education and Citizenship Forum	Activist, academic, NGO, and social movement network	Institutionalization and implosion stage (2003–to date)	First attempt to forge links between university experts and solidarity organizations in CDCS in Spain
Onda Color Radio	Community radio	Community media have been present during the three stages, so it would be a cross-sectional case	Community media are paradigmatic representatives of CDCS at an international level
Comunicambio	Facebook group linked to communication, development cooperation, and solidarity organizations	Institutionalization and implosion stage (2003–to date)	Social network groups emerging from the "Indignados" or 15-M Movement (2011)

Source: Own elaboration

historiographical and bibliometric analysis. For the theoretical fundamentals of the case studies, we have resorted to Coldevin (1986, 2008), Eisenhardt (1989), Yin (2009), and Snow and Trom (2002).

The four case studies (see Table 1) meet the following requirements: (1) each one of the three stages in which the history of CDSC is divided in Spain is represented; (2) the diversity of the social actors that have promoted this field of research in the country (development research institutes, activist networks, scholars and NGOs, community media, and, more recently, the "Indignados" or 15-M Movement) is described; and (3) the communication practices fostered by the case studies encompass the publication of books (IEPALA), the production of radio programs (Onda Color), website design (Communication, Education and Citizenship Forum), and techno-communicative practices in the digital space (Grupo Comunicambio on Facebook).

We will now take a look at the research results, setting out the different analytical techniques that have been implemented in chronological order.

74.3 Historiographical Analysis: The Stages of CDSC in Spain

For the sake of brevity, rather than offering a detailed description of the characteristic features of each one of the three stages of CDSC in Spain – (1) pioneers (1980–1994); (2) emergence (1994–2002); and (3) institutionalization and implosion (2003–to date) – we will provide a tentative overview (Table 2):

1. Firstly, the accent must be placed on the subsidiary nature gradually assumed by those institutions, fields of knowledge, and disciplines not directly associated with the epistemological framework created by CDSC at a global level, as in the case of institutes dedicated to issues such as international cooperation or education for development (including Hegoa, Etea, and IEPALA).
2. Secondly, what is remarkable is the *décalage temporal* of the introduction of the field of CDSC in Spain, compared with other geographically (Europe) or culturally closer (Latin America) contexts.
3. The *absence of academic spaces per se* in which to review international debates on CDSC, as well as the *dearth of solid research* offering an overview with which to reconstruct the field's history in the Spanish context.
4. The *tardy, sporadic*, and *weak ties with researchers and social activists promoting CDSC at an international level.* Notwithstanding the multiple ties that would have made it relatively easy for Spanish research to accommodate the most representative authors in the field of Latin American communicology (Kaplún 1998; Luis Ramiro Beltrán 1980; Juan Díaz Bordenave 1976; among others), the truth is that their presence and influence have been tardy and marginal. The same can be said about the field's key authors such as Servaes, Tufte, and Tacchi Servaes (2012, 2015).

Table 2 Stages of CDSC in Spain

Pioneer stage (1980–1994)	Emergence stage (1994–2002)	Institutionalization and implosion stage (2003–to date)
The NGO phenomenon: from marginality to popularization	Creation of the Institute of Communication (InCom) (Autonomous University of Barcelona – UAB, 1997) involving some of the field's international experts (Thomas Tufte, for example)	Basic research into the field of CDCS in Spain (Javier Erro, 2003, 2004)
Emergence of the first CDCS experiences, albeit fragmented and disperse	First rigorous scientific communication and NGO studies (Sampedro, Ariel Jerez, López-Rey)	Founding of the Communication, Education and Citizenship Forum (2006)
First studies and publications dealing with solidarity communication	Galvanizing role of the EMA-rtv network (local citizen radio and television broadcasters of Andalusia) in promoting congresses and actions consistent with CDCS	Institutionalization of CDCS in the field of research (EMA-rtv Congress 25th Anniversary McBride, 2005; Congress of the Spanish Association of Communication Research, 2011, dedicated to these issues)

Source: Own elaboration

74.4 A Brief Look at the Bibliometric Analysis of CDSC in Spain

Based on the aforementioned criteria, and after performing an analysis of the 3782 papers comprising the sample, it can be observed that a total of 24 papers meeting the established requirements, plus another 19 only partially so, have been published to date. Of these 24 papers, 19 have appeared in 4 journals: *Comunicar* (3), *Revista Latina de Comunicación Social* (3), *Telos* (5), and *CIC* (8). Only the first three have a name for publishing research papers on CDSC. None of them have an editorial line exclusively or preferentially focusing on CDSC, although they do indeed show a great affinity for the subject, due to the fact that their key thematic areas are associated with fields close to it, as in the case of media literacy or political communication.

In relation to the authors of papers on CDSC published in Spain (see list in Annex 1), first and foremost, the scant presence of international researchers who have served as touchstones for the field's construction can be observed. Furthermore, these authors include a significant number (7) of researchers linked to universities and institutions that have promoted and strengthened the field of CDSC in the country.

Bearing in mind the work of Jiménez and Arriola (2016), nonetheless, specific authors have indeed broached the field of development in papers on CDSC published in Spain. Based on the bibliometric research conducted by (Marí and Ceballos 2015), Jiménez and Arriola have performed a quantitative and qualitative study aimed at

Table 3 Development researchers cited more than once in papers on CDSC published in Spain

Author(s)	No. of citations	No. of cited publications	No. of publications citing them
Jan Servaes	9	7	6
Rosa María Alfaro	8	2	8
Raquel Martínez and Mario Lubetkin	6	3	4
Javier Erro	4	4	2
Jo Ellen Fair	4	3	1
Gilbert Rist	4	1	4
Amartya Sen	4	2	4
Alberto Acosta and Esperanza Martínez	3	1	3
Anthony Bebbington	3	3	1
Serge Latouche	2	2	2
Paolo Mefalopulos	2	2	2

Source: Arriola and Jiménez (2016)

identifying which development theories and reference authors dealing with these subjects have been used by communication researchers to support their claims. One of their most noteworthy conclusions is that, as a rule, a sufficiently solid state of the question still has not been established. Most of the papers published to date have lapsed into what we could call a *post-modern fragmentation* when framing their objects of study in the field's historical and research context. On the whole, such a contextualization has not been performed or has been fragmentary and deficient at the very most.

When reviewing the authors working in the field of development listed in Table 3, it is possible to detect a clear bias toward those linked to the post-development (Rist 2008), degrowth (Latouche 2010), or good living (Acosta 2011) currents, while at the same time, other theoretical positions and perspectives forming part of the field's history have been disregarded. For example, these omissions include the dependency theory originally developed by authors like Cardoso and Faletto (1979) and Baran (1957), to name but a few. Similarly, although the most cited authors include eminent international researchers (Servaes), there are also important oversights (Tufte 2006; Tacchi 2013; Wilkins 2013; and more besides).

74.5 The Four Case Studies

Lastly, the third research technique that we have employed to analyze CDSC in Spain has been the case study. The most significant results obtained from the four case studies performed here are as follows:

– On the whole, in all four cases, the pioneering social actors have implemented communication practices and initiatives with a far-reaching social impact in terms of promoting development and social change, although the theoretical frameworks in which they are grounded (journalism, dissemination, or public relations, for instance) have not allowed their inherent transforming potential to be fully leveraged.

- In two of the case studies (Communication, Education and Citizenship Forum and the community radio station Onda Color), however, it has indeed been possible to identify a number of communication, development, and social change theories associated with the field's tradition (Servaes, Beltrán, Bordenave, etc.), to wit, as a social process affected by diverse mediations and linked to culture, with a marked pedagogical-liberating dimension (Freire 1970).
- All the cases have had strong ties with local and specific historical contexts of social change. In this connection, it could be claimed that, following the scheme of McQuail (1994), they are cases closer to *social centrism* than to *media centrism*. Thus, the centrality of the process in the history of each case has been crucial, this being understood in communicative, political, and educational terms (Barbas and Postill 2017).
- The Internet has been used in all four cases, although the strategies of technological appropriation employed have been patchy. This has meant that it has been impossible to reach intensive use levels of what Hamelink (2000) calls "informational capital" in all the cases. For him, this technological appropriation culminates when ICT and digital communication practices lead to the creation of knowledge that serves to transform reality. From this perspective, the highest levels of informational capital have been observed in the case of the community radio station Onda Color which has leveraged the Internet to foster media literacy processes, as well as establishing channels for dialogue and citizen and audience participation.

74.6 Discussion and Conclusions

In light of the above, the time has now come to discuss the research results and draw a series of conclusions. To this end, in relation to the objective of identifying the relevance of CDSC in Spanish research, we believe that it is more convenient to talk about an *implosion* than a *boom*. According to the *Oxford Dictionary*, an implosion is "an instance of something collapsing violently inwards," which to our mind better describes the current situation in which the field's internal dynamics are weaker than its external ones.

Specifically, we have used the term "implosion" here on several occasions to refer to the theoretical inconsistency and shortcomings of the field of CDSC in Spain. Consequently, now at a moment of relative expansion, its present and future substantiation and consolidation lack firm foundations. This can be seen in several aspects of which we will highlight two. On the one hand, there is the structural weakness of the elements defined by Martínez Nicolás (2016) which would put a field in the process of being institutionalized on a firmer footing: the institutional context of production (universities, research groups, research networks) and the epistemological context (a field's substantiation and delimitation).

On the other, the rationales of academic production prevailing in the field of CDSC in Spain involve the subjugation of scientific production to commodity logic and to the intensive exploitation of the productivity of researchers in the short term. This second trend has a negative impact on knowledge construction processes and the theories sustaining them, since it ultimately favors those elements of reality that are easily measurable, in line with mainstream thought. In pragmatic terms, from a quantitative point of view, this had led to the predominance of metadata analyses in the case of bibliometric studies, which are necessary but insufficient if the intention is to identify the reading, comprehension, and citation rationales that we have analyzed here. Moreover, these dominant dynamics in contemporary academic production ultimately favor the use of functionalist theoretical frameworks in which the empiricist principle of analyzing what is "easily measurable and observable" displaces and marginalizes that of theories and methodologies that are more useful for critically analyzing the contradictions of reality, power structures, or the rationales of domination or submission in a specific academic field.

Furthermore, CDSC poses a great paradox. On the one hand, it is an approach whose terminology (*development* and *social change*) is insufficient to deal with problems of a sociopolitical nature. However, despite these difficulties, the theoretical and practical background of CDSC enables us to claim that this framework offers more possibilities to think about communication sociopolitically than other approaches to the study of third sector communications, such as social marketing and public relations, understood at least in conventional terms.

As has been seen, by comparing the structure of the field of CDSC in Spain with international trends in this regard, it can be noted that there has been a *décalage temporal* with respect to the introduction of the discipline's key authors and concepts. It has also been observed that this albeit tardy introduction has been patchy. Key international authors in the field do not appear in the theoretical references of the studies on CDSC that have been conducted in the Spanish context.

In short, CDSC has reached a critical point in Spain: even though it has developed, this growth is not sufficiently grounded in scientific production, institutional advancement, and epistemological development. How this critical situation is tackled through research, theorization, and practical action will have a crucial impact on the future success or failure of CDSC in Spain. There is potential for change: the ongoing cycle of social mobilizations commencing with the "Indignados" or 15-M movement in 2011, and worldwide with movements such as Occupy Wall Street, provides an opportunity to build significant bridges between researchers and activists – that is, an opportunity for praxis.

Acknowledgments This chapter forms part of the R&D project "Evaluación y monitorización de la Comunicación para el Desarrollo y el Cambio Social en España" (MINECO, España) (CSO2014-52005-R) (2015-2017), funded by the Spanish Ministry of Economy, Industry and Competition.

List of Pertinent Authors and Universities Publishing Papers on CDSC in the Top Ten Spanish Journals During the Period 2000–2015

Journal	Author	University	Vol.	Year
1. *Revista Comunicar*	Thomas Tufte	University of Copenhagen, Denmark	26	2006
2. *Revista Comunicar*	Alejandro Barranquero Carretero	University of Malaga (UMA)	29	2007
3. *Revista Comunicar*	Sherri H. Culver, Thomas Jacobson	University of Philadelphia, Pennsylvania, USA	39	2012
4. *Revista Latina de Comunicación Social*	Francisco Sierra Caballero	University of Seville (US)	26	2000
5. *Revista Latina de Comunicación Social*	Chiara Sáez Baeza	Autonomous University of Barcelona (UAB)	64	2009
6. *Revista Latina de Comunicación Social*	Txema Ramírez de la Piscina Martínez	University of the Basque Country (UPV/EHU)	65	2010
7. *Telos. Cuadernos de Comunicación en Innovación*	José Marques de Melo	Methodist University of São Paulo, Brazil	51	2002
8. *Telos. Cuadernos de Comunicación en Innovación*	Luis Ramiro Beltrán Salmón	Researcher, Bolivia	72	2007
9. *Telos. Cuadernos de Comunicación en Innovación*	Manuel Chaparro Escudero	University of Malaga (UMA)	74	2008
10. *Telos. Cuadernos de Comunicación en Innovación*	Manuel Chaparro Escudero	University of Malaga (UMA)	81	2009
11. *Telos. Cuadernos de Comunicación en Innovación*	Manuel Chaparro Escudero	University of Malaga (UMA)	94	2013
12. *Cuadernos de Información y Comunicación*	Jan Servaes	University of Massachusetts, Amherst, USA	17	2012
13. *Cuadernos de Información y Comunicación*	Ana Fernández Viso	Institute of Communication – Autonomous University of Barcelona (InCom-UAB)	17	2012
		Charles III University (UC3M), Madrid	17	2012

(continued)

Journal	Author	University	Vol.	Year
14. *Cuadernos de Información y Comunicación*	Alejandro Barranquero Carretero			
15. *Cuadernos de Información y Comunicación*	Raquel Martínez-Gómez, Pinar Agudiez Calvo	Complutense University of Madrid (UCM)	17	2012
16. *Cuadernos de Información y Comunicación*	María Cruz Alvarado López	University of Valladolid (UVa)	17	2012
17. *Cuadernos de Información y Comunicación*	Eloísa Nos, Luís Amador Iranzo Montés, Alessandra Farné	Institute of Social Development and Peace (IUDESP) – Jaume I University (UJI), Castellón	17	2012
18. *Cuadernos de Información y Comunicación*	Julio César Herrero, Ana Toledo Chávarri	Camilo José Cela University (UCJC), Madrid	17	2012
19. *Cuadernos de Información y Comunicación*	Isidoro Arroyo	Rey Juan Carlos University (URJC), Madrid	18	2013
20. *Revista de Estudios en Comunicación ZER*	Víctor Marí	University of Cadiz (UCA)	22	2007
21. *Revista Ámbitos*	Francisco Collado Campaña	Pablo de Olavide University (UPO), Seville	17	2008
22. *Revista Comunicación Sociedad*	Summer Harlow, Dustin Harp	University of Texas at Austin, School of Journalism; University of Texas at Arlington, Department of Communication	26	2013
23. *Revista Estudio del Mensaje Periodístico*	María José Gámez Fuentes, Eloísa Nos Aldás	Jaume I University (UJI), Castellón	18	2012
24. *Revista Estudio del Mensaje Periodístico*	Concepción Travesedo de Castilla	University of Malaga (UMA)	19	2013

References

Acosta A (2011) Buen Vivir: Today's tomorrow. Development 54(4):441–447

Baran P (1957) The political economy of growth. Monthly Review Press, New York

Barbas Á, Postill J (2017) Communication activism as a School of Politics: lessons from Spain's indignados movement. J Commun 67(5):646–664

Beltrán LR (1980) A farewell to Aristotle: horizontal communication. Communication 5(1):5–41

Cardoso FH, Faletto E (1979) Dependency and development in Latin America. University of California Press, Berkeley

Castoriadis C (1987) The imaginary institution of society. MIT Press, Cambridge
Coldevin G (1986) Evaluation in rural development communications. A case study from West Africa. Media Educ Dev 19(3):112–118
Coldevin G (2008) Making a difference through development communication: some evidence-based results from FAO field projects. In: Servaes J (ed) Communication for development and social change. Sage, London, pp 232–253
De Certeau M (1984) The practice of everyday life. University of California Press, Berkeley
Díaz Bordenave J (1976) Communication of agricultural innovations in Latin-America: need for new models. Commun Res 3(2):135–154
Eisenhardt KM (1989) Building theories from case study research. Acad Manag Rev 14(4):532–550
Enghel F (2011) Communication, development and social change: future alternatives. Presented at the ICA congress, 26–30 May 2011, Boston
Erlandson D (1993) Doing naturalistic inquiry: a guide to methods. Sage, London
Erro J (2003) Comunicación, desarrollo y ONGD. Hegoa. Bilbao, España
Erro J (2004) El trabajo de comunicación en las ONGD del País Vasco. Hegoa. Bilbao, España
Freire P (1970) Pedagogy of the oppressed. Continuum, New York
Gumucio-Dagron A (2001) Making waves. Stories of participatory communication for social change. The Rockefeller Foundation, New York
Gumucio-Dagron A, Tufte T (eds) (2006) Communication for social change anthology: historical and contemporary readings. Communication for Social Change Consortium, South Orange
Hamelink C (2000) The ethics of cyberspace. Sage, London
Jiménez M, Arriola J (2016) Cómo se concibe el desarrollo económico y social en los textos de C4D en España. Communication presented at the international conference on regional science, "Treinta años de integración en Europa desde la perspectiva regional: balance y nuevos retos", 16–18 Nov 2016 (Santiago de Compostela, Spain) [How economic and social development are conceived in C4D texts in Spain]
Latouche S (2010) Farewell to growth. Polity Press, Cambridge
Lennie J, Tacchi J (2013) Evaluating communication for development. A framework for social change. Routledge, London
Lie R, Servaes J (2015) Disciplines in the field of communication for development and social change. Commun Theory 25(2):244–258
Marí VM (2005) Communication, networks and social change. In: Gumucio-Dagron A, Tufte T (eds) Communication for social change anthology. Historical and contemporary readings. Communication for Social Change Consortium, South Orange, pp 1009–1014
Marí, V and Ceballos G (2015) Bibliometric analysis of the articles published in 'communication, development and social change' in the top ten journals of communication in Spain. Cuadernos. info 37:201–212
Martínez-Nicolás M, Saperas E (2016) Research focus and methodological features in the recent Spanish communication studies (2008–2014): an analysis of the papers published in Spanish specialized journals. Rev Lat Comun Soc 71:1365–1384
McDonald K, Tipton T (1993) Using documents. Sage, London
McQuail D (1994) Mass communication theory. An introduction. Sage, London
Mills W (1959) The sociological imagination. Oxford University Press, Oxford
Rist G (2008) The history of development: from western origins to global faith. Zed Books, London
Ruiz Olabuénaga, José e ispizua, María Antonia (1989) La decodificación de la vida cotidiana: Métodos de Investigación Cualitativa. Bilbao: Publicaciones Universidad de Deusto
Sáez VMM (2016) Communication, development, and social change in Spain: a field between institutionalization and implosion. Int Commun Gaz 78(5):469–486
Servaes J (2002) Approaches to development communication. UNESCO, Paris
Servaes (2008) (ed.) Communication for development and social change. Sage, London

Servaes J, Polk E, Shi S, Reilly D, Yakupitijage T (2012) Towards a framework of sustainability indicators for "communication for development and social change" projects. Int Commun Gaz 74(2):99–123
Snow DA, Trom D (2002) The case study and the study of social movements. In: Klandermans B, Straggenborg S (eds) Methods of social movement research. University of Minnesota Press, Minneapolis, pp 146–172
Stake RE (1995) The art of case study research. Sage, Thousand Oaks
Tufte T (2017) Communication and social change. A citizen perspective. Polity Press, Cambridge
Valles Martínez MS (1999) Técnicas cualitativas de investigación social: reflexión metodológica y práctica profesional. Síntesis, Madrid
Wilkins KG, Enghe F (2013) The privatization of development through global communication industries: living proof? Media Cult Soc 35(2):165–181
Yin RK (2009) Case study research: design and methods. Sage, London

A Sense of Community in the ASEAN

75

Pornpun Prajaknate

Contents

Abstract

The ASEAN Community was established in 2015 to achieve the three pillars of sustainable development: political security, economic, and sociocultural development. However, this is only the beginning of establishing a sense of community there. The question remains whether the ASEAN can reach its goals and whether the people will feel this sense of community. Understanding the factors that predict the development of a sense of community in the ASEAN can help answer this question. The survey results from 1200 citizens of the ASEAN member states provide insight for future regional communication campaigns, which may help to strengthen emotional connections, trust, and a sense of belonging to the ASEAN Community.

P. Prajaknate (✉)
Graduate School of Communication Arts and Management Innovation, National Institute of Development Administration, Bangkok, Thailand
e-mail: pornpun2@hotmail.com; pornpun.p@nida.ac.th

J. Servaes (ed.), *Handbook of Communication for Development and Social Change*,
https://doi.org/10.1007/978-981-15-2014-3_50

Keywords

Sense of community · ASEAN community · Multimedia campaigns

75.1 Introduction

One Vision, One Identity, One Community

This motto was stated by the Chairman of the 11th ASEAN Summit on December 12, 2005. Ten years after this statement, the ASEAN Community was formally established on December 31, 2015 (The ASEAN Secretariat Community Relations Division 2017a). Member states include Brunei Darussalam, Cambodia, Indonesia, Lao PDR, Malaysia, Myanmar, the Philippines, Singapore, Thailand, and Vietnam (The ASEAN Secretariat Community Relations Division 2017b).

Following the community's motto, three blueprints reflecting the three pillars of development were issued: (1) the ASEAN Economic Community Blueprint 2025 (AEC); (2) the ASEAN Political-Security Community Blueprint 2025 (APSC); and (3) the ASEAN Socio-Cultural Community Blueprint 2025 (ASCC) (Aekaputra 2011). By 2025, according to the ASEAN secretariat (2015), member states should be working together to achieve the vision of these three blueprints, including the following: (1) to strengthen an integrated economy, (2) to ensure peace and a strong shared sense of togetherness, and (3) to promote a sustainable community and a shared sense of ASEAN identity.

However, there is ample evidence that these visions have not yet been accomplished. Regarding the first pillar, the ASEAN Economic Community Blueprint 2025 (AEC), Rillo (2017) states that the ten member states have faced difficulties in achieving economic integration. The existing literature reveals that about half of the respondents in the extant studies engaged in small- and medium-sized enterprises (SMEs) in Cambodia and about 20% of the respondents working in manufacturing in Indonesia appeared to lack awareness of the AEC (Thangavelu et al. 2017; Anas et al. 2017). The SMEs in Lao were not seen to be closely linked to regional economic integration (Kyophilavong et al. 2017), and most of the SMEs in the Philippines have not recognized the impact of the AEC on their business performance (Aldaba 2017). Additionally, the Gen Y Malaysian population reported a strong sense of economic nationalism and at the same time the possibility of disagreement with the idea of regional economic integration in the ASEAN (Benny 2016).

As for the second pillar, the ASEAN Political-Security Community Blueprint 2025 (APSC), some member states have not yet achieved its vision of peace and security. Thailand's southern insurgency has also contributed to a sense of insecurity and distrust between the leaders of Thailand and Malaysia (Zha 2016, 2017; na Thalang 2017). With the last pillars, the ASEAN Socio-Cultural Community Blueprint 2025 (ASCC), the issues of humanitarian crises that have systematically crushed the ethnic minority and religious identity of the Rohingya, for example, have impeded the process of promoting ASEAN Community integration (Paik 2016).

As pointed out above, various problems still occur within this community. Therefore, the question remains whether the ASEAN Community can reach its visions and whether the people of the ten member states will feel this sense of community. To date, only one study has examined the sense of belonging to the ASEAN Community among its citizens, and it was discovered that about 75% of the respondents from the ten member states reported a moderate to strong sense of belonging to the ASEAN Community. Out of 2322 respondents, only 69 reported a very low sense of ASEAN citizenship and belonging (Intal and Ruddy 2017). However, no study has looked into the influence of communication factors on developing a sense of community in the ASEAN (Dawson 2006). Section 75.2 provides a comprehensive review of the literature and empirical research on the concepts and definitions of community across the disciplines of sociology, community psychology, and development communication. The roles of communication tools in community development are also discussed in this section. Section 75.3 explores the factors that predict a sense of community in a general context. The crucial factors that can enhance a sense of community in the ASEAN are identified in Sect. 75.4. Section 75.5 presents a concise summary of the key factors predicting a sense of community in the ASEAN. Finally, the future directions for member states in developing a sense of community in the ASEAN are proposed in Sect. 75.6.

75.2 Sense of Community and Communication for a Sustainable Community

75.2.1 Definition of Community

The definition of "community" is often ambiguous (Giddens and Sutton 2017). For several decades, sociologists, along with psychologists, have been working to construct a definition of "community" (Hutchison 2008). Attempts have been made to define "community" in the field of sociology. Hillery (1955), for example, examined 94 studies and found that the definition of community included three common elements: geographic area, social interaction, and common ties.

Community psychologists also have highlighted the importance of the sense of community (Hughey and Speer 2012; McMillan and Chavis 1986; Sarason 1976). Among the concepts discussed in this connection, the most prominent framework of the community is the idea of the "psychological sense of community" developed by McMillan and Chavis (Flaherty et al. 2014). The sense of community comprises four dimensions: (1) membership, (2) influence, (3) integration and fulfillment of needs, and (4) a shared emotional connection. The first dimension, *membership*, refers to a sense of belonging or personal relations, which has five elements: (1) boundaries, (2) emotional safety, (3) a sense of belonging and identification, (4) personal investment, and (5) a common symbol system. The second dimension, *influence*, is the perception of individuals of their contribution to the community and a sense of cohesion with people in the community. The third dimension, *integration and fulfillment of needs*, refers to the perceived fulfillment of an individual's needs through the resources available in the community, such as membership status and the reputation of the community. The last dimension, *shared emotional connection*, is a bonding between people or a group with a shared emotional connection, history, place, time, experience, as well as spirituality (McMillan and Chavis 1986).

75.2.2 Communication for a Sustainable Community

The community concept can also be seen to have roots in the development of the communication approach. Servaes (2008) pointed out, for example, that communication for sustainable social change has become an influential paradigm, and from his standpoint, it is crucial to examine the "changes from bottom-up, from the self-development of the local community." Servaes and Malikhao (2016) further proposed five levels of communication strategies for development and social change as listed below:

(a) *Behavior change communication* (BCC): mainly interpersonal communication
(b) *Mass communication* (MC): community media, mass media, online media, and ICT
(c) *Advocacy communication* (AC): interpersonal and/or mass communication
(d) *Participatory communication* (PC): interpersonal communication, community media, and social media
(e) *Communication for structural and sustainable social change* (CSSC): interpersonal communication, participatory communication, mass communication, and ICT

In particular, Servaes and Malikhao (2014) have stated that CSSC should come to the forefront for the social and sustainable development of the ASEAN Community, and sustainable development should be accomplished via a combination of communication and other dimensions, including structural, organizational, cultural, demographic, sociopolitical, socioeconomic, and physical environmental factors. Furthermore, Servaes and his students (2016a) formulated sustainability indicators for evaluating the influence of communication on development projects. One of the indicators suggests investigating a variety of media, including face-to-face communication, print, radio, television, ICT, and mobile phone. Therefore, the next section will provide a review of the literature on the disadvantages and advantages of the media regarding community development and the role of each type.

The mass media, for example, have been criticized for being one-way communication oriented and for overestimating the importance of audience exposure to the mass media without considering the content or sources of messages (Melkote and Steeves 2015). Similarly, one of the disadvantages of community media is that they are small-scale organizations which fail to reach a large audience (Carpentier et al. 2003).

In contrast, social media and face-to-face communication have several advantages when compared to mass media and community media. Social media have been seen to foster the relationships between citizens and government organizations (Criado et al. 2017) and to provide opportunities for civic engagement and building trust among citizens, which leads to high trust in an institution (Warren et al. 2014). Likewise, face-to-face communication that focuses on receiver-oriented and horizontal communication is crucial for building a sustainable community. According to the communication for social change paradigm, it is important to build communication networks for sharing knowledge and information among key stakeholders, including academic scholars, communication practitioners, field-specific professional, NGOs, and community members (Servaes and Lie 2015). During their interaction period within networks, mutual trust and honesty have increased, and this has triggered information flows and true participatory action, which eventually leads to the success of the community building effort (Servaes and Malikhao 2008). Thus, focusing on the local community's viewpoint and building trust among the people in the community through collaboration between local and international stakeholders are crucial for reaching community sustainability (Servaes 2016b).

75.3 Factors Predicting a Sense of Community

The first predictor begins with the perceived physical environment, and several studies have demonstrated an association between this and the perceived sense of community. For example, it has been suggested that perceived neighborhood safety is positively associated with a sense of belonging to the community (Gonyea et al. 2017), and a safe physical environment, pedestrian facilities, and recreation areas have been seen to promote a sense of community (French et al. 2014). In particular, the sense of belonging among the citizens of the ten ASEAN member states has been seen to be derived from the geographical setting (Intal and Ruddy 2017).

Apart from the physical environment factor, there is also a social interaction factor that plays an important role in the prediction of a sense of community. Jabareen and Zilberman (2017) developed a social interaction methodological framework that emphasized the physical environment, transportation, and social interaction as predicting factors in promoting a sense of community. Evidence from previous studies also supports this framework; that is, the availability of public spaces can promote face-to-face interaction among people in communities (van den Berg et al. 2014). Likewise, Wood et al. (2010) found that communities with pedestrian-friendly streets strengthened the social interaction among residents.

Reid (2015) also found that the perception of living in a safe physical environment with common facilities increased opportunities for social interaction and enhanced the sense of community among the residents of multi-owned properties in Australia. Likewise, Seo and Lee (2017) demonstrated that the Crime Prevention

Through Environment Design project in Korea, which renovated the pedestrian environment for safety and established crime protection devices, increased the residents' perceived quality of the environment by 13.7–18.5%. As a result, the residents have increased their participation in several types of social activities, including social gathering and joining a children's playground. Consequently, their perceived sense of community rose by 5.4–7.9%. Thus, a high-quality physical environment, or the perception of living in one, can be seen to promote a sense of community through the availability of opportunities for formal and informal interaction between the members in the community.

The role of social interaction itself is also important. Drawing on McMillan and Chavis's sense of community concept, Garrett et al. (2017) proposed co-constructing a sense of community model which highlights the importance of the relationship between participating in community activities and the sense of community. According to this model, community members develop their sense of community through their interaction during daily routines and social events. For young people, engaging in sports and recreational and religious activities strengthens a sense of community and identification with the community (Cicognani et al. 2008).

In communities where various social activities have been organized, community members have plenty of opportunities to spend time together. Consequently, they are able to develop social connectedness, neighborhood-based friendships, and social relationships with other people in the community (Lenzi et al. 2013). Along the same lines, the results of previous study have demonstrated that leisure activities, sports, and cultural activities have been seen to significantly predict the social connectedness of older people through social gatherings with family and friends from within and outside the neighborhood (Toepoel 2013).

A more specific social gathering has been discussed by Rosenthal et al. (2007), who stated that people that report spending time with others through religious activities have a higher level of perceived connectedness to the people in the city than those that have not participated in any activities because they know their source of emotional support. Similarly, Sohi et al. (2017) found that the more that people participate in ritual activities, the more likely they are to have a greater sense of community. Because the members of a ritual community have worshiped together, they are bonded together. Thus, their sense of membership and a shared emotional connection are developed.

Another factor that previous studies have highlighted as a possible determinant of the sense of community has to do with the information from different sources – there is evidence that community members develop their sense of community through the information gathered from the various sources of media in the community. For instance, a case study of the success a local initiative against the carbon dioxide capture and geological storage in a Canadian community demonstrated the importance of information sources in developing a sense of community. Boyd (2017) investigated this case and found that strong community attachment was one element of a sense of community that was derived from the communication networks in the community. Sharing cooperative connectedness, gathering information through online media and mass media, and disseminating the information via face-to-face communication helped the community members succeed in their opposition to the carbon dioxide capture system in this Canadian community.

Information sources such as face-to-face communication, participatory communication, ICT, and mass communication have been viewed as crucial tools for sustainable social change (Servaes 2017). Out of these tools, online media, such as digital media and mobile phones, have become prominent forms of communication for social change projects (Servaes and Lie 2015). ICTs help people develop social contact with friends and family, enhance economic performance, and increase the sense of security and peace (Baqir et al. 2011). It was for this reason that Hoffman (2017) suggested distributing information concerning community activities across social media channels (e.g., Twitter and Facebook) in order to enhance the perception of the importance of community service activities and to strengthen the sense of connectedness with the community. Another tool that has been posited to serve community development is community media (Carpentier et al. 2003), which are accessible sources of information, education, and entertainment for the local community, according to Berrigan (1979).

The role of information sources in community development is crucial in many ways, not only for sustainable social change but also for fostering a sense of community. Friends and radio were the desired sources of information that were able to establish a sense of community among Trinidadians and Jamaicans living in Washington, D.C. The more that information was gathered from friends at work, meetings, and with the telephone, the more were the immigrants likely to have a greater sense of community. Thus, exposure to online media, community media, mass media, and face-to-face communication are one of the factors that predict a sense of community (Regis 1988).

In order to understand the sense of community in the ASEAN, a prediction model is proposed. As shown in Fig. 1, the perceived sense of community among the citizens of the ASEAN Community may be a result of the perceived physical environment, social interaction, and exposure to information sources.

Fig. 1 Predictors of a sense of community in the ASEAN

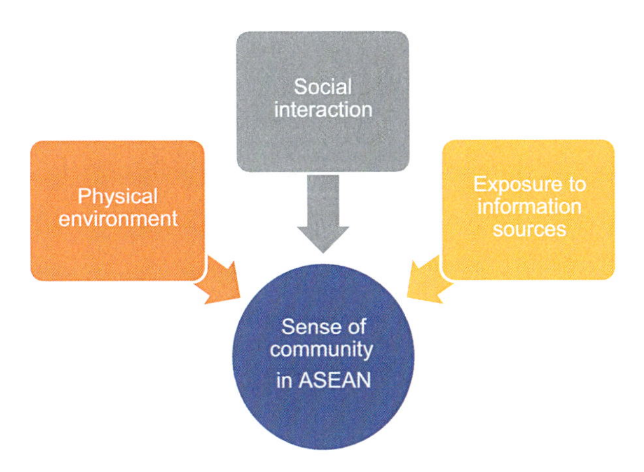

75.4 Sense of Community in ASEAN and Its Predictors

To explore the factors that predict a sense of community among the citizens of the member states (Fig. 1), a survey was conducted in six provinces across three geographical regions of Thailand: the northeast, north, and south. These setting are unique and located along three special border economic zones, including the Thailand-Laos border, the Thailand-Malaysia border, and the Thailand-Myanmar border. According to the Official Statistics Registration System of Department of Provincial Administration (2015), the total size of the populations in the 6 selected provinces was 5,358,866, comprising Chiang Rai (1,282,544), Tak (631,965), Nakhon Phanom (716,873), Nong Khai (520,363), Songkhla (1,417,440), and Narathiwat (789,681). A total of 1200 respondents residing in 6 selected provinces were recruited to fill out the survey, employing the quota sampling method with a proportional allocation of 400 samples for each region. Each of the predictors is explored. An exploration of perceived sense of community in the ASEAN and its predictors are demonstrated below.

75.4.1 Perceived Sense of Community in the ASEAN

The citizens of the member states have a strong sense of community in the ASEAN (Mean = 3.54). Further examining the sense of community dimensions, the highest average score went to “membership” with a mean score of 3.65. The membership questionnaire items that measured the feeling of who is a part of the ASEAN community and a recognition of the common symbol system received a higher mean score than all others. The lowest average score went to the integration and fulfillment of needs dimension with a mean score of 3.46. The lowest-scoring question item assessed whether the ASEAN community would succeed in fulfilling the needs of its citizens. (Appendix 1, Table 2).

75.4.2 Factor Predicting Sense of Community in the ASEAN

The perception of the citizens of the ASEAN member states toward border environmental features, such as cross-border connecting roads, public transportation, and commercial areas, was high, with a mean value of 3.54, 3.49, and 3.48, respectively (Fig. 2).

Regarding their social interaction with people that were residing in ASEAN member states, the citizens of the ASEAN Community were more likely to exchange greetings (e.g., “hello” and “smile”) than to engage in cross-border spiritual activities, have conversations, and get together with friends (Fig. 3). In addition, the citizens of the member states reported that their main sources of information about the ASEAN Community were online media such as websites, Google, Line, Facebook, and YouTube. The second most common information source for exposure

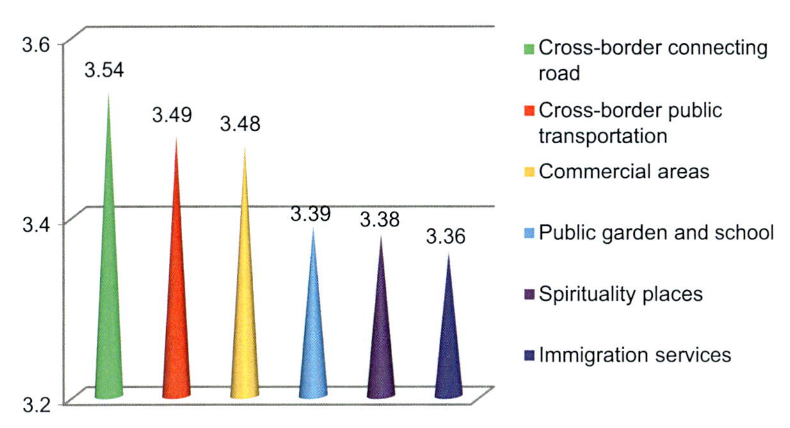

Fig. 2 Perceived physical environment in the ASEAN. (Data were collected for this study. The respondents were asked to rate the items using a five-point Likert scale, measured on the basis of a 7-item questionnaire derived from Cerin et al. (2009).)

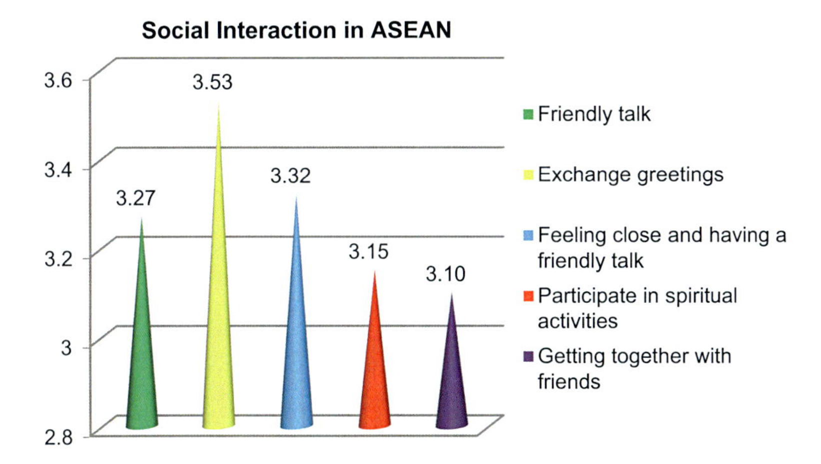

Fig. 3 The level of social interaction among the citizens of the ASEAN Community. (Data were collected for this study. The respondents were asked to rate the items using a five-point Likert scale. The 5-item questionnaire on perceived social interaction was adapted from Mendes de Leon et al. (2009).)

was the mass media, followed by community media. The least-used information source was face-to-face communication, which scored on average 3.35 (Table 1).

As shown in Fig. 4, the model for predicting a sense of community in the ASEAN was developed using the stepwise regression technique. The analysis showed that the combination of perceived physical environment, social interaction, exposure to face-to-face communication, and exposure to online media explained 58.7% of the perceived sense of community ($R^2 = .587$) (Appendix 1, Tables 2, 3). The

Table 1 Exposure to information sources among the citizens of the ASEAN Community

Information sources	Mean
Online media (websites, Google, Line, Facebook, and YouTube)	3.81
Mass media (television, radio, newspapers, and magazines)	3.55
Community media (cable television, community radio, and local newspapers)	3.51
Face-to-face communication (family, teachers, government officers, and community leaders)	3.35

Source: Data were collected for this study. The respondents were asked to rate their level of exposure to information sources on a five-point Likert scale ranging from "highest level of exposure" to "lowest level of exposure"

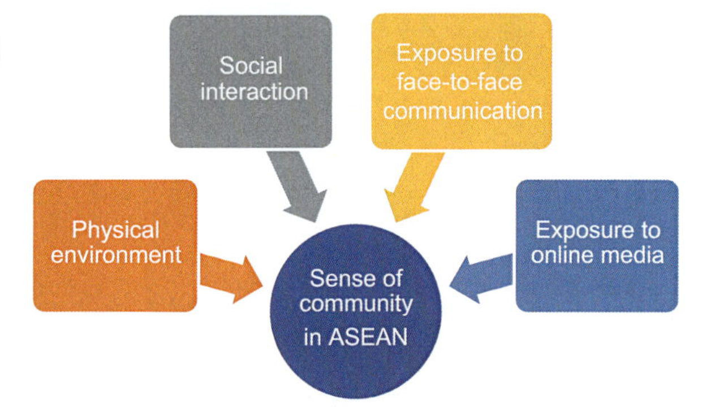

Fig. 4 Predictors of a sense of community in the ASEAN

standardized coefficients for perceived physical environment (β = .360; $p < .001$), social interaction (β = .309; $p < .001$), exposure to face-to-face communication (β = .171; $p < .001$), and exposure to online media (β = .136; $p < .001$) were significant. This explained that these factors were variables that predicted the perceived sense of community in the ASEAN. The strongest predictive contribution was the perceived physical environment variable ($R^2 = .459$; $R = .678$); however, exposure to community media and mass media did not significantly predict the perceived sense of community.

75.5 Summary

This is just the early period of establishing the ASEAN Community. Therefore, it is crucial to understand the sense of what the ASEAN Community is and the social interaction and media exposure among the citizens of the community so that the ten member states can accomplish the three-pillar visions. This study provides an understanding of the sense of community in the ASEAN.

The perceived physical environment was the strongest predictor of the perceived sense of community in the ASEAN, accounting for 45.9%. The citizens of the ASEAN Community perceived that they could easily commute across the borders of the ASEAN member states using individual and public transportation for the purposes of education, business, and ritual practices. Thus, living in a community with a "walkable" space, a recreation area, and high availability of public transportation engenders a real sense of community (Francis et al. 2012; Gonyea et al. 2017; Wood et al. 2010; French et al. 2014). The member states of the ASEAN Community are obviously situated in close proximity, and this leads to an increase in the sense of belonging (Intal and Ruddy 2017).

In particular, public spaces and facilities can promote social interaction among the people in communities (van den Berg et al. 2014). A sense of community is developed after engaging in social interaction and social activities in the community (Garrett et al. 2017; Cicognani et al. 2008). The more people engage in spiritual activities, the more likely they are to have a greater sense of community (Rosenthal et al. 2007; Sohi et al. 2017). The physical environment, together with the social interaction variables, explained 54.2% of the perceived sense of community: greater accessibility to transportation, infrastructure, and public facilities led to a higher level of engagement in social activities on the part of the ASEAN Community citizens. Social interaction such as friendly talk with people from member states, exchange greetings, and participating in spiritual practices can enhance their sense of community in the ASEAN.

Exposure to face-to-face communication and online media were also found to predict the sense of community in the ASEAN. As emphasized by Servaes and Malikhao (2014), the communication for sustainable social change (CSSC) should be central to the sustainability development initiatives of the ASEAN Community, and in particular, the current study demonstrated that face-to-face communication is one of the crucial information sources for strengthening a sense of belonging to this greater community. Moreover, Servaes and Lie (2015) have stated that communication networks including academic scholars, NGOs, and community members are drivers of social change: as information regarding community development is exchanged during social interactions, mutual trust is established, and, consequently, a community is developed sustainably (Servaes and Malikhao 2008; Servaes 2016b). Additionally, social media channels (e.g., Twitter and Facebook) and ICTs are crucial tools for fostering relationships between citizens and government agencies (Criado et al. 2017; Warren et al. 2014) and for strengthening a sense of connectedness with the community and developing a sense of security (Hoffman 2017; Baqir et al. 2011).

In spite of the fact that the citizens of the member states have a high level of exposure to mass media and community media, these factors failed to predict the perceived sense of community within the ASEAN. This may be explained by the weaknesses of these types of media: the mass media focus too heavily on one-way communication but ignore content (Melkote and Steeves 2015), and the community media have low capability to reach large audiences (Carpentier et al. 2003).

According to the idea of Servaes and Malikhao (2014), sustainable social change is not only the result of exposure to the media; it can also be achieved through the integration of communication factors and other factors such as the perceived physical environment. The model that is the most successful predictor of the perceived sense of community must include the perceived physical environment and exposure to online and other media.

The citizens of the member states recognized the beginning of the establishment of the ASEAN Community through their exposure to face-to-face communication and online media. In dealing with community issues, the members of the community tended to obtain information from different sources. They then passed this information through face-to-face interaction, and this in turn created strong community attachment and a sense of community (Boyd 2017). Thus, the more frequently that the respondents were exposed to these information sources (e.g., face-to-face communication and online media), the more was information regarding the ASEAN Community distributed and the more this led to social interaction. Additionally, the respondents live close to other ASEAN member states along the borders, and this close proximity increases opportunities for social interaction and eventually develops a sense of belonging to the greater community.

75.6 Future Directions

The ten member states of the ASEAN Community have faced challenges in accomplishing the visions of the three pillars. It is important to gain insight into the key factors that can predict the ASEAN citizens' sense of community. As mentioned, the geographical proximity advantage was seen to be a major predictor of this sense of community; however, the communication aspects, such as the social interaction among the citizens of member states, along with exposure to online media and face-to-face communication, were also seen to be crucial keys in fostering a sense of community.

This study suggests future directions for exploration. Given the partial roles of exposure to online media and face-to-face communication, emphasis should be placed upon the use of online media (e.g., websites and Facebook Fan Pages) as a major tool for the promotion of the ASEAN Community's visions in fostering integration and trust among its citizens. Along with online media, the ten member states should convey information regarding these visions through government officers, community leaders, teachers, and local people because they are the most important sources of information in many border communities in the ASEAN Community. Although in this study exposure to the mass and community media was not seen to effectively predict the perceived sense of community, these cannot be dismissed as unimportant channels for the ASEAN Community Awareness Campaigns. As the final level of strategic communication for sustainable development is communication for sustainable social change (CSSC), this suggests incorporating interpersonal communication, participatory communication, mass communication, and ICT into development communication campaigns (Servaes and Malikhao 2014).

Appendix 1: Survey from a Citizen of ASEAN's Member States

Table 2 Mean for the perceived sense of community in the ASEAN

Perceived sense of community in the ASEAN	Mean
Membership	3.65
Influence	3.50
Integration and fulfillment of needs	3.46
A shared emotional connection	3.50
Total	3.54

Sources: Data were collected for this study. The respondents were asked to rate the items using a Likert scale. The questionnaire on the perceived sense of community was adopted from Chavis et al. (2008). Permission to use this questionnaire was obtained from Community Science.

Table 3 Model summary of linear stepwise regression analysis

	SOC	PE	SI	O	F	M	b	β	SR^2 incremental
PE	0.678						0.347***	0.360	0.459
SI	0.674	0.683					0.270***	0.309	0.084
O	0.369	0.304	0.265				0.130***	0.171	0.027
F	0.493	0.383	0.487	0.247			0.082***	0.136	0.016
M	0.364	0.328	0.299	0.227	0.592		–	–	–
C	0.335	0.341	0.353	0.329	0.515	0.611	–	–	–
							Intercept		0.710***
Mean	3.541	3.446	3.278	3.811	3.354	3.512			
SD	0.669	0.694	0.766	1.106	0.878	0.846			
							R^2		0.587
							R^2_{adj}		0.585
							R		0.766***

SOC sense of community, *PE* physical environment, *SI* social interaction, *E* exposure to information sources, *F* face-to-face communication, *O* online media
***$p < 0.00$

References

Aekaputra P (2011) Report on the ASEAN economic cooperation and integration. In: Herrmann C, Terhechte JP (eds) European yearbook of international economic law 2011. Springer Berlin Heidelberg, Heidelberg, pp 375–388

Aldaba RM (2017) Philippine SME participation in ASEAN and East Asian regional economic integration. J SE Asian Econ (JSEAE) 34(1):39–76

Anas T, Mangunsong C, Panjaitan NA (2017) Indonesian SME participation in ASEAN economic integration. J SE Asian Econ (JSEAE) 34(1):77–117

Baqir MN, Palvia P, Nemati HR, Casey K (2011) Defining ICT and socio-economic development. In: AMCIS 2011, Detroit

Benny G (2016) Attitude, challenges and aspiration for the ASEAN Community 2015 and beyond: comparative public opinion in Malaysia and Thailand. Soc Sci 11(22):5488–5495

Berrigan FJ (1979) Community communications: the role of community media in development. reports and papers on mass communication no. 90. ERIC, New York

Boyd AD (2017) Examining community perceptions of energy systems development: the role of communication and sense of place. Environ Commun 11(2):184–204. https://doi.org/10.1080/17524032.2015.1047886

Carpentier N, Lie R, Servaes J (2003) Community media: muting the democratic media discourse? Continuum: J Media Cult Stud 17(1):51–68

Cerin E, Conway TL, Saelens BE, Frank LD, Sallis JF (2009) Cross-validation of the factorial structure of the neighborhood environment walkability scale (NEWS) and its abbreviated form (NEWS-A). Int J Behav Nutr Phys Act 6(1):32. https://doi.org/10.1186/1479-5868-6-32

Chairman's Statement of the 11th ASEAN Summit (2005) One vision, one identity, and one community statement

Chavis DM, Lee KS, Acosta J (2008) The sense of community (SCI) revised: the reliability and validity of the SCI-2 paper presented at the 2nd International Community Psychology Conference, Lisboa, Portugal

Cicognani E, Pirini C, Keyes C, Joshanloo M, Rostami R, Nosratabadi M (2008) Social participation, sense of community and social well being: a study on American, Italian and Iranian university students. Soc Indic Res 89(1):97–112. https://doi.org/10.1007/s11205-007-9222-3

Criado JI, Criado JI, Rojas-Martín F, Rojas-Martín F, Gil-Garcia JR, Gil-Garcia JR (2017) Enacting social media success in local public administrations: an empirical analysis of organizational, institutional, and contextual factors. Int J Public Sect Manag 30(1):31–47

Dawson S (2006) A study of the relationship between student communication interaction and sense of community. Internet High Educ 9(3):153–162

Department of Provincial Administration (2015) A number of population in Thailand. Official Statistics Registration System, Bangkok

Flaherty J, Zwick RR, Bouchey HA (2014) Revisiting the sense of community index: a confirmatory factor analysis and invariance test. J Community Psychol 42(8):947–963

Francis J, Giles-Corti B, Wood L, Knuiman M (2012) Creating sense of community: the role of public space. J Environ Psychol 32(4):401–409

French S, Wood L, Foster SA, Giles-Corti B, Frank L, Learnihan V (2014) Sense of community and its association with the neighborhood built environment. Environ Behav 46(6):677–697

Garrett LE, Spreitzer GM, Bacevice PA (2017) Co-constructing a sense of community at Work: the emergence of community in coworking spaces. Organ Stud 38(6):821–842. https://doi.org/10.1177/0170840616685354

Giddens A, Sutton PW (2017) Essential concepts in sociology. Wiley, Cambridge

Gonyea JG, Curley A, Melekis K, Lee Y (2017) Perceptions of neighborhood safety and depressive symptoms among older minority urban subsidized housing residents: the mediating effect of sense of community belonging. Aging Ment Health: 1–6. https://doi.org/10.1080/13607863.2017.1383970

Hillery GA (1955) Definitions of community areas of agreement. Rural Sociol 20:111–123

Hoffman AJ (2017) Millennials, technology and perceived relevance of community service organizations: is social media replacing community service activities? Urban Rev 49(1):140–152. https://doi.org/10.1007/s11256-016-0385-6

Hughey J, Speer PW (2012) Community, sense of community, and networks. In: Fisher AT, Sonn CC, Bishop BJ (eds) Psychological sense of community: research, applications, and implications. Springer US, New York

Hutchison ED (2008) Dimensions of human behavior: person and environment. Sage Publications, California

Intal P, Ruddy L (2017) Voices of ASEAN what does ASEAN mean to ASEAN peoples? Economic Research Institute for ASEAN and East Asia, Jakarta

Jabareen Y, Zilberman O (2017) Sidestepping physical determinism in planning: the role of compactness, design, and social perceptions in shaping sense of community. J Plan Educ Res 37:1):18–1):28

Kyophilavong P, Vanhnalat B, Phonvisay A (2017) Lao SME participation in regional economic integration. J SE Asian Econ (JSEAE) 34(1):193–220

Lenzi M, Vieno A, Santinello M, Perkins DD (2013) How neighborhood structural and institutional features can shape neighborhood social connectedness: a multilevel study of adolescent perceptions. Am J Community Psychol 51(3–4):451–467. https://doi.org/10.1007/s10464-012-9563-1

McMillan DW, Chavis DM (1986) Sense of community: a definition and theory. J Community Psychol 14(1):6–23

Melkote S, Steeves HL (2015) Place and role of development communication in directed social change: a review of the field. J Multicult Discourses 10(3):385–402. https://doi.org/10.1080/17447143.2015.1050030

Mendes de Leon CF, Cagney KA, Bienias JL, Barnes LL, Skarupski KA, Scherr PA, Evans DA (2009) Neighborhood social cohesion and disorder in relation to walking in community-dwelling older adults: a multi-level analysis. J Aging Health 21(1):155–171. https://doi.org/10.1177/0898264308328650

na Thalang C (2017) Malaysia's role in two South-East Asian insurgencies:'an honest broker'? Aust J Int Aff 71(4):389–404

Paik W (2016) Domestic politics, regional integration, and human rights: interactions among Myanmar, ASEAN, and EU. Asia Europe Journal 14(4):417–434. https://doi.org/10.1007/s10308-016-0458-x

Regis HA (1988) Communication and the sense of community among the members of an immigrant group. J Cross-Cult Psychol 19(3):329–340. https://doi.org/10.1177/0022022188193003

Reid S (2015) Exploring social interactions and sense of community in multi-owned properties. Int J Hous Mark Anal 8(4):436–450

Rillo AD (2017) Monitoring the ASEAN Economic Community. In: De Lombaerde P, Saucedo Acosta EJ (eds) Indicator-based monitoring of regional economic integration: fourth world report on regional integration. Springer International Publishing, Cham, pp 287–297. https://doi.org/10.1007/978-3-319-50860-3_13

Rosenthal DA, Russell J, Thomson G (2007) Social connectedness among international students at an Australian university. Soc Indic Res 84(1):71–82. https://doi.org/10.1007/s11205-006-9075-1

Sarason SB (1976) Community psychology, networks, and Mr. Everyman. Am Psychol 31 (5):317–328

Seo SY, Lee KH (2017) Effects of changes in neighbourhood environment due to the CPTED project on residents' social activities and sense of community: a case study on the Cheonan Safe Village Project in Korea. Int J Urban Sci 21:1–18

Servaes J (2008) Communication for development and social change. SAGE Publications India, New Delhi

Servaes J (2016a) How 'sustainable' is development communication research? Int Commun Gaz 78 (7):701–710

Servaes J (2016b) Sustainable development goals in the Asian context. Springer, Singapore

Servaes J (2017) The resiliency of social change. In: Tumber H, Waisbord S (eds) The Routledge companion to media and human rights. Routledge, London, p 136

Servaes J, Lie R (2015) New challenges for communication for sustainable development and social change: a review essay. J Multicult Discourses 10(1):124–148. https://doi.org/10.1080/17447143.2014.982655

Servaes J, Malikhao P (2008) Development communication approaches in an international perspective. In: Servaes J (ed) Communication for development and social change. SAGE Publications India, New Delhi, pp 158–179

Servaes J, Malikhao P (2014) The role and place of communication for sustainable social change (CSSC). Int Soc Sci J 65(217–218):171–183. https://doi.org/10.1111/issj.12080

Servaes J, Malikhao P (2016) Communication is essential for global impact. Procedia Eng 159 (Supplement C):316–321. https://doi.org/10.1016/j.proeng.2016.08.187

Sohi KK, Singh P, Bopanna K (2017) Ritual participation, sense of community, and social well-being: a study of seva in the Sikh community. J Relig Health. https://doi.org/10.1007/s10943-017-0424-y

Thangavelu SM, Oum S, Neak S (2017) SME participation in ASEAN and East Asian integration: the case of Cambodia. J SE Asian Econ (JSEAE) 34(1):175–192

The ASEAN Secretariat (2015) ASEAN 2025: forging ahead together. The Association of Southeast Asian Nations, Jakarta

The ASEAN Secretariat Community Relations Division (2017a) Celebrating ASEAN: 50 years of evolution and progress. The Association of Southeast Asian Nations, Jakarta

The ASEAN Secretariat Community Relations Division (2017b) Towards ASEAN Economic Community 2015: monitoring ASEAN economic integration. ASEAN Community, Jakarta

Toepoel V (2013) Ageing, leisure, and social connectedness: how could leisure help reduce social isolation of older people? Soc Indic Res 113(1):355–372. https://doi.org/10.1007/s11205-012-0097-6

van den Berg P, Kemperman A, Timmermans H (2014) Social interaction location choice: a latent class modeling approach. Ann Assoc Am Geogr 104(5):959–972

Warren AM, Sulaiman A, Jaafar NI (2014) Social media effects on fostering online civic engagement and building citizen trust and trust in institutions. Gov Inf Q 31(2):291–301. https://doi.org/10.1016/j.giq.2013.11.007

Wood L, Frank LD, Giles-Corti B (2010) Sense of community and its relationship with walking and neighborhood design. Soc Sci Med 70(9):1381–1390

Zha W (2016) Trans-border ethnic groups and interstate relations within ASEAN: a case study on Malaysia and Thailand's southern conflict. Int Relat Asia-Pacific 17(2):301–327

Zha W (2017) Ethnic politics, complex legitimacy crisis, and intramural relations within ASEAN. Pac Rev: 1–19. https://doi.org/10.1080/09512748.2017.1391866

Part XII

Case Studies

76 Entertainment-Education in Radio: Three Case Studies from Africa

Kriss Barker and Fatou Jah

Contents

Abstract

This chapter presents the findings from three radio dramas in Africa (Nigeria, Burundi, and Burkina Faso) developed using Population Media Center's unique entertainment-education methodology.

Each case study is presented in detail, including background on the country or region where the program was implemented, a description of the program design, and highlights of the results obtained by the program in question.

Keywords

Entertainment-education · Social and behavioral change communication · Behavior change · Case studies · Africa · Nigeria · Burundi · Burkina Faso

K. Barker (✉) · F. Jah
International Programs, Population Media Center, South Burlington, VT, USA
e-mail: krissbarker@populationmedia.org

J. Servaes (ed.), *Handbook of Communication for Development and Social Change*,
https://doi.org/10.1007/978-981-15-2014-3_68

76.1 Introduction

This chapter presents the findings from radio programs in three African countries that were developed using a specific entertainment-education methodology, applied by Population Media Center (PMC). Over its 20-year history, PMC has developed over 50 entertainment-education serial dramas in over 54 countries, reaching over 500 million people worldwide.

76.2 Background

Radio serial dramas using entertainment-education have been shown in numerous settings to increase use of family planning, increase spousal communication about fertility and family planning, motivate HIV testing, and change attitudes toward girls and women (Piotrow et al. 1990; Nariman 1993; Rogers et al. 1999; Ryerson 2010; Bertrand and Anhang 2006; Singhal et al. 2004; Smith et al. 2007; Vaughan and Rogers 2000; Westoff and Bankole 1997; Barker 2012). Though "the mass media alone seldom effect individual change" (Papa et al. 2000, p. 33), it can influence social norms and stimulate community dialogue and discussion on important social and health topics (Papa et al. 2000; Rogers et al. 1999; Vaughan and Rogers 2000). As a result, even those who have not tuned into a program can examine previously held beliefs, consider options, and identify steps to adopt novel behaviors and attitudes attained through interactions and discussions with the primary audience of an entertainment-education drama (Rogers 2004).

76.2.1 Theoretical Underpinnings

Audiences across the globe have experienced such transformations thanks in part to the use of PMC's entertainment-education methodology which is based on an integrated multidisciplinary theoretical framework that uses a long-running serial drama format (Nariman 1993). A long-running format lends to audience behavior change because it allows time for (1) audience members to identify with characters, (2) characters to change their own attitudes at a believable pace, and (3) audience members to test the attitudinal and behavior changes themselves (Ryerson 2010). Each aspect of a PMC methodology drama is developed according to a theoretical and empirical research-based formula in order to reinforce a coherent set of interrelated values that is tied to specific prosocial behaviors. Social cognitive theory (SCT) and other theories in entertainment-education (Barker 2012; Sood et al.2004) play a prominent role in the PMC methodology via the use of role-modeled behaviors. According to SCT theory, learning can occur via mediated role models who inspire increased confidence or self-efficacy to perform a specific behavior. Not only can new behaviors be learned from such models, but many cognitive and affective responses to a behavioral situation can be learned as well (Bandura 1986).

76.2.2 Program Development

The PMC methodology for developing dramas for social change is a well-tested entertainment-education methodology, based on role modeling rather than dictating. PMC-style dramas consist of positive, negative, and transitional characters to model behaviors, with the latter characters being crucial to the promotion of positive behavior change. Characters may begin these long-running series exhibiting the antithesis of the values being promoted, but through interaction with other characters, twists, and turns in the plot, and sometimes even outside intervention, they come to see the value of the program's underlying messages. The audience observes the positive benefits enjoyed by these characters resulting from their new attitudes and behaviors and is motivated by their emotional ties with these characters to adopt similar values. The high entertainment value of such dramas leads to widespread popularity and significant audience share. The resulting large audience following along each week, adjusting their behaviors and values, make this approach the most cost-effective method of behavior change communications, on a per-behavior change basis.

Good dramas have to be local and relatable. PMC hires and trains local producers, writers, directors, and actors to create social and behavior change communication (SBCC) programs that are culturally sensitive and appropriate on radio or TV in the form of a serial drama. The issues addressed in each program are based on the concerns and policies of the host country and the findings of thorough research on the values, norms, and customs of the people of that country.

76.2.3 Research-Based Programs

PMC's methodology for developing social content serial dramas is grounded in thorough research (qualitative and quantitative research), which is conducted prior to (formative), during (monitoring), and after the program (summative evaluation). PMC contracts with independent local research firms to conduct all research activities for its programs.

76.2.4 Formative Research

Cognizant of countries' diversity in culture, values, languages, media markets, and constantly changing contexts, PMC dramas begin with extensive formative research based on qualitative approaches. Focus group discussions and individual in-depth interviews are conducted within the community to identify health and social processes of concern and to assess potential audience members' attitudes toward these issues. The formative research also helps to identify country laws and policies and the availability of health, social, and environmental services that are critical to any social behavior change program. It also assesses media habits and ideal listening/viewing times of potential audience members and includes a pretest of preliminary versions of the first four episodes of the drama. Thus the formative research enables the creative team to design appropriate, realistic, and credible characters, settings, and story lines that are realistic and engaging to members of the target audience. Local staff establishes a production advisory committee to assure high-quality and appropriate content of the program.

76.2.5 Service Points Monitoring

Quantitative monitoring surveys of service points around the final quarter of a drama broadcast are one of several monitoring activities that PMC employs. The objectives of the clinic monitoring research are to evaluate (1) whether the drama program influenced listeners in the broadcast areas to seek services (health, social, or environmental) that were promoted by the program and (2) the popularity of the drama. The contracted local research firm conducts the survey with a convenience sample of service points drawn from both urban and semi-urban/rural segments of the broadcast areas, including government-operated, private, and nongovernmental organization (NGO) service points. Clients are randomly selected as they exit the facilities and interviewed on (1) their motivation to seek services that day, (2) whether they listen to the drama or not, and (3) whether people in their community/village listen to the drama or not.

76.2.6 Summative Evaluation

The summative or end line evaluation is conducted at the end of broadcast and draws on a mixed-methods (quantitative and qualitative) approach to assess the impact of the drama. Primary data collection for the quantitative end line survey includes individual face-to-face interviews with a statistically representative sample of the target audience. The purpose of the end line survey is to generate information on socioeconomic, demographic and health, and environmental indicators that can be used for measuring progress in achieving project objectives and to evaluate the overall impact at the end of the project period. An additional exposure module that contains questions about frequency of listening, appeal, and character recognition is included in the survey instrument. The impact of the drama is determined by making statistical comparisons between respondents who report listening to the program versus those who did not on key indicators controlling for potentially influential sociodemographic factors. Exposure to a drama is the independent variable and is derived from respondents that listened to a drama one or more times weekly.

76.2.7 Case Studies

76.2.7.1 Nigeria: *Ruwan Dare* (Midnight Rain)

Nigeria is one of the most unique, and most challenging, countries in the world. It is the most populous country in Africa (Population Reference Bureau 2018), and its borders encircle peoples of many different ethnicities, cultures, and languages. The states of northern Nigeria, which are remote and desolate, have historically been a difficult place to work, especially for social service and development agencies. However, it is here that one finds the poorest health and socioeconomic indicators in Nigeria (NPC and ICF Macro 2009). PMC produced *Ruwan Dare* (Midnight Rain) in Nigeria. This 208-episode radio serial drama aired July 2007 through July 2009 in four northern states in Hausa, the most widely spoken and understood language in that area. It was also rebroadcast from July 2009 through November 2010. In addition to the radio serial drama, PMC used its whole society strategy to create a radio talk show aimed at engaging listeners. The radio talk show allowed people to phone in and talk about the issues addressed in *Ruwan Dare*.

The northwest region of the country, including Kano, Kaduna, Katsina, and Sokoto states, was identified as the target region for broadcast because this region has the lowest levels of knowledge of contraceptive methods, ever use of contraceptives, education, and exposure to family planning messages in Nigeria. These states also have the highest fertility rates, largest desired family size (between six and ten children), highest population growth rates, and highest rates of unmet need for contraception in Nigeria (NPC and ICF Macro 2009).

Clinic Monitoring

To monitor the effects of the program on listeners' behaviors, PMC established 11 clinic research sites in the 4 states and conducted 4 rounds of clinic exit interviews with a purposive sample of new clinic clients. Results found *Ruwan Dare* increasingly motivated new clients to seek family planning and reproductive health services with the last round of interviews indicating that 67% of new clients named *Ruwan Dare* as the direct or indirect source of influence to seek family planning and/or reproductive health services for the first time.

Impact Evaluation

Multivariate analysis that controlled for influencing factors of age, marital status, urban/rural location, education, and sex revealed that exposure to the program had an impact on listeners.

Figure 1 presents results of the evaluation in adjusted odds ratios derived from multivariate analysis. Results found that *Ruwan Dare* impacted several of the indicators measured in the evaluation, including knowledge, attitudes, and behaviors. In terms of behaviors, listeners were 2.4 ($p \leq 0.05$) times as likely as non-listeners to say they "currently use something to delay or avoid pregnancy" and listeners were 1.7 ($p \leq 0.05$) times as likely as non-listeners to say they "discuss family planning with others." Similarly, attitudes relating to attitudes were impacted by the drama, especially the indicators measuring gender attitudes/women's empowerment. Thus, attitudinal indicators (four in all) were the most impacted followed by behaviors (two indicators), with knowledge indicators being the least impacted. These results are promising given that a change in attitudes precedes any form of behavior change.

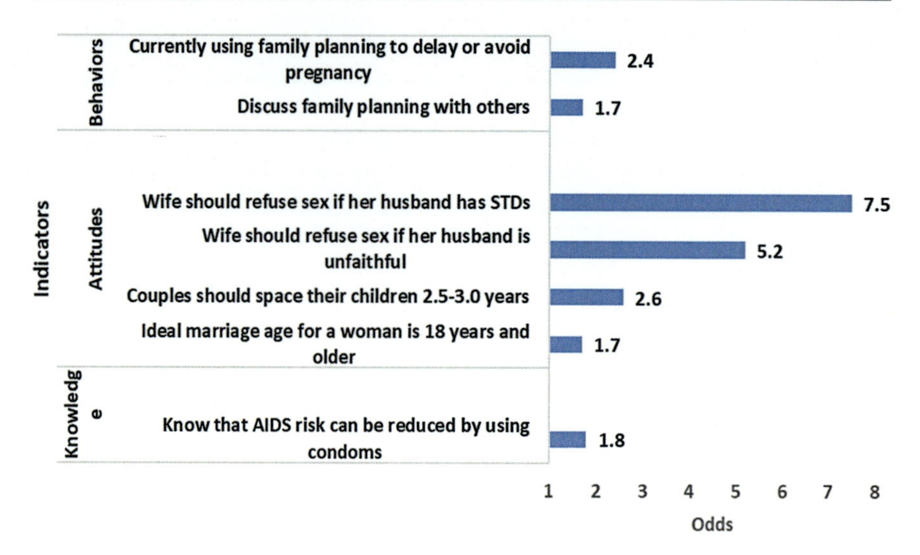

Fig. 1 Multivariate effects of listening to *Ruwan Dare*, 2009 adjusted odds ratios comparing listeners and nonlisteners at least $p \leq 0.05$

Ruwan Dare was not only successful in changing behaviors and attitudes, but it was also one of PMC's most cost-effective dramas. With 70% of the population in the broadcast area listening, *Ruwan Dare* proved to be a remarkably popular series. After adjusting for age (using 2007 state-level census estimates for Nigeria), it was determined that approximately 12.3 million people in the four states listened to *Ruwan Dare*. It catalyzed 1.1 million new family planning users with 60% of listeners agreeing that *Ruwan Dare* was both "entertaining and educational." The cost per listener was calculated to be $0.08 US cents, and the cost per behavior change (new adopter of family planning) was only $0.30 US cents.

76.2.7.2 Burundi: *Agashi* (Hey, Look Again!)

Burundi is a small landlocked country in East Africa, about two-thirds the size of Switzerland with a population of about ten million (PRB 2017). Today, Burundi is one of the five poorest countries in the world (World Bank 2017), and the Population Institute has ranked Burundi the fourth most vulnerable state in their 2015 Demographic Vulnerability Report. Burundi's annual population growth rate is 3.21%, which, if unchanged, means its population is projected to double in 22 years. PMC produced *Agashi* ("Hey! Look Again!") in Burundi. This 208-episode radio serial drama aired from January 2014 to January 2016 in Kirundi, Burundi's national language.

Clinic Monitoring

Figure 2 shows the results of two rounds of exit surveys that were conducted in randomly selected facilities in September 2014 (Round 1) and from September to October 2015 (Round 2). Sixty percent of 1,224 clinic clients surveyed reported

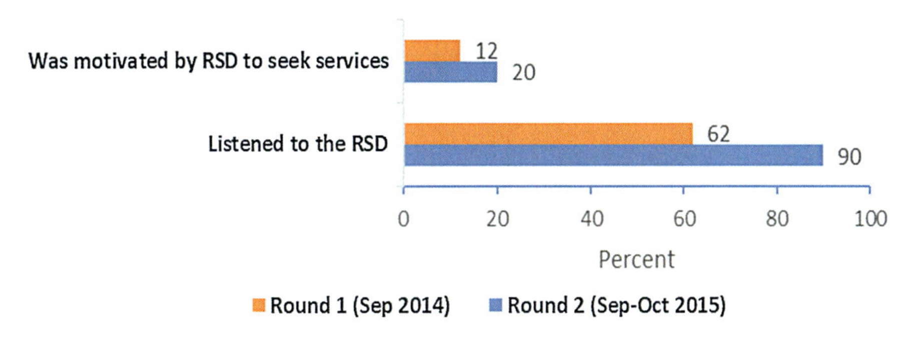

Fig. 2 Clinic monitoring survey of FP/RH services Burundi radio serial drama, *Agashi* 1.0

listening to *Agashi*; this figure increased to 90% of 1,236 clients surveyed in 2015. In 2014, 12% of the 1,244 clients surveyed cited the program as their primary motivation for attending the clinic; this figure increased to 20% of clients in 2015, meaning that 1 in 5 clients sought health services as a result of the program (see Fig. 2).

Impact Evaluation

Figure 3 shows the effects of the Burundi radio serial drama *Agashi* in adjusted odds ratios from multivariate analysis. Like *Ruwan Dare*, *Agashi* impacted many indicators measured in the evaluation, including indicators not only in the knowledge, attitudes, and behaviors categories but also an indicator in the self-efficacy category (see Fig. 3).

In terms of behaviors, listeners were 2 ($p \leq 0.007$) times more likely than non-listeners to say they treated childhood diarrhea with homemade oral rehydration solution (ORS) recommended by government and 1.7 ($p \leq 0.025$) times more likely than non-listeners to say they visited a health facility for information on the secondary effects of family planning in the past 12 months. Regarding self-efficacy, listeners were 1.7 ($p \leq 0.007$) times more likely than non-listeners to say they can confidently negotiate condom use with their sexual partner. Unlike *Ruwan Dare* where attitudinal indicators were the most strongly impacted, the most strongly impacted indicators by *Agashi* fall under the knowledge category (bottom of Fig. 3).

The evaluation also found *Agashi* to be very popular and cost-effective. Eighty-one percent listened to the series at least once weekly, which translates into an estimated national audience size of 2.25 million Burundians. The cost per listener of $0.74 cents (US) is higher than found for *Ruwan Dare* but still cost-effective, and the average cost per behavior change (over five behaviors) attributable to *Agashi* is $6.70 (US). Cost per adopter of voluntary HIV testing is $1.13 (US) and per adopter of ORS is $2.11 (US).

76.2.7.3 Burkina Faso: *Yam Yankré* (The Choice) and *Here S'ra* (The Road to Happiness)

Burkina Faso is a country approximately the size of New Zealand with a population of almost 16 million located in West Africa. The fertility rate in Burkina Faso is the

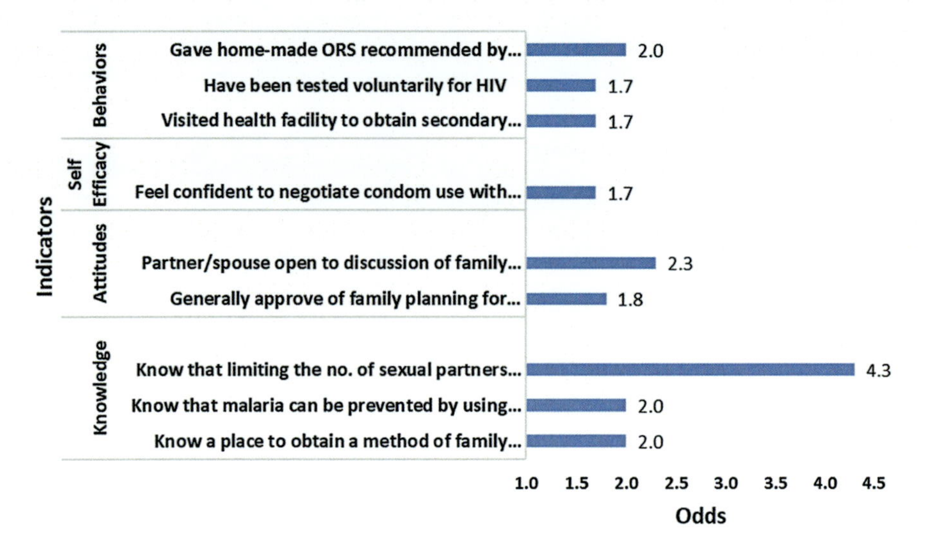

Fig. 3 Multivariate effects of listening to *Agashi*, 2015 adjusted odds ratios comparing listeners and nonlisteners at least $p \leq 0.05$

seventh highest in the world (PRB 2017), driven by large family size. In general, men and women want large families, but 24% of married women ages 15–49 reported that they would like to space or limit births, but they're not currently using any method of family planning (INSD and ICF 2012). The top reasons married women cited for not using a modern method of contraception are the desire for more children (18%); personal, partner, or religious opposition (17%); fear of health effects (10%); and not knowing a method or a source (10%). Cost is cited by 2.5%, and lack of access is cited by 0.7% (StatCompiler 2017). To address these issues, PMC produced *Yam Yankré* (The Choice) in Mooré and *Here S'ra* (The Road to Happiness) in Djoula, two unique radio serial dramas of 156 episodes each, which aired from September 2012 through March 2014.

Clinic Monitoring

Clinic monitoring results show an increased uptake in family planning and reproductive health services during program broadcast, and the program was a contributing factor to the observed demand for services. Results found that 25% of new family planning and reproductive health clients surveyed said they sought services because of a radio program.

Impact Evaluation

Results of the summative evaluation in adjusted odds ratios using multivariate analysis are presented in Fig. 4 above. Results reveal that the program succeeded in positively changing audience knowledge, attitudes, intentions, and behaviors regarding family planning, child health/nutrition, female genital mutilation (FGM), HIV/AIDS, and sustainable environmental practices. With respect to knowledge, the

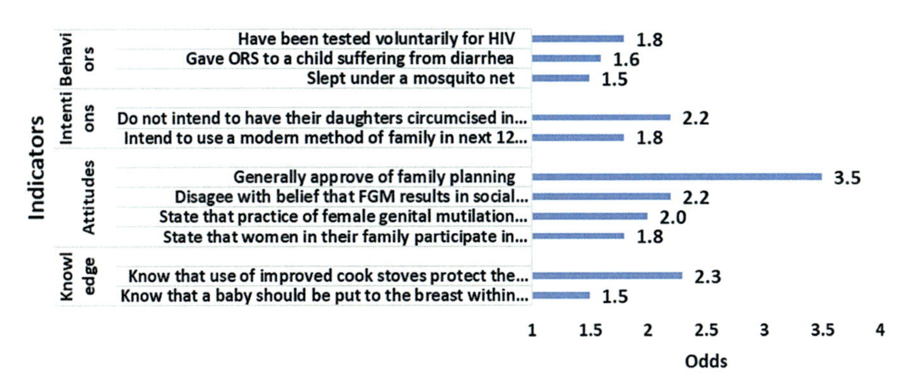

Fig. 4 Multivariate effects of listening to *Yam Yankre* or *Hɛrɛ S'ra,* 2014 adjusted odds ratios comparing listeners and nonlisteners at least $p \leq 0.05$

dramas impacted two new thematic indicators not observed in the Nigeria and Burundi case studies, including the environment and child health/nutrition. Listeners were 2.3 ($p \leq 0.004$) times more likely than non-listeners to say that the use of improved cook stoves can protect the environment and 1.5 ($p \leq 0.004$) times more likely than non-listeners to know that a baby should be put to the breast within an hour of birth. Similarly, under the intentions category, two thematic indicators not impacted by *Ruwan Dare* nor *Agashi* were found to be positively impacted: listeners were 2.2 ($p \leq 0.038$) times more likely than non-listeners to say that they do not intend to have their daughters circumcised in the future and 1.8 ($p \leq 0.005$) times more likely than non-listeners to say they intend to adopt a modern method of contraception to limit the number of children. Regarding behaviors, while two (voluntary testing for HIV and treating diarrhea with a homemade solution) of the three indicators influenced by the dramas are not new, the third is. It relates to maternal health where listeners were observed to be 2.2 ($p \leq 0.016$) times more likely than non-listeners to say that they slept under a mosquito net the night before. As observed with *Ruwan Dare*, the effects of the dramas are strongest under the attitudes domain, with four indicators impacted. For instance, listeners were 3.5 ($p \leq 0.007$) times more likely than non-listeners to state that they generally approve of family planning. The other three indicators pertain to gender and reproductive health issues. On these indicators, results indicate that listeners, when compared to non-listeners, were 2.2 ($p \leq 0.046$) times more likely to state that they disagree with the belief that FGM results in social acceptance, 2 ($p \leq 0.014$) times more likely to state that the practice of FGM should disappear, and 1.8 ($p \leq 0.014$) times more likely to state that women in their family participate in decisions regarding the education of children.

Clinic monitoring estimated listenership to the two programs at approximately six million people. Regarding cost-effectiveness, the evaluation research on the impact of *Yam Yankré* and *Here S'ra* revealed that the cost per listener was $2.00 (US). While this cost per listener is higher than those found for *Ruwan Dare* and *Agashi*, the cost per adoption of behavior attributable to one or both dramas is relatively low.

It is $3.00 (US) for treating childhood diarrhea with ORS and $4.00 (US) each for voluntary HIV testing and for sleeping under a mosquito net.

76.3 Conclusion

As has been proven in other parts of the developing world, the above three case studies (Nigeria and Burkina Faso from West Africa and Burundi from East Africa) indicate that PMC's serial drama methodology is replicable. The case studies also indicate that while each drama follows the same production process, the outcomes are anything but formulaic, yielding unique and powerful stories capable of generating social and behavior change. This is because the linchpin of the methodology is its thorough qualitative and quantitative research of the local context prior to program development. The methodology places the audience at the heart of the program, using formative audience research prior to the first broadcast and monitoring research with audience participation during the entire broadcast. Thus, the creative team is able to gain insights into audience responses to the characters and story lines during the creative process, resulting in a feedback loop that allows for adjustments by the creative team, which ultimately leads to audience engagement, realism, and authenticity. Of note is the methodology's ability to address very sensitive issues that are usually shrouded in secrecy as evidenced by the indicators being investigated here. Finally, the observed level of cost-effective impact proves the efficiency of such an extensive communication campaign designed to promote long-term behavior changes.

References

Bandura, A. (1986). Social foundations of thought and action. Englewood Cliffs, NJ: Prentice-Hall.

Barker K (2012) Sex, soap and social change: an examination of the elements underlying the successful application of entertainment-education. University of Cape Town, Cape Town

Bertrand JT, Anhang R (2006) The effectiveness of mass media in changing HIV/AIDS-related behavior among young people in developing countries. (Preventing HIV/AIDS in young people: a systematic review of the evidence from developing countries). WHO technical report series. WHO, Geneva

Institut National de la Statistique et de la Démographie (INSD) et ICF International (2012) Enquête Démographique et de Santé et à Indicateurs Multiples du Burkina Faso 2010. INSD et ICF International, Calverton

Nariman H (1993) Soap operas for social change. Praeger, Westport

National Population Commission (NPC) and ICF Macro (2009) Nigeria Demographic and Health Survey 2008. National Population Commission and ICF Macro, Abuja

Papa MJ, Singhal A, Law S, Pant S, Sood S, Rogers E, Shefner-Rogers CL (2000) Entertainment-education and social change: an analysis of parasocial interaction, social learning, collective efficacy, and paradoxical communication. J Commun 50(4):31–55

Piotrow PT, Rimon JG, Winnard K, Lawrence Kincaid D, Huntington D, Convisser J (1990) Mass media family planning promotion in three Nigerian cities. Stud Fam Plan 21(5):265–274

Population Reference Bureau (PRB) (2017) Available at: http://www.prb.org/pdf17/2017_World_Population.pdf. Downloaded on 27 Dec 2017

Population Reference Bureau (2018) 2018 World Population Data Sheet. Washington DC, Population Reference Bureau
Rogers EM (2004) A prospective and retrospective look at the diffusion model. J Health Commun 9(Suppl 1):13–19
Rogers EM, Vaughan PW, Swalehe RM, Rao N, Svenkerud P, Sood S (1999) Effects of an entertainment-education radio soap opera on family planning behavior in Tanzania. Stud Fam Plan 30(3):193–211
Ryerson WN (2010) The effectiveness of entertainment mass media in changing behavior. Population Media Center, Shelburne
Singhal A, Cody MJ, Rogers EM, Sabido M (2004) Entertainment education and social change. Lawrence Erlbaum Associates Publishers, London
Smith RA, Downs E, Witte K (2007) Drama theory and entertainment education: exploring the effects of a radio drama on behavioral intentions to limit HIV transmission in Ethiopia. Commun Monogr 74(2):133–153
Sood S, Menard T, Witte K (2004). The theory behind entertainment-education. In: Singhal, A., Cody, M.J., Rogers, E.M. and M. Sabido (Eds.) Entertainment-Education and Social Change: History, Research, and Practice. (pp. 117–149). Mahwah, NJ: Lawrence Erlbaum Associates.
StatCompiler. Available at https://www.statcompiler.com/en/. Downloaded on 27 Dec 2017
Vaughan PW, Rogers EM (2000) A staged model of communication effects: evidence from an entertainment-education radio drama in Tanzania. J Health Commun 5(3):203–227
Westoff CF, Bankole A (1997) Mass media and reproductive behavior in Africa. Demographic and health surveys analytical reports no. 2. Macro International Inc, Calverton
World Bank. Available at http://www.worldbank.org/en/country/burundi. Downloaded on 30 Dec 2017

The Role of Participatory Communication in Strengthening Solidarity and Social Cohesion in Afghanistan

77

Hosai Qasmi and Rukhsana Ahmed

Contents

Abstract

Participatory communication is an important consideration in the development process (Tufte and Mefalopulos 2009). Communication plays an important role in empowering people to influence the decisions that affect their lives and to strengthen solidarity and social cohesion among them (Noorzai 2006; Tufte and Mefalopulos 2009). The purpose of this chapter is to discuss the role of participatory communication in strengthening solidarity and social cohesion through development programs in Afghanistan. Primary interview data facilitate discussion of how participatory communication approach of development programs can be effective in increasing unity and solidarity among members of the Afghan society. Nevertheless, of note is that ethnic diversity and power relations at the community level could affect solidarity among people, both in negative and positive ways. As such, people's participation in development activities can

H. Qasmi
Institute of Feminism and Gender Studies, University of Ottawa, Ottawa, ON, Canada
e-mail: hqasm075@uottawa.ca

R. Ahmed (✉)
Department of Communication, University at Albany, SUNY, Albany, NY, USA
e-mail: rahmed4@albany.edu

J. Servaes (ed.), *Handbook of Communication for Development and Social Change*,
https://doi.org/10.1007/978-981-15-2014-3_63

contribute to achieving equality, providing them the ownership of their development can help reduce conflict and thus increase collectiveness.

Keywords
Afghanistan · Community development · Development communication · Ownership · Participatory communication · Solidarity · Social cohesion · Trust

77.1 Introduction

After decades of conflict and fall of the dark regime of the Taliban in 2001, Afghanistan entered the period of development, reconstruction, and rehabilitation. To succeed in the process of reconstruction, rehabilitation, and development, the newly established government of the Islamic Republic of Afghanistan had to reach rural population of the country, expand opportunities to marginalized communities, and promote a sense of solidarity and unity among people. To achieve the goals, the government of Afghanistan, and its international partners, developed a community-driven development (CDD) program. As Beath et al. (2018) state, community-based development or community-driven development programs stress on community members' participation and bottom-up approach to development that allows people to specify and prioritize their needs and address them. Such projects are effective in terms of increasing participation and building social capital among community members by allowing them to select and manage projects based on their needs and priorities (Beath et al. 2018). Participatory development empowers beneficiaries that previously had limited control over their development decisions (Cooke and Kothari 2001). Considering the importance of people's participation in their development process, the National Solidarity Program (NSP) was implemented in Afghanistan to grant community members the ownership of their development (National Solidarity Program 2012) and to increase the state's administrative reach to the rural population (Beath et al. 2018).

Since the NSP is based on community-driven development approach that emphasizes two-way communication and participation, this chapter discusses the role of communication for development strategy through the community-driven development program NSP in strengthening social cohesion among people. Likewise, Beath et al. (2018) argue that CDD programs increase participation and build social capital among community members; it is valuable to explore the case of National Solidarity Program in Afghanistan that is considered as one of the largest and most successful programs in the country (National Solidarity Program 2012) and second largest in the world (Kakar 2005). Hence, the purpose of this chapter is to discuss the less studied role of interpersonal communication and participation in social change. Specifically, this chapter focuses on the case of Afghanistan's community-based program that used communication for development through a participatory approach to strengthen solidarity among people. Against this backdrop, preceded by an overview of the country and the program contexts, this chapter discusses the role of participatory communication in strengthening solidarity and social cohesion in Afghanistan informed by thematic analysis of primary interview findings within a development communication theory framework. The chapter ends with concluding remarks.

77.2 Country Context

Afghanistan is known in the international media either for the Soviet invasion during the 1970s and 1980s or for the Taliban regime. Focus on one part of the history of the country has resulted in a biased view toward the country and its people (Emadi 2005). However, Afghanistan, like any other country on the globe, has a diverse geography, people, and civilization (Emadi 2005).

Afghanistan is a landlocked country that depends on agriculture and livelihoods. A major part of the population lives in rural parts of the country (Beath et al. 2018; Torabi 2007). The rural life of the country has a collective structure (Noorzai 2006). Afghanistan is a diverse state where different ethnic groups, religions, and tribal communities live together treasuring its unique tradition (Emadi 2005; Murtazashvili 2009).

Due to the decades of conflict in the country, the central government has had weaker intervention in the rural areas of the country; therefore, rural communities function on the bases of customary local governance mechanism called *Jirga* or *Shura*, i.e., a participatory council led by village male elders (Beath et al. 2018). For villages to connect with the central government, there is mostly a mediator termed Malik, Arbab, or Qaryadar who connects the village with the provincial and central authorities (Beath et al. 2018). However, Jirga or Shura is a participatory mechanism; it is still not fully participatory because it does not include women (Boesen 2004). Likewise, throughout the history, reforms and development in Afghanistan had a top-down approach which most of the times conflicted with the values and norms of people (Noorzai 2006). Reforms were mostly centralized; development programs and initiatives were designed and planned at top levels without involving local perspectives (Noorzai 2006).

77.2.1 Program Context

In 2001, after the fall of the Taliban the then established transitional government in Afghanistan and its international partners acknowledged the need to be visible and to gain the trust of the rural population, which encompasses approximately 80% of the country's population; thus, the National Solidarity Program (NSP) was designed with the objective of reaching the rural population, gaining their support and trust, and involving them in the rehabilitation process of their country (Torabi 2007). The NSP, being facilitated by nongovernmental organizations (NGOs), allows the program to profit from knowledge, experience, and expertise of the NGO sector in Afghanistan buildup during more than the two decades of conflict in the country (Beath et al. 2018). Adapting the bottom-up approach to development, the NSP, as the largest development program in Afghanistan, has been able to introduce CDD in all 34 provinces of Afghanistan and thereby overcome challenges such as insecurity, dominant gender norms, and mistrust of the central government (Beath et al. 2018). As argued by Murtazashvili (2009), the Afghan Ministry of Rural Rehabilitation and Development (MRRD) was successful to gain the international donors support for the program by highlighting the need of building the social capital among people in rural areas as the NSP is based on the Afghan tradition of "Ashar" that refers to working together voluntarily.

The NSP is a community-driven development program in Afghanistan that gives local communities the ownership of their development. The National Solidarity Program [in Dari, Hambastagi Milli; in Pashto, Milli Paiwastoon] was developed in mid-2003 (Torabi 2007). MRRD executes the program, and the World Bank funds the program through bilateral donors funding pool and is facilitated by national and international NGOs (Beath et al. 2018). The program involves two main steps, first the election of Community Development Council (CDC) that should involve both men and women, and second, provision of block grants that is a specific amount of fund allocated to each villages based on the number of households living there (Beath et al. 2018). The block grant is utilized for community development projects selected by CDCs in consultation with village members (Beath et al. 2018). The process of the NSP implementation in each village is a three-year-long process during which CDCs select projects, design projects in consultation with villagers, submit their proposals to receive funds, and once funding is received, implement the projects (Beath et al. 2018). Such programs have been designed and implemented in many other countries for the reconstruction purposes. In Rwanda and East Timor, this model was implemented as a participatory peace-building program (Torabi 2007). However, Torabi (2007) believes that the East Timor project was not as successful due to its structure that opposed social and local norms of the communities. Torabi (2007) further explains that the Afghan National Solidarity Program (NSP) had strong input from local authorities throughout its design and implementation. The NSP is the Afghan government's only program that functions almost all over the country (Nagl et al. 2009).

77.3 Role of Participatory Communication in Strengthening Solidarity and Social Cohesion

In order to understand the role of participatory communication in strengthening solidarity and social cohesion through development programs in Afghanistan, semi-structured in-depth interviews were conducted with ten participants from the Khost province of Afghanistan, which is located in the Eastern part of the country bordering with the Northern Waziristan in Pakistan. These participants constituted three groups, one nonprofit organization as a facilitating partner (CARE International) that implements the NSP in the village, three CDC members, and six community members. The interview findings show that participatory communication approach helped strengthen solidarity and social cohesion among the Afghan people by instilling in them a sense of acceptance, ownership, and trust.

Connection between social cohesion and sustainable development, in recent years, has received augmented attention in both development theory and practice (King et al. 2010). Some studies (for example, Easterly et al. 2006; Ritzen et al. 2000) have exhibited a strong connection between social cohesion and development (King et al. 2010). Solidarity is visible among people when they show a desire to work together for achieving a common goal. Paulo Freire (as cited in Tufte and Mefalopulos 2009) believed in giving voice to a marginalized population of society by giving them power, time, voice, and space to define their problems and find solutions for them. Free and open dialogue is the main element of participatory communication (Tufte and Mefalopulos 2009); similarly, the NSP, through its participatory approach, has given marginalized populations an opportunity to identify their needs and problems by allowing them to come together and communicate. Maynard (2007) argues that the NSP is designed in a way that promotes solidarity among people. The NSP's participatory approach encourages community members to be involved in their own development projects by working together. The NSP requires peoples' involvement and support in all the steps of the project implementation process. Their collective action is visible by their participation in different activities of the NSP. Collective action and involvement starts from the first step of the program that is the election process of CDC members as representatives of their community. Community members collectively make decisions on their community development projects, including project implementation, monitoring and evaluation, and audit. The NSP requires community members to meet regularly, interact, and communicate frequently.

The participatory approach does not only promote collective action among people it also likely reduces conflict. In some cases, the NSP has shown that people put aside their conflict and issues to achieve their development goals and they likely give importance to their collective good more than their conflicts for the sake of community development and betterment. Community members demonstrated their desire and willingness to work together that would benefit everyone.

Participatory communication ensures inclusion of all the stakeholders, especially, the most marginalized by providing them a voice (Tufte and Mefalopulos 2009). To have a dialogic communication, a catalyst is required to facilitate the process (Tufte and Mefalopulos 2009), and in the NSP program, the CDC is the catalyst that facilitates the process. Tufte and Mefalopulos (2009) further argue that in addition to dialogue and reflection, participatory communication is also strongly action-oriented, which is demonstrated in the NSP program by encouraging and authorizing communities to execute every step to facilitate development of their own community.

Colletta and Cullen (2000) consider social cohesion an important element for sustainable peace through strengthening interpersonal and intergroup networks, trust, and reciprocity. As such, two-way communication is necessary to reunite different perceptions and prioritize development plans (Tufte and Mefalopulos 2009). Similarly, the participatory approach of communication adopted by the NSP likely increases interaction among community members by requiring them to have frequent meetings. Furthermore, the dialogic approach of participatory communication values communication in revealing oppressive conditions to encourage collective action (Wilkins 2008). Communication articulates social relations between people, and therefore participation is not possible without communication (Serveas 1999, as cited in Msibi and Penzhorn 2010). According to Msibi and Penzhorn (2010), there can be true participation only when there is involvement of people and development planners throughout the decision-making process, a genuine dialogue, and when people are empowered to control the actions taken.

The NSP not only gives communities the ownership of their development but also responsibility for their development. Communities decide and take the control of their decisions. One of the components of development communication theory is to allow people to take ownership of their own development. Development communication allows individuals to make the decisions that affect their lives. Siddique (2001) states that social cohesion is built when people consider themselves the stakeholders of their society. Tufte and Mefalopulos (2009) also assert that genuine participation and engagement by local people increases a sense of ownership among them. According to them, participatory communication empowers people to engage in their communities, feel committed to, and be able to have the ownership of problems (2009). Development communication claims that ownership of development projects by participants help sustain the project in the longer run. Similarly, the participatory approach of the NSP allows local people to take responsibility and ownership of their development projects. By achieving the ownership, they take the responsibility to sustain and protect their projects from failure or any risks.

Social cohesion and solidarity also require a sense of acceptance among people. Solidarity requires cooperative action, and cooperative action requires people to work together and respect each other's views (USATAA 2013). According to the Council of Europe (2001), "acceptance of diversity and the interaction between cultures foster harmonious relations between people, enrich their lives and provide them with creativity to respond to new challenges. It is not the denial, but rather, the recognition of differences that keeps a community together" (p. 11). Through active participation, people in a society will most likely to develop a sense of acceptance, resolve tensions, and respect each other's views. Further, working together and interacting with each other more often increases the understanding among people. A participatory approach will likely not only promote working together for a common objective but also resolve issues that arise when implementing projects. The NSP, through its participatory approach, was able to increase communication among community members. However, conflict resolution in a collective manner is part of Afghan culture and etiquette. Afghan cultural values are carefully integrated into the design of the NSP program that is also a contributing factor of its success and acceptance of this program. The NSP's bottom-up participatory approach enables people to communicate and interact more frequently, which more likely increase a sense of acceptance among them. Fearon et al. (2007) also argue that through the participatory approach of a community-driven program with a successful community mobilization, it is possible to reduce tensions present in a community and encourage collective action. Through the establishment of the CDCs, community members are now connected to their CDC representatives besides having different roles and responsibilities in their particular community. Community members are bound together for their community's development. Regular meetings held by the CDCs with community members increase interaction among them.

Furthermore, trust is an important factor in building and strengthening solidarity among people. According to the opinion survey by the National Unity and Reconciliation Commission (2005–2007), a meaningful relationship is developed as a result of trust among people. Trust helps to encourage people to work together for achieving mutual goals and increase the effectiveness of community projects, and reciprocity, responsibility, and moral obligations help to develop trust among people. Wlech et al. (2005) believe that trust is important in mobilizing individuals, encouraging them to work toward a common goal, and making community projects more effective. Likewise, the World Bank, Communication Initiative, and Food and Agriculture Organization (FAO) of the United Nations report (2007) posit that development communication is about "seeking change at different levels, including listening, building trust, sharing knowledge and skill, building policies, debating and learning for sustainable and meaningful change" (p. xxxiii). Thus, trust is an important component in development communication. Trust is required among people to work together for the sustainable development of their society.

Individual honor, loyalty to colleagues and friends, tolerance to others, and directness are some cultural qualities that most Afghan share (Dupree 2002). Accordingly, trust, loyalty, and respect are important fabrics of the Afghan culture (Emadi 2005). The first step of the NSP is to elect community councils, i.e., CDC members, through voting. Community members elect those whom they trust to represent them accurately, understand their needs, and those who are transparent. However, rural communities in Afghanistan have Jirga or Shora, an informal institution that come to gather at times to resolve a conflict in the community. Jirga or Shora members are not elected but are elders of the community. However, CDC members are elected in a democratic voting. Election of CDC members is a new shift that allows community members to designate those they trust and want to represent them. Additionally, people also have a role in monitoring the work; therefore, CDC members have to be truthful while reporting to people. Community members are involved in every step of the project; therefore, there is no room for misleading reporting. As mentioned earlier, honesty and integrity are important cultural values of the Afghan society and particularly among rural population that is dominantly characterized by communal and tribal life style. Additionally, people trust each other when there is transparent information sharing among them, and thus truthfulness can strengthen the bond among people. Therefore, communication can play an important role in building trust among people.

The NSP program's social audit approach has the purpose of reducing corruption and increasing the transparency, where monitoring and auditing are performed transparently with the involvement of community members. Furthermore, a CDC member would want to display his/her honesty and transparency to community members and be truthful to them and to maintain their respect, as dignity is one of the more visible cultural traits in the Afghan society (Dupree 2002).

People's involvement and engagement is crucial in community-based projects, and development communication emphasizes participation and members' engagement in their own development process. One of the objectives of the NSP, through its participatory approach, is to increase Afghan rural populations' involvement and engagement in their development. Through the NSP, people are not only engaged in the development of their community but also developing trustworthy relations among each other to work together. Communication and interaction foster trust and thereby promotes cooperation. In some communities, the participatory communication approach of the NSP has been able to reduce the chances of conflict, increase the level of trust among community members and their CDCs as well as a sense of acceptance among people. However, the NSP still needs to reach the goal of strengthening 'intercommunal' solidarity and social cohesion, i.e., solidarity at a larger level. Although the NSP's success can be seen in building solidarity and social cohesion among people in small and homogenous groups and communities, the goal of strengthening solidarity and social cohesion at the national level is yet to be achieved.

77.4 Conclusion

In summary, reconstruction and development through participation by the NSP appear to be a promoter of solidarity and social cohesion among people in a small community of the Khost province of Afghanistan. In general, the process of participatory communication seems to reinforce the consciousness and willingness of communities regarding cooperation and contribution to reconstruction and development. The process of participation and communication can help to reduce conflicts and promote cooperation among people. Participatory communication, through the NSP, seems to promote intercommunity cooperation among people in the villages in Afghanistan. This participatory process of the NSP is a key to empowerment of rural population. Participatory communication through the NSP has the potential to reduce conflicts and increase solidarity and social cohesion among people in Afghanistan. CDCs play an important role in encouraging people to participate in community development and work in collaboration for their community development.

The establishment of the NSP in 2003 was an initiative to develop the rural side of Afghanistan, where the majority of the population resides, and provide an opportunity for people to work together for their development. Through the NSP, the rural population can decide and design their development projects according to their needs and priorities by working together. The NSP is one of the most successful programs in Afghanistan that is functioning all over the country. One of the reasons behind the success of the NSP is its design. People have accepted it because the notion of working voluntarily as "Ashar" and consulting each other as "Jirga" is present in their culture. The NSP is a community-driven development program that allows development from a bottom-up approach. The NSP has been able to communicate its message of strengthening solidarity and social cohesion through its participatory communication approach. For Afghanistan, the reconstruction of social damages of war and conflict is as important as rebuilding physical infrastructure.

Social cohesion and development have strong ties. Development brings people together and motivates them to achieve a mutual goal. Working together for a common goal is an indicator of strong bond among people. In the meanwhile, meaningful development requires people's participation at every step, the involvement of all groups, and providing an opportunity to voiceless groups. People should be the agents of change in their communities and societies. People's engagement and interaction with each other is crucial for reducing tensions, building trust, developing social cohesion, and establishing social relationships.

77.5 Cross-References

- ▶ A Community-Based Participatory Mixed-Methods Approach to Multicultural Media Research
- ▶ A Threefold Approach for Enabling Social Change: Communication as Context for Interaction, Uneven Development, and Recognition
- ▶ Asian Contributions to Communication for Development and Social Change
- ▶ Communication for Development and Social Change: Three Development Paradigms, Two Communication Models, and Many Applications and Approaches
- ▶ Development Communication as Development Aid for Post-Conflict Societies
- ▶ Multiplicity Approach in Participatory Communication: A Case Study of the Global Polio Eradication Initiative in Pakistan
- ▶ Multidimensional Model for Change: Understanding Multiple Realities to Plan and Promote Social and Behavior Change
- ▶ Participatory Development Communication and Natural Resources Management

References

Beath A, Christia F, Enikolopov R (2018) The national solidarity programme: assessing the effects of community-driven development in Afghanistan. In: Gisselquist MR (ed) Development assistance for peacebuilding. Routledge, New York, pp 20–38

Boesen IW (2004) From subjects to citizens: local participation in the National Solidarity Programme. AREU, Kabul. Available via reliefweb. https://reliefweb.int/sites/reliefweb.int/files/resources/3A038948365DBD7DC1256F020047544D-areu-afg-31aug.pdf

Colletta NL, Cullen ML (2000) Violent conflict and the transformation of social capital: lessons from Cambodia, Rwanda, Guatemala, and Somalia (English). Washington, D.C.: The World Bank. http://documents.worldbank.org/curated/en/799651468760532921/Violent-conflict-and-the-transformation-of-social-capital-lessons-from-Cambodia-Rwanda-Guatemala-and-Somalia.

Cooke B, Kothari U (2001) Participation: the new tyranny? Zed Books, London

Council of Europe (2001) Promoting the policy debate on social cohesion from a comparative perspective. Council of Europe Publications, USA. Available at. http://www.coe.int/t/dg3/socialpolicies/socialcohesiondev/source/Trends/Trends-01_en.pdf

Dupree N (2002) Cultural heritage and national identity in Afghanistan. Third World Q 23 (5):977–989

Easterly W, Ritzen J, Woolcock M (2006) Social cohesion, institutions, and growth. Econ Polit 18:103–120. https://doi.org/10.1111/j.1468-0343.2006.00165.x

Emadi H (2005) Culture and customs of Afghanistan (culture and customs of Asia). Greenwood Press, Westport

Fearon J, Humphreys M, Weinstein J (2007) Community-driven reconstruction in Lofa County: baseline survey preliminary report. Available via http://www.columbia.edu/~mh2245/FHW/FHW_baseline.pdf

Kakar P (2005) Fine-tuning the NSP: discussions of problems and solutions with facilitating partners (working paper series). AREU, Kabul. Available via AREU. https://areu.org.af/wp-content/uploads/2015/12/530E-Fine-Tuning-the-NSP-WP-print.pdf

King E, Sami C, Snilstveit B (2010) Interventions to promote social cohesion in sub-Saharan Africa. J Dev Eff 2:336–370. https://doi.org/10.1080/17449057.2010.504552

Maynard KA (2007) The role of culture, Islam and tradition in community driven reconstruction: the international rescue committee's approach to Afghanistan's national solidarity program. (Evaluation report). International Rescue Committee, Kabul

Msibi F, Penzhorn C (2010) Participatory communication for local government in South Africa: a study of the Kungwini local municipality. Inf Dev 26:225–236. https://doi.org/10.1177/0266666910376216

Murtazashvili JB (2009) The micro foundations of state building: informal institutions and local public goods in Rural Afghanistan. Doctoral dissertation, University of Wisconsin-Madison

Nagl JA, Exum AM, Humayun AA (2009) A pathway to success in Afghanistan: the national solidarity program (policy brief). Center for New America Security. Available at http://www.cnas.org/files/documents/publications/CNAS%20Policy%20Brief%20-%20Supporting%20Afghanistans%20NSP%20March%202009.pdf

National Solidarity Program (website) (n.d.) National Solidarity Program. http://www.nspafghanistan.org

Noorzai R (2006) Communication and development in Afghanistan: a history of reforms and resistance. Master's thesis. Available via Ohiolink. http://rave.ohiolink.edu/etdc/view?acc_num=ohiou1154640245

Ritzen J, Easterly W, Woolcock M (2000) On 'good' politicians and 'bad' policies – social cohesion, institutions, and growth. The World Bank, Washington, DC

Siddique S (2001) Social cohesion and social conflict in Southeast Asia. In: Colleta N, Lim TG, Kelles-Viitanen A (eds) Social cohesion and conflict prevention in Asia. The World Bank, Washington, DC, pp 17–42

Torabi Y (2007) Assessing the national solidarity program: the role of accountability in reconstruction. Tiri, London

Tufte T, Mefalopulos P (eds) (2009) Participatory communication: a practical guide (electronic resource). The World Bank, Washington, DC

USATAA (2013) Building community through cooperation. In United States of America transactional analysis association (Joomla Website Design). Available at. http://www.usataa.org/articlesand-links/articles/24-building-community-through-cooperation

Welch MR, Rivera RE, Conway BP, Yonkoski J, Lupton PM, Giancola R (2005) Determinants and consequences of social trust. Sociol Inq 75:453–473. https://doi.org/10.1111/j.1475-682X.2005.00132.x

Wilkins K (2008) Development communication. In W. Donsbach (ed.) The International Encyclopedia of Communication. Oxford: Wiley-Blackwell, pp. 1229–1238

World Bank, Communication Initiative, Food and Agriculture Organization of the United Nations (2007) World congress on communication for development: lessons, challenges, and the way forward. The World Bank Publications, Washington, DC

Sinai People's Perceptions of Self-Image Portrayed by the Egyptian Media: A Multidimensional Approach

78

Alamira Samah Saleh

Contents

Abstract

The media's marginalization of the Sinai Bedouins in Egypt could explain the popular image of them as being considered second-class citizens or just as smugglers, arms dealers, or human traffickers.

Little or no benefits have reached the local inhabitants from the development of the Sinai's tourism industry, and Bedouin communities are largely prevented from working in these luxury resorts. A career in the army is also prohibited, given the Egyptian military's tendency to restrict Bedouin conscription. Any response to the Sinai's current troubles must however address the socioeconomic inequalities affecting the local inhabitants and avoid repeating past mistakes that have essentially tended toward securitization of the issue. Although the officials of the Egyptian authorities introduced many symbols and discourses of unity to Sinai people, quick observations show that the development of a sense of belonging at their significant level is almost nonexistent. By drawing upon the established social identity theories, risk perception scales, this chapter takes the

A. S. Saleh (✉)
Faculty of Mass Commination, Cairo University, Giza, Egypt
e-mail: Samahsaleh2002@cu.edu.eg

J. Servaes (ed.), *Handbook of Communication for Development and Social Change*,
https://doi.org/10.1007/978-981-15-2014-3_40

investigation back to basics. How Sinai people interpret mediated messages about themselves and how these messages can help enhance their feeling of risk in a country they even don't have legislation to be the owners of their homes, anxiety, and belonging? Is the media involved in getting some citizens of the Egyptians feel indeed Egyptians and others do not?

Keywords

Sinai · Marginalization · Self-image · Egyptian media · Development

78.1 Introduction

Having suffered for many years from miserable, risky, and isolating socioeconomic challenges, the Sinai's 61,000 km^2 peninsula is one of Egypt's least developed regions, with the highest rates of unemployment and poverty.

Major bombings occurred in 2004, 2005, and 2006 and repeatedly in severe and terrifying terroristic attacks in the last few years, ended by the most brutal attack in November 2017 on one of the most prominent mosques there (Al-Rawda Mosque) and in a praying time which resulted in killing more than 300 persons.

More than 2500 people were imprisoned as a consequence of the repeated attacks, and many of them were detained for years without charge while being prevented visits by their family members or lawyers. However, each time, the political regime responds with a security-first strategy, sending in the army to conduct punitive sweeps and mass arrests (Dessì 2012).

These harsh measures only added to the widespread marginalization of the Sinai people among the Egyptian society, being considered as second-class citizens or just as smugglers, arms dealers, or human traffickers (Abdel-Meguid 2012).

And once again, comprehensive development is forgotten, every time, only for an argument to be revived when the next terrorist attack occurs!

Rarely done in the context of the social identity theory, or in the context of broader research of media effects on self-image and others' image, the first-hand analysis of this study can explain the difficulties with the formation of the Sinai people's identity, their self-image, their mediated image, or their feeling of risk or security.

However, such analysis will not solely depend on what's in the media but what's around the media that matters to Sinai people. Meso- and microlevel analysis that affects the making of their image, their understanding of messages in programs, and ultimately their sense of who they are will come to play a crucial role in contextualizing the Sinai's people image. On the meso-level of the Sinai people self-image, some factors shape the process: military, economic, social, and long years of marginalization.

The microlevel of analysis refers to the Sinai people being Egyptians: what they do, with whom they engage in everyday life, and how they feel about their safety in a country they even don't have legislation to be the owners of their homes!!

Only after that, we could ask about the media they watch on media, how they interpret mediated messages about themselves, and how these messages can help enhance their feeling of risk, anxiety, and/or belonging!

However, the rigorous question for the current analysis will be based on the media involved in getting some citizens of the Egyptians feel Egyptians indeed, and others do not.

78.2 The Beginnings That Matter

To fully understand the meaning of the events in Sinai, we need to look back elsewhere.

> The fate of the Sinai's people as 'nomad' was sealed for good with the signing of the Sykes-Picot agreement in 1916. The agreement divided the Arab provinces of the Ottoman Empire outside the Arabian Peninsula into areas of British and French control or influence. (Serhan 2012)

As a roaming people whose livelihood depended on seasonal movement from pasture to another, cementing the border left them with no choice but to become sedentary. This cutoff from "fundamental elements in their economic, commercial and social universe" (Dawn 1986, 30) exposed Sinai to a whole new level of poverty. The fact that the people were excluded from the Egyptian army at the time, a major force for upward mobility, did not help their case either. But worst of all, they were denied identification cards long after their compatriots had received them (Cole 2003, 250).

All these combining factors are not just pushing toward underestimating the Sinai's people and not just to obliterate their way of life, but it created a sense of erasing them from existence altogether. The 400,000 Bedouins of Sinai thus never fitted into the Egyptian nation-building project, from which they were marginalized from the outset.

However, what is considered, very generally, as "Bedouin" is a mosaic of populations that reflect the complicated settlement history of the peninsula, with clear distinctions regarding origins, traditions, economic activities, and even language (Middle East/North Africa Report 2007, 10).

From 1979 to 1981, the years in which Egypt was under the presidency of Anwar Al-Sadat and regained control of Sinai from Israel and during which Egypt also faced its first – and arguably most threatening – Islamist insurgency, another phase began. These two processes are elaborately linked, which didn't end by the assassination of Al-Sadat in 1981 (Goldberg 2015).

Under the 1979 peace treaty with Israel, it was supposed that Egypt had got back its control over the Sinai Peninsula. This was a time during which Sinai was actively introduced into Egyptian public knowledge as a symbol of pride and dignity. The return of the Sinai was presented to the Egyptians as military and social will's victory, something that the Mubarak regime was keen on tapping into. On an almost daily basis, national television aired footage of then President Hosni Mubarak raising the Egyptian flag on "Taba," the last point handed over to the Egyptian army on April 25, 1982. In the background, the voice of the Egyptian famous singer "Shadya" rang out, "All of Sinai has returned to us. Egypt is rejoicing today and forever after!"

Sinai was the focus of attention during those years, but despite all the media hype, it was still unrecognized terra for most of the Egyptians. Sinai existed in its national symbolism rather than in its developmental or cultural continuity with the homeland. It was hardly an attractive place for either strategic inhabitation except for a few nomadic Bedouin tribes with whom most Egyptians rarely had any direct contact with or through well-established industries. It has kept on being far away, dryish, and poorly connected to the rest of the mainland (Hosni 2013).

The tribe is the building unit of Bedouin life. It can include any number between 20 and 20,000 people. Social roles are determined by age, sex, and seniority and work best in the absence of social differentiation. "The more similar people are in other respects, the more fully kinship, sex and age can differentiate their roles") Sivini 2007, 447).

The fighting capabilities of the tribe are fundamental to maintain the strength, necessary for survival and loyalty. It is of topmost importance. For this reason, they are commonly viewed not as individuals but as groups.

These tribes have their political structure and laws known as *Ourf*, meaning what is known, or what has become a custom. In the case of any disgraceful behavior, for example, the Bedouin sheikhs convene and try the perpetrator/attacker in a Bedouin court. Unlike civil penal law, which is based on punishments, Bedouin law is based on retribution. "An eye for an eye, a tooth for a tooth" is the gist of their sense of justice.

Sinai's people feel that their customs and tribal laws have been largely overlooked in the development of the peninsula. This has been exacerbated by the government's heavy-handed, security-focused approach to the region in response to terrorist attacks since the mid-2000s (Stothard 2014).

From 1982 onward, Egypt thought of the Sinai question as above all a matter of population settlement. Increasing the population was an essential element in the way to control and integrate the peninsula. The "Egyptians," as they are called locally (even by themselves), come from the entire Nile Valley, including the Delta and Upper Egypt cities. They comprise distinct groups, distinguished by accent and economic activity, and most often settle together in clusters or associations according to their village or governorate of origin. For example, in North Sinai, families from Monufia (a governorate in the Delta) are essential. The government greatly encouraged people of this governorate (from where current president Sisi and former President Mubarak and even the late president Sadat originate) to migrate, including by offering different incentives like attractive salaries and fixed public sector employment (International Crisis Group 2007, 11).

Although the government has undertaken development initiatives in the Sinai, they had been poorly funded, of limited scope, and managed by disengaged Cairo-based residents. The lack of real work opportunities in the Sinai is such that many would inevitably make a living through illegal activities or the shadow economy.

The development in the Sinai has centered primarily on the tourism industry in the south. South Sinai with about 160,000 people is lightly populated, but it has world-class beaches, scuba diving, and hotels.

Settlements grew in tandem with the tourism industry from the 1990s onward. Although some of South Sinai's people benefited from tourism activities, employers primarily recruited Egyptians from the Nile Valley. Bedouin communities are largely prevented from working in these luxury resorts which considered open discrimination against Sinai residents (The tourism sector accounts for approximately 11% of Egypt's gross domestic product).

Foreign mass tourism in Egypt is a post-Sadat phenomenon, and it has provided regime opponents with a soft target. The most famous and still the most deadly incident is the 1997 attack on Hatshepsut's Temple in Upper Egypt in which 68 people (including 6 attackers) died. The attack was carried out by members of the Islamic Group who opposed a truce some of their leaders had arranged with the Mubarak government. In 2004, 2005, and 2006, hotels and tourist attractions in Taba, Sharm el-Sheikh, and Dahab in South Sinai were attacked, and more than 150 people killed in total.

However, tourism has not done well since 2011 as revenues have shrunk from over $14 billion to under $ 5 billion a year due to the disturbing political scene and deteriorating security circumstances.

With no stable source of income and exclusion from fundamental and indigenous rights in Sinai, such as the right to register land and property they own that thus remain without any legal guarantees, the peninsula could become a safe haven for foreign jihadists. If South Sinai remains a tourist destination, attacks there may be a "red line" for the Egyptian government, and the tourism industry may further collapse.

North Sinai has never been the object of much investment by Egyptian governments. It has suffered an even more catastrophic economic collapse than the south. Although North Sinai is also lightly populated, with 420,000 people and it is much larger than the south.

The region's economy would inevitably orient toward being a large market for Palestinian people. Moreover, the North Sinai's economy includes human trafficking. According to some estimates, tens of thousands of Eritrean and other African citizens have attempted to enter Israel through the Sinai border crossings illegally.

Since the 2011 revolution against the Mubarak's regime in Egypt, there has been increasing instability in the Sinai Peninsula. Sinai's population retaliated against the security state, but the cause of violence in the peninsula soon transitioned from Bedouin grievances to Salafi jihadism. Hard-line militant factions used Sinai as a launch point for attacks against the Egyptian state for its way of handling the struggle with their fellow Muslims. After the July 2013 events against former President Mohamed Morsi, who is a Muslim Brotherhood leader, attacks in and from Sinai noticeably increased. Even when the Egyptian military launched large, sustained operations to counter the threat, Sinai's Salafi jihadists expanded their fight: by attacking Egypt's "mainland" with car bombs, carrying out an assassination campaign against Egyptian security officials, and using advanced weapons to take down a military helicopter.

Until recently, Sinai's Salafi jihadist groups' primary targets were Egypt's security establishment. The growing concern that their targets would continue to shift, however, became more justified with a February 2014 attack on tourists. Such a trend puts Western interests, including tourists, foreign embassies, and perhaps even ships in the Suez Canal into their sights.

Unfortunately, the comprehensive reactive approach from the different Egyptian political regimes and their practices usually aims merely to prevent the next attack, instead of finding profound solutions to the underlying problems that fuel militancy and terrorism inter-generationally (Gold 2014, 19).

Now, it seems more apparent that an overarching theme of Egyptian policy toward Sinai has been the focus on Sinai as a "security" issue.

The successive periods have differed in the intensity of Cairo's response to Sinai's impact on national security, but there has been a little and judicious attention of Sinai in any other context. "Since Mubarak's ouster, successive governments have spoken of the need to address Sinai's economic, developmental and unemployment problems, but no major projects have been implemented" (Gold 2014).

78.3 Development That Is Sung For

Although three times the land mass of the Nile Valley and Delta, Sinai's residents comprise less than 1% of Egypt's population (Afifi 2014).

For the sake of immediate necessity, it is supposed that the increasing size of the Egyptian population would primarily rely on relocating a sizeable portion of the population from the Nile Valley and Delta and Suez Canal governorates to Sinai.

The size of the population in North and South Sinai (including those originally from other governorates which are working in Sinai) is less than 600,000 people or less than 0.7% of Egypt's population, and it has a population density of fewer than 10 people per square kilometer (Al-Najjar 2014).

The number of schools in Sinai is 767 schools distributed among 533 schools in North Sinai, which accommodate 98,235 male and female students and 234 schools in South Sinai, which serve 21,881 male and female students.

Thus, it could be noted that in the Sinai Peninsula, which is characterized lack of population density per region, the rate of learners per class decreases as does the percentage of learners per teacher and the proportion of teachers covering the teaching load compared to the target rates. A phenomena that affects providing educational service with appropriate investment and operational efficiency.

Despite the success of educational policies in their ability to absorb, still, we find that the retention of students until the completion of the three stages of education still needs improvement. The dropout rates in Sinai exceeded the average dropout rate in the other governorates of the Republic. Moreover, it increases among female students compared to male students of the same stage as well as in Bedouin communities compared to urban ones. The solution to eliminating the phenomenon of dropout from education requires a package of interventions based on geographical targeting, conditional cash transfers that combine financial support for families, literacy programs, education and training in the skill-building programs needed by the labor market, as well as awareness of the importance of education, especially for females.

The low dropout rates not only decrease the waste of resources, but it also represents a drying of the sources of illiteracy. The illiteracy rate in North Sinai reaches 19.1%, and the rate of illiteracy among males reaches 13.2% while 25.4% among female. However, the illiteracy rate in South Sinai reaches 17.1% as it reaches 14.5% among men, whereas 22.6% among women and the majority of illiterates are in Bedouin areas and communities. This problem could be due to reasons related to shared cultural and religious values. Moreover, it is relevant to the nature of business distribution within the family in the desert and the degree of demand for them, as well as due to the high poverty rates (over 50% of Sinai Bedouin live in poverty). Needless to say, the majority (80%) of those who did not join schools are mainly girls in Bedouin areas.

Therefore, the status of education in the Sinai is still suffering from various problems, both at the level of lack of schools and shortage of teachers and the qualified characters. In addition to that, the government does not support education in remote places in Sinai where it is difficult to reach out the schools due to the lack of transportation that helps the teachers to reach their schools daily.

The reality of health services in different areas of the Sinai does not differ too much from their previous conditions in education. The absence of the efficient medical staff working in these institutions and the lack of access to medicines and diagnostic devices profoundly contributed to the low level of existing health services and even its total absence from some Bedouin communities.

Numerous declarations on investments in infrastructure and aid for the marginalized Sinai's population have been heard recently, and meetings have been held between the Egyptian interior minister and Sinai's sheikhs representing their tribes. But it takes more to regain the trust of people who – for decades – have been treated as if they were opponents working against the state.

A lack of legitimate economic opportunities, underdevelopment, and the denial of land claims has firmly established long-standing grievances with the Egyptian government. Combining these issues with high levels of unemployment has meant that smuggling and trafficking are, for many, the only source of income. Within the security vacuum that has developed in North Sinai, this and other serious forms of organized crime will increase.

The expansion of tourism has influenced the local Bedouins in that they increasingly interact with other cultures and struggle for their livelihood as they transform from a pastoral community to a residential one. This transformation has emerged in the type of jobs they acquire at present, and therefore they have started to move to a more modern lifestyle.

During the Mohamed Morsi's former presidency, a mixed approach was again employed. Morsi indicated his readiness to respond to the socioeconomic grievances of Sinai people and initiated talks with the Sinai's armed groups, mediated by Salafists and former Salafist jihadists. He also supported a relaxing of the overground movement of people and goods into Gaza. Despite these efforts, however, fighting in the Sinai continued, prompting a return to a military response.

However, the political transformation that had occurred in Egypt recently and the new roles assigned and connected to youth challenged many patriarchal powers in Sinai and even more the state authorities. Accordingly, the influence of the old sheikhs is not what it used to be. The old sheikhs no longer have their traditional patriarchal control over the various segments of their tribes. These further alterations left the new social powers in Sinai more attracted to radical Islam (Tuastad 2013).

Unlike other equally impoverished areas of Egypt, Sinai in the wake of the 2011 uprising did have one important economic sector: illegal trade. It was precisely its illegality that made it rewarding. Illegal business (in the north) and tourism (in the south) are important drivers of the economy and politics in these two areas.

It was briefly highlighted here that Sinai's people don't have the right to own their lands. However, what is worst is the recent displacement policies that the government decided to implement after the recent severe terroristic attacks from Sinai's area. Here we come with "Loss of Home" new grievances!!

In 1963, thousands of Nubian families were displaced for the sake of building the High Dam in Aswan. At the time, the events were framed as a sacrifice for the nation, a small price Nubians had to pay for the sake of a substantial national project that was going to change Egypt's future.

Before 1979, when the Egyptian government re-controlled a narrow strip of land adjacent to the Suez Canal, the governor of Sinai began to plan for development. These plans reflected the input of many interest groups with the military on the top, the most famous industrialist families, the ministers, and members of the governing political party. (Lavie, 75).

In 1975, the office of the Sinai governor was filled with petitions, requests, and demands from who claim property and other interest groups in Sinai that the governor published an open letter in one of the most Egyptian influential newspapers (Al-Ahram) asking for moderation.

However, once a consensus was reached about a particular project, the governor unilateral announces the plan for it without any public debate or agreement.

In 1980, "Law 104" casted the interplay of elite interest groups by the government's consistent behavior in publicizing and implementing its plans. Further, steps were made including plans which allowed the state ownership of desert land and thus making the whole Sinai a governmental property and changing the permission of private ownership totally for the benefit of some elite groups. The law had some devastating effects on the Bedouins. Their land claims were not legally recognized, and they were subsequently displaced with no government compensation (Lavie, 76). In their place, the land was repopulated with peasants to solve the unemployment problem in the urban center (Serhan 2012).

In 2014–2015, the same discourse repeated once again to justify the displacement of over 1,000 families in Rafah/Sinai to create a buffer zone at the border, which the state insists is necessary for national security. However, the correlative damage – which has severe social and, potentially, security consequences – is largely ignored (Afify 2014).

The people of Rafah have been living under a stifling 4 p.m. curfew for more than 3.5 years now. They have been living in the midst of military operations that locals report have taken civilian casualties.

Debates about whether this is a good move or not are almost nonexistent. A campaign widely backed by a media establishment moved in lockstep with official policy line. Reports in mainstream media emphasized that the people of North Sinai are willing to face relocation for the good of the country (Esterman 2014).

However, demolition began, and pictures were circulated. Photos of buildings splintered in the air great clouds of smoke appeared under headlines declaring, "Sinai sacrifices for the sake of Egypt," as the privately owned newspaper *Al-Masry Al-Youm* put it.

The human cost of these state measures is continuously represented as a piddling "price that has to be accepted, even overlooked, for a higher purpose. Accordingly, the media coverage of the recent Rafah operation has blatantly ignored the human factor, only covering the state's security perspective" (Afifi 2014).

Newspapers this week led with reports of President Abdel Fattah al-Sisi's meeting with Rafah community leaders, where he expressed his desire to compensate the recently displaced residents promptly. This is in addition to his pledge to plan development projects for the area.

Meanwhile, Sinai residents have complained in the Egyptian media about human rights abuses and the feeling of being treated with suspicion by the rest of the country. On 25 January 2013, 13 domestic human rights groups wrote an open letter to the Egyptian government, advocating a comprehensive approach which states that addressing terrorism requires that a more comprehensive vision should be adopted, a vision that should take into consideration the economic, social, and political circumstances in which terrorism emerges and spreads.

From a developmental perspective, the history of Sinai is a series of interconnected programs that hold many promises by statements of needs and field reports. The first was a report conducted by the firm Dames and Moore between 1980 and 1983 for the Ministry of Housing and Development. There was also the Ministry of Reconstruction Plan of 1981 and the National Plan for the Development of Sinai issued by the Ministry of Planning in 1992)Hosni 2013).

Then the European Union funded South Sinai Regional Development Program (SSRDP) in 2001. Then, following the ousting of Mubarak in 2011, the troubled security situation in North Sinai revived the development discourse. In 2012, the government submitted a draft bill that should lead to the foundation of a "permanent" Sinai development agency. A step that hadn't been finalized yet!!

What these development plans had in common was a commitment to tourism, both as a means for development and a dream in itself. The overall futurity was fit out toward the turning of Sinai into a tourist destination of international import. Large strips of coastland in the southeast were sold to investors for low prices. The Gulf of Aqaba coast was labeled the "Red Sea Riviera." Each one of these development plans was an update of the previous one (Hosni 2013).

We could argue that much of the violence and disorders that erupted since 2011's revolution is more of a reaction against the government's attempts to bind the underground economy (by destroying the tunnels) than an ideologically motivated plan as it is usually presented. However, the peninsula is compounded by a long history of marginalization and poverty; so it looks like a time bomb, even without taking into account the complex politics on the other side of the border. The situation in the peninsula is a reflection of an endemic illness and oppression which is not limited to the border region (Hosni 2013).

These circumstances bring us to our main argument: Who is responsible for producing these images? And are they intrinsic to Sinai's nature, or carefully constructed by external forces to serve specific interests?

78.4 The Politics of Image

Based on the assumptions of social identity theory, promoting certain media representations which feature particular aspects of specific groups and ignoring others plays a role in creating shared norms and activating the use of these constructs in subsequent evaluations for this group/s (Mckinley et al. 2014, 2). "In particular, media messages have the potential to (1) influence the importance/relevance of, and ability to prime, different group memberships; (2) contribute to viewers' perceptions about the features/dimensions that characterize different groups; (3) provide norms of treatment for different groups; (4) define the status and standing of different groups; and ultimately (5) normalize these notions by suggesting that media representations are consensually accepted" (Mckinley et al. 2014, 3). Similarly, the current contribution argues that the Egyptian media was an essential tool for self-perceptions and others' perceptions of Sinai people. But before getting into the various self-perceptions that were gathered throughout the current contribution (The current data and statistics were gathered from (343) inhabitants of North Sinai's governorate between 1 December 2015 and 22 January 2016), it is worth passing by some of the literature that dealt with the Egyptian media's representations of Sinai's people in more details.

In 1987 Egyptian public television aired the soap opera Sonbol Ba'd al-Milyon (Sonbol after the Million), a comically exaggerated series of sorts starring the famous Egyptian comedian Mohamed Sobhy. It came as a continuation of a popular first part titled The Million Pound Journey, in which Sonbol, the series' main character, rises the social ladder, despite his poor origins. Sonbol was an ambitious self-made young man from the Nile Delta who decided to invest his savings by reclaiming desert land in the Sinai Peninsula. Sonbol bought a piece of land from the government and set out to build a farm) (Hosni 2013).

Sonbol's journey was the first and most prominent representation of Sinai in Egyptian popular culture. There is a reason why Sobhy chose Sinai and the theme of land reclamation as the center of his plot.

Looked down upon with fear and suspicion, Sonbol symbolized this tense relationship (or rather non-relationship) with the land. It was arguably the first time the figure of the Bedouin is represented with some detail, albeit not favorably most of the time.

Eventually, the two groups reconcile; Sonbol won the tribe's confidence and gained their approval to marry Ziba and settle on the land.

In 1987 Sinai seemed so distant, but Sonbol showed it could be otherwise. He appeared as a man of the frontier, who brought the territory under his dominion and tamed its unruly and insurgent population.

In the case of Sinai, the notion of collective identity was the key to producing many of its people's stereotypes. By grouping all Bedouins into one undifferentiated lot, the *Badu*, many beneficiaries were able to generalize and thereby sidestep the particularities that make the Bedouin a living being. It allowed abstraction and left no room for other representations or renegotiation.

Egyptian media coverages of Sinai give a perfect example that encounters with this old-new other: the other land and its other peoples. It played out the old typical and ideal archetype of Egypt versus its age-old desert strangers. The desert is the avoided unknown, and its people, the Bedouins, are semi-mythical.

Generally, the image of the Bedouin as an "outlaw" became an official government coin, with no way of climbing the social or the economic ladder.

The logic was, if the state treated us as outsiders, then we might as well exist outside the law, thus supporting Sinai's reputation as a "lawless frontier."

After a while, the "outlaw" image temporarily took the backseat, and a new image topped the scene, driven by Israeli interests in South Sinai. To win the Bedouins' loyalty, Israel built a whole new platform for communication based on the Bedouins as "Children of Israel." According to the Hebrew Biblical reference (Serhan 2012).

In spite of all the Israeli efforts, Israel was still viewed as an occupier by the Sinai. One way for the Sinai to mark their territory was to come up with an image that would help identify and differentiate them. Here we came, the "Muslim Bedouin." The issue of self-identification became an urgent one when relations with outsiders conducted exceptionally through sheikhs and Bedouins came into increasing contact with the West. They felt that all Westerners, whether tourists or soldiers, Israelis or Europeans, Jews or Christians, invaded their privacy and threatened their traditions and customs (Lavie 1990, 72).

In response, the Bedouins encouraged an Islamic revival of a very paradoxical nature. They still work in tourism and came into contact with tourists every day, but all the money made was "purified" by bestowing generously on mosques and shrines of saints and extreme manifestations of religious enthusiasm. "'We are Muslims;' (they said) 'they are the Jews'" (Lavie 1990, 68).

Returning back to the current study's results, the Islamic identity came at the first response by Sinai people when asked to arrange their core identity as they feel it; 27.9% of the respondents confirmed such a choice and then came the Egyptian identity 26.6% and the Arab identity at the third rank 19.5%, while both the Sinai identity and the tribal one came at the end by 15.5% and 10.2%, respectively.

However, a decade later, this "Muslim" image faced a new set of challenges as the whole of South Sinai was once again part of the Egyptian land after 15 years under Israeli occupation.

One could imagine that an Islamic spirit in the Sinai would have helped bridge the gap between the Egyptian state and the Bedouins – but that was not the case.

Although President Sadat was popularly known as the Muslim president, "state-supported Muslim institutions, such as Al-Azhar University, invested this official policy with an Islamic sanction" (Smadar, 74).

"The fact that Egypt signed a peace treaty with Israel did not help bridge the gap either. Were the Bedouins to be viewed as fellow Egyptian returning from exile or were they, treacherous collaborators?" (Serhan 2012). More importantly, which of these images was more beneficial to the state?

To realize its development plans in South Sinai, the state decided to make the best use of the latter. Once again, the image of the "lawbreaker" was popped up and given a face-lift, and "the villain" was reborn, an all-encompassing figure who stood for many deficits all at once (Serhan 2012).

They were portrayed as the uncivilized, lawbreakers, treacherous, and dangerous. The most important thing for the state was to cater to the economic interests of Cairo's elite in the Sinai.

In the face of all these depictions, judgments, and marginalization, one image had come to simultaneously "embody and subvert all Bedouin stereotypes, all the paradoxes of the local South Sinai hybridization: the traditional Muslim nomad pastoralist ideology and the western culture imposed on (them) by international interventions: The Sinai man is the 'Jack of all trades'," (Serhan 2012).

The question now is: will the ever held negative image of Sinai's people held by different parties found its resonance by the Sinai people themselves?!

Previous studies indicated that every Bedouin stereotype out there has been readily absorbed and exploited by the Sinai people themselves. They have become these stereotypes. They exploited "the sheikh" to make connections with the government, they have exploited "the lawbreaker" to get rich, and they line the coast with fake palm-frond huts to export an "exotic" image to backpacking tourists!! (Serhan 2012).

On the contrary, the current study results showed that the majority of the sample asserted the responsibility of the Egyptian media representations through most of its news reports and programs for transferring and exporting such images to the rest of Egyptians.

More than 80% of the sample had beliefs that the Egyptian media coverage depicts negative images of them. Moreover, they went further, giving reasons for that. As Table 1 shows below, the majority of the sample confirmed that the brutal terrorists' attacks on Egypt's security forces, as well as the connections of Sinai's violent groups to international Salafi jihadist networks, including Al-Qaeda (AQ), are behind most of the exported image of Sinai.

This begs many questions concerning governmental transparency that expose the facts about what happens in Sinai and what is being done to confront the threat of terrorism there. Information monopolizing about Sinai and its inhabitants limits the credibility of media coverage and provides a fearful hole for inaccurate speculations (Akl 2015).

Table 1 Sinai people's justifications for/perceptions of self-image portrayed by the Egyptian media

Reasons	Frequency
The coverage focuses on the destructive actions of the terrorists	57.4
The coverage relies on the governmental news sources only	34.7
The coverage is influenced by the state's high policies	25.7
The coverage rekindles sedition and rumors against us	24.8
The coverage discloses the real ugly reality of Sinai	9

However, it was astonishing that 84.5% of the study's sample mentioned that they have friends and relatives all over Egypt and they don't find any kind of problems contacting and living with them. The main severe obstacle that resulted from such circulated negative images is the economic consequences that hit their incomes and everyday life activities. 86.9% confirmed that they feel paralyzed after the stifling enforcement and security campaigns. It is worth noting here that the government had imposed the emergency law procedures at the peninsula for more than 5 years now.

They complained the government ignorance of providing them with the promised financial aid to people affected by military operations:

> Since our news had been continuously stigmatized in the media, our ethnic and tribal dimensions remain underrated and incorrectly handled, and we've been treated as less-fledged compatriots under different political regimes, latent anger and frustration, grievances, and low self-esteem are escalating.

For the people here, reaching out to the tribes of Sinai never exceeded a ritual process or a protocol of gatherings of state officials with tribe leaders on an irregular basis. The ongoing results of such meetings usually never surpassed flashy statements and news headlines to the media highlighting future avenues of cooperation, ones that never came to be accomplished!! (Akl 2015).

Moreover, even as the Egyptian government launched large, sustained operations to counter the terroristic threats in Sinai, 79.3% of the respondents reported that they fear of being victims personally or any of their families because of the violence around them.

The growing and great concern that their interests and living conditions would get worse lets 64.7% of the sample say that they feel collectively insecure and they do fear significantly about the future of their children.

When it comes once again to the role of Egyptian media portrayals of Sinai, the research responses revealed many negative and annoyed attitudes of Sinai people toward the media.

The first criticism of the media was that it lacks enough, sustainable, and multiple angles of Sinai life and its people. 78.4% bitterly said that the media don't genuinely know the real importance of the Sinai territories.

While that was the first attitude, the worst one was mentioned by 68.5% of the sample who said that "The Egyptian media always oppressing us by the repeated accusations of spying, infiltration, and treason"!!

They elaborated further, blaming the Egyptian media for not hosting Sinai's people themselves to talk about their real problem, living conditions, customs, etc., instead of depending on the official sources only that always work only when there is a catastrophe that hit the peninsula!!. "Our news is mostly bloody." "Egyptian media coverage of Sinai is a mere reflection of the official marginalization, irresponsibility, and ignorance of Sinai!!"

It is worth noting here that Egypt has six local TV channels that cover the country's prominent regions like Alexandria channel, Delta channel, Upper Egypt channel, and Tiba channel. However, no single media outlet is explicitly concerned with Sinai region despite the exceptional social, environmental, political, and cultural circumstances that surround the peninsula.

"The media always present us as caring only for our tribal matters and belonging without paying any sincere devotion or belonging to our country as a whole," confirmed 60.9% of the sample.

Egyptian media seems to continue and strengthen the isolation strategy of Sinai's people to the extent that pushed 58.9% of them to say: "When we watch the Egyptian media, we feel like being second-class citizens who lack the simplest rights to live and to respect our privacy and customs"!

We could end addressing Sinai people status quo by the significant positive relationship between watching media covering their life and both their feeling of pessimism, anxiety, and collective insecurity and thinking of taking new cautious steps toward their future and their sons also ($p < .01$).

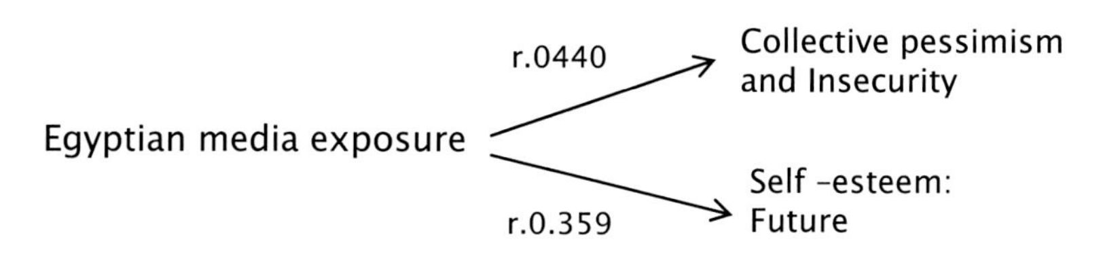

78.5 Conclusion

This chapter sought to analyze the portrayals of Sinai people in the Egyptian media as it pertains to the people themselves. It started by giving the reader an overview of the current Sinai's social, security, and economic status quo and structures. Then, it examined how image building could be or not an essential tool for political control over Sinai; before moving into how and why specific people's images are constructed and deconstructed by the mass media and how this is affecting and affected by the different and multiple aspects of development.

While many Arabic and Western studies tackled the Sinai Peninsula issues, yet almost no single research focused on the people themselves: How they perceive their own identity?! How they evaluate their image in other Egyptian's eyes?! And how they judge what is being presented to them via Egyptian media?!

A quick glance at the development measures that could boost the integration of Sinai's people into the formal Egyptian economy and provide alternatives to smuggling, we would easily realize that "the poor state of Egypt's public finances and the myriad challenges facing the country as a whole could mean that the needs of the Sinai may continue to be neglected despite the promise to invest" (Watanabe 2015).

Bright promises only follow almost every terroristic attack and then fade out. Although the officials quickly introduce many symbols and discourses of unity to the people there, quick observations more quickly show that the development of a sense of belonging at their popular level is slow.

References

(2007) Egypt's Sinai question, Middle East/North Africa Report N°61 – 30 January 2007. In: Files.ethz.ch. https://www.files.ethz.ch/isn/28095/061_egypts_sinai_question.pdf. Accessed 10 Nov 2017
(2014) Rights groups condemn terrorist explosions, concerned by increasing violence and excessive force by security forces – Cairo Institute for Human Rights Studies. In: Cairo Institute for Human Rights Studies. http://www.cihrs.org/?p=7971&lang=en. Accessed 21 Dec 2017
Abdel-Meguid W (2012) Egypt's Sinai: development versus security – opinion – Ahram online. In: English.ahram.org.eg. http://english.ahram.org.eg/NewsContent/4/0/51615/Opinion/Egypts-Sinai-Development-versus-security.aspx. Accessed 18 Aug 2017
Afifi H (2014) Egypt's population to reach 91 Million in June, Up from 90 in December. https://goo.gl/XPdkoD. Accessed 19 Oct 2017
Afify H (2014) When home is lost. In: Mada Masr. https://www.madamasr.com/en/2014/11/06/opinion/u/when-home-is-lost/. Accessed 12 Nov 2017
Akl Z (2015) Countering terrorism in Sinai: towards a comprehensive strategy – Opinion – Ahram Online. In: English.ahram.org.eg. http://english.ahram.org.eg/NewsContent/4/0/136695/Opinion/-Countering-terrorism-in-Sinai-Towards-a-comprehen.aspx. Accessed 14 Oct 2017
Al-Naggar A (2014) Sinai: forgotten development and present terrorism – Opinion – Ahram Online. In: English.ahram.org.eg. http://english.ahram.org.eg/NewsContent/4/0/114451/Opinion/Sinai-Forgotten-development-and-present-terrorism.aspx. Accessed 17 Sept 2017
Cole D (2003) Where have the Bedouin gone? Anthropol Q 76:235–267. https://doi.org/10.1353/anq.2003.0021
Dawn C (1986) From camel to truck: the Bedouin in the modern world. Vantage Press, New York
Dessì A (2012) Shifting Sands: security and development for Egypt's Sinai, Op-Mid. In: http://www.gmfus.org/publications/shifting-sands-security-and-development-egypt%E2%80%99s-sinai. Accessed 11 Sept 2017
Esterman I (2014) A history of forced relocations. In: Mada Masr. https://www.madamasr.com/en/2014/11/05/feature/politics/a-history-of-forced-relocations. Accessed 23 Oct 2017
Gold Z (2014) Securing the Sinai: present and future. International Center for Counter-Terrorism. https://www.icct.nl/download/fle/ICCT-Gold-Security-In-Te-Sinai-March-2014.pdf. Accessed 16 Jul 2017
Goldberg E (2015) Sinai: war in a distant province. In: Jadaliyya – جدلية. http://jadaliyya.com/Details/32327/Sinai-War-in-a-Distant-Province. Accessed 15 Nov 2017
Hosni A (2013) Sonbol in Sinai: a narrative of Territorialization. In: Jadaliyya – جدلية. http://jadaliyya.com/Details/29898/Sonbol-in-Sinai-A-Narrative-of-Territorialization. Accessed 26 Oct 2017 https://newmeast.wordpress.com/2012/04/23/paying-the-price-for-marginalizing-bedouins-in-sinai/. Accessed 20 Sept 2017
Lavie S (1990) The poetics of military occupation. University of California Press, Berkeley
McKinley C, Mastro D, Warber K (2014) Social identity theory as a framework for understanding the effects of exposure to positive media images of self and other on intergroup outcomes. Int J Commun 8:1049–1068
Serhan M (2012) The politics of image: the Bedouins of South Sinai. In: Jadaliyya – جدلية. http://jadaliyya.com/Details/26332/The-Politics-of-Image-The-Bedouins-of-South-Sinai. Accessed 17 Sept 2017
Sivini G (2007) Resistance to modernization in Africa: journey among peasants and nomads. Transaction Publishers, Edison
Stothard R (2014) A cycle of insecurity in Egypt's North Sinai. In: Jadaliyya – جدلية. http://jadaliyya.com/Details/30414/A-Cycle-of-Insecurity-in-Egypt%60s-North-Sinai. Accessed 21 Oct 2017
Tuastad D (2013) Paying the price for marginalizing Bedouins in Sinai. In: The New Middle East Blog. https://newmeast.wordpress.com/2012/04/23/paying-the-price-for-marginalizing-bedouins-in-sinai/
Watanabe L (2015) CSS analyses in security policy. In: Css.ethz.ch. http://www.css.ethz.ch/publications/pdfs/CSSAnalyse168-EN.pdf. Accessed 26 Sept 2017

Protest as Communication for Development and Social Change in South Africa 79

Elizabeth Lubinga

Contents

Abstract

South Africa has one of the highest rates of protest in the world. The country experiences a strong protest culture, with more than two million people protesting every year (Plaut, Behind the Marikana massacre. New Statesman. Retrieved from https://www.newstatesman.com/blogs/world-affairs/2012/08/behind-marikana-massacre, 2012). Such frequent occurrences of protest action make some critics assert that the country may be the protest capital of the world (Bhardwai 2017, Runciman, SA is the protest capital of the world. Pretoria News, 22 May 2017. Retrieved 26 Oct 2017 from https://www.iol.co.za/pretoria-news/sa-is-protest-capital-of-the-world-9279206, 2017).

E. Lubinga (✉)
Department of Strategic Communication, School of Communication, University of Johannesburg, Johannesburg, South Africa
e-mail: elizabethl@uj.ac.za

J. Servaes (ed.), *Handbook of Communication for Development and Social Change*,
https://doi.org/10.1007/978-981-15-2014-3_133

The development and social challenges confronting post-1994 South Africa, after the country achieved democracy, are mostly responsible for fuelling protest action with mass mobilization for social change often taking the form of visible street protests. Yet protest as a tool for communication is not new to the South African arena. It is rooted in a history of mass action against the pre-1994 government as attested to by among others, the 1976 Youth Uprising against the then proposed education policies. There is an increased use of protest as communication for change as evidenced by the sheer volume of protests, with three protests and labor strikes recorded per day (Institute for Security Services, At the heart of discontent: measuring public violence in South Africa. Retrieved from https://issafrica.org, 2016).

This chapter identifies some of the core development and social challenges confronting the country and outlines how protest has been used to draw the attention of stakeholders including policy makers to the problems. Exemplars of protest action used in the areas of health, housing, education and literacy, gender and other human rights, social justice, and access to social service provision in communities are provided. Finally, the chapter offers a critique of protest action in South Africa, by discussing their success stories and limitations.

Keywords

Protest action · Development and social challenges · Service delivery · Communication for change · Protest communication strategies

79.1 Development and Social Challenges in South Africa

South Africa has a deeply entrenched protest culture. In order to gain insight into how protest action has been used in the country to address development and social change in many areas, it is relevant to begin with a discussion of some of the prevailing developmental and social challenges in South Africa. The discussion premises from the fact that many of the developmental and social challenges responsible for protest action in South Africa today are legacies of the pre-1994 apartheid era. This view does not negate the possibility that other protests may have emanated from public discontent towards post-1994 government inadequacies. In fact, Bedasso and Obikili (2016) posit that unfulfilled expectations with respect to one's own human capital accumulation have had the biggest effect on the propensity for militant protest, not only at the individual level but at the municipal level too. Mpofu (2017) points out that in disrupting the world around them, the poor and marginalized majority in South Africa seek for authorities to address their grievances through strategic and effective communication.

From a sociological point of view, Burawoy (2017) proposes two types of social movements. Movements based on unequal inclusion in major institutions of society, such as labor, and those based on forcible exclusion from the same institutions, manifesting in service delivery deficiencies electricity, water, healthcare, and sanitation. The latter type breeds the kind of protest action that this chapter addresses.

A brief contextualization of prevailing challenges, followed by exemplars in areas responsible for most of the recent protest action namely: settlement (housing), health-related issues, local government service provision, access to education; unemployment, human rights – gender, racial, labor, LGBTQI rights, and economic equity, follows.

Poverty, unemployment, and inequality. During the first quarter of 2017, the unemployment rate rose to a 14-year high of 27.7% – the highest figure since September 2003 later decreasing to 26.7% at the end of the year (Stats SA 2018). Such a high rate inevitably translates into competition for scarce resources between citizens-vs-citizens, but mostly, citizens-vs-foreign nationals, with the latter perceived as threats to access to socioeconomic resources.

In 1994, the government identified poverty as one of the greatest challenges to socioeconomic development. A Stats SA (2017) Report on trends in South Africa reveals that poverty is on the increase in the country. In 2015, 55.5% or more than half of the country's population, equivalent to 30.4 million people were living in poverty in 2015, rising from 53.2% in 2011. Over the years, the government has devised various policies in an attempt to alleviate poverty, from the Reconstruction and Development Programme to the more recent, controversial proposed Radical Economic Transformation Policy.

Access to land. Land and agricultural-related labor are some of the important areas of economic development in any society. Historically, The Natives Land Act of 1913 and 1936, which set aside 13% of the land in South Africa for citizens categorized as blacks, also restricted them to areas of their cultural origin, with minority whites having access to 87%. By 2012, postapartheid land reform had transferred 7.95 million hectares into black ownership, equivalent to 7.5% according to the University of the Western Cape's Institute for Poverty, Land, and Agrarian Studies (Naki 2017). The consequences of these legislations then, were: many families lost their land and faced restrictions to providing labor in areas lying outside of their "homelands." Today, land expropriation remains a controversial "bone" of contention among various demographic groups in South Africa, with the December 2017 ANC National Conference provoking a flurry of divergent discussions on proposing that land should be expropriated without compensation.

Provision of basic services. Since 1994, government has made significant headway in the provision of basic services to previously disadvantaged communities through the RDP strategy, which not only emphasizes sustainable but also equitable access to social services. Between 1993 and 2000, local governance underwent fundamental transformations in a bid to tum it into an effective and efficient service delivery tool. Previously, disadvantaged sections of the population appear to have benefitted from the provision of basic services by the post-1994 government. A 2016 General Household Survey indicates that households with inadequate or severely inadequate access to food decreased from 23.9% in 2010 to 22.3% in 2016. The percentage of households that experienced hunger decreased from 23.8% to 11.8%, while the percentage of individuals who experienced hunger decreased from 29.3% to 13.4% over the same period (Stats SA 2017).

Table 1 Percentages of households with access to basic services

Year	Piped or tap water	Electricity	Housing (fully owned)
2004	86.9	80.9	54.7
2006	88.9	80.7	59.2
2008	88.8	81.9	61.4
2010	90.0	82.9	57.5
2012	90.8	85.3	54.5
2013	90.8	85.4	

Table 1 shows that access to basic services has generally increased over the years. In late 1993 and early 1994, 53.6% of households had access to electricity (Wilkinson 2015), although in 2012, the then Minister of Energy Dipuo Peters allegedly claimed that only 30% of households had access to electricity in 1994. To date, more than 85% of households in the country have access to the service.

In terms of the improvement of sanitation, while the Western Cape (94.8%) and Gauteng (90.2%) have greatly improved, some rural provinces such as Limpopo (50%), Mpumalanga (62%), Eastern Cape (71.2%), and Kwazulu Natal (73.9%) still fall short of the national rate of improvement of 77.9% (Stats SA 2017).

For some communities, the use of bucket toilets as a form of sanitation prevails. The Nelson Mandela Bay Government in the Eastern Cape Province, for instance, in 2013 made progress in reducing the number of consumers using bucket toilets from a high of 30,202 to 16,317 in 2016, yet the system remains a repugnant reality for many residents. In May 2016, frustration over bucket toilets erupted into a violent service delivery protest by Missionvale residents within the city following in the wake of bucket-system protests by residents of Khayelitsha in the Western Cape in 2013 and Soweto Township in Gauteng Province in 2014, among others. A 2016 census of municipalities by Stats SA (2017) reveals that of the 68, 028 consumer units in South Africa using a bucket toilet in 2016, 24% were located in Nelson Mandela Bay. Possibly, the country has not witnessed the end to such protests, given that by 2016, 45 of South Africa's 278 municipalities, including Free State (44%), Eastern Cape (33%), and Northern Cape (15%), were still providing communities with the bucket system instead of portable or permanent flush toilets (PFTs) (Fig. 1).

Health services. The government has invested a lot of money into redressing health inequality in the country. Healthcare was the fourth largest item of government expenditure (totalling R157 billion – 11 cents out of every rand, spent by government) in 2014/15. HIV/AIDS and related diseases such as tuberculosis remain responsible for consuming a large portion of the health budget. For example, the number of people living with HIV in South Africa increased from an estimated 4.94 million in 2002 to 7.06 million by 2017 (Stats SA 2017). As Venter et al. (2017) reveal:

> ... for 2014–2015, USD350 million was spent on ART for just <3 million people living with HIV, most of it on first-line treatment. From mid-2016, the number of people on ART increased to 3.4 million, with approximately 145,000 of them on second-line treatment and just 700 on third-line treatment. The increase in public spending is expected to continue...the national Department of Health (DoH) announced a drop of the previous

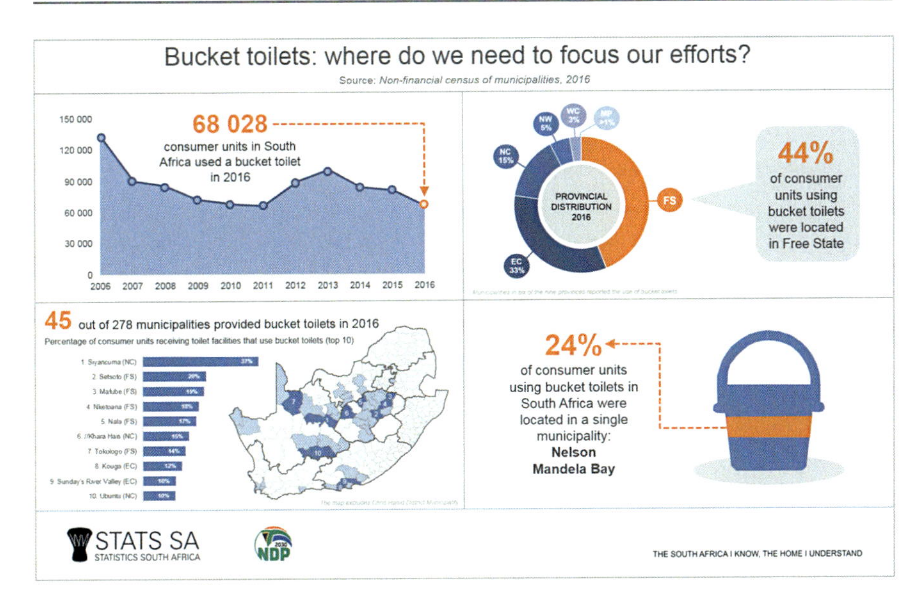

Fig. 1 2016 statistics showing the prevalence of the bucket system in South Africa

> CD4 thresholds for ART initiation from September 2016, theoretically doubling the number of people eligible for ART to >6 million people (p. 28).

In South Africa, the burden of the provision of health services largely falls onto the public health system, with as many as 45 million, or 82 out of every 100 - South Africans, falling outside the medical aid net, and are therefore largely dependent on public healthcare. Only 17 in 100 South Africans have medical insurance, guaranteeing access to private healthcare (Stats SA 2017). Yet, as Mayosi and Benatar (2014) point out, many of the state-run hospitals in South Africa are in a state of crisis, with much of the public health care infrastructure run down and dysfunctional as a result of underfunding, mismanagement, and neglect. Such existing health challenges serve to underscore that health equity and provision of adequate health services are yet to be achieved in South Africa. Schiavo (2014) conceptualizes health equity as a guiding principle to eliminating disparity in health, by minimizing or reducing differences in the well-being and health status of diverse populations and groups.

Racial inequality and participation: transformation/access to university education. Historically, access to tertiary education, especially at elite universities also referred to as historically advantaged institutions (HAIs) was limited to wealthy white minority and the middle class. Although under post-1994 governance, higher education continues to be transformed, a number of students from poor and low-income backgrounds continue to face serious challenges in pursuing and successfully completing higher education degrees. The National Student Financial Aid Scheme Act, 1999 (Act No. 56 of 1999) was established to provide for the granting of loans and bursaries to eligible students attending public higher education and training, yet to-date tertiary institutions still hemorrhage students.

Protests in the field of education, especially higher education, have not solely focused on the lack of access by economically disadvantaged masses, or related financial constraints evidenced by the #FeesMustFall (#FMF) movement, but have also addressed the decolonization of education through the #RhodesMustFall (#RMF) movement that began at the University of Cape Town (UCT) in March 2015.

Economic racial inequality and participation: economic empowerment. Economically, attempts to redress the legacies of economic inequality in South Africa were made through the Broad-Based Black Economic Empowerment (B-BBEE) Act, amended in 2013, put into place to achieve greater participation in and control of the economy by black South Africans. The policy has since been criticized for benefitting a small elite group of black middle class, with the envisaged economic transformation not filtering down to the poor masses. Southall (2007) points out that B-BBEE has become highly controversial, that it serves as a block to foreign investment, encourages a re-racialization of the political economy and promotes the growth of a small but remarkably wealthy politically connected "empowerment" elite.

Gender-related issues. Socially, gender issues may not have been a direct legacy of the apartheid system; however, they too have been responsible for a number of protests in society. Protests have resulted from gender violence including societal rape of vulnerable groups such as children and women, institutionalized sexual abuse at institutions such as universities, as well as hate crimes against the formerly Lesbian, Gay, Bisexual, and Transgender (LGBT) now LGBTQIA (adding Queer, Intersex, and Asexual/allied) community. Accurate rape statistics in South Africa are not readily available. Publicly available police crime statistics may reveal how many rapes are reported but not how many are committed. A 2017 Press Release by the South African Institute of Race Relations (SAIRR) argues that "the data is certainly an undercount as many women are too afraid to report rapes... in 2015/16 alone, 20 254 children were reported to have been victims of rape and sexual assault (p. 1)."

Hate crimes against the LGBT community occur even though the 1996 South African Constitution globally recognized as being the first jurisdiction in the world to provide constitutional protection to LGBT people, under Section 9(3) disallows discrimination on race, gender, sexual orientation among others. A countrywide survey among over 2000 members in South Africa by Out LGBT Wellbeing (2016) reveals they experienced hate crimes ranging from murder, corrective rape and sexual abuse to being verbally insulted.

In sum, it is impossible to map out the entire gamut of the causes of protests in South Africa. In what follows, some exemplars of the use of protest as communication for social and development change are highlighted.

79.2 Using Protest as Communication for Change

A variety of protest communication strategies have been used to raise the awareness of society and policymakers such as government and other institutions, to various areas of discontent by citizens in South Africa. Bond and Mottiar (2013) suggest that different protesters prefer to use specific tactics. In community protests related to service delivery and political accountability, burning tires and barricading roads seem to be the most favored tactics. For workers, strike action often includes, marching, demonstrating, and picketing. Student protests favor strikes and boycott, as well as marching, demonstrating, and picketing, including destruction of property and vandalism, in addition to intimidation and disruptions. Protest action is evolving. Recently, protest action increasingly encompasses destruction of public infrastructure such as schools, public libraries; destruction of local government representatives' property and possessions such as houses and cars; dumping of rubbish and human waste as well as nudity or baring of body parts in public places.

Yet as Brown (2015) argues, whether the sites that citizens use to express their political agency and to express their demands for social citizenship or effective democracy are situated within, outside of the state, or at grassroots, the scope revolves around articulation of claims for proper healthcare, access to education, housing, and basic services such as water, electricity, or sewage removal.

79.3 Service Delivery Protests: The Case of Ses'khona People's Rights Movement in the Western Cape and Vuwani in Limpopo Provinces

Occurrence of service delivery protests in the country has been both localized in communities, but has in some cases, been taken to the public arena, rendering it debatable as to which protest communication modality is more visible. For instance, in June 2013, members of the Ses'khona People's Rights Movement in the Western Cape engaged in an internationally visible protest against being provided with portable rather than flushing toilets by the Provincial government. Protesters dumped human waste and raw sewerage at the Cape Town International Airport disrupting domestic and international services. The group had previously engaged in a similar "feces war," dumping human waste on the steps of the Western Cape Provincial legislature in Wale Street, outside provincial offices in Greenmarket Square and on a convoy of vehicles in which the then Premier of the Province, Helen Zille was travelling to a Green Economy event (Dano and Barnes 2013).

At the local community level, it would be a misnomer to classify the 2015–2016 Vuwani community protests, responsible for a complete shutdown of services in the area as service delivery protests. This is because residents revolted against municipal boundary demarcation. However, what was prominent about these protests is that they led to a complete shutdown of services in the community, leading some to classify them as service delivery protests. Relevant to the discussion, is that they can be indirectly traced back to demarcations imposed by the pre-1994 government, on the black population. As Buccus (2016) notes, the struggle over the demarcation was shaped by old Bantustan borders and identities and therefore became an ethnic conflict. In contextualization, the Municipal Demarcation Board (MDB) in 2016 made a decision to incorporate Vuwani into a new Malamulele Municipality (a different tribal group based on apartheid restrictions) from the community-favored Makhadho Municipality.

79.4 Protests for the Delivery of Health Services: The Treatment Action Campaign (TAC)

Civil Society Group, Treatment Action Campaign (TAC) often credited with ensuring health policy change in South Africa, was cofounded by Zackie Achmat in 1998, with the primary aim of campaigning for equitable treatment access for all people living with HIV/AIDS (Sabi and Rieker 2017). The organization engaged in demonstrations in Cape Town, Durban, and Johannesburg in March 1999, specifically advocating for Prevention of Mother-to-Child Transmission (PMTCT) by provision of AZT (an antiretroviral – ARV) at birth. Organization members went to various public places such as malls, hospitals, clinics, and schools, to solicit signatures from people from different occupations, in support of the cause. On South Africa's Human Rights Day (21 March 1999), another campaign "Fast to save lives," which spread countrywide, was held at one of South Africa's largest hospitals, Chris Hani Baragwanath in Johannesburg. It attracted people living with HIV/AIDS, medical doctors, traditional healers, other civil society, and political groups such as the Congress of South Africa Trade Unions (COSATU) and the Johannesburg branch of the South African Communist Party who pledged to support the movement (TAC 2010). At a personal level, TAC cofounder Zackie Achmat who is also HIV positive, pledged to abstain from using ARVs until the treatment was available to the public (Achmat 2004).

Today, the organization engages in activism intended to strengthen the South African health system, monitors TB and AIDS response, advocates for access to quality and affordable medicine, and builds local activism (TAC 2017). For instance, in September 2016, 500 members of the TAC held a picket outside a hospital in Vosloorus in the Gauteng Province to highlight problems such as negligence, poor staff attitudes, overcrowded and dirty facilities, long waiting times, and even patients being turned away. On 24 October 2017, over 1000 TAC and 36 patient groups marched to the Department of Trade and Industry (DTI) in support of government's efforts, in a #FixThePatentLaws protest. The campaign focused on advocating for access to cheaper generic cancer medicine.

79.5 Gender-Related Protests: Societal, Institutionalized, Corrective Rape, and Gender Violence

The use of the body, specifically semi/nudity, has become popular in South Africa as a strategy of protest mostly against but not excluded to gender-related issues. In May 2017, over 100 female students from the University of Pretoria held a topless protest against rape and sexual harassment allegedly taking place at the University campus and on their commute between the University campus and their residences (Chauke and Keppler 2017). In a similar protest in 2016, students at Rhodes University protested while topless against the rape culture at the institution and rape in general.

Nudity resurfaced as a protest tactic in November 2017, when some female students at UCT protested in full nudity during #FMF protests (Fig. 2).

Fig. 2 2017 #FeesMustFall protest at UCT. (Image: Twitter/VernacNews)

Thompson (2017) argues that globally, South Africa included, bare breasts have for long been sexualized, and feminists have long known that there are few things more powerful and important than their own bodies in the fight against oppression and patriarchy. He notes that many women have pushed to desexualize the breast and female body adding that one spin-off of this is the increasing prevalence of topless protests in countries throughout the world. However, in South Africa, not every protest against gender-related issues such as rape has used (semi)nudity as a strategy or solely involved female participants.

In August 2017, a mixed-gender one-day silent protest against victims raped and murdered through gender-based violence was held at the UCT (as well as the University of Witwatersrand). Organized by the Aids Healthcare Foundation, UCT survivors and UCT Sexual Assault Response Team, protesters taped their mouths and marched in silence through the University campus and broke the silence with shouts demanding for justice (Jacobs 2017). On 20 May 2017, a large group mostly made up of men and some women named hashtags #NotInOurName and #StopItNow, organized by several NGOs and civil society groups, marched in Pretoria to protest against killings of women and children in the country.

South Africa has also witnessed protests against homophobic violence, especially the corrective rape of lesbians. Corrective rape involves men raping lesbian women to “turn” them straight or “cure” them of their sexual orientation (Koraan and Geduld 2015). On 5 January 2010, South African volunteer anti-rape advocacy group, Luleki Sizwe from the townships of Cape Town petitioned the Ministry of Justice to address corrective rape. They collected more than 100,000 signatures from 163 countries asking the then Justice Minister Jeff Radebe to combat “corrective rape,” which they felt was on the increase. The signatures were sent by email circulation through international platform for change: change.org (Nkalane 2011).

79.6 Protests for Affordable Access to Tertiary Education: "#MustFall Movements"

Student-led protests are not new to South Africa, dating as far back as the 1976 protests. Post-1994 student protests have been occurring at various institutions, especially at historically disadvantaged institutions (HDIs). For example, at the Tshwane University of Technology (TUT Soshanguve Campus), it is estimated that more than once every year between 2005 and 2010, 28 student protests took place against academic exclusion, access to basic services, insufficient funding and financial exclusion (Vilakazi 2017). Critics argue that from 2015, university student protests gained visibility only because, they were engineered from HAIs, but that they had for long been taking place at HDIS (Malabela 2017a; Nyamnjoh 2016).

Different from previous episodes, October 2015 to 2017 University protests were nationwide, yet they were not homogenous to all the universities. For example, even though #FMF protests were geared towards achieving a 0% increment in University fees, the actual issues of contention differed across and were specific to individual universities, or were even campus-specific. While, the #FMF protests stemmed from the University of Witwatersrand, some institutions such as the University of Zululand (UNIZULU) and the University of Limpopo (UL) did not experience any/as massive protests as some HAIs and HDIs (Malabela 2017a; Edwin 2017). Furthermore, issues within institutions as Malabela (2017b) points out that at the University of the Witwatersrand (WITS), the 2016 movement although not as prominent as it was in 2015, was more about challenging the status quo and it centered around the decolonization of the university rather than on free fees.

At some institutions, campus-specific causes ensured that the level of violence varied between campuses. At the University of KwaZulu-Natal (UKZN), Kujeke (2017) notes that Durban-Westville's violent protests were linked to its history as a historically black university. Students interviewed at Durban-Westville strongly felt that resources allocated to their campus were inferior compared to those allocated to the Howard College campus, which is historically a former white campus under the former University of Natal. At some of the merged institutions, students viewed the inequalities to be highly entrenched and resistant to any transformative initiatives. Post-1994 mergers between former white and black universities thus did not seem to resolve the racial and class inequalities that existed prior to 1994 (Robus and Macleod 2006).

In relation, the #RMF movement attempted to fight against physical and epistemological colonial symbols, such as statues, names, resources, and curricula. The #RMF spread across many universities too, addressed a different aspect of tertiary education: decolonization of education, and initially included other stakeholders such as academics, alumni among others. It targeted the desecration and/or destruction of symbols associated with colonialism, sentiments of which are captured below (UCT #RhodesMustFall mission statement 2015):

> This movement is not just about the removal of a statue... Its removal will not mark the end but the beginning of the long overdue process of decolonising this university. In our belief,

> the experiences seeking to be addressed by this movement are not unique to an elite institution such as UCT, but rather reflect broader dynamics of a racist and patriarchal society that has remained unchanged since the end of formal apartheid. (p. 6)

Like #FMF, the #RMF decolonization movement played out differently at different institutions and different campuses. The focus may have been on destruction/ removal of colonial statues at UCT and UKZN, yet at Rhodes University, protesters demanded for change of a name viewed to be a legacy of the colonial past. At Stellenbosch University, the focus was on the (alienating) role of Afrikaans as the language of instruction, with some of the black students talking about their lived-experiences of racism in lecture halls and residences in a video gone viral, *Luister*. Again, at some HDIs such as UNIZULU and UL, students were reluctant to participate in the countrywide student movements. For example, students from UL criticized "the perceived arrogance of previously white universities like Wits, which have the power to capture news headlines," adding that the struggles waged in previously black universities do not grab headlines, unless there is a death or something of that nature (Malabela 2017a).

79.7 A Critical Analysis of Protest Action in South Africa: Successes and Limitations

The fruits of protest action in South Africa post-1994 have been several and varied. Protests have produced success stories, but they have also highlighted limitations to its use as a tool for communicating development and social change.

79.7.1 Successes

79.7.1.1 Opens Channels of Communication

Some researchers have argued that the very fact that protests take place reveals that formal channels of communication are blocked or non-existent. Municipal IQ (2016) points out that a large part of the problem sparking protests in South Africa has been very poor communication between representatives of metros and communities, essentially a task of ward councillors and local officials. Protest action thus gives protesters a voice to highlight their grievances with a possibility of opening avenues for policy makers or government to participate in finding solutions to problems.

79.7.1.2 Creates Awareness About Existing/Underlying Problems

Protest action creates awareness about existing social and developmental problems. The #RMF Movement, by illustration, which started at UCT unearthed deeper underlying problems at university level. As Chaudhuri (2016) argues, while the protest action may have started with demands to remove the statue of Cecil Rhodes from prominence at UCT, what was at stake was much deeper. At the center of the widespread protest, as Chaudhuri (2016) aptly states, was an ethos that gives "academic" space and even preeminence to such a figure (Cecil Rhodes) and hesitates to interrogate Rhodes's legacy.

79.7.1.3 Partial/Full Achievement of Goals of Protest Action

When protest action takes place, it is often able to achieve some or most of the goals that initially inspired the action. For example, under #RMF, students and some members of staff at Stellenbosch University demanded for change regarding the use of Afrikaans as the language of instruction. On 12 November 2015, the University of Stellenbosch's Rector's Management Team recommended a new language policy in line with Open Stellenbosch's (protest) demands; specifically, the adoption of English as a lingua franca (De Villiers 2015). At UCT, #RMF's origin, members of the university community were invited to motivate their positions on whether the Rhodes statue should stay or be removed. On 9 April 2015, the UCT Convocation, senate, and council authorized the removal of the Cecil Rhodes statue created in 1934.

In the health sector, the DoH's success in reducing the mother-to-child transmission (MTCT) of HIV is partially attributed to the TAC. The TAC set up a court case against government regarding provision of ARVs for PMTCT. A Constitutional Court monumental judgement compelled the DoH to make the ARV Nevirapine available to pregnant women in the public healthcare sector in 2002 and paved the way for the introduction of ARV treatment for all South Africans. In 2015, HIV transmission from mother-to-child in South Africa fell to just 1.5% in 2015, down from 30% in the early 2000s, exceeding the projected national target of 1.8% (AVERT 2015).

In local governance, Von Holdt et al. (2011) state that at times, protests would result into high visibility responses by the ruling parties' senior African National Congress (ANC) officials or even Cabinet Ministers as was the case of the protest in Vuwani in Limpopo Province and probes into municipal management and/or suspensions of municipal staff would ensue.

The #FMF and #RMF movements that swept across universities in South Africa may not have achieved free education for university students or immediate decolonization of education, as the goals were. In the short-term, at a communicative level, they collectively brought the issues of decolonization and transformation back into South African discourse (Malabela 2017b), in addition the President established a Commission, following the 2015 and 2016 #FMF. While free education for all was not yet feasible in 2017, the free university education matter was still under consideration by the Presidency, with a view to finding a feasible long-term solution. In December 2017, then President Jacob Zuma announced that government would provide subsidized tertiary education for the poor.

79.7.1.4 Creates Legacies to the Cause of Protest

Results are not always restricted to local sites of protest action, they have shown the potential to spread from local to national or even international level. The #RMF movement, which started at UCT, spread to other South African Universities, became as far-reaching as Oxford and Edinburg universities in the UK and the

University of California, Berkeley, in the USA (News24 2015). According to the Herts and Essex Observer (2015) at the University of Oxford, students called for a statue of Rhodes to be removed from Oriel College. They also started a movement at the university to better represent non-white culture in the curriculum as well as to combat racial discrimination and insensitivity (Aftab 2015; Rhoden-Paul 2015).

The legacies of one problem may spread to other social issues. At TUT Soshanguve, a new student movement, Black First Land First, not aligned to the Student Representative Council or political parties emerged during #FMF. In addition, whereas the #RMF movement emanated from an academic environment, it spread into local governance, with South African political party Economic Freedom Front (EFF) calling for the removal of other apartheid-era statues situated in various municipalities. At UCT, subsequent to #RMF, UCT fallists effected various other programs including protests and art installations on 16 August 2015 to mark the third anniversary of the Marikana massacre, demonstration under #PatriarchyMustFall, which confronted UCT's institutionalized misogyny, queerphobia, and transphobia as well as demands for the insourcing of outsourced workers (Langa et al. 2017).

79.7.1.5 Fast-Tracks Crafting of Solutions to Longstanding Social and Developmental Problems

Many of the legacies of apartheid have been long-standing and have/will take time to solve. However, as a result of some of the protests, there has been fast-tracking of the crafting the necessary solutions. As Duncan (2016, p. 1) points out, protests hold the capacity to "wake society up from its complacent slumber, make it realize that there are problems that need to be addressed urgently and hasten social change."

79.7.2 Limitations

79.7.2.1 May Deviate Attention from the Legitimacy of/or the Cause

The use of shock tactics such as semi/nudity may be legitimate in protest; however, they may distract attention from the genuine causes that initiated the protest. In November 2017, for example, some of the UCT students who were arrested as part of the #FMF protests, meant to demand free university education, instead appeared in court on charges of public indecency. The media conducted intensive coverage of nudity used by some female students, while videos circulated on the social media focusing on the use of police brutality in arresting the naked students.

In addition, criminal elements have often been known to "hijack" legitimate social protests, turning them into xenophobic destruction "movements." Bond and Mottiar (2013) refer to these as "popcorn" protests in view of their tendency to flare up and settle down immediately. "While 'up in the air', protesters are often subject to the prevailing winds, and if these were from the right, the protests could – and often did – become xenophobic" (p. 289).

79.7.2.2 Overturning the Gains of Democracy/Failure to Achieve Long-Term Goals

The destruction of public roads, schools, libraries, clinics, and other infrastructure characteristic of many of the service delivery protests may reverse the gains of democracy. The lengthy Vuwani Community protests witnessed the burning down of more than 30 schools, with books and school records. Section 27, a non-governmental organization (NGO) that had been working for several years in the Vuwani area to ensure the construction of schools, provision of textbooks among others, pointed out that the strike negatively affected the rights of 42,664 pupils' access to education. The costs for the reconstruction of the schools was estimated at over R400 million, which only covers the buildings, almost half of the total Limpopo Department of Education's 2016/2017 budget that caters for all the province's 4062 schools is R930 million.

79.7.2.3 Danger of Becoming the "Dreaded Vice" that Originated the Protest

Social movements often face the danger of becoming the very "thing" they are fighting against, as was the case of the #RMF student movement where "alienation of the other" arose (Nyamnjoh 2016), in a protest against alienation of black students and institutional racism. At UCT, members of the #RMF Movement (fighting against discrimination amongst others) took over the food service in Fuller Hall's dining room and discriminated against white, colored, and Indian students by barring them from entering the hall, where they served food to black students only. One of the students is quoted as saying that the protesting students "took something that was pure and good and turned it into a fight: Black against white. . ." (Huisman 2016).

In December 2017, Rhodes University banned two student females for life after they were convicted of committing acts of criminality by dragging male students (alleged to be rapists) out of their rooms, during a protest against rape at the university.

Nyamnjoh also criticizes measures proposed to solve decolonization of the currirulum at Universities by protesters:

> When one claims that black lecturers will channel the 'politics of being black' in their disciplines. . .It seems to suggest an experience particular to an identity that all others who identify with said identity possess. It gives the impression that more black representation in the academic staff is sufficient for the decolonisation of an otherwise alienating curriculum. . .One could reasonably end up with their worst nightmare – a prospective black lecturer whose neurotic obsession with Kant precludes possibilities of Africanisation (p. 263).

79.7.2.4 Co-constructed Violence

Violence during the student protests at UCT over the 2015/16 period was co-constructive (Langa et al. 2017) whereby ". . .violence reproduced violence in a game of disruption and destruction between the students and the university, the latter through the security apparatus of the police services as well as private security

(p. 74)". The students used symbolism, damage to property, art installations and the disruption of physical space as modalities of protests, punctuated with traditional modes of protest such as mass toyi-toying animated by dancing and the singing of struggle songs, stemming from pre-1994 era (Bond and Mottiar 2013).

Violence turned personal during the #MustFall movements, when students retaliated against representatives of institutions perceived to be perpetrators of violence. Langa et al. (2017) report that in 2016, a private security officer was seriously injured when students dropped a rock on his head from an elevated position. Students at UCT physically assaulted the vice-chancellor on two occasions. In 2015 during a senate meeting, a student threw a bottle filled with water that hit Price on his forehead, while in 2016 outside the Bremner Building, a student reportedly struck Price on his abdomen while he was addressing the students' demands.

But media's coverage of protests often depends on who is involved and whether or not the protest is violent. Media tend to focus mainly on violent protests, which creates a perhaps erroneous impression that protests in South Africa are inherently violent. Von Holdt et al. (2011), in a Centre for the Study of Violence and Reconciliation (CSVR) Report on Collective Violence, quote violent protesters:

> Violence is the only language that our government understands. Look we have been submitting memos, but nothing was done. We became violent and our problems were immediately resolved. It is clear that violence is a solution to all problems (p. 28).

79.8 Conclusion

The development and social challenges as well as exemplars provided in the chapter are but a meagre representation of the numerous protests that take place on a daily basis in South Africa. For instance, prominent protest action towards contemporary problems, such as the enduring citizen noncompliance towards payment of tollgate levies in Gauteng Province and countrywide 2017 #BlackMonday protests against growing farm attacks and murders have not been covered.

Also, whereas protest as a tool for CDSC in South Africa has been largely successful in some instances, its successes and limitations indicate that successful achievement of long-term goals to protest action cannot solely attributed to protest action. Violent protest, if or when used as a means of communication, should work in conjunction with other tools, not only in order to achieve short-term but also long-term goals. With the massive volume of protests, that become increasingly more violent in order to gain visibility in the media and in society, the possibility exists that different, yet effective communication strategies may have to be crafted. Peaceful means could be a solution. They too, though rarely used in South Africa, have been shown to be successful in drawing the attention of stakeholders such as policy makers, responsible for effecting the necessary social change, a case in point being the 100,000 email campaign against corrective rape.

References

Achmat Z (2004, November 10) John Foster lecture. Speech. HIV/AIDS and human rights: a new South African struggle

Aftab A (2015, June 19) Oxford University students call for greater 'racial sensitivity' at the institution and say it must be 'decolonised'. The Independent

Allan K, Heese K (2016) Understanding why service delivery protests take place and who is to blame. Retrieved from https://www.municipaliq.co.za/publications/articles/sunday_indep.pdf

AVERT (2015) South Africa exceeds national mother-to-child transmission target. Retrieved from https://www.avert.org/news/south-africa-exceeds-national-mother-child-transmission-target

Bedasso BE, Obikili N (2016) A dream deferred: the microfoundations of direct political action in pre- and post-democratisation South Africa. J Dev Stud 52(1):130–146

Bhardwai V (2017) Are there 30 service delivery protests a day in South Africa? Africa Check. Retrieved 8 Oct 2017 from https://africacheck.org/reports/are-there-30-service-delivery-protests-a-day-in-south-africa-2/

Bond P, Mottiar S (2013) Movements, protests and a massacre in South Africa. J Contemp Afr Stud 31(2):283–302

Brown J (2015) South Africa's insurgent citizens: on dissent and the possibility of politics. Zed Books, Chicago

Buccus I (2016, May 19) Understand the burning of schools in #Vuwani. The Mercury. Retrieved from https://www.iol.co.za/mercury/understand-the-burning-of-schools-in-vuwani-2023671

Burawoy M (2017) Social movements in a neoliberal age. In: Paret M, Runciman C, Sinwell L (eds) Southern resistance in critical perspective: the politics of protest in South Africa's contentious democracy. Routledge, Abingdon, pp 21–35

Chaudhuri A (2016, March 16) The real meaning of Rhodes must fall: after the nation's long retreat from multiculturalism and the return of a rose-tinted memory of empire, it is no accident that the Rhodes Must Fall movement has come to Britain. The Guardian. Retrieved from https://www.theguardian.com/uk-news/2016/mar/16/the-real-meaning-of-rhodes-must-fall

Chauke N, Keppler V (2017, May 17) Naked protest against UP campus rapes. The Citizen. Retrieved from https://citizen.co.za/news/south-africa/1516239/naked-rage-at-up-campus-rapes

Dano Z, Barnes C (2013, June 26) Faeces fly at Cape Town airport. Retrieved from https://www.iol.co.za/news/crime-courts/faeces-fly-at-cape-town-airport-1537561

De Villiers W (2015) Professor. politicsweb

Duncan J (2016) Protest nation: the right to protest in South Africa. UKZN Press, Scottsville

Edwin Y (2017) South African higher education at a crossroads: The Unizulu Case Study. In M. Langa (Ed.) *#Hashtag An analysis of the #FeesMustFall Movement at South African universities*. Johannesburg: Centre for the Study of Violence and Reconciliation (CSVR)

Herts and Essex Observe (2015, April 13) University students remove statue of Stortford's Cecil Rhodes from campus

Huisman B (2016, February 21) Students question #RhodesMustFall. City Press

Jacobs Y (2017, August 29) LOOK: silent protest against sexual violence speaks volumes. Retrieved from https://www.iol.co.za/news/south-africa/look-silent-protest-against-sexual-violence-speaks-volumes-10987142

Koraan R, Geduld A (2015) "Corrective rape" of lesbians in the era of transformative constitutionalism in South Africa. Potchefstroomse Elektroniese Regsblad 18(5):1931–1952. https://doi.org/10.4314/pelj.v18i5.23

Kujeke M (2017) Violence and the #FeesMustFall movement at the University of Kwazulu Natal. In M. Langa (Ed.) *#Hashtag An analysis of the #FeesMustFall Movement at South African universities*. Johannesburg: Centre for the Study of Violence and Reconciliation (CSVR)

Malabela M (2017a) We are already enjoying free education: Protests at the University of Limpopo (Turfloop). In M. Langa (Ed.) *#Hashtag An analysis of the #FeesMustFall Movement at South African universities*. Johannesburg: Centre for the Study of Violence and Reconciliation (CSVR)

Malabela M (2017b) We are not violent but just demanding free decolonised education: University of the Witwatersrand. In M. Langa (Ed.) *#Hashtag An analysis of the #FeesMustFall Movement at*

South African universities. Johannesburg: Centre for the Study of Violence and Reconciliation (CSVR)
Mayosi BM, Benatar SR (2014) Health and health care in South Africa – 20 years after Mandela. N Engl J Med 371(14):1344–1353
Mpofu S (2017) Disruption as a communicative strategy: the case of #FeesMustFall and #RhodesMustFall students'protests in South Africa. J Afr Media Stud 9(2):355–372
Municipal IQ (2016) Index of crime incidence in municipalities. Retrieved from http://www.municipaliq.co.za/index.php?site_page=cim.php
Naki E (2017, January 10) Land expropriation now top of ANC agenda because of EFF, says expert. The Citizen. Retrieved from https://citizen.co.za/news/south-africa/1391231/anc-must-take-land-issue-seriously-says-expert
News24 (2015, March 26) American students support #RhdesMustFall Campaign. Retrieved from http://pp.m.news24.com/news24/southafrica/news/american-students-support-rhodesmustfall-campaign-20150326?mobile=true
Nkalane M (2011, January 6) Protest against 'corrective rape'. The Sowetan. Retrieved from https://www.sowetanlive.co.za/news/2011-01-06-protest-against-corrective-rape
Nyamnjoh A (2016) The phenomenology of Rhodes must fall: student activism and the experience of alienation at the University of Cape Town. Strateg Rev South Afr 39(1):256–277
Out LGBT Wellbeing (2016) Hate crimes against lesbian, gay, bisexual and transgender (LGBT) people in South Africa. Retrieved from https://www.gala.co.za/resources/docs/Free_Downloads/hatecrimes.pdf
Rhoden-Paul A (2015, June 18) Oxford Uni must decolonise its campus and curriculum, say students. The Guardian
Robus D, Macleod C (2006) "White excellence and black failure": the reproduction of racialized education in everyday talk. S Afr J Psychol 36(3):463–480
Sabi SC, Rieker M (2017) The role of civil society in health policy making in South Africa: a review of the strategies adopted by the treatment action campaign. Afr J AIDS Res 16(1):57–64
Schiavo R (2014) Health communication: from theory to practice, 2nd edn. Jossey-Bass, San Francisco
South African Institute of Race Relations (2017) Rape plague terrorises SA women, children – IRR report 13 June 2017. Retrieved from http://irr.org.za/reports-and-publications/media-releases/rape-plague-terrorises-sa-women-children-2013-irr-report/view
Southall R (2007) Ten propositions about black empowerment in South Africa. Rev Afr Political Econ 34(111):67–84
Stats SA (2017) Poverty trends in South Africa: an examination of absolute poverty between 2006 and 2015; Quarterly Labour Force Survey – QLFS Q1:2017; General Household Survey, 2016. Retrieved from http://www.statssa.gov.za/?p=9960.pdf
Stats SA (2018) South Africa's jobless rate falls to 26.7% in Q4. Retrieved from https://tradingeconomics.com/south-africa/unemployment-rate
TAC (2010) Fighting for our lives: the history of the treatment action campaign 1998–2010. Treatment Action Campaign, Cape Town
TAC (2017) Working for access to quality public healthcare in South Africa since 1998. Retrieved from https://tac.org.za
Thompson A (2017) How topless south African women are embodying resilience. Retrieved from https://theculturetrip.com/africa/south-africa/articles/how-topless-south-african-women-are-embodying-resilience/
University of Cape Town (UCT) (2015) #RhodesMustFall mission statement. Retrieved from http://jwtc.org.za/resources/docs/salon-volume-9/RMF_Combined.pdf
Venter WDF et al (2017) Cutting the cost of South African antiretroviral therapy using newer, safer drugs. S Afr Med J 107(1):28–30
Vilakazi M (2017) Tswane University of Technology: Shoshanguve Campus protest cannot be reduced to #FeesMustFall. In M. Langa (Ed.) *#Hashtag An analysis of the #FeesMustFall Movement at South African universities*. Johannesburg: Centre for the Study of Violence and Reconciliation (CSVR)

Von Holdt K, Langa M, Molapo S, Mogapi N, Ngubeni K, Dlamini J, Kirsten A (eds) (2011) The smoke that calls: insurgent citizenship, collective violence and the struggle for a place in the New South Africa: eight case studies of community protest and xenophobic violence. Centre for the Study of Violence and Reconciliation (CSVR) and the Society, Work and Development Institute (SWOP), University of the Witwatersrand

Wilkinson K (2015, May 14) Did 34% of households have access to electricity in 1994? Retrieved from https://mg.co.za/article/2015-05-14-did-only-32-of-households-have-access-to-electricity-in-1994

Case Study of Organizational Crisis Communication: Oxfam Responds to Sexual Harassment and Abuse Scandal

80

Claudia Janssen Danyi

Contents

Abstract

The #MeToo movement has put a spotlight on sexual abuse in a wide range of industries and organizations, including the aid sector. In February 2018, Oxfam GB faced a severe organizational crisis after news reports alleged a cover-up of sexual abuse and harassment perpetrated by its staff in Haiti from 2010–2011. This case study applies crisis communication theory to analyze Oxfam's response to the scandal. The analysis shows that Oxfam's initial response focused on explaining and justifying its past decisions fell short to address the public issues at hand. As the organization quickly shifted strategies toward full apology and corrective action, inconsistencies and misstatements continued to undermine its messages. The case study points to unique crisis communication challenges for global NGOs in the #MeToo era and the importance of crisis preparedness and training for NGOs.

C. Janssen Danyi (✉)
Department of Communication Studies, Eastern Illinois University, Charleston, IL, USA
e-mail: cijanssen@eiu.edu

J. Servaes (ed.), *Handbook of Communication for Development and Social Change*,
https://doi.org/10.1007/978-981-15-2014-3_41

Keywords

Crisis communication · #MeToo movement · Oxfam · Scandal · Sexual abuse and harassment · Strategic communication

80.1 Introduction

NGOs provide vital services to communities in crises. Yet, sometimes they face crises of their own (see Gibelman and Gelman 2004). On February 9, 2018, Oxfam GB came under intense public scrutiny after *The Times* (O'Neill 2018a, b) alleged that the organization had covered up sexual harassment and abuse perpetrated by its staff in Haiti from 2010 to 2011. The crisis led to parliamentary hearings, official investigations, resignations of Oxfam executives, and the loss of 7,000 regular donors, among others (BBC 2018).

The #MeToo movement has gained momentum after prominent women broke their silence about sexual harassment and abuse in the film industry in October 2017. Following, women across the world took to social media to share their stories and named abusers (Zacharek et al. 2017). In consequence, beside powerful individuals, corporations, and public authorities, NGOs have also been at the center of a wave of scandals (Beaumont and Ratcliffe 2018). These crises critically undermine trust in NGOs, which commonly provide services to the most vulnerable populations, including women and children. What's more, because they depend on the support of multiple stakeholder groups, including governments, donors, partner organizations, and volunteers, organizational crises caused by misconduct can be particularly severe for NGOs.

While organizational wrongdoing ultimately demands change and corrective action, effective and ethical crisis communication is essential. The practice of crisis management focuses on alleviating impacts of crises on the organization and others. This includes crisis prevention and preparedness, crisis response, and postcrisis management (Institute for Public Relations 2007). This case study applies crisis communication theory, particularly the Situational Crisis Communication Theory (SCCT), to analyze how Oxfam addressed the recent sexual harassment and abuse scandal. More specifically, it focuses on the organization's response and provides insights into important considerations for NGOs in crisis.

80.2 Crises and the Situational Crisis Communication Theory (SCCT)

Coombs (2012) defined crises as "the perception of an unpredictable event that threatens important expectancies of stakeholders and can seriously impact an organization's performance and generate negative outcomes" (pp. 2–3). According to the premise that crises are perceptual, they arise when enough stakeholders view a situation as problematic, whether the organization agrees or not. Other scholars have focused on crises as revolving around so-called legitimacy gaps (Sethi 1977), which occur when publics perceive that the organization's actions are not aligned with their expectations, norms, and values. The wave of recent scandals around past cases of sexual abuse and harassment indicates a change in public expectations toward organizational behavior and policies. Once legitimacy gaps arise, organizations need to find ways to realign themselves with norms and values they are perceived to have violated (for instance, Hearit 1995).

The Situational Crisis Communication Theory (SCCT) developed by Coombs (2007) continues to be the most influential crisis communication framework to date (see Ma and Zhan 2016). SCCT holds that if the level of attribution of responsibility toward the organization for a crisis is high, then the reputational threat to the organization is high and vice versa. In other words, if the organization is viewed as at fault, the crisis will have a bigger impact on its reputation. SCCT then clusters crises into three crisis types on a continuum from low to high levels of responsibility: victim, accident, and preventable crises (Coombs 2012). When attacked by hackers, for instance, the organization may be viewed as bearing little responsibility for the situation (victim). A tragic fire may not have been caused by wrongdoing or neglect (accident), and when organizational members commit fraud, the crisis would have been preventable.

In any crisis, SCCT advises organizations to prioritize messages that (a) prevent harm to people and the environment and (b) explain how and why the situation occurred and what the organization is doing to prevent similar situations from occurring in the future. To restore an organization's reputation, however, the theory asserts that response strategies be selected according to crisis type to be effective (see Coombs 2007). Specifically, Coombs' (2012) drew on established rhetorical strategies of image restoration (see Benoit 1997) and categorized them according to the level of responsibility they accept. First, *denial postures* include messages aimed at attacking the accuser, denial, and scapegoating. These messages accept the least amount of responsibility and are most appropriate for crises caused by unfounded rumors, among others. Second, *diminishment postures* focus on excusing and justifying, which includes "minimizing the organization's responsibility" and "the perceived damage" (p. 155) and are best suited for accident crises. Third, *rebuilding postures* focus on apology and corrective action. They accept the highest amount of responsibility and are most appropriate for preventable crises. Finally, *bolstering postures* remind stakeholders of the value of the organization's work, praise stakeholders, and/or emphasize how the organization is also a victim. These strategies can be used in combination with other strategies. However, victimage postures are only appropriate when the organization faces a victim crisis (Coombs 2012).

Finally, SCCT established two intensifying factors – crisis history and prior reputation – that also influence the attribution of responsibility toward an organization. Accordingly, if an organization has a negative reputation and/or a history of similar crises, it will be less likely perceived as bearing low levels of responsibility, even in a genuine victim or accident crisis. In such cases, stakeholders tend to treat a victim crisis as an accident crisis and an accident crisis as a preventable crisis. Thus, depending on their prior reputation and crisis history, organizations have more or less latitude when crafting crisis responses (Coombs 2012). Indicating the complexity of crisis situations, scholars have further identified additional variables that impact the effect of crisis messages, such as communication channels (Liu et al. 2011) or individual's locus of control (Claeys et al. 2010).

Overall, empirical studies have provided strong evidence for the tenets of SCCT (for instance, Coombs and Holladay 2002; Jeong 2009). Recent research, however, indicates that even well-matched crisis responses have comparably lower effects on organizational reputation than the type of crisis (see Ma and Zhan 2016). Yet, a poor crisis response, or lack thereof, can cause severe additional damage (see Coombs 2016; Grebe 2013). While SCCT thus provides important guidelines, initial crisis communication tactics should not be viewed as magic bullets that replace long-term efforts to rebuild trust.

80.3 Case Study: Oxfam in Crisis

80.3.1 Organizational Background

In 1995, Oxfam was founded by several NGOs to increase their "impact on the international stage to reduce poverty and injustice" (Oxfam n.d., para 1). Beside its International Secretariat (or Oxfam International) headquartered in Oxford, its member organizations, including Oxfam GB, operate independently in 20 countries. The organizations' work focuses on combatting poverty and hunger through empowerment, providing disaster relief and emergency response, as well as advocacy for social justice, including equality, sustainability, and women's rights.

Oxfam GB employs approximately 5,000 people (Oxfam 2017). In 2015/2016, more than 31,000 volunteers supported its work (Oxfam n.d.). Donations and retail operations each made up approximately 28% of its total 2016/2017 income of £406.8 million. 43% was generated from governments and public authorities (Oxfam 2017). At the onset of the crisis, Oxfam GB's most visible spokespersons included CEO Mark Goldring and the Chair of the Board of Trustees Caroline Thompson. Chief Executive Director Winnie Byanyima represented Oxfam International.

Headlines had come to question Oxfam's legitimacy before. In October 2017, after the release of Oxfam's annual report, several news articles (Bacchi 2017; Watt 2017) reported the dismissal of 22 Oxfam staff members "over allegations of sexual abuse" (para 1) and "87 claims of sexual exploitation and abuse involving its workers" (para 5). These reports, however, did not create a crisis for the organization back then. In 2014, Oxfam GB's former counter-fraud director was convicted of fraud. He had enriched himself with payments adding up to £65,000 (BBC 2014). Fraud has also been a repeated issue as several of its annual reports showed lost funds of over £400,000 due to fraudulent activities (for instance, Austin 2018). The crisis history likely created a perception of a potentially problematic organizational culture at the onset of this crisis.

80.3.2 The Crisis Situation and the Consequences

On October 5, 2011, Oxfam International issued a little-noticed press release. The organization reported the results of an internal investigation, which had found that "a small number of its staff members working in Haiti" (para 1) had violated the organization's code of conduct, undermined the organization's reputation, abused power, and engaged in bullying. While the press release did not specify the type of misconduct, it stated that these employees had already left Haiti and no longer worked for Oxfam (Oxfam International 2011).

Seven years later, on February 9, 2018, *The Times*' cover read "Top Oxfam staff paid Haiti survivors for sex. Charity covered up scandal in earthquake zone. Girls at 'Caligula orgy' may have been underage" (O'Neill 2018b). Below the headline, the newspaper placed a large picture of the director of Oxfam's operations in Haiti, Roland Van Hauwemeiren, who was at the center of the scandal. Oxfam thus faced a preventable crisis caused by organizational decisions and actions on two levels: first, sexual misconduct by employees and, second, mishandling of the misconduct by the organization.

Initial consequences included the resignation of Oxfam's Deputy Chief Executive Penny Lawrence, official investigations by the British Charity Commission and Haiti's government, and threats from the European Commission to withdraw its funding. Haiti suspended Oxfam GB from operating in the country for the duration of its investigation (Almasy 2018), and Oxfam's leadership agreed to abstain from applying for public grants in the UK until it could show significant improvements (Slawson 2018). What's more, Desmond Tutu and Minnie Driver publicly cut all ties with the organization, Oxfam's leadership testified in front of the International Development Committee, and the organization lost approximately 7,000 regular donors within just 11 days (BBC 2018). While the Charity Commission's inquiry is still ongoing at this time, Oxfam CEO Mark Goldring announced his resignation by the end of the year in May 2018 (Rawlinson 2018).

80.3.3 Oxfam Responds

Phase 1: Explanations and Justifications. Best practices of crisis communication include timeliness, honesty, and consistency (Coombs 2012). Oxfam GB and Oxfam International responded within hours of the publication of *The Times*' article. The early responses, however, indicate that Oxfam's leadership may have underestimated how the public and the press viewed this crisis. They also showed a lack of consistency between Oxfam International's and Oxfam's GB's statements. Oxfam GB's (2018a, b) first statements condemned the misconduct as "totally unacceptable" and focused on explaining and justifying how the organization had handled the incidents in 2011. In addition, the statements informed the public about measures, which had been put in place as a direct result of the 2011 investigation; a safeguarding team and a whistle-blower hotline. Stating that Oxfam GB (2018b) "hopes that they (our supporters) will be reassured by the steps we have taken" (para 7), the organization indicated that it believes it had sufficiently addressed any past shortcomings. Oxfam International's (2018a) February 9 press release applied a less defensive strategy and acknowledged failure, which undermined Oxfam GB's message. After condemning sexual misconduct and reiterating the organization's stance against exploitation and abuse, the statement admitted that "[. . .] we have not done enough to change our own culture and to create the strongest possible policies to prevent harassment and protect people we work with around the world" (para 4). In addition, it mentioned that more cases of sexual harassment have recently surfaced at Oxfam.

Its global structure composed of Oxfam International and its independent national organizations resulted in several spokespersons from two structurally separate entities engaging in a public response at the onset of the crisis. Adding to inconsistencies and a multitude of voices was Oxfam GB's former CEO Dame Barbara Stocking who appeared on BBC Newsnight (2018) on February 10. Stocking had headed the organization in 2011 and fiercely defended the decisions made at the time without showing a hint of regret to the visible disbelief of the interviewer. Meanwhile, Oxfam's current CEO Goldring also gave press interviews. Still refusing to acknowledge that the misconduct had been addressed insufficiently, he issued a first narrow apology that mirrored sentiments of the statements from the prior day. "What I'm apologizing for," Goldring (cited from Guardian News 2018, 0:17) stated in off-the-cuff remarks, "is that nine Oxfam staff behaved in a way that was totally unacceptable and contrary to our values."

These first statements did not settle the situation. The public and press primarily seemed to judge this crisis as preventable, based on the perception that Oxfam had failed to handle sexual harassment and abuse appropriately and seriously enough. Oxfam GB's early focus on justifications and excuses – strategies most appropriate for accident and victim crises – did not address these concerns. Goldring (cited from Aitkenhead 2018) himself reflected on this dynamic in a now infamous interview; "What I felt really clearly is many people haven't wanted to listen to explanations" (para 4).

Crises can be fast-paced, dynamic, and unpredictable. Oxfam soon found itself facing a constantly shifting rhetorical situation. During the following two days, the organization issued two more statements (Oxfam International 2018b, c) in response to news reports, which had cast further doubt on the organization's handling of sexual harassment and abuse. *The Times* (O'Neill 2018c) reported that the employees who had resigned and were fired in 2011 simply took jobs with other aid agencies. News also surfaced that Van Hauwemeiren had been accused of similar misconduct on Oxfam's premises in Chad in 2006 (Ratcliffe and Quinn 2018). Adding to these revelations, the Charity Commission for England and Wales (2018), which registers and regulates charities, publicly responded to the emerging scandal; "Our approach to this matter would have been different had the full details that have been reported been disclosed to us at the time" (para 3). This statement cast further doubt on Oxfam GB's assurances that it had handled the incidents appropriately. Under mounting pressure, Oxfam shifted its response from a diminishment to a rebuilding posture.

Phase 2: Corrective Action, Apologies, and Resignations. On February 11, the Chair of Oxfam GB's Board of Trustees (cited from Oxfam 2018g) admitted in a press release that further improvement was necessary. Thompson apologized "unreservedly" (para 2) and assured that Oxfam would "fully learn the lessons of events in 2011" (para 3) by strengthening staff vetting and training and by extending reviews of its practices, among others. Oxfam International Executive Director Winnie Byanyima published a video statement that mirrored the acknowledgement of shortcomings and announced new initiatives (Oxfam International 2018e). These were later spelled out in a "comprehensive action plan to stamp out abuse" (Oxfam 2018d) aimed at increasing transparency and accountability as well as improving Oxfam's policies, practices, and culture (Oxfam International 2018f). Among others, these measures included an independent commission headed by women's rights activists (Oxfam International 2018g). The acknowledgments and repentant tone were promptly supported with the resignation of Oxfam GB's Deputy Chief Executive. "As programme director at the time," Penny Lawrence was quoted in two identical releases (Oxfam 2018c; Oxfam International 2018d), "I am ashamed that this happened on my watch and I take full responsibility" (para 4).

The organization followed up on the high-profile resignation with an open letter of apology signed by Goldring and Thompson and published in *The Guardian* and *The Times* on February 17, 2018 (see Sampson 2018). The letter was also posted in Oxfam stores (Ferguson 2018). The leaders explicitly apologized to "Oxfam supporters, friends and volunteers, [...]" as well as "to the people of Haiti and other places where the conduct of Oxfam staff has been reprehensible" (para 1) for misconduct and for the organization's failure to report the incidents accurately. Stating in bold letters that "We are listening" (para 4), they assured readers that Oxfam continues to work "hard to rebuild" (para 5) trust and reiterated the major initiatives to prevent similar wrongdoing in the future. In closing, the letter applied a strategy of bolstering by praising the "amazing, brave, committed staff and volunteers who are making remarkable life-saving, life-changing work happen in desperate situations" (para 9) and reminding readers of Oxfam's good work.

Placing a full-page ad in a high-profile newspaper is a risky choice for an NGO in a crisis. An ad provides full control over the content. However, spending a charity's money on what may be perceived as "just PR" could lead to further backlash during crises. Oxfam addressed possible concerns with a disclaimer on the bottom of the letter; "Private donors who wish to remain anonymous have kindly paid for this message" (para 13). What's more, Liu et al. (2011) have shown that publics are least likely to accept accommodative (or rebuilding) crisis messages when they are delivered via traditional media.

According to SCCT, rhetorical strategies of apology paired with corrective action are more appropriate to address this preventable crisis. Inconsistencies, however, continued to undermine Oxfam's attempts to restore its reputation. Particularly, Goldring's (cited from Aitkenhead 2018) statements in an interview with *The Guardian*, which preceded the open letter of apology by a day, further fueled the crisis:

> The intensity and the ferocity of the attack makes you wonder, what did we do? We murdered babies in their cots? Certainly, the scale and the intensity of the attacks feels out of proportion to the level of culpability. I struggle to understand it. You think, 'My God, there's something going on there.' (para 11)

This unfortunate quote quickly made headlines (for instance, Swinford and Bird 2018). It distracted from the official apology, and the posture of victimage and articulated lack of understanding for the public outrage directly contradicted any official expression of remorse. It hence cast a shadow on the sincerity of Oxfam's public apologies, and Goldring's interview had created an additional preventable issue that now needed to be addressed.

Despite the public outcry over this statement, Oxfam continued its rebuilding strategy. Leading up to the parliamentary hearing in London, the organization published its 2011 report (Oxfam 2018f), apologized to Haiti's government, and met with government officials in the country. The report revealed that Oxfam staff members under investigation had physically threatened witnesses. A few days later, Haiti suspended Oxfam's operations in the country for the duration of its own investigation (Almasy 2018).

When Goldring, Byanyima, and Thompson faced the parliamentary commission on February 20 (see International Development Committee 2018), the organization was still under intense public scrutiny and criticism. The three leaders reiterated their apologies, committed to measures of improvement, and shared recently reported information on additional cases of sexual harassment within the organization. Prompted by a committee member, Goldring also apologized for his prior remarks in *The Guardian*. This part of the hearing captured significant media attention (for instance, Reuters 2018; Smout 2018).

While public attention to the scandal peaked on February 12, the intensity slowly decreased after the parliamentary hearing on February 20 (see Fig. 1) and remained

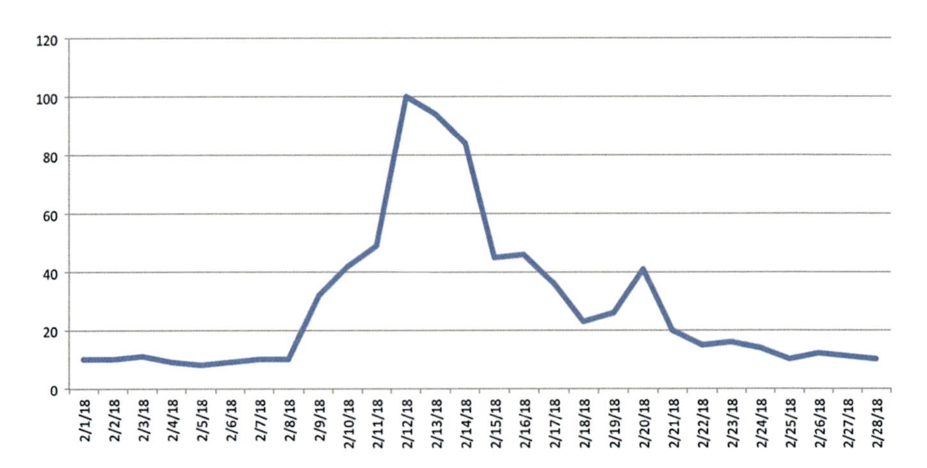

Fig. 1 Interest over time relative to the maximum search interest (100) based on google searches for "Oxfam" in the United Kingdom from February 1 to 28, 2018

low. This, however, did not mean that Oxfam could go back to business as usual as its reputation had significantly suffered. What's more, other NGOs had now also come under scrutiny for similar issues (Beaumont and Ratcliffe 2018). Addressing the sector-wide issue, Goldring signed an open letter of apology with leaders from 21 other UK aid agencies, including Save the Children, UNICEF, and World Vision. The letter, published in the Huffington Post, stated:

> As aid agencies we will take every step to right our wrongs and eradicate abuse within our industry. We are truly sorry that at times our sector has failed. We must and will do better. (Goldring et al. 2018, para 1–2)

The organizations further announced joint efforts to improve safeguarding, referencing systems, and reporting to authorities and agencies. In addition, the leaders reiterated their intentions to work with regulators to overcome "barriers to rigorous background checks in the UK." (para 7). Joining a large group of affected NGOs and communicating jointly allowed Oxfam to transcend its isolated status in the public arena.

Oxfam now also had to follow up on its promises. In March, Oxfam International (2018h) appointed the co-chairs and members of the Independent Commission charged with developing recommendations for creating "a culture of zero tolerance for any kind of sexual harassment, abuse or exploitation" (para 8) within the organization and industry. In addition, both Oxfam International and Oxfam CB continue to report updates about ongoing efforts on their websites (Oxfam International 2018f, i; Oxfam 2018e).

While April and March remained relatively calm, Goldring announced his resignation on May 16 by the end of 2018 (Rawlinson 2018). Indeed, while he had joined Oxfam two years after the incidents and internal investigation in question, the few but prominent errors at the onset of the crisis damaged his credibility. This, in turn, likely undermined his ability to lead efforts to restore Oxfam GB's reputation as one of its most visible spokespersons. Whether remaining in this role until the end of the year will be beneficial for the organization's recovery in the public eye or not remains an open question.

80.4 Conclusion

In our current times, many organizations will need to revisit their past and present policies, actions, and decisions regarding sexual harassment and abuse. The case of Oxfam's crisis shows that NGOs can face high-stakes crises even around issues that they may view as matters of the past. Oxfam GB initially got overwhelmed with the sudden scandal over its handling of sexual abuse and harassment in 2011. Its early messages focused on excusing and justifying past decisions and fell short to address the core of the preventable crisis. Under growing pressure, the organization eventually had to apologize and focus on corrective action.

Early inconsistencies caused by shifting strategies, multiple spokespersons and postures of victimage, however, continued to undermine the organization's ability to get a handle on the situation and regain trust. The early statements came to provide a backdrop upon which any apology would be interpreted. Further, with public interest peaking three days after the onset of the scandal, stakeholders might remember Oxfam's earlier messages more than its official apology. Had the organization been able to accurately assess the situation as a preventable crisis and responded with a consistent rebuilding strategy right away, its crisis communication might have been more effective.

Oxfam's structure with many independent affiliate organizations, not too different from other global NGOs, increased challenges for consistent crisis responses. At the same time, stakeholders will be unlikely to distinguish separate entities under the same "brand." A crisis for one member organization might quickly undermine the reputation of all affiliates and require each to communicate within their areas of operation. NGOs are thus well advised to develop strong crisis protocols and invest in crisis preparedness beyond the boundaries of each individual member organization.

Oxfam's errors also underline necessary crisis communication expertise. NGOs, particularly charities, are more constrained when seeking external crisis counsel, as investments in "public relations" may be perceived negatively and further fuel the crisis. As part of proper pre-crisis planning and preparedness (see Coombs 2006), NGOs should ensure that their leadership and communication staff are well-versed, trained, and experienced in crisis communication.

Finally, crises also provide opportunities for renewal and social change. Transitioning into a phase of postcrisis communication (Ulmer et al. 2007), Oxfam can now better itself as it continues to regain trust. To prevent future crises and beyond, other NGOs should take the wave of recent scandals as an impetus for reflection on their own cultures and policies.

References

Aitkenhead D (2018) Oxfam boss Mark Goldring: 'Anything we say is being manipulated. We've been savaged'. https://www.theguardian.com/world/2018/feb/16/oxfam-boss-mark-goldring-anything-we-say-is-being-manipulated-weve-been-savaged. Accessed 15 Apr 2018

Almasy S (2018) Haiti suspends Oxfam operations over sexual misconduct scandal. CNN. https://www.cnn.com/2018/02/22/americas/oxfam-report-haiti-threats-intl/index.html. Accessed 15 Apr 2018

Austin J (2018) Oxfam scandal: Under fire charity is losing half a million pounds a year to fraud. https://www.express.co.uk/news/uk/918425/OXFAM-SCANDAL-charity-losing-half-a-million-pounds-A-YEAR-fraud. Accessed 15 Apr 2018

Bacchi U (2017) British charity Oxfam dismisses 22 staff over sexual abuse. https://www.reuters.com/article/us-britain-charity-sexual-exploitation/british-charity-oxfam-dismisses-22-staff-over-sexual-abuse-idUSKBN1D01UI. Accessed 15 Apr 2018

BBC (2014) Oxfam ex-fraud chief jailed for scamming charity. http://www.bbc.com/news/uk-england-oxfordshire-27588466. Accessed 15 Apr 2018

BBC (2018) Oxfam Haiti allegations: How the scandal unfolded. http://www.bbc.com/news/uk-43112200. Accessed 15 Apr 2018

BBC Newsnight (2018) Former Oxfam boss knew of sexual misconduct claims. https://www.youtube.com/watch?v=VwlH0XtmA3Y. Accessed 15 Apr 2018

Beaumont P, Ratcliffe R (2018) #MeToo strikes aid sector as sexual exploitation allegations proliferate. https://www.theguardian.com/global-development/2018/feb/12/metoo-strikes-aid-sector-as-sexual-exploitation-allegations-proliferate. Accessed 15 Apr 2018

Benoit WL (1997) Image repair discourse and crisis communication. Pub Rel Rev 23:177–186

Charity Commission for England and Wales (2018) Charity Commission statement on Oxfam. https://www.gov.uk/government/news/charity-commission-statement-on-oxfam. Accessed 15 Apr 2018

Claeys AS, Cauberghe V, Vyncke P (2010) Restoring reputations in times of crisis: an experimental study of the Situational Crisis Communication Theory and the moderating effects of locus of control. Pub Rel Rev 36:256–262

Coombs WT (2006) Code red in the boardroom: crisis management as organizational DNA. Greenwood, Westport. https://books.google.com/books/about/Code_Red_in_the_Boardroom.html?id=V8UdG3xANxAC

Coombs WT (2007) Protecting organization reputations during a crisis: the development and application of situational crisis communication theory. Corp Reputation Rev 10:163. https://doi.org/10.1057/palgrave.crr.1550049

Coombs WT (2012) Ongoing crisis communication: planning, managing, and responding. SAGE, Thousand Oaks

Coombs WT (2016) Reflections on a meta-analysis: crystallizing thinking about SCCT. J Public Relat Res 28(2):120–122

Coombs WT, Holladay SJ (2002) Helping crisis managers protect reputational assets: initial tests of the situational crisis communication theory. Manag Commun Q 16:165–186

Ferguson S (2018) Oxfam apology letters appear in windows of the charity's Bath stores in wake of recent Haiti scandal. https://www.bathchronicle.co.uk/news/oxfam-apology-letters-appear-windows-1239657. Accessed 15 Apr 2018

Gibelman M, Gelman SR (2004) A loss of credibility: patterns of wrongdoing among nongovernmental organizations. Voluntas 15:355–381. https://doi.org/10.1007/s11266-004-1237-7

Goldring M et al (2018) As aid agencies, we will take every step to right our wrongs and eradicate abuse within our industry. https://www.huffingtonpost.co.uk/entry/aid-exploitation_uk_5a901de7e4b0ee6416a271ed?guccounter=1. Accessed 15 Apr 2018

Grebe SK (2013) Things can get worse: how mismanagement of a crisis response strategy can cause a secondary or double crisis: the example of the AWB corporate scandal. Corp Commun: An Int J 18:70–86. https://doi.org/10.1108/13563281311294137

Guardian News (2018) 'I am deeply ashamed' says Oxfam CEO of the Haiti sex scandal. https://www.youtube.com/watch?v=tDO9bgnnr2M. Accessed 15 Apr 2018

Hearit KM (1995) "Mistakes were made": organizations, apologia, and crises of social legitimacy. Comm Stud 46:1–17. https://doi.org/10.1080/10510979509368435

Institute for Public Relations (2007) Crisis management and communications. https://instituteforpr.org/crisis-management-and-communications/. Accessed 15 Apr 2018

International Development Committee (2018) Sexual exploitation in the aid sector. https://parliamentlive.tv/event/index/d01474b7-9989-4db3-8183-eda3ecdfd449. Accessed 15 Apr 2018

Jeong SH (2009) Public's responses to an oil spill accident: a test of the attribution theory and situational crisis communication theory. Public Relat Rev 35:307–309

Liu BF, Austin L, Jin Y (2011) How publics respond to crisis communication strategies: the interplay of information form and source. PR Rev 37:345–353

Ma L, Zhan M (2016) Effects of attributed responsibility and response strategies on organizational reputation: a meta-analysis of situational crisis communication theory research. J PR Res 28:102–119. https://doi.org/10.1080/1062726X.2016.1166367

O'Neill S (2018a) Top Oxfam staff paid Haiti survivors for sex. The Times February 8:1. London
O'Neill S (2018b) Oxfam in Haiti: 'It was like a Caligula orgy with prostitutes in Oxfam T-shirts'. https://www.thetimes.co.uk/article/oxfam-in-haiti-it-was-like-a-caligula-orgy-with-prostitutes-in-oxfam-t-shirts-p32wlk0rp. Accessed 15 Apr 2018
O'Neill S (2018c) Oxfam sex scandal: sacked staff found new aid jobs. https://www.thetimes.co.uk/article/new-shame-for-oxfam-h5nq8lmfn. Accessed 15 Apr 2018
Oxfam GB (2017) 2016/17 Oxfam annual report & accounts. http://www.oxfamannualreview.org.uk/wp-content/uploads/2017/11/oxfam-annual-report-2016-17-v2.pdf. Accessed 15 Apr 2018
Oxfam GB (2018a) Oxfam response to The Times story – 9 February 2018. https://www.oxfam.org.uk/media-centre/press-releases/2018/02/oxfam-response-to-the-times-story. Accessed 15 Apr 2018
Oxfam GB (2018b) Oxfam's reaction to The Times article (9 Feb 2018). https://www.oxfam.org.uk/media-centre/the-times-statement. Accessed 15 Apr 2018
Oxfam GB (2018c) Oxfam announces resignation of Deputy Chief Executive. https://www.oxfam.org.uk/media-centre/press-releases/2018/02/oxfam-announces-resignation-of-deputy-chief-executive. Accessed 15 Apr 2018
Oxfam GB (2018d) Oxfam announces comprehensive action plan to stamp out abuse. https://www.oxfam.org.uk/media-centre/press-releases/2018/02/oxfam-announces-comprehensive-action-plan-to-stamp-out-abuse. Accessed 15 Apr 2018
Oxfam GB (2018e) Stamping out abuse: Information and updates. https://www.oxfam.org.uk/what-we-do/about-us/stamping-out-abuse. Accessed 15 Apr 2018
Oxfam GB (2018f) Oxfam releases report into allegations of sexual misconduct in Haiti. https://www.oxfam.org.uk/media-centre/press-releases/2018/02/oxfam-releases-report-into-allegations-of-sexual-misconduct-in-haiti. Accessed 15 Apr 2018
Oxfam GB (2018g) Oxfam commits to improvements in aftermath of Haiti reports. https://www.oxfam.org.uk/media-centre/press-releases/2018/02/oxfam-commits-to-improvements-in-aftermath-of-haiti-reports. Accessed 15 Apr 2018
Oxfam GB (n.d.) Oxfam's CEO, Directors and Trustees. https://www.oxfam.org.uk/what-we-do/about-us/our-trustees. Accessed 15 Apr 2018
Oxfam International (2011) Internal investigation confirms staff misconduct in Haiti. https://www.oxfam.org/en/pressroom/pressreleases/2011-09-05/internal-investigation-confirms-staff-misconduct-haiti. Accessed 15 Apr 2018
Oxfam International (2018a) Oxfam's reaction to sexual misconduct story in Haiti. https://www.oxfam.org/en/pressroom/reactions/oxfams-reaction-sexual-misconduct-story-haiti. Accessed 15 Apr 2018
Oxfam International (2018b) Statement responding to allegations in today's Times that Oxfam failed to warn other NGOs not to re-employ those guilty of sexual misconduct in Haiti. https://www.oxfam.org/en/pressroom/reactions/statement-responding-allegations-todays-times-oxfam-failed-warn-other-ngos-not. Accessed 15 Apr 2018
Oxfam International (2018c) Statement on Haiti including response to allegations that Oxfam staff used sex workers in Chad in 2006. https://www.oxfam.org/en/pressroom/reactions/statement-haiti-including-response-allegations-oxfam-staff-used-sex-workers-chad. Accessed 15 Apr 2018
Oxfam International (2018d) Oxfam Great Britain announces resignation of Deputy Chief Executive. https://www.oxfam.org/en/pressroom/pressreleases/2018-02-12/oxfam-great-britain-announces-resignation-deputy-chief-executive. Accessed 15 Apr 2018
Oxfam International (2018e) Oxfam International Executive Director Winnie Byanyima responds to Haiti and Chad reports: "We are committed to our work and to our values". https://www.oxfam.org/en/multimedia/video/2018-oxfam-international-executive-director-winnie-byanyima-responds-haiti-and-chad. Accessed 15 Apr 2018
Oxfam International (2018f) Immediate response actions: sexual misconduct. https://www.oxfam.org/en/immediate-response-actions-sexual-misconduct. Accessed 15 Apr 2018

Oxfam International (2018g) Oxfam asks women's rights leaders to carry out urgent independent review. https://www.oxfam.org/en/pressroom/pressreleases/2018-02-16/oxfam-asks-womens-rights-leaders-carry-out-urgent-independent. Accessed 15 Apr 2018

Oxfam International (2018h) Oxfam announces Zainab Bangura and Katherine Sierra to co-lead Independent Commission on Sexual Misconduct. https://www.oxfam.org/en/pressroom/pressreleases/2018-03-16/oxfam-announces-zainab-bangura-and-katherine-sierra-co-lead. Accessed 15 Apr 2018

Oxfam International (2018i) How we are working to rebuild your trust. https://www.oxfam.org/en/how-we-are-working-rebuild-your-trust. Accessed 15 Apr 2018

Oxfam International (n.d.) History of Oxfam International. https://www.oxfam.org/en/countries/history-oxfam-international. Accessed 15 Apr 2018

Ratcliffe R, Quinn B (2018) Oxfam: fresh claims that staff used prostitutes in Chad. https://www.theguardian.com/world/2018/feb/10/oxfam-faces-allegations-staff-paid-prostitutes-in-chad. Accessed 15 Apr 2018

Rawlinson K (2018) Oxfam chief steps down after charity's sexual abuse scandal. https://www.theguardian.com/world/2018/may/16/oxfam-head-mark-goldring-steps-down-sexual-abuse-scandal. Accessed 18 May 2018

Reuters (2018) Oxfam chief apologises for "babies in cots" comment on sex abuse scandal. https://www.reuters.com/article/britain-oxfam/oxfam-chief-apologises-for-babies-in-cots-comment-on-sex-abuse-scandal-idUSL8N1QA2TV. Accessed 15 Apr 2018

Sampson J (2018) Oxfam publishes apology in newspapers. https://www.newsworks.org.uk/news-and-opinion/oxfam-publishes-apology-in-newspapers/194094. Accessed 15 Apr 2018

Sethi P (1977) Advocacy advertising and large corporations: social conflict, big business image, the news media, and public policy. D.C. Heath, Lexington

Slawson N (2018) Oxfam government funding cut off after Haiti scandal. https://www.theguardian.com/world/2018/feb/16/oxfam-government-funding-cut-off-after-haiti-scandal. Accessed 15 Apr 2018

Smout A (2018) Oxfam chief apologizes for 'babies in cots' comment as more abuse reported. https://www.reuters.com/article/us-britain-oxfam/oxfam-chief-apologizes-for-babies-in-cots-comment-as-more-abuse-reported-idUSKCN1G41DH. Accessed 15 Apr 2018

Swinford S, Bird S (2018) Oxfam boss: what did we do? Murder babies in cots? https://www.telegraph.co.uk/news/2018/02/16/oxfam-boss-baffled-ferocious-criticism-claiming-critics-gunning/. Accessed 15 Apr 2018

Ulmer RR, Seeger MW, Sellnow TL (2007) Post-crisis communication and renewal: expanding the parameters of post-crisis discourse. PR Rev 33:130–134. https://doi.org/10.1016/j.pubrev.2006.11.015

Watt H (2017) Oxfam says it has sacked 22 staff in a year over sexual abuse allegations. https://www.theguardian.com/world/2017/oct/31/oxfam-says-it-has-sacked-22-staff-in-a-year-over-sexual-abuse-allegations. Accessed 15 Apr 2018

Zacharek S, Dockterman E, Sweetland Edwards H (2017) Person of the year: the silence breakers. http://time.com/time-person-of-the-year-2017-silence-breakers/. Accessed 15 Apr 2018

Communication and Culture for Development: Contributions to Artisanal Fishers' Wellbeing in Coastal Uruguay

81

Paula Santos and Micaela Trimble

Contents

Abstract

The combination of the concepts of Wellbeing in Developing Countries (WeD) and social representations offers a relevant methodological framework to investigate how the exercise of the capacity of voice by vulnerable populations can alter positively aspects related to their wellbeing and aspirations. These alterations can contribute directly to development since they relate to people's "capacity of voice," in the sense of Arjun Appadurai. Using the First Artisanal Fisheries Festival in Piriápolis (coastal Uruguay, 2012) as a case study – an activity of communication for development arisen from a participatory action research group – we investigated how the exercise of fishers' capacity of voice promoted a

P. Santos (✉)
Universidad Católica del Uruguay, Montevideo, Uruguay
e-mail: paulasantosvizcaino@gmail.com

M. Trimble (✉)
South American Institute for Resilience and Sustainability Studies (SARAS), Bella Vista-Maldonado, Uruguay
e-mail: mica.trimble@gmail.com

J. Servaes (ed.), *Handbook of Communication for Development and Social Change*, https://doi.org/10.1007/978-981-15-2014-3_97

change in the figurative nucleus of their self-representation. Semi-structured interviews with 12 informants were performed. The results show that the Fisheries Festival tackled at least four material aspects of wellbeing: catch, boat, fishing gear, and knowledge. Subjective wellbeing implications were also found as the Festival confirmed fishing as a way of life and added value to the existence of the fishers by recirculating positive values about them. In terms of relational wellbeing (the third dimension), the Festival approach to fishers' representation transcended to other sectors of society while enriching their original connection with the environment. In conclusion, there was a change in the fishers' terms of recognition, placing them in a more advantaged position to negotiate their own interests. This modification to existent cultural consensus about fishers in Piriápolis will have a positive effect in their future existence, thus contributing to development.

Keywords

WeD approach · Social representations · Aspirations · Capacities · Participatory research · Fisheries

81.1 Introduction

Artisanal fishing communities stand out for their knowledge of the sea. Their oral practices transmitted from generation to generation directly contribute to environmental sustainability and food security, among others. For different reasons, in Uruguay (and the world), these communities historically experience situations of vulnerability, and the richness of their practices remains ignored, particularly from the perspective of fishers themselves.

Communication and culture can play a key role in projects aimed at improving their wellbeing. Giving them "voice" (Appadurai 2004:62) could improve the terms in which they are recognized in society, thus favoring their repositioning in order to better negotiate their interests. This work offers an exploratory model for inquiring about the social change produced in a particular moment of a communication for development initiative in a specific time and space, at a given cultural scenario such as fisheries. The model was applied to the case study of the First Artisanal Fisheries Festival in Piriápolis (February 11–12, 2012), organized by the POPA Group (For Artisanal Fishing), a participatory research group formed by multiple stakeholders.

In general, results indicated that the fishers raised their "voice" to promote a new condition of "different" (Appadurai 2004:62) of themselves in direct relation with their own interests. This increased their ability to "navigate" in society more fluidly. In short, they were able to publicly express their visions and promote transformations in relation with their manifested aspirations of wellbeing. However, the persistence of the transformative process initiated in the social representation of fishers as of the Festival and its scope in the fisher population are presented as challenges to these transformations.

81.1.1 Social Representations as Determinants of Wellbeing

Social representations are a socially elaborated and shared form of knowledge. They are also defined as "common sense knowledge" or "natural knowledge" to differentiate them from scientific knowledge (Moscovici, in Jodelet 1997:53). Through them humans attempt to understand and explain the phenomena of everyday life. They contain a pragmatic or functional dimension, not only in terms of behavior but also in the transformation of the environment in which these behaviors take place. They allow an individual or group to take a stand against different situations, events, objects, and communications that concern them and to orient their action according to that position. Social representations are forms of constituted thought, as they constitute sociocultural products that intervene in social life as preformed structures that serve as a framework for interpretation. Meanwhile, they are forms of constituent thought, in the sense that they intervene in the elaboration or conformation of the same object that they represent, and in this way, they contribute to shape the social reality of which they are a part, determining in a different measure their effects in the daily life. This allows for understanding social representations as processes of construction of reality (Ibáñez 1988:37 in Araya 2002:30).

The social wellbeing approach was developed by the Economic and Social Research Council Research Group on Wellbeing in Developing Countries (WeD) at the University of Bath. The key lines of thought and research relating to social wellbeing arise within the approaches of economics of happiness, poverty and development, capabilities, gender, human rights, sustainable livelihoods, vulnerability, and social capital (Weeratunge et al. 2013), defined by the WeD group as the "state of being with others, which arises where human needs are met, where one can act meaningfully to pursue one's goals, and where one can enjoy a satisfactory quality of life." Viewed as a social process, wellbeing is composed of material, relational, and subjective dimensions that occur simultaneously in a specific time and space. The material dimension includes assets, social security (welfare), and quality of life. The relational dimension is divided into two spheres: the social sphere, social relations and access to public goods, and the human sphere, which refers to aspects-capabilities and attitudes in relation to life and personal relationships. Subjective aspects of wellbeing also have two dimensions: on the one hand, people's perception of their position (material, social, and human), and on the other, people's cultural values, ideologies, and beliefs. These different aspects of wellbeing are interlinked and cannot exist without the others. Wellbeing arises in the constant interaction of externally observable and verifiable aspects of reality, with their "subjective" perceptions and evaluations by subjects. Through this process, people build the meaning of their lives (White 2009:9). Wellbeing is then socially and culturally constructed. Therefore, if socially and culturally constructed knowledge is determinant of wellbeing, changes in social representations will produce changes in wellbeing.

The combination of these two concepts (social representations and wellbeing) is rooted in at least four common elements that regard the conceptualization of reality and the centrality of subjects in that reality (see Table 1). In particular, both concepts

Table 1 Common elements to the concepts of social wellbeing and social representation

	Reality		The subject	
A combination of two concepts:	A process (. . .)	(. . .) in constant change	Capable subjects (meaning having capacities) (. . .)	(. . .) act collectively/ intersubjectively
Social representation (knowledge)	Seen as processes of construction of reality	They contain a pragmatic dimension that makes the transformation of the environment possible	Through social representations, humans attempt to understand and explain the phenomena of everyday life	They contribute to shape the social reality of which they are a part, determining in a different measure their effects in the daily life
Social wellbeing (a state)	It is viewed as a social process with material, relational, and subjective dimensions that occur in a specific time and space	It is realized through the "work" people put daily into making meaning out of their lives	People act meaningfully to pursue their goals and enjoy a satisfactory quality of life	Understandings of wellbeing are socially and culturally constructed

consider reality as a process in constant transformation. Subjects, as capable beings, are active builders of reality. To do so, they rely on their subjectivity and on their capacity to objectivize aspects of reality intersubjectively, socially, and culturally.

Both concepts approach change, but at different levels. Social representations are abstract bodies of knowledge circulating and recirculating indefinitely in society in general, whereas the concept of wellbeing, defined as "state," shapes and re-shape itself throughout a specific time and space. It is in this subtle combination that micro-generation of social common sense matters for the collective conceptualization of wellbeing (under construction in a macro space and time).

A last justification to this combination of methodological frameworks is the possibility of investigating social change provided by the concept of social representation. The theory of the figurative nucleus of the social representation explains how discourse is structured and objectivized within the representation in a figurative synthetic and concrete scheme of lived clear images. These structured images are the figurative nucleus, a concentrated and graphic conceptual core that captures the essence of the concept. These simplified visions are what enables people to talk and understand things and other people more easily and become natural facts. This scheme or nucleus is the more stable and solid part of the representation. It organizes the representation as a whole and provides meaning to the rest of elements present in the field of the representation (Moscovici, 1979, 1981, 1984 a, b, in Araya 2002:35).

The theory of the figurative scheme has implications in social change, because only those actions intending to modify a social representation that are addressed to modify the nucleus will be successful, as the global meaning of the representation depends on it (Araya 2002:41).

Thus, the combination of both concepts provides a methodological approach to investigate change in social wellbeing as of the study of changes in social representations.

81.1.2 Raising the Voice for Development

In turn, the direct relation between the concept of social wellbeing and that of "capacity of voice" (Appadurai 2004:59–84), a cultural capacity originated in the concept of developmental capabilities of Amartya Sen (1985a, 1999 in Appadurai 2004:63), indicates that alterations in social wellbeing affect development.

Based on Hirschman (1970 in Appadurai 2004:63), Appadurai claims that the poor oscillate between the exit from society and the loyalty to norms that reproduce their vulnerability. Between these extremes, there appears the capacity of voice, a cultural capacity, as a means for overcoming this oscillation. This can be achieved because cultural consensus cannot be taken for granted (Fernandez, 1965, 1986, in Appadurai 2004:64). Appadurai finds also that future is implicit in factors central to society and culture (understood as specific and multiple designs of social life), such as norms, beliefs, and values that circulate in society, where cultural consensus is produced.

Therefore, according to Appadurai (2004), the capacity of vulnerable populations to exercise their voice to debate, challenge, and oppose directions is vital for living a collective social life according to their own interests. Not only because the capacity of voice is virtually a definition of inclusion and participation in any democracy but also because it is the only way they will be able to find possible local forms of altering what he calls "the terms of recognition" in any cultural regime. To strengthen voice as a cultural capacity, vulnerable populations should find metaphors, rhetorics, and ways of organizing themselves and acting publicly that better function in their cultural worlds. Appadurai has seen in various movements in the past that, when these work, a change takes place in the terms of recognition and in the cultural framework. This will increase their capacity to navigate in society more fluidly and consequently project their existence in time, to express their visions publicly and obtain results adapted to their own wellbeing and for development in general (Appadurai 2004:66).

Just as the combination of the concepts of social representation and social wellbeing is rooted in the coincidence of at least four conceptual elements, the concept of capacity of voice shares the consideration of reality as a process in constant transformation, where subjects, as capable beings, are responsible for building it socially and culturally and produce change.

An additional theoretical observation is made for methodological purposes regarding a particular contribution from the concept of social representation to the study of the capacity of voice. Based on Durkheim, Appadurai recognizes that "there is no self outside the social frame, setting, mirror." He says that "wellbeing aspirations are never simply individual but formed in interaction and in the thick of social life" (2004:67). On his side, Moscovici had several years before proposed a step forward in the classic conception of "collective representations" of Durkheim, putting the focus on the individual (rather than the collective approach) and in the elaboration of representations in a process of intersubject exchange (Moscovici, cfr. Banchs 2000:8–9 in Araya 2002). According to this work, the concept of social representations offers the possibility of digging deeper in methodological analysis of the exercise of the voice capacity for wellbeing, finely tackling the very centrality of the subject in the process, and building upon the intersubject level.

In sum, the interweaving of these three concepts offers the possibility of investigating how the exercise of the capacity of voice can alter social wellbeing and how these alterations can be investigated through the study of the figurative nucleus of social representations to assess social change at the microlevel for further upscaling.

81.2 Case Study with Artisanal Fisheries in Piriápolis, Uruguay

In this research, we investigated the First Artisanal (or Small-scale) Fisheries Festival in Piriápolis (coastal Uruguay, 2012) as a case study (Stake, 1995 in Cresswell 2003, 2013:15).

Piriápolis is a tourist city in Maldonado Department (Uruguay), founded in 1890. Around 10,000 people live in Piriápolis throughout the year, but this number increases to 40,000 during the austral summer. At the time of this research, there were 50 fishing boats in the Piriápolis area, although some were disused or used only seasonally. Two to four fishers work on board boats that range in length from 4 to 8 m and use motors that vary from 8 to 60 horsepower (see Fig. 1). The fishing gear most commonly employed consists of bottom-set longlines and gill nets of varied mesh sizes to catch different fish species. Most of the catch is sold via middlemen, although some fishers have a fish stall (Trimble 2013).

Over the past 25 years, Piriápolis has received migrant fishers from other coastal localities during the high fishing season. At the time of this research, most artisanal fishers in Piriápolis were mobile ("nomads" or "migratory" in their own words): they move along the coast (either sailing or carrying their boats on a truck) in response to whitemouth croaker movements (one of the main commercial species). The number of fishers therefore varies greatly throughout the year (e.g., from 30 to 150 fishers) mainly due to resource availability. Some migrant fishers also have additional occupations seasonally. Their most common additional occupations are as construction workers, as crew on industrial trawler vessels based in Montevideo, as vendors in the fish market, as woodcutters, and as professional gardeners (especially during the summer).

81.2.1 The POPA Group of Piriápolis and the First Fisheries Festival

The POPA Group (POPA stands for "Por la Pesca Artesanal en Piriápolis" in Spanish, which is equivalent in English as For Artisanal Fisheries in Piriápolis) is

Fig. 1 (**a**) Fishing boat used by artisanal fishers (POPA photo-art show inaugurated at the First Fisheries Festival in Piriápolis). (**b**) Fishing gear (Gill Net) used in Piriápolis (POPA Group)

a participatory research group which originated in 2011 in the context of the PhD thesis of the coauthor of this chapter. The creation of a participatory action research group was proposed to analyze whether the foundations for fisheries co-management could be laid in that area (Trimble and Berkes 2013). To that end, in May 2011, artisanal fishers from Piriápolis, representatives of the state (National Fisheries Agency), academia, and civil society were invited to participate to contribute collectively to the solution of problems of artisanal fisheries in that location (Trimble and Lazaro 2014). Over the months, these different actors formed the so-called POPA Group. During the first year, the Group brought closely together 13 people from the local fishery, academia, civil society, and the state. The first author of this work was a member at the time of creation of the Group and design of the festival.

When discussing the problems of the fisheries, fishers and other Group members acknowledged that, despite representing a relevant economic activity and the way of life of many families, artisanal fishing was not valued by a large part of the inhabitants, who sometimes identify this activity as detrimental to a growing tourism (when in fact tourists show interest in learning more about fishing in the area). The time of greatest sale of fish is the austral summer (December to February), when Piriápolis receives many tourists (from Uruguay and abroad). In winter time, the fish sale decreases, since the permanent population in Piriápolis is much lower than in the summer (and Uruguayans have traditionally preferred beef).

Around 2010, the sale of fish caught by the artisanal fishing sector in Piriápolis began to be negatively affected by the increasing sale of Pangasius (also known as basa). This catfish began to be imported into Uruguay in 2008, coming from hatcheries in Vietnam at very low cost (being an aquaculture product). The sale of Pangasius in Piriápolis (and in the rest of the country) increased from that moment onward, be it in local markets, supermarkets, or restaurants, where it is often sold as if it were a species caught locally, deceiving the consumer. This problem of the sale of Pangasius in Piriápolis, together with the low social valorization of the fishery, was diagnosed by the fishers themselves, who were eager to be part of POPA Group. During a process of monthly workshops, the team discussed strategies to address this social-environmental problem. As a result, the intersectoral and interdisciplinary group conformed opted for the organization of a communication activity.

Organized by POPA, the First Artisanal Fisheries Festival (Festival or Show) took place from February 11 to 12, 2012 in Piriápolis. It involved an organization process of approximately 9 months. The variety of actors that made up the group in the process favored the support of local, national, and international actors from the public, private, academic, and civil society sectors to the activity implementation.

The Festival offered five simultaneous proposals:

- Art show including varied manifestations: a photographic exhibition of 25 figures of the local fishery (taken by Group members), titled “A Day in the Life of an Artisanal Fisher”; an exhibition of 27 drawings by children of the state primary school of Piriápolis; posters with specific information on artisanal fishing also contributed content. In addition, national artists exhibited their works of art inspired by artisanal fishing and performed live oil painting of works in dialog with the public.
- An exhibition of fishing gear by artisanal fishers allowed the public to know the gill nets and longlines used locally and participate in the preparation (see Fig. 2).
- Health education workshops by a medical doctor focused on disseminating the nutritional properties of fish species caught in Uruguay.
- A gastronomy area, with participation of local chefs, a group of students of local international cuisine, an artisanal fisher, and an international chef, offering marine specialties.
- National music and dance groups offered their shows at specially prepared stage.

The Festival had the participation of approximately 3000 people. Several media disseminated it (before, during, and after its realization): television channels, radios, local newspapers, information center of Piriápolis, social networks on the Internet, and websites of the institutions that make up POPA. Local media and national media supported the proposal. Artisanal fishers were the sole spokespersons of the Group to announce, cover, and inform about the Festival. In general, artisanal fishers in Uruguay appear in the national and international media agenda in relation to fisheries problems. The Festival, however, allowed for an innovative vision of artisanal fishers in the media, repositioning them as agents in control of their condition, not without problems, but addressing them in a positive and constructive manner (see Fig. 2).

81.2.2 Fishers Communicating for Development

The Festival can be described as a communication for development activity. Tufte and Mefalopulos (2009:8) identify three main conceptual approaches to development communication: the diffusion model (one-way/monologic communication), the life skills model, and the participatory model (two-way/dialogic communication). The last two models are presented in contrast to the diffusion one. The life skills model is considered

Fig. 2 (**a**) Artisanal fishers interviewed by the media during the Festival (POPA Group). (**b**) Artisanal fisher sharing their practices with visitors (POPA Group)

an intermediate model of communication for development (Hendricks 1998, in Tufte and Mefalopulos 2009:9). It focuses on the development of personal skills. It originates in adult education, but in development it works connected to the exercise of rights, as well as to address the structural conditions that impede the development of skills. On the other hand, there is the participatory model based on the liberating pedagogy of Paulo Freire of the 1960s. It is characterized by being dialogical and horizontal. Instead of communicating the "correct" information to specific audiences, just as the diffusion model proposes, it is about articulating collective action processes and the reflection of relevant stakeholders. The center of attention is the empowerment of citizens through the active involvement in the identification of their own problems, the search for solutions, and the implementation of problem-solving strategies.

In relation to these conceptual models, POPA Group designed a communication activity spontaneously combining elements of the three models. On the one hand, through a positive approach to artisanal fishers, it addressed the structural problem of their terms of recognition in society, which prevent them from developing their full environmental, economic, and cultural potential as social actors. On the other hand, the strategy of action, which was set a priori by the participatory research model put in practice by POPA, was to empower the fishers through dialogic and horizontal exchange to solve a priority problem in an intersectoral and interdisciplinary manner. Lastly, and in contrast, the Group opted for a diffusion model of communication to tackle the problem identified.

The Festival applied a multitrack model of communication for development, which considers communication fundamentally as a horizontal and participatory process (Tufte and Mefalopulos 2009:14). This approach divides communication in two categories, monological and dialogic communication. Monological communication includes one-way communication approaches, such as information dissemination, media campaign, and other diffusion approaches, while dialogical communication refers to bidirectional exchange. This last category foresees open-ended processes and results and enables the exploration of topics and the generation of knowledge and problem-solving initiatives, instead of just transmitting information. The Festival combined both categories.

81.2.3 Methods and Analysis

The Festival, a communication for development activity, was an exercise of the capacity of voice of artisanal fishers from Piriápolis intended to reposition them in society. Evidence of change in the figurative nucleus of the social representation of the fishers in these three aspects would mean change toward repositioning themselves in society.

The proposed framework combining these different concepts allowed for qualitative analysis of the case by means of semi-structured interviews (Cresswell 2003:17–19), associative charts (Abric, 1994 in Araya 2002:60–61), and field participant observation. The reconstruction of the figurative nucleus of the representation of the fishers was facilitated by the open questions of semi-structured interviews and the use of associative charts. Spontaneity was the key feature provided by these tools to catch the core structural scheme of the representation. Closed questions double-checked findings.

Fieldwork for this research was conducted in Piriápolis from May 2012 to September 2014. The perceptions of (12) informants were registered. During the interviews with fishers (eight, four of whom were members of the POPA Group) and other fisheries stakeholders (three were members of the POPA Group representing the academy, the civil society, and the state and one external local journalist), the following topics were tackled:

- What ideas, values, and beliefs the Festival attributed to the artisanal fishers from Piriápolis?
- How these relate to ideas, values, and beliefs about artisanal fishers that circulate in the area of Piriápolis?
- Did the Festival contribute to improve the wellbeing of fishers from Piriápolis in any manner?

The data gathered was subject of two types of analysis. A primary analysis was based on the recovery of unit of analysis containing perceptions about what is an artisanal fisher and what wellbeing aspirations they have. This first part of the work gave rise to general categories, each of which was assigned to one of the three aspects of wellbeing they relate to. A secondary analysis on the categorized content assigned the wellbeing categories to one of three stages established by the research questions (see Table 2).

81.3 Results and Discussion

Results indicate that, as a consequence of the First Fisheries Festival, the figurative nucleus of social representation of the artisanal fisher of Piriápolis experienced a change in the number and content of social wellbeing categories (see Fig. 3). In some categories, their content was confirmed, whereas in others, innovations were incorporated.

Table 2 Stages of analysis of the figurative nucleus of the social representation of fishers from Piriápolis within the framework of the First Artisanal Fisheries Festival taking into consideration the three dimensions of social wellbeing

Stage	Description
1. Social representation of artisanal fishers from Piriápolis	Identification of material, subjective, and relational aspects of the figurative nucleus of the representation Categorized information from interviews with 8 fishers
2. Social representation of artisanal fishers from Piriápolis within the framework of the First Artisanal Fisheries Festival	Identification of material, subjective, and relational aspects of the figurative nucleus of the representation: Which ones were confirmed, which ones did not appear; comparative study of content Categorized information from interviews with 12 informants (including the 8 fishers)
3. Wellbeing aspirations of artisanal fishers from Piriápolis	Identification of material, subjective, and relational aspects of the wellbeing aspirations of the fishers Categorized information from 8 fishers

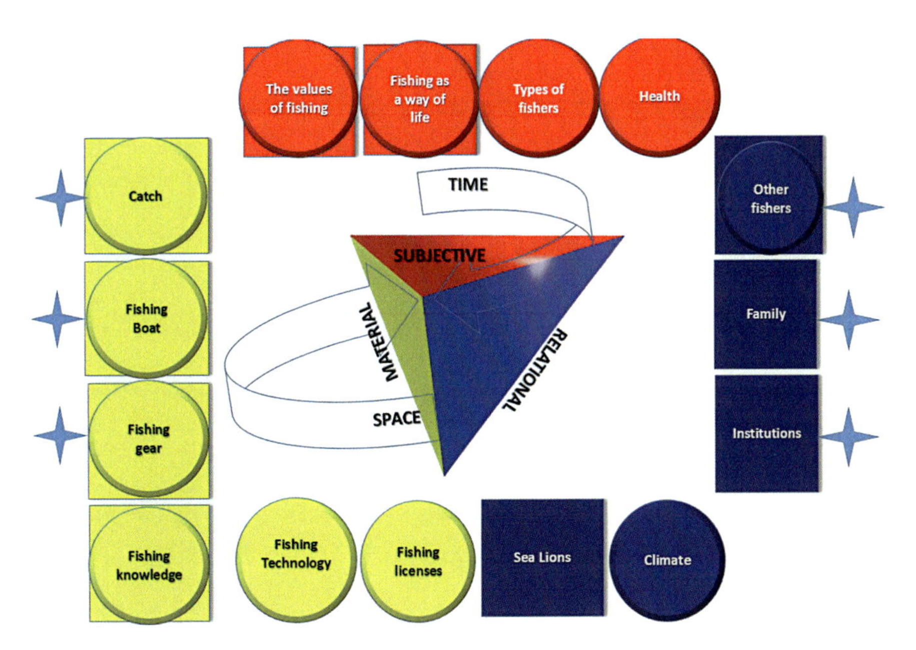

Fig. 3 Figurative nucleus of the social representation of the artisanal fisher from Piriápolis following White's presentation of the triadic dimensions of tridimensional wellbeing (2009). Circles indicate information categories of the self-representation of fishers. Squares indicate information categories of the representation of fishers incorporated by informants as of the Festival. A square around the circles indicates that the Festival confirmed the information category of the self-representation of fishers. Stars indicate those categories related to the personal aspirations of fishers

Regarding fishers' material wellbeing, the Festival confirmed the categories of catch, boat, fishing gear, and fishers' knowledge that were present in the nucleus of general self-representation of the fisher. On the contrary, technology used on the boats (such as the echo sounder) and the fishing licenses issued by the government were not mentioned when asked about the Festival.

Regarding aspects of fishers' subjective wellbeing identified within the nucleus, the Festival described fishers as of two categories, already present in their self-representation: fishing as a way of life (in contrast of a job) and the values associated with the fishing activity. However, a turning point was the change in their content, as it will be explained in the following paragraphs.

Lastly, when referring to relational aspects of wellbeing, the Festival added categories to the nucleus of the representation originally retrieved, namely, the family and the institutions (the state, the municipality, the military, and the media, among others, all of which supported the Festival in different modalities). These two categories do not appear in the general self-representation of fishers, who explicitly name themselves a Quixote or pirates, in contact almost exclusively with peers. The middlemen, who play a central role in the commercialization of the catch, did not appear in fishers' relations in any of the two moments analyzed (neither in the self-representation nor in the representation at the Festival). Nature was present in fishers' self-representation through their reference to the climate and the sea, but marine fauna was only mentioned when reflecting upon the Festival, through the presence of information about sea lions in particular.

Regarding the content of the categories found, the obvious centrality of the material category of catch appeared in both moments. However, the difficulties to obtain it, and a perceived decrease in the resource (noticeably reiterated in the self-description), did not arise when reflecting upon the Festival. The Festival recreated the boat and the fishing gear; informants remembered them positively as a possibility offered to visitors for boarding the boats (mainly kids) and for learning how to prepare the longlines for fishing.

With respect to subjective aspects of wellbeing, fishing as a way of life, as an adventure, as a legend, and as an act of freedom was also perceived at the Festival, but from a positive vision (on the contrary to the self-representation). As of the Festival, innovative content was retrieved. Fishers stated that they started to "exist" for society as of the Festival, that they "felt included" (in society), and even described themselves as being able to make others feel happy. When analyzing how fishers value themselves, the general self-representation pictures them as Quixote, pirates, as mentioned above, and even thieves or drug addicts. According to a non-fisher informant, the Festival achieved "a space for valorization, or at least a step towards a new value of fishers that not everyone recognizes." Informants also stated that the Festival brought forward the human aspect of fishing, the values of dignity, sacrifice, and honesty, as characteristics of the fishers' way of life.

Lastly, the Festival innovated also in the relational contents of the nucleus of the representation of Piriápolis fishers. It depicted them as having families and adding value to their representation through their presence. It showed "a different fisher," "one who has a family integrated in the fishing practices," according to a fisher informant.

Regarding their relationship with nature, the sea lions' category appeared when referring to the Festival, not in their self-representation in general. Although there is a conflictive interaction between the fishery and the sea lions, the Festival innovated in showing the fisher concerned and investigating about how to tackle the problem.

At a final stage, when comparing material, subjective, and relational categories of the nucleus of the self-representation in both instances, with the wellbeing aspirations of fishers, the communication process poses challenges and opportunities to the fishers. The Festival strengthened material categories present in the self-representation of fishers that are in direct relation with fishers' wellbeing aspirations.

"Progress," conceived as material growth by the fishers, is an aspiration for most of them, even though an informant defended the existence of a "subsistence fisher" that even gives away exceeding catch. Material growth can be "making a good wage," "being able to fish full time and not having to switch occupations." Both aspects are closely linked to the improvement of material assets: catch, boats, and fishing gear.

"Comfort," which stems from economic growth, is also an aspiration of fishers, together with the education of their children, housing, and social welfare. Neither the concept of "progress" nor the concept of "comfort" was mentioned when reflecting upon the Festival.

Regarding relational categories, there is evidence that the "destination" of fishers' aspirations is their families, but they also feel responsible for their peers. The Festival appears as a scenario that clearly contributed to the positive presence of the family and the importance of their peers, but also widens their relation to others (including institutions), which is also part of their aspirations.

Finally, the interviews showed a noticeable gap between the richness of subjective aspects in the self-representation and the absence of spontaneous reference to subjective aspirations. From the theoretical point of view, subjective aspects of wellbeing were confirmed as a component of the figurative nucleus of the representation and, therefore, matter for social wellbeing and development. Developing subjective aspects of fishers' representation should be an aspiration in itself for them as a means to improve their capacities and capabilities for achieving the "progress" and "comfort" they desire.

81.4 Final Considerations

This research indicates that valorizing the dimensions of wellbeing in the self-representation of vulnerable populations is key to repositioning them in society. However, this work calls the attention on lack of subjective aspects of fishers' wellbeing (the need for improving the society's perception of their position, for instance) when asked about their aspirations in general. This is a contradiction between their manifested material and relational aspirations of "progress" and "comfort," their aspiration to valorize their way of living (which motivated the Festival), and the need for valuing their own representation in order to achieve the state of triadic wellbeing. This aspect is particularly worrying, considering recent research in the area of Piriápolis concludes that "they feel neglected by state" and that "they need to be made to feel that they are valued partners in the governance of fisheries and in coastal development" (Trimble and Johnson 2013:7).

Regarding the social and cultural process of Piriápolis fishers in space and time, and the historical milestone the Festival constituted to them, the first challenge was whether the positive transformations in fishers' representation would remain over time. In this respect, fishers and non-fisher informants stated that the Festival was a point in time. Therefore, despite this research shows a positive change in fishers' representation as of the communication activity, the question that arose was: would material, subjective, and relational dimensions valorized in the nucleus of Piriápolis fishers' representation finally have an impact in the material and relational aspirations they have? In 2017, the POPA Group turned 5 years of uninterrupted activity, having implemented two development and research projects with funding from the Ministry of Cattle, Agriculture, and Fisheries of Uruguay (approx. US$ 100.000):

- "Mitigation of the impact of interactions between sea lions and artisanal fisheries: a participatory research for the evaluation of fish traps as alternative fishing gear." (2013–2014)
- "Innovations for family production in artisanal fisheries of Piriápolis: improvement and evaluation of fishtraps as complementary fishing gear." (2015–2016)

Among the activities developed involving natural and social sciences research, both projects foresaw the continuity of communication for development components with the fishers as protagonists. When informally exchanging with the Group members on the success of the fund raised, they expressed that this can largely be attributed to the mobilization initiated at the Festival.

A second challenge was whether the effect of the change in the nucleus would transcend to other fishers from Piriápolis or abroad or even reach other sectors of society. Both projects abovementioned involved spontaneously fishers from other localities of Canelones, Maldonado, and Rocha (three of the six coastal departments of Uruguay), who approached specific project activities of the Group attracted by the results of its action. A milestone of the reach of the Group was the invitation received by the National Union of Sea Workers (SUNTMA, "Sindicato de Trabajadores del Mar") to the Parliament of Uruguay to discuss artisanal fisheries' priority issues with the National Congress Commission for Cattle, Agriculture, and Fisheries (Act of the Committee Session, July 8, 2014). Fishers of the Group accepted the invitation to present their views of fisheries situation in Piriápolis. However, it was also the opportunity for them to undertake the presentation of the project under implementation. Finally, the Group directly participates upon invitation or request at the meetings of the Fisheries Council of Piriápolis, a governance structure created by the Ministry of Cattle, Agriculture, and Fisheries in fishing locations of Uruguay. The last presence of the Group in the Council dates back to June 2, 2016.

Finally, at the general level, the value of this investigation is centered in the study of the transformations that the exercise of the capacity of voice by a vulnerable population can promote in the self-representation of subjects in a situation of vulnerability and how these changes relate to their wellbeing aspirations. This research proposed a theoretical and methodological model integrating the concepts of capacity of voice, a cultural capacity, with the Wellbeing in Developing Countries (WeD), along with the concept of social representation, which helps assess alterations in cultural consensus. The research process could serve as an example or guide for evaluating communication for development actions in the field or actions focused on promoting the wellbeing of vulnerable populations.

References

Appadurai A (2004) The capacity to aspire: culture and terms of recognition. In: Rao W, Walton M (eds) Culture and public action. Stanford University Press, California, pp 59–84

Araya S (2002) Las representaciones sociales: ejes teóricos para su discusión. Cuaderno de Ciencias sociales. Facultad Latinoamericana de Ciencias Sociales. Sede Académica, Costa Rica. ISSN:1409-3677

Cresswell J (2003) Research design. Qualitative, quantitative and mixed methods approaches. Sage, California. ISBN 0-7619-2441-8

Jodelet D (ed) (1997) Les représentations sociales, 5th edn. Presses Universitaires de France, Paris ISBN 2 13 048570 7

Trimble M (2013) Towards adaptive co-management of artisanal fisheries in coastal Uruguay: analysis of barriers and opportunities, with comparisons to Paraty (Brazil). Doctorate thesis, Universidad de Manitoba, Canadá

Trimble M, Berkes F (2013) Participatory research towards co-management: lessons from artisanal fisheries in coastal Uruguay. J Environ Manag 128:768–778. https://doi.org/10.1016/j.jenvman.2013.06.032

Trimble M, Johnson D (2013) Artisanal fishing as an undesirable way of life? The implications for governance of fishers' wellbeing aspirations in coastal Uruguay and southeastern Brazil. Mar Policy 37:37–44. https://doi.org/10.1016/j.marpol.2012.04.002

Trimble M, Lazaro M (2014) Evaluation criteria for participatory research: insights from coastal Uruguay. J Environ Manag 54(1):122–137. https://doi.org/10.1007/s00267-014-0276-0

Tufte T, Mefalopulos P (2009) Participatory communication. A practical guide. World Bank working paper, vol 170. The International Bank for Reconstruction and Development/The World Bank, Washington, DC

Weeratunge N et al (2013) Small-scale fisheries through the wellbeing lens. Fish and fisheries. Blackwell Publishing. https://doi.org/10.1111/faf.12016

White S (2009) Bringing wellbeing into development practice. Working paper. University of Bath/Wellbeing in Developing Countries Research Group, Bath. (Wellbeing in Developing Countries (WeD) Working Papers; WeD Working Paper 09). https://purehost.bath.ac.uk/ws/portalfiles/portal/334487/WeDWP_09_50.pdf

Fostering Social Change in Peru Through Communication: The Case of the Manuani Miners Association

82

Sol Sanguinetti

Contents

Abstract

Much has been done to formalize miners in Peru; since its inception in 2006 the Ministry of Environment passed laws to align mining policies with international standards. However, the combination of the necessity to have a source of income and the rapid acquisition of money that illegal and informal gold mining provides has proven a serious impasse for the Peruvian government; hence these mining practices keep growing. As part of the Initiative for the Conservation of the Andean Amazon II (ICAA II), the Manuani Miners Association in Madre de Dios, one of the regions more prone to illegal mining and bad environmental practices in Peru, started a land restoration, remediation, and reforestation process of the degraded rainforest. This chapter analyzes how a program within ICAA II used communication to promote this significant change in behavior, what communicational tools were used, and what is their possible further applicability in similar scenarios to foster positive change.

S. Sanguinetti (✉)
Universidad de Lima, Lima, Peru

Programme, Environment and Technology Institute, Lima, Peru
e-mail: solsanguinetti@eti-ngo.org; ssanguin@ulima.edu.pe

J. Servaes (ed.), *Handbook of Communication for Development and Social Change*,
https://doi.org/10.1007/978-981-15-2014-3_96

Keywords

Environmental communication · Informal gold mining · Amazon rainforest

82.1 Introduction

Historically Peru has always been a mining country, and extractive mining activities are an important part of its economy, currently accounting for "about 65% of Peru's export earnings" (mining.com 2017). Worldwide, Peru is the third largest producer of silver and copper, and seventh in gold (MINEM 2017), just to name some of the most significant metal products.

Informal economic activities are very common in Peru; as a CEPLAN Report informs, in 2013, 19% of the national gross product came from informal sources (2016, p. 12). It also states that Peru remains as part of the five countries with the most informal non-agrarian labor market in the Latin American and Caribbean regions, even though there was a reduction from 75% in 2004 to 64% in 2013 (p. 5).

As a combination of both of these tendencies, Peru faces informal mining activities throughout its territory. Despite the fact that mining is one of the two sectors that pay better (CEPLAN 2016, p. 22), there is still significant informal and illegal mining activities in Peru, where there is no application of any kind of labor benefits or taxation to revenues. Many efforts have been made by the government to change this, and new laws have been approved that are in par with international standards, but their actual implementation is still far from a reality.

Informal gold mining in the Amazon basin is increasing and constitutes a serious environmental threat; activities involved in this type of mining modify the landscape extensively, as waterways are shifted or blocked and hills and holes are created as a result of digging for gold, depleting forest cover, polluting land and water sources, destroying ecosystems, threatening local communities, and turning swaths of rainforest into barren waste sites (Aljazeera 2016). "Global demand for gold has led to a massive increase in mining activity around the world. During the last decade, gold mining grew significantly in the Amazon becoming a major driver for land degradation and heavy metal contamination" (VerticalNews 2016). Currently alluvial gold mining is the second anthropogenic driver after the expansion of the agricultural frontier due to settlement in forest areas.

In Peru, the Region of Madre de Dios is one of the most harshly affected by informal and illegal gold mining; it rests right in the Andes-Amazon transition, making it an area of remarkable biodiversity which still maintains significant primary forest cover (UF et al. 2010, p. 2):

> Madre de Dios, until the year 2011, had lost an area of 50,000 ha due to degraded soils by mining (Asner et al. 2013); to that we must add the daily utilization of approximately 175,000 gallons of diesel and gasoline and the shedding of about 1,500 liters of machinery oil into the rivers. This excessive depletion by mining brings direct detriment to our

> biodiversity and natural landscapes such as wetlands, swamps, primary-forests, among others, affecting the conservation of both forest and aquatic ecosystems. (MDDC 2016, p. 5)

As a response to this and other threats to the Amazon basin's biodiversity, the Initiative for the Conservation of the Andean Amazon II (ICAA II), from the US Agency for International Development (USAID), carried on the work started in its first stage since 2006. ICAA II was composed of several consortia, one of which was the Madre de Dios Consortium (MDDC) composed of five institutions that worked jointly with University of Florida (UF) as its prime partner. From 2011 to 2016 the MDDC worked directly on informal alluvial gold mining as one of its four thematic programs, focusing their work on capacity building on Peruvian mining legislation, adoption of clean technologies, and recuperation of degraded mining areas through reforestation (UF et al. 2010, pp. 8–9), as well as providing the foundation for training the beneficiaries to reduce environmental impact (MDDC 2016, p. 5).

One of the most successful experiences within the mining program was the work with the "Asociación de Mineros Artesanales y Agricultores de la Cuenca del Río Manuani (AMAAM)" (Artisanal Miner and Agricultural Association of the Manuani River basin) with which the MDDC worked on reforestation to mitigate and remediate the impact of alluvial gold mining in the Tambopata National Reserve Buffer Zone where AMAAM is located.

The remediation work started with ICAA II:

> The MDD Consortium, through its Mining Program, has been installing and monitoring experimental plots to develop a proposal for recovering degraded soils by mining with emphasis on native and pioneer species and in the application of bio-fertilizers in the Manuani Sector. (MDDC 2016, p. 5)

More specifically, the work the MDDC carried out in Manuani was within experimental plots to test the capability of four native tree species, *Apeiba membranacea, Ochroma pyramidale, Ceiba pentandra*, and *Erythrina ulei*, to remediate degraded soils after alluvial gold mining. They worked with two planting methods, bare root and transplantation, and the periodic application of bio-fertilizer in three different concentrations. Additionally, the MDDC worked with the Manuani Association for the adoption of a comprehensive technical proposal for restoration both socially and institutionally (MDDC 2016, pp. 51–52).

The scientific research method is well documented by Dr. Francisco Román Dañobeytia and his research team; the objectives of the study were:

> (1) assess the level of soil degradation after informal mining operations, (2) evaluate the survival and growth of four native tree species using two planting methods and three levels of biofertilization, (3) detect possible accumulation of Hg in plant tissues of planted species, and (4) estimate the cost of reforestation in abandoned gold mined areas. (Román-Dañobeytia et al. 2015, p. 40)

82.2 How Communication Played In

This was highly scientific work, and paired with that, it applied new technologies not thoroughly documented, particularly not in Peru. The key achievement of these experimental plots has been the ability of the MDDC team to include the AMAAM as actual partners of this scientific experience; they avoided the passive beneficiaries' model and worked for an more active approach to the processes.

How did the MDDC achieve this? In a word: communication.

As mentioned before, informal activities are a common practice in Peru; hence, there is a natural resistance from informal miners to formalize or work within established parameters that could make their work, in their view, less economically efficient. This was the first and most significant obstacle that the MDDC had to face when first starting operations in 2011 as part of ICAA II; they had to change the attitude of these informal miners and try to modify their mining practices, so they would be more in line with the law and more friendly both with the environment and with human health.

Peru is a mega-diverse country with significant biodiversity, containing approximately 10% of the worldwide species (WB 2013); its geography is an important factor for this wealth both in animal and plant species. The Andes-Amazon transition, where Madre de Dios is located, is particularly biodiverse; however, as most of the illegal and informal miners in Madre de Dios, AMAAM is composed of a migrant population from Andean backgrounds.

The ecosystems where these migrant miners come from are significantly different from the lush rainforest that is this region; therefore they lack key ancestral knowledge to manage these landscapes, and they keep applying their Andean landscape management skills, from agriculture practices that include cutting and burning to non-rotational crops, to cutting down trees without managing soil coverage in a highly erosion-prone landscape.

Additionally to that, this population view themselves as transient dwellers of a land that they do not even own. Therefore, their relationship toward the land is fleeting, and without respect, as whatever they inflict on the forest is not viewed as a permanent detriment to themselves and their lives, it is just a temporary hardship that they must endure.

This was the second gap that the MDDC needed to bridge in order to achieve an efficient management of the forest striking a balance between livelihood improvement and biodiversity conservation.

The MDCC considered several tools to use in order to change the behavior of these informal miners, and bearing in mind that "Communication is as old as homo sapiens and has always been a powerful force in all cultures for sharing ideas and knowledge, and for influencing values and behaviour to bring about social change" (Balit 2012, p. 105), they decided to have a strong emphasis on communicational tools.

Further to this, effective communication is crucial to achieve the involvement of stakeholders at all levels, and its lack thereof functions as a detriment for this participation, as Silvia Balit (2012) says "many development programs have failed because they failed to communicate with the intended beneficiaries and other stakeholders" (p. 106); therefore the MDDC focused on high involvement from the beneficiaries and they worked with them as partners.

Finally, considering that "Communication is used to address global environmental issues of general public interest. Within this framework, very often communication, education, participation, and public awareness approaches are used in an integrated manner to reach out to key groups" (IBRD/WB 2007, p. 73).

Therefore, the first step toward the envisioned change of behavior was to effectively communicate what the project intended to achieve with the involved stakeholders, in particular with AMAAM members and chiefly its leaders as the direct beneficiaries/partners of the project.

The primary message to communicate was that the MDDC and AMAAM needed to work together, jointly and as partners, in order to achieve a better management of the forest that could be as economically appealing to this Association as gold mining.

> ... the MDD Consortium coordinated the implementation of awareness-raising, training and research with Manuani Miners Association, Asociación de Mineros Artesanales y Agricultores de la Cuenca del Río Manuani AMAAM (in a forestry concession for reforestation in the buffer zone of the Bahuaja Sonene National Park/Tambopata National Reserve).... (MDDC 2016, p. 11)

The MDDC did this in a series of interpersonal activities which included workshops, training activities, talks, presentations, and particularly an open and constant dialogue between MDDC staff members and the directives of AMAAM. This allowed for both the MDDC and AMAAM to be on the same page from the start until the end of the project.

But that was just the first step. Soil remediation is still not a widely applied science, and not much has been researched about it, as we mentioned before from Dr. Román Dañobeytia and his research group's work. Additionally to that, the members from AMAAM were not certified scientists but informal miners and farmers, most of whom actually lacked formal education. Thus, the MDDC needed to train these beneficiaries on a subject matter well beyond their scope of knowledge.

The MDDC (2016) decided to use targeted audiovisual support tools (training videos) within a plan:

> CMDD produced two videos: "Siempre hay orito... no es que no puede haber" [There is always a little gold... there is no way there cannot be; named "Manuani" in the MDDC webpage (translation and note added)] and "Viveros volantes" [Movable nurseries (translation added)]. They show the transition from mining to land recovery in different manners. Once validated, these videos were used to support decision making among migrant miners, indigenous peoples and authorities, to generate a reflection on the meaning of what it is to recover degraded areas and how they can change a destructive activity into a environmentally responsible response. (p. 22)

With these two videos they raised awareness to a problem that initially was not even considered as such by AMAAN nor by the regional or local authorities. Interviews and workshop activities were part of the production of the videos, and

AMAAM leaders participated giving their testimonies, furthering the strength of the message as it comes from some of their own members.

Particularly with "Viveros volantes," the MDDC was able to use it as a training tool, as it is a video developed to sensitize and teach gold miners how to build, maintain, and use forestry nurseries. It was an important method used for capacity building with a population with no previous forestry production or management background. AMAAM members were first trained (workshop style), and then the whole process was recorded, step by step, demonstrating each procedure as part of the nursery building and its management, with a voice-over commentary on the views from different AMAAM members and how these changed throughout their learning process.

This is a significant break from the previously used methods in Madre de Dios, where the regional government was completely in charge of the management of nurseries and of handing out the tree seedlings to the final beneficiaries. With the introduction of these movable nurseries, which are smaller and more easily manageable, the idea was to transfer the technical knowledge to the beneficiaries and thus make them partners in the process from start to end (Sanguinetti, personal interview, October 10, 2017).

Additionally the video "Biodigestores" (Bio-digesters) was also used as a capacity building instrument. This video is more structured as a training tool and uses graphics to depict chemical and physical processes that take place within these bio-digesters resulting in their final products of gas, used for cooking, and bio-fertilizer used for soil enrichment as it introduces nutrients in the soil as well as improving its edaphological characteristics for better root growth of the transplanted seedling, both at the experimental plots and later on the land they will be using as final planting sites for the saplings.

The importance of the application of this video comes from the use of bio-fertilizer as an additional resource applied specially to degraded soils. It is an economically accessible, as well as a low-tech, alternative to commercially available fertilizers, and it can be produced directly by the final users from their daily waste materials.

By the end of the project, the MDDC had started working on the introduction and adoption of cleaner technologies with AMAAM, such as the "shaking table," to change the popular practice of using mercury (Hg), which has been proven toxic to human health and which, as of January 2016, was illegal in Peru as the Minamata Convention was signed and ratified by the Peruvian government to be implemented by August of that same year (Minamata Convention 2013).

AMAAM members were willing to explore this new technology and thus improve their health and also have a lower impact on the environment. This willingness was possible due to a process that included significant activities of communication, becoming a clear indicator of how the adequate application of these tools had made it possible for AMAAM members to change their behavior.

Further to the communication work the MDDC carried out directly with AMAAM, they developed policy briefs, targeted to a different tier of project partners: decision makers. Particular to this case, they developed a "policy brief related to the cost structure of mining and its implications for both the regional economy and the policy proposals of control and remediation activity," called "Evaluation of the production chain of gold" (MDDC 2016, p. 22).

Additionally the MDDC also worked on different task forces, with interpersonal communication activities, related to gold mining and restoration of degraded landscapes, with the regional and local governments as an effort to foster information exchange among decision makers. For instance the MDDC:

> ...sponsored the Regional Forum 'Canon en Madre de Dios: Situación actual y perspectivas' [Madre de Dios Mining Tax: Current situation and perspective (translation added)] to propose the creation of the 'Mesa Multisectorial para el Canon de Madre de Dios [Task Force for Mining Tax in Madre de Dios (translation added)]'. This event was attended by civil society, Regional and Central government including the Madre de Dios Congressman. (MDDC 2016, p. 10)

82.2.1 How They Structured Communication

The MDDC worked with AMAAM and the related authorities in three differentiated communicational stages:

82.2.1.1 Information

As explained above, all activities included an informational component to start the process. This was centered on expounding, through a clear message, what was the objective and how it could be achieved in a jointly fashion.

For this the MDDC mainly used interpersonal communication, such as in the informational talks and awareness-raising activities (MDDC 2016, p. 11) with AMAAM, or through its participation in the different task forces with regional and local governments; thus the MDDC worked with different kinds of stakeholders participating in linked or related processes.

82.2.1.2 Education

The next stage was when the MDDC provided the stakeholders with theoretical knowledge imparted on interpersonal communication activities, such as workshops. But this went beyond just informing, they shared more detailed information, explaining more thoroughly how the objectives could be achieved.

Additionally the MDDC used audiovisual tools, for instance, the use of the video "Biodigestores" is a good example, as it disaggregates the process of waste becoming gas and bio-fertilizer, and it explains chemical and physical changes that take place within the bio-digesters, a highly technical concept put into context in a simple and understandable manner.

82.2.1.3 Training

And finally the MDDC worked with AMAAM so they had to apply what they had been informed and educated about, by putting it into practice on the field and developing actual skills such as building bio-digesters, building forestry nurseries, and planting saplings on the final planting sites.

This final stage was implemented with the AMAAM and not with decision-making stakeholders, since the nature of the work carried out with these two beneficiaries was intrinsically different.

This three-stage process is not per se very complicated; it is in fact simple and very intuitive, but its simplicity has proven to be rather effective. It is important to point out that the project manager of the MDDC, Bruno Sanguinetti, comes from a Food and Agriculture Organization (FAO) background, where this methodology has been widely applied (Sanguinetti, personal interview, October 10, 2017).

82.3 What They Achieved

Even though the scientific findings after all this work on scientific research and social change are not particularly encouraging or definitive, particularly the scientific component, this process has been significant to determine that "More experimental reforestation and remediation studies are needed to improve the science and practice of forest restoration in gold mined areas" (Román-Dañobeytia et al. 2015, p. 45).

Thus, this project, and the specific experience with AMAAM, can be viewed as opening the door for other projects and initiatives to focus on restoring gold-mined landscapes, particularly in the Amazon basin, and creating further knowledge on this subject matter. This is the case with the current CINCIA project (http://cincia.wfu.edu/en/), which is also partnered by USAID and where the chief scientist is Dr. Román Dañobeytia, who worked directly with AMAAM as part of the MDDC.

On the policy side, as a direct result of this experience in Manuani, the MDDC reached an agreement with the Ministry of Energy and Mines (MINEM) to promote recovery of areas degraded by alluvial gold mining and to promote the process for legalization of informal miners (ISU 2016, p. 33). Prior to this experience, restoration of mined areas was not at all the focus of the MINEM. It is important that as a result of this project this has changed, because many people will continue to practice informal and illegal mining activities, as these informal miners are a symptom of the lack of opportunities in the formal work chain and will continue to emerge until many other policy reforms, not only related to mining, are in place and operational.

However, the most important and lasting impact this project has had has been the involvement and appropriation of land restoration by the beneficiary population, an actual social change experience. As the ICAA-wide Final Report (ISU 2016) clearly states, the MDDC's achievements were significant:

> . . . with local land-owners to develop restoration techniques for the recovery of degraded soils. Based on participatory research, the Manuani community has developed technologies that are now being adopted by governmental agencies in other outmined areas in the Tambopata [National Reserve (added)] buffer zone. The Manuani people now are able to transfer technology for the implementation of ecological restoration in degraded areas and act as service providers. (p. 27)

> The MDD consortium developed field research on restoration in outmined degraded areas with members of the Manuani community. Its success in terms of restored area and people

able to apply this technique in other areas was nationally recognized by winning the Premio Ambiental Antonio Brack Egg. (November 2015) (p. 41)

These indicators clearly show that the installed capacity that AMAAM attained through the ownership of the nurseries, as well as the bio-digesters, has been put to use and has elevated the social value the AMAAM can deliver, not only internally but also as external service providers. But particularly it demonstrates that these *beneficiaries* breached the gap between receivers, to actual partners, and finally implementers of their own change and that this experience could and should be replicated in similar scenarios, because its feasibility has already been proven.

As discussed in the World Congress on Communication for Development (IBRD/ WB 2007, p. 74), "Communication about the environment shouldn't be so difficult because people do care about the environment – it is important to give people a reason to care, to act." The people in Manuani definitely cared and decided to act, and because the communication tools were effective, the transfer of technological knowledge that they needed to apply to their daily life was effective as well.

Once ICAA II came to a closing by 2016, the AMAAM continued with their restoration work, both at their experimental plots and offering their services to other mining associations. This did not played out particularly well, however, since they lacked the institutional leverage the MDDC, as part of the ICAA II-wide initiative, offered.

As reported by Vanessa Romo (El Comercio, 2014), the problem of illegal and informal miners invading AMAAM is a harsh and constant reality; therefore, they need institutional support to be able to carry on with the work they have started.

References

Aljazeera (2016) Peru's dirty gold. Retrieved from: http://www.aljazeera.com/programmes/techknow/2016/07/peru-dirty-gold-160719111106642.html. Accessed 14 Sept 2017

Asner G, Llactayo W, Tupayachi R, Luna E (2013) Elevated rates of gold mining in the Amazon revealed through high-resolution monitoring. Proc Natl Acad Sci USA 110(46):18454–18459. Retrieved from http://www.jstor.org/stable/23757565

Balit S (2012) Communication for development in good and difficult times the FAO experience. Nordicom Rev 33:105

Centro Nacional de Planeamiento Estratégico CEPLAN (2016) Economía informal en Perú: Situación actual y perspectivas. Available at: http://pardee.du.edu/sites/default/files/Econom%C3%ADa%20informal%20en%20PerúV11%2015-03-2016.pdf. Accessed 8 Oct 2017

CINCIA Project. http://cincia.wfu.edu/en/. Accessed 10 Oct 2017

ICAA Support Unit ISU (2016) ICAA – WIDE FINAL REPORT. Available at: http://pdf.usaid.gov/pdf_docs/PA00M1TT.pdf. Accessed 29 Sept 2017

Madre de Dios Consortium MDDC (2016) Consolidating regional environmental management capacity in Madre de Dios for the 21st century: final report FY 2012 – FY 2016. USAID, Lima

Minamata Convention (2013). Signatories and ratifiers list. http://www.mercuryconvention.org/Countries/tabid/3428/language/en-US/Default.aspx. Accessed 29 Sept 2017

mining.com (2017) Peru's copper output slightly up. http://www.mining.com/perus-copper-output-slightly-up/. Accessed 15 Sept 2017

Ministerio de Energía y Minas MINEM (2017). Producción Minera, por Principales Productos. http://www.minem.gob.pe/_estadistica.php?idSector=1&idEstadistica=11876
Multiple videos. http://mddconsortium.org. Accessed on multiple dates of 2017
Román-Dañobeytia, Francisco, Mijail Huayllani, Anggela Michi, Flor Ibarra, Raúl Loayza-Muro, Telésforo Vázquez, Liset Rodríguez, Mishari García (2015). Reforestation with four native tree species after abandoned gold mining in the Peruvian Amazon. Ecol Eng, Elsevier, 39, 46, 85
Romo V. (2014) Los mineros artesanales de Tambopata quieren de nuevo su bosque. El Comercio. Retrieved from: https://elcomercio.pe/peru/madre-de-dios/mineros-artesanales-tambopata-quieren-nuevo-bosque-285995. Accessed 5 Oct 2017
Sanguinetti B (2017) MDDC Project Manager. Personal interview
The International Bank for Reconstruction and Development/The World Bank IBRD/WB (2007) World congress on communication for development, lessons, challenges, and the way forward. World Bank, Washington, DC
The World Bank (2013) Peru: a mega-diverse country investing in National Protected Areas. http://www.worldbank.org/en/news/feature/2013/06/06/peru-pais-megadiverso-que-invierte-en-areas-naturales-protegidas-gpan-pronanp. Accessed 26 Dec 2017
University of Florida UF, Proyecto Especial Madre de Dios, Universidad Nacional Amazónica de Madre de Dios, Woods Hole Research Center and Futuro Sostenible (2010) Proposal document: consolidating regional environmental capacity in Madre de Dios for the 21st century
VerticalNews (2016) Ecology research; researchers from university peruana Cayetano Heredia Report recent findings in ecology research (Reforestation with four native tree species after abandoned gold mining in the Peruvian Amazon). Ecology, Environment & Conservation. Atlanta. http://fresno.ulima.edu.pe/ss_bd00102.nsf/RecursoReferido?OpenForm&id=PROQUEST-41716&url=/docview/1751280387?accountid=45277

Communicative Analysis of a Failed Coup Attempt in Turkey

83

Zafer Kıyan and Nurcan Törenli

Contents

Abstract

A failed coup attempt occurred in Turkey on 15 July 2016. This incident has generated three significant results in terms of communication. Firstly, it has proved that there is a symbiosis, rather than a contrast, between the traditional and new media. Secondly, it has demonstrated that different communication technologies can assume various roles during the stages of a movement. Thirdly, it has revealed that mosques can serve as a unique communication network. This chapter attempts, on one side, to explore why the failed coup attempt has a communicative character and, on the other side, to investigate how different communication technologies played various roles during the stages of this event. The role of mosques in the coup attempt is examined under a separate heading. Finally, both the coup attempt and protests against it are considered in their own social context since each protest movement occurs within a particular context.

Z. Kıyan (✉) · N. Törenli
Department of Journalism, Ankara University, Ankara, Turkey
e-mail: zafkiyan@gmail.com; zkiyan@media.ankara.edu.tr; torenli@ankara.edu.tr

J. Servaes (ed.), *Handbook of Communication for Development and Social Change*,
https://doi.org/10.1007/978-981-15-2014-3_11

Keywords

Turkey · Coup attempt · Traditional and new media · Sala · FaceTime · Recep Tayyip Erdoğan

83.1 Introduction

Turkey has a history of coups in which the military has intervened several times in country's politics since 1960. Not surprisingly, the country again witnessed an attempt, starting in the evening of 15 July 2016 and continuing until the early hours of the next day, in which a section of the Turkish military calling itself "Peace at Home Council" launched a movement against the government and President Recep Tayyip Erdoğan. The attempt did not succeed, but the cost was heavy. During the attempt, many government buildings including the Turkish Parliament were bombed by fighter jets and helicopters, resulting in over 300 people being killed and more than 2000 being injured.

This incident has generated three significant consequences from the communication point of view while demonstrating a distinctive characteristic in terms of its onset, progress, and ending. First, it has shown that there is a symbiosis, rather than a contrast, between the traditional and new media. Many commentators have emphasized how social media, especially FaceTime, a popular video chat service, helped to quell the coup attempt. However, the role of traditional media in preventing the attempt has remained unexplored. In fact, these two forms of media coexisted and complemented each other in the coup attempt. The picture of President Erdoğan talking via FaceTime to a privately owned television channel has been not just one of the popular images of the event but also an iconic scene showing the interaction between the traditional and new media. Second, it has demonstrated that different communication technologies can assume different roles during the various stages of a movement. Indeed, during the event, while television served to meet the institutional information needs of the people, social media played a major role in coordinating and organizing the protests on the streets. Third, it has revealed that mosques, which are unique to Muslim societies, can be used as a means of communication. The Salas recited by mosques across the country not just encouraged people to take to the streets but also played a significant role in galvanizing those who were already on the streets to resist the coup attempt.

The role of the media in protests has become a popular research subject among scholars in recent years. In this context, social movements that have emerged over the last few years in diverse regions of the world (e.g., Arab Spring in MENA, Occupy Wall Street in the USA, Indignadas in Spain, Gezi Park in Turkey) have been well documented and analyzed. For example, Castells (2012) has examined the movements from both communicative and sociological perspectives. Other authors (Gerbaudo 2012; Khondker 2011; Park et al. 2015; Rosa 2014; Salem 2015; Tudoroiu 2014) have directly focused on the role of the media in the movements.

Although there have been many attempts to explain the role of the media in social movements, the majority of existing research has focused largely on the role of new media, especially social networking sites such as Twitter, Facebook, and YouTube (see, e.g., Anduiza et al. 2014; Harlow 2011; Kharroub and Bas 2016; Theocharis et al. 2015; Thorson et al. 2013). What has often remained invisible is the role of alternative forms of media in protests. For instance, little research (Ahy 2016; Boyle and Schmierbach 2009; Rane and Salem 2012) has explored whether the traditional media has effects on protests.

In this chapter, it is argued that protest movements are not solely linked to new or social media; instead, different media platforms may coexist with, interact with, and complement each other in a protest movement. The outline of the chapter is as follows:

First, it is revealed why this coup attempt has a communicative character by examining the communication infrastructure and media usage in Turkey. Then, the roles that different communication technologies played during different stages of the coup attempt are investigated. The role of mosques in this event is analyzed under a separate heading. Finally, both the coup attempt and protests against it are considered in their own social context since each protest movement occurs within a particular context.

Three semi-structured face-to-face interviews were conducted. The first interview was conducted with the General Manager of CNN Türk, the second with the Domestic News Manager of Anadolu Agency (AA), and the third with the Vice President of Religious Affairs (RA). The importance of these three institutions stems from their key roles regarding the coup attempt. CNN Türk, the CNN International Turkish affiliate, is the media outlet that President Erdoğan spoke to via FaceTime on the night of the event to urge people to take to the streets. AA is the official news agency of Turkey that met the information needs of many media outlets due to its wide news network covering the coup attempt. RA is the official religious body of Turkey that gave the order for the Salas to be read from mosques across the country. The interviews enabled to collect critical data concerning the event.

83.2 The Communicative Character of the Recent Coup Attempt in Turkey

In the 15 July coup attempt, various media platforms were not just intensively used by the government and protestors to foil the coup attempt but also by plotters to manage the attempt. Hence, it is important to primarily investigate the existing telecommunication infrastructure and media usage in Turkey.

Telecommunication services in Turkey are currently widely provided by Türk Telekom, which has a monopoly in the sector. There are also 460 other operators providing service in the telecommunication sector (e.g., internet service providers, mobile operators, satellite communication operators, and cable platform services providers). With more than 300,000 km of fiber infrastructure, Turkey has 4.5G technology for mobile communication. In Turkey, there are nearly 70 million broadband subscribers, and the number of mobile subscribers is about 78 million in this country of 80 million people (Information and Communication Technologies Authority 2017).

In Turkey, social media takes first place among the activities for Internet users. According to a latest report released by We Are Social (Kemp 2017), there are

48 million active social media users in Turkey, and of these, 42 million connect to social media via their mobile devices. The report reveals that 87% of Internet users go online every day, and 36% of web traffic takes place using laptop and desktop computers, 61% via mobile devices, and 3% through tablets. PC and tablet users spend an average of 6 h 46 min per day; however, this activity falls to 2 h 59 min on mobile devices. Internet users who use both channels to access the Internet spend an average of 3 h 01 min on social media platforms. A survey conducted by the Turkish Statistical Institute (2017) reports that 83.7% of the individuals use the Internet to participate in social media.

Television is the leading outlet for traditional media usage in Turkey. Among the 20 million households in the country, over 19 million has one or two televisions. According to the Turkish Statistical Institute (2015), the most popular entertainment and cultural activity of 94.6% of people aged 10 years and above is watching television. During the daily 5.5 h spent in front of television, magazine programs and series are the most watched (Deloitte 2014). Television is mostly watched between the hours 21:01 and 24:00 (Radio and Television Supreme Council 2016). Interestingly, the starting time of the coup attempt fell within this time slot.

The data presented above offers important clues concerning why the 15 July coup attempt has a communicative character. First, it shows that Turkey has a relatively advanced communication infrastructure. This infrastructure paved the way for an uninterrupted communication between protestors who took part in anti-coup demonstrations. The same infrastructure also facilitated the connection between President Erdoğan and multiple television channels via FaceTime and Skype. The data also demonstrates that social media has become an important platform for daily communication between individuals in Turkey. The intensive social media usage between individuals in Turkey resulted in the fast dissemination of vital information and rapid mobilization of protests. Finally, this data reveals that the traditional media has not yet lost any ground. Television is still a popular medium of communication in Turkey; thus, the people watching television at the start of the coup attempt were able to observe the events via the news, especially the breaking news.

83.3 The Sequence of Events on the Night of 15 July

It was an ordinary day, just like any other, on 15 July 2016. Yet, around 10:00 p.m. local time, military forces closed off the main routes over the Bosphorus and Atatürk Airport in Istanbul, Turkey's largest city. They also established strategic checkpoints in different parts of the capital city, Ankara. Then, above both cities, Turkish F-16 jets roared across the sky at a low level, just a few hundred feet above the rooftops. Shortly after, the first images of the uniformed soldiers began to appear over social media, urging people to go to their homes. Meanwhile, some national television channels reported clashes between the police and the army on the streets of several major cities. People were astonished and were unable to grasp the situation. Something was happening, but what? A terror attack? A coup? There was total confusion and uncertainty.

Only one example is enough to give an idea about what people were thinking at that moment. V.M.A. (full name concealed) was sitting in a park in Ankara when the jets buzzed across the skyline of the city showing red and blue lights and leaving a white light beam behind them. V.M.A, who was as astonished by the situation as anyone else, tried to understand what was happening. The red, blue, and white color spectrum caused by the jets brought the French flag to mind. V.M.A. concluded that it was a fly past to show of support for the French people who had been hit by an ISIS attack the day before (V.M.A., personal communication, 18 July 2016).

When Prime Minister Binali Yıldırım broadcast live on the Turkish NTV television channel at 11:05 p.m. local time to say that they were investigating the possibility of an uprising, then the military maneuvers became clear. Only 1 h after the Prime Minister's statement, the coup plotters released a statement announcing the coup attempt on Turkish Radio and Television (TRT), the state-run television broadcaster in Ankara. This was not unexpected as it was what the military had done in the previous coups in 1960, 1971, and 1980, when there had been one main radio and television station in the country.

President Erdoğan was on holiday in Marmaris, on the southern Turkish coast, when the coup attempt unfolded; thus, he was unable to appear on any media platform in the early hours of the uprising. At about 11:44 p.m. local time, immediately before the statement from the coup plotters had been read out on TRT, the local press in Marmaris received a message via WhatsApp stating that the President was going to make a statement. Press members who received the message immediately went to the hotel where President Erdoğan was staying. He gave a brief statement on the doorstep of the hotel and made his first call for people to take to the streets in protest.

Interestingly, the statement was unable to be broadcast on any national television channel due to a "technical" problem. The statement was relayed from the Facebook account of one of the journalists who was present (Irmak 2016). Yet, whatever had happened, this statement never had the desired effect in the social media. At that time, President Erdoğan was involved in a situational assessment with his aides and, at the same time, watching the coup statement being read out on television. The presenter was saying that the Turkish military had completely taken over the administration of the country. Although President Erdoğan has faced significant challenges since he came to power in 2002 including Gezi Park protests starting with environmental protests and turning into a general riot against his regime in 2013, this situation was considerably different and was probably the greatest challenge in his political career.

83.4 Erdoğan on FaceTime, FaceTime on CNN Türk: The Traditional and New Media Are Hand in Hand

President Erdoğan's unconventional appearance on television via FaceTime was the pivotal moment of the coup attempt. It was a marvelous moment to watch since it was a merging of the old and new media. When President Erdoğan appeared on CNN Türk at 12:24 a.m. local time, it was Hande Fırat, the Head of the Ankara office of the channel, who held the smartphone that facilitated his appearance. According to her own account (Fırat 2016), Fırat called the Presidential Principal Clerk and requested a live broadcast. Upon learning that his previous statement had not been broadcast on national television channels, President Erdoğan promptly welcomed the offer as he needed to show the public he was in safe hands and not being held hostage. Soon after, all preparations were made for the broadcast. Then, President Erdoğan sat in front of a white curtain in order not to disclose his location and gave his message: "... I call on our people to gather in squares and airports... Let us gather as a nation in city squares... Those who attempted a coup will pay the highest price..." (BBC Türkçe 2016).

This broadcast is important for at least two reasons. Firstly, it proved that every protest movement has a breaking point. This is especially so in the case of this movement, because it was the main trigger of the protests against the coup attempt. Although people were used to watching President Erdoğan on a podium almost every day, seeing him framed on the tiny screen of a smartphone was both astonishing and terrifying. Even the President, who was sitting atop the state, seemed deprived of conventional means of communication. Yet, the scene motivated those who had doubts to take to the streets and protest.

Secondly, it reminded that the capacity of an infrastructure which allows communication is as important as the content of the communication. Indeed, President Erdoğan successfully made his broadcast because Turkey had launched its fourth-generation mobile telecommunications technology, called 4.5G, just a few months before the event. The former 3G technology did not allow smooth video chat due to spectrum use limitations. Obviously, as indicated by Nini (2016), it will never be known what would have happened if President Erdoğan had been unable to make his broadcast.

Becky Anderson from CNN International conducted an interview with President Erdoğan after the event, in which she reminded President Erdoğan of the importance of the moment he took to the air on CNN Türk, using FaceTime, by asking him this question: "...do you have an appreciation, to a certain extent, of the free press and social media since your recent experience?" (CNN International 2016). The implication behind this question referred to President Erdoğan's previous adverse attitudes toward social media. During the Gezi Park protests in 2013, President Erdoğan had called it "the worst menace to society" (Guardian 2013). At a campaign rally in 2014, he had talked about eradicating Twitter (Rayman 2014). Moreover, he had compared social media to a "knife in the hand of a murderer" in 2014 (Arab News 2014). Thus, once President Erdoğan resorted to FaceTime in an attempt to rally people against the attempted coup, the irony immediately appeared. He probably had no other choice; yet, the situation was deliciously reminiscent of a famous phrase by a prominent figure of Turkish politics, former President Süleyman Demirel, who said: "Yesterday was yesterday, today is today."

83.5 The Role of the Media in the 15 July Coup Attempt

The role of the media in the coup attempt has been a controversial issue. Not only journalists but also scholars have emphasized that social media, especially FaceTime, played a key role in the failure of the attempt (see, e.g., Boyle 2016;

Srivastava 2016). However, one of the more interesting remarks on the role of the media in the coup attempt has come from Unver and Alassaad (2016), and a closer look at their study offers a profitable ground in the context of this chapter.

The study of Unver and Alassaad, *How Turks mobilized against the coup*, aims to show that mobilization against the coup attempt was first initiated online, then gained momentum through the network of mosques, and finally continued to the grass roots. Their work provides an insight into how people mobilized against the coup attempt through empirical data. However, it seems that the authors are wrong in their claims concerning the reasons that motivated protesters in the coup attempt.

On the one hand, Unver and Alassaad argue that President Erdoğan played a belated role in mobilizing protesters against the attempted coup. Despite the fact that the authors do not ignore the triggering effect of President Erdoğan on the protesters, they tend to evaluate the mobilization as a natural political reflex of people who already were on the streets when the coup attempt started. One can wholeheartedly agree with the notion that the initial protests began sprouting up simultaneously in various cities in the first hours of the event. Yet, after President Erdoğan's appearance on television over FaceTime, the protests intensified and turned into a nation-wide movement across the country. There are at least two reasons to support this.

First, there were problems in accessing Twitter, Facebook, and YouTube since the Internet was blocked through throttling when the coup attempt started. Social media was only accessible through a virtual private network program which allows users to circumvent limitations. Turkey Blocks, a watchdog group that monitors Internet bans, confirmed that access to these three social media platforms was restricted in Turkey after 10:50 p.m. local time (Turkey Blocks 2016). Once the government recognized that the Internet and social media could act in their favor against the attempted coup, they immediately lifted the restrictions. Although restrictions lasted a little less than 2 h, it is important to note that protesters were able to use neither social media nor other online digital tools during this period. Therefore, they were unable to organize a massive protest over these platforms in the first hours of the coup attempt. Second, it is important to indicate that a portion of the people who went onto the streets when the coup attempt began had formed long queues at many shops, ATMs, and gas stations in the cities. Thus, the mention of a natural reflex could be considered as the survival instinct of worried people rather than being political as Unver and Alassaad emphasized.

On the other hand, Unver and Alassaad claim that the mosques and digital media had the more important role in mobilizing people against the coup attempt. According to the authors, both these elements created a two-way channel for political communication. Clearly, Unver and Alassaad are correct in asserting the importance of mosques and digital media, but for a better understanding of the role of the media in the coup attempt, a more nuanced analysis is needed. Within this context, instead of answering the question of which means of communication played a greater role, it would be more accurate to emphasize that different communication technologies assumed different roles during the various stages of the coup attempt.

The results of surveys conducted by three leading Turkish research companies are a gauge of what Turkish citizens thought of the event. The importance of these surveys stems from the fact that they included people who took to the streets on the night of 15 July. Interestingly, all three surveys revealed that more than one means of communication played a significant role in mobilizing protesters against the coup attempt.

The first survey was carried out by Andy-Ar (2016) on 19 July 2016 with 1496 respondents. The participants were asked "Did you watch President Erdoğan's statement on television on the night of the coup attempt?," and 83.9% responded "Yes." The second question was "Did you or any of your relatives go out on the street after President Erdoğan's call for people to take to the streets?," to which 65.7% responded "Yes." The second survey was conducted by SETA (Miş et al. 2016) between 18 and 24 July 2016 through a semi-structured interview with 146 respondents. The researchers detected that three factors motivated the participants to take to the streets. First was President Erdoğan's call to take to the streets, second was the coup declaration being read on TRT, and third was the Salas that were broadcast from the mosques. Finally, the third survey was administered by Konda (2016) on 26 July 2016 and comprised 1875 respondents. One of the questions posed in the survey was "Through which source did you receive the initial news about the coup attempt?" The responses were as follows: 62% of the participants stated television, 24% friend or acquaintance, 9% social media, 3% news websites, 1% SMS, and another 1% mosques. A further question was "At what point of the events did you decide to take to the streets?," to which 53% responded that they did so after President Erdoğan's call to go into the street, 27% went before this call and 20% took to the streets after 16 July.

To examine the role of the media in the coup attempt, this chapter proposes that this role can be classified under three distinct stages (Khamis et al. 2012): The first stage is when the attempt unfolded, the second stage occurred when the anti-coup protests began, and the third happened when the attempt ended. Different means of communication played different roles in all these three stages.

First and foremost, television and FaceTime played a major role in the events of the coup attempt. The reason was twofold; the attempt began at the peak hours of the television in the country, which was partly why television served as a vital source of news and information in the first hours of the event. The second and more important reason was that President Erdoğan made his crucial statement on television via FaceTime. The ratings provide evidence of the role played by television and FaceTime in the coup attempt. Table 1 shows the ratings of the four leading news channels (NTV, CNN Türk, Habertürk, and A Haber) in Turkey on the night of 15 July and the relation to the crucial moments of the coup attempt and the time slice of President Erdoğan's FaceTime connection.

A semi-structured interview was conducted on 30 November 2016 with Erdoğan Aktaş, the CNN Türk General Manager. According to Aktaş, since the ratings of news channels in Turkey are measured in 15-minute slices, it becomes difficult to determine the rating of the FaceTime connection that started at 12:24 and ended at 12:31 a.m. local time. The FaceTime video chat was relayed not just through many national television stations in Turkey, but also via several international channels around the world. Hence, it is even more difficult to determine precise ratings since the ratings of other television

Table 1 Ratings of the Turkish television news channels on the night of 15 July

		NTV	CNN Türk	Habertürk	A Haber
	12:00–12:15 a.m.	12.43%	12.81%	5.45%	9.80%
FaceTime broadcast 12:24 a.m. (start)	12:15–12:30 a.m.	15.35%	16.30%	6.78%	9.67%
FaceTime broadcast 12:31 a.m. (end)	12:30–12:45 a.m.	18.28%	14.48%	6.53%	11.15%
	12:45–01:00 a.m.	18.64%	12.83%	5.80%	13.60%

Note: The ratings data was obtained with special permission from Erdoğan Aktaş, CNN Türk General Manager

channels need to be included in these measurements. However, the data in Table 1 shows that total ratings of the four news channels reached nearly 50% at the most critical time slice of the coup attempt and remained stable at this level. Even more importantly, since the connection was broadcast live on more than one television channel, this created a multiplier effect and reached a large audience as with in the social media. Television preserved its effectiveness before and after the broadcast. For example, TRT who broadcast the moment when the coup declaration was read out and FOX TV who broadcast the moment when the fighter jets bombed the Turkish Parliament live peaked in the ratings in Turkey (Medya Tava 2016).

In addition to television and FaceTime, news agencies played an essential role in the attempted coup with AA. During the event, AA became a vital source of news and information due to having a wide-ranging news network in the country. A semi-structured interview was carried out on 1 February 2017 with Zekeriya Kaya, AA Domestic News Editor, in which he stated that nearly 800 AA reporters were active on the night of 15 July. Furthermore, Kaya emphasized that many television stations and Internet news sites obtained their first news and images about the event from AA. According to Kaya, this explained the sixfold increase in the volume of news distribution of AA on the night of 15 July.

Interestingly, radio was another conventional medium which played an active role on the night of the coup attempt. When President Erdoğan appeared on television to make his call for people to take to the streets, the call was heard by people outside via people in their cars turning up the volume of their radios.

Mobile phones also featured prominently in the coup attempt. That night, GSM operators sent text messages signed by President Erdoğan to their subscribers asking them to stand up for democracy. Additionally, they provided extra talk and Internet packages to their subscribers. Text messaging also helped the ruling Justice and Development Party (AKP) to mobilize and coordinate its membership. That night, AKP sent text messages to its nearly ten million registered members, urging them to take to the streets in a show of support for the government. Mustafa Ataş, the AKP Deputy Chairman and the Head of party organization, stated that about 30 million text messages were sent from the headquarters of the party to the members on the night of 15 July. According to Ataş, party members were successfully mobilized through text messaging (Selvi 2016).

The role of social media surfaced after people took to the streets, and this platform facilitated the protests against the coup attempt. By using social networking sites such as Twitter and Facebook, protesters were not only able to just communicate with each other, but they could also quickly organize and mobilize in the cities where the most chaotic events happened. The role of social media, especially Twitter, becomes clear in the light of empirical data; on the night of 15 July, between 10:00 p.m. and 12:00 a.m. local time, 71,938 tweets were sent. In the same time interval on an average day, 16,500 tweets would be sent. This corresponds to a fourfold increase. Furthermore, between 12:00 and 4:00 a.m. local time, the number of tweets related to the attempted coup was 495,000, which corresponds to a 35-fold increase compared to the average figure of 14,142 over a normal day (Can 2016). In total, seven million tweets were sent on the night of 15 July under popular hashtags such as #DarbeyeHayır (no to coup), #VatanİçinNöbetteyiz (on guard for the motherland), and #TekYürek (single heart) (Demir 2016). Not surprisingly, the number of tweets continued to increase after 10:00 p.m. local time and reached a peak after President Erdoğan's FaceTime broadcast.

What is particularly interesting is that the coup plotters used online digital tools to organize themselves, and this is why the plotters did not intervene in the Internet although they destroyed the country's satellite communication and cable TV operator, Türksat. The plotters tried, on the one hand, to cut the flow of information by taking out traditional means of communication, for example, trying to seize the major Turkish broadcasters such as TRT and CNN Türk. On the other hand, they allowed the Internet to stay active as they were using instant messaging applications, especially WhatsApp, which enabled them to send encrypted messages over the Internet. Therefore, if they had shut down the Internet, it would have resulted in a collapse in their internal communication networks. Clearly, the plotters made a strategic mistake because although Türksat was experiencing a blackout and satellite platforms like D-Smart and Digiturk were also unable to broadcast, these platforms could present themselves on the Internet. The irony was that while the plotters, on the one hand, were trying to seize control of the traditional means of communication, the Internet and especially social media, on the other hand, were experiencing one of their busiest moments in the country. Actually, the plotters were trying to stage "an analogue coup in a digital age" (Unver and Alassaad 2016).

In the process of the coup attempt, the two most popular live-streaming services, Twitter's Periscope and Facebook Live, were effectively generating information about the attempt. Around 2:00 a.m. local time, "more than 80 streams were active on the Periscope platform in Istanbul, compared to 30 in London and two in San Francisco" (Frier et al. 2016). Meanwhile, Facebook's real-time map demonstrated that a large group of users were active in Turkey (Couts 2016). Both these platforms allowed users to either stream the instant events of the coup attempt or show protesters on the street resisting the plotters. The Google search engine was also active. While tanks were rolling down the street, the protestors were searching on Google for ways to stop tanks. According to Google Trends (https://trends.google.com.tr), the online searches for "How to stop a tank" skyrocketed on the night of 15 July. This was in a sense the embodiment of the "search engine society" argument that Halavais (2009) argued years ago.

The coup attempt was suppressed in less than 24 h; however, the so-called "democracy watch" protests continued for 29 days to protect the country against another potential attempt. The media played an essential role in this process as well. For instance, television channels, newspapers, magazines, and Internet portals covered the coup attempt with all its details. Furthermore, governmental and non-governmental organizations prepared publications (e.g., reports and photograph albums), which brought to the forefront the dramatic side of the event. These publications, while enabling the formation of a wide consensus against the coup attempt, also led to the building of a societal memory regarding the event.

83.6 Mosques as a Communication Network

One of the interesting features concerning the 15 July coup attempt was the involvement of mosques and the use of Sala. Mosques are a religious institution where Muslims gather for prayers, and Sala has its etymological roots in Arabic and means a kind of an Islamic call to prayer. Traditionally, Salas are read out to announce Friday prayers or funerals at the mosques. Hence, the main meaning of the word Sala is "annunciation." It can be argued that the Salas that were read out on the night of 15 July had two purposes: first to announce the coup attempt in line with the meaning of the word and the second to rally people against this attempt.

Historically, Sala is not a new practice; it has been used, for example, during the Turkish War of Independence 1919–1922. During the Arab Spring, although not specifically the Sala, mosques and prayers played a particular role in the protests (Aslam 2017). That the same thing has not been seen in the protest movements in the West indicates that the cultural features and religious beliefs of societies may be determinant in the decision about which medium could be used in protests.

A semi-structured interview was conducted on 14 November 2016 with Mehmet Emin Özafşar, Vice President of RA. He explained that on the night of 15 July, the imams, the highest-ranking religious officers in mosques, began reading the Salas without having received any order from RA. Özafşar then continued to state that to avoid any confusion, RA sent a text message to its 110,000 imams, asking them to read the Sala and call for Tekbeer (a kind of Islamic call) in their mosques. There was an immediate positive effect; the people on the streets passionately replied with the slogan, Allahu Akbar (God is greatest), and this is why the reading of Sala continued for weeks even after the coup attempt had been suppressed.

Viewed closely, it can be argued that the network of mosques in Turkey is not complicated. At the top, there is the RA under which there are the offices of the mufti in all 81 cities and 919 counties in the country, who represent the highest religious authority in their locations and under them are the mosques. According to Özafşar, there are 86,762 mosques in Turkey. In terms of the distribution across the cities, in Istanbul and Ankara where the most intensive anti-coup protests took place, there are 3317 and 2994 mosques, respectively. Given their widespread presence in the cities, mosques have been relatively effective communication networks in Muslim societies.

Considering mosques as means of communication, the open to communication structure of these religious institutions resembles the structure of traditional means of communication. Through the loudspeakers connected to their minarets, mosques allow only one-way communication just like in means of mass communication. However, what makes mosques unique is not their communication structure but much rather the fact that they turn communication into a religious character. Mosques played an important role during the coup attempt when on the night of 15 July, the imams read the Salas, calling for the people to take to the streets, which became an irresistible religious order. Thus, the mosques were able to effectively play their role in the events that unfolded. This was partly because of the political identity of the protesters, of whom nearly 80% had voted for AKP in the general elections of 2015, which is known for its conservatism (Konda 2016). Thus, the Salas called from the mosques showed a total parallelism with the religious beliefs of the protesters known for their conservative identities. In this respect, the Salas surrounded these protesters with their holy ambiance and motivated them to stand against the coup attempt even risking death.

83.7 Conclusion

An advanced communication network provides possibilities not only for a movement to become visible but also for a counter movement to parry it. However, this does not mean that communication technologies alone have the ability to start and end a protest movement and only focusing on technology inevitably turns into a discussion of determinism. When protest movements are evaluated with slogans such as "Facebook revolution" or "the revolution will not be tweeted," the discussion inevitably is locked in the question of "what the determinant is" (see, e.g., Gladwell 2010; Shirky 2011). Technology cannot be considered as the main cause of protest movements as many other social factors (e.g., political, social, economic, psychological, cultural, and religious) are at play in these movements (Ainger 2016), neither can it be interpreted as completely ineffective as it can facilitate protests (Comunello and Anzera 2012).

The 15 July coup attempt was a movement that aimed to usurp the government and unseat President Erdoğan. While the plotters attempted to do this by cutting the interaction of the government with social powers, the opposite movement endeavored to suppress the coup attempt by keeping communication channels open. Therefore, the common feature of the both movements was their communication style; however, the social factors that motivated people to take to the streets were different. Those who took to the streets on the night of 15 July believed that the goal of the coup was to capture and divide Turkey, which would result in chaos and civil war (Konda 2016). Consequently, both movements started and ended as a chain of numerous overlapping events in the same social context. In this process, different communication technologies played their own roles at various stages of the coup attempt. As the stages changed, so did the means of communication and the role they played.

83.8 Cross-References

▶ Online Social Media and Crisis Communication in China: A Review and Critique

References

Ahy MH (2016) Networked communication and the Arab spring: linking broadcast and social media. New Media Soc 18(1):99–116. https://doi.org/10.1177/1461444814538634

Ainger K (2016) The social fabric of resilience: how movements survive, thrive or fade away. In: Price S, Sabido RS (eds) Sites of protests. Rowman&Littlefield, London, pp 37–53

Anduiza E, Cristancho C, Sabucedo JM (2014) Mobilization through online social networks: the political protest of the indignados in Spain. Information. Commun Soc 17(6):750–764. https://doi.org/10.1080/1369118X.2013.808360

Andy-Ar (2016) Darbe teşebbüsü: Türkiye siyasi gündem darbe araştırması. http://www.andy-ar.com/wp-content/uploads/2016/07/Darbe-Ara%C5%9Ft%C4%B1rmas%C4%B1-Temmuz-2016.pdf. Accessed 30 Aug 2016

Arab News (2014) Erdogan says social media like "murderer's knife". http://www.arabnews.com/news/middle-east/616176. Accessed 15 Aug 2016

Aslam A (2017) Salat-al-Juma: organizing the public in Tahrir Square. Soc Mov Stud 16 (3):297–308. https://doi.org/10.1080/14742837.2017.1279958

BBC Türkçe (2016) Erdoğan: Milletimi meydanlara davet ediyorum. https://www.youtube.com/watch?v=7LEfGo0uN-o. Accessed 20 Dec 2017

Boyle D (2016) How faceTime stopped the Turkish coup: President Erdogan's last-ditch mobile phone call to privately owned TV station mobilised mass support by harnessing social media. Daily Mail. http://www.dailymail.co.uk/news/article-3693987/How-FaceTime-stopped-Turkish-coup-President-Erdogan-s-ditch-mobile-phone-call-privately-owned-TV-station-mobilised-mass-support-harnessing-social-media.html. Accessed 17 Jul 2016

Boyle MP, Schmierbach M (2009) Media use and protest: the role of mainstream and alternative media use in predicting traditional and protest participation. Commun Q 57(1):1–17. https://doi.org/10.1080/01463370802662424

Can A (2016) Darbe gecesi 35 kat tweet attık! Hürriyet. http://www.hurriyet.com.tr/darbe-gecesi-35-kat-tweet-40172748. Accessed 27 Jul 2016

Castells M (2012) Networks of outrage and hope: social movements in the internet age. Polity, Cambridge

CNN International (2016) Turkish president describes night of coup attempt. http://edition.cnn.com/videos/world/2016/07/18/turkey-erdogan-interview-becky-anderson.cnn/video/playlists/turmoil-in-turkey/. Accessed 27 Jul 2016

Comunello F, Anzera G (2012) Will the revolution be tweeted? A conceptual framework for understanding the social media and the Arab spring. Islam and Christian–Muslim Relations 23(4):453–470. https://doi.org/10.1080/09596410.2012.712435

Couts A (2016) Turkey's military coup is playing out in real time on facebook live. http://www.dailydot.com/layer8/turkey-military-coup-facebook-live/. Accessed 15 Jul 2016

Deloitte (2014) World's most colourful screen TV series sector in Turkey. https://www2.deloitte.com/tr/en/pages/technology-media-and-telecommunications/articles/turkish-tv-series-industry.html. Accessed 30 Aug 2014

Demir ST (2016) 15 Temmuz darbe girişiminde medya. https://www.setav.org/15-temmuz-darbe-girisiminde-medya/. Accessed 30 Aug 2016

Fırat H (2016) 24 saat: 15 Temmuz'un kamera arkası. Doğan Kitap, İstanbul

Frier S, Wang S, Ackerman G (2016) Turkey's Erdogan used social media he despised to stay in power. Bloomberg. https://www.bloomberg.com/news/articles/2016-07-19/turkey-s-erdogan-used-social-media-he-despised-to-stay-in-power. Accessed 19 Jul 2016

Gerbaudo P (2012) Tweets and the streets: social media and contemporary activism. Pluto, London

Gladwell M (2010) Small change: why the revolution will not be tweeted. The New Yorker. http://www.newyorker.com/magazine/2010/10/04/small-change-malcolm-gladwell. Accessed 4 Oct 2010

Guardian (2013) Social media and opposition to blame for protests, says Turkish PM. https://www.theguardian.com/world/2013/jun/02/turkish-protesters-control-istanbul-square. Accessed 30 Jul 2016

Halavais A (2009) Search engine society. Polity, Cambridge

Harlow S (2011) Social media and social movements: Facebook and an online Guatemalan justice movement that moved offline. New Media Soc 14(2):225–243. https://doi.org/10.1177/1461444811410408

Information and Communication Technologies Authority (2017) Electronic communications market in Turkey. https://www.btk.gov.tr/File/?path=ROOT%2f1%2fDocuments%2fPages%2fMarket_Data%2f2017_Q3_Eng.pdf. Accessed 12 Dec 2017

Irmak T (2016) Cumhurbaşkanı Erdoğan açıklama yapıyor. https://www.facebook.com/temel.irmak1/videos/1116233715087074/?pnref=story. Accessed 10 Sep 2016

Kemp S (2017) Digital in 2017: a study of internet, social media, and mobile use throughout the region. https://www.slideshare.net/wearesocialsg/digital-in-2017-western-asia. Accessed 30 Jan 2017

Khamis S, Gold PB, Vaughn K (2012) Beyond Egypt's "Facebook revolution" and Syria's "YouTube uprising": comparing political contexts, actors and communication strategies. Arab Media and Society 15:1–30. http://www.arabmediasociety.com/?article=791. Accessed 25 Oct 2016

Kharroub T, Bas O (2016) Social media and protests: an examination of twitter images of the 2011 Egyptian revolution. New Media Soc 18(9):1973–1992. https://doi.org/10.1177/1461444815571914

Khondker HH (2011) Role of the new media in the Arab spring. Globalizations 8(5):675–679. https://doi.org/10.1080/14747731.2011.621287

Konda (2016) Democracy watch research: the profile of the squares. http://konda.com.tr/en/rapor/democracy-watch-research/. Accessed 26 Jul 2016

Medya Tava (2016) 15 Temmuz 2016 Cuma tarihli rating tablosu. http://www.medyatava.com/rating/2016-07-15#. Accessed 17 Jul 2016

Miş N, Gülener S, Coşkun İ, Duran H, Ayvaz ME (2016) 15 Temmuz darbe girişimi toplumsal algı araştırması. SETA, Ankara. https://setav.org/assets/uploads/2016/08/15-temmuz-darbe1.pdf. Accessed 17 Sep 2016

Nini A (2016) How facetime saved the Turkish president from his country's attempted coup. Vice. https://www.vice.com/en_us/article/erdogan-facetime-coup-attempt. Accessed 17 Jul 2016

Park SJ, Lim YS, Park HW (2015) Comparing twitter and youTube networks in information diffusion: the case of the "Occupy wall street" movement. Technol Forecast Soc Chang 95:208–217. https://doi.org/10.1016/j.techfore.2015.02.003

Radio and Television Supreme Council (2016) Medya okuryazarlığı araştırması. RTÜK, Ankara. https://www.rtuk.gov.tr/assets/Icerik/AltSiteler/medya-okuryazarligi-arastirmasi.pdf. Accessed 30 Dec 2016

Rane H, Salem S (2012) Social media, social movements and the diffusion of ideas in the Arab uprisings. J Int Commun 18(1):97–111. https://doi.org/10.1080/13216597.2012.662168

Rayman N (2014) Turkey's Erdogan now says he'll shut down twitter, too. Time. http://time.com/32339/turkey-erdogan-twitter/. Accessed 30 Jul 2016

Rosa A (2014) Social media and social movements around the world: lessons and theoretical approaches. In: Pătruţ B, Pătruţ M (eds) Social media in politics: case studies on the political power of social media, vol 13. Springer, New York, pp 35–47

Salem S (2015) Creating spaces for dissent: the role of social media in the 2011 Egyptian revolution. In: Trottier D, Fuchs C (eds) Social media, politics and the state. Routledge, New York, pp 171–188

Selvi A (2016) "AK party able to bring millions to the street in one hour". Hürriyet Daily News. http://www.hurriyetdailynews.com/ak-party-able-to-bring-millions-to-the-street-in-one-hour.aspx?pageID=449&nID=105310&NewsCatID=581. Accessed 25 Oct 2016

Shirky C (2011) The political power of social media: technology, the public sphere, and political change. Foreign Affairs. https://www.foreignaffairs.com/articles/2010-12-20/political-power-social-media. Accessed 30 Jul 2016

Srivastava M (2016) How Erdogan turned to social media to help foil coup in Turkey. Financial Times. https://www.ft.com/content/3ab2a66c-4b59-11e6-88c5-db83e98a590a. Accessed 16 Jul 2016

Theocharis Y, Lowe W, Van Deth JW, García-Albacete G (2015) Using twitter to mobilize protest action: online mobilization patterns and action repertoires in the Occupy Wall street, indignados, and Aganaktismenoi movements. Information, Communication & Society 18(2):202–220. https://doi.org/10.1080/1369118X.2014.948035

Thorson K, Driscoll K, Ekdale B, Edgerly S, Thompson GL, Schrock A, ... Wells C (2013) YouTube, twitter and the occupy movement: connecting content and circulation practices. Info Commun Soc 16(3):421–451. https://doi.org/10.1080/1369118X.2012.756051

Tudoroiu T (2014) Social media and revolutionary waves: the case of the Arab spring. New Political Science 36(3):346–365. https://doi.org/10.1080/07393148.2014.913841

Turkey Blocks (2016) Confirmed: twitter, facebook & youTube blocked in #Turkey at 10:50PM after apparent military uprising in #Turkey. https://twitter.com/TurkeyBlocks/status/754081725970468865. Accessed 16 Jul 2016

Turkish Statistical Institute (2015) Time use survey. http://www.turkstat.gov.tr/PreTablo.do?alt_id=1009. Accessed 18 Nov 2016

Turkish Statistical Institute (2017) Information and communication technology usage survey on households and individuals. http://www.tuik.gov.tr/PreTablo.do?alt_id=1028. Accessed 20 Dec 2017

Unver HA, Alassaad H (2016) How Turks mobilized against the coup: the power of the mosque and the hashtag. Foreign affairs. https://www.foreignaffairs.com/articles/2016-09-14/how-turks-mobilized-against-coup. Accessed 14 Sep 2016

84 Plurality and Diversity of Voices in Community Radio: A Case Study of Radio Brahmaputra from Assam

Alankar Kaushik

Contents

Abstract

Community radio is a powerful medium, which serves to voice the voiceless and is at the heart of communication and democratic processes within societies. Considering the diversity of the northeastern region of India, it is not possible for state-run broadcasting agencies to reach all the communities through its programming content. Radio Brahmaputra, located in Dibrugarh, eastern part of Assam, is the first and only grassroot community radio of Northeast India. Radio Brahmaputra produces content in various languages and local dialects, which create impact in the community engagement of the respective areas. Radio Brahmaputra assists in production of plural voices from the margins of the country against the backdrop of the politics of community radio in India. This chapter will also deal with the process of selecting themes for the radio program by the community reporters and the importance of narrowcasting over broadcasting for a visible impact of the stakeholders.

A. Kaushik (✉)
EFL University, Shillong Campus, Shillong, India
e-mail: alankar@eflushc.ac.in

J. Servaes (ed.), *Handbook of Communication for Development and Social Change*,
https://doi.org/10.1007/978-981-15-2014-3_9

Keywords

Assamese · *Bhojpuri* · *Bodo* · *Chadri* · Community radio · Democratic space · Diverse voice · Indigenous communities · *Mishing* · Narrowcasting · Radio Brahmaputra · Radio program · Spiral of silence

84.1 Introduction

Community radio is a powerful medium that gives voice to the voiceless and is at the heart of communication and democratic processes within societies. The growth of community radio in India is remarkable in the last decade with university campuses, some NGOs, and other civil society groups running their own community radio stations. The community radio stations play a pivotal role in making the people aware about their basic rights, entitlements, and duties while providing a strong platform to freely disseminate ideas among the community. Though from the late 1970s, community radio gained momentum in India as an alternative medium to reach the marginalized, but the irony is, despite having the maximum diversity of culture and communities in the country, the northeastern region of India have only three community radio stations till 2017 with addition of two other community radio station in Manipur in 2018.

More than two decades after the historic Supreme Court of India ruling which declared airwaves as public property, the Government of India has been cautious in releasing broadcasting news and current affairs but restricted its use only to entertainment. Even under the new policies, the private commercial FM radio stations have not been permitted to broadcast news and current affairs programs.

On the other hand, considering the diversity of the northeastern region in India, it is not possible for state-run broadcasting agencies to reach all the communities through its programming content. Therefore, it is important for us to understand the diversity and the quality of information that is increasingly helping people from the margins to voice out their concerns by experimenting with participatory communication for social change. Radio Brahmaputra is the first grassroot community radio station of Northeast India situated at Dibrugarh district of Assam. Dibrugarh district of Assam is situated 439 km east of the capital city of Guwahati, Assam. Dibrugarh is home to sprawling tea gardens, a significant stretch of the Brahmaputra River, with Island villages as well as a range of ethnic groups, cultures, and languages among the districts located in the Upper Assam.

84.2 Radio Brahmaputra (90.4 MHz)

Radio Brahmaputra has been established to cater the communication needs of the marginalized communities living in the media dark zones. The media dark zone includes islands, tea gardens, and other villages where majority of the population speak local languages and dialects. These local languages and dialects are hardly represented in the mainstream media as a part of their regular programs.

The vision of Radio Brahmaputra is to build a network of community reporters who will report, inform, and broadcast in their local languages and dialects. Radio Brahmaputra produces programs in five different languages and mother tongue such as *Assamese, Chadri, Bhojpuri, Mishing, and Bodo.* The communities speaking the abovementioned languages and dialects are the most marginalized communities of the districts of Dibrugarh, Dhemaji, and partly Lakhimpur. Out of these five languages, *Chadri* is a dialect of the tea community easily comprehensible for the majority of Assamese community. There are 2,044,776 *Chadri* speakers in India. When the British tea planters brought the adivasis to Assam, as laborers, in the nineteenth century, *Chadri* as a link language came along with them. Radio Brahmaputra caters to an approximate population of 1 lakh *Chadri* speakers through its broadcast coverage near Dibrugarh. According to the Government of India of 2001 census, there are 33,099,497 *Bhojpuri* language speakers in Assam. The migrants from Bihar and northern India speak the *Bhojpuri* language. Among the mentioned languages, *Assamese* is widely spoken with 13,168,484 persons as per the Government of India census of 2001. For *Mishing* language in Assam, there are 517,170 persons and 1,350,478 *Bodo* speakers in Assam. *Assamese* and *Bodo* languages are a part of the 8 schedule of the Indian constitution, whereas *Mishing* is a dialect of the *Mishing* tribe who inhabit in 11 districts of Assam and its neighboring Arunachal Pradesh.

Presently Radio Brahmaputra is located in an 80-year-old revamped building at Maijan Borsaikia village near Maijan Ghat, Dibrugarh. The radio station is near to the temporary port where the vessels take off for different destinations every day. Radio Brahmaputra is covering nearly 180 mainland villages, 12 island villages, and 31 tea garden and subdivisions including Dibrugarh municipality area. A variety of radio formats from interviews, chat shows, information-based programs, drama, folk music, quiz shows for school children, special programs for women, and public service announcements can be heard when tuned into Radio Brahmaputra 90.4 FM. The team of Radio Brahmaputra includes community reporter, field community reporter, and radio producers. They have been working as radio broadcasters to produce programs relating to their own community in their own languages. For the management section, a station manager is placed at the radio station to look after the regular programmatic and other logistical issues.

Radio Brahmaputra reports, documents, and broadcasts on issues of local concern, ranging from economic and governance to social and cultural issues. It also aims to create awareness of social challenges, document cultural diversity and social practices, and report on the physical environment in which people inhabit in rivers, floods, islands, farms, etc. in Dibrugarh district. The radio station also showcases local talent for the listening public by encouraging entertainment, competition, and information. The primary goal of Radio Brahmaputra has been to empower people by making them aware of the issues that affect them directly or indirectly, enabling their concerns to be shared over the air, bringing it to the notice of government and administration, and providing the government with information and opportunities to address the challenges in an open and democratic manner.

84.3 Content Production and Management

Considering the contemporary ideas of participatory democracy, Radio Brahmaputra is contributing to the creation of a new relationship between media and the grassroots. In this process, Radio Brahmaputra is playing a constructive role to enable rural people to know about the process of development in their own areas.

Participatory communication advocates that media technology can become a full partner in the development process only when the ownership of both the message and the medium – the content and the process – resides with the communities (Pavarala and Malik 2007). The primary issue in participation is the facilitation of dialogue and debate among members of a community so that they become ultimate arbiters and negotiators of the development norms suitable for them. Participatory communication must be transactional in nature wherein sender and receiver of messages interact over a period of time "to arrive at shared meanings" (Nair and White 1987; White 1999). Participatory development communication that reflects a transactional process is defined as "a two-way, dynamic interaction between grassroots receiver and the information source, mediated by development communicators, which facilitates participation of the target group in the process of development" (Nair and White 1987: 37).

Unlike the traditional content format of All India Radio Dibrugarh, Radio Brahmaputra brings different segment for one program. For example, in a 1 h program, it produces programs of health, a public service message, a related folk song, etc. which combine education, awareness, and entertainment in one package and get broadcasted. The team of Radio Brahmaputra also tries to integrate important messages and deliver through various innovative programs (Personal Interview with Bhaskar J Bhuyan, Station Manager, 3 October, 2017, at Radio Brahmaputra).

Format of the program schedule of Radio Brahmaputra in different languages and dialects

THU	Assamese	Chadri	Bhojpuri	Chadri	Assamese
	Opening Presentation Bani(3A+C +B) Bhokti Sagar Promo Ajir Deha Hathor Dinlipi [Live	Signature Tune Presentation Aijker Subh Bichar Local Artist Song Mon Ker Awaz/ **ADDA** **new rept** Local Song Radio Promo Quiz 2 and Ans Station	Opening Music Opening Presentation Darpan (Health Programme) Folk Song Bhojpuri Community Hum Hai Sitare (Children Programme) Folk Song Bhojpuri Community News Paper Updates/Quiz Ending	Repeat CHADRI (Daily News Paper Update and Ending (Presentation)	Presentation Song Hathor Ajir ATITHI Amar Dictionary 2nd Repat Listener Feedback 2nd 2 no's Presentation New (Livelihood Women) Beli Mar Gol Ending Presentation

(*continued*)

		Stinger Tips (Education) Folk Song Radio Promo Quiz 2 and Ans Drama	Presentation Ending Music		
FRI	Assamese	Chadri	Assamese	Chadri	Assamese
	Opening Presentation Bani(3A+C +B) Bhokti Sagar Promo Ajir Deha Hathor Dinlipi [Live]	Signature Tune Presentation Aijker SuhhBichar Folk Song Local Song Radio Promo Quiz 3 and Ans Station Stinger Tips (House) Local Artist Song Kahani ker Duniya Radio Promo Quiz 3 and Ans	Presentation Rang Rupalir Hathor Kesa Matir Hubah Aapunar Swasthy Aamar Diha 2nd (Repeat) GaneGane	Repeat (Daily News Paper) Update and Ending (Presentation)	Presentation Hathor Krikhokor kotha 2nd Ajir ATITHI Song Apunar Sasthy amar Diha(Rpt) Song New(Village Based) Presentation Beli Mar Gol Ending Presentation

The content of the radio programs are primarily broadcasted in five different languages and dialects in the coverage area. Radio Brahmaputra caters to around 6 lakhs (six hundred thousand) listeners of their coverage area. The coverage provides opportunities for the media dark regions, viz., Dibrugarh tea garden areas, Dhemaji, and some part of Lakhimpur district, to listen to the radio program in their own dialects. Though Assamese is the lingua franca for the state of Assam, yet celebration of different local dialects by Radio Brahmaputra in its various program proves the potential of a community radio station to engage and result in participation of people from different communities and tongues.

Apart from *Bhojpuri, Chadri, and Assamese* language programs at Radio Brahmaputra, the other two languages *Bodo* and *Mishing* could not sustain in the programming because of the difficulties of production cost and accessibility to various resources. Through years of experience at Radio Brahmaputra, it was felt that, in the community learning process, the dialect of the communities seem utterly important, but when they attempt to bring out community voice, a common language like Assamese is always encouraged so that the message get disseminated to a large section of people.

The most important element of a community radio station is to bring out the community voice. Considering this idea into consideration, Radio Brahmaputra embarked on the idea of producing community programs in different languages and dialects to be made available for people who resides within the coverage area of the station. The term "voice," as used here, does not derive from a particular view of economic processes (consumer "voice") or even mechanisms of political representation (political "voice") but from a broader account of how human beings are. The value of voice articulates some basic aspects of human life that are relevant *whatever* our views on democracy or justice, so establishing common ground between contemporary frameworks for evaluating economic, social, and political organization (Nick Couldry 2010).

Charles Taylor in his work on *Self Interpreting Animals* describes man as "a self-interpreting animal." According to him, what we do beyond a basic description of how our limbs move in space already comes embedded in narrative, our own and that of others. This is why to deny value to another's capacity for narrative to deny her potential for voice is to deny a basic dimension of human life. A form of life that systematically denied voice would not only be intolerable but also barely be a culture at all (Taylor 1986).

Henceforth, to allow a community voice to run in a radio station becomes the prime prerogative of Radio Brahmaputra. But as mentioned earlier, the two languages and dialect *Bodo* and *Mishing* could not sustain due to lack of resources, monitoring of content, high production cost, and also accessibility of participants for the programs. This is due to the fact that the access to the community radio station is tough and also not cost-effective considering the practical aspects of running a community radio station. The celebration of diverse voice from different communities is not an easy task to perform. In order to bring a particular program to be aired in the station, one needs to travel across the river to Dhemaji district where the population of *Mishing* tribes are in large number. In addition to it, in order to broadcast a program of 1 h in a particular language, viz., *Mishing*, one would require a preproduction time of at least 25 h, which makes it more challenging to sustain the economic model of a community media station. If someone travels from across the river to get trained at the community radio station and spend time during the training, he or she needs to go back to their respective villages for their work in the agricultural field, fisheries, and so forth. While doing so, they tend to completely forget the basics of technical training in their second visit as they do not have access to any equipment and electricity while going back to the villages. Therefore, the earlier training stands no value, and finally the product doesn't meet the required expectations for broadcast. Though the radio programs broadcasted from Dibrugarh are more popular in the said region, yet it is difficult for the people from that region to train them at the station for regular programs in *Mishing* dialect. (Personal Interview with Bhaskar Bhuyan, 6 October, 2017).

The possible solution to this issue could be to create capacities of community correspondent of that area. They could participate either voluntarily or as paid correspondents and contribute regular stories through the use of mobile devices.

There has been a lot of academic and policy discussion on the role and potential of a community radio in developing countries. Many experiments also suggested immense value to grassroot communities in improving their lives especially when they have been channelized to a more participatory model of communication.

The range of programs that are being broadcasted over Radio Brahmaputra on a daily basis are on health issues, colloquial riddles, folk song, radio promo, health tips, SMS quiz, dictionary word, stories of local people, agriculture- and farming-related programs, old songs, village-based program, career-related information, youth-based program, children issues, songs from local artist, local newspaper updates, and women issue-based programs.

Apart from regular planned programs, Radio Brahmaputra also reported during the flood in August 2017 in the Dibrugarh and Dhemaji district. The news of the people affected due to flood were never paid due attention by the mainstream press. One of the community reporters Ms. Rumi Naik (in the picture), who usually report in *Chadri* language, went with a boat along with her friend to report in the flood-affected areas of Mothola block of Maijan village, Dibrugarh district of Assam.

Seen in the pic is Rumi Naik, community reporter of Radio Brahmaputra reporting during the August 2017 flood in Dibrugarh district

Given the media's role in crystallizing public opinion, media access becomes crucial for those who desire to shape the public mood. The phrase "spiral of silence" actually refers to how people tend to remain silent when they feel that their views are in the minority.

> *People ... live in perpetual fear of isolating themselves and carefully observe their environment to see which opinions increase and which ones decrease. If they find that their views predominate or increase, then they express themselves freely in public; if they find that their views are losing supporters, then they become fearful, conceal their convictions in public and fall silent. Because the one group express themselves with self-confidence whereas the others remain silent, the former appear to be strong in public, the latter weaker than their numbers suggest. This encourages others to express themselves or to fall silent, and a spiral process comes into play.* (Noelle-Neumann 1981)

Individuals, who notice that their own personal opinion is spreading and is taken over by others, will voice this opinion self-confidently in public. On the other hand, individuals who notice that their own opinions are losing ground will be inclined to adopt a more reserved attitude (Noelle-Neumann 1977).

This creates a spiral process, where the *majority* viewpoint becomes more powerful and the *minority* viewpoint becomes less dominant over time. Therefore, a community radio station like Radio Brahmaputra through its regular programming on various languages and dialects provides space for the minority viewpoints to get communicated. The right to communicate and to have affordable access to the means of communication is increasingly being acknowledged throughout the world as a fundamental human right. In the struggle to attain the rights of indigenous people from the margins, Radio Brahmaputra is benefitting various linguistic communities of Assam to have access to deliver voice and participate in communication. Radio Brahmaputra is also strengthening the cultural rights of cultural minorities in Assam such as migrants from Bihar and tea garden laborers by providing access to the means of communication.

Mere access to the means of communication does not necessarily benefit the process of integrated development goals set up by the Radio Brahmaputra and its mission to create awareness of social challenges. Therefore, to initiate an inclusive development model, the radio station supported education and training to develop critical understanding among the people of the communities. Here, solely the people are seen as producers and contributors to information. This experiment with participatory communication for social change has found Radio Brahmaputra to be the facilitator in breaking the barriers of isolation. A Rockefeller Foundation report asserts that community radio is "one of the best ways to reach excluded or marginalized communities in targeted, useful ways" and in the giving them a 'voice' that matters most in development communication (Dagron 2001).

84.4 Narrowcasting over Broadcasting

The facilitation of dialogue and debate among the members of a community is the primary goal of participatory communication. Therefore, it is very important to understand the real participation of the communities who not only express their viewpoints but also share their knowledge with their fellows. In this aspect, Radio Brahmaputra with the experience of broadcasting since mid-2015 understood that taking the radio to the communities facilitate more dialogues on issues that concerns them and also create more awareness in the blocks, talukas, and villages.

Henceforth, the team of Radio Brahmaputra is planning for a long-term effectiveness by building capacities of the people and target groups and training them to understand and analyze issues of their concern. Ms. Sandhya, one of the community reporters who have joined Radio Brahmaputra in 2012, works primarily with women and adolescent girls group of a slum in Dibrugarh district, thereby bringing out their priorities and issues to be addressed. For example, in raising issues about health, nutrition, or hygiene, Sandhya discovered that women and girl groups generally lack knowledge about issues related to health and have very less information about government schemes. They have also negligible amount of information on how to access government facilities in mitigating the problem faced by them. Basic issues like the importance of taking iron supplements, folic acid by a female, and knowledge about the importance of checking levels of hemoglobin are discussed in various groups over a period of time with the women and girl folks (Personal Interview with Ms. Sandhya Sharma).

Though the slum area in which Sandhya work as a community reporter is under the coverage area of the radio station, yet due to the accessibility issues of mobile devices by the female members unlike male members of the family, they do not get an opportunity to listen to the radio programs focused on women issues. Therefore, Radio Brahmaputra felt the need of narrowcasting a particular issue, which carries the potential to bring more awareness and raise critical question to the forefront. *Sandhya Didi*, as she is popularly called in the area, added that narrowcasting help people of the area to come together and discuss important issues. According to her, if Radio Brahmaputra can show them the path where to get the resources provided by the government, that itself is a big achievement for the team of Radio Brahmaputra. The housewife turned community reporter Sandhya concluded by saying in Hindi *Hum Rahe ya na rahe, lekin yeh to rahega* which means "Irrespective of my presence or absence, the Radio Station should sustain".

The process of narrowcasting a program by Radio Brahmaputra involves the following steps:

(a) Capacity-building exercise of community reporters by bringing in experts on a particular issue or topic

(b) Selecting villages, blocks, and number of households by undertaking a preliminary survey for the formation of target groups to be studied
(c) Group formation (at present there are around 94 groups of adolescent girls from the age group of 10 to 19 years old who are being trained in the field of menstrual hygiene)
(d) Building capacity of few girls from the group and select them as community representatives
(e) And the community representatives work in a radius of 1 to 2 km from their area to bring in people for discussion
(f) Have regular narrowcasting and discussion on particular issues that concern the group
(g) Organize regular community quiz for updates and revisions
(h) Take the results to the district administration regarding the gaps in communication campaigns of the government public service messages on crucial issues

Score Sheet
Community Quiz
Brahmaputra Community Radio Station, Dibrugarh

Venue: .... Date: ....

Score Name: ....

Question Number with Set Number (Ex. Set 1 Q 1)	Answered or Not Answered (If someone able to give correct answer please mark below as Y and if nobody able to give correct answer then mark as X)	Answered By (If someone able give correct answer then write his/her name)
1		
2		
3		
4		
5		
6		
7		

Format of the community quiz score sheet

Report format of Community Counseling
Brahmaputra Community Radio Station, Dibrugarh, Assam

Date: ______//______//______	***Name of Location:*** ______
Name of Venue: ______	***Type of Group:*** ______
Age group: ______	***Total Number of Attendees:*** ______
Name of Radio Program: ______	***Episode Number:*** ______
Name of Facilitator: ______	

Pre Discussion Topic:

______

Views of Participants on discussion Topic: (at least two important views of participants)

______

Feedback of Participants on Radio Program/Program Issue (at leasst one feedback of a participant with name, age, educational qualification):

______

Quarries of participants on any relevant issues: (with age and educational qualification)

______

Format for community counseling in field level intervention by Radio Brahmaputra

It was discovered through the field-level intervention by the team of Radio Brahmaputra that there is lack of in-depth study on basic issues such as health, hygiene, and nutrition, and therefore the introduction of community-based participatory action research with the help of the community reporters and community representatives shall strategize a methodology to collect data and use proper communication channel to bridge the gap in communication. According to the station manager, the goal is to create an army of local community-based researchers who shall be able to analyze issues within communities and thereby contribute to the root of the solution to any issues of the community. This process will also enhance their knowledge of exercising the democratic rights every 5 years during elections. It is expected to generate demand for development of the communities by the local administrators (Personal Interview, Bhaskar Bhuyan, 6 Oct 2017).

Sl No	Sample No	Number of Children of the Family		No of Never Enrolled Children		Reason for Not Enrolled in School	No of School Dropout Student		Reason for Drop Out	
		Girls	Boys	Girls	Boys		Girls	Boys		
1	1	0	4	0	3	Code 4	0	1	Code 18	Mothola
2	2	0	2	0	1	Code 19	0	1	Code 3	Mothola
3	3	3	2	2	1	Code 3	1	1	Code 19	Mothola
4	4	3	1	2	1	Code 5	1	0	Code 22	Mothola
5	5	0	2	0	0		0	2	Code 19	Mothola
6	6	1	2	0	0		0	2	Code 1	Mothola
7	7	1	3	0	2	Code 7	0	1	Code 1	Mothola
8	8	2	1	0	0		1	1	Code 2	Mothola

Format of Collecting Data from Blocks for Community Based Participatory Action Research

84.5 Conclusion

As per the UNESCO document, "*Grand realities- Community Radio in India, 2011*," a large gap remains between policy and practice in India. It further comments that communities from the dark regions of India continue to struggle to get their voices heard and to receive critical and locally relevant information (Jethwaney 2016).

Since 2009, Radio Brahmaputra has been producing programs to narrowcast and since 2015 started its regular broadcast. Over time few of the community reporters have received training from consulting training firms based on New Delhi, and consequently the community reporters started local training programs for the fellow community representatives. The community reporters strongly believe that there are a lot of issues to be addressed of their communities, which remain unreported in local and regional media. Rumi Naik, who reported the flood of August 2017, made people aware of the department of local disaster management plan and their failure in mitigating flood. There is a huge acceptance of Radio Brahmaputra within the communities because of its ability to involve people through participation at all levels of decision-making. By providing a proper platform for various languages and dialects of different communities to exchange views, this initiative of Radio Brahmaputra is a struggle to retain cultural and linguistic rights of indigenous communities of Assam.

Such is the popularity of the station that when the transmitter of Radio Brahmaputra was facing issues during October 2017, it could not transmit more than 3 km radius, and the Dhemaji district therefore could not receive the signals. An octogenarian lady of Dhemaji district came to a radio mechanic and asked him to repair the radio and told him to tune into Radio Brahmaputra. When the mechanic said the radio has no technical issues, the old lady asked gently, then *why the Radio Brahmaputra is not working*? (Personal interview Pinku Gohain, Community Reporter, Radio Brahmaputra).

To build up a meaningful democratic space, it is very important for community spaces like Radio Brahmaputra to create a harmonious listening environment in which everyone claims the Radio Brahmaputra as "our radio."

References

Couldry N (2010) Why voice matters: culture and politics after neoliberalism. Sage, London

Dagron AG (2001) Making waves: stories of participatory communication for social change. The Rockefeller Foundation, New York

Noelle-Neumann E (1981) Mass-media and social change in developed societies. In: Katz E, Szecsko T (eds) Mass media and social change. Sage, London, p 139

Jethwaney J (2016) Social sector communication in India: concepts, practices, and case studies. Sage, New Delhi

Nair KS, White SA (1987) Participation is key to development communication. Media Dev 34 (3):36–40

Noelle-Neumann E (1977) Turbulences in the climate of opinion: methodological applications of the spiral of silence theory. Public Opin Q 41(2):143–158. https://doi.org/10.1086/268371

Pavarala V, Malik K (2007) Other voices: the struggle for Community Radio in India. Sage, Thousand Oaks

Personal Interview, Bhaskar J Bhuyan, 3 Oct, 2017, Radio Brahmaputra, Dibrugarh, Assam

Personal Interview, Pinku Gohain, 6 Oct, 2017, Radio Brahmaputra, Dibrugarh, Assam

Personal Interview, Rumi Naik, 6 Oct, 2017, Radio Brahmaputra, Dibrugarh, Assam

Personal Interview, Sandhya Sharma, 6 Oct, 2017, Radio Brahmaputra, Dibrugarh, Assam

Taylor C (1986) Self-interpreting animals. In: Philosophical papers, vol 1. Cambridge University Press, Cambridge, pp 45–76

White R (1999) The need for new strategies of research on the democratisation of communication. In: Jacobson TL, Servaes J (eds) Theoretical approaches to participatory communication. Hampton Press, Cresskill, pp 229–262

Part XIII

Conclusion

Communication for Development and Social Change: Conclusion

85

Jan Servaes

Contents

Abstract

This concluding chapter attempts to answer the question whether "the glass is half empty or half full." It summarizes the main findings presented in this Handbook of Communication for Development and Social Change and highlights some of the major questions for the future – What about the Sustainable Development Goals? What about children, women, and seniors rights? What about worldwide inequality? What about the Global South? What about digital rights and digital cooperation? – to conclude that communication for development and social change is crucial to effectively tackle the major problems of today.

J. Servaes (✉)
Department of Media and Communication, City University of Hong Kong, Hong Kong, Kowloon, Hong Kong

Katholieke Universiteit Leuven, Leuven, Belgium
e-mail: jan.servaes@kuleuven.be; 9freenet9@gmail.com

J. Servaes (ed.), *Handbook of Communication for Development and Social Change*,
https://doi.org/10.1007/978-981-15-2014-3_116

Keywords

Advocacy communication · Behavior change communication · Digital cooperation · Human development index · Mass communication · Participatory communication · Sustainable development goals (SDGs) · Transdisciplinarity · Worldwide inequalities

> Without a vision, movement-building is fragmented and slow. Without a movement, visions are narrow and competing, rather than shared. (Julian Agyeman 2013: 168)

What are the main conclusions we can draw from all the chapters in this *Handbook of Communication for Development and Social Change*? Though we dare not claim that this volume is the unique and final overview of our field, we are confident that it will stand out for its comprehensive and multidimensional approaches and nontraditional perspectives, which can be summarized as follows:

- Human and environmental *sustainability* must be central in development and social change activities. Besides political-economic approaches, we need socio-cultural approaches to guarantee acceptable and integrated levels of sustainability and to build resilience. *Building resilient communities* should be a priority issue in the field of communication for development and social change.
- We should recognize that development problems are *complex. Complex or so-called wicked problems*, such as the existence of climate change, conflict and war, overpopulation, HIV/AIDS, and malaria, are problems that do not have one single solution that is right or wrong, good or bad, or true or false.
- These problems need negotiating or "social dialogue" from *a rights-based perspective*. A rights-based perspective is founded on the belief that every person has an inherent dignity and right to well-being and self-determination.
- *Communication and information* play a strategic and fundamental role by (a) contributing to the interplay of different development factors, (b) improving the sharing of knowledge and information, and (c) encouraging the participation of all concerned. It works by:
  - Facilitating participation: giving a voice to different stakeholders to engage in the decision-making process.
  - Making information understandable and meaningful. It includes explaining and conveying information for the purpose of training, exchange of experience, and sharing of know-how and technology.
  - Fostering policy acceptance: enacting and promoting policies that increase people's access to services and resources.
- Though participatory approaches have gained some visibility among mainstream development agencies, big data and quantitative assessments remain popular. *New creative techniques and methodologies* need further attention, exploration, and experimentation.
- There is a need for *transdisciplinarity* to rethink and reorder the relationships between communication academics, communication professionals, technical field-specific professionals, policy-makers, civil society members, and local people. There

is a need for building knowledge and communication networks and to attach importance to stakeholder interactions and knowledge system approaches.
- *Mainstreaming communication for development* is firmly grounding the field of communication for development and social change in thematic and nonthematic sub-disciplines of the communication sciences, like strategic communication, participatory communication, crisis communication, health communication, journalism, or environmental communication. These sub-disciplines provide a foundation by underpinning the work of development communication professionals and academics and giving them a solid basis to work from.
- *Communication strategies for development and social change* can be distinguished at five levels: (a) *behavior change communication* (BCC) (mainly interpersonal communication), (b) *mass communication* (MC) (community media, mass media, online media, and ICTs), (c) *advocacy communication* (AC) (interpersonal and/or mass communication), (d) *participatory communication* (PC) (interpersonal communication, community media, and social media), and (e) *communication for structural and sustainable social change* (interpersonal communication, community mobilization, participatory communication, mass communication, and ICTs). Interpersonal communication and mass communication form the bulk of what is being studied in the mainstream discipline of the communication sciences. Behavior change communication is mainly concerned with short-term individual changes in attitudes and behavior. It can be further subdivided in perspectives that explain individual behavior, interpersonal behavior, and community or societal behavior.
- There is a need to *connect communication to learning, education, and knowledge exchange* to better understand processes of transformative learning, social learning, experiential learning, organizational learning, and double-loop learning in order to better understand processes of change by looking at how people learn and change their attitudes and behavior.
- Four specific areas of focus for sustaining development momentum are (a) *enhancing equity*, including on the gender dimension; (b) *enabling greater voice and participation of citizens*, including women, youth, seniors, and indigenous peoples; (c) *confronting environmental pressures*; and (d) *managing demographic change.*

In other words, *communication is viewed as a social process that is not just confined to the media or to messages*. Communication for development and social change needs to deal with the complex issues of sustainable development in order to:

- Improve access to knowledge and information to all sectors of society and especially to vulnerable and marginalized groups
- Foster effective management and coordination of development initiatives through bottom-up planning
- Address equity issues through networking and social platforms influencing policy-making
- Encourage changes in behavior and lifestyles, promoting sustainable consumption patterns through sensitization and education of large audiences

- Promote the sustainable use of natural resources considering multiple interests and perspectives and supporting collaborative management through consultation and negotiation
- Increase awareness and community mobilization related to social and environmental issues
- Ensure economic and employment opportunities through timely and adequate information
- Solve multiple conflicts ensuring dialogue among different components in a society

If the above appropriately summarizes the assumptions underpinning most of the chapters in this handbook, some of the contributions go even further and problematize some of the complex issues in more detail.

85.1 What About the Sustainable Development Goals?

Jan Vandemoortele (2018: 1), the co-architect of the Millennium Development Goals, who was the Director of the Poverty Group at UNDP in 2001–2005, concludes that "respectable progress was made towards the Millennium Development Goals (MDGs) between 2000 and 2015." The challenge that remains, he contends, is twofold: "environmental sustainability and high inequality." However, the Sustainable Development Goals (SDGs), to be reached by 2030, dodge these challenges because the relevant targets lack precision and ambition. The Agenda 2030 is not universal in scope because the few targets that are verifiable – those that contain conceptual clarity, numerical outcomes, and specific deadlines – apply primarily to developing countries. For instance, the omission of targets for overweight and breastfeeding exemplifies the reluctance of developed countries to commit themselves to specific, quantitative, and time-bound targets.

Most SDG targets that are verifiable are actually not dissimilar from the MDGs. They clearly constitute a difficult intergovernmental compromise and made extra arduous by the deepening North–South divide, a return of East–West tensions, and a resurging sense of nationalism among several countries. To a large extent, the context of weak multilateralism explains why the SDGs are not fit for purpose to address the dual challenge of environmental sustainability and high inequality. Vandemoortele (2018: 12) proposes two vital steps to help realize the transformative potential of the SDGs: "At the national level, each country must select from among the SDG items those that are most relevant to the local context. ... At the global level, the important step is to choose fitting indicators to help fix several of the flawed targets. ... It seems that governments are not ready to accept indicators that could reveal politically sensitive dimensions of reality."

This observation applies not only to countries in the South but also to those in the North. In the South, serious questions are recently raised regarding rising population trends threatening the SDGs in Asia and Africa (Deen 2019). In the North, a detailed analysis finds that no European capital city or large metropolitan area has fully achieved the SDGs. Nordic European cities – Oslo, Stockholm, and Helsinki – are closest to the SDG targets but still face significant challenges in achieving one or several SDGs. Overall, cities in Europe perform best on SDG 3 (Health and Well-Being), SDG 6 (Clean Water and Sanitation), SDG 8 (Decent Work and Economic Growth), and SDG 9 (Industry, Innovation and Infrastructure). By contrast, performance is lowest on SDG 12 (Responsible Consumption and Production), SDG 13 (Climate Action), and SDG 15 (Life on Land). Further efforts are needed to achieve zero net carbon dioxide (CO_2) emissions or very close to zero net emissions by 2030 (Lafortune et al. 2019).

Therefore, the four questions we raised in Servaes (2017: 164) regarding the future of the SDGs seem to be still relevant today:

- First, how can we bring together the right stakeholders at the right time in the right place?
- Second, how do we make difficult trade-offs?
- Third, how do we build in accountability and transparency for action?
- Fourth, how to organize this in a participatory and democratic way?

85.2 What About Children Rights?

This can be further illustrated by looking at the SDGs and children rights. One of the latest United Nations Children's Fund (UNICEF)'s reports shows that, despite some progress, the most vulnerable people and countries still suffer the most and that the global response so far is not ambitious enough (https://data.unicef.org/children-sustainable-development-goals/). The evaluation includes quality education (SDG 4), inequality (SDG 10), migration (SDG 10.7), climate change (SDG 13), and all forms of violence against children (SDG 16). According to UNICEF, there remains a lot of room for improvement.

For instance, on *SDG 4*: ensure equal access to quality education and promote lifelong learning for all. Around 262 million children and adolescents in the world do not have the ability to go to school or finish school. When they are in primary school, an estimated 250 million children do not even learn basic skills. The quality and availability of data for this purpose remain an obstacle. But the data is not sufficient: the most important phase of human development takes place during the first years of life, and therefore early childhood must form the core of sustainable development. In addition, school success is measured not only in terms of school enrollment rates but also by what the child learns effectively. It is therefore essential to increase the capacity for measuring and monitoring learning outcomes.

On *SDG 10*: reduce inequality in and between countries. The vast majority of children do not have real social coverage. Half of the world's poor are under the age of 18. After a long fall, hunger seems to be rising again. Conflicts, droughts, and natural disasters linked to climate change are the most important factors in reversing this trend. Inequality starts at birth and influences the equal opportunities of every child.

Target 10.7 relates to migration and is particularly important for children. Half of the refugees are children. Almost 50 million children around the world live far from home, displaced within and beyond the borders of their own country. Their stories show us how painful it is for children to leave their country, sometimes without parents, to escape from conflict, violence, or poverty. Ensuring that migration runs smoothly is not only a challenge; it is also a great opportunity to protect children and thus to remind that migrant and refugee children are first and foremost children and enjoy all the rights of the Convention on the Rights of the Child. Well-managed migration, tackling its causes and mitigating its risks, is at the heart of the priorities of an equality-driven development.

On *SDG 13*: take urgent action to combat climate change and its impact. Globally, 93% of children under 15, or 1.8 billion children, breathe so much polluted air every day that their health and development are seriously threatened. Today, more than 500 million children live in areas where the risk of flooding is extremely high and nearly 160 million where the risk of drought reaches a high or extreme level. The Committee on the Rights of the Child recognizes that "environmental damage is an urgent human rights challenge" that has an impact on children's lives today and in the future. Violations of rights as a result of environmental damage can have irreversible, long-term, and even transgenerational consequences.

On *SDG 16*: promote peaceful and inclusive societies for sustainable development; ensure access to justice and develop effective, responsible, and accessible institutions at all levels. Worldwide, one billion children fall victim to some form of violence every year, and every 5 min a child dies of it. Three million girls run the risk of genital mutilation every year. It is therefore essential to prevent this violence and to ban all violence against children (regardless of context) and to strengthen child protection systems. To enable children to assert their rights, it is necessary that they have access to appropriate legal systems and that their right to legal identity (including through birth registration) is respected. Unfortunately, a substantial number of children, deprived of their freedom, have largely been forgotten in the statistics.

We cannot achieve sustainable development by excluding part of the population. Inequality starts at birth and influences the equal opportunities of every child. Uneven patterns are passed on from generation to generation. That is why there is an urgent need for early intervention and priority investments for all children, but especially for those children most affected by poverty and social exclusion. Therefore, at the High-Level Political Forum (HLPF) of July 2019, an important follow-up study will be conducted that is of particular interest to children http://sdg.iisd.org/events/high-level-political-forum-on-sustainable-development-hlpf-2019/.

85.3 What About Women?

Indeed, and what about women? To name just one authoritative report, the 2013 Human Development Report singled out the importance of women education: "It has often been stressed that improving education for women helps raise their levels of health and nutrition and reduces fertility rates. Thus, in addition to its intrinsic value in expanding women's choices, education also has an instrumental value in enhancing health and fertility outcomes of women and children" (Malik 2013: 33). In other words, educating women through adulthood is the closest thing to a "silver bullet" formula for accelerating human development. However, "important as education and job creation for women are, they are not enough. Standard policies to enhance women's income do not take into account gender differences within households, women's greater burden of unpaid work and gender division of work as per cultural norms. Policies based on economic theory that does not take these factors into account may have adverse impacts on women, even though they create economic prosperity. Key to improving gender equity are political and social reforms that enhance women's human rights, including freedom, dignity, participation, autonomy and collective agency" (Malik 2013: 45).

The report concluded clearly that "unless people can participate meaningfully in the events and processes that shape their lives, national human development paths will be neither desirable nor sustainable" (Malik 2013: 18). And, unfortunately, that continues to be the case. In its 2019 Human Development Report, UNDP observed that though gender gaps in early years are closing, inequalities persist in adulthood. One key source of inequality within countries is the gap in opportunities, achievements, and empowerment between women and men. Worldwide the average Human Development Index (HDI) for women is 6% lower than for men, due to women's lower income and educational attainment in many countries. Although there has been laudable progress in the number of girls attending school, there remain big differences between other key aspects of men and women's lives. Women's empowerment remains a particular challenge.

85.4 What About Worldwide Inequality?

Wide inequalities in people's well-being cast a shadow on sustained human development progress (UNDP 2019). According to the latest Human Development Index (http://hdr.undp.org/en/2018-update), people living in the very high human development countries can expect to live 19 years longer, and spend 7 more years in school, than those living in the group of low human development countries.

Movements in the HDI are driven by changes in health, education, and income. Health has improved considerably as shown by life expectancy at birth which has increased by almost 7 years globally, with sub-Saharan Africa and South Asia showing the greatest progress, each experiencing increases of about 11 years since 1990. And, today's school-age children can expect to be in school for 3.4 years longer than those in 1990. Disparities between and within countries continue to stifle progress. Average HDI levels have risen significantly since 1990 – 22% globally and 51% in least developed countries – reflecting that on average people are living longer, are more educated, and have greater income. But there remain massive differences across the world in people's well-being. In sum, there is tremendous variation between countries in the quality of education, healthcare, and many other key aspects of life. And many of these issues are interrelated. Take, for instance, the issue of migration. United Nations Population Fund (UNFPA)'s World Population Prospects 2019 (UNFPA 2019) describes how international migration has become an important determinant of population growth and change in some parts of the world. "Ironically, many countries which have low population growth are shutting their door to immigrants who could take on economic roles which would benefit those countries as well as the immigrants who are moving away for a better life and to leave behind the political and economic instability in their own countries," observed Purnima Mane, the former UN Assistant Secretary-General at the UN Population Fund (in Deen 2019).

85.5 What About the Global South?

The same 2013 Human Development Report assessed that, for the first time in 150 years, the combined output of the developing world's three leading economies – Brazil, China, and India – was about equal to the combined GDP of the long-standing industrial powers of the North, Canada, France, Germany, Italy, the United Kingdom, and the United States. This represented a dramatic rebalancing of global economic power.

The middle class in the South is growing rapidly in size, income, and expectations. The South is now emerging alongside the North as a breeding ground for technical innovation and creative entrepreneurship. Not only the larger countries have made rapid advances, notably Brazil, China, India, Indonesia, Mexico, South Africa, Thailand, and Turkey; but substantial progress has also been made in smaller economies, such as Bangladesh, Chile, Ghana, Mauritius, Rwanda, and Tunisia (for more details, see UNDP 2014, 2015).

The ways of engagement are changing from multilateralism to bi- or unilateral relations. The "America First" policy is an extreme example of this shift in global relations. But also the policy of other major players seems to have changed. For instance, the purpose of Russia's engagement with the African states is to cement its influence in the continent through political and military assistance in the same way as China has done through infrastructure and economic initiatives. Chinese and Russian engagement, characterized by the geopolitical convergence of Chinese and Russian interests, has seen the development of a Sino-Russian relationship based on competitive cooperation which has become a feature of the geopolitics of the African continent and beyond, concludes Tom Harper (2019), researcher at the University of Surrey specializing in China's relations with the developing world.

Fortunately, some promising news as well is the recent resurgence of South–South cooperation, which has moved once again onto the center stage of world politics and economics, leading to a renewed interest in its historic promise to transform world order (Gray and Gills 2016).

85.6 What About Digital Rights and Digital Cooperation?

From 2016 to 2018, there were 371 instances worldwide in which governments restricted Internet service or mobile apps, according to the digital rights advocacy group, Access Now (Bengali 2019). The vast majority of shutdowns – 310 – occurred in Asia, home to emerging economies with large numbers of new Internet users and where the free flow of information often poses a direct challenge to authoritarian governments. China remains the model for Internet censorship and surveillance, but India, which prides itself as the world's largest democracy, has been the quickest to cut off Internet service.

An expert panel, co-chaired by American philanthropist Melinda Gates and Chinese e-commerce magnate Jack Ma, convened by the UN has urged greater cooperation among governments, civil society, and the private sector, saying that while the rise of digital technologies has resulted in "unprecedented advances," it has also posed "profound new challenges" to people's human rights (for more general assessments Servaes 2014; Servaes and Hoyng 2017; UNDP 2012).

New forms of "digital cooperation" are needed to ensure that digital technologies are "built on a foundation of respect for human rights and provide meaningful opportunity for all people and nations," said the panel in its report, The Age of Digital Interdependence (2019).

In view of growing threats to human rights and safety posed by social media, the panel called on social media enterprises to work with governments, international and local civil society organizations, and human rights experts around the world "to fully understand and respond to concerns about existing or potential human rights violations." These concerns do also feature prominently on the European agenda. See, for instance, two recent policy documents from the High Level Group on Fake News (The European Commission 2018a) and the one on Tackling Online Disinformation (2018b) or Martens et al. (2018).

Companies like Facebook or Google have often reacted slowly and inadequately to complaints and reports that their technologies are being used in ways that undermine human rights. "We need more forward-looking efforts to identify and mitigate risks in advance: companies should consult with governments, civil society and academia to assess the potential human rights impact of the digital technologies they are developing," the report argued. It added: "From risk assessment to ongoing due diligence and responsiveness to sudden events, it should be clarified what society can reasonably expect from each stakeholder, including technology firms."

The report discussed the ways digital technologies have improved people's lives and "revolutionized the ability to communicate with others and to share access and knowledge." However, at the same time, more and more people are "increasingly – and rightly – worried that our growing reliance on digital technologies has created new ways for individuals, companies, and governments to intentionally cause harm or to act irresponsibly." "Virtually every day brings new stories about hatred being spread on social media, invasion of privacy by businesses and governments, cyberattacks using weaponized digital technologies, or states violating the rights of political opponents."

The panel was asked to consider questions of "digital cooperation" – defined as ways "to address the social, ethical, legal, and economic impact of digital technologies to maximize their benefits and minimize their harm." It was asked to look at how digital cooperation can contribute to the achievement of the Sustainable Development Goals (SDGs) and to consider models of digital cooperation "to advance the debate surrounding governance in the digital sphere." Digital cooperation, the panel said, must be grounded in the principles of inclusiveness, respect, human-centeredness, human rights, international law, transparency, and sustainability.

Universal human rights apply equally online as offline. For it to be effective, it must be multilateral and multi-stakeholder – involving not only governments but a diverse spectrum of civil society, academics, technologists, and traditionally marginalized groups such as women, youth, indigenous people, rural populations, and older people.

The panel also argued that digital technologies will only advance toward the full sweep of the SDGs if it goes more broadly than the issue of access to the Internet and digital technologies. There must be cooperation on "broader ecosystems that enable digital technologies to be used in an inclusive manner." This will require policy frameworks that directly support economic and social inclusion, special efforts to bring traditionally marginalized groups to the fore, investments in both human capital and infrastructure, smart regulatory environments, and significant efforts to assist workers facing disruption from technology's impact on their livelihoods.

"Those who lack safe and affordable access to digital technologies are overwhelmingly those from already marginalized sectors – women, elderly people, and those with disabilities; Indigenous groups; and those who live in poor, remote, or rural areas." This has reinforced many existing and widening inequalities – in wealth, opportunity, education, and health – it added.

The panel also recommended the following as priority actions:

- By 2030, every adult should have affordable access to digital networks, as well as digitally enabled financial and health services, as a means to make a substantial contribution to achieving the SDGs. Provision of services should guard against abuse by building on best practices, including the ability to opt in and opt out, and by encouraging informed public discourse.
- The creation of a broad, multi-stake alliance, involving the UN, for sharing digital public goods, engaging talent, and pooling data assets, "in a manner that respects privacy, in areas related to attaining the SDGs."
- The adoption by the private sector, civil society, national governments, multilateral banks, and the UN of policies to support full digital inclusion and digital equality for women and marginalized groups. International bodies, such as the World Bank and the UN, should strengthen research and promote action on barriers women and marginalized groups face to digital inclusion and digital equality.
- A set of metrics for digital inclusiveness should be urgently agreed, measured worldwide, and detailed with sex-disaggregated data.

- The UN Secretary-General must institute an agencies-wide review of how existing international human rights accords and standards apply to new and emerging digital technologies. Civil society, governments, and the private and the public sector should be invited to submit their views on how to apply existing human rights instruments in the digital age.
- The development of a Global Commitment on Digital Trust and Security to shape a shared vision, identify attributes of digital stability, and strengthen implementation of norms for responsible uses of technology and propose priorities for action.
- That autonomous intelligent systems be designed in ways that enable their decisions to be explained and humans to be accountable for their use. Audits and certification schemes should monitor compliance of AI systems with engineering and ethical standards, which should be developed using multi-stakeholder and multilateral approaches. Life and death situations should not be delegated to machines.

Reacting to the findings of the report, WACC General Secretary Philip Lee (2019) said, "individually, these challenges are not new, but their pace and interconnection in the digital age pose huge and complex obstacles to people and communities seeking to participate fully and equally as responsible citizens in democratic societies." He added that while government action is needed, "it must be informed and supported by civil society and based on a strong foundation of human rights, social justice and democratic principles. To counter and control the digital transformation of societies requires an equally transformative movement of people."

All things considered, hang in there!

References

Agyeman J (2013) Introducing just sustainabilities. policy, planning and practice. ZED Books, London

Bengali S (2019) An increasingly popular authoritarian tool: shutting down the Internet. https://www.latimes.com/world/la-fg-myanmar-internet-shutdown-20190627-story.html?fbclid=IwAR0twj18RlkFXf4ccRkC5gYN_QtCqb-EOjEeFSOrvEyFG37sWnoa5S8B2EA. Accessed 27 June 2019

Deen T (2019) Rising population trends threaten UN's Development Goals in Asia & Africa. Inter Press Service, 19 June 2019. www.ipsnews.net/2019/06/rising-population-trends-threaten-uns-development-goals-asia-africa. Accessed 24 June 2019

European Commission (2018a) A multi-dimensional approach to disinformation. Report of the independent high-level group on fake news and online disinformation. March 2018. https://ec.europa.eu/digital-single-market/en/news/final-report-high-level-expert-group-fake-news-and-online-disinformation

European Commission (2018b) Communication on tackling online disinformation. March 2018. https://eur-lex.europa.eu/legal-content/EN/TXT/?uri=CELEX:52018DC0236

Gray K, Gills BK (2016) South–South cooperation and the rise of the Global South. Third World Q 37(4):557–574. https://doi.org/10.1080/01436597.2015.1128817

Lafortune G, Zoeteman K, Fuller G, Mulder R, Dagevos J, Schmidt-Traub G (2019) The 2019 SDG Index and Dashboards report for European cities (prototype version). Sustainable Development

Solutions Network (SDSN) and the Brabant Center for Sustainable Development (Telos). http://unsdsn.org/wp-content/uploads/2019/05/Full-report_final-1.pdf

Lee P (2019). http://www.waccglobal.org/articles/un-appointed-panel-calls-for-global-cooperation-to-ensure-safe-inclusive-digital-economy-and-society-for-all?fbclid=IwAR2FXYWt2pLh2daDp4CnKtltNNMbMMbTvVseLcqkmiEUDEnnW62C5qVcyKM

Malik K (ed) (2013) The rise of the South: human progress in a diverse world. Human development report. UNDP, New York

Martens B, Aguiar L, Gomez-Herrera L, Mueller-Langer F (2018) The digital transformation of news media and the rise of disinformation and fake news. An economic perspective; Digital economy working paper 2018-02; JRC technical reports. European Commission, Sevilla

Servaes J (ed) (2014) Technological determinism and social change. Communication in a tech-mad world. Lexington Books, Lanham

Servaes J (ed) (2017) The Sustainable Development Goals in an Asian context. Springer, Singapore, 174pp. http://www.springer.com/in/book/9789811028144

Servaes J, Hoyng R (2017) The tools of social change: a critique of techno-centric development and activism. New Media Soc 19(2):255–271. https://doi.org/10.1177/1461444815604419

Tom Harper T (2019) Competitive cooperation: the connection between Chinese and Russian initiatives in Africa and beyond. Asia Dialogue. https://theasiadialogue.com/2019/06/26/competitive-cooperation-the-connection-between-chinese-and-russian-initiatives-in-africa-and-beyond/

UN (2019) The age of digital interdependence. Report of the UN Secretary-General's High-level Panel on Digital Cooperation. June 20. https://digitalcooperation.org/

UNDP (2014) Human Development Report: sustaining human progress: reducing vulnerabilities and building resilience. http://hdr.undp.org/en/content/human-development-report-2014

UNDP (2015) Human Development Report – rethinking work for human development. http://hdr.undp.org/en/rethinking-work-for-human-development

UNDP (2019) Focusing on inequality. http://hdr.undp.org/en/towards-hdr-2019

UNFPA (2019) World Population Prospects 2019. https://population.un.org/wpp/

United Nations Development Programme (UNDP) (2012) Mobile technologies and: enhancing human development through participation and innovation. http://www.undp.org/content/undp/en/home/librarypage/democratic-governance/access_to_informationande-governance/mobiletechnologiesprimer.html

Vandemoortele J (2018) From simple-minded MDGs to muddle-headed SDGs. Dev Stud Res 5(1):83–89. https://doi.org/10.1080/21665095.2018.1479647

Index

J. Servaes (ed.), *Handbook of Communication for Development and Social Change*,
https://doi.org/10.1007/978-981-15-2014-3

B

D

F

Y